D0084953

McGraw-Hill Higher Education

A Division of The ***McGraw-Hill*** *Companies*

FOUNDATIONS OF ENGINEERING, SECOND EDITION

Published by McGraw-Hill, a business unit of The McGraw-Hill Companies, Inc., 1221 Avenue of the Americas, New York, NY 10020.

This book is printed on acid-free paper.

International 1 2 3 4 5 6 7 8 9 0 VNH/VNH 0 9 8 7 6 5 4 3 2
Domestic 3 4 5 6 7 8 9 0 VNH/VNH 0 9 8 7 6

ISBN-13: 978-0-07-248082-5
ISBN-10: 0-07-248082-3
ISBN-13: 978-0-07-119561-4 (ISE)
ISBN-10: 0-07-119561-0 (ISE)

Publisher: *Elizabeth A. Jones*
Sponsoring editor: *Kelly Lowery*
Developmental editor: *Maja Lorkovic*
Executive marketing manager: *John Wannemacher*
Senior project manager: *Jill R. Peter*
Production supervisor: *Kara Kudronowicz*
Lead media project manager: *Audrey A. Reiter*
Senior media technology producer: *Phillip Meek*
Coordinator of freelance design: *Rick D. Noel*
Cover designer: *Jamie E. O'Neal*
Cover image: *©Getty Images, image number ab62750, Pyramids in Egypt*
Senior photo research coordinator: *Lori Hancock*
Photo research: *Randall Nicholas/Nicholas Communications*
Compositor: *Lachina Publishing Services*
Typeface: *10/12 Times Roman*
Printer: *Von Hoffmann Press, Inc.*

The credits section for this book begins on page 709 and is considered an extension of the copyright page.

Library of Congress Cataloging-in-Publication Data

Holtzapple, Mark Thomas.
Foundations of engineering/ Mark T. Holtzapple, W. Dan Reece.—2nd ed.

p. cm.

Includes bibliographical references and index.
ISBN 0-07-248082-3 — ISBN 0-07-119561-0 (ISE)
1. Engineering. I. Reece, W. Dan. II. Title.

TA145.H59 2003
620—dc21

2002022660
CIP

INTERNATIONAL EDITION ISBN 0-07-119561-0

www.mhhe.com

Foundations of ENGINEERING

SECOND EDITION

Mark T. Holtzapple
W. Dan Reece
Texas A&M University

Boston Burr Ridge, IL Dubuque, IA Madison, WI New York San Francisco St. Louis
Bangkok Bogotá Caracas Kuala Lumpur Lisbon London Madrid Mexico City
Milan Montreal New Delhi Santiago Seoul Singapore Sydney Taipei Toronto

CONTENTS

Foundations of ENGINEERING

SECTION THREE: ENGINEERING FUNDAMENTALS

CHAPTER 10 NEWTON'S LAWS

CHAPTER 11 INTRODUCTION TO THERMODYNAMICS

CHAPTER 15

INTRODUCTION TO STATICS AND DYNAMICS

CHAPTER 16

INTRODUCTION TO ELECTRICITY

SECTION FOUR:

ENGINEERING ACCOUNTING

CHAPTER 17

ACCOUNTING

CHAPTER 18

ACCOUNTING FOR MASS

CHAPTER 19

ACCOUNTING FOR CHARGE

CHAPTER 20

ACCOUNTING FOR LINEAR MOMENTUM

CHAPTER 21

ACCOUNTING FOR ANGULAR MOMENTUM

CHAPTER 22

ACCOUNTING FOR ENERGY

CHAPTER 23

ACCOUNTING FOR ENTROPY

CHAPTER 24

ACCOUNTING FOR MONEY

APPENDIX A

APPENDIX B

APPENDIX C

APPENDIX D

TO THE PROFESSOR

Traditional engineering courses—such as courses on heat transfer, circuits, and fluids—are fairly well defined. In contrast, there is no general agreement on the content of freshman engineering courses. Current freshman engineering texts choose from a range of topics including professionalism, creativity, ethics, design, technical writing, graphing, systems of units, engineering science, and problem solving. All of these topics are important aspects of the freshman engineering experience, but we found no one text that adequately encompassed them all. Therefore, we decided to write our own text to fill the void.

Many freshman engineering texts describe specific engineering disciplines, such as mechanical or electrical engineering, and give sample problems involving statics or electrical circuits. Given the increasing number of new engineering disciplines (e.g., biochemical engineering) and the increasingly interdisciplinary nature of engineering (e.g., mechatronics), we feel this discipline-specific approach is inadequate. Instead, we feel a more unified approach is required, with less emphasis on traditional disciplines. The goals of our text are listed here:

- *Excite the student about engineering.* Most practicing engineers find their work to be very exciting and creative. However, freshmen must struggle with the rigors of their science and mathematics classes, so they may be unaware of the pleasures that await them. We hope to stimulate the students' interest in engineering by describing engineering history, challenging them with "brain teaser" problems, and explaining the creative process.
- *Provide a strong foundation in engineering fundamentals.* Engineering has grown beyond the traditional disciplines (e.g., civil, mechanical, and electrical engineering) and now includes nontraditional disciplines (e.g., biomedical, environmental, and nuclear engineering). The common threads through all these disciplines are fundamental physical and mathematical laws.
- *Cultivate problem-solving skills.* The most important engineering skill is the ability to solve problems. We describe many heuristic approaches to creative problem solving as well as a systematic approach to solving well-defined engineering problems.
- *Challenge advanced students.* Students who have good high school backgrounds will have been exposed to calculus and physics. To stimulate their interest in engineering, advanced topics are sprinkled throughout the book.

- *Integrate computing with other engineering topics.* This book contains numerous sample computer programs illustrating a variety of engineering applications. This will help the student realize that computing is not a separate topic, but is a tool used by engineers to solve problems.
- *Provide reference material.* Most students will not purchase handbooks until later in their engineering careers. This book provides unit conversion factors and material properties so that students have the resources to solve real-world problems.
- *Provide information the student is unlikely to encounter elsewhere.* Often, important engineering information that does not fit neatly into advanced courses is put into a freshman engineering course. Thus, this text includes information such as statistics, grammatical rules for the SI system, and graphing rules.
- *Connect with their high school experience.* Many students may be concerned about possible gaps between their actual knowledge and the knowledge college professors expect of them. Touching upon topics with which they are already familiar will ease their anxiety and improve their confidence.
- *Review high school mathematics.* Most freshman engineering students no longer have their high school mathematics textbooks, nor is high school mathematics discussed in college calculus textbooks. For students who need to refresh their mathematics skills, the book's website, http://www.mhhe.com/holtzapple, offers a mathematics supplement complete with practice problems.
- *Connect with their freshman science and mathematics courses.* Some students may perceive that their freshman science and mathematics classes are a hazing process, and may not understand that these courses form the backbone of engineering. We purposely incorporate topics they see in other courses to show the connection with engineering.
- *Provide "soak time" for difficult topics.* Learning is a process that requires repetition. A few difficult topics that students will encounter in later engineering courses (e.g., thermodynamics, rate processes) are introduced here at a very simple level. This allows them to become acquainted with the ideas, so their next detailed exposure is easier.
- *Introduce the design process.* To help freshmen experience the joy of engineering, we think it is necessary to assign a design problem during their first semester. To support this notion, early in the text, we introduce design.
- *Emphasize the importance of communication skills.* Too often, engineers are criticized for lacking communication skills. To help overcome this problem, we provide information on both oral and written communication that will be immediately useful to freshmen during their design project.

The topics in *Foundations of Engineering* are presented in a sequential manner, so it can be read from front cover to back cover with each new topic building on previously presented topics. Although the book is designed so that it **can** be read from cover to cover, this does not imply that it **must** be read from cover to cover. The accompanying figure indicates how the chapters fit together.

The "road map" in the accompanying figure shows that Chapters 1 through 9 are independent; if you decide to skip these chapters, it will not seriously affect the students' understanding of later chapters. In contrast, Chapters 10, 11, 13, and 14 are interdependent and must be covered in sequence. Chapters 12 and 15 are optional, but if covered, they must

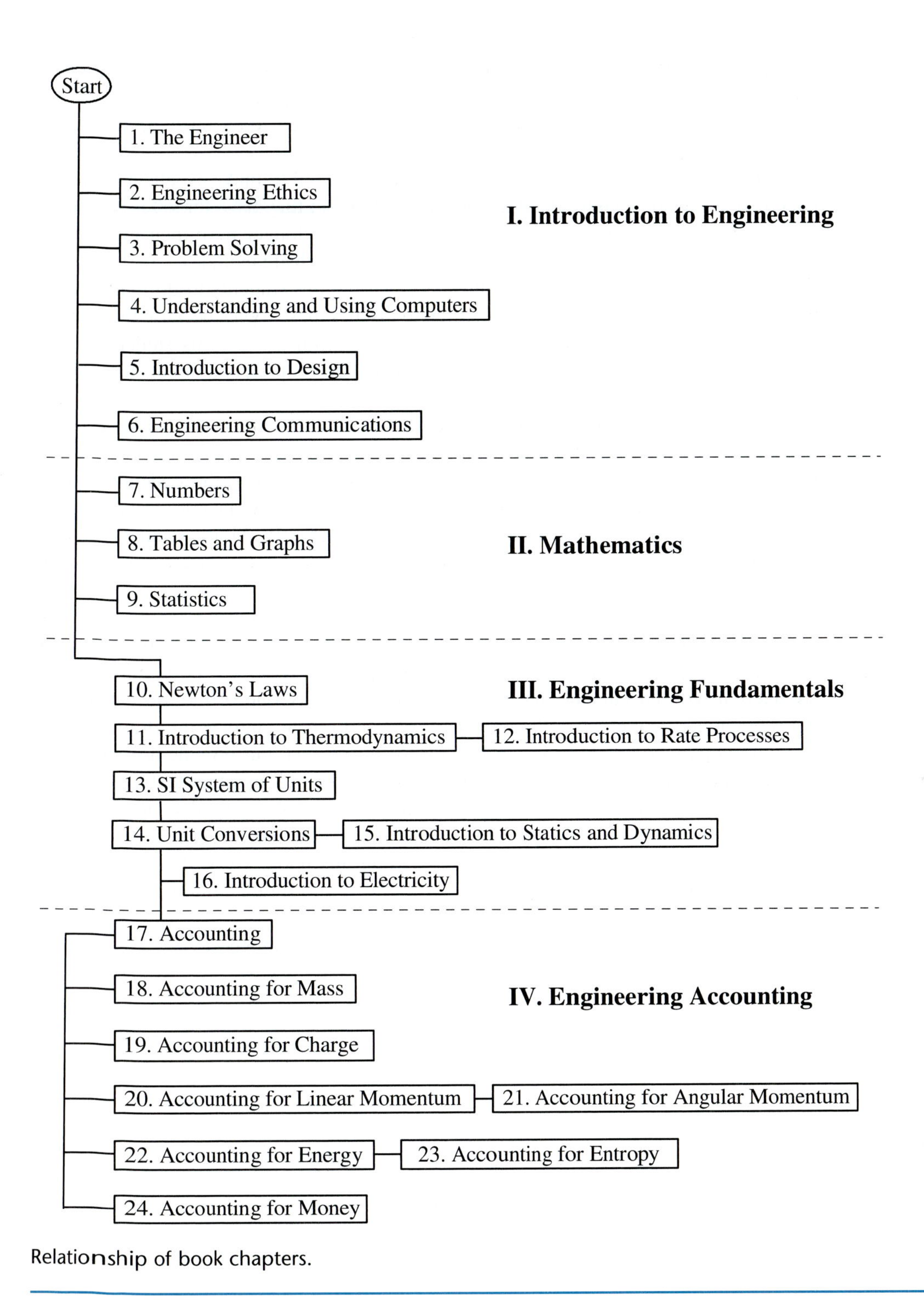

Relationship of book chapters.

and detailed. In fact, it is unlikely that you will be able to cover the entire book in a single semester. Your professor will decide which of the many topics will be covered in your particular course. However, your professor's decisions should not preclude you from reading on your own. All of the topics in this text should be covered at some point in your studies.

We have divided the book into four sections:

- *Introduction to Engineering:* This is an overview of the engineering professions and the skills required to become a good engineer.
- *Mathematics:* We touch on a few mathematical concepts that you are not likely to encounter in your calculus class.
- *Engineering Fundamentals:* We feel the topics discussed here are absolutely fundamental to engineering education. You will be introduced to topics such as thermodynamics, rate processes (e.g., heat transfer, electricity), and Newton's laws. Unit conversions are given particular attention because this topic is so important.
- *Engineering Accounting:* We have cast the basic conservation laws (e.g., conservation of energy or mass) as a simple "accounting" procedure. We feel that accounting is a unifying concept that transcends the individual engineering disciplines. Here, you have the opportunity to apply your new skills to a variety of problems. The fundamental accounting principles are applied to such quantities as mass, energy, linear momentum, and angular momentum.

In case your high school mathematics is rusty, the book's website, at http://www.mhhe.com/holtzapple includes a mathematics supplement which reviews topics such as algebra, mathematical notation, probability, geometry, trigonometry, logarithms, polynomials, zeros of equations, and calculus. Each chapter with mathematical content informs you of the mathematical prerequisites needed to fully understand the chapter, and directs you to the appropriate section on the website.

The website also contains useful supplemental learning materials. Please visit the site; we're sure you'll find it useful.

We think of our book as a smorgasbord of delightful delicacies. There are so many delicacies, it is impossible for you to eat them all in a single sitting. However, with many sittings, it is possible for you to enjoy them all.

As many topics as we cover in this book, we still do not attempt to cover everything you will need to know. For several topics of major importance to engineers, particularly engineering graphics and the details of computing, we expect that you are receiving training from other texts. Both topics are essential to the practicing engineer. Even a simple engineering drawing passes more information than several volumes of words alone. Computers have revolutionized engineering. What took hours of drudgery just 20 years ago can now be done in seconds by using personal computers and software.

As shown in the "pyramid of learning" depicted earlier, all engineering disciplines use knowledge gained in mathematics and science courses. In addition, an important foundation of engineering is communications. One of the most important functions of engineers is to present their findings clearly and succinctly, both orally or in writing. It is no accident that English and technical writing are included in your engineering studies! The ability to convey ideas well comes only with hard work, practice, and constructive feedback; this may be the most important skill you have to learn.

We recommend that you hold onto this book. It has many useful charts, tables, conversion factors, and formulas that you will find invaluable in your later studies. Also, the topics are covered in a friendly, unified approach. If you are having troubles grasping a concept in your later studies, we hope you will take this book off your shelf and read—or reread—the appropriate chapters.

Mark T. Holtzapple W. Dan Reece

ACKNOWLEDGMENTS

A project of this size cannot be completed without assistance from many individuals. Dan Turner, the undergraduate dean at the Dwight Look College of Engineering at Texas A&M, initiated this project to provide a text for our introductory engineering course *Engineering Problem Solving and Computing.* He provided financial support and organized internal reviews of the manuscript. Karan Watson, who followed Dan Turner, provided invaluable support and encouragement for this project.

Curtis Johnson, an undergraduate nuclear engineering student at Texas A&M, provided assistance with computers and graphics. We are deeply indebted to him. The secretarial support from Brenda Mooney is much appreciated. Also, we thank Seth Adelson, a graduate chemical engineering student, for providing some historical descriptions of engineering and for creating some key computer programs.

Bill Bassichis is commended for giving a very thorough and helpful review of our physics chapters. Mike Rabins and Ed Harris, who jointly work on an NSF-supported engineering ethics project, were enormously helpful in their critical review of the ethics chapter. John Fleming deserves our praise for his very careful review of the manuscript. We also appreciate Larry Piper's and Jim Morgan's efforts in coordinating many of the activities that went into the book.

Charles Glover is thanked for the countless hours we spent discussing the accounting principles used widely in this text. He is a key player in an NSF-sponsored project to create a unifying framework for all engineering disciplines. His insight and deep thinking are essential to this book.

As mentioned earlier, Dan Turner organized a review of the manuscript by the following individuals: Lee Carlson, Glen Williams, Mac Lively, Alberto Garcia, Larry Piper, Skip Fletcher, Ray James, Tom Tielking, Mike McDermott, Ron Hart, Richard Griffin, Gerald Miller, Pierce Cantrell, Aaron Cohen, Vincent Sweat, and Kaylan Annamalai. We are grateful for their helpful comments.

It took thousands of hours to write this book, mostly during the evenings and weekends. We thank our families for graciously providing us with this time.

LIST OF REVIEWERS

Second Edition

Sven Bilén	The Pennsylvania State University
Jerome N. Borowick	California State Polytechnic University, Pomona
John T. Demel	The Ohio State University
Lawrence J. Genalo	Iowa State University
Robert J. Gustafson	The Ohio State University
Mark Hernandez	University of Colorado–Boulder
William E. Howard	Milwaukee School of Engineering
Jean C. Malzahn Kampe	Virginia Polytechnic Institute
Andrew Lau	The Pennsylvania State University
Gary A. Pertmer	University of Maryland
Raymond H. Russell	University of Texas–Austin
David R. Thompson	Oklahoma State University
Ronald L. Thurgood	Utah State University

First Edition

Barry Crittenden	Virginia Polytechnic Institute
John T. Demel	The Ohio State University
James Garrett	Carnegie Mellon University
Jeff Kantor	University of Notre Dame
Rajiv J. Kapdia	Mankato State University

James L. Kelly	University of Virginia
Hillel Kumin	University of Oklahoma
James Morgan	Texas A&M University
William Park	Clemson University
Joey Parker	University of Alabama
Harry J. Ploehn	University of South Carolina
Larry G. Richards	University of Virginia
David N. Rocheleau	University of South Carolina
Sheryl Sorby	Michigan Technological University
Linda L. Vahala	Old Dominion University
Gretchen L. Van Meer	Northern Illinois University
Thomas Walker	Virginia Polytechnic Institute
Daniel White	University of Alaska Fairbanks
Steve Yurgartis	Clarkson University

ABOUT THE AUTHORS

Mark T. Holtzapple

Mark T. Holtzapple is Professor of Chemical Engineering at Texas A&M University. In 1978, he received his BS in chemical engineering from Cornell University. In 1981, he received his PhD from the University of Pennsylvania. His PhD research focused on developing a process to convert fast-growing poplar trees into ethanol fuel.

After completing his formal education, in 1981 Mark joined the U.S. Army and helped develop a portable backpack cooling device to alleviate heat stress in soldiers wearing chemical protective clothing.

After completing his military service, in 1986 Mark joined the Department of Chemical Engineering at Texas A&M University. It quickly became apparent that he had a passion for teaching: within a 2-year period he won nearly every major teaching award offered at Texas A&M, including Tenneco Meritorious Teaching Award, General Dynamics Excellence in Teaching Award, Dow Excellence in Teaching Award, and two awards offered by the Texas A&M Association of Former Students. Mark particularly has a passion for teaching freshman engineering students. He wrote this book to excite students about engineering and to help lay a solid foundation for their future studies.

In addition to his role as an educator, Mark is a prolific inventor. He is developing an energy-efficient, ecologically friendly air-conditioning system that uses water instead of Freons as the working fluid. He is also developing a high-efficiency, low-pollution Brayton cycle engine suitable for automotive use. In addition, he is developing technologies for converting waste biomass into useful products, such as animal feeds, industrial chemicals, and fuels. To recognize his contributions in biomass conversion, in 1996 he received the Presidential Green Chemistry Challenge Award offered by the president and vice president of the United States.

W. D. Reece

Dr. Reece is an Associate Professor in the Nuclear Engineering Department and Director of the Nuclear Science Center at Texas A&M University. He received his Bachelor of Chemical Engineering, Master of Science in Nuclear Engineering, and PhD in Mechanical Engineering all at the Georgia Institute of Technology. He has worked as an analytical chemist, a chemical engineer, and a staff scientist at the Pacific Northwest National Laboratory, before his current positions at Texas A&M.

Much of Dr. Reece's research is in the area of radiation monitoring, novel uses of radiation in medicine, and the health effects of radiation. Like Dr. Holtzapple, he has a passion for teaching and has won a Distinguished Teaching Award from the Texas A&M Association of Former Students. Dr. Reece teaches many topical courses in dosimetry and health physics, has an active consulting business, and, whenever his schedule allows him free time, enjoys backpacking, playing tennis, and running. His greatest enjoyment comes from his children, his students, and the advances in medicine and worker protection he has helped to make.

SECTION ONE

INTRODUCTION TO ENGINEERING

This book is divided into four sections. This first section addresses the question of what exactly are engineers, and what do they do? In this section we will explore the various disciplines within engineering, some history of engineering, and what characteristics are usually present in good engineers. Next, we examine engineering professionalism and engineering ethics. Lastly, we will look at the most basic activities of engineering: solving problems, using computers, designing things, and communicating findings.

Mathematical Prerequisite

- Geometry (Appendix I, Mathematics Supplement)

CHAPTER 1

The Engineer

Nearly all the manmade objects that surround you result from the efforts of engineers. Just think of all that went into making the chair upon which you sit. Its metal components came from ores extracted from mines designed by mining engineers. The metal ores were refined by metallurgical engineers in mills that civil and mechanical engineers helped build. Mechanical engineers designed the chair components as well as the machines that fabricated them. The polymers and fabrics in the chair were probably derived from oil that was produced by petroleum engineers and refined by chemical engineers. The assembled chair was delivered to you in a truck that was designed by mechanical, aerospace, and electrical engineers, in plants that industrial engineers optimized to make best use of space, capital, and labor. The roads on which the truck traveled were designed and constructed by civil engineers.

Obviously, engineers play an important role in bringing ordinary objects to market. In addition, engineers are key players in some of the most exciting ventures of humankind. For example, the Apollo program was a wonderful enterprise in which humankind was freed from the confinement of earth and landed on the moon. It was an engineering achievement that captivated the United States and the world. Some pundits say the astronauts never should have gone to the moon, simply because all other achievements pale in comparison; however, we say that even more exciting challenges await you and your generation.

1.1 WHAT IS AN ENGINEER?

Engineers are individuals who combine knowledge of science, mathematics, and economics to solve technical problems that confront society. It is our practical knowledge that distinguishes engineers from scientists, for they too are masters of science and mathematics. Our emphasis on the practical was eloquently stated by the engineer A. M. Wellington (1847–1895), who described engineering as "the art of doing . . . well with one dollar, which any bungler can do with two."

Although engineers must be very cost-conscious when making ordinary objects for consumer use, some engineering projects are not governed strictly by cost considerations. President Kennedy promised the world that the Apollo program would place a man on the moon prior to 1970. Our national reputation was at stake and we were trying to prove our technical prowess to the Soviet Union in space, rather than on the battlefield. Cost was a secondary consideration; landing on the moon was the primary consideration. Thus, engineers can be viewed as problem solvers who assemble the necessary resources to achieve a clearly defined technical objective.

Engineer: Origins of the Word

The root of the word *engineer* derives from *engine* and *ingenious*, both of which come from the Latin root *in generare*, meaning "to create." In early English, the verb *engine* meant "to contrive" or "to create."

The word *engineer* traces to around A.D. 200, when the Christian author Tertullian described a Roman attack on the Carthaginians using a battering ram described by him as an *ingenium*, an ingenious invention. Later, around A.D. 1200, a person responsible for developing such ingenious engines of war (battering rams, floating bridges, assault towers, catapults, etc.) was dubbed an *ingeniator.* In the 1500s, as the meaning of "engines" was broadened, an engineer was a person who made engines. Today, we would classify a builder of engines as a mechanical engineer, because an engineer, in the more general sense, is "a person who applies science, mathematics, and economics to meet the needs of humankind."

1.2 THE ENGINEER AS PROBLEM SOLVER

Engineers are problem solvers. Given the historical roots of the word engineer (see box above), we can expand this to say that engineers are *ingenious* problem solvers.

Fulfilling President Kennedy's promise, the United States landed on the moon in 1969.

In a sense, all humans are engineers. A child playing with building blocks who learns how to construct a taller structure is doing engineering. A secretary who stabilizes a wobbly desk by inserting a piece of cardboard under the short leg has engineered a solution to the problem.

Early in human history, there were no formal schools to teach engineering. Engineering was performed by those who had a gift for manipulating the physical world to achieve a practical goal. Often, it would be learned through apprenticeship with experienced practitioners. This approach resulted in some remarkable accomplishments. Appendix D summarizes some outstanding engineering feats of the past.

Current engineering education emphasizes mathematics, science, and economics, making engineering an "applied science." Historically, this was not true; rather, engineers were largely guided by intuition and experience gained either personally or vicariously. For example, many great buildings, aqueducts, tunnels, mines, and bridges were constructed prior to the early 1700s, when the first scientific foundations were laid for engineering. Engineers often must solve problems without even understanding the underlying theory. Certainly, engineers benefit from scientific theory, but sometimes the solution is required before the theory can catch up to the practice. For example, theorists are still trying to fully explain high-temperature superconductors while engineers are busy forming flexible wires out of these new materials that may be used in future generations of electrical devices.

1.3 THE NEED FOR ENGINEERING

Appendix D describes how humankind's needs have been met by engineering throughout history. As you prepare for a career in engineering, you should be aware of the problems you will face. Here, we look briefly at some of the challenges in our future.

1.3.1 Resource Stewardship and Utilization

The history of engineering can be viewed as "humans versus nature." Humans made progress when they overcame some of nature's terrors by redirecting rivers, paving land, felling trees, and mining the earth. In view of our large population (about 6 billion), we can claim victory.

The Trebuchet: An Engine of War

The trebuchet (pronounced *tray-boo-shay*) pictured below is an ancient "engine" of war. It consists of a long beam that rotates about a fixed fulcrum. One end of the beam has a cup or sling into which the projectile is placed. At the other end is a counterweight that, when released, causes the beam to rotate and throw the projectile into the air.

The trebuchet was invented in China about 2200 years ago and reached the Mediterranean about 1400 years ago. It could throw objects weighing up to 1 ton great distances; in fact, it was used even after the invention of the cannon, because its range was greater than that of early artillery. A modern trebuchet constructed in England could throw a 476-kg car (without engine) 80 meters using a 30,000-kg counterweight. Ancient machines threw stones, dead horses, and even diseased human corpses as a form of biological warfare.

As is often the case, practice preceded theory; trebuchets were constructed and used long before their theory was understood. Many modern concepts, such as force vectors and work (a force exerted over a distance), are thought to have been developed by engineers seeking to improve trebuchet performance. The trebuchet is an example of military necessity causing advances in scientific understanding, a process that is still occurring.

Adapted from: P. E. Chevedden, L. Eigenbrod, V. Foley, and W. Soedel, "The Trebuchet," *Scientific American,* July 1995, pp. 66–71.

The rising wave of environmentalism results from our recognition that a fundamental change is now required. We can no longer be nature's adversary, but must become its caretaker. We have become so powerful, we literally can eliminate whole ecosystems either deliberately (e.g., by felling rain forests) or inadvertently (e.g., by releasing pollution into the water and air). Many scientists are also concerned that human activity may result in changing weather patterns due to the release of "greenhouse gases" such as carbon dioxide, methane, chlorofluorocarbons, and nitrogen oxides. Some chlorine-containing gases are implicated in the destruction of the ozone layer, which protects plants and animals from damaging ultraviolet light.

Although we humans have become extremely powerful, we still depend upon nature to provide the basics of life, such as food and oxygen. These basics do not come easily. NASA has spent millions of dollars to develop regenerative life support systems for use on the moon or Mars that allow people to live independently of earth's life support system. The research continues because the problem is so challenging.

"Sustainable development" is a recent economic philosophy that recognizes humans' right to live and improve their standard of living, while simultaneously protecting the environment. This philosophy attempts to reshape our economy to achieve sustainability. For example, basing our energy sources on fossil fuels is not sustainable. Eventually they will run out, or the pollution resulting from their use will make the planet uninhabitable. Sustainable development would require the use of renewable energy sources such as solar, wind, and biomass fuels, or "infinite" energy sources such as fission (with breeder reactors) or fusion. Resource conserving, recycling, and nonpolluting technologies are also essential to sustainable development.

In modern times, many resources are used once and then thrown away. This "one-pass" approach is increasingly unacceptable, because of the finite nature of our resources and because discarded resources cause pollution. Instead, engineers must develop a cyclical approach in which resources are reused. Some products are now designed to be dismantled when their useful life is completed. They are constructed of metals and polymers that can be reformed into new products.

All processes, including the cyclical processes developed by future engineers, are driven by energy. Because energy production expends resources and causes pollution, it is incumbent upon engineers to develop energy-efficient processes. Many of our current processes use energy inefficiently and can be greatly improved by future engineers.

Unavoidably, all processes produce waste. In the future, many engineers will be required to design processes that minimize wastes, produce wastes that can be converted to useful products, or convert the wastes to forms that can be safely stored.

1.3.2 Global Economy

During World War II, while much of the world economy was destroyed, the U.S. economy remained intact. For a few decades immediately following the war, the U.S. economy was very strong with high export levels. Foreign nations wanted our goods—not because they were of superior quality, but because there were few alternatives. In fact, the quality of many U.S. goods actually deteriorated due to sloppy manufacturing practices, adopted because our industry was not challenged by competition.

Today, the world economy is completely different. The economies of the world have long since recovered from the war. Many nations are capable of producing goods that are equal or superior to the quality of U.S. goods. After the war, a product labeled "Made in Japan" was assumed to be of poor quality; today, this label is an indication that the product is well made and affordable.

In a free market, consumers are able to buy products from all over the world. When they select products made in other countries, it represents a loss of jobs for the United States. American industry is meeting this challenge by instituting "quality" into the corporate culture. A company that is committed to quality must identify their customers, learn their requirements, and transform its manufacturing and management practices to create products that meet the customers' needs and expectations.

Because labor is generally less expensive overseas, many labor-intensive products cannot be economically manufactured in the United States using current technology. However, if engineers develop manufacturing methods that use machines to replace labor, then many of these products can be made in the United States.

Another way for the United States to compete is by developing high-technology products. A major U.S. competitive advantage is our very strong science base. We have a very healthy scientific enterprise in this nation. By translating the latest scientific research into consumer products, we can maintain a competitive edge.

1.4 THE TECHNOLOGY TEAM

Modern technical challenges are seldom met by the lone engineer. Technology development is a complex process involving the coordinated efforts of a technology team consisting of:

A Few Words on Diversity

To fully describe a person, the list of traits might include intellectual ability, personality, creativity, educational level, hobbies, hair color, skin color, body weight, height, age, physical strength, gender, religion, ethnic background, sexual orientation, nationality, language, parental upbringing, and so forth. The list is long, and there are so many variations within each trait that certainly every person is unique.

Because humanity is so diverse, you can be assured that the teammates on your technology team will be different from you. This diversity will be a source of either strength or weakness, depending upon how you respond to it.

Diversity is a source of strength when people with various backgrounds and abilities all work together on the technical problem. The benefits of diversity have long been recognized; hence the expression "two heads are better than one." This simple statement recognizes the fact that a single person may not have all the skills necessary to solve a complex problem, but collectively, the needed skills are there. Also, a diverse team has a useful variety of viewpoints. For example, although traditional automobile design teams have been strictly male, women have recently joined these teams. The female teammates have introduced a new perspective to automobile design, making the cars safer and more appealing to women, who constitute about 50% of the car-buying public.

Diversity is a source of weakness if teammates are so different that they cannot communicate, or they mistrust each other and cannot work together toward a common end. This potential weakness results from two common human tendencies: tribalism and overgeneralization. *Tribalism* refers to the fact that during most of human history, people have lived in tribes composed of similar members. When outsiders entered the tribal land, they were often treated with suspicion because they were potential enemies. *Overgeneralization* refers to the fact that in their attempt to understand the world, humans make generalizations from specific observations—but sometimes the generalizations go too far. For example, if Laura were watching a basketball game, she would observe that the team is composed primarily of tall people. After the game, if Laura were to meet Greg, who happens to be seven feet tall, she might assume that Greg plays basketball, when, in fact, he has no interest in the game. Tribalism and overgeneralization prevent people from dealing with each other as individuals; instead, perceived attributes of a group are automatically assigned to an individual. Not acknowledging the true character of a co-worker makes a working relationship impossible.

To gain strength from diversity and avoid potential pitfalls, it is important that the technology team share a common set of core values that allow it to work together. Some sample core values are shown below:

- Teammates are rewarded on the basis of hard work, not politics.
- Teammates are treated with respect.
- Teammates are treated as unique persons with their own skills, talents, abilities, and perspectives.

Adopting these core values, and others, will allow the team to function in harmony and gain strength from diversity.

- *Scientists,* who study nature in order to advance human knowledge. Although some scientists work in industry on practical problems, others have successful careers publishing results that may not have immediate practical applications. Typical degree requirement: BS, MS, PhD.
- *Engineers,* who apply their knowledge of science, mathematics, and economics to develop useful devices, structures, and processes. Typical degree requirement: BS, MS, PhD.
- *Technologists,* who apply science and mathematics to well-defined problems that generally do not require the depth of knowledge possessed by engineers and scientists. Typical degree requirement: BS.
- *Technicians,* who are generally supervised by engineers and scientists to accomplish specific tasks such as drafting, laboratory procedures, and model building. Typical degree requirement: two-year associate's degree.
- *Artisans,* who have the manual skills (welding, machining, carpentry) to construct devices specified by scientists, engineers, technologists, and technicians. Typical degree requirement: high school diploma plus experience.

Elijah McCoy: Mechanical Engineer and Inventor

Elijah McCoy was born in the early 1840s in Colchester, Ontario, Canada. His parents were former slaves who escaped from Kentucky via the Underground Railroad, a network of individuals who helped slaves reach freedom.

At that time, educational opportunities for blacks were limited, so at age 15, McCoy's parents sent him to study in Scotland, where he achieved the title "master mechanic and engineer." He returned to North America and settled in Detroit, Michigan. During the 1860s, it was difficult for blacks to obtain jobs in the professions, so his first job was a fireman/oilman on the Michigan Central Railroad. As a fireman, he shoveled coal into the firebox. As an oilman, he lubricated the machinery, which had to be stopped for that purpose, causing delays and reducing efficiency. This experience inspired his first patent (U.S. Patent 129,843, issued July 12, 1872), for a device that lubricated machinery while in motion. This lubricating device was so superior to the competition that some engineers would ask if machinery was equipped with *the real McCoy,* a popular American expression meaning *the real thing.* Interestingly, this expression originated in an 1856 advertising slogan *the real MacKay,* used to promote a Scottish brand of whiskey.

During his life, McCoy developed 57 patents. They were issued in the United States, Great Britain, Canada, France, Germany, Austria, and Russia. Among them were an ironing board and a lawn sprinkler.

In 1920, he established the Elijah McCoy Manufacturing Company to manufacture and sell his numerous inventions. He died nine years later in 1929. To honor his achievements as an inventor, he was inducted into the National Inventors Hall of Fame in 2001.

Adapted from the following websites:

www.princeton.edu/~mcbrown/display/mccoy.html
web.mit.edu/www/inventorsI-Q/mccoy.html
www.inventorsmuseum.com/elijahmccoy.htm
www.invent.org/book/book-text/mccoy.htm
www.uselessknowledge.com/word/mccoy.shtml

Successful teamwork results in accomplishments larger than can be produced by individual team members. There is a magic when a team coalesces and each member builds off of the ideas and enthusiasm of teammates. For this magic to occur and to produce output that surpasses individual efforts, several characteristics must be present:

- Mutual respect for the ideas of fellow team members.
- The ability of team members to transmit and receive the ideas of the team.
- The ability to lay aside criticism of an idea during early formulation of solutions to a problem.
- The ability to build on initial or weakly formed ideas.
- The skill to accurately criticize a proposed solution and analyze for both strengths and weaknesses.
- The patience to try again when an idea fails or a solution is incomplete.

1.5 ENGINEERING DISCIPLINES AND RELATED FIELDS

At this point in your engineering career, you may not have selected a major. Does your future lie in mechanical engineering, chemical engineering, electrical engineering, or other engineering fields? Once you have made your selection, you will have decided upon your engineering *discipline.* To help in this decision, we briefly describe the major engineering disciplines and some related fields.

Josephine Garis Cochrane: Inventor of the Dishwasher

In 1839, Josephine Garis was born into an industrious family. Her father, John Garis, was a civil engineer who supervised mills along the Ohio River and drained swamps to develop Chicago during the 1850s. Her great-grandfather, John Fitch, built a steamboat of his own design that served Philadelphia in 1786.

In 1853, at age 19, Josephine Garis married William Cochran, a handsome 27-year-old man who became wealthy in the dry goods business. Josephine was an independent woman—although she took her husband's name, she insisted on ending it with an *e*.

The young socialite couple was popular and had many friends, whom they entertained frequently with elaborate dinner parties using family heirloom china. The servants who washed the china were careless and broke too many plates, so Josephine decided to wash and dry the dishes herself. She soon concluded that this activity wasted her precious time, so she resolved to design a machine that would wash the dishes for her. Within a half hour, she decided that the machine should hold the dishes in a rack and high-pressure water would scrub them clean.

Shortly thereafter, in 1883, Josephine's husband died. Although he was a wealthy man, he had spent more than he earned, leaving her destitute. Nonetheless, with the help of mechanic George Butters, she built the first dishwasher in the shed behind her home, a site now marked with a historical marker. Powered by a hand pump, it cleaned dishes using streams of soapy water. Friends and neighbors came to see the contraption. They were delighted and encouraged her to pursue it further. On December 28, 1886, Mrs. Cochrane received her first patent on the dishwasher.

Because it was expensive, she decided to market her dishwasher to institutions, rather than homes. The Palmer House, a famous Chicago hotel, was her first customer. Her dishwasher could wash and dry 240 dishes within 2 minutes.

Because she had no capital, she hired a contractor to build the units. Relations were strained; the contractor would often ignore her ideas because she had no formal mechanical training and because she was a woman. Further, the contractor took most of the profits, even though she had the brains, patents, entrepreneurial talent, and sales orders. She was not able to raise capital from investors because they refused to invest in a company headed by a woman.

In 1893, nine of her Garis-Cochran dishwashers cleaned dirty dishes at the World's Columbian Exposition, a large fair held in Chicago. Judges awarded her machine the highest prize, stating it had the "best mechanical construction, durability and adaptation to its line of work." The resulting publicity generated more orders. By 1898, Mrs. Cochrane had saved enough money to open her own manufacturing facility and no longer depend on a contractor to build her machines. George Butters became the foreman and oversaw the three employees. Finally, she could work with people who respected her and did not challenge her ideas. Her dishwashers were acclaimed by hotels and other institutions because they saved labor, reduced breakage, and sanitized the dishes. Josephine had succeeded and lived to see her business thrive.

After her death in 1913, the company continued to manufacture dishwashers of her design. In 1926, the company was purchased by Hobart, a manufacturer of well-engineered appliances. Hobart changed the name of the dishwasher subsidiary to KitchenAid, which finally introduced a home dishwasher in the 1940s. Later, KitchenAid was acquired by Whirlpool, a major home appliance manufacturer.

Adapted from: J. M. Fenster, "The Woman Who Invented the Dishwasher," *American Heritage of Invention & Technology,* vol. 15, no. 2, pp. 54–61, Fall 1999.

Figure 1.1 shows when the major engineering disciplines were born. Nearly all disciplines are thought to have evolved from civil engineering. Note that all engineering disciplines require extensive knowledge of physics, whereas chemical and materials engineering require extensive knowledge of physics and chemistry. Some recent disciplines (biochemical and biomedical) require extensive knowledge of physics, chemistry, and biology.

1.5.1 Civil Engineering

Civil engineering is generally considered the oldest engineering discipline—its works trace back to the Egyptian pyramids and before. Many of the skills possessed by civil engineers (e.g., building walls, bridges, roads) are extremely useful in warfare, so these engineers worked on both military and civilian projects. To distinguish those engineers who work on civilian projects from those who work on military projects, the British engineer John Smeaton coined the term *civil engineer* in about 1750.

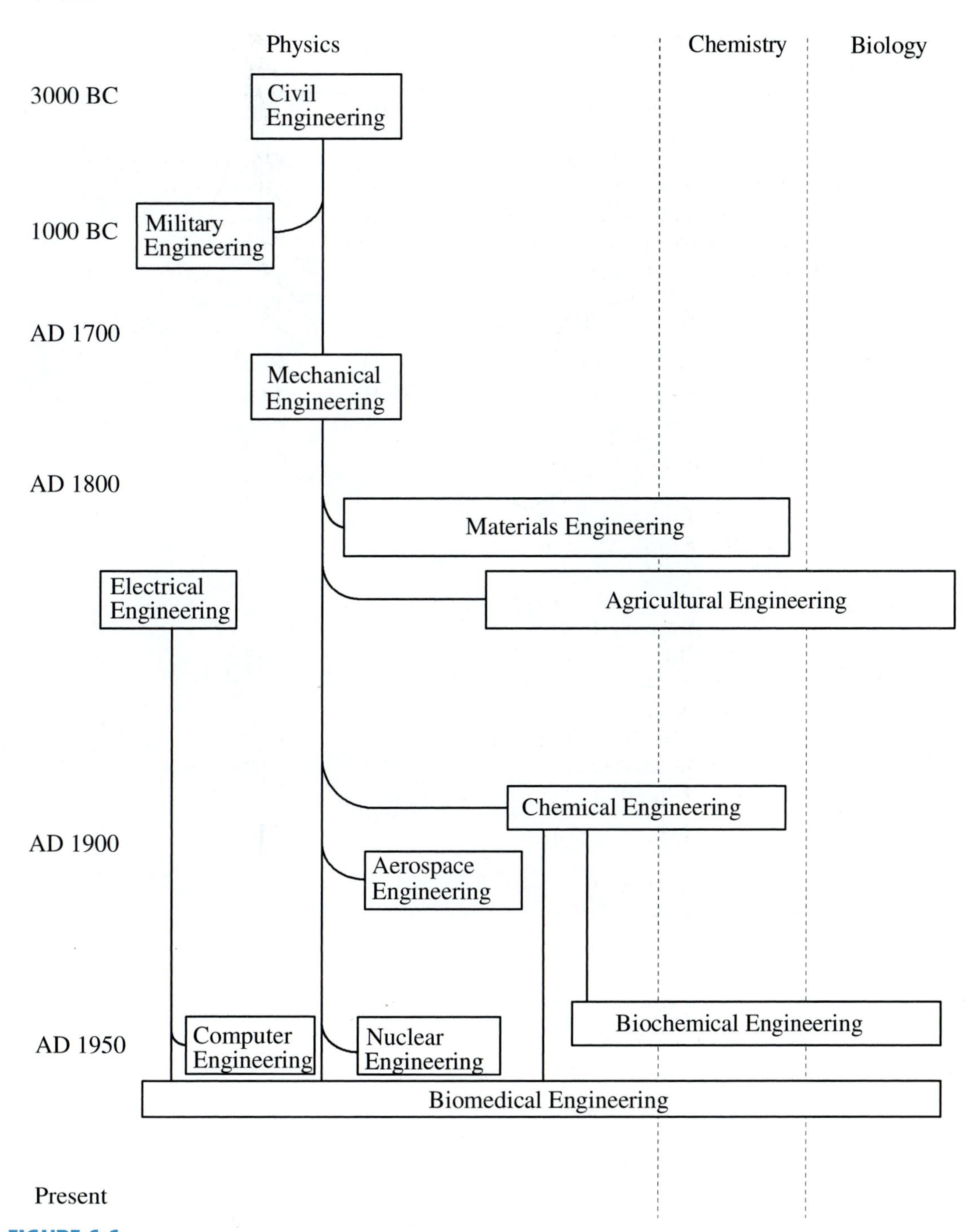

FIGURE 1.1
Birth of engineering disciplines (birth dates are approximate).

Ancient Egypt: From Engineer to God

Egyptian civilization ascended from the Late Stone Age, around 3400 B.C., with vigorous advancements in several engineering fields. While we can still see the spectacular construction feats of the Pyramid Age (3000–2500 B.C.), the ancient Egyptians also pioneered other engineering fields. As hydraulic engineers, they manipulated the Nile River for agricultural and commercial purposes; as chemical engineers, they produced dyes, cement, glass, beer, and wine; as mining engineers, they extracted copper from the Sinai Peninsula for use in the bronze tools that built the pyramids.

One of the key players of this period was Imhotep, known today as "The Father of Stone Masonry Construction." Imhotep served the pharaoh Zoser as chief priest, magician, physician, and head engineer. Most archaeologists credit Imhotep with designing and building the first pyramid, a stepped tomb for Zoser at Sakkara, around 2980 B.C. This pyramid consists of six stages, each 30 feet high, built from local limestone, and hewn with copper chisels. While only 200 feet high (the height of an 18-story building), this unique structure served as a prototype for the Great Pyramid at Giza, constructed 70 years later, which covers four city blocks in area and originally stood 480 feet high.

Imhotep acquired an extensive reputation as a sage, and in later centuries was recognized as the Egyptian god of healing. Although Egyptian civilization saw great engineering progress during the Pyramid Age, 2000 years of stagnation and decline followed.

Hypostyle Hall of Karnak Temple in Luxor, Egypt.

Courtesy of: Seth Adelson, graduate student.

Civil engineers are responsible for constructing large-scale projects such as roads, buildings, airports, dams, bridges, harbors, canals, water systems, and sewage systems.

1.5.2 Mechanical Engineering

Mechanical engineering was practiced concurrently with civil engineering because many of the devices needed to construct great civil engineering projects were mechanical in nature. During the Industrial Revolution (1750–1850), wonderful machines were developed: steam engines, internal combustion engines, mechanical looms, sewing machines, and more. Here we saw the birth of mechanical engineering as a discipline distinct from civil engineering.

Mechanical engineers make engines, vehicles (automobiles, trains, planes), machine tools (lathes, mills), heat exchangers, industrial process equipment, power plants, consumer items (typewriters, pens), and systems for heating, refrigeration, air conditioning, and ventilation. Mechanical engineers must know structures, heat transfer, fluid mechanics, materials, and thermodynamics, among many other things.

1.5.3 Electrical Engineering

Soon after physicists began to understand electricity, the electrical engineering profession was born. Electricity has served two main functions in society: the transmission of power and of information. Those electrical engineers who specialize in power transmission design and build electric generators, transformers, electric motors, and other high-power equipment. Those who specialize in information transmission design and build radios, televisions, computers, antennae, instrumentation, controllers, and communications equipment.

Electronic equipment can be **analog** (meaning the voltages and currents in the device are *continuous* values) or **digital** (meaning only *discrete* voltages and currents can be attained by the device). As analog equipment is more susceptible to noise and interference than digital equipment, many electrical engineers specialize in digital circuits.

Modern life is largely characterized by electronic equipment. Daily, we rely on many electronic devices—televisions, telephones, computers, calculators, and so on. In the future, the number and variety of these devices can only increase. The fact that electrical engineering is the largest engineering discipline—comprising over 25% of all engineers—underscores the importance of electrical engineering in modern society.

1.5.4 Chemical Engineering

By 1880, the chemical industry was becoming important in the U.S. economy. At that time, the chemical industry hired two types of technical persons: mechanical engineers and industrial chemists. The chemical engineer combined these two persons into one. The first chemical engineering degree was offered at the Massachusetts Institute of Technology (MIT) in 1888.

Chemical engineering is characterized by a concept called *unit operations.* A unit operation is an individual piece of process equipment (chemical reactor, heat exchanger, pump, compressor, distillation column). Just as electrical engineers assemble complex circuits from component parts (resistors, capacitors, inductors, batteries), chemical engineers assemble chemical plants by combining unit operations together.

Chemical engineers process raw materials (petroleum, coal, ores, corn, trees) into refined products (gasoline, heating oil, plastics, pharmaceuticals, paper). Biochemical engineering is a growing subdiscipline of chemical engineering. Biochemical engineers combine biological processes with traditional chemical engineering to produce food and pharmaceuticals and to treat wastes.

1.5.5 Industrial Engineering

In the late 1800s, industries began to use "scientific management" techniques to improve efficiency. Early pioneers in this field did time-motion studies on workers to reduce the amount of labor required to produce a product. Today, industrial engineers develop, design, install, and operate integrated systems of people, machinery, and information to produce either goods or services. Industrial engineers bridge engineering and management.

Industrial engineers are famous for designing and operating assembly lines that optimally combine machinery and people. However, they can also optimize train or plane schedules, hospital operations, banks, or overnight package delivery services. Industrial engineers who specialize in human factors design products (e.g., hand tools, airplane cockpits) with the human user in mind.

1.5.6 Aerospace Engineering

Aerospace engineers design vehicles that operate in the atmosphere and in space. It is a diverse and rapidly changing field that includes four major technology areas: aerodynamics, structures and materials, flight and orbital mechanics and control, and propulsion. Aerospace engineers help design and build high-performance flight vehicles (e.g., aircraft, missiles, and spacecraft) as well as automobiles. Also, aerospace engineers confront problems associated with wind effects on buildings, air pollution, and other atmospheric phenomena.

1.5.7 Materials Engineering

Materials engineers are concerned with obtaining the materials required by modern society. Materials engineers may be further classified as:

- *Geological engineers,* who study rocks, soils, and geological formations to find valuable ores and petroleum reserves.
- *Mining engineers,* who extract ores such as coal, iron, and tin.
- *Petroleum engineers,* who find, produce, and transport oil and natural gas.
- *Ceramic engineers,* who produce ceramic (i.e., nonmetallic mineral) products.
- *Plastics engineers,* who produce plastic products.
- *Metallurgical engineers,* who produce metal products from ores or create metal alloys with superior properties.
- *Materials science engineers,* who study the fundamental science behind the properties (e.g., strength, corrosion resistance, conductivity) of materials.

1.5.8 Agricultural Engineering

Agricultural engineers help farmers efficiently produce food and fiber. This discipline was born with the McCormick reaper. Since then, agricultural engineers have developed many other farm implements (tractors, plows, choppers, etc.) to reduce farm labor requirements. Modern agricultural engineers apply knowledge of mechanics, hydrology, computers, electronics, chemistry, and biology to solve agricultural problems. Agricultural engineers may specialize in: food and biochemical engineering; water and environmental quality; machine and energy systems; and food, feed, and fiber processing.

1.5.9 Nuclear Engineering

Nuclear engineers design systems that employ nuclear energy, such as nuclear power plants, nuclear ships (e.g., submarines and aircraft carriers), and nuclear spacecraft.

Some nuclear engineers are involved with nuclear medicine; others are working on the design of fusion reactors that potentially will generate limitless energy with minimal environmental damage.

1.5.10 Architectural Engineering

Architectural engineers combine the engineer's knowledge of structures, materials, and acoustics with the architect's knowledge of building esthetics and functionality.

1.5.11 Biomedical Engineering

Biomedical engineers combine traditional engineering fields (mechanical, electrical, chemical, industrial) with medicine and human physiology. They develop prosthetic devices (e.g., artificial limbs), artificial kidneys, pacemakers, and artificial hearts. Recent developments will enable some deaf people to hear and some blind people to see. Biomedical engineers can work in hospitals as clinical engineers, in medical centers as medical researchers, in medical industries designing clinical devices, in the FDA evaluating medical devices, or as physicians providing health care.

1.5.12 Computer Science and Engineering

Computer science and engineering evolved from electrical engineering. Computer scientists understand both computer software and hardware, but they emphasize software. In contrast, computer engineers understand both computer software and hardware but emphasize hardware. Computer scientists and engineers design and build computers ranging from supercomputers to personal computers, network computers together, write operating system software that regulates computer functions, or write applications software such as word processors and spreadsheets. Given the increasingly important role of computers in modern society, computer science and engineering are rapidly growing professions.

1.5.13 Engineering Technology

Engineering technologists bridge the gap between engineers and technicians. Engineering technologists typically receive a 4-year BS degree and share many courses with their engineering cousins. Their course work evenly emphasizes both theory and hands-on applications, whereas the engineering disciplines described above primarily emphasize theory with less emphasis on hands-on applications. Engineering technologists can acquire specialties such as general electronics, computers, and mechanics. With their skills, engineering technologists perform such functions as designing and building electronic circuits, repairing faulty circuits, maintaining computers, and programming numerically controlled machine shop equipment.

1.5.14 Engineering Technicians

Engineering technicians typically receive a 2-year associate's degree. Their education primarily emphasizes hands-on applications with a minimum of theory. Their work is often directed by engineers. Because they have little theoretical background, their assigned tasks must be well defined, such as drafting, taking laboratory data, analyzing data according to prescribed procedures, and constructing electronic circuits designed by someone else.

1.5.15 Artisans

Artisans often receive no formal schooling beyond high school. Typically, they learn their skills by apprenticing with experienced artisans who show them the "tricks of the trade." Artisans have a variety of manual skills such as machining, welding, carpentry, and equipment operation. Artisans are generally responsible for transforming engineering ideas into reality; therefore, engineers often must work closely with them. Wise engineers highly value the opinions of artisans, because artisans frequently have many years of practical experience.

1.5.16 Engineering Employment Statistics

Table 1.1 shows the number of engineers employed in the United States. Approximately 1.4 percent of all employees are engineers.

TABLE 1.1
Number of engineers and other professions in the United States

Engineers	Men	Women	Total
Electrical and electronic	420,471	46,552	467,023
Civil	235,162	17,646	252,808
Mechanical	176,092	9,780	185,872
Industrial	151,859	24,474	176,333
Aerospace	131,786	11,648	143,434
Chemical	57,163	7,157	64,320
Petroleum	22,908	1,657	24,565
Metallurgical and materials	17,021	2,209	19,230
Nuclear	10,108	693	10,801
Mining	6,063	415	6,478
Agricultural	2,012	136	2,148
Marine and naval architecture	12,776	493	13,269
Other	308,540	33,423	341,963
Total	1,551,961	156,283	1,708,244
Other professionals			
Lawyers	564,332	182,745	747,077
Physicians	465,468	121,247	586,715
Pharmacists	114,949	66,849	181,798
Architects	133,212	23,662	156,874
Dentists	135,588	19,941	155,529
Scientists			
Chemists	102,505	38,750	141,255
Biologists	36,207	25,930	62,137
Physicists	24,238	3,604	27,842
Total employed	62,704,579	52,976,623	119,550,000
Total U.S. population	121,172,379	127,537,494	248,709,873

Adapted from: 1990 Census.

1.6 ENGINEERING FUNCTIONS

Regardless of their discipline, engineers can be classified by the functions they perform:

- *Research engineers* search for new knowledge to solve difficult problems that do not have readily apparent solutions. They require the greatest training, generally an MS or PhD degree.
- *Development engineers* apply existing and new knowledge to develop prototypes of new devices, structures, and processes.
- *Design engineers* apply the results of research and development engineers to produce detailed designs of devices, structures, and processes that will be used by the public.
- *Production engineers* are concerned with specifying production schedules, determining raw materials availability, and optimizing assembly lines to mass produce the devices conceived by design engineers.
- *Testing engineers* perform tests on engineered products to determine their reliability and suitability for particular applications.
- *Construction engineers* build large structures.
- *Operations engineers* run and maintain production facilities such as factories and chemical plants.
- *Sales engineers* have the technical background required to sell technical products.
- *Managing engineers* are needed in industry to coordinate the activities of the technology team.
- *Consulting engineers* are specialists who are called upon by companies to supplement their in-house engineering talent.
- *Teaching engineers* educate other engineers in the fundamentals of each engineering discipline.

To illustrate the roles of engineering disciplines and functions, consider all the steps required to produce a new battery suitable for automotive propulsion. (The probable engineering discipline is in parentheses and the engineering function is in italics.) A *research engineer* (chemical engineer) performs fundamental laboratory studies on new materials that are possible candidates for a rechargeable battery that is lightweight but stores much energy. The *development engineer* (chemical or electrical engineer) reviews the results of the research engineer and selects a few candidates for further development. She constructs some battery prototypes and tests them for such properties as maximum number of recharge cycles, voltage output at various temperatures, effect of discharge rate on battery life, and corrosion. If the development engineer lacks expertise in corrosion, the company would temporarily hire a *consulting engineer* (chemical, mechanical, or materials engineer) to solve a corrosion problem. When the development engineer has finally amassed sufficient information, the *design engineer* (mechanical engineer) designs each battery model that will be produced by the company. He must specify the exact composition and dimension of each component and how each component will be manufactured. A *construction engineer* (civil engineer) erects the building in which the batteries will be manufactured and a *production engineer* (industrial engineer) designs the production line (e.g., machine tools, assembly areas) to mass produce the new battery. *Operations engineers* (mechanical or industrial engineers) operate the production line and ensure that it is properly maintained. Once the production line is operating, *testing engineers* (industrial or electrical engineers) randomly select batteries and test them to ensure that they meet company specifications. *Sales engineers* (electrical or mechanical engineers) meet with automotive companies to explain the advantages of their company's battery and answer technical questions. *Managing engineers* (any discipline) make decisions about financing plant expansions, product pricing, hiring new personnel, and setting company goals. All of these engineers were trained by *teaching engineers* (many disciplines) in college.

In this example, the engineering disciplines that satisfy each function are unique to the project. Other projects would require the coordinated efforts of other engineering disciplines. Also, the disciplines selected for this project are an idealization. A company might not have the ideal mix of engineers required by a project and would expect its existing engineering staff to adapt to the needs of the project. After many years, engineers become cross trained in other disciplines, so it becomes difficult to classify them by the disciplines they studied in college. An engineer who wishes to stay employed must be adaptable, which means being well acquainted with the fundamentals of other engineering disciplines.

1.7 HOW MUCH FORMAL EDUCATION IS RIGHT FOR YOU?

Knowledge is expanding at an exponential rate. It is impossible to fully grasp engineering in a 4-year BS degree. Although you will continue learning on the job, your experience there will tend to be narrowly focused on the needs of the company.

As you proceed through your engineering studies, you should ask yourself, How much more formal education do I need? The answer depends upon your ultimate career objectives. Many of the job functions described above can be performed adequately with a BS degree. However, others—like the research engineer and the development engineer—generally require an MS or a PhD. These individuals are engaged in the early stages of product development. More education is required because they must solve more challenging technical problems.

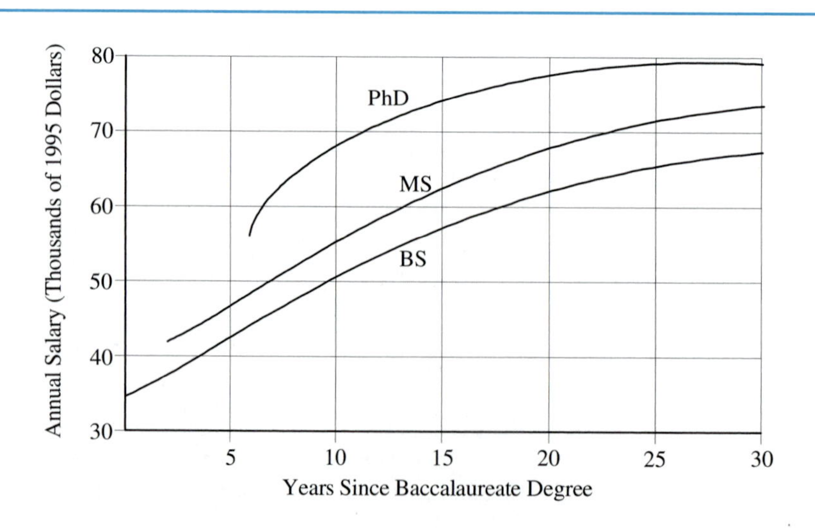

FIGURE 1.2
Median salaries for engineers with different levels of education.
Source: Engineering Workforce Commission of the American Association of Engineering Societies, *Engineers: A Quarterly Bulletin on Careers in the Profession* 1, no. 3 (July 1995).

If you think that you would enjoy the technical challenges met by advanced-degree engineers, do not let the educational costs dissuade you. Most graduate schools provide financial assistance to their students in the form of a stipend. Although the stipend does not equal the pay received in industry, it is usually enough to live a comfortable life. Because people with advanced degrees generally earn higher salaries (Figure 1.2), the short-term financial loss may eventually be recouped. Financial gain should not be your primary motivation for obtaining an advanced degree, however. You should consider it only if you would enjoy a job with greater technical challenges.

Some BS engineering students decide to continue formal education in other fields such as law, medicine, or business. The engineering curriculum provides an excellent background for these other fields because it develops excellent discipline, work habits, and thinking skills.

1.8 THE ENGINEER AS A PROFESSIONAL

Historically, a professional was simply a person who professed to be "duly qualified" in a given area. Often, these professionals professed adherence to the monastic vows of a religious order. So, being a professional meant not only mastering a body of knowledge, but also abiding by proper standards of conduct.

In the modern world, our concept of a professional has become more formalized. We consider a **professional** to have the following traits:

- *Extensive intellectual training*—all professions require many years of schooling, at the undergraduate or post-graduate level.
- *Pass qualifying exam*—professionals must demonstrate that they master a common body of knowledge.
- *Vital skills*—the skills of professionals are vital to the proper functioning of society.
- *Monopoly*—society gives professionals a monopoly to practice in their respective fields.
- *Autonomy*—society entrusts professionals to be self-regulated.
- *Code of ethics*—the behavior of professionals is regulated by self-imposed codes.

Engineering, architecture, medicine, law, dentistry, and pharmacy are examples of professions; they are some of the most prestigious occupations in our society.

1.8.1 Engineering Education

Since 1933, engineering education has been accredited by the Accrediting Board for Engineering and Technology (ABET). The primary purpose of accreditation is to ensure that graduates from engineering programs are adequately prepared to practice engineering. Although schools can offer nonaccredited engineering programs, graduates from these schools may have difficulty finding employment.

When an engineering program is evaluated by ABET, the evaluation team assesses the quality of the students, faculty, facilities, and curriculum. The curriculum must include (1) general education courses, (2) 1 year of college-level mathematics and basic sciences, and (3) $1\frac{1}{2}$ years of engineering science and design. The curriculum must culminate with a major design experience that employs realistic constraints from the following list: economic, environmental, sustainability, manufacturability, ethical, health and safety, social, and political.

The Roman Republic and Empire: Paving the World

Over a period of 800 years, the city-state of Rome grew from a crowded Latin settlement to the nerve center of an empire stretching from present-day Scotland to Israel. To maintain the stability of their vast realm, the Romans implemented many public works, using contemporary technology to supply water, remove sewage, allow transportation, traverse rivers, and provide entertainment.

Scientific achievement under Roman rule was minimal; the Romans were not interested in theory. Applying simple principles with plenty of cheap materials and slave labor gave satisfactory results. For example, early Roman builders used the semicircular arch, an architectural concept developed by the Etruscans (a non-Indo-European people from northern Italy), to construct the magnificent aqueducts that supplied Rome with water. Although many earlier peoples had used concrete, Roman engineers manufactured an improved mixture, yielding a building material as hard and as waterproof as natural rock. With this improved concrete, they built well-planned cities with apartment buildings, or *insulae* ("islands"), that rose five stories high and provided central heating.

The Roman network of roads, beginning with the famed Via Appia in central Italy and then expanding outward into the Empire, was originally intended for military use. To defend its borders and continue its expansion, the Empire required rapid transportation of soldiers over a hard surface with sure footing. Roman engineers built their roads to last; they used simple instruments with plumb bobs to keep the surfaces level, and they often laid down four or five layers, 4 feet thick and 20 feet wide.

Roman aqueduct in Segovia, Spain.

Courtesy of: Seth Adelson, graduate student.

Rather than prescribing a list of courses, ABET allows each engineering department to design its own curriculum that allows students to meet specified goals. During the evaluation of an engineering program, ABET determines if the graduates have the following skills:

a. An ability to apply knowledge of mathematics, science, and engineering.
b. An ability to design and conduct experiments, as well as to analyze and interpret data.
c. An ability to design a system, component, or process to meet desired needs.
d. An ability to function on multidiscipline teams.
e. An ability to identify, formulate, and solve engineering problems.
f. An understanding of professional and ethical responsibility.
g. An ability to communicate effectively.
h. The broad education necessary to understand the impact of engineering solutions in a global and societal context.
i. A recognition of the need for, and an ability to engage in, life-long learning.
j. A knowledge of contemporary issues.
k. An ability to use techniques, skills, and modern engineering tools necessary for engineering practice.

1.8.2 Registered Professional Engineer

Each state has the power to license and register professional engineers. The purpose is to protect the public by ensuring minimum standards through testing, experience, and letters of recommendation. In 1907, the need to license engineers was made evident in Wyoming. Chaos resulted when homesteaders surveyed their own water rights and declared themselves as the surveying engineer. Today, all states have an engineering board that licenses and registers engineers.

An engineer does not need a license to practice engineering, but those who do have licenses have more career opportunities. Many industrial and government positions can only be filled by licensed engineers.

Although each state has its own licensing regulations, the procedure is generally as follows:

1. Obtain a degree from an institution recognized by the state engineering board. This requirement is automatically satisfied if the institution is accredited by ABET.
2. Successfully complete the Fundamentals of Engineering examination. This is an 8-hour exam on discipline specifics, as well as fundamentals in chemistry, mathematics, structures, electronics, economics, and other subjects. The title "Engineer in Training" (EIT) is given to engineering graduates who pass the exam.
3. Work 4 years as an engineer.
4. Obtain letters of recommendation.
5. Successfully complete the Principles and Practice examination, which is another 8-hour exam on the engineer's discipline.

Both exams are prepared by the National Council of Examiners for Engineering and Surveying (NCEES) and are offered throughout the country at about the same time. If you wish to become a registered professional engineer, you should plan to take the Fundamentals exam during your last semester of college, when the knowledge is still fresh in your mind.

China Through the Ages: Walls, Words, and Wells

No discussion of ancient engineering feats is complete without mention of the Great Wall of China. Construction began in the third century B.C., under the rule of the brutal emperor Qin Shi Huang Di (a title meaning "First Divine Autocrat of the Qin Dynasty"). The emperor's goal was to secure China from the murderous Huns of northern Asia. To this end, he forced hundreds of thousands of Chinese peasants, men and women, to leave their homes and fields and join the building effort. Though not completed during Qin's lifetime, the Wall eventually grew to a length of over 2200 miles, including the spurs and branches. If placed in America, it would stretch from New York City to Des Moines, Iowa. Materials and dimensions vary over the entire length, but the Wall is largely constructed of clay bricks, 25 feet thick at the base and rising 30 feet high. Watchtowers are spaced every few hundred yards. The Wall has been rebuilt by various rulers throughout history, even into the 19th century.

Great Wall of China.

Later Chinese engineering accomplishments, though not quite as large, are equally remarkable. In the 1st century A.D., the courtly eunuch Cai Lun concocted paper from tree bark, hemp, rags, and fishnets. Later, development of the printing press in the 9th through the 12th centuries made the Chinese the first publishers as well as the first to circulate printed currency.

Pioneering the chemical industry, engineers of the landlocked Sichuan province in central China collected brine from deep wells for salt production as early as the 11th century A.D. Salt accounted for the bulk of the local economy for over 800 years. Using bamboo cables to drill and bamboo pipes to collect, the wells grew from 100 to 1000 meters deep as technology improved. As early as the 16th century, the Sichuanese learned to store the natural gas that also came from the wells, and used it to fire the brine boilers.

Courtesy of: Seth Adelson, graduate student.

1.8.3 Professional Societies

Most professions have professional societies. The American Medical Association (for physicians) and the American Dental Association (for dentists) serve the interests of those professions. Similarly, we engineers have professional societies that serve our interests. The first engineering professional society was the Institute of Civil Engineers, founded in Britain in 1818. The first American professional society was the American Society of Civil Engineers, founded in 1852. Since then, many other professional societies have been founded (Table 1.2).

TABLE 1.2
Internet addresses of major professional societies

AAES	American Association of Engineering Societies	www.aaes.org
NSPE	National Society of Professional Engineers	www.nspe.org
IEEE	The Institute of Electrical and Electronics Engineers	www.ieee.org
ASCE	American Society of Civil Engineers	www.asce.org
ASME	The American Society of Mechanical Engineers	www.asme.org
AIChE	American Institute of Chemical Engineers	www.aiche.org
IIE	Institute of Industrial Engineers	www.iienet.org
AIAA	American Institute of Aeronautics and Astronautics	www.aiaa.org
ACM	Association for Computing Machinery	www.acm.org
AIME	American Institute of Mining, Metallurgical and Petroleum Engineering	www.aimeny.org
ASAE	American Society of Agricultural Engineers	www.asae.org
ANS	American Nuclear Society	www.ans.org
BMES	Biomedical Engineering Society	www.mecca.org/BME/BMES/society
MAES	Society of Mexican American Engineers and Scientists	www.maes-natl.org
NSBE	National Society of Black Engineers	www.nsbe.org
SWE	Society of Women Engineers	www.swe.org

The primary function of professional societies is to exchange information between members. This is accomplished in such ways as publishing technical journals, holding technical conferences, maintaining technical libraries, teaching continuing education courses, and providing employment statistics (salaries, fringe benefits) so members can assess their compensation. Some professional societies will assist members to find jobs or advise government in technical matters related to their profession.

As a student, you are highly encouraged to get involved with student chapters of professional societies in your discipline. They provide many benefits, such as group meetings that allow you to interact with industry, fellow students, and faculty. If you become a student officer, the leadership experience will be invaluable to your future success. Many student chapters arrange for plant trips so you can learn about the "real world" of engineering. Also, student chapters have social gatherings where you can become better acquainted with your peers.

1.9 THE ENGINEERING DESIGN METHOD

In high school, you probably have been exposed to the **scientific method:**

1. Develop *hypotheses* (possible explanations) of a physical phenomenon.
2. Design an experiment to critically test the hypotheses.
3. Perform the experiment and analyze the results to determine which hypothesis, if any, is consistent with the experimental data.
4. Generalize the experimental results into a law or theory.
5. Publish the results.

Although engineers use knowledge generated by the scientific method, they do not routinely use the method; that is the domain of scientists. The goals of scientists and engineers are different. Scientists are concerned with discovering what *is,* whereas engineers are concerned with designing what *will be.* To achieve our goals, engineers use the **engineering design method,** which is, briefly stated:

1. Identify and define the problem.
2. Assemble a design team.
3. Identify constraints and criteria for success.
4. Search for solutions.
5. Analyze each potential solution.
6. Choose the "best" solution.
7. Document the solution.
8. Communicate the solution to management.
9. Construct the solution.
10. Verify and evaluate the performance of the solution.

This method is described in much greater detail in Chapter 5, "Introduction to Design."

Your engineering education will focus primarily on **analysis.** The hundreds (or thousands) of homework and exam problems you will work during your studies are all designed to sharpen your analytical skills.

In their analysis of physical systems, engineers use **models.** A model represents the real system of interest. Depending upon the quality of the model, it may, or may not, be an accurate representation of reality.

1.9.1 Qualitative Models

A **qualitative model** is a simple relationship that is easily understood. For example, if you were designing a grandfather clock, the *period* of the pendulum—the time it takes to swing back and forth—would be a critical design issue because the pendulum regulates the clock (Figure 1.3). By observing a swinging rock tied to a string, you may notice that longer strings lengthen the period. A simple relationship such as this is very useful to the engineer; however, it is generally insufficient for rigorous analysis. We usually require more quantitative information. To build the clock, we need to know the exact period for a given pendulum length.

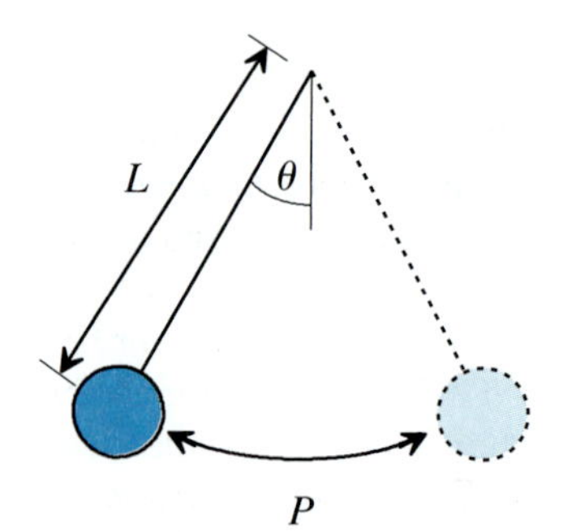

FIGURE 1.3
Pendulum.

1.9.2 Mathematical Models

Because engineering usually needs quantitative values, we transform these qualitative ideas about string length into mathematical formulas. For small displacement angles θ (less than about 15°), physics tells us that the period P of the pendulum (the time it takes to return to its original starting position) can be calculated by the simple formula

$$P = 2\pi\sqrt{\frac{L}{g}} = \frac{2\pi}{\sqrt{g}}\sqrt{L} = k\sqrt{L} \tag{1-1}$$

where L is the pendulum length (measured from the pivot point to the center of the pendulum mass), g is the acceleration due to gravity (9.8 m/s^2), and k is a proportionality constant. This relationship tells us "exactly" how the period changes with length.

Actually, this mathematical relationship is not exact; it applies only for small angles θ. Even at small angles, there is some error in the model. This simple mathematical model neglects factors such as air drag, friction at the pivot point, and the buoyancy of the swinging mass in air. Because air density changes with height above the earth, a complete model would have to account for this effect, even though the mass changes height by only a few centimeters. This simple model assumes that g, the acceleration due to gravity, is a constant. In fact, g decreases at distances farther from the center of the earth. Again, a complete model would have to account for the slight changes in g as the pendulum swings back and forth. A complete model should include the effects of electrical eddy currents generated in the metal pendulum as it swings through the earth's magnetic field. Because light exerts a slight pressure on objects, the complete model would have to account for the effect of light pressure.

You can see from this discussion that a complete model of the pendulum is hopelessly complex. Engineers rarely are able to develop complete mathematical models. However, even incomplete mathematical models may be extremely useful for design purposes, so we use them. A good engineer designs the final product so adjustments can be made to correct for minor effects not considered in the model, or to accommodate slight variations in the manufacturing process. In the case of the grandfather clock, the pendulum could have an adjustment screw that slightly changes its length.

Once a mathematical model of the system has been developed, the complete power of mathematics is at the disposal of the engineer to manipulate the mathematical description of the system. Insofar as the mathematical model is a reasonably accurate description of reality, the mathematical manipulations will also result in equations that approximate reality.

1.9.3 Digital Computer Models

Mathematical models may be programmed and solved using digital computers. In our pendulum example, we could write a computer program that calculates the position of the pendulum as time progresses. At each position, we could calculate the air density, the buoyancy forces, the acceleration due to gravity, the light-pressure forces, the air drag, and the pivot-point friction. The computer model would use all of this information to calculate the next position. All of this information would then be recalculated, allowing the next position to be determined. This may sound like a lot of work. It is. The amount of modeling effort expended depends upon how accurately the period must be known. Perhaps it would be better to use the simpler model and, using an adjustment screw on the pendulum, calibrate it against an electronic clock.

1.9.4 Analog Computer Models

Electronic circuits can be configured to simulate physical systems. Before digital computers became widely available, analog computers were frequently used. Today, they are rarely used because digital computers are more versatile and powerful.

1.9.5 Physical Models

Some systems are extremely complex and require physical models. For example, wind tunnel models of the space shuttle were constructed to determine its flight characteristics. Engineers use a physical model of the Mississippi River to understand the effect of silt deposits and rainfall on its flow rate. Chemical engineers build a pilot plant to test a chemical process before the industrial-scale plant is constructed.

1.10 TRAITS OF A SUCCESSFUL ENGINEER

All of us would like to be successful in our engineering careers, because it brings personal fulfillment and financial reward. (For most engineers, financial reward is not the highest priority. Surveys of practicing engineers show that they value exciting and challenging work performed in a pleasant work environment over monetary compensation.) As a student, you may feel that performing well in your engineering courses will guarantee success in the real engineering world. Unfortunately, there are no guarantees in life. Ultimate success is achieved by mastering many traits, of which academic prowess is but one. By mastering the following traits, you will increase your chances of achieving a successful engineering career:

- *Interpersonal skills.* Engineers are typically employed in industry where success is necessarily a group effort. Successful engineers have good interpersonal skills. Not only must they effectively communicate with other highly educated engineers, but also with artisans, who may have substantially less education, or other professionals who are highly educated in other fields (marketing, finance, psychology, etc.).
- *Communication skills.* Although the engineering curriculum emphasizes science and mathematics, some practicing engineers report that they spend up to 80% of their time in oral and written communications. Engineers generate engineering drawings or sketches to describe a new product, be it a machine part, an electronic circuit, or a crude flowchart of new computer code. They document test results in reports. They write memos, manuals, proposals to bid on jobs, and technical papers for trade journals. They give sales presentations to potential clients and make oral presentations at technical meetings. They communicate with the workers who actually build the devices designed by engineers. They speak at civic groups to educate the public about the impact of their plant on the local economy, or address safety concerns raised by the public.
- *Leadership.* Leadership is one of the most desired skills for success. Good engineering leaders do not follow the herd; rather, they assess the situation and develop a plan to meet the group's objectives. Part of developing good leadership skills is learning how to be a good follower as well.
- *Competence.* Engineers are hired for their knowledge. If their knowledge is faulty, they are of little value to their employer. Performing well in your engineering courses will improve your competence.
- *Logical thinking.* Successful engineers base decisions on reason rather than emotions. Mathematics and science, which are based upon logic and experimentation, provide the foundations of our profession.
- *Quantitative thinking.* Engineering education emphasizes quantitative skills. We transform qualitative ideas into quantitative mathematical models that we use to make informed decisions.
- *Follow-through.* Many engineering projects take years or decades to complete. Engineers have to stay motivated and carry a project through to completion. People who need immediate gratification may be frustrated in many engineering projects.
- *Continuing education.* An undergraduate engineering education is just the beginning of a lifetime of learning. It is impossible for your professors to teach all relevant current knowledge in a 4-year curriculum. Also, over your 40-plus-year career, knowledge will expand dramatically. Unless you stay current, you will quickly become obsolete.

- *Maintaining a professional library.* Throughout your formal education, you will be required to purchase textbooks. Many students sell them after the course is completed. If that book contains useful information related to your career, it is foolish to sell it. Your textbooks should become personalized references with appropriate underlining and notes in the margins that allow you to quickly regain the knowledge years later when you need it. Once you graduate, you should continue purchasing handbooks and specialized books related to your field. Recall that you will be employed for your knowledge, and books are the most ready source of that knowledge.
- *Dependability.* Many industries operate with deadlines. As a student, you also have many deadlines for homework, reports, tests, and so forth. If you hand homework and reports in late, you are developing bad habits that will not serve you well in industry.
- *Honesty.* As much as technical skills are valued in industry, honesty is valued more. An employee who cannot be trusted is of no use to a company.
- *Organization.* Many engineering projects are extremely complex. Think of all the details that had to be coordinated to construct your engineering building. It is composed of thousands of components (beams, ducting, electrical wiring, windows, lights, computer networks, doors, etc.). Because they interact, all those components had to be designed in a coordinated fashion. They had to be ordered from vendors and delivered to the construction site sequentially when they were required. The activities of the contractors had to be coordinated to install each item when it arrived. The engineers had to be organized to construct the building on time and within budget.
- *Common sense.* There are many commonsense aspects of engineering that cannot be taught in the classroom. A lack of common sense can be disastrous. For example, a library was recently built that required pilings to support it on soft ground. (A *piling* is a vertical rod, generally made from concrete, that goes deep into the ground to support the building that rests on it.) The engineers very carefully and meticulously designed the pilings to support the weight of the building, as they had done many times before. Although the pilings were sufficient to hold the building, the engineers neglected the weight of the books in the library. The pilings were insufficient to carry this additional load, so the library is now slowly sinking into the ground.
- *Curiosity.* Engineers must constantly learn and attempt to understand the world. A successful engineer is always asking, Why?
- *Involvement in the community.* Engineers benefit themselves and their community by being involved with clubs and organizations (Kiwanis, Rotary, etc.). These organizations provide useful community services and also serve as networks for business contacts.
- *Creativity.* From their undergraduate studies, it is easy for engineering students to get a false impression that engineering is not creative. Most courses emphasize **analysis,** in which a problem has already been defined and the "correct" answer is being sought. Although analysis is extremely important in engineering, most engineers also employ **synthesis,** the act of creatively combining smaller parts to form a whole. Synthesis is essential to design, which usually starts with a loosely defined problem for which there are many possible solutions. The creative engineering challenge is to find the best solution to satisfy the project goals (low cost, reliability, functionality, etc.). Many of the technical challenges facing society can be met only with creativity, for if the solutions were obvious, the problems would already be solved.

1.11 CREATIVITY

Imagination is more important than knowledge.
Albert Einstein

If the above quotation is correct, you should expect your engineering education to start with Creativity 101. Although many professors do feel that creativity is important in engineering education, creativity *per se* is not taught. Why is this?

- Some professors feel that creativity is a talent students are born with and cannot be taught. Although each of us has different creative abilities—just as we have different abilities to run the 50 yard dash—each of us *is* creative. Often, all the student needs is to be in an environment in which creativity is expected and fostered.
- Other professors feel that because creativity is hard to grade, it should not be taught. Although it is important to evaluate students, not everything a student does must be subjected to grading. The students' education should be placed above the students' evaluation.
- Other professors would argue that we do not completely understand the creative process, so how could we teach it? Although it is true we do not completely understand creativity, we know enough to foster its development.

Rarely is creativity directly addressed in the engineering classroom. Instead, the primary activity of engineering education is the transfer of knowledge to future generations that was painstakingly gained by past generations. (Given the vast amount of knowledge, this is a Herculean task.) Further, engineering education emphasizes the proper manipulation of knowledge to correctly solve problems. Both these activities support analysis, not synthesis. The "analysis muscles" of an engineering student tend to be well developed and toned. In contrast, their "synthesis muscles" tend to be flabby due to lack of use. Both analysis and synthesis are part of the creative process; engineers cannot be productively creative without possessing and manipulating knowledge. But it is important to realize that if you wish to tone your "synthesis muscles," it may require activities outside the engineering classroom.

Table 1.3 lists some creative professions, of which engineering is one. Although the goals of authors, artists, and composers are many, most have the desire to communicate. However, the constraints placed upon their communication are not severe. The author e.e. cummings is well known for not following grammatical conventions. We have all been to

TABLE 1.3
Creative professions

Profession	Goals	Constraints
Author	Communication, exploration of emotions, development of characters	Language
Artist	Communication, creation of beauty, experimentation with different media	Visual form
Composer	Communication, creation of new sounds, exploration of potential of each instrument	Musical form
Engineer	Simplicity, increased reliability, improved efficiency, reduced cost, better performance, smaller size, lighter weight, etc.	Physical laws and economics

art galleries in which a blob passes for art. The musician John Cage composed a musical piece entitled 4′ 33″ in which the audience listens to random ambient noise (e.g., the air handling system, coughs, etc.) for 4 minutes and 33 seconds.

The goals of engineers differ from those of the other creative professions (Table 1.3). To achieve these goals, we are constrained by physical laws and economics. Unlike other creative professions, we are not free to ignore our constraints. What success would an aerospace engineer achieve by ignoring gravity? Because we must work within constraints to achieve our goals, engineers must exhibit tremendous creativity.

Of those engineering goals listed in Table 1.3, one of the most important is simplicity. Generally, a simple design tends to satisfy the other goals as well. The engineer's desire to achieve simplicity is known as the *KISS* principle: "Keep It Simple, Stupid."

Although the creative process is not completely understood, we present here our own ideas about the origins of creativity. People can crudely be classified into *organized thinkers, disorganized thinkers,* and *creative thinkers.* Imagine we tell each of these individuals that "paper manufacture involves removing lignin (the natural binding agent) from wood, to release cellulose fibers that are then formed into paper sheets." Figures 1.4 through 1.6 show how each thinker might store the information.

The organized thinker has a well-compartmentalized mind. Facts are stored in unique places, so they are easily retrieved when needed. The papermaking fact is stored under "organic chemistry," because lignin and cellulose are organic chemicals.

The disorganized thinker has no structure. Although the information may be stored in multiple places, his mind is so disorganized that the information is hard to retrieve when needed. The disorganized thinker who needed to recall information about papermaking would not have a clue where to find it.

The creative thinker is a combination of organized and disorganized thinkers. The creative mind is ordered and structured, but information is stored in multiple places so that when the information is needed, there is a higher probability of finding it. When creative people learn, they attempt to make many connections, so the information is stored in different places and is linked in a variety of ways. In the papermaking example, they might store the information under "organic chemistry" because they are organized, but also under "biochemistry" (because lignin and cellulose are made by living organisms) and under "art prints" (because high-quality prints must be printed on "acid-free" paper, which uses special chemistry to remove the lignin).

When an engineer tries to solve a problem, she works at both the conscious and subconscious level (Figure 1.7). The subconscious seeks information that solves a qualitative model of the problem. As long as it finds no solution, the subconscious mind keeps searching the information data banks. Here, we see the advantage of the creative thinker. With information stored in multiple places and connected in useful ways, there is a greater probability that a solution to the qualitative model will be found. When the subconscious finds a solution, it emerges into consciousness. You have certainly experienced this. Perhaps you went to bed with a problem on your mind, and when you woke up, the solution seemingly "popped" into your head. In actuality, the subconscious worked on the problem while you were sleeping, and the solution emerged into your consciousness when you awoke. For engineers, generally what emerges from the subconscious is a potential solution. The actual solution won't be known until the potential solution is analyzed using a quantitative model. If analysis proves the solution, then the engineer has cause for celebration; she has solved the problem.

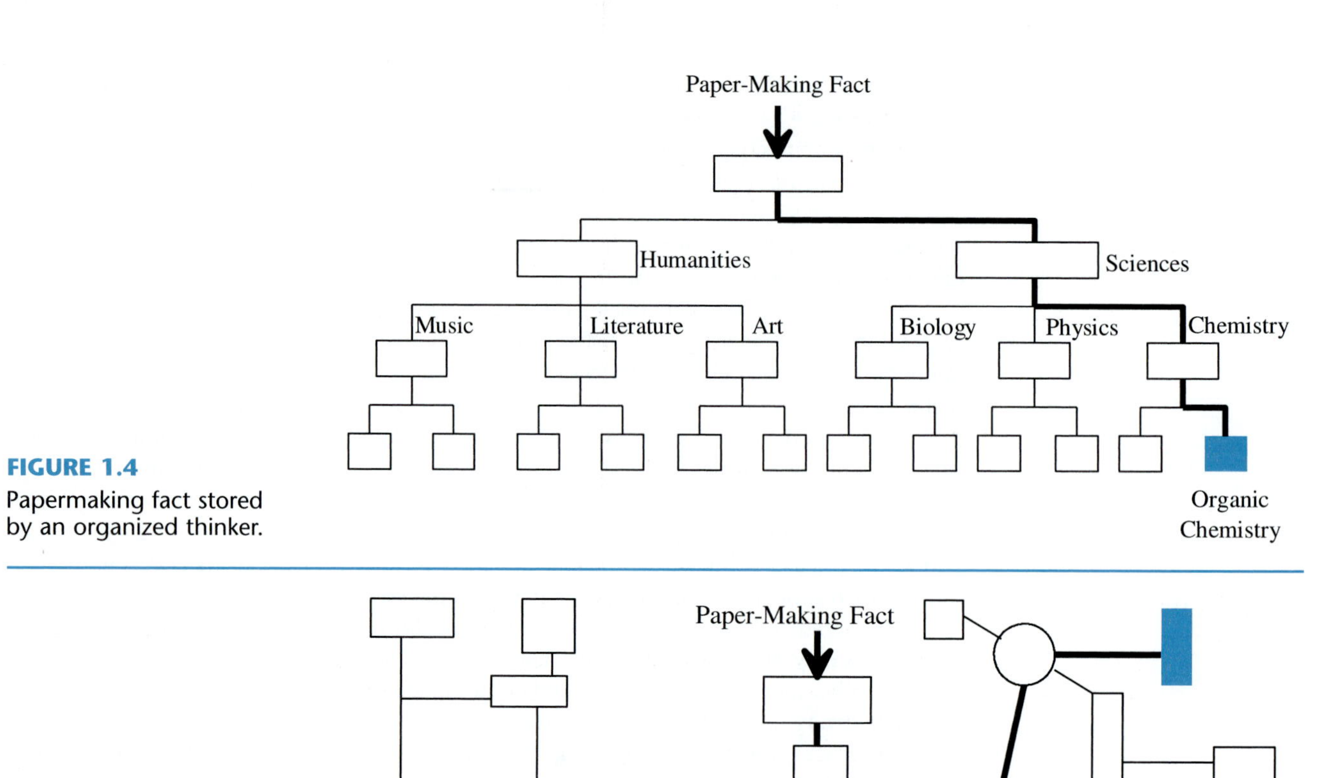

FIGURE 1.4
Papermaking fact stored by an organized thinker.

FIGURE 1.5
Papermaking fact stored by a disorganized thinker.

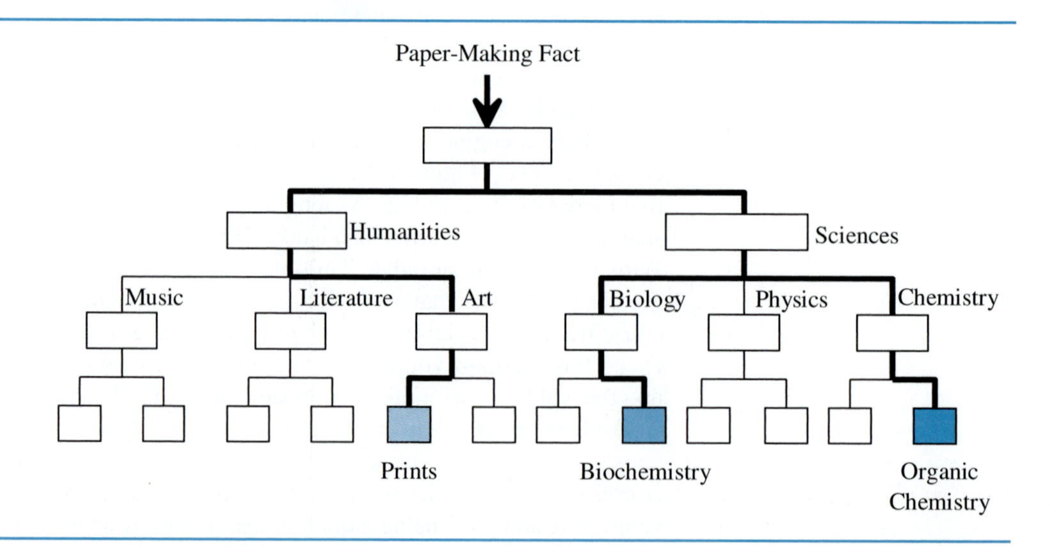

FIGURE 1.6
Papermaking fact stored by a creative thinker.

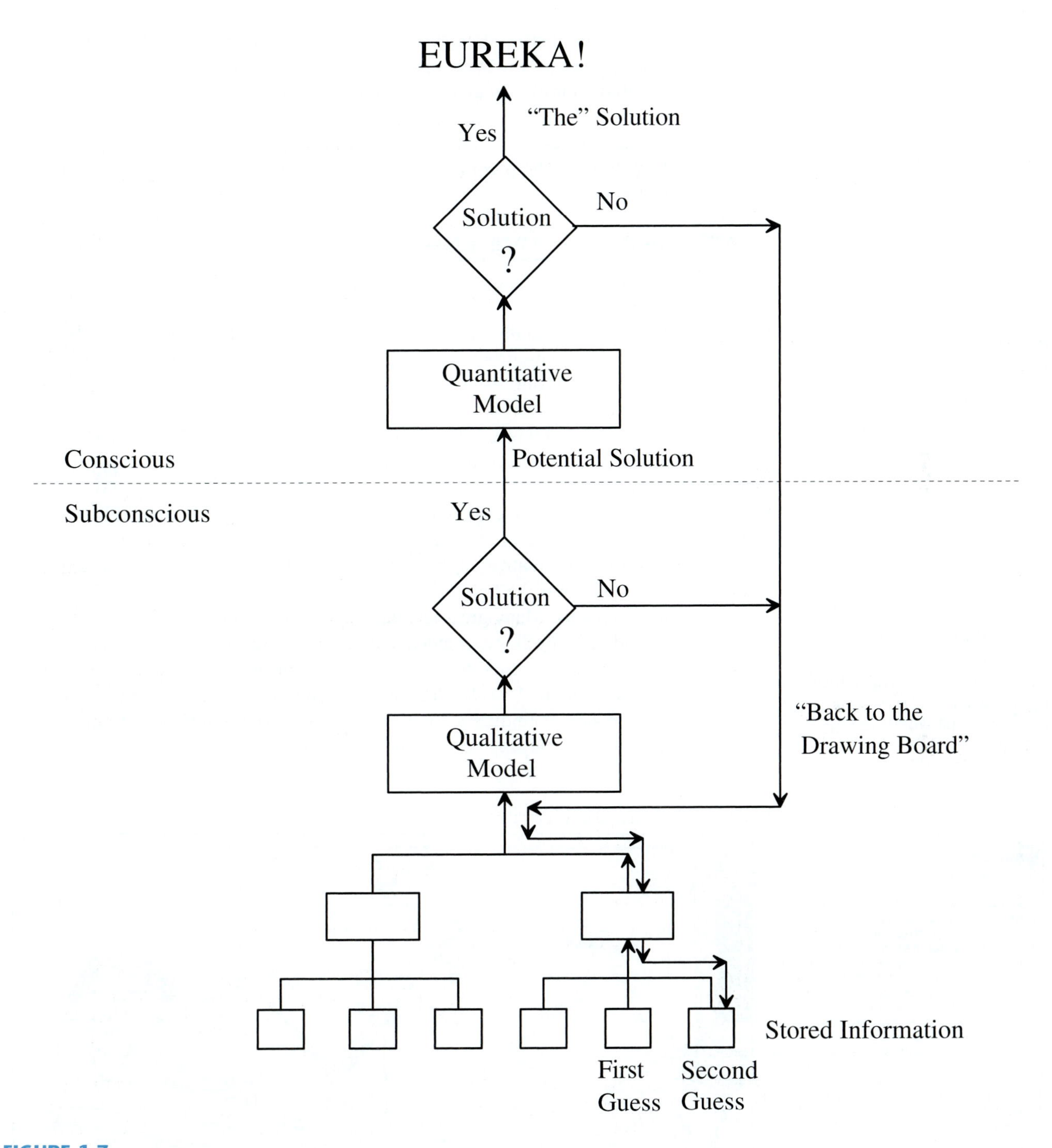

FIGURE 1.7
The problem-solving process.

Most of your engineering education will focus on analysis, the final step in the problem-solving process. However, unless your subconscious is trained, you won't have good potential solutions to analyze. Notice that the subconscious requires a qualitative model. A good engineer develops a "feeling" for numbers and processes and often does not have to feed mathematical formulas to get answers. Developing a feeling for numbers will also help your analysis skills, as it provides an essential check on your calculated answers.

1.12 TRAITS OF A CREATIVE ENGINEER

The following list describes some traits of a creative engineer:

- *Stick-to-it-iveness.* Producing creative solutions to problems requires unbridled commitment. There are always problems along the way. A successful creative engineer does not give up. Thomas Alva Edison said that "Genius is 1 percent inspiration and 99 percent perspiration."
- *Asks why.* A creative engineer is curious about the world and is constantly seeking understanding. By asking why, the creative engineer can learn how other creative engineers solved problems.
- *Is never satisfied.* A creative engineer goes through life asking, How could I do this better? Rather than complaining about a stoplight that stops his car at midnight when there is no other traffic, the creative engineer would say, How could I develop a sensor that detects my car and turns the light green?
- *Learns from accidents.* Many great technical discoveries were made by accident (e.g., Teflon). Instead of being single-minded and narrow, be sensitive to the unexpected.
- *Makes analogies.* Recall that problem solving is an iterative process that largely involves chance (Figure 1.7). By having rich interconnections, a creative engineer increases the chance of finding a solution. We obtain rich interconnections by making analogies during learning so information is stored in multiple places.

The cartoons of Rube Goldberg are well-known for accomplishing a simple task in an overly complex manner, thus violating the KISS principle.

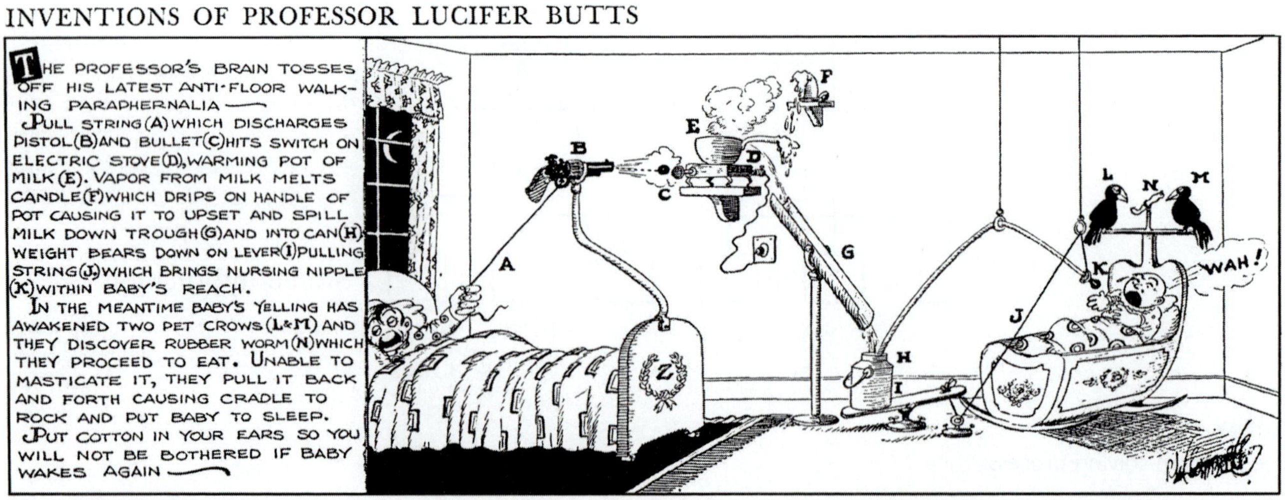

INVENTIONS OF PROFESSOR LUCIFER BUTTS

- *Generalizes.* When a specific fact is learned, a creative engineer seeks to generalize that information to generate rich interconnections.
- *Develops qualitative and quantitative understanding.* As you study engineering, develop not only quantitative analytical skills, but also qualitative understanding. Get a feeling for the numbers and processes, because that is what your subconscious needs for its qualitative model.
- *Has good visualization skills.* Many creative solutions involve three-dimensional visualization. Often, the solution can be obtained by rearranging components, turning them around, or duplicating them.
- *Has good drawing skills.* Drawings or sketches are the fastest way by far to communicate spatial relationships, sizes, order of operations, and many other ideas. By accurately communicating through engineering graphics and sketches, an engineer can pass her ideas easily and concisely to her colleagues, or with a little explanation, to non-engineers.
- *Possesses unbounded thinking.* Very few of us are trained in general engineering. Most of us are trained in an engineering discipline. If we restrict our thinking to a narrowly defined discipline, we will miss many potential solutions. Perhaps *the* solution requires the combined knowledge of mechanical, electrical, and chemical engineering. Although it is unreasonable that we be expert in all engineering disciplines, each of us should develop enough knowledge to hold intelligent conversations with those in other disciplines.
- *Has broad interests.* A creative engineer must be happy. This requires balancing intellectual, emotional, and physical needs. Engineering education emphasizes your intellectual development; you are responsible for developing your emotional and physical skills by socializing with friends, having a stimulating hobby (e.g., music, art, literature), and exercising.
- *Collects obscure information.* Easy problems can be solved with commonly available information. The hard problems often require obscure information.
- *Works with nature, not against it.* Do not enter a problem with preconceived notions about how it must be solved. Nature will often guide you through the solution if you are attentive to its whispers.
- *Keeps an engineering "toolbox."* An engineering "toolbox" is filled with simple qualitative relationships needed by the qualitative model in the subconscious. These simple qualitative relationships may be the distilled wisdom from quantitative engineering analysis. The following sections describe a few "tools." As you progress through your career, you will need a large toolbox to hold all the tools you acquire from your experience.

1.12.1 Cube-Square Law

An example of information an engineer may store in her toolbox is the **cube-square law.** The cube-square law says that as an object gets smaller, its volume decreases much faster than its area. Therefore, the surface-area-to-volume ratio increases with smaller objects.

To illustrate this law, imagine our object is a sphere (Figure 1.8). The surface area A is

$$A = 4\pi r^2 \tag{1-2}$$

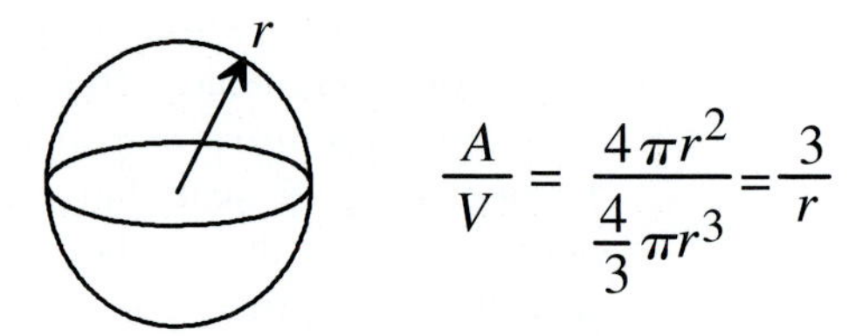

FIGURE 1.8
Illustration of the cube-square law using a sphere.

and the volume V is

$$V = \frac{4}{3}\pi r^3 \tag{1-3}$$

The surface-area-to-volume ratio is

$$\frac{A}{V} = \frac{4\pi r^2}{\frac{4}{3}\pi r^3} = \frac{3}{r} \tag{1-4}$$

This equation says that the area-to-volume ratio increases as the radius decreases.

The cube-square law is one of the most overarching laws in nature, as shown by the following examples:

Example 1.1. Imagine you work in a cannonball factory that casts cannonballs from molten metal and cools them in air. From the cube-square law, you know that smaller cannonballs will cool much faster than large cannonballs, because the rate of heat loss is affected by the surface area but the total amount of heat loss is determined by the cannonball volume.

Example 1.2. Imagine that you must select the most energy-efficient method to fly 500 passengers from New York to Paris. You could charter five 100-passenger planes or one 500-passenger plane. Fuel is primarily required to overcome air drag, which is dictated by the plane surface area. Passenger capacity is determined by the plane volume. To improve fuel economy, you want a small surface area relative to the volume, so one large plane is better than five smaller planes.

Example 1.3. You want to store 50,000 gallons of diesel fuel. You are contemplating purchasing one 50,000 gallon tank or five 10,000 gallon tanks. Tank vendors charge for the metal, not the air in the tank; therefore, tank cost is mostly determined by the surface area. Because a large tank has less surface area per volume, it will be less expensive to purchase one large tank rather than five smaller tanks.

Example 1.4. A whale has limited surface area relative to its volume. Metabolic energy generated within the whale's volume must be eliminated to the surrounding fluid via the whale's surface area. When swimming in water, there is no problem transferring the heat, because water has good heat transfer properties. But when a whale becomes beached on land, it is surrounded by air, which has poor heat transfer properties. The whale literally cooks, because it cannot transfer metabolic energy to the surroundings.

Example 1.5. Normally, coal burns safely at a controlled rate within a furnace. However, in mining operations, fine coal dust is produced, which can burn explosively because there is so much surface area relative to its volume. This presents a significant safety hazard against which numerous precautions are taken.

Example 1.6. The mass of an animal is dictated by its volume, but leg strength is determined by its cross-sectional area. If a deer were scaled up to the size of an elephant, its long slender legs would snap. That is why an elephant must have short, thick legs.

1.12.2 Law of Diminishing Returns

Many engineering systems exhibit the behavior shown in Figure 1.9. With small inputs, the output increases linearly, but with larger inputs, the output levels off. There are "diminishing returns" in the nonlinear region, meaning there is not as much output per unit of input.

Example 1.7. The more time an engineer spends designing a product, the better the product becomes. In this example, time is the input and product quality is the output. In the early stages of the design, product quality improves dramatically with each additional hour spent on the project. However, at some point, an additional hour spent does not improve the design much more. Once this occurs, the engineer is getting diminishing returns on his effort.

Example 1.8. Crop production improves as fertilizer is applied to a farmer's field. In this example, fertilizer is the input and crops are the output. Low fertilizer additions improve crop production dramatically *per unit of fertilizer added.* Although high fertilizer additions do increase crop production, the improvement *per unit of fertilizer added* is low. An agricultural engineer designing a crop production system would realize diminishing returns at high fertilizer additions.

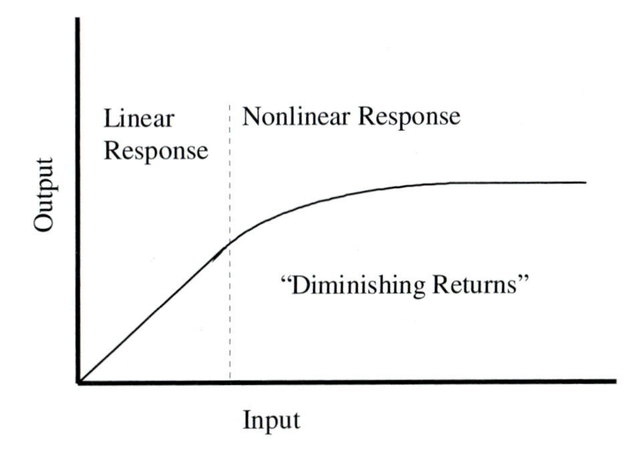

FIGURE 1.9 The **law of diminishing returns.**

1.12.3 Put Material Where Stresses Are Greatest

When designing a structure or product, engineers minimize the amount of materials used. This is an important objective for many reasons:

- The cost of most products is directly related to the amount of material in them: therefore, by reducing the amount of material, the product will be less expensive.
- The size of the support columns on the lower levels of a building are proportional to the mass of the materials used in the upper levels. If the upper levels are heavy, the support columns become so large that there is little usable space available.
- Fuel requirements increase as a vehicle becomes more massive. Reducing materials from an automobile improves fuel economy. Similarly, a very powerful engine is required to accelerate a heavy automobile, whereas a less powerful engine can accelerate a light automobile.
- The cost of lifting objects into earth orbit is about $5000 to $10,000 per pound. Lightening the load reduces launch costs.
- Portable products (e.g., notebook computers) must be lightweight to be accepted by the public.

The weight of a product can be reduced by putting the materials where the stresses are, as illustrated by the following examples:

Example 1.9. Figure 1.10 shows the stresses in a rectangular beam that is loaded in the middle and supported on the two ends. The material on the top edge is in compression, the material on the bottom edge is in tension, and the material on the centerline has no stress. By removing material from the center area (where there is little stress) and adding it to the edges, the beam becomes much stronger. This is the design principle for an I-beam, which is commonly used in large buildings.

Example 1.10. Figure 1.11 shows a dam holding back a lake. As the water depth increases, so does the pressure, which puts more stress at the bottom of the dam. To counteract these greater stresses, the dam is thicker at the bottom.

1.13 SUMMARY

Engineers are individuals who combine knowledge of science, mathematics, and economics to solve technical problems that confront society. As civilization has progressed and become more technological, the engineers' impact on society has increased.

Engineers are part of a technology team that includes scientists, technologists, technicians, and artisans. Historically, various disciplines within engineering have evolved (e.g., civil, mechanical, industrial). Regardless of their discipline, engineers fulfill many functions (research, design, sales, etc.). Because of engineers' importance to society, their education is regulated by ABET and professional licenses are granted by states. There are many engineering professional societies serving a variety of roles, such as providing continuing education courses and publishing technical journals.

In meeting the needs of society, engineers use the engineering design method. An important step in the engineering design method is to formulate models of reality. These models can range from simple qualitative relationships to detailed quantitative codes in digital computers.

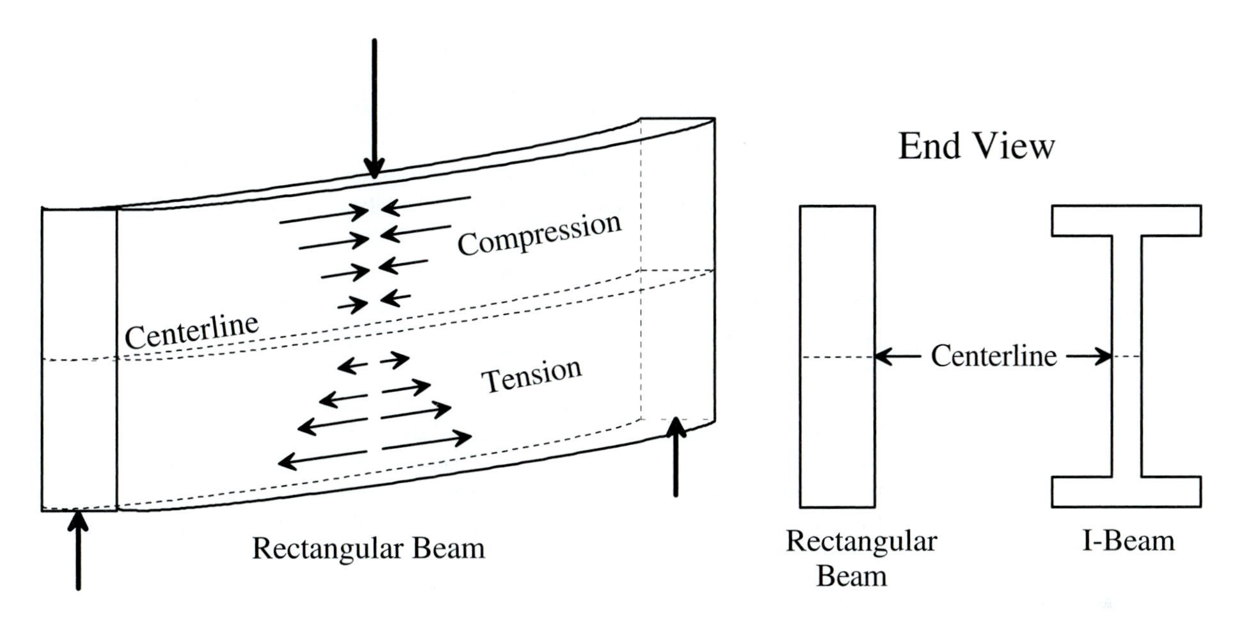

FIGURE 1.10
Loading of a rectangular beam. End view of a rectangular beam and I-beam.

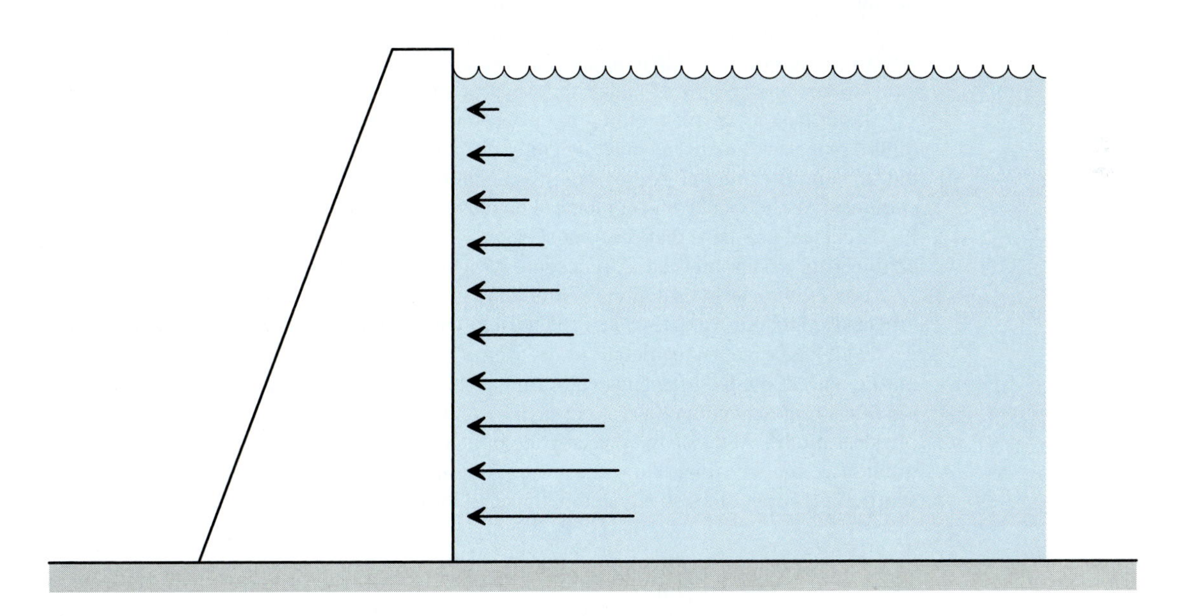

FIGURE 1.11
Stresses on a dam.

To be successful, engineers need to cultivate many traits, such as competence and communication skills. Among the more important skills is creativity, which is needed to solve the more difficult problems faced by society. The creative process involves an interplay between qualitative models that are understood by the subconscious and quantitative models understood by the conscious. These qualitative models may be viewed as tools that engineers keep in their "toolbox" to help guide their creativity in productive directions.

Further Readings

Beakley, G. C., and H. W. Leach. *Engineering: An Introduction to a Creative Profession.* 4th ed. New York: Macmillan, 1983.

Eide, A. R.; R. D. Jenison; L. H. Mashaw; and L. L. Northup. *Engineering Fundamentals and Problem Solving.* 3rd ed. New York: McGraw-Hill, 1997.

Florman, S. C. *The Existential Pleasures of Engineering.* New York: St. Martin's Press, 1976.

Hickman, L. A. *Technology as a Human Affair.* New York: McGraw-Hill, 1990.

Petroski, H. *Beyond Engineering.* New York: St. Martin's Press, 1986.

———. *To Engineer Is Human: The Role of Failure in Successful Design.* New York: Vintage Books, 1992.

Smith, R. J.; B. R. Butler; and W. K. LeBold. *Engineering as a Career.* 4th ed. New York: McGraw-Hill, 1983.

Wright, P. H. *Introduction to Engineering.* New York: Wiley, 1989.

Glossary

analog An electronic device that uses continuous values of voltages and currents.

analog computer model An analog electronic circuit that simulates a physical system.

analysis The process of defining and seeking an answer to a problem.

cube-square law A law of nature that says as an object gets smaller, its volume decreases much faster than its area; therefore, the surface-area-to-volume ratio increases with smaller objects.

digital An electronic device that uses only discrete voltages and currents.

digital computer model A program in a digital computer that simulates a physical system.

engineer An individual who combines knowledge of science, mathematics, and economics to solve technical problems that confront society.

engineering design method A procedure of synthesis, analysis, communication, and implementation used to design solutions to problems.

law of diminishing returns The concept that with small inputs, the output increases linearly, but with larger inputs, the output levels off.

mathematical model Mathematical formulas that describe a physical system.

model A representation of a real physical system.

physical model A physical representation of a complex system.

qualitative model A non-numerical description of a physical system.

synthesis The act of creatively combining smaller parts to form a coherent whole.

Mathematical Prerequisite

- Probability (Appendix G, Mathematics Supplement)

CHAPTER 2

Engineering Ethics

Engineering is a *profession,* similar to law, medicine, dentistry, and pharmacy. A distinguishing feature of all these professions is that their practitioners are highly educated. Engineers are hired by clients (and employers) specifically for their specialized expertise. Generally, the client knows less about the subject than the engineer. Therefore, engineers have ethical obligations to the clients, because the client often cannot assess the quality of the engineer's technical advice. These obligations are part of **engineering ethics,** the set of behavioral standards that all engineers are expected to follow. Engineering ethics are an extension of the ethical standards we all have as human beings.

Engineers have a long tradition of ethical behavior that is widely recognized. Public opinion polls consistently list engineering among the most ethical professions.

2.1 INTERACTION RULES

Engineers rarely work as lone individuals; we generally work in teams. Further, the products of our labor—automobiles, roads, chemical plants, computers—impact society as a whole. Therefore, we need a set of **interaction rules** outlining the expected sets of behavior between the engineer, other individuals, and society as a whole (Figure 2.1). The interaction rules go both ways: the engineer has obligations to society (e.g., to be honest, unbiased, hardworking, careful) and society has obligations to the engineer (e.g., to pay for work performed, to protect intellectual property).

Interaction rules can be classified as etiquette, law, morals, and ethics.

2.1.1 Etiquette

Etiquette consists of codes of behavior and courtesy. It addresses such issues as how many forks to place on the dinner table, proper dress at weddings, seating arrangements, and invitations to parties. Although we generally learn these rules from our everyday experience, they have been codified in various books.

The rules of etiquette are often arbitrary, and they evolve rapidly. For example, in the past, it was common for women to wear white gloves at formal functions; now, this is rarely done. The consequences of violating rules of etiquette are generally not severe. Although a faux pas may cause those "in the know" to snicker, it does not result in jail time. In some cases, etiquette can have important impacts. The peace talks during the

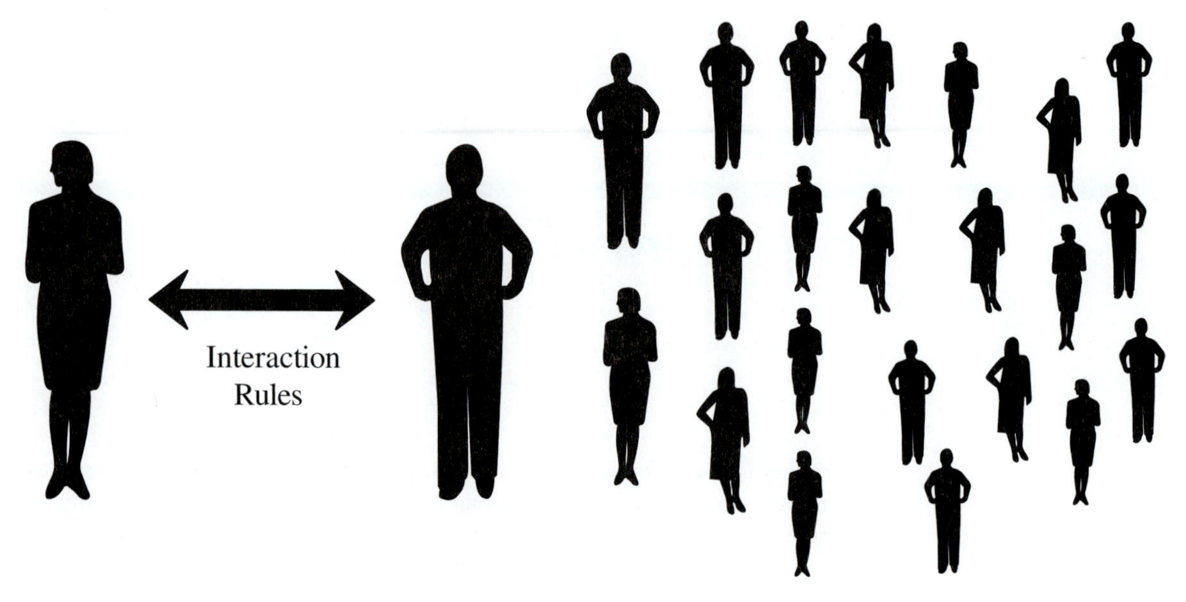

FIGURE 2.1
Interaction rules between other individuals and society.

Vietnam War were mired down in a discussion about the proper shape of the negotiating table; many lives were lost during the delay.

Within the engineering world, proper etiquette is manifested by showing proper respect to employers and clients, not embarrassing colleagues, answering the phone in a professional manner, and so forth.

2.1.2 Law

Law is a system of rules established by authority, society, or custom. Unlike etiquette, violations of law carry penalties such as imprisonment, fines, community service, death, dismemberment, or banishment. Each society has its own consequences for law violations. In Middle Eastern societies, robbers may have a hand amputated, whereas Western society favors imprisonment.

Because severe penalties may result from violating the law, the law must be clearly identifiable so everyone can know the boundaries separating legal and illegal behavior. In some cases, this requirement may lead to seemingly arbitrary laws. For example, it is illegal for a 15-year-old to drive in most states, whereas 16-year-olds can. This is somewhat arbitrary, as a mature 14-year-old may be capable of driving but an immature 18-year-old may not. Because we have no test to quantify the maturity level needed for driving, society relies on age as a nonsubjective (although imperfect) measure of maturity. Similarly, the United States specifies 21 as the responsible drinking age. Legal driving ages and drinking ages vary between countries, another indication that the exact age is somewhat arbitrary.

Legal rights are "just claims" given to all humans within a government's jurisdiction. Most governments grant their citizens rights through a constitution. For example, Amendment VIII of the U.S. Constitution protects citizens from "cruel and unusual punishments." A citizen need not perform acts of kindness to earn this right, nor can it be removed if he engages in crime.

2.1.3 Morals

Morals are accepted standards of right and wrong that are usually applied to personal behavior. We derive moral standards from our parents, religious background, friends, and the media (television, movies, books, and music). Many moral codes are recorded in religious writings. Despite the wide variety of cultures and religions in the world, there is agreement on many moral standards. Most cultures consider murder and stealing to be immoral behavior. From the view of cultural evolution, we can say there is a strong selective pressure against these behaviors. Societies that did not develop moral codes against these behaviors degenerated into anarchy and disappeared.

For some behaviors, there is not universal agreement as to whether they are immoral. Activities such as gambling, dancing, and consumption of alcohol, meat, coffee, and cigarettes are considered immoral in some cultures and religions, but not others. From the view of cultural evolution, we can say there is not a strong selective pressure against these activities. Societies can maintain viability while tolerating these activities, although some people would argue that a society that bans them is stronger.

Moral rights are "just claims" that belong to all humans, regardless of whether these rights are recognized by government. Civilization recognizes that simply being a human endows us with rights; we need not do anything to earn these rights. For example, most of the civilized world believes that because prisoners are human beings, they should not be tortured, regardless of the cruelty of their crime. This moral right has been codified as a legal right in the U.S. Constitution (Amendment VIII).

Moral rights can be a source of controversy. For example, do we all have a right to health care? If so, how much? Some people believe that moral rights should be extended to nonhuman species. If this concept is widely adopted, it will limit some engineering activities. For example, the construction of dams and drainage of swamps can dramatically impact the survival of animal and plant species.

2.1.4 Ethics

Ethics consists of general and abstract concepts of right and wrong behavior culled from philosophy, theology, and professional societies. Because professions draw their members from many cultures and religions, their ethical standards must be secular. Most professional societies have a formal code of ethics to guide their members. Appendix B shows the National Society of Professional Engineers Code of Ethics.

2.1.5 Comparison of Interaction Rules

From the above discussion, you can see we have a complex web of interaction rules governing our behavior. In some cases, all the interaction rules agree. For example, murder is illegal, immoral, unethical, a violation of human rights, and certainly bad etiquette.

In general, lawmakers try to formulate laws that are consistent with morality. However, there can be conflicts between the law and morality for the following reasons:

- The legal system has not considered the situation.

Example 2.1. A chemical company develops a new process that has a waste by-product. Its own internal studies show this by-product to be extremely carcinogenic; however, it is not on the government list of banned chemicals because it is so new. If this company releases large amounts of this carcinogen into the environment, it would not violate the law. However, we would all agree this is immoral behavior.

- Encoding some moral standards into law would be unenforceable.

Example 2.2. Some moral codes ban alcohol. During Prohibition, this ban was enacted into U.S. law. However, it was not enforceable and it created more problems than it solved.

- Laws must be impartial and treat everyone the same.

Example 2.3. Government self-regulations require that all purchases be made through purchasing agents. A government engineer wishes to purchase a used alternator at the local junkyard to perform a quick experiment. Obtaining this part through the purchasing agent is unworkable—it is not readily identifiable by a catalog number, and it takes too long to process the order. The engineer decides to purchase the used alternator with his own funds and reimburse himself with computer disks of equivalent value.

This is not immoral behavior because no theft was involved; however, it is illegal. The government cannot make a regulation stating that "honest employees are entitled to reimburse themselves with computer disks; dishonest employees are required to use the purchasing agent." The law must be impartial and treat every government employee as though he or she is dishonest.

- Laws must govern observable behavior.

Example 2.4. In some moral codes, to think a bad thought is equivalent to having performed it. How could a law be written that bans bad thoughts?

- Laws may be enacted by immoral regimes.

Example 2.5. Nazi law forbade hiding Jews during World War II, but many considered it their moral obligation to break the law and hide innocent Jews.

Morality may also conflict with legal rights. U.S. citizens have the legal right to free speech. Therefore, we have the legal right to tell racist jokes; however, many people consider this to be immoral. Similarly, although the legal rights of pornography publishers have been protected by freedom of the press, many consider this to be immoral activity.

2.2 SETTLING CONFLICTS

A major purpose of interaction rules is to avoid conflicts between members of society. For example, a law tells us on which side of the road to drive. Without it, there would be many lethal conflicts.

Inevitably, human interactions result in conflicts. To settle a conflict, it is necessary to discern its source, which may result from moral issues, conceptual issues, applications issues, and factual issues.

2.2.1 Moral Issues

A **moral issue** is involved if the issue can be resolved only by making a moral decision. For example, when automobiles were first introduced onto roads, a moral decision had to be made whether limits should be placed upon their speeds. One side of the issue would argue that motorists should go as fast as they wish either for their own pleasure or to save time for their business. The other side of the issue would argue that excessive speeds place other motorists and pedestrians at risk. Clearly, moral considerations favor speed limits, for peoples' lives are more valued than the pleasures or business interests of speed-seeking motorists.

2.2.2 Conceptual Issues

A **conceptual issue** arises when the morality of an action is agreed upon, but there is uncertainty about how it should be codified into a clearly defined law, rule, or policy. For example, society has agreed that innocent motorists should be protected from other motorists who go too fast. The conceptual issue is: What speed is too fast? To resolve the conceptual issue, *too fast* may be defined as "highway speeds that exceed 55 mph under favorable driving conditions or speeds likely to result in an accident under adverse driving conditions such as fog, snow, ice, or rain."

2.2.3 Application Issues

An **application issue** results when it is unclear if a particular act violates a law, rule, or policy. Suppose a motorist is going 50 mph on a road posted for 55 mph. During a light rain, the motorist skids off the road and has an accident. The police officer who arrives at the accident scene must decide if she should cite the motorist for excessive speed; in other words, the application issue is whether the light rain qualifies as an adverse driving condition.

2.2.4 Factual Issues

A **factual issue** arises when there is uncertainty about morally relevant facts. It can usually be resolved by acquiring more information. If a motorist is stopped by a police officer for going 60 mph in a 55-mph zone and the motorist claims she was going 55 mph, the conflict can be resolved by getting more data. For example, the motorist might be able to show that the police radar gun is out of calibration by 5 mph.

Notice that the preceding issues were arranged from the most abstract to the most concrete. Factual issues are more clearly defined and can generally be resolved regardless of upbringing and cultural background. In contrast, moral issues are often hard to define and their resolution may depend upon upbringing and cultural background. As a result, moral issues can be difficult to resolve. They are addressed by applying moral theories to help the decision-making process.

2.3 MORAL THEORIES

As much as we would like to have a "moral algorithm" that always leads us to the correct answer, such an algorithm does not exist. If it did, we could program computers to make moral and ethical decisions for us. Instead, we have **moral theories** that provide a framework for making moral and ethical decisions. Sometimes these different theories lead to different answers, but often they lead to the same answer.

To illustrate moral theories, consider this example: A civil engineer works for the city as a building inspector. As a large building is being erected, he is responsible for ensuring it is built according to the city code. (The code protects the public by specifying proper construction materials and methods suitable for a particular city. For example, buildings in San Francisco are constructed according to a code that allows them to withstand earthquakes.) The building inspector is offered a $10,000 bribe to overlook some shoddy construction that would cost the contractor $50,000 to correct. Should the engineer accept the bribe?

It does not take a great moral theorist to determine that the answer is no. Each moral theory arrives at this answer through slightly different paths.

2.3.1 Ethical Egoism

Ethical egoism is a moral theory stating that an act is moral provided you act in your enlightened self-interest. For example, if a mugger were to attack you with a knife and you killed him in self-defense, this would not be immoral.

Societies that structure themselves to harness our natural desire to act in self-interest are more successful. (The recent collapse of communism attests to this.) However, ethical egoism is not a license for selfish behavior. In the long term, selfish behavior is not rewarded; selfish people have few friends and are not likely to be promoted in a company.

In our building-inspector example, if he were to take the bribe, there is always a chance that he would be caught. Imprisonment and the loss of his job are certainly not worth $10,000. Therefore, he can argue it is in his self-interest not to take the bribe.

Not all ethical and moral issues involve a single individual. In cases where many people or a society are involved, we must consider the broader moral theories of utilitarianism and rights analysis.

2.3.2 Utilitarianism

According to **utilitarianism,** moral activities are those that create the most good for the most people. This moral theory attempts to optimize the **happiness objective function,** such that

$$\text{Happiness objective function} = \sum_i (\text{benefit})_i(\text{importance})_i - \sum_j (\text{harm})_j(\text{importance})_j$$

Those actions that increase benefits and reduce harm are considered best.

To perform utilitarian analysis:

1. Determine the target audience (e.g., an individual, a company, or a society).
2. For each action, determine the harms, benefits, and importance to the target audience.
3. Evaluate the happiness objective function for each action.
4. Select the action that maximizes the happiness objective function.

In our building-inspector example, if he were to apply this moral theory, he would count the $10,000 bribe as a benefit, the $50,000 savings to the contractor as a benefit, but the deaths from the building collapse as a harm. The harm so overwhelms the benefits that the correct action is clear.

Utilitarianism is very logical and appealing to engineers. Nonetheless, it has some problems: (1) it implies we have enough knowledge to evaluate the happiness objective function, (2) value judgments are required to assess the importance of each harm/benefit, and (3) it may lead to injustice for individuals.

EXAMPLE 2.6

Problem Statement: Boonville is a small agricultural community located along a major highway. Because farming is not very profitable, there is a lot of poverty. The schools are poor; many children drop out and never amount to much. There is no hospital in Boonville, so health care is poor.

A major trucking company decides it would like to locate a distribution center in Boonville because it is strategically located on the highway. The town council must decide whether to allow the company to locate there. During the public debate, the yea-sayers emphasize that the company will bring many jobs to the community. The resulting tax base will allow them to improve the school system and build a hospital. The nay-sayers observe that the increased truck traffic will probably cause an additional highway death every 5 years. They also point out that there will be a lot of noise and pollution.

Calculate the happiness objective function to determine an ethical course of action.

Solution: The happiness objective function is

$$\begin{aligned}\text{Happiness objective function} = {} & (\text{good schools})(\text{importance}) \\ & + (\text{hospital})(\text{importance}) \\ & + (\text{employment})(\text{importance}) \\ & - (\text{death})(\text{importance}) \\ & - (\text{noise})(\text{importance}) \\ & - (\text{pollution})(\text{importance})\end{aligned}$$

When applied to the community, the equation is

$$\begin{aligned}\text{Happiness objective function} &= (8)(10) + (10)(7) + (5)(20) - (50)(2) \\ &\quad - (3)(1) - (7)(2) \\ &= +133 \Rightarrow \text{Do it!}\end{aligned}$$

When applied to the family whose child was killed by the truck, the equation is

$$\begin{aligned}\text{Happiness objective function} &= (8)(10) + (10)(7) + (5)(20) \\ &\quad - (50)(1{,}000{,}000) - (3)(1) - (7)(2) \\ &= -49{,}999{,}767 \Rightarrow \text{Don't do it!}\end{aligned}$$

Strictly speaking, utilitarian analysis is applied only to the entire community affected by the decision, and not subsets of the community. However, here we applied this analysis to a subset of the community (the family) to show that utilitarianism can benefit society as a whole at the expense of individuals. Rights analysis, explained next, attempts to correct this failing of utilitarianism.

2.3.3 Rights Analysis

According to **rights analysis,** moral actions are those that equally respect each human being. This is often summarized in the *Golden Rule:* Do unto others as you would have them do unto you. Many cultures use the Golden Rule; however, it does not work in every case. If it were strictly followed, a manager (who himself would not want to be laid off) could not lay off workers even if it were required for the health of the company.

As another example of Golden Rule failure, consider an Italian foreman who likes to tell Polish jokes. His Polish subordinates are offended and complain to him. He counters that he doesn't mind Italian jokes, and proceeds to tell one.

To solve this problem, we could formulate the *Revised Golden Rule:* Do unto others as *they* would have done unto *them.* The Revised Golden Rule would ask the foreman to put himself in the shoes of his subordinates. Feeling the pain his subordinates experience from the jokes, he would stop his offensive behavior even though he personally is not offended by the jokes. Even the Revised Golden Rule cannot be applied universally. If it were, a judge would be unable to sentence criminals to jail, because the criminals certainly do not want to go to jail.

Because not all rights are equally important, a hierarchy has been established. They are listed below from most important to least important:

1. Right to life, physical integrity, and mental health.
2. Right to maintain one's level of purposeful fulfillment (e.g., right not to be deceived, cheated, robbed, or defamed).
3. Right to increase one's level of purposeful fulfillment (e.g., right to self-respect, to nondiscrimination, and to acquire property).

To perform rights analysis:

1. Determine the target audience.
2. Evaluate the seriousness of the rights infringement according to the above list.
3. Choose the course of action that imposes the least serious rights infringement.

In our example of the building inspector who was offered a bribe, he would know the correct action to take via rights analysis. His accepting the $10,000 bribe may lead to a more fulfilling life for him, but this is subordinate to the rights of those persons who may be killed if the building collapses.

Example 2.7. Suppose a kidnapper has taken a person hostage and threatens to kill him. The police become involved and lay a trap that involves deception. Here we have a conflict between the rights of the hostage not to be killed and the rights of the kidnapper not to be lied to. Clearly, the rights of the hostage take precedence over the rights of the kidnapper.

2.3.4 Making Moral Decisions When Moral Theories Diverge

In our building-inspector example, we determined that he should not take the bribe regardless of the moral theory that we applied. This is an example of **convergence.** However, this is not always the case. Sometimes, the moral theories do not agree—they **diverge.**

When applied to society, utilitarianism represents one extreme: Do the most good for society regardless of the consequences to the individual. Rights analysis represents the other extreme, in which individual rights are protected regardless of the impact on society. Society must determine how it will strike a balance between these two extremes.

To illustrate how moral theories can diverge, consider highway construction. Engineers decide on the most efficient route between two points that reduces construction costs and allows motorists to efficiently travel between population centers. Often, the most efficient route goes through some homes. The government will condemn those homes under **"eminent domain"** and reimburse the home owners according to fair market price. The rights of the individual home owners are violated. Perhaps they have strong emotional ties to their home and do not want to sell. However, society benefits by constructing the road, because people can move more quickly between population centers, trucking costs are reduced, and fuel is saved. In this case, society has chosen the utilitarian approach.

As another example of diverging moral theories, consider a situation where a sickly brother has a rare disease that will certainly be fatal if he does not receive a kidney transplant. His healthy brother has a closely matching tissue type, so the transplant would be successful. All other relatives do not have closely matching tissue types, so their transplanted kidneys would fail. The healthy brother never liked his sickly brother and refuses to give him the kidney. The utilitarian approach would forcibly demand that the healthy brother donate a kidney to his sickly brother, as total happiness is greater with this option. The sickly brother would be helped much more than the healthy brother would be harmed. In contrast, rights analysis would honor the right of the healthy brother not to be dismembered. In this case, society has chosen the rights analysis approach.

Although there are no algorithms to tell us exactly what to do, a reasonable approach to making moral decisions when moral theories diverge is to use utilitarianism unless an individual's rights are seriously violated.

2.4 THE ETHICAL ENGINEER

Most of the professional societies have prepared ethical codes for their members. The purpose of these codes is to provide guidance to engineers on ethical behavior. A distillation of these codes provides the following guidelines:

1. Protect the public safety, health, and welfare.
2. Perform duties only in areas of competence.
3. Be truthful and objective.
4. Behave in an honorable and dignified manner.
5. Continue learning to sharpen technical skills.
6. Provide honest hard work to employers or clients.
7. Inform the proper authorities of harmful, dangerous, or illegal activities.
8. Be involved with civic and community affairs.
9. Protect the environment. [Only in a few codes.]
10. Do not accept bribes, or gifts that would interfere with engineering judgment.
11. Protect confidential information of employer or client.
12. Avoid conflicts of interest.

A **conflict of interest** is a situation in which an engineer's loyalties and obligations may be compromised because of self-interest or other loyalties and obligations. This can lead to biased judgments. Suppose an engineer were responsible for selecting bearings for an engine her employer is constructing. It so happens that her father owns a bearing company that the engineer will inherit when her father dies. This situation makes it very difficult for the engineer to make an unbiased selection of bearings because she has an obvious conflict of interest. She and her family would benefit by selecting her father's bearings. Even if her father's bearings are the best ones for the job, the selection of these bearings gives the *appearance* of impropriety. Therefore, the situation must be avoided. The engineer should inform her boss that she has a conflict of interest and that another engineer should make the bearing selection.

Informing authorities of harmful, dangerous, or illegal activities is often called **whistle blowing.** An engineer who is involved with an organization that is doing these activities has a conflict of interest. He has obligations to protect society, but he also has obligations to his fellow workers and employers. Clearly, the need to protect the public is paramount. However, when performing his public duty, the whistle blower must be prepared to pay the consequences. He may lose his job or be given a flunky job. He may find it difficult to find new employment because potential employers may be unwilling to hire a whistle blower. If the engineer has a family, the effects of lost income could be devastating. If he keeps his job, he is likely to be ostracized by his coworkers. Before whistle blowing, the engineer should try every method possible to persuade the wrongdoers to correct their ways. It takes a lot of strength and courage to do the right thing. Although the consequences of being a whistle blower may be severe, the consequences of knowing you are unable to do the right thing can also be severe. It is best to avoid this situation as much as possible and work with an honorable company.

No code of ethics can cover every possible ethical situation. Perhaps the simplest guideline is to imagine that you have been selected by the *New York Times* as Engineer of the Year. A reporter follows you around and records all your activities, which are then published daily. Because most of us have an inherent sense of right and wrong, this should lead us to correct ethical behavior.

2.5 RESOURCE ALLOCATION

One of the greatest challenges to society is the proper allocation of resources. When resources are misallocated, it costs lives.

Prisoner's Dilemma

Moral codes exist to enable members of society to cooperate with each other. If each member of society follows the code, behavior is predictable and both the individual and society benefit. However, for his personal benefit, an individual can defect from cooperative behavior and take advantage of a code-follower's predictable behavior.

According to ethical egoism, actions are correct that result in self-interest. Thus, should a person seeking maximum self-interest cooperate or defect?

The answer to this question can be determined by playing *games.* A classic game is called the "Prisoner's Dilemma," which goes as follows: The police have just captured two suspects at the scene of a crime. Both suspects are guilty, but the police do not know that. The only way the police can determine guilt is to separate the prisoners and separately interrogate them. While sitting in their separate cells, each prisoner contemplates his strategy. Should he confess or blame the other prisoner for the crime?

If a prisoner confesses, this is cooperative behavior. If both confess, then the guilt can be shared between the two of them and they can plead for a lesser sentence. If a prisoner blames the other for the crime, this defects from cooperative behavior because he is attempting to be set free while having the other prisoner take a harsher sentence.

The situation is not so simple. If both confess, they will share a lesser sentence. If one confesses and the other blames, then the confessor will get a particularly harsh sentence while the blamer is released. However, if each blames the other, the police will know that both are guilty and each will receive a harsh sentence for his crime; their pleadings for a lesser sentence will go unheeded because they both lied to the police. The following *truth table* shows the benefits to each prisoner depending on which actions are taken.

Action			Benefit	
Me	Him		Me	Him
Coop	Coop		3	3
Coop	Defect	⇒	0	5
Defect	Coop		5	0
Defect	Defect		1	1

The strategy that maximizes self-interest—a strategy endorsed by ethical egoism—is the subject of **game theory.** In the early 1980s, Robert Axelrod (a political science professor at the University of Michigan) organized a test of computer programs to determine which strategy results in the most self-interest. For a single round of play, defecting always is the best strategy. [The above truth table shows this. Assuming the other prisoner randomly cooperates or defects, my average benefit by defecting is 3 (i.e., (5 + 1)/2). If I cooperate, my average benefit is only 1.5 (i.e., (3 + 0)/2), assuming the other prisoner randomly cooperates or defects.)

In real life, we generally interact with other individuals multiple times. In Axelrod's computer contest, the winning strategy for multiple interactions was "tit-for-tat." This strategy always cooperated in the first round and simply repeated what the opponent did in subsequent rounds, as shown:

1st Round			2nd Round
Me	Him		Me
Coop	Coop	⇒	Coop
Coop	Defect		Defect

The tit-for-tat strategy, which places a premium on cooperative behavior, shows convergence between ethical egoism, utilitarianism, and rights analysis.

In the early 1990s, Martin Nowak (University of Oxford) and Karl Sigmund (Vienna University) showed that when occasional mistakes are made (as occurs in real life), the tit-for-tat strategy degenerates into backbiting. They determined that an alternative strategy was more successful. The "Pavlov" strategy repeats a response if it was successful (i.e., obtains a benefit of 3 or 5) and changes the response if it was unsuccessful (i.e., obtains a benefit of 0 or 1). The truth table for the Pavlov strategy follows.

1st Round				2nd Round
Me	Him	My Benefit		Me
Coop	Coop	3		Coop
Coop	Defect	0	⇒	Defect
Defect	Coop	5		Defect
Defect	Defect	1		Coop

In a population of *generally* cooperative individuals, the Pavlov strategy also leads to cooperation. Thus, in this case, the Pavlov strategy also shows convergence of ethical egoism, utilitarianism, and rights analysis. However, if the population consists of *always* cooperative individuals, Pavlov will exploit these "suckers" by defecting.

In summary, game theory is a tool for evaluating behavioral strategies that conform with ethical egoism. Interestingly, many of the most successful strategies are consistent with the morality of utilitarianism and rights analysis. However, the Pavlov strategy warns that a moral person should be on guard against being exploited as a sucker by less moral individuals.

Adapted from: T. Beardsley, *Scientific American,* October 1993, p. 22.

As an example, consider well-meaning legislation designed to reduce the amount of carcinogens released by the chemical industry. Carcinogens can be reduced to arbitrarily small levels, but always at a cost. If the chemical industry were to spend $1 billion for pollution-control equipment that removes enough carcinogens to save one person's life, would that be a proper allocation of resources? To the person whose life was saved, the answer is yes; but perhaps that $1 billion would be better spent on cancer research or improved prenatal health care. In that case, perhaps hundreds or thousands of lives could be saved.

The big question is, What is the value of a human life? You might reply, "Human life is invaluable; you cannot quantify it with dollars." Although this is a wonderful sentiment, it does not help society to allocate resources. If life has an infinite value, then the chemical industry should install $10 billion of pollution control equipment to save a life, or $100 billion, or $1000 billion. When the chemical industry spends these billions, it loses the opportunity to use those finite resources for other fruitful endeavors. And, of course, the cost is ultimately passed on to the consumer.

In a sense, each of us makes a decision about the value of our own life based upon the car we drive. Every year, about 43,000 people die in traffic accidents; it is the most prevalent cause of accidental death by far. In any year, each of us has about one chance in 7000 of being killed in an automobile accident, assuming we drive an average car in an average manner for an average number of miles. Suppose we are deciding about the purchase of a car we expect to last for 10 years. During the 10-year life of the car, we have one chance in 700 of being killed in an automobile accident if we drive an average car. Fortunately, some cars are built with safety as a high priority (e.g., BMW, Volvo, Lexus, Mercedes Benz). Imagine we have narrowed our car purchase decision to two models:

Car	Cost	Probability of Fatal Accident During 10-year Life of Car
Average	$15,000	1/700
Safe	$40,000	1/1400

If 1400 people spend $40,000 for the safe car, they will have spent $56 million and one of them will be dead in 10 years because of a car accident. If 1400 people spend $15,000 for the average car, they will have spent $21 million and two of them will be dead in 10 years because of car accidents. Thus, by spending an extra $35 million, one life can be saved. If you are aware of this safety/cost trade-off and you still select the $15,000 car, you have valued your life at less than $35 million, because you could be that extra person who dies.

Suppose people insisted that a life has infinite value and that engineers must design automobiles that are perfectly safe. The design solution might look something like a tank, cost about $2 million, and have two-gallons-per-mile fuel economy. If we all drove tanks, our roads must be stronger, which means more of us would be killed making stronger roads. Also, the pollution and environmental damage resulting from burning the tremendous amounts of fuel would cost additional lives.

The conclusion is inevitable: Life has a value, and engineers must know what it is so we can make intelligent design decisions. There is no accepted method for determining the value of a human life. Some approaches include determining the amount of lost wages due to an untimely death, determining ransom payments made to kidnappers, or the amount of extra wages workers demand for extra-risky jobs. Valuations of human life range from about \$200,000 to \$8 million. In his book *Fatal Tradeoffs: Public and Private Responsibilities for Risk,* W. Kip Viscusi argues that the average American assesses his life at \$6 million. (Of course, wealthy individuals assess their lives at a higher figure.)

EXAMPLE 2.8

Problem Statement: A highway engineer has \$1 million to spend on guardrails for roads. Guardrails cost \$13/ft, so 14.5 miles of guardrails can be installed. They have a service life of 20 years. She is considering spending the money on two roads. One is a scenic two-lane road through the mountains that has very steep shoulders. If a car falls off the shoulder, there is certain death for the passenger. The other is a four-lane highway on fairly level ground. If a car drives off the shoulder, there is only a 10% chance of death for the passenger. The scenic two-lane road traffic is 20 cars per day and the four-lane highway traffic is 22,000 cars per day. To help make her decision, she consults *Benefit-Cost Analysis of Road Side Safety Alternatives* (1986, Transportation Research Record 1065). According to this document, at a rate of 20 cars per day on a two-lane road, there will be only 0.01 "encroachments" per mile per year whereas at a rate of 22,000 cars per day on a four-lane road, there will be 3 "encroachments" per mile per year. (An "encroachment" is when a car drives off the road.) Which road should have the guardrails? How much money is spent per life saved? (Assume there is only one occupant per car, and that the presence of the guardrail prevents death from the automobile accident.)

Solution: For the two-lane scenic road,

$$\text{Cost} = \frac{\$1{,}000{,}000}{20\text{ yr}} \times \frac{\text{yr}\cdot\text{mi}}{0.01\text{ Encroachments}} \times \frac{\text{Encroachment}}{1\text{ Life saved}} \times \frac{1}{14.5\text{ mi}} = \frac{\$344{,}827}{\text{Life saved}}$$

For the four-lane highway,

$$\text{Cost} = \frac{\$1{,}000{,}000}{20\text{ yr}} \times \frac{\text{yr}\cdot\text{mi}}{3\text{ Encroachments}} \times \frac{\text{Encroachment}}{0.1\text{ Life saved}} \times \frac{1}{14.5\text{ mi}} = \frac{\$11{,}494}{\text{Life saved}}$$

Obviously, the money is better spent on erecting guardrails along the four-lane highway.

Engineers must always keep safety in mind when designing products. The amount of money engineers spend to reduce risk depends on (1) whether the product user voluntarily accepts the risks and (2) the amount of benefit the user derives from the product. For example, driving a car is *much* riskier than living near a nuclear power plant. However, the driver voluntarily accepts these risks and feels that the benefits outweigh the risks. Because radiation release from a power plant accident could be widespread, anyone could be exposed. Therefore, the risk is not voluntary. Also, a person may not feel that electricity derived from nuclear energy has particular value, because it could just as well be derived from other sources. With these considerations, the safety levels demanded in nuclear power plants far exceed those of automobiles.

Risk is something we must deal with every day. Table 2.1 puts many of these risks in perspective.

2.6 CASE STUDIES

We use case studies to illustrate the points made in our discussion. As you read the case studies, place yourself in the role of the engineer and imagine what you would do under the same circumstances.

2.6.1 Challenger Explosion

The *Challenger* was one of three space shuttles launched by NASA (Figure 2.2). A space shuttle consists of a reusable delta wing orbiter that contains the main engines, a cargo bay, living quarters, and the cockpit. The main engines are powered by liquid hydrogen and oxygen supplied from an expendable external tank. At liftoff the craft is extremely heavy,

TABLE 2.1
Risk of death for an "average" American†

Cause of Death	Probability $\left(\frac{\text{deaths}}{\text{person} \cdot \text{year}}\right)$	Cause of Death	Probability $\left(\frac{\text{deaths}}{\text{person} \cdot \text{year}}\right)$
From any cause	8585×10^{-6}	Accidents	354×10^{-6}
Disease	7266×10^{-6}	Motor vehicle	173×10^{-6}
Cardiovascular	3622×10^{-6}	Falls	50×10^{-6}
Cancer	2040×10^{-6}	Poison	26×10^{-6}
Diet*	714×10^{-6}	Fire	16×10^{-6}
Tobacco*	612×10^{-6}	Drowning	16×10^{-6}
Sexual behavior*	143×10^{-6}	Choking	13×10^{-6}
Occupation*	82×10^{-6}	Medical complications	10×10^{-6}
Alcohol*	61×10^{-6}	Firearms	6×10^{-6}
Pollution*	41×10^{-6}	Water transport	3×10^{-6}
Industrial products*	$<20 \times 10^{-6}$	Railroad	3×10^{-6}
Food additives*	$<20 \times 10^{-6}$	General aviation	3×10^{-6}
Suicide	120×10^{-6}	Electrocution	2×10^{-6}
Murder (white male)	93×10^{-6}	Commercial aviation	0.8×10^{-6}
Murder (black male)	720×10^{-6}		
Murder (white female)	30×10^{-6}		
Murder (black female)	142×10^{-6}		

*Causes of cancer according to the best estimates published in L. A. Sagan, "Problems in Health Measurements for the Risk Assessor," in *Technological Risk Assessment.* The Hague, Netherlands: Martinus Nijhoff, 1984.
†Calculated as the number of occurrences in 1991 divided by the U.S. population.
Adapted from: U.S. Department of Commerce, *Statistical Abstract of the United States,* 1994.

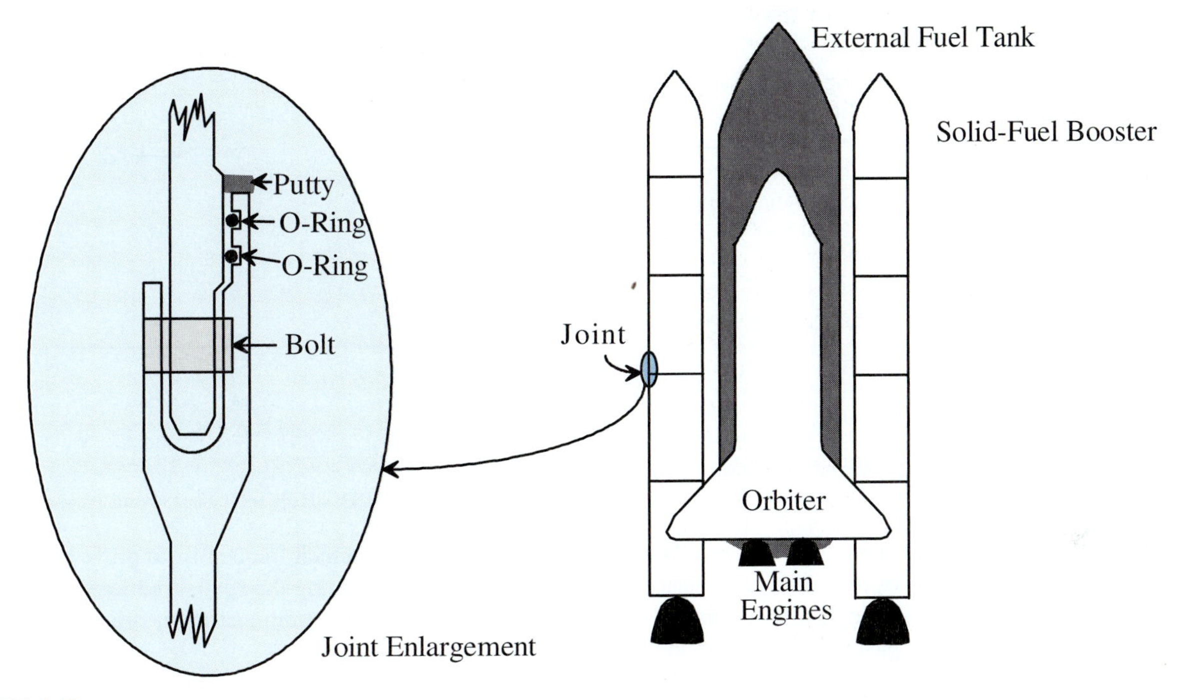

FIGURE 2.2
Challenger space shuttle.

because the filled tanks weigh many millions of pounds. Therefore, two solid-fuel booster rockets are needed during the first 2 minutes of flight. The booster rockets are jettisoned early enough in flight that they do not burn up in the earth's atmosphere and can be recovered for potential reuse.

The booster rockets are extremely large (150 feet long, 12 feet in diameter), so they are difficult to transport. They are assembled from smaller sections that are sealed with vulcanized rubber O-rings and zinc chromide putty (Figure 2.2). The joined sections are then filled with a million pounds of aluminum/potassium chloride/iron oxide propellant. The booster rockets for all three shuttles were manufactured by Morton-Thiokol.

The booster rocket seals had been giving NASA trouble. Inspection of the recovered boosters showed that combustion gases can blow by the O-ring seals, causing them to char. From experience, the engineers knew this problem to be particularly acute during cold weather, when the putty and O-rings are less pliable. Therefore, a program was under way to improve the design.

The space shuttle is a marvelous engineering achievement that brought new capabilities to NASA, such as the ability to repair near-earth satellites. However, it was a program born in controversy. As it was extremely expensive, many other programs were canceled, including heavy-lift expendable rockets. NASA claimed the space shuttle would be an inexpensive "space truck," but it required frequent launches to amortize the devel-

Risk on the Road

As shown in the figure, automobile driving became safer in the United States between 1975 and 1991. Engineers added safety features to both the automobile (e.g., seat belts, air bags, crumple zones) and roads (break-away signs, crash cushions, guardrails) that contributed to making driving safer, even though the number of cars on the road increased each year.

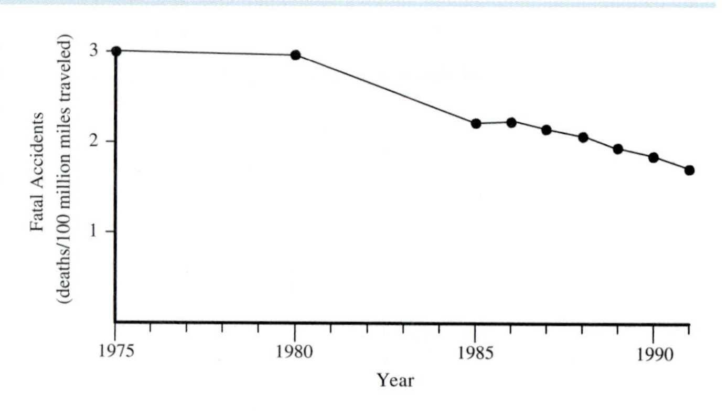

Adapted from: U.S. Department of Commerce, *Statistical Abstract of the United States,* 1994.

opment costs. Thus, NASA was under pressure to launch frequently, to prove the viability of its enormous investment. In fact, NASA was claiming the space shuttle could generate enough revenue to be self-supporting. A few Reagan administration officials even suggested it be privatized and sold to a commercial airline.

Any large engineering undertaking, like the space shuttle, should be viewed as an experiment. Although some components can be independently tested (e.g., electronics, engines, pumps), the only way to test the entire system is to launch it. Each launch experiences problems, such as cracked pump blades or some blow-by in the booster seals. Each time a decision is made to proceed with a launch in spite of equipment concerns, it is viewed as a test of the system. Eventually, as the limits of the system are explored, the system will break.

On January 28, 1986, NASA inadvertently tested the temperature limits of the space shuttle *Challenger.* The evening temperature had been below freezing and the morning temperature was 36°F. Prior to all launches, discussions are held between NASA management and engineers about the viability of the upcoming launch. This time, engineers expressed concern that ice at the launch site might damage the orbiter or its fuel tank. Violent waves had forced booster rocket recovery ships to return to the coast. Further, there was engineering concern about the booster rocket O-rings. Roger Boisjoly, a Morton-Thiokol engineer, recommended that no launch should occur below 53°F, based on his previous launch experience. He and his associate, Arnold Thompson, expressed their concerns to Morton-Thiokol and NASA management. Because Morton-Thiokol was negotiating for future booster rocket contracts, they did not want to portray their product as being of poor quality. NASA was interested in an aggressive launch schedule to prove space shuttle economic viability. Thus, the many engineering concerns were overruled by management. The decision was made to launch the space shuttle at 11:38 A.M. After 76 seconds of flight, at 50,000 feet, the booster rocket joints failed, allowing hot escaping gases to ignite the hydrogen fuel. All seven crew members, including Christa MacAuliffe ("teacher in space"), died, as millions watched on television.

Ethical points for discussion:

1. Should the Morton-Thiokol engineers have blown the whistle and announced to the press that NASA management was endangering the lives of the crew members and risking destruction of the space shuttle?
2. No launch is completely safe; space travel is inherently risky. The astronauts accepted this risk when they volunteered for the job. Should they have been informed that the risks were higher for this particular launch?
3. Of the three main moral theories—ethical egoism, utilitarianism, and rights analysis—which moral theory were the Morton-Thiokol and NASA management using when they made the decision to launch over the objections of the engineers? In retrospect, we could judge their action to be wrong. What moral theory or theories are we applying?
4. Is it unfair to place blame on NASA managers? After all, every launch has risks and someone has to make the decision to launch.
5. Should the engineers be faulted for developing an inferior design? Perhaps they should have incorporated heating tape into the joints if low temperatures were known to be a problem.
6. Should an escape mechanism be installed on space shuttles even though it imposes a severe weight penalty and reduces the shuttle payload capability?

Challenger explosion.

2.6.2 Missouri City Television Antenna Collapse

In 1982, a television antenna was erected in Missouri City, Texas. An engineering company designed it, and a separate rigging company erected it. The rigging company approved the detailed engineering drawings and said they could erect the antenna as designed.

The design consists of a three-legged tower constructed from premade sections (Figure 2.3). A movable crane sits at the top of the already-erected sections. It lifts a new section into place, which is then attached to the tower by the riggers. The crane is designed so it can crawl up the tower, allowing it to install the next section at the top. In a sense, the tower is lifted up by its boot straps. All the tower sections are identical except for the top one, which has the microwave basket attached.

The bottom sections of the three-legged tower were successfully erected by the rigging company. The final step was to hoist the top section. When the riggers attempted to attach the cables to attachment points on the top section of the tower, they found that the microwave basket interfered with the cables. They called the engineering company to see if the microwave basket could be removed, allowing the top section to be lifted just like all the others. The microwave basket would be lifted independently and attached after the tower was completely erected. The engineering company refused; their past experience with this

FIGURE 2.3
The Missouri City TV tower disaster.

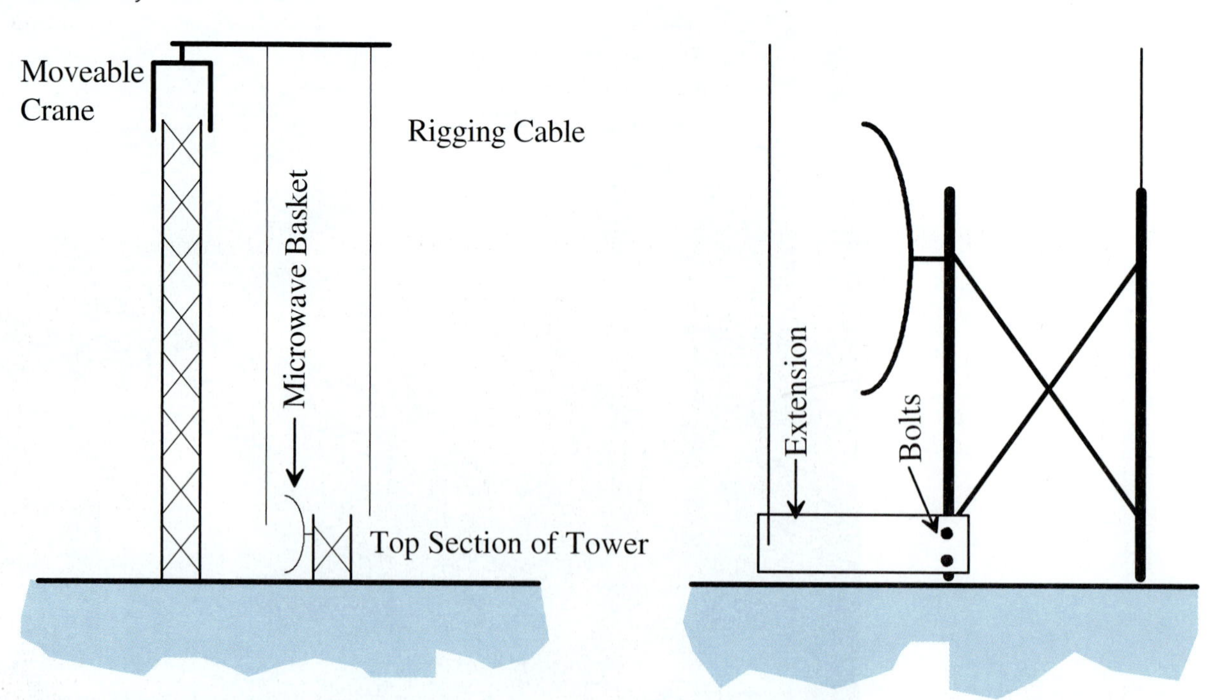

method was that the microwave basket became damaged, and they were held liable for the damage, so they lost money when they repaired it. The engineers informed the riggers that if they removed the microwave basket, it would void the warranty.

Later, the rigging company contacted the engineering company with a proposal to add an extension to the top section that kept the rigging cables away from the microwave basket. The rigging company did not have in-house engineering expertise and wanted the engineering company to look at a drawing of their proposed modification. Although the engineer wanted to help the riggers, his boss was firmly against the idea. If the engineering company approved the modifications, they would be held liable if anything went wrong. The engineering company reminded the riggers that they had already approved the original drawings; it was the riggers' responsibility to determine how to raise it.

The riggers knew the weight of the top section, so they bought bolts that were rated sufficiently strong to carry the weight. They installed the extension and attached the cables to lift the top section. Six riggers rode with the top section because they were needed to install it once it got to the top. They never made it, because the bolts sheared, causing them all to fall to their deaths. The riggers had neglected to allow for the extra moment arm of the extension, which increased the forces on the bolts by about 12 times. (*Note*: You will learn more about moment arms in the statics chapter.)

Ethical points for discussion:

1. Did the engineering company have a faulty design? Should they have anticipated the problem and designed accordingly?
2. Were the riggers operating outside their area of expertise? Should they have hired an engineering consultant to calculate the necessary bolt size?
3. Should the engineering company have been more helpful and less legalistic?

Wreckage from the Missouri City television antenna collapse. The movable crane from the top of the tower is lying in the foreground.

4. Where did the responsibility of the engineers end and the riggers begin?
5. Of the three major moral theories (ethical egoism, utilitarianism, rights analysis), which theory was the boss of the engineering company using?
6. Elaborate on the legal and moral conflicts faced by the engineer who was contacted by the rigging company.

2.6.3 Kansas City Hyatt Regency Walkway Collapse

In 1976, Crown Center Redevelopment Corporation decided to build a Hyatt Regency Hotel in Kansas City, Missouri. G.C.E. International, Inc. was selected as the structural engineering firm, and Havens Steel Company did the steel fabrication. The building was to have 750 rooms and a large atrium measuring 117 × 145 feet on the floor and 50 feet high. Three suspended walkways through the atrium connected the second, third, and fourth floors. From these walkways, patrons could look down into the atrium, adding to the dramatic effect.

In spring 1978, construction began. In January 1979, Havens contacted Daniel M. Duncan (a G.C.E. engineer) about a proposed design change. The original design specified the suspended walkways for floors two and four to be connected to the ceiling by a single 46-foot threaded rod (Figure 2.4). Havens determined that the single rod was simply too

FIGURE 2.4

Detail of walkway support structure. In the original design, each nut has a load of $\frac{1}{2}F$. In the revised design, the nuts holding the fourth-floor walkway have load F, which is twice their anticipated load.

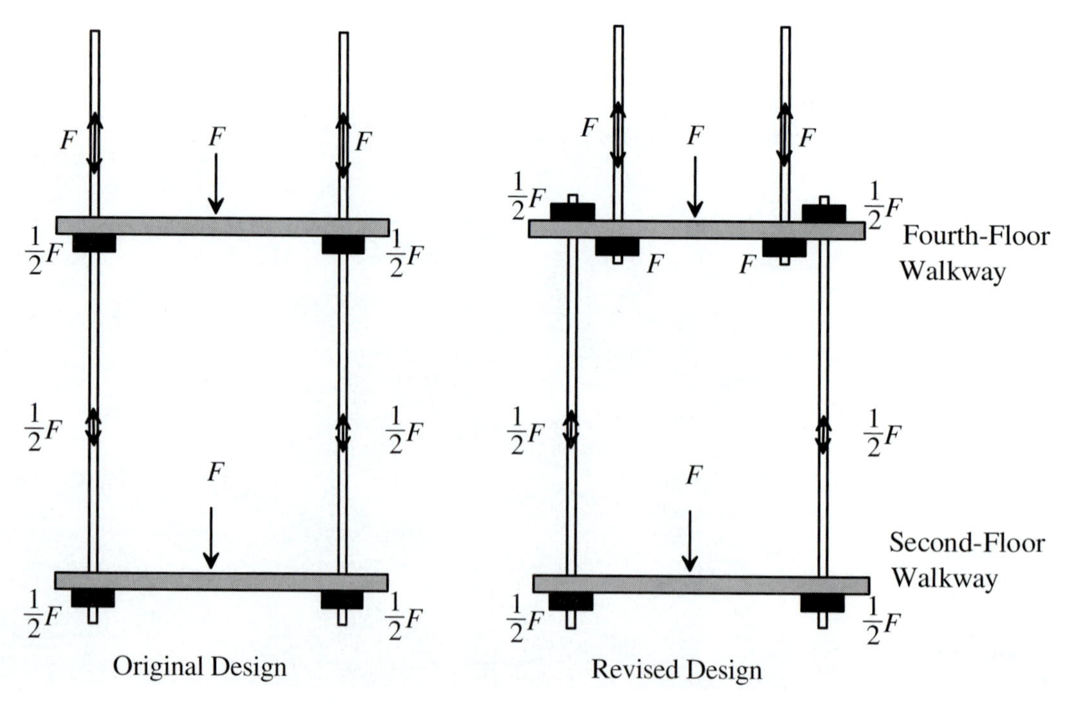

long to build; so they proposed that the threaded rod be separated into two. The design changes were incorporated into shop drawings that received the engineering seal of approval by Jack D. Gillum, president of G.C.E.

In July 1980, the hotel was open for business. One year later, on July 17, 1981, during a party held at the hotel, many party-goers were dancing, standing, and walking on the atrium walkways. The fourth-floor walkway collapsed onto the second-floor walkway, killing 114 and injuring over 200. It was the most deadly structural failure in U.S. history.

In November 1985, Gillum and Duncan were found guilty of gross negligence, misconduct, and unprofessional conduct in the practice of engineering. The collapse brought great scrutiny to their work and revealed many irregularities. Duncan's design did not employ adequate stiffeners at the joints between the threaded rod and walkway. Critical welds near the joints were not specified, and the material selected for the threaded rods was too weak. Duncan failed to review the shop drawings that described the new two-rod support system. Most importantly, Duncan failed to perform engineering calculations of either the original design or the modification. Had he done so, he would have learned that the modification doubled the load on the fourth-floor walkway support bolt. This exceptionally high load caused the joint to fail and led to the walkway collapse. Although Gillum had delegated responsibility to Duncan for performing the walkway design, because Gillum placed his engineering seal of approval on the drawings and had not adequately checked Duncan's work, he was held "vicariously liable" and was subjected to the same penalties as Duncan.

Although Duncan and Gillum lost their Missouri licenses, they are practicing in other states.

Ethical points for discussion:

1. Should these engineers be allowed to practice engineering in other states?
2. Should the engineers be held responsible for a simple error such as this? After all, they had to make thousands of decisions and everyone is entitled to make a mistake.
3. Engineering is a discipline requiring attention to detail. Is it ethical for professors to give students partial credit for wrong answers that are sort of right?
4. In court testimony, G.C.E. claims it was never called by Havens about the proposed design change, yet their seal was affixed to the revised drawings. Does it appear G.C.E. was telling the truth?
5. Should the fabricator be held responsible for overloading the support nut?
6. What can be done to prevent such catastrophes in the future? Should the government review all drawings? If procedural changes are implemented, what is their impact on efficiency and cost?
7. Should the engineers have lost their licenses?
8. Upon further investigation it was found that the atrium roof had collapsed during construction. G.C.E. claims that the owner (Crown Center Redevelopment Corporation) was unwilling to pay for onsite inspection, even though G.C.E. requested it three times. How much responsibility does the owner have to ensure that the engineers are funded well enough to properly perform their job?
9. Even the original walkway design was questionable. Were the engineers living up to their obligations to be technically competent?

Wreckage from the walkway collapse of the Kansas City Hyatt Regency.

10. Because Gillum had delegated responsibility to Duncan to do the design, should Gillum be held liable for Duncan's mistakes?
11. How can competent engineers protect themselves from the impact of incompetent engineers?
12. Should stronger certification laws be enacted to eliminate incompetent engineers?

2.7 SUMMARY

As professionals, engineers are expected to behave in an ethical manner. Ethical rules are among the interaction rules that govern the relationship between individuals and society; other interaction rules are classified as etiquette, law, and morals. Generally, there is consistency between these various interaction rules, but occasionally, they conflict.

The purpose of interaction rules is to eliminate conflict; however, it is impossible to eliminate all conflicts, so the source of the conflict must be identified. Conflicts can result from moral issues, conceptual issues, application issues, or factual issues. Factual issues are very concrete and can generally be settled regardless of a person's upbringing and background. In contrast, moral issues tend to be abstract. Resolution of moral conflicts depends upon a person's upbringing and background.

When attempting to settle moral issues, various moral theories may be employed. Ethical egoism states that actions are moral if you act in your enlightened self-interest. Utilitarianism states that the most moral action is the one that brings the most good to the most people. Rights analysis states that actions are moral if they do not violate the rights of individuals. Often, these three moral theories converge, indicating that one action is clearly correct. However, sometimes they diverge, making it difficult to select a moral action. Each society deals with this problem differently. Some philosophers recommend using utilitarianism unless an individual's rights are seriously impaired.

A major issue confronting society is the proper allocation of resources. Although we all would like to live in a risk-free society, there are insufficient resources to achieve that. Often, engineers must make difficult decisions regarding acceptable levels of risk and safety.

Further Readings

Harris, C. E., Jr.; M. S. Pritchard; and M. J. Rabins. *Engineering Ethics: Concepts and Cases.* Belmont, CA: Wadsworth, 1995.

Johnson, D. G. *Ethical Issues in Engineering.* Englewood Cliffs, NJ: Prentice Hall, 1991.

Martin, M. W., and R. Schinzinger. *Ethics in Engineering.* 2nd ed. New York: McGraw-Hill, 1989.

PROBLEMS

2.1 In the following examples, determine whether the source of the conflict is a moral issue, conceptual issue, application issue, or factual issue.

(a) John is a newly hired engineer for the JMT Electronics Company. He is responsible for selecting vendors that supply components for his company's products. One of the vendors invites John to lunch at Chez Pierre Restaurant so they can become better acquainted. When John returns from lunch, his boss is angry with him. He explains that John could be fired for accepting a bribe, because a typical lunch at Chez Pierre costs $50.

(b) In response to the incident described in Part (a), John's boss issues a memo stating that "engineers are not permitted to accept lunch from vendors if the value of the meal exceeds $10." A vendor takes John to another restaurant and spends $9 for the food and leaves a $2 tip. John is concerned that he might be fired for not complying with the memo, because the total spent by the vendor was $11, a figure exceeding the policy guidelines. Jane, a fellow engineer, says not to worry, because the food was only $9 and falls within the allowable limit.

(c) During World War II, engineers were important players in the Manhattan Project, the U.S. program that built the first atomic bomb. Mark and Edward were two engineers who debated whether they should participate in the project. Edward argued that they should not become involved because the atomic bomb would mean a level of mass destruction never before unleashed on the world. In his view, humankind is not wise enough to control such power. Mark argued that millions of lives, both U.S. and Japanese, would be sacrificed if conventional weapons were used to end the war. He felt that although the atomic bomb was certainly destructive, many fewer lives would be lost because the war would end much earlier if the atomic bomb were used.

(d) Fred is a quality control engineer in an automotive plant that makes drive shafts on a high-speed lathe requiring specially hardened tool bits. He notices that the automotive drive shafts are no longer meeting specifications on a critical dimension, causing 1000 of them to be rejected. Because his bonus depends on a low rejection rate, Fred is angry at this development. He calls the lathe operator into his office and accuses him of installing the wrong tool bit in the high-speed lathe. The lathe operator swears that he used the right tool bit.

(e) Larry is an aerospace engineer employed by Boeing to design planes. He is a member of the Quaker religion, which is famous for nonviolence. (During wars, Quakers often refuse to serve as fighting soldiers, but will serve as medics.) Larry was hired by Boeing to design passenger airplanes; however, his boss has recently reassigned him to design military fighters. Larry must decide whether to accept the new assignment, or quit and find a new job.

(f) Sally is a mechanical engineer employed by General Motors to design automotive gas tanks. According to U.S. Government safety standards, the automobile must survive a moderate impact with no chance of the gas tank catching fire. In recent tests, cars that crashed at 35 mph had no fires, whereas 20% of cars that crashed at 45 mph had fires. She is considering whether to redesign the gas tank.

(g) A new government law requires that the lead content of drinking water be less than 1.0 ppb (parts per billion). Melissa is a safety engineer who has tested her company's drinking water by two methods. Method A gives a reading of 0.85 ppb, whereas Method B gives a reading of 1.23 ppb. She must fill out a government report describing the quality of her company's water. If the lead content exceeds 1.0 ppb, her company will be fined. She is contemplating whether to report the results from Method A or Method B.

2.2 The environmental organization Greenpeace wants to stop "toxic colonialism," in which developed nations send their hazardous waste to less developed countries with lax environmental laws. The United Nations has proposed regulations entitled "The Basel Convention on the Control of Transboundary Movements of Hazardous Wastes & Their Disposal." The regulations seek to ban the export of hazardous waste from developed to lesser developed countries. In 1995, 91 nations were signatories to the agreement.

Mary is an engineer who works for Copper Recyclers, Inc. They ship bales of used copper wire from the United States to Mexico, where they have a processing plant. Because the copper is used, it is contaminated with small amounts of tin-lead solder. She is concerned that the Basel Convention will affect her company's business because of the lead content in their bales. She must make a recommendation to her boss whether to relocate the processing plant in the United States.

From this scenario, identify a moral issue, a conceptual issue, an application issue, and a factual issue.

2.3 Volvo recently introduced an automobile to the U.S. market in which the headlights are on whenever the engine is on; thus, the lights are on even during the day. In Sweden, the law requires all motorists to drive with their lights on even during the day. Since enacting this law, motorist deaths have been reduced by 11%. Because of the extra energy used by the headlights and the need for more frequent replacement of the headlights, this practice adds an expense of approximately $0.002/mi. Use the following assumptions: There are about 180 million automobiles and 250 million people in the United States. Each vehicle is driven about 15,000 miles per year. If the United States adopted the Swedish law,

(a) How many lives would be saved?

(b) What is the cost of saving each life?

(c) Assuming a human life is valued at $7 million, is this a legitimate expense?

2.4 A retired engineer is contemplating whether to fly or drive to a destination that is 1000 miles away. He has heard that flying is safer than driving and decides to calculate the probability of being killed while traveling.

(a) Calculate the probability of the retired engineer being killed on a 1000-mile one-way trip taken by car and by plane. Use the following assumptions: There are about 180 million automobiles in the United States and 250 million people. Each vehicle is driven about 15,000 miles per year. An average American travels 1800 miles by plane each year. The engineer is in good health and has driving skills comparable to the average American.

(b) The direct cost of driving (fuel, oil, wear-and-tear) is about $0.25/mi. (*Note*: This cost does not include fixed costs such as insurance, registration fees, automobile depreciation, and interest. These direct costs must be paid regardless of whether he drives or not; therefore, they will not be charged to the trip. If both fixed and direct costs are included, the typical cost of driving an automobile is about $0.45/mi.) A round-trip airplane ticket for the 1000-mile trip costs $700. By flying rather than driving, how much money is spent to save a life? The engineer places no value on his time because he is retired, but he does place a $7 million value on his life. Is the extra cost of the airplane ticket worth it to the retired engineer?

(c) The above calculation shows flying to be much safer than driving on a per-mile basis. The retired engineer spends two hours in the car getting to the airport and two hours on the plane. Is it valid for him to feel much safer during the airplane trip and much less safe during the automobile trip? In making your calculation, use the following assumptions: Automobiles average about 35 mph and airplanes average about 500 mph.

2.5 Prepare truth tables for the Prisoner's Dilemma game using the following strategies. Play four rounds using each strategy. Report your "benefit" after each round. Compute your average benefit for each strategy after the four rounds are completed. Be sure to evaluate all possible initial conditions.

Problem	My Strategy	His Strategy
a.	Random*	Random*
b.	Random	Always cooperate
c.	Random	Always defect
d.	Tit-for-tat	Random
e.	Tit-for-tat	Always cooperate
f.	Tit-for-tat	Always defect
g.	Tit-for-tat	Tit-for-tat
h.	Pavlov	Random
i.	Pavlov	Always cooperate
j.	Pavlov	Always defect
k.	Pavlov	Tit-for-tat
l.	Pavlov	Pavlov

**Random* means that in an unpredictable manner, about half the time the prisoner cooperates and about half the time the prisoner defects.

2.6 Write a computer program that allows you to play against the computer. Program the computer to follow the random, tit-for-tat, or Pavlov strategy.

2.7 Write a computer program that calculates the average benefit to you after playing 100 rounds of one strategy listed in Problem 2.5.

2.8 Write a computer program that calculates the average benefit to you after playing 100 rounds of one strategy listed in Problem 2.5. Include in your program a feature that allows for occasional mistakes. For example, rather than always cooperating, perhaps the other prisoner cooperates 90% of the time and defects 10% of the time.

Glossary

Σ The Greek symbol for sum.
application issue The lack of clarity as to whether a particular act violates a law, rule, or policy.
conceptual issue The morality of an action is agreed upon, but there is uncertainty about how it should be codified into clearly defined law, rule, or policy.
conflict of interest A situation in which an engineer's loyalties and obligations may be compromised because of self-interest or other loyalties and obligations.
convergence To tend toward a common solution.
diverge To differ.
eminent domain The right of the government to take possession of private property for public use.
engineering ethics The set of behavioral standards that all engineers are expected to follow.
ethical egoism A moral theory stating that an act is moral provided you act in your enlightened self-interest.
ethics The general and abstract concepts of right and wrong behavior culled from philosophy, theology, and professional societies.
etiquette The codes of behavior and courtesy.
factual issue Uncertainty about morally relevant facts.
game theory A tool for evaluating optimal behavior strategies.
happiness objective function Σ (benefit) (importance) $-$ Σ (harm) (importance)
interaction rules Expected sets of behavior (etiquette, law, morals, and ethics) between the engineer, other individuals, and society as a whole.
law The system of rules established by authority, society, or custom.
legal rights The "just claims" given to all humans within a government's jurisdiction.
moral issue An issue that can be resolved only by making a moral decision.
moral rights The "just claims" that belong to all humans, regardless of whether these rights are recognized by government.
moral theories A framework for making moral and ethical decisions.
morals The accepted standards of right and wrong that are usually applied to personal behavior.
rights analysis A moral theory that equally respects each human being.
ulitarianism A moral theory that seeks to create the most good for the most people.
vulcanized To improve the strength and resiliency of a polymer such as rubber or plastic by combining it with sulfur (or some other type of agent).
whistle blowing The act of informing authorities of harmful, dangerous, or illegal activities.

Mathematical Prerequisites

- Algebra (Appendix E, Mathematics Supplement)
- Geometry (Appendix I, Mathematics Supplement)

CHAPTER 3

Problem Solving

Engineers are problem solvers; employers hire them specifically for their problem-solving skills. As essential as problem solving is, it is impossible to teach a specific approach that will always lead to a solution. Although engineers use science to solve problems, this skill is more art than science. The only way to learn problem solving is to do it; thus, your engineering education will require that you solve literally thousands of homework problems.

In the modern world, computers are often used for problem solving. The novice student may think that the computer is actually solving the problem, but this is untrue. Only a human can solve problems; the computer is merely a tool.

3.1 TYPES OF PROBLEMS

A *problem* is a situation, faced by an individual or a group of individuals, for which there is no obvious solution. There are many types of problems that we confront:

- *Research problems* require that a hypothesis be proved or disproved. A scientist may hypothesize that CFCs (chlorofluorocarbons) are destroying the earth's ozone layer. The problem is to design an experiment that proves or disproves the hypothesis. If you were confronted with this research problem, how would you approach it?
- *Knowledge problems* occur when a person encounters a situation that he does not understand. A chemical engineer may notice that the chemical plant produces more product when it rains. The cause is not immediately obvious, but further investigation might reveal that heat exchangers are cooled by the rain and hence have more capacity.
- *Troubleshooting problems* occur when equipment behaves in unexpected or improper ways. An electrical engineer may notice that an amplifier has a 60-cycle hum whenever the fluorescent lights are turned on. To solve this problem, she determines that extra shielding is required to isolate the electronics from the 60-cycle radiation emitted by the lights.
- *Mathematics problems* are frequently encountered by engineers, whose general approach is to describe physical phenomena with mathematical models. If a physical phenomenon can be described accurately by a mathematical model, the engineer unleashes the extraordinary power of mathematics, with its rigorously proved theorems and algorithms, to help solve the problem.

- *Resource problems* are always encountered in the real world. It seems there is never enough time, money, people, or equipment to accomplish the task. Engineers who can get the job done in spite of resource limitations are highly prized and well rewarded.
- *Social problems* can impact engineers in many ways. A factory may be located where there is a shortage of skilled labor because the local schools are of poor quality. In this environment, an engineer running a training program for factory workers must design the program to accommodate the low reading abilities of the trainees.
- *Design problems* are the heart of engineering. To solve them requires creativity, teamwork, and broad knowledge. A design problem must be properly posed. If your boss said, "Design a new car," you would not know whether to design an economy car, a luxury car, or a sport/utility vehicle. A well-posed design problem must include the ultimate objectives of the design project. If the boss said, "Design a car that goes from 0 to 60 miles per hour in 6.0 seconds, gets 50 miles-per-gallon fuel economy, costs less than $10,000, meets government pollution standards, and appeals to aesthetic tastes," then you could begin the project—even though it has difficult objectives.

3.2 PROBLEM-SOLVING APPROACH

The approach to solving an engineering problem should proceed in an orderly, stepwise fashion. The early steps are qualitative and general, whereas the later steps are more quantitative and specific. The elements of problem solving can be described as follows:

1. *Problem identification* is the first step toward solving a problem. For students, this step is done for them when the professor selects the homework problems. In the real engineering world, this step is often performed by a manager or creative engineer.

As an example, the management of an automotive firm may be painfully aware that the firm is losing market share. They challenge the engineering staff to design a revolutionary automobile to gain back lost sales.

2. *Synthesis* is a creative step in which parts are integrated together to form a whole.

For example, the engineers may determine that they can meet the design objectives for the new car (high fuel economy and rapid acceleration) by combining a highly efficient engine with a sleek, aerodynamic body.

3. *Analysis* is the step where the whole is dissected into pieces. Most of your formal engineering education will focus on this step. A key aspect of analysis is to translate the physical problem into a mathematical model. Analysis employs logic to distinguish truth from opinion, detect errors, make correct conclusions from evidence, select relevant information, identify gaps in information, and identify the relationship between parts.

For example, the engineers may compare the drag of a number of different body types and determine if the engine can fit under the hood of each body.

4. ***Application*** is a process whereby appropriate information is identified for the problem at hand.

For example, the engineers determine that a key question is to find the required force needed to propel the automobile at 60 mph at sea level, knowing the car has a projected frontal area of 19 ft^2 and a drag coefficient of 0.25.

5. ***Comprehension*** is the step in which the proper theory and data are used to actually solve the problem.

For example, the engineers determine that the drag force F on the automobile may be calculated using the formula

$$F = \tfrac{1}{2} C_d \rho A v^2$$

where C_d is the drag coefficient (dimensionless), ρ is the air density (kg/m^3), A is the projected frontal area (m^2), v is the automobile velocity (m/s), and F is the drag force (N). From the data, the force required to overcome air drag is

$$F = \tfrac{1}{2}(0.25)\left(1.18\,\frac{\text{kg}}{\text{m}^3}\right)\left[19\ \text{ft}^2 \times \left(\frac{\text{m}}{3.281\ \text{ft}}\right)^2\right]$$

$$\times \left(60\,\frac{\text{mi}}{\text{h}} \times \frac{\text{h}}{3600\ \text{s}} \times \frac{5280\ \text{ft}}{\text{mi}} \times \frac{\text{m}}{3.281\ \text{ft}}\right)^2 \times \frac{\text{N}}{\frac{\text{kg}\cdot\text{m}}{\text{s}^2}}$$

$$= 190\ \text{N} \times \frac{\text{lb}_\text{f}}{4.448\ \text{N}} = 42\ \text{lb}_\text{f}$$

The required force to overcome air drag is 190 newtons (in the metric system) or 42 pounds-force (in the American Engineering System). (*Note*: The above calculation involves many **conversion factors** from Appendix A. If you are uncomfortable with the conversions, do not worry; they will be discussed in more detail later in the book. You may wish to look at the box "A Word About Units" to review some basics regarding units.)

A Word About Units

Undoubtedly, you have been exposed to units of measure in high school. Here, our purpose is to refresh your memory. You are certainly familiar with the formula relating distance d, speed s, and time t,

$$d = st$$

Suppose your speed is 60 miles per hour and your trip takes 2 hours. We can calculate the distance traveled as

$$d = \frac{60\ \text{miles}}{\cancel{\text{hour}}} \times 2\ \cancel{\text{hour}} = 120\ \text{miles}$$

The hours cancel, leaving units of miles. Some students prefer to show their calculations as follows:

$$d = \frac{60\ \text{miles}\ \big|\ 2\ \cancel{\text{hour}}}{\cancel{\text{hour}}\ \big|} = 120\ \text{miles}$$

Either approach is correct as long as you are careful with your units.

If you wish to express the distance in kilometers, then a conversion factor is required. There is about 0.6 mile in a kilometer. Use this relationship to convert miles to kilometers:

$$d = 120\ \cancel{\text{miles}} \times \frac{\text{kilometer}}{0.6\ \cancel{\text{mile}}} = 200\ \text{kilometers}$$

When a unit is raised to a power, the conversion factor also must be raised to a power. For example, if a large land area were 120 square miles, then the area would be converted to square kilometers as follows:

$$A = 120\ \cancel{\text{miles}^2} \times \left(\frac{\text{kilometer}}{0.6\ \cancel{\text{mile}}}\right)^2 = 333\ \text{kilometers}^2$$

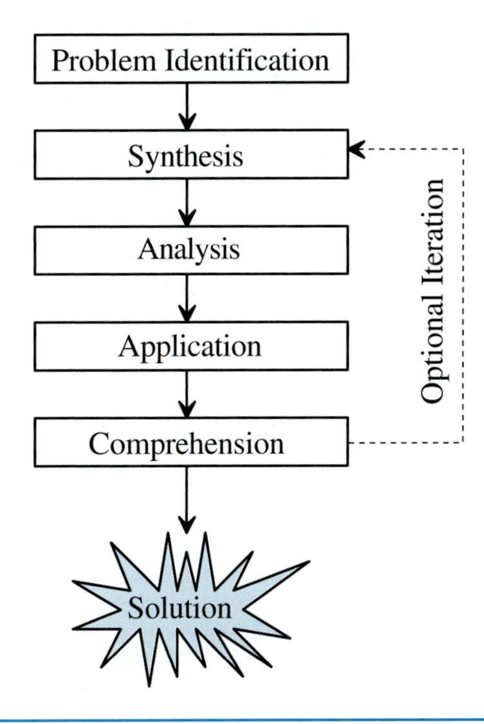

FIGURE 3.1
Problem-solving approach.

Although we would like to believe that these five steps can be followed in a linear sequence that always leads us to the correct solution, such is not always the case. Often, problem solving is an **iterative procedure,** meaning the sequence must be repeated because information learned at the end of the sequence influences decisions early in the sequence (Figure 3.1). For example, if the engineers determine that the force calculated in Step 5 is too high, they would have to return to earlier steps and try again.

3.3 PROBLEM-SOLVING SKILLS

Problem solving is a process in which an individual or a team applies knowledge, skills, and understanding to achieve a desired outcome in an unfamiliar situation. The solution is constrained by physical, legal, and economic laws as well as by public opinion.

To become a good problem solver, the engineer must have the following:

- Knowledge (first acquired in school, but later on the job).
- Experience to wisely apply knowledge.
- Learning skills to acquire new knowledge.
- Motivation to follow through on tough problems.
- Communication and leadership skills to coordinate activities within a team.

Table 3.1 compares skilled and novice problem solvers. Among the most important capabilities of a skilled problem solver is **reductionism,** the ability to logically break a problem into pieces. (Question: How do you eat an elephant? Answer: One bite at a time.)

TABLE 3.1
Comparison of skilled and novice problem solvers

Characteristic	Skilled Problem Solver	Novice Problem Solver
Approach	Motivated and persistent	Easily discouraged
	Logical	Not logical
	Confident	Lacks confidence
	Careful	Careless
Knowledge	Understands problem requirements	Does not understand problem requirements
	Rereads problem	Relies on a single reading
	Understands facts and principles	Cannot identify facts and principles
Attack	Breaks the problem into pieces*	Attacks the problem all at once
	Understands the problem before starting	Tries to calculate the answer right away
Logic	Uses basic principles	Uses intuition and guesses
	Works logically from step to step	Jumps around randomly
Analysis	Organized	Disorganized
	Thinks carefully and thoroughly	Hopes the answer will come
	Clearly defines terms	Uncertain about the meaning of symbols
	Careful about relationships and meaning of terms	Jumps to unfounded conclusions about the meanings of terms
Perspective	Has a feel for the correct magnitude of answers	Uncritically believes the answers produced by the calculator or computer
	Understands the differences between important and unimportant issues	Cannot differentiate between important and unimportant issues
	Uses rule of thumb to estimate the answer	Cannot estimate the answer

* Very important
Adapted from: H. S. Fogler, "The Design of a Course in Problem-Solving," in *Problem Solving,* AIChE Symposium Series, vol. 79, no. 228, 1983.

Reductionism contrasts with *synthesis,* the creative process of putting pieces together. If your problem were to design an airplane, you would use reductionism to design subsystems (engines, landing gear, electronic controls, etc.) and synthesis to combine the pieces together.

3.4 TECHNIQUES FOR ERROR-FREE PROBLEM SOLVING

All students hope to do error-free problem solving, because it assures them of excellent grades on homework and exams. Further, error-free calculations will be required when the student enters the engineering workforce. The truth is that engineers can never be certain that their answer is correct. A civil engineer who goes through elaborate calculations to design a bridge cannot be certain of her calculations until a heavy load is placed on the bridge and the deflection agrees with her calculations. Even then there is uncertainty, because there may be errors in the calculations that tend to cancel each other.

Although we can never be certain our answer is correct, we can increase the probability of calculating a correct answer using the following procedure:

Given:

1. Always draw a picture of the physical situation.
2. State any assumptions.
3. Indicate all given properties on the diagram ***with their units.***

Find:

4. Label unknown quantities with a question mark.

Relationships:

5. From the text, write the *main equation* that contains the desired quantity. (If necessary, you might have to derive the appropriate equation.)
6. Algebraically manipulate the equation to isolate the desired quantity.
7. Write *subordinate equations* for the unknown quantities in the main equation. Indent to indicate that the equation is subordinate. You may need to go through several levels of subordinate equations before all the quantities in the main equation are known.

Solution:

8. After all algebraic manipulations and substitutions are made, insert numerical values ***with their units***.
9. Ensure that units cancel appropriately. Check one last time for a sign error.
10. Compute the answer.
11. Clearly mark the final answer. ***Indicate units.***
12. Check that the final answer makes physical sense!
13. Ensure that all questions have been answered.

Notice that the first step in solving an engineering problem is to draw a picture. We cannot overstate the importance of graphics in engineering. Solving most engineering problems requires good visualization skills. Further, communicating your solution requires the ability to draw.

The above procedure is a process of working backward. You start from the end with the unknown, desired quantity and work backward using the given information. Not every problem fits into the paradigm described above, but if you use it as a guide, it will increase the likelihood of obtaining a correct answer. Notice that units are emphasized. Most calculation errors result from a mistake with units.

It goes without saying that the solution should be as neat as possible so it can be easily checked by another person (a co-worker or boss). Calculations should be performed in pencil so changes can be made easily. It is recommended that you use a mechanical pencil because it does not need to be sharpened. A fine (0.5-mm diameter), medium-hardness lead works best. Any erasures need to be complete.

Most engineering professors prefer that students use **engineering analysis paper,** because it has a light grid pattern that can be used for graphing or drafting. It also has headings and margins that allow the calculations to be labeled and documented. Use one side of the paper only. If at all possible, complete the solution on a single page; this allows for easy checking. Sample Problems 1, 2, and 3 present examples of well-solved homework problems.

	3 OCT '95	HOMEWORK # 3	JOE ENGINEER 378-15-5432 ENGR 109 SECTION 521	1 OF 3

1.

$L_1 = ?$ $\quad L = 40.0\text{ cm}$ $\quad r$

$\rho_{wood} = 0.6 \frac{g}{cm^3}$ $\quad \rho_{water} = 1.00\ g/cm^3$

L_2

$m_{wood} = m_{water}$ (ARCHIMEDES PRINCIPLE)

$$m_{wood} = \rho_{wood} V_{wood}$$

$$V_{wood} = \pi r^2 L$$

$$m_{wood} = \rho_{wood} \pi r^2 L$$

$$m_{water} = \rho_{water} V_{water}$$

$$V_{water} = \pi r^2 L_2$$

$$L = L_1 + L_2$$

$$L_2 = L - L_1$$

$$V_{water} = \pi r^2 (L - L_1)$$

$$m_{water} = \rho_{water} \pi r^2 (L - L_1)$$

$$\rho_{wood} \pi r^2 L = \rho_{water} \pi r^2 (L - L_1)$$

$$\rho_{wood} L = \rho_{water} (L - L_1) = \rho_{water} L - \rho_{water} L_1$$

$$\rho_{water} L_1 = \rho_{water} L - \rho_{wood} L = L(\rho_{water} - \rho_{wood})$$

$$L_1 = L \frac{\rho_{water} - \rho_{wood}}{\rho_{water}} = 40.0\text{ cm} \frac{1.00\ g/cm^3 - 0.600\ g/cm^3}{1.00\ g/cm^3}$$

$$\boxed{L_1 = 16.0\text{ cm}}$$

Sample Problem 1:

Archimedes' principle states that the total mass of a floating object equals the mass of the fluid displaced by the object. A 40.0 cm log is floating vertically in the water. Determine the length of the log that extends above the water line. The water density is 1.00 g/cm^3 and the wood density is 0.600 g/cm^3.

	3 OCT '95	HOMEWORK #3	JOE ENGINEER 378-15-5432 ENGR 109 SECTION 521	2 OF 3

2. $L_1 = ?$ $L_2 = ?$

$m_1 = 30.0$ kg $m_2 = 20.0$ kg

F_1 F_2

$L = 5.00$ m

ASSUME MASS OF TEETER-TOTTER IS NEGLIGIBLE

$F_1 L_1 = F_2 L_2$ @ STATIC EQUILIBRIUM

$F_1 = m_1 g$

$F_2 = m_2 g$

$m_1 g L_1 = m_2 g L_2$

$L_1 = \frac{m_2}{m_1} L_2$

$L = L_1 + L_2$

$= \frac{m_2}{m_1} L_2 + L_2$

$= \left(\frac{m_2}{m_1} + 1\right) L_2$

$$L_2 = \frac{L}{\frac{m_2}{m_1} + 1} = \frac{5.00 \text{ m}}{\frac{20.0 \text{ kg}}{30.0 \text{ kg}} + 1} = \boxed{3.00 \text{ m}}$$

$$L_1 = L - L_2 = 5.00 \text{ m} - 3.00 \text{ m} = \boxed{2.00 \text{ m}}$$

Sample Problem 2:

An object is in static equilibrium when all the moments balance. (A moment is a force exerted at a distance from a fulcrum point.) A 30.0 kg child and a 20.0 kg child sit on a 5.00 m long teeter-totter. Where should the fulcrum be placed so the two children balance?

	3 OCT '95	HOMEWORK #3	JOE ENGINEER 378-15-5432 ENGR 109 SECTION 521	3 OF 3

3.

$V = 1.5\ \text{V}$, $i = ?$

$R_1 = 5.00\ \Omega$, $R_2 = 10.0\ \Omega$, $R_3 = 15.0\ \Omega$

$$V = iR$$

$$i = \frac{V}{R}$$

$$\frac{1}{R} = \frac{1}{R_1} + \frac{1}{R_2} + \frac{1}{R_3}$$

$$R = \left[\frac{1}{R_1} + \frac{1}{R_2} + \frac{1}{R_3}\right]^{-1}$$

$$= \left[\frac{1}{5.00\ \Omega} + \frac{1}{10.0\ \Omega} + \frac{1}{15.0\ \Omega}\right]^{-1}$$

$$= 2.73\ \Omega$$

$$i = \frac{1.5\ \text{V}}{2.73\ \Omega} \times \frac{\Omega}{\text{V/A}}$$

$$\boxed{i = 0.55\ \text{A}}$$

Sample Problem 3:
The voltage drop through a circuit is equal to the current times the total resistance of the circuit. When resistances are placed in parallel, the inverse of each resistance sums to the inverse of the total resistance. Three resistors (5.00 Ω, 10.0 Ω, and 15.0 Ω) are placed in parallel. How much current flows from a 1.5-V battery?

3.5 ESTIMATING

The final step in solving a problem is to check the answer. One can accomplish this by working the problem using a completely different method, but often there is not time for this. Instead, a very valuable approach is to estimate the answer.

The ability to estimate comes with experience. After many years of working similar problems, an engineer can "feel" if the answer is in the right ballpark. As a student, you do not have experience, so estimating may be difficult for you; however, your inexperience is not an excuse for failing to learn how. If you cultivate your estimating abilities, they will serve you well in your engineering career. It has been our experience that much business is conducted over lunch, using napkins for calculations and drawings. An engineer who has the ability to estimate will impress both clients and bosses.

A question we often hear from students is, How do I make reasonable assumptions during the estimation process? There is no quick and simple answer. One of the reasons that "has broad interests" and "collects obscure information" are included among the traits of a creative engineer (Chapter 1) is that these traits will supply you with raw data and cross checks for reasonableness when making estimations. With practice and experience you will become better. Notice how much your skills improve with the few examples you do in the classroom and for homework. After working in a particular engineering field for a while, many estimations will become second nature.

To get you started as an expert estimator, we have prepared Table 3.2. It lists relationships that we find to be very useful during our business lunches. Use this table as a starting point; as you mature in your own engineering discipline, you will undoubtedly learn other relationships and rules of thumb that will serve you well.

The following examples describe five approaches to estimation.

EXAMPLE 3.1 Simplify the Geometry

Problem Statement: Estimate the surface area of an average-sized man.

Solution: Approximate the human as spheres and cylinders.

$$A = A_{\text{head}} + A_{\text{torso}} + 2A_{\text{leg}} + 2A_{\text{arm}}$$

$$= 4\pi r_{\text{head}}^2 + (2\pi rL)_{\text{torso}} + 2(2\pi rL)_{\text{leg}} + 2(2\pi rL)_{\text{arm}}$$

$$= 4\pi(3.75\text{ in})^2 + [2\pi(5\text{ in})25\text{ in}] + 2[2\pi(3\text{ in})32\text{ in}] + 2[2\pi(1.5\text{ in})24\text{ in}]$$

$$= 2620\text{ in}^2 \times \left(\frac{2.54\text{ cm}}{\text{in}}\right)^2 \times \left(\frac{\text{m}}{100\text{ cm}}\right)^2 = 1.69\text{ m}^2$$

This is very close to the accepted value of 1.7 m^2.

TABLE 3.2
Useful relationships for estimation purposes

Conversion Factors

1 ft = 12 in
1 in = 2.54 cm
1 mi = 5280 ft
1 km ≈ 0.6 mi
1 ft^3 = 7.48 gal
1 gal = 3.78 L
1 bbl = 42 gal
1 mi^2 = 640 acre
1 kg ≈ 2.2 lb_m
1 atm = 14.7 psi
1 atm ≈ 1×10^5 N/m^2
1 atm = 760 mm Hg
1 atm ≈ 34 ft H_2O
1 min = 60 s
1 h = 60 min
1 d = 24 h
1 year = $365\frac{1}{4}$ d
1 Btu ≈ 1000 J
1 Btu ≅ $\frac{1}{4}$ kcal
1 hp = 550 ft·lb_f/s
1 hp ≈ 0.75 kW

Temperature Conversions

[K] = [°C] + 273.15
[°F] = 1.8 [°C] + 32
[°R] = [°F] + 459.67
[°C] = ([°F] −32)/1.8

Ideal Gas

Standard temperature = 0°C
Standard pressure = 1 atm
Molar volume (@ STP) = 22.4 L/gmole
Molar volume (@ STP) = 359 ft^3/lbmole

Molecular Weights

H = 1
C = 12
N = 14
O = 16
Air = 29 (21 mol % O_2, 79 mol % N_2)

Geometry Formulas

Circle area = πr^2
Circle circumference = $2\pi r$
Cylinder volume = $\pi r^2 L$
Cylinder area (without end caps) = $2\pi rL$
Sphere area = $4\pi r^2$
Sphere volume = $\frac{4}{3}\pi r^3$
Triangle area = $\frac{1}{2}$ (base)(height)

Temperature References

Absolute zero = 0 K
He boiling point = 4 K
N_2 boiling point = 77 K
CO_2 sublimation point = 195 K = −78°C
Mercury melting point = −39°C
Freezer temperature = −20°C
H_2O freezing point = 0°C
Refrigerator temperature = 4°C
Room temperature = 20°C to 25°C
Body temperature = 37°C
H_2O boiling point = 100°C
Lead melting point = 327°C
Aluminum melting point = 660°C
Iron melting point = 1540°C
Flame temperature (air oxidant) = 2100°C
Flame temperature (pure oxygen) = 3300°C
Carbon melting point = 3700°C

Physical Constants

Speed of light ≈ 3×10^8 m/s
Speed of sound (air, 20°C, 1 atm) ≈ 770 mph
Avogadro's number ≈ 6×10^{23}
Acceleration from gravity ≈ 9.8 m/s^2 ≈ 32.2 ft/s^2

Mathematical Constants

$e \approx 2.718$
$\pi \approx 3.14159$

Water Properties

Density = 1 g/cm^3 = 62.4 lb_m/ft^3 = 8.34 lb_m/gal
Latent heat of vaporization ≈ 1000 Btu/lb_m
Latent heat of fusion ≈ 140 Btu/lb_m
Heat capacity = 1 Btu/(lb_m · °F) = 1 kcal/(kg · °C)
Melting point = 0°C
Boiling point = 100°C

Densities

Air (0°C, 1 atm) = 1.3 g/L
Wood ≈ 0.5 g/cm^3
Gasoline = 0.67 g/cm^3
Oil = 0.88 g/cm^3
Concrete = 2.3 g/cm^3
Aluminum = 2.6 g/cm^3
Steel = 7.9 g/cm^3
Lead = 11.3 g/cm^3
Mercury = 13.6 g/cm^3

Earth

Circumference ≈ 40,000 km
Mass ≈ 6×10^{24} kg

People

Male	Female
Avg mass = 168 lb_m	Avg mass = 139 lb_m
Avg height = 68 in	Avg height = 63 in

Sustained work output of man = 0.1 hp

Energy

Natural gas: 1000 standard ft^3 ≈ 10^6 Btu
Crude oil: 1 bbl ≈ 6×10^6 Btu
Gasoline: 1 gal ≈ 125,000 Btu
Coal: 1 lb_m ≈ 10,000 Btu

Ultimate Strength of Materials

Steel = 60,000 to 125,000 lb_f/in^2
Aluminum = 11,000 to 80,000 lb_f/in^2
Concrete = 2000 to 5000 lb_f/in^2

Automobiles

Typical mass = 2000 to 4000 lb_m
Typical fuel usage = 600 gal/year
Typical fuel economy = 20 to 30 mi/gal
Typical power = 50 to 500 hp
Engine speed = 1000 to 5000 rpm

Typical Appliance Energy Usage

Fluorescent lamp = 40 W
Incandescent lamp = 50 to 100 W
Refrigerator-freezer = 500 W
Window air conditioner = 1500 W
Heat pump = 12,000 W
Cooking range = 12,000 W
Avg continuous home electricity usage ≈ 1 kW

EXAMPLE 3.2 Use Analogies

Problem Statement: Estimate the volume of an average-sized man.

Solution: Assume the density of a man is 0.95 that of water. (People are mostly water, but they do float slightly when swimming, so the density must be slightly less than water.)

$$V = \frac{m}{\rho} = \frac{168\ \text{lb}_\text{m}}{0.95(1.0\ \frac{\text{g}}{\text{cm}^3})} \times \frac{\text{kg}}{2.2\ \text{lb}_\text{m}} \times \frac{1000\ \text{g}}{\text{kg}} \times \left(\frac{\text{m}}{100\ \text{cm}}\right)^3 = 0.080\ \text{m}^3$$

EXAMPLE 3.3 Scale Up From One to Many

Problem Statement: How many bed pillows can fit in the back of a tractor trailer?

Solution: One pillow measures 3 in thick by 16 in wide by 21 in long. The cargo bed of a tractor trailer measures roughly 8 ft wide by 10 ft tall by 35 ft long.

$$\text{Number of pillows} = \frac{V_\text{truck}}{V_\text{pillow}} = \frac{(8\ \text{ft})(10\ \text{ft})(35\ \text{ft})}{(3\ \text{in})(16\ \text{in})(21\ \text{in})} \times \left(\frac{12\ \text{in}}{\text{ft}}\right)^3 = 4800$$

This calculation neglects the compression of the pillows as they stack on top of each other (which would increase the number carried). It also neglects the volume occupied by packing materials (which would decrease the number carried). Hopefully, the errors due to the simplifying assumptions will tend to cancel each other out.

EXAMPLE 3.4 Place Limits on Answers

Problem Statement: Estimate the mass of an empty tractor trailer.

Solution: It seems reasonable that the tractor trailer mass should be more than five automobiles, but less than 30 automobiles.

$$\text{Lower bound} = 5 \times 3000\ \text{lb}_\text{m} = 15{,}000\ \text{lb}_\text{m}$$

$$\text{Upper bound} = 30 \times 3000\ \text{lb}_\text{m} = 90{,}000\ \text{lb}_\text{m}$$

Because automobiles typically have a mass between 2000 and 4000 lb_m, an average value of 3000 lb_m was used. The estimated mass of the tractor trailer is reasonable; a literature search revealed that heavy trucks weigh about 42 tons (84,000 lb_m).

EXAMPLE 3.5 Extrapolate From Samples

Problem Statement: How much fuel is burned by Texas A&M students for the Thanksgiving visit home?

PROBLEM 3.3

Finally, using six equilateral triangles with finite thickness (plane or paper-thin triangles are not the correct shape), make each touch:

Two others (two completely different ways)
Three others (two completely different ways)

As with rectangular parallelepipeds, corner touches do not count.

3.6.2 Classifying Problem-Solving Strategies

A good engineer's time is always spoken for. And because good engineers enjoy engineering, it is indeed an act of love of the profession to take time from their work to write books and articles for the general public. It is even rarer for engineers to analyze how they approach engineering problems and write this down for the student. In 1945, George Polya, a mathematician, published the book *How to Solve It* (Polya, 1945), describing in general terms methods to solve mathematical problems. These methods are well suited to engineering problems. (Alas, the book is out of print. If you can find a used copy, it is well worth reading.)

Polya summarized his method as the "How to Solve It List":

First. You have to *understand* the problem.

What is the unknown? What are the data? What is the condition?

Is it possible to satisfy the condition? Is the condition sufficient to determine the unknown? Or is it insufficient? Or redundant? Or contradictory?

Draw a figure. Introduce suitable notation.

Separate the various parts of the condition. Can you write them down?

Second. Find the connection between the data and the unknown. You may be obliged to consider auxiliary problems if an immediate connection cannot be found. You should obtain eventually a *plan* of the solution.

Have you seen it before? Or have you seen the same problem in a slightly different form?

Do you know a related problem? Do you know a theorem that could be useful?

Look at the unknown! And try to think of a familiar problem having the same or a similar unknown.

Here is a problem related to yours and solved before. Could you use it? Could you use its results? Could you use its method? Should you introduce some auxiliary element in order to make its use possible? Could you restate the problem? Could you restate it still differently? Go back to definitions.

If you cannot solve the proposed problem, try to solve first some related problem. Could you imagine a more accessible related problem? A more general problem? A more special problem? An analogous problem? Could you solve a part of the problem? Keep only a part of the condition, drop the other part; how far is the unknown then determined, how can it vary? Could you derive something useful from the data? Could you think of other data appropriate to determine the

TABLE 3.3
Problem-solving strategies

Polya	Woods et al.	Bransford and Stein	Schoenfeld	Krulik and Rudnick
Understand the problem	Define the problem	Identify the problem	Analyze the problem	Read the problem
	Think about it	Define and represent it	Explore it	Explore it
Devise a plan	Plan	Explore possible strategies	Plan	Select a strategy
Carry out the plan	Carry out the plan	Act on the strategies	Implement	Solve
Look back	Look back	Look back and evaluate the effects of your activities	Verify	Look back

unknown? Could you change the unknown or the data, or both if necessary, so that the new unknown and the new data are nearer to each other?

Did you use all the data? Did you use the whole condition? Have you taken into account all essential notions involved in the problem?

Third. *Carry out* your plan.

Carrying out your plan of the solution, *check each step.* Can you see clearly that the step is correct? Can you prove that it is correct?

Fourth. *Examine* the solution obtained.

Can you *check the result?* Can you check the argument? Can you derive the result differently? Can you see it at a glance? Can you use the result, or the method, for some other problem?

There are many variations of this basic scheme; Table 3.3 summarizes strategies suggested by other authors whose work you will find in the bibliography. Most notable is that they all offer roughly the same advice, and that all of them emphasize *checking your results* as the last step. To err is human, but most human of all is to screw everything up in the final steps after the fun of solving the problem is gone and deadlines gallop closer. Many an engineer wishes he had taken an extra 5 minutes to check units or to see if the result really made sense.

Many problem-solving methods have been categorized. The following examples are by no means a complete list—but the particular problem one is pondering has a vanishingly small chance of being categorized anyway. The object of these examples is to get you to think about general methods of attack on problems and, hopefully, to ponder your own methods of tackling problems. Moving from abstract discussions of problem solving to concrete examples will help you to grasp each strategy.

EXAMPLE 3.6 Exploit Analogies or Explore Related Problems

Exploiting analogies is one of the most common approaches to solving a problem. In the old days of engineering (20 years ago), machines called analog computers existed. By using resistors to model fluid friction, capacitors to model holdup tanks, and batteries for pumps, the flow of fluid through a piping system could be modeled with electronics. As you practice engineering, your list of solved problems will grow and become your list against which to test new problems, looking for similarities and analogies.

Problem Statement: Freight carriers charge by rate proportional to the length, depth, and height of a package. What are the minimum lengths, depths, and heights of a rectangular parallelepiped to ship a rod of length q, if the diameter of the rod is negligible? (Adapted from Polya, 1945, and others.)

Solution:

What is the unknown? The lengths of the sides of the parallelepiped, say a, b, and c.
What is the given? The length of the rod q, and that the diameter of the rod is negligible.
Draw a figure.

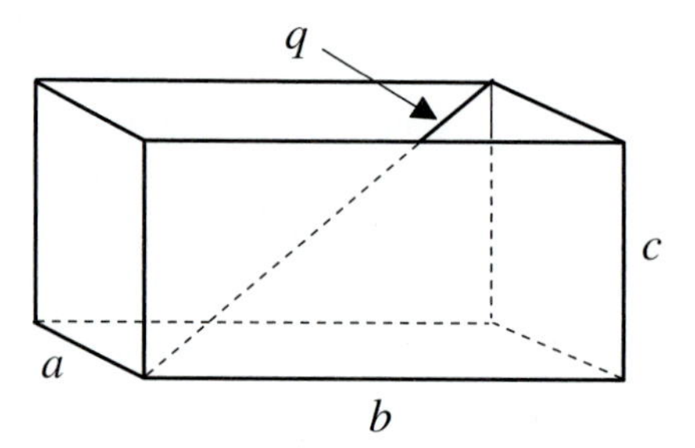

What is the unknown? The diagonal of the parallelepiped.
Do you know any problems with a similar unknown? Yes, sides of right triangles!
Devise a plan.

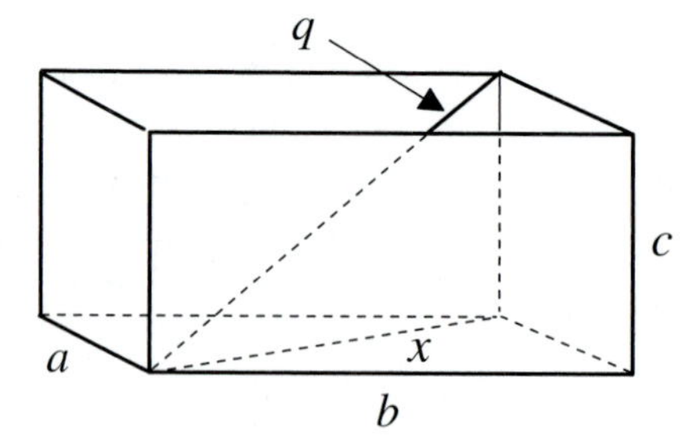

Carry out the plan. If we could find x, then we could relate c and x to q by the **Pythagorean theorem,** that is,

$$q^2 = x^2 + c^2$$

But x^2 is equal to $b^2 + a^2$, so that

$$q = \sqrt{a^2 + b^2 + c^2}$$

Look back. Did we use the appropriate data? Does the answer make sense? (What happens if a goes to zero? Check special cases.) Is there symmetry? (Because a, b, and c can be arbitrarily rotated to any edge, is the form of the answer symmetrical?)

EXAMPLE 3.7 Introduce Auxiliary Elements; Work Auxiliary Problem

Sometimes a problem is too difficult (or it's too early in the morning) to tackle head on. An approach that allows progress and often lights the path to the real insight to unknot the problem is to tackle the problem with one or more of the constraints lifted. By solving the easier problem, we can sometimes maneuver the parameters of the simpler problem until the lifted constraint is also satisfied and we're done. As in this problem, it's spectacular when it works.

Problem Statement: Inscribe a square in any given triangle. Two vertices of the square should be on the base of the triangle, the other two vertices of the square on the other two sides of the triangle, one on each. (From Polya, 1945.)

Solution:

What is the unknown? A method to inscribe a square in any triangle.
What is the given? An arbitrary triangle. Directions for placing square in triangle.

Draw a figure.

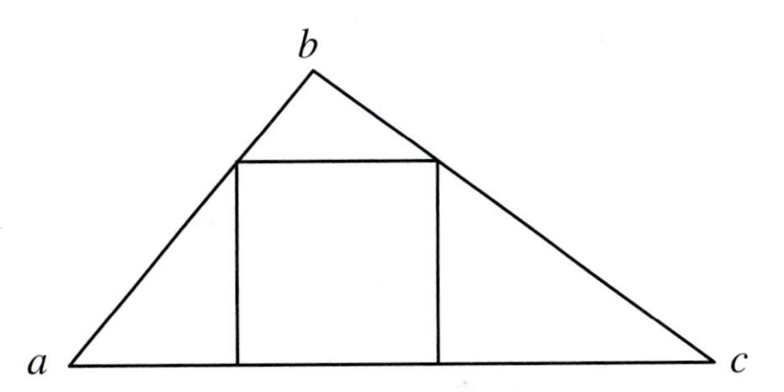

Ouch! This is tough. Even using a graphics program to do the hard part, it still takes us 10 minutes to jockey a square about the right size into a triangle, and then it's only an approximation. With odd-shaped triangles (short height with a long base, or tall with a narrow base), the whole thing gets worse.

Hmmm . . . What if we let squares sit on the base, but they only have to touch the triangle on side *ab*? If we draw a bunch of squares, maybe the lights will come on.

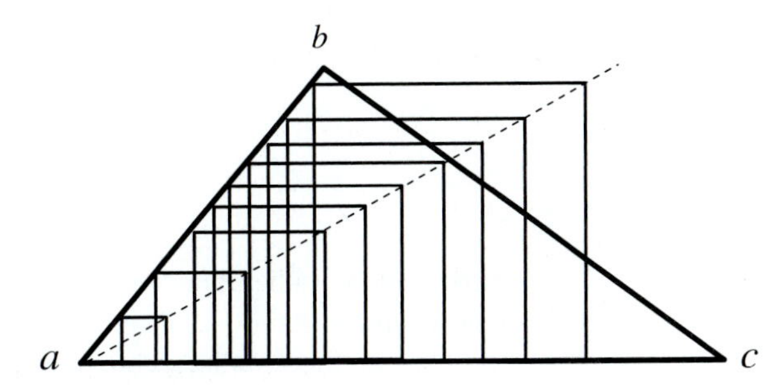

Notice anything unusual about the free vertices? They all lie along a single line! From there, algorithm construction is simple.

Carry out the plan. Construct an arbitrary square on the base of the triangle and draw a line from point a on the triangle through the free vertex of the square. Where this line intersects side bc on the triangle is where the free vertex of the desired square will coincide with side bc of the triangle. Construct a perpendicular line to the base of the triangle through this point. This is one side of the square. Mark off the same distance along the base of the triangle; this is the second side of the square. Construct a perpendicular from this point on the base to side ab on the triangle. This locates all four vertices of the square.
Look back. With the clarity of hindsight, why do the vertices lie on a single line? Is this right? Knowing this, can an easier algorithm be developed rather than constructing an arbitrary square on the base of the triangle and drawing a line from point a on the triangle through the free vertex of the square?

EXAMPLE 3.8 Generalizing: The Inventor's Paradox

Sometimes it is easier to solve a more general version of a given engineering problem and put in the specific parameters at the end, rather than to solve the particular problem straight away. Much work has been done to solve general problems, so the efforts of others can sometimes be brought to your aid by enlarging your problem. In the study of differential equations, for example, the trick is to identify which general class of equations the particular equation you're working with belongs. Once you've done this, the answer is obvious, because general solutions are given for each class of equations.

Problem Statement: A satellite in circular orbit about the earth, 8347.26 miles from the center of the earth, moves outward 6.1 ft. How much new area is enclosed in its new orbit?

Solution:

What is the unknown? The extra area inside the orbit.
What is the given? Circular orbit with a radius of 8347.26 miles. Move out 6.1 ft.
Draw a figure.

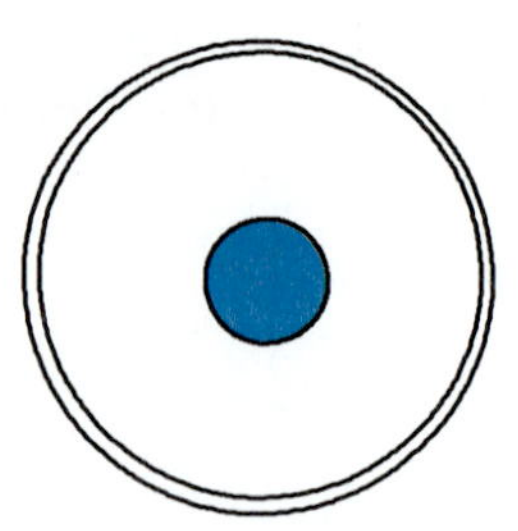

Make a plan. Rather than launch in and crunch numbers, let's do a bit of algebra for a second.
Carry out the plan. Let R be the beginning radius, and let q be the distance added to the radius. For the smaller orbit, the area enclosed was πR^2. For the new orbit, the area enclosed is $\pi(R + q)^2$. The difference is the new area enclosed, or

$$\text{New area} = \pi(R + q)^2 - \pi R^2 = \pi(R^2 + 2Rq + q^2) - \pi R^2$$

$$= \pi(2Rq + q^2) = \pi q(2R + q)$$

$$= 3.14159\left(6.1 \text{ ft} \times \frac{\text{mile}}{5280 \text{ ft}}\right)\left[2(8347.26 \text{ mi}) + \left(6.1 \text{ ft} \times \frac{\text{mile}}{5280 \text{ ft}}\right)\right]$$

$$= 61 \text{ square miles}$$

Look back. Is this reasonable? How could this result be checked? Are the units right? Are the conversion factors right? If q went to zero, does the formula work? If R went to zero?

If you are not convinced that this algebraic approach is better, start right out with numbers and see how you do. Besides that, by solving this a general way, several avenues open up for checking the result.

EXAMPLE 3.9 Specializing; Specializing to Check Results

Specialization is our all-time favorite tool for problem solving. For checking results as the final step in problem solving, specialization is the tool of choice. Results are often known for special cases, and this offers a wonderful way to check a general algorithm. Another standard approach is to push the algorithm's parameters to their limits and assure that the algorithm fails gracefully or scoots to infinity if it's supposed to.

Problem Statement: In a triangle, let r be the radius of the inscribed circle, R the radius of the circumscribed circle, and H the longest altitude. Then prove

$$r + R \leq H$$

(In Polya, 1945, quoting from *The American Mathematical Monthly* 50, 1943, p. 124 and 51, 1944, pp. 234–236.)

Solution:

What is the unknown? How to show that the sum of the radii of inscribed and circumscribed circles is less than or equal to the greatest height of the triangle.
What is the given? The formula and directions for constructing the figure.
Draw a figure.

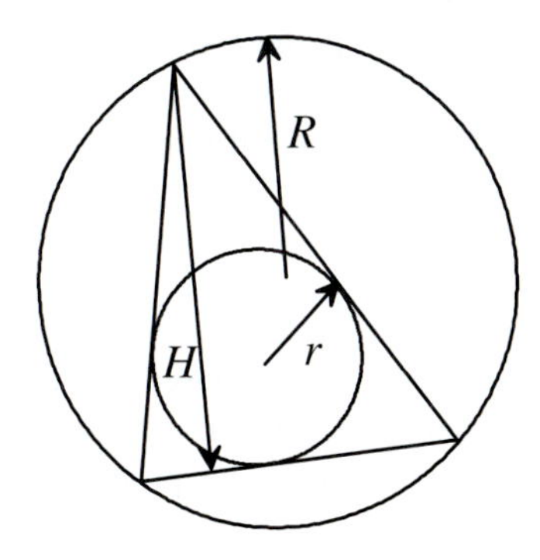

Solution:

What is the unknown? How to construct the circle.
What is the given? Two intersecting lines and one point of tangency.
Draw a figure.

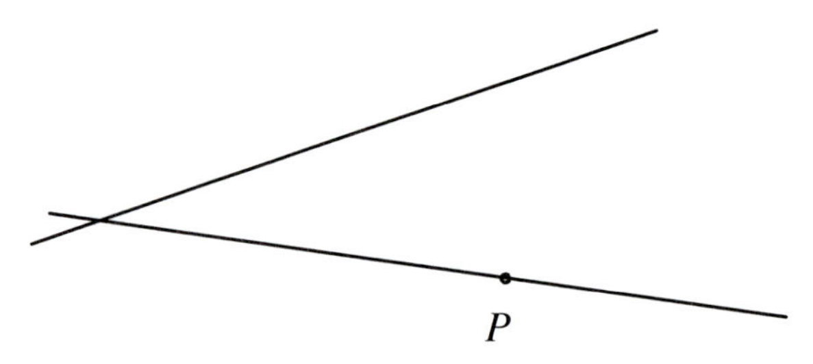

Devise a plan. Let's sketch in something that's close to a circle and close to the right size and see what happens.

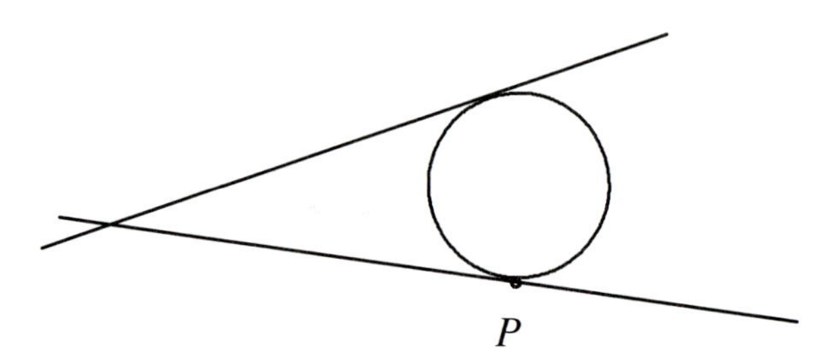

Well, not too bad for a sketch, but now what? The center of the circle looks straight up from point P. But it shouldn't be, right? Aren't tangents at right angles to the radius of the circle at the point of tangency? Well, that helps. I can construct a perpendicular at point P, but I still can't draw the circle because I don't know where the center is. Hmmm . . .

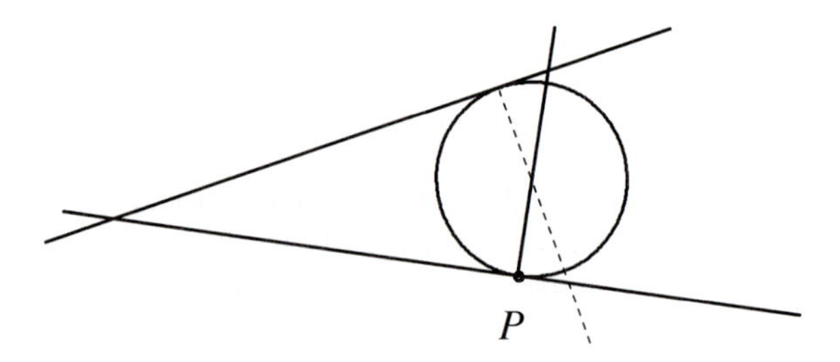

If I only knew where to place the other perpendicular from the other intersecting line (the dotted line in the figure), I'd be done. But I don't. Rats. . . . But wait a minute! If that line *were* there, then the radius from the center to point P is the same distance as the radius to the tangency on the other line, right?

That means that the center of the circle is on the angular bisector, thus

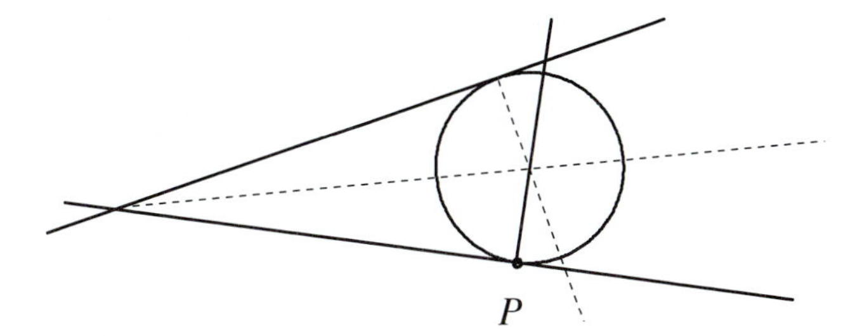

So now the scheme is clear. At point *P*, construct a perpendicular. Bisect the angle between the two intersecting lines. The point where the bisector and perpendicular meet is the center of the circle.

Look back. Is this true for any two intersecting lines? How about obtuse angles? Can we do that? How about as the lines intersect at angles near zero (i.e., as the lines approach being parallel)?

EXAMPLE 3.12 Working Forward / Working Backward

One of our students informed us that learning this trick alone was worth all the effort of the entire course. His work is synthesizing organic compounds, and this is his standard trick. He looks at what he wants to make and then mentally breaks the compound down to simpler components that he can make or, better yet, has sitting on his shelf. Another, less esoteric, application of this method is in solving maze puzzles. Start at the end and work back toward the start. Often the puzzle is easier this way. What usually happens in practice is that one starts from the beginning and works forward until stymied, then starts at the end and works backward. With luck one meets oneself somewhere in the middle.

Problem Statement: Measure exactly 7 ounces of liquid from a large container using only a 5-ounce container and an 8-ounce container. (Practiced with beer at many student hangouts.) (From Polya, 1945, and many others.)

Solution:

What is the unknown? How to measure seven ounces.
What is the given? Unlimited liquid, 5-oz and 8-oz containers.
Draw a figure.

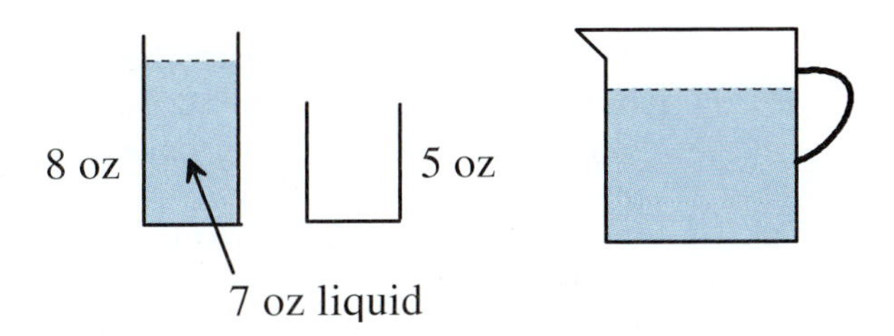

This is, of course, the final state we seek: the 7 oz of liquid in the 8-oz container. (It's hard to put that much liquid into the 5-oz container.)
Devise a plan. The obvious comes to mind first. We can get 3 oz by filling the 8-oz container and then pouring from it into the 5-oz container until the 5-oz is full, leaving 3 oz of liquid in the 8-oz container. Also, we can get 2 oz by filling the 5-oz container, pouring it into the 8-oz container, filling the 5-oz container again, and emptying it into the 8-oz container until the 8-oz container is full, leaving 2 oz in the 5-oz container. Well, so what? Let's start at the end and see if we can meet in the middle.
Carry out the plan. If we back up one step from the final state we would have:

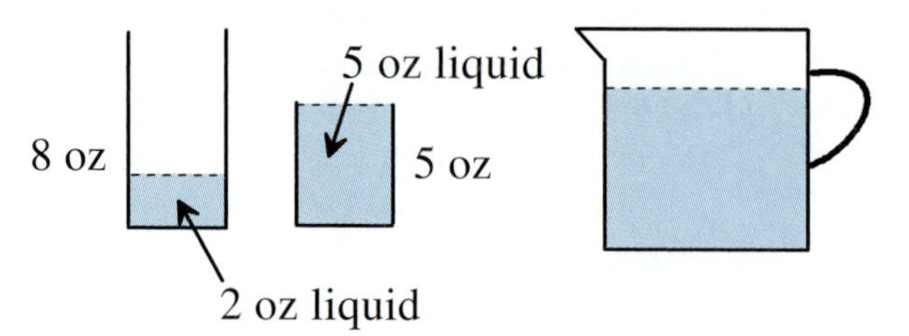

Bingo! I know how to get 2 oz. So now we perform the forward sequence to get 2 oz, then empty the full 8-oz container, pour in the 2 oz, refill the 5-oz container, and empty it into the 8-oz container.
Look back. Is each step correct? Anything special about 8-, 7-, and 5-oz measurements?

EXAMPLE 3.13 Argue by Contradiction; Using Reductio ad Absurdum

Occasionally, nature is kind enough to present a problem that must have only one of a small number of answers. It may be impossible to prove directly that one particular answer is correct. However, if you can prove that all others are not the answer, you prove indirectly that the remaining choice is correct.

Problem Statement: Write numbers using each of the 10 digits (0 through 9) exactly once such that the sum of the integers is exactly 100. For example, $29 + 10 + 38 + 7 + 6 + 5 + 4 = 99$ uses each digit once, but adds to 99 instead of 100. And then there's $19 + 28 + 31 + 7 + 6 + 5 + 4 = 100$, but . . . no zero. (From Polya, 1945.)

Solution:

What is the unknown? How to create numbers using the digits 0 through 9 so that the sum is 100.
What is the given? The digits 0 through 9, examples.
Draw a figure. For once, this doesn't help.
Devise a plan. After a lot of trial and error, one is convinced that the answer is not obvious, if it exists. However, proof that it is impossible is not very obvious either. We'll bet it has something to do with which digits are in the tens place. Let's analyze how the sums can be made and, assuming it can be done, look for a contradiction.
Carry out the plan. First, the sum of the 10 digits, each taken singly, is:

$$0 + 1 + 2 + 3 + 4 + 5 + 6 + 7 + 8 + 9 = 45$$

If we add "0" to any of the digits, for example we change 1 to 10, the sum increases by 10 times whichever digit we used, less the digit itself. For 10, the sum is $45 + 10 - 1 = 54$. For 40, the sum is $45 + 40 - 4 = 81$. For 60, the sum is 99; for 70 the sum is 108, too large as are 80 and 90. If digits are combined with numbers other than 0, for example combining the 4 and the 2 to make 42, the sum is $45 - 4 + 40 = 81$. Ahhh, only the digits used in the tens place are "lost" in the sum. We don't know how many digits will be lost (at most three, right?), but we can write an equation about their sum:

$$10 \times (\text{tens\_sum}) + 45 - \text{tens\_sum} = 100$$

Now we're cooking. But wait—solving the equation for tens_sum, we get 55/9???? What kind of junk is this? Exactly the kind of junk we hoped for. Because there is no way to add integers and get 55/9, then the conjecture that the digits can sum to 100 must be false.

Look back. Is there some way to prove our method is correct? How about if we relax the requirement that the digits be integers? Can we use the digits and get 100?

3.7 SUMMARY

During the 40 or so years you will practice engineering, technology will change dramatically. Therefore, it is impossible for your professors to teach you every fact you will need; today's state-of-the-art fact will become obsolete tomorrow. Instead, to prepare you for the future, we can help train your mind to think and solve problems. These skills never become obsolete.

Engineers face a variety of problems, including research, troubleshooting, mathematics, resource, and design problems. The problem-solving process can start once the problem is identified. To solve problems, engineers apply both synthesis (where pieces are combined together into a whole) and analysis (where the whole is dissected into pieces).

During your schooling, you will be confronted with thousands of problems. So you can check your work, the solution to these problems is provided to you. If you make a mistake, the consequences are not severe—the loss of a few points on a homework assignment or exam. In the real world, there are no answers in the back of the book, and the consequences of making a mistake can be catastrophic. You are well-advised to develop a systematic problem-solving strategy that leads you to the correct answer. We offer a suggested approach in the section on "error-free" problem solving.

One of the most valuable skills in engineering is the ability to estimate answers from incomplete information. Because we almost never have all the information we need to precisely solve a problem, nearly every engineering problem can be viewed as an estimation problem; they differ only in the degree of uncertainty in the final answer.

Many authors agree that problem solving can be broken into four or five steps: (1) understand the problem, (2) think about it, (3) devise a plan, (4) execute the plan, and (5) check your work. While thinking about the problem, a number of **heuristic** approaches (i.e., hints) can lead you to the solution. For example, you can exploit analogies, introduce auxiliary elements, generalize, specialize, decompose, take the problem as solved, work forward/backward, or argue by contradiction.

Perhaps you enjoy solving puzzles in the Sunday newspaper. Although these may seem like trivial games, these puzzles can actually train your mind to solve engineering problems. Just as a boxer may train with a jump rope for an upcoming fight, an engineer can train with puzzles to prepare for the main event: solving engineering problems.

Further Readings

Albrecht, K. *Brainpower: Learning to Improve Your Thinking Skills.* New York: Prentice Hall, 1987.

Bransford, J. D., and B. S. Stein. *The Ideal Problem Solver.* New York: W. H. Freeman, 1984.

de Bono, E. *The Five-Day Course in Thinking.* New York: Signet, 1967.

———. *New Think.* New York: Avon, 1968.

———. *Lateral Thinking: Creativity Step by Step.* New York: Harper, 1970.

Epstein, L. C. *Thinking Physics.* San Francisco: Insight Press, 1990.

Gardner, M. *Mathematical Puzzles and Diversions.* New York: Simon & Schuster, 1959.

———. *Mathematical Carnival.* New York: Alfred A. Knopf, 1975.

———. *Mathematical Circus.* New York: Alfred A. Knopf, 1979.

Graham, L. A. *The Surprise Attack in Mathematical Problems.* New York: Dover Publications, 1968.

Krulik, S., and J. A. Rudnick. *Problem Solving: A Handbook for High School Teachers.* Boston: Allyn and Bacon, 1989.

Miller, J. S. *Millergrams: Some Enchanting Questions for Enquiring Minds.* Sydney: Ure Smith, 1966.

Moore, L. P. *You're Smarter Than You Think.* New York: Holt, Rinehart & Winston, 1985.

Polya, G. *How to Solve It.* Princeton: Princeton University Press, 1945.

Row, T. S. *Geometric Exercises in Paper Folding.* New York: Dover Publications, 1966.

Sawyer, W. W. *Mathematician's Delight.* London: Penguin Books, 1943.

Schoenfeld, A. H. *Mathematical Problem Solving.* San Diego: Academic Press, 1985.

Schuh, F. *The Master Book of Mathematical Recreations.* New York: Dover Publications, 1968.

Stewart, I. *Game, Set, and Math: Enigmas and Conundrums.* Oxford: Basil Blackwell, 1989.

Witt, S. *How to Be Twice as Smart: Boosting Your Brainpower and Unleashing the Miracles of Your Mind.* West Nyack, NY: Parker Publishing, 1983.

Woods, D. R.; J. D. Wright; T. W. Hoffman; R. K. Swartman; and I. D. Doig. "Teaching Problem Solving Skills." *Engineering Education,* December 1975, pp. 238–243.

PROBLEMS

3.1 Mary Mermaid is taking swimming lessons in a circular pool. Mary starts at one edge of the pool and swims in a straight line for 12 meters, where she hits the edge of the pool. She turns and swims another 5.0 meters and again hits the edge of the pool. As she is examining her various scrapes, she realizes that she is exactly on the opposite side of the pool from where she started. What is the diameter of the pool?

3.2 What is the shortest path for an ant crawling on the surface of a unit cube from the starting point to the ending point shown:

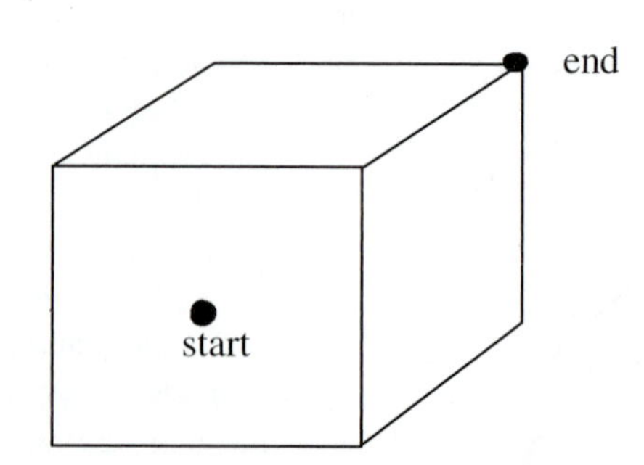

3.3 What is the shortest path for an ant crawling on the surface of a unit cube from the starting point to the ending point shown:

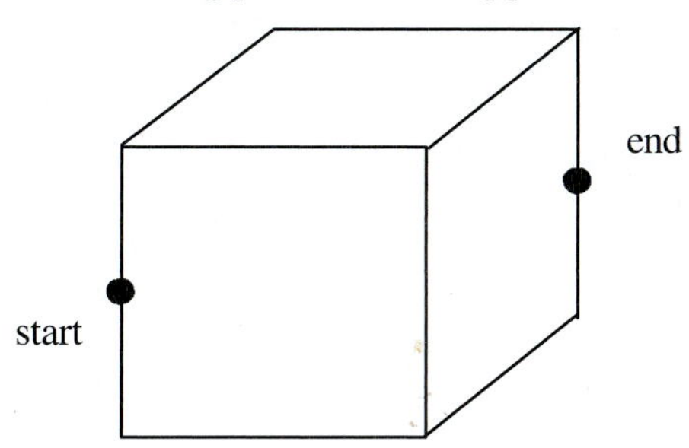

3.4 A string is fit snugly around the circumference of a spherical hot air balloon. More hot air is added (probably by a prominent scientist's lecture) and it now takes an additional 12.4 feet of string to fit around the circumference. What is the increase in diameter?

3.5 The poet Henry Wadsworth Longfellow, in his novel *Kavenaugh,* presented the following puzzle:

When a water lily stem is vertical, the blossom is 10 cm above the water. If the blossom is pulled to the right keeping the stem straight, the blossom touches the water 21 cm from where the stem came through the water when vertical. How deep is the water? (*Hint*: See figure.)

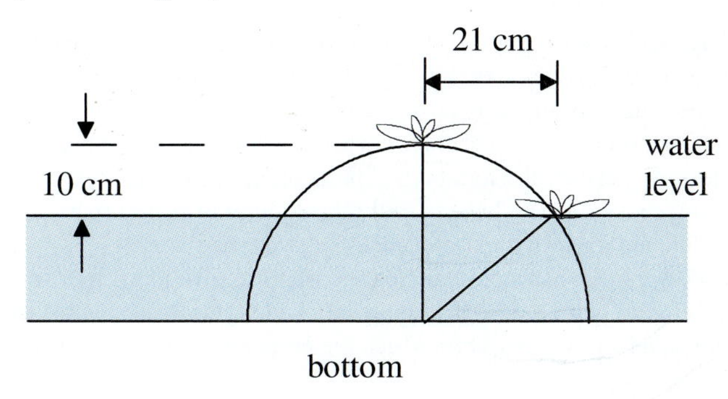

3.6 A farmer goes to market with 100 dollars to spend. Cows cost 10 dollars each, pigs cost 3 dollars each, and sheep cost half a dollar each. The farmer buys from the cattle dealer, the pig dealer, and the sheep dealer. She spends exactly 100 dollars and buys exactly 100 animals. How many of each animal does she buy?

3.7 The king and his two children are imprisoned at the top of a tall tower. Stone masons have been working on the tower and have left a pulley fixed at the top. Over the pulley runs a rope with a basket attached to either end. In the basket on the ground is a stone like the ones used to build the tower. The stone weighs 35 kg_f (75 lb_f). The king figures out that the stone can be used as a counterbalance—provided that the weight in either basket does not differ by more than 7 kg_f (15 lb_f). The king weighs 91 kg_f (195 lb_f), the princess weighs 49 kg_f (105 lb_f), and the prince weighs 42 kg_f (90 lb_f). How can they all escape from the tower? (They can throw the stone from the tower to the ground!) (Attributed to Lewis Carroll.)

3.8 From Problem 3.7, add a pig that weighs 28 kg_f (60 lb_f), a dog that weighs 21 kg_f (45 lb_f), and a cat that weighs 14 kg_f (30 lb_f). There is an extra limitation: there must be one human at the top and bottom of the tower to put the animals in and out of the basket. How can all six escape?

3.9 Calculate the ratio of the area to the volume for a unit cube, a unit sphere inscribed inside the cube, and a right cylinder inscribed inside the cube.

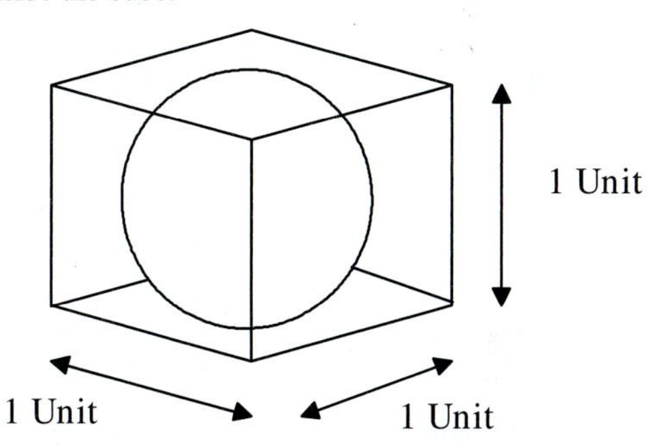

Next, for each having a unit volume (i.e., all three solids have the same volume) calculate the area-to-volume ratios for a sphere, cube, and cylinder.

3.10 Dr. Bogus, a close friend of ours, during a long session at the Dixie Chicken restaurant, was doodling on a napkin and "proved" that all numbers are equal. This came as quite a surprise to us, and we have delayed his calling the president only until you can review his proof. Here is a translation of the napkin doodles:

> Pick two different numbers, a and b, and a nonzero number c that is the difference between a and b, thus:
>
> $$a = b + c \qquad c \neq 0 \tag{1}$$
>
> Multiply both sides by $a - b$
>
> $$a(a - b) = (b + c)(a - b) \tag{2}$$
>
> or
>
> $$a^2 - ab = ab - b^2 + ac - bc \tag{3}$$
>
> Subtract ac from both sides,
>
> $$a^2 - ab - ac = ab - b^2 - bc \tag{4}$$
>
> Factor out a from the left side of the equation, and b from the right,
>
> $$a\,(a - b - c) = b\,(a - b - c) \tag{5}$$
>
> Eliminate the common factor from both sides, and
>
> $$a = b \tag{6}$$
>
> Because a and b can be any number, c can be positive or negative; thus, all numbers are equal to each other. Exactly which step(s) are wrong in the proof above, and why?

3.11 Dr. Bogus, now on a roll, also presented us with a proof that all our attempts to measure area are wrong. Again, we were somewhat startled, but he showed us that by simply rearranging the space within an area, one could get different answers. He presented us with the following drawing:

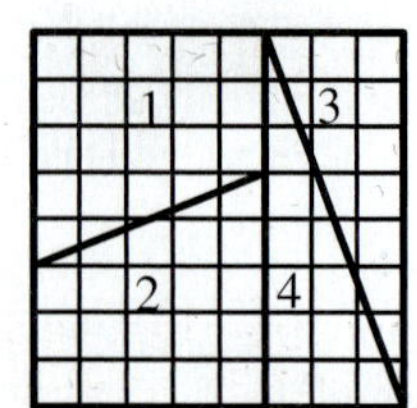

A square 8 units on a side is cut into 4 pieces. The pieces are then rearranged into a 5×13 rectangle.

Wait a minute!

$8 \times 8 = 64$
$5 \times 13 = 65$

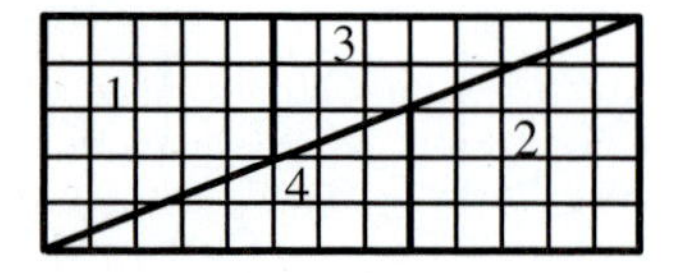

Where does the extra square come from? Help us, or Dr. Bogus will get a Nobel Prize before we do.

3.12 What is the plane angle between Line A and Line B, drawn on a unit cube?

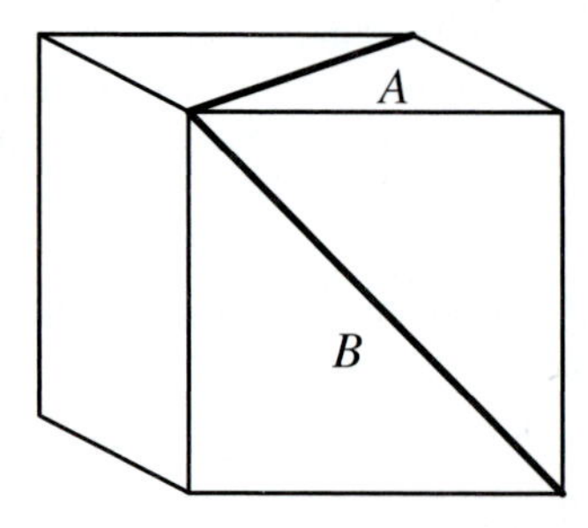

3.13 A fly is at the midpoint of the front edge of a unit cube as shown in the figure. What minimum distance must it crawl to arrive at the midpoint of the opposite top edge?

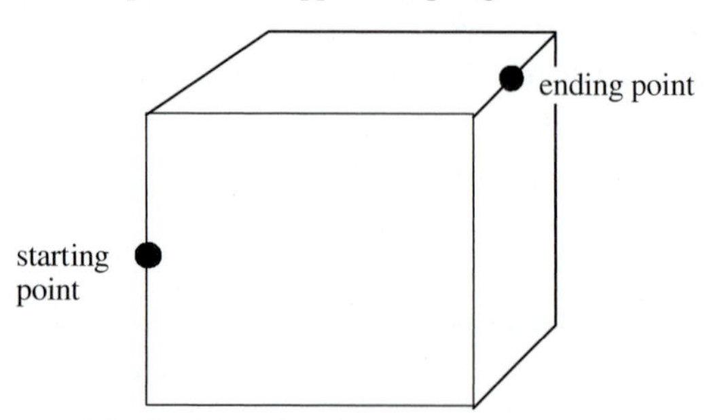

3.14 Suppose you wish to average 40 mph on a trip and find that when you are half the distance to your destination you have averaged 30 mph. How fast should you travel in the remaining half of the trip to attain an overall average of 40 mph?

3.15 A ship leaves port at 12:00 P.M. (noon) and sails east at 10 miles per hour. Another leaves the same port at 1:00 P.M. and sails north at 20 miles per hour. At what time are the ships 50 miles apart?

3.16 At a certain Point A on level ground, the angle of elevation to the top of a tower is observed to be 33°. At another Point B, in line with A and the base of the tower and 50 feet closer to the tower, the angle of elevation to the top is observed to be 68°. Find the height of the tower.

3.17 Using a straight track with 10 spots, set four pennies in the leftmost spots and four dimes in the rightmost spots, leaving the center two spots blank, as shown in the figure:

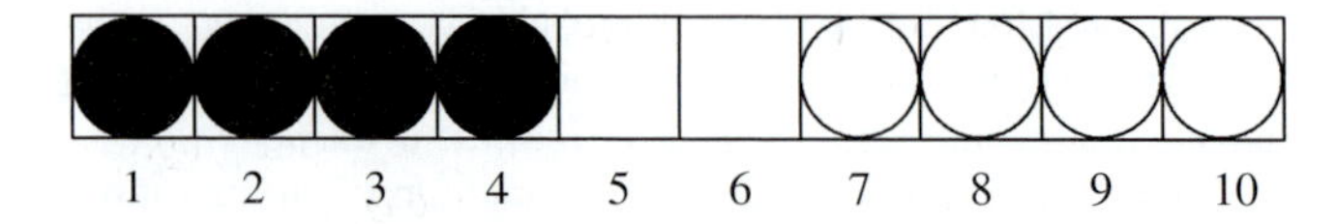

The object of the game is to move the pennies to the rightmost spots and the dimes to the left. You can move coins only by jumping a single coin into a vacant spot or by moving forward one to a vacant spot. No backward moves are allowed: all pennies must move to the right; all dimes to the left. Number the squares as shown to describe your solution.

3.18 Jack Jokely got up late one morning and was trying to find a clean pair of socks. He has a total of six pairs (three maroon and three white) in his drawer. Jack is not a very good housekeeper and his socks were just thrown in the drawer and, to make the morning perfect, the lightbulb is burned out so he can't see. How many socks does he have to pull out of the drawer before he gets a matched pair (two maroon socks or two white socks)?

3.19 A farmer has a piece of land he wants to give to his four sons (see figure below). The land must be divided into four equal-sized and equal-shaped pieces. How can the farmer accomplish his task?

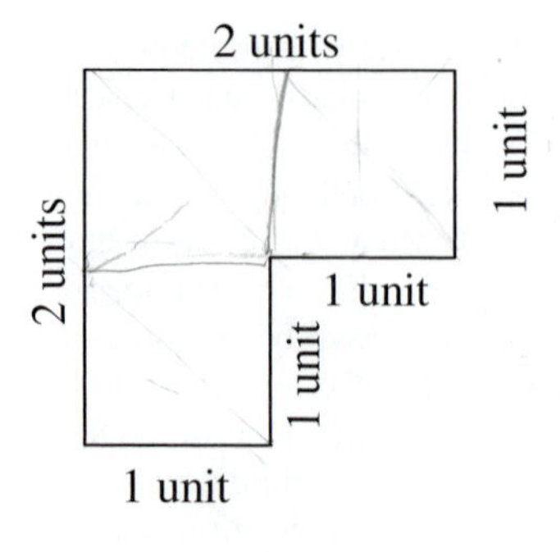

3.20 A peasant had the wonderful good fortune to save the life of the King of Siam. The King, in turn, granted her payment in any form she wished. "I am a simple woman," said the peasant; "I wish merely to never be hungry again." The peasant asked to have payment made onto a huge chessboard painted on the floor. Payment was to be as follows: one grain of rice on the first square, two grains on the second square, four grains on the third square, eight onto the fourth, etc., until all 64 squares were filled. How many grains of rice will the peasant receive from the King? Roughly what volume of rice is this?

3.21 In a field there are cows, birds, and spiders. Spiders have four eyes and eight legs each. In the field there are 20 eyes and 30 legs. All three animals are present, and there is an odd number of each animal. How many spiders, cows, and birds are present? (From Gardner, 1978.)

3.22 Florence Fleetfoot and Larry Lethargy race on a windy day: 100 yards with the wind, they instantly turn around, and race 100 yards back against the wind. Larry is unaffected by the wind, but Florence goes only 90% as fast running against the wind as when running with no wind. Running with the wind improves Florence's no-wind speed by 10%. On a day with no wind, Florence and Larry tie in a 100-yard race. Who wins the windy-day race, and by how much?

3.23 Given a fixed triangle T with base B, show that it is always possible to construct with a straightedge and compass, a straight line parallel to B dividing the triangle into two parts of equal area. (From Schoenfeld, 1985.)

3.24 A pipe one mile in length and with 1-inch inside diameter is set in the ground such that it is optically straight, that is, a laser beam will pass down the centerline of the whole pipe. The centerline of the pipe is exactly level, that is, it is at right angles with a line passing through the center of the earth. Now, by means of a funnel and rubber tubing, water is poured into one end of the pipe until it flows from the other end. The funnel and tubing are removed and water is allowed to flow freely from the open ends. Neglect the surface tension of the water (it has little effect on this problem anyway) and calculate how much water is left in the pipe (say to within 10%). (*Hint*: The answer is not zero, and take the radius of the earth to be exactly 4000 miles should you need that information. *Further hint*: This problem is tougher than the rest of 'em put together.)

3.25 Estimate the number of toothpicks that can be made from a log measuring 3 ft in diameter and 20 ft long.

3.26 Estimate the number of drops in the ocean. Compare your estimate to Avogadro's number.

3.27 Estimate the maximum number of cars per hour that can travel down two lanes of a highway as a function of speed (i.e., at 50, 60, and 70 mph). For safety reasons, the cars must be spaced one car length apart for each 10 mph they are traveling. For example, if traffic is moving at 50 mph, then there must be five car lengths between each car. A city councilman proposes to solve traffic congestion by increasing the speed limit. How would you respond to this proposal?

3.28 If electricity costs \$0.07/kWh (kilowatt-hour), how much money does a typical household spend for electricity each year?

3.29 Estimate the number of books in your university's main library.

3.30 Estimate the amount of garbage produced by the United States each year.

3.31 Estimate the amount of gasoline consumed by automobiles in the United States each year. If this gasoline were stored in a single tank measuring 0.1 mi by 0.1 mi at the base, what is its height?

3.32 A vertical 10-ft column must support a 1,000,000-lb_f load located at the top of the column. An engineer must decide whether to construct it from concrete or steel. She will select the column that is lightest. Assume the column will be designed with a safety factor of 3, meaning the constructed column could support a load three times heavier, but no more. Estimate the mass of a steel column and a concrete column.

3.33 Estimate the number of party balloons it would take to fill your university's football stadium to the top.

3.34 Estimate the amount of money students at your university spend on fast food each semester.

3.35 Estimate the mass of the air on planet earth. What fraction of the total earth mass is air?

3.36 Estimate the mass of water on planet earth. What fraction of the total earth mass is water?

3.37 Estimate the length of time it would take for a passenger jet flying at Mach 0.8 ($\frac{8}{10}$ the speed of sound) to fly around the world. Make allowances for refueling.

3.38 Using Archimedes' principle, estimate the mass that can be lifted by a balloon measuring 30 ft in diameter. The temperature of the air in the balloon is 70°C, and its pressure is 1 atm.

Glossary

application A process whereby appropriate information is identified for the problem at hand.

Archimedes' principle The total mass of a floating object equals the mass of the fluid displaced by the object.

comprehension The step in which the proper theory and data are used to actually solve the problem.

conversion factor A numerical factor that, through multiplication or division, converts a quantity expressed in one system of units to another system of units.

engineering analysis paper A paper with a light grid pattern that can be used for graphing or drafting.

used is discussed next, but for those students who haven't encountered base-2, -8, or -16 number systems, here's how they work.

As you learned in grade school, it makes a difference where the digits are placed within a number; namely, for each digit left of the decimal point, the digit is multiplied by a higher order of 10. Thus, we interpret the number 6945 as

$$\begin{array}{lcr} 6 \times 10^3 & = & 6000 \\ 9 \times 10^2 & = & 900 \\ 4 \times 10^1 & = & 40 \\ 5 \times 10^0 & = & 5 \\ \hline \text{Sum} & = & 6945 \end{array}$$

The same scheme holds true for the binary number system, except that the number increases by a power of 2 rather than of 10 as one moves left of the rightmost digit. The binary system is easy to use in a mechanical or electronic system because there are only two digits: 1 and 0. The first 12 numbers in binary are:

Base-10	Binary
0	0000
1	0001
2	0010
3	0011
4	0100
5	0101
6	0110
7	0111
8	1000
9	1001
10	1010
11	1011

To demonstrate the position/power-of-two relationship, consider the number 11 in binary notation:

$$\begin{array}{lcr} 1 \times 2^3 & = & 8 \\ 0 \times 2^2 & = & 0 \\ 1 \times 2^1 & = & 2 \\ 1 \times 2^0 & = & 1 \\ \hline \text{Sum} & = & 11 \end{array}$$

Because only two digits are needed to specify any whole number, the binary system is the easiest to model in a mechanical or electronic computing machine. Only two *states* are needed in the computing machine, for example, a pin or lever up for 1 and down for 0 in a mechanical computer, or a voltage above 5 volts for 1 and a voltage near zero for 0 in electronic systems.

This "volts/no volts" system is, of course, the basis for essentially all computers in use today. Whereas the binary system is easy to use electronically, it takes a large number of powers-of-two positions (or **bits**) to name a number. The numbers 2 and 3 take 2 bits, the numbers 4 through 7 take 3 bits, the numbers 8 through 15 take 4 bits, and so on. Because it takes many bits to specify a number, even the earliest electronic computers grouped bits together for routing numbers within the memory and operating system of the computer. Eight bits bundled together as part of a number is called a **byte.**

The largest integer that can be represented by a single byte is

$$
\begin{aligned}
1 \times 2^7 &= 128 \\
1 \times 2^6 &= 64 \\
1 \times 2^5 &= 32 \\
1 \times 2^4 &= 16 \\
1 \times 2^3 &= 8 \\
1 \times 2^2 &= 4 \\
1 \times 2^1 &= 2 \\
1 \times 2^0 &= \underline{1} \\
& 255
\end{aligned}
$$

It's easy to show that, in general, the largest integer that can be represented by n bits is $2^n - 1$, so for 4-byte integers, the largest would be

$$2^{32} - 1 = 4{,}294{,}967{,}295$$

Actually, a number only half that large can be stored because most computers use the leading bit as a sign indicator so that negative and positive integers can be stored in the computer. Real numbers are stored in a variety of ways depending on the operating system, computer architecture, and so forth, but often the leading bit sets the number to positive or negative, the next 7 bits are used to determine the value of the exponent, and the last 24 bits are the mantissa, or fractional part of the number.

When listing numbers using binary notation, even moderate-sized numbers require many digits. But converting from decimal to binary is awkward for large numbers. To circumvent these problems, computer programs often used octal- and hexadecimal-based systems. It is easy to convert between binary and these two systems of numbers, and both are much more compact than binary. To convert from binary to octal, group the binary number into groups of 3 bits starting with the least significant bit, then convert each 3-bit group to a number between 0 and 7. For example,

$$100 \;\; 010 \;\; 001 \;\; 111 \Rightarrow 4217_8$$

where the subscript signifies base-8 numbers. To convert to hexadecimal, group the bits into groups of 4 and convert each bit to a number between 0 and 15. Because 10 is the number 16 in hexadecimal notation, the following convention is used in "hex" numbers:

Decimal	0	1	2	3	4	5	6	7	8	9	10	11	12	13	14	15	16
Hexadecimal	0	1	2	3	4	5	6	7	8	9	A	B	C	D	E	F	10

For the same number, the hex conversion would be

$$1000 \qquad 1000 \qquad 1111 \Rightarrow 88F_{16}$$

For the same number, the decimal conversion would be

$$
\begin{aligned}
1 \times 2^{11} &= 2048 \\
0 \times 2^{10} &= 0 \\
0 \times 2^9 &= 0 \\
0 \times 2^8 &= 0 \\
1 \times 2^7 &= 128 \\
0 \times 2^6 &= 0 \\
0 \times 2^5 &= 0
\end{aligned}
$$

$$
\begin{array}{rcl}
0 \times 2^4 & = & 0 \\
1 \times 2^3 & = & 8 \\
1 \times 2^2 & = & 4 \\
1 \times 2^1 & = & 2 \\
1 \times 2^0 & = & 1 \\
\hline
 & & 2191
\end{array}
$$

4.2.2 Algorithms

Throughout this chapter and your work with computers, you will be expected to find and use an appropriate **algorithm** to solve a particular problem. What specifically is an algorithm? An algorithm is a sequence of actions that will bring about the solution to a particular problem. A recipe book, for example, is filled with "algorithms" on how to prepare various parts of a meal. The word itself comes from the name of a famous Persian mathematician, Mûsâ al-Khowârizm, who in 825 A.D. wrote a widely read book on mathematics, *Kitab al jabr w'al-muqabala.* From the book's title we derive the word *algebra.*

Algorithms were described and used much earlier. One widely known algorithm for finding the largest common factor between two integers is attributed to Euclid from about 300 B.C. His algorithm can be stated as:

1. From the two integers for which the largest common factor is sought, divide one by the other.
2. Discard the integer quotient but retain the remainder.
3. Divide the previous divisor by the remainder, discard the integer quotient, but retain the new remainder.
4. Repeat the above step until the remainder is zero.
5. The divisor that produced a remainder of zero is the largest common multiple.

To test this algorithm, let's see how it works with two specific numbers: 1798 and 2666.

2666/1798 has a remainder of 868.
1798/868 has a remainder of 62.
868/62 has a remainder of 0, so the largest common factor is 62.

Knowing this is a common factor, then it's easy to show that $1798 = 62 \times 29$ and $2666 = 62 \times 43$, and indeed 62 is the largest common factor because 43 and 29 have no common factors other than 1.

Alan Turing, during World War II, proposed a model of all computing machines. His idealized model consists of a machine capable of reading and writing 0s and 1s on an infinite tape that bases its actions on the marks on the tape. (Actually Turing proposed using more complicated marks than 0s and 1s, but the marks can be restricted to these two without any loss of generality.) By performing thought experiments with this machine, Turing almost single-handedly wrote the beginning chapters of computer science. He was able to show that any finite algorithm could be reduced to a series of 0s and 1s on the tape, along with a finite set of instructions for his machine.

From all this, we can state a rough summary of how engineers solve problems:

1. From the universe, the engineer chooses those variables important to the problem at hand.
2. She models the interrelation of the significant variables mathematically.
3. The math is converted to an algorithm.

4. If computers are used, the algorithm is converted to a series of binary instructions (as Turing showed).
5. The results are compared to what happens in the universe.
6. If successful, the problem is solved; if not, repeat all steps.

4.2.3 The Basic Building Blocks in a Computer

Today's computers are well modeled as electronic Turing machines, in the sense that they can implement a wide range of algorithms stored as 0s and 1s in the machine, operating on data also stored as 0s and 1s. How this is done is well beyond what you need to know about computer languages, but a general knowledge of computer architecture will help you understand why programming languages share many common features.

A computer can be built with remarkably few kinds of components that perform the basic functions within the computer. Basically, it needs a number storage device, a method to store and retrieve numbers, a method to add two numbers, a method to move nonsequentially along a string of operations, and a method to communicate with the outside world. From this simple list, the complicated algorithms the computer performs can be synthesized.

Memory.

A basic need in the computer is, of course, the ability to "store" a number. The physical device that allows this storage is called a **flip-flop** because it "flips" up to a high voltage (modeling a 1) or "flops" to zero volts (modeling a 0). The unique feature of a flip-flop is that until it is signaled to change, it stays set at high or low voltage. Figure 4.1 shows the diagram of a flip-flop.

Most flip-flops operate such that when the latch voltage is low, the output exactly follows the input: when input is high so is output, and when input is low so is output. Thus, when the latch voltage is low, the flip-flop is *transparent* and signals are passed directly from the input to the output. Then, when the latch voltage becomes high, it *freezes* the output. The output will stay high if the input was high, and the output will stay low if the input was low. *In either case, the output will stay that way no matter what the input does.* The ability of the flip-flop to freeze the output regardless of the input allows the storage of numbers in the computer.

Figure 4.2 shows a 4-bit computer storage device that passes the binary number 1010 to the output for storage. In Figure 4.2(a), the latch voltage is low, so the input bits are passed directly to the output. In Figure 4.2(b), the latch voltage becomes high, which freezes the output (i.e., the output becomes stored). In Figure 4.2(c), the latch voltage is still high so the output stays the same even though the inputs have changed. Therefore, the number in the output is permanently stored until the latch voltage becomes low again.

FIGURE 4.1
Flip-flop.

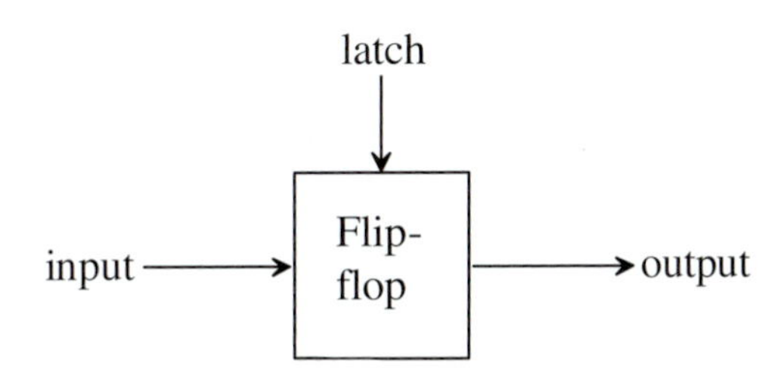

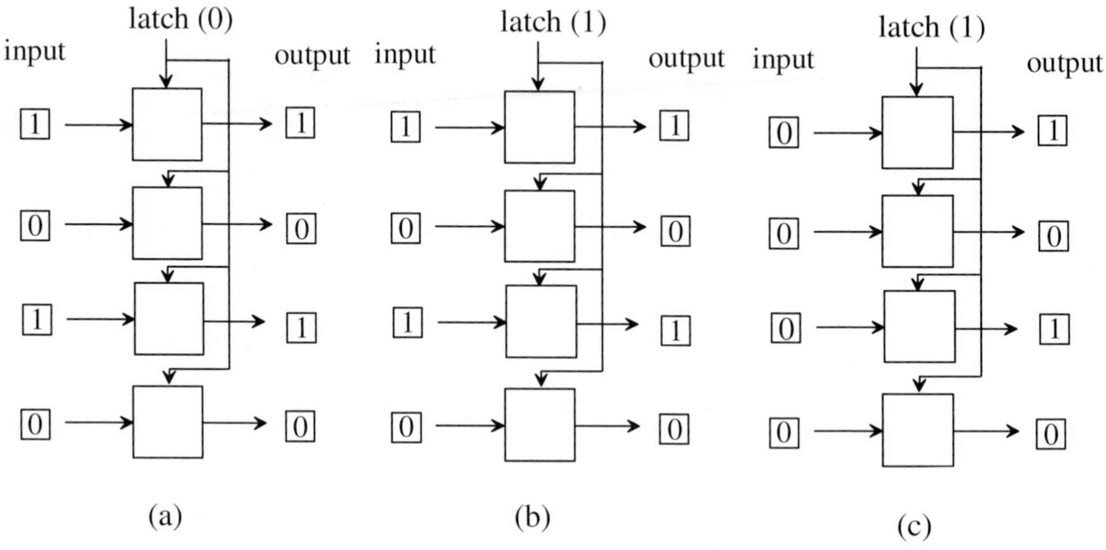

FIGURE 4.2
Schematic of 4-bit memory storing the binary number 1010 in the output.

In a computer, the signal to latch a number into a gang of flip-flops is regulated by the computer *clock.* A computer can move a number from one gang of flip-flops (memory location) to another at about the same rate as the **clock cycle,** the speed at which the computer performs a single operation. The clock cycle speed is an important specification used to describe the speed of a computer.

The Central Processing Unit (CPU).

Binary machine instructions are stored by flip-flops and can direct the computer to retrieve numbers from various memory locations (each identified by a unique binary number); to do simple additions of binary numbers; to test two numbers to determine if one is larger, smaller, or equal to the other; and, based on the results of number comparisons, to jump to a different place in the sequence of instructions. The part of the computer that performs these operations is called the **central processing unit** or CPU. From the simple list of operations given above, impressive algorithms can be built.

To understand how the CPU accesses numbers in memory, consider the schematic of a switching tree shown in Figure 4.3. The switching tree is analogous to a train switch yard in which there are numerous branch points. At each point, the train can be directed to go left or right. In the case of computers, an individual bit (either 0 or 1) is directed to go left or right according to the directions given by the **address,** a numerical designation for a memory location. As shown, a single bit (indicated by an x, which designates either a 0 or 1) is directed to memory location 1010 because the address appropriately sets the switches (indicated as black squares). For 16-bit computer memories, a total of 16 trees are used in parallel to route each 16-bit number to computer memory.

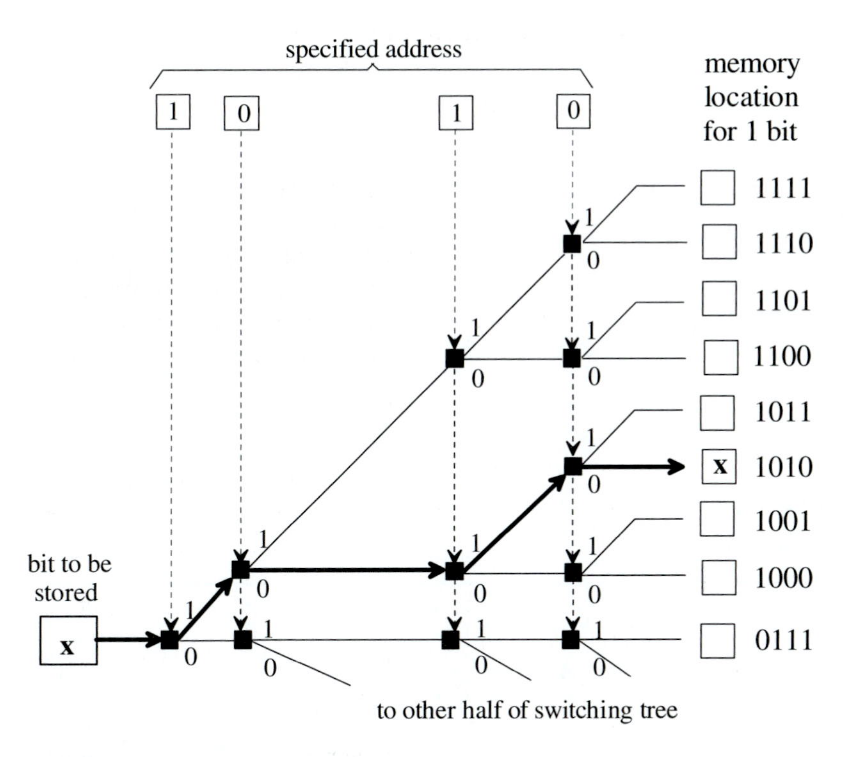

FIGURE 4.3
Schematic of a switching tree for storing a single bit. (*Note:* x indicates a single bit, either 0 or 1. Each black square indicates a switch.)

Computer Arithmetic.

Figure 4.4 shows logic circuits that take an input (*A* or *B*) and produce an output (*Y*). The *truth tables* give the output for a given input.

Binary addition is not difficult to understand—it follows most of the same rules you learned for decimal arithmetic. The following gives all the necessary conditions to add two binary numbers with a sum less than 2:

```
  0     0     1     1     A
 +0    +1    +0    +1    +B
 00    01    01    10    XY
```

To perform the additions shown above, the truth table must be as follows:

A	B	X	Y
0	0	0	0
0	1	0	1
1	0	0	1
1	1	1	0

This truth table can be obtained by using the combination of logic circuits shown in Figure 4.5.

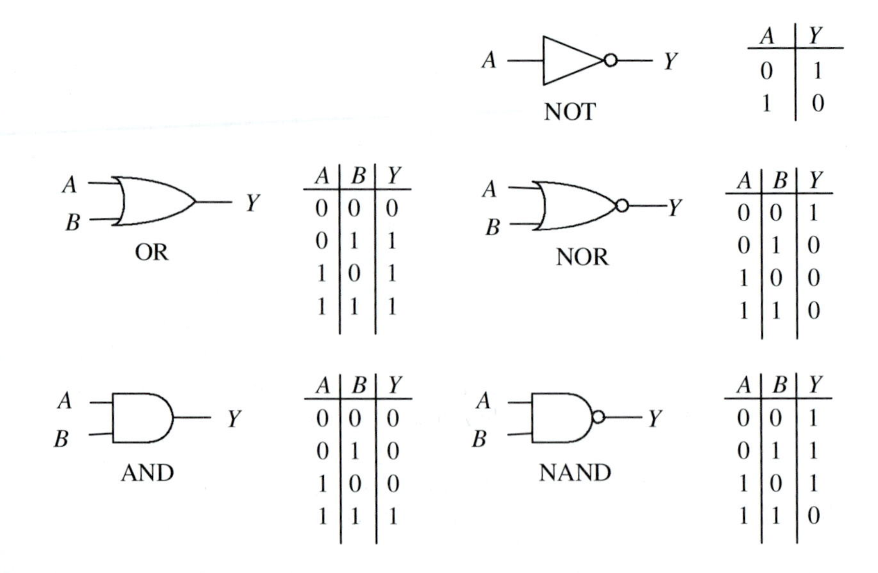

FIGURE 4.4
Logic circuits.

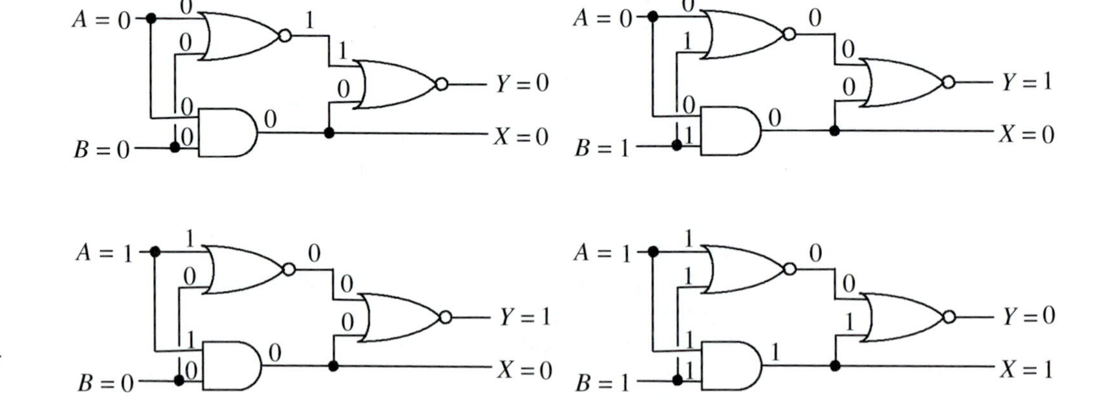

FIGURE 4.5
Circuit for computer addition.

Binary multiplication is identical to decimal multiplication by hand. For each higher power of 2, the product is shifted to the left and all of the partial products are summed. To multiply 5 by 3 in binary:

```
  101
 ×011
  101
 101
000
01111  →  15₁₀
```

Computer multiplication is usually implemented by using lookup tables.

Another important feature of CPUs that allows nonsequential execution of instructions is the **program counter.** The program counter stores the memory location of the next instruction and, if necessary, it can be changed by the program itself. For example, a condition may be tested (e.g., "is $a > 1000$?"). If the condition is true, the counter is changed to start sequencing at a distant memory location. If the condition is false, then instead of jumping, the next instruction in the sequence is executed. These simple manipulations, and a handful of others, comprise the basic operating instructions for the CPU.

Levels of Computer Languages.

Within a computer, the CPU shuffles numbers around, does simple arithmetic, and routes numbers in and out of the computer according to instructions stored in memory. These instructions, called **machine language,** are a series of binary numbers that direct the CPU. Because it is hideously slow to program a computer in machine language, even computer scientists use a higher-level system called **assembly language,** which uses Englishlike commands to generate machine language instructions. A simple interpreter changes a sequence of assembly commands into machine language. For mere mortals trying to use a computer, computer scientists have developed high-level languages to allow a user to run a computer without having to understand very much of the internal processes. These high-level languages include computer languages such as Fortran, C, and Basic. The computer **operating system** is also a high-level language. The operating system is the fundamental interface through which the user loads instructions onto the CPU, routes output to a printer, or saves work for later use.

4.3 GETTING STARTED

Now that we have established a history that explains how computers have evolved and generally how they work, we present the information necessary for you to understand today's computers. We are presenting only the bare essentials here; you will learn more during your programming exercises. Here are some of the key words and concepts you need to begin.

4.3.1 Hardware

Hardware is a term for the computer and the associated peripherals, such as the printer, disk drives, and networks. This brief overview will help orient you to the resources available in a typical computer laboratory.

Personal Computers.

Many different types of computers are used in nearly every industry in the world. These computers range from supercomputers to handheld calculators. Near the middle of the spectrum, as far as price and power are concerned, are the class of computers known as *personal computers* (PCs).

Two of the most important parts of the computing machine are the CPU (discussed earlier) and the **memory,** a specific part of the machine used solely to store numbers. Memory in personal computers has risen from a few thousand bytes in the early 1970s to many million bytes today.

Personal computers became a market success in the early 1980s with the introduction of the Intel-IBM 8088 CPU-based systems. Over the past decade or so, the PC CPU has continually been upgraded, advancing the PC considerably in functionality

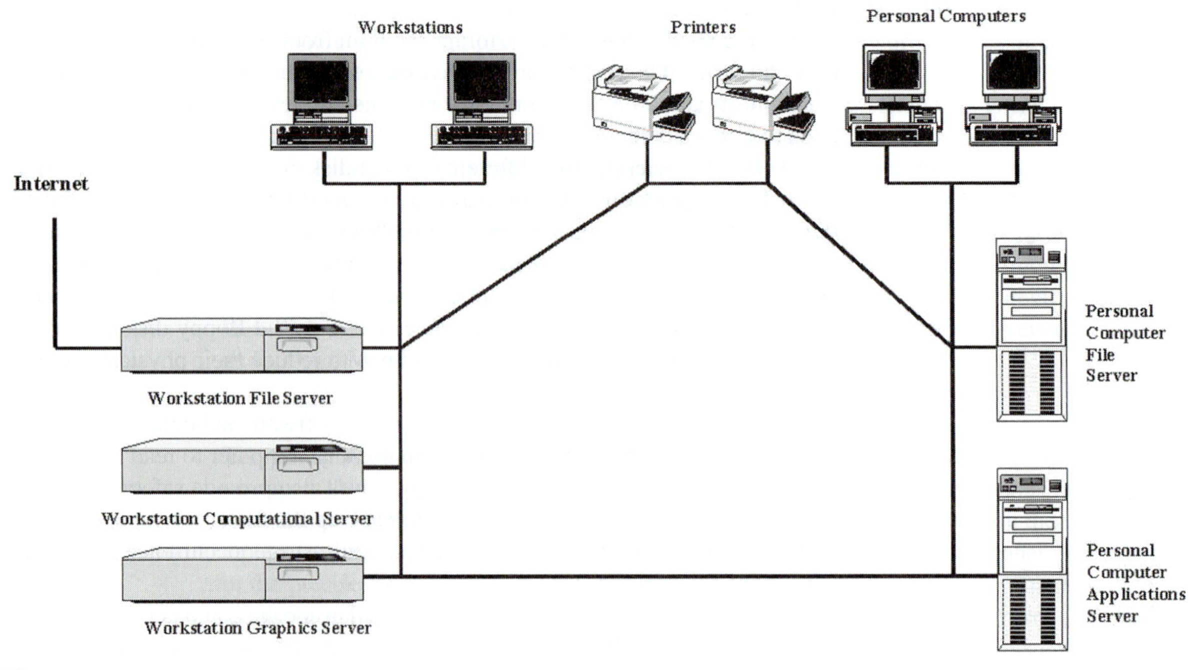

FIGURE 4.7
Today, computers are routinely part of a larger computing environment allowing them to share information with local users and the world via the Internet. The computer network illustrated here has both personal computers and workstations, which are powerful computers used for computationally intensive applications. The servers store data and application programs, and also support graphics and computations.

output (I/O) devices. A **local area network** (LAN) allows one PC to be used as a **server** to other computers (Figure 4.7). The server is usually used only for routing communication between PCs and as a centralized repository for files and programs. The PCs still act independently, but the LAN enables them to pass files, send mail, and share equipment with greater flexibility and speed. This allows software, printers, and plotters to be used by multiple PCs while maintaining the advantages of a PC-based system.

Such a back-and-forth move by the computer industry may at first seem absurd, yet the end result is a rapidly advancing technology that allows users all the power of personal computing while maintaining the group management positives of the mainframe—a union previously unthinkable.

4.3.2 Software

Software is a collection of instructions that directs computer hardware to perform specific tasks. Common examples of software are word processors, spreadsheets, graphics packages, and games.

Files.

Often, software will generate information to be stored on a disk. For example, if you were to compose a letter using a word processor, the letter would be stored on a disk, either a hard disk or a floppy disk. Both software and the information they generate are stored as a long series of 1s and 0s termed a *file.* To aid the retrieval of this series of 1s and 0s, it is given a *filename.* A file that contains a letter to your mother might be called "MOM." To further describe the file, an *extension* is appended to the end of the filename that serves as an adjective that describes the type of file. Your letter to your mother might have an extension ".TXT," indicating that it is a text file. Thus, the complete name for the file that contains the letter to your mother is "MOM.TXT." In some computers, the filename has eight letters or less, and the extension has three letters.

Every file must have a unique filename. **Note well**—saving a file using the same name as an existing file will overwrite the old file. If this is what you intended, fine; if not, you just lost your old file. A word to the wise: save your work often, and preserve intermediate work, deleting working files only when you are sure that you are done.

Disk Operating Systems and Other Operating Systems.

DOS is the Disk Operating System used to control the PC's most basic functions. If you sit down at a computer and see prompts such as

```
C:\>
```

then what you see is called the *DOS prompt,* and the computer is awaiting a DOS command from the keyboard. DOS commands can be used to start a program (e.g., word processor), copy files from one disk to another, delete files, print files, and so forth. Mainframe computers use more sophisticated operating systems (e.g., VMS, UNIX, and JCL), which serve similar functions as DOS.

The Windows Environment.

Although the DOS system is relatively simple to use compared to mainframe operating systems, the average PC user is not enthusiastic about learning a couple of dozen arcane commands and the associated concepts to operate his computer. *Windows* is a user-friendly environment that replaces typed DOS commands with point-and-click commands from a pointing device such as a *mouse.* The various applications, files, and operations are represented by small pictorial symbols known as *icons.* By moving the mouse to the desired icon and double clicking (pressing twice in rapid succession) on the left button, you may start the application. Once you have opened an application, you can control the program applications by clicking the mouse in the dialogue boxes on the screen. Unlike DOS, a Windows environment allows you to see and work with more than one file at a time. There are only a few essential commands that must be learned; all other commands and functions can be derived from the basic commands.

Writing Equations in Computers.

So far, your interactions with computers may have been limited to communication tools, such as word processors, Internet browsers, and e-mail. They are all fairly recent applications; computers were originally developed to perform numerical calculations. As you progress in your engineering studies, increasingly you will use the computer as a calculational tool.

In order to use a computer for computations, it is necessary to enter mathematical formulas into the computer. In an algebra textbook, you might see a formula like this:

$$y = \frac{x(x + 3)^2}{\sqrt[3]{x} - 7}$$

When entered into the computer, the formula would look like this:

```
y = x * (x + 3) ^2/(x - 7)^(1/3)
```

or

```
y = x * (x + 3.)**2/(x - 7.)**(1./3.)
```

Compared with the way the formula appears in the textbook, we observe some important differences:

- The computer formula appears on a single line.
- Division is indicated by /.
- Multiplication is indicated by *.
- Exponentiation is indicated by ^ or **.
- Roots are indicated by exponentiation (1/2 for square root, 1/3 for cube root, etc.).
- Parentheses are used to indicate the order of mathematical operations.

Also, some computer languages distinguish between **real numbers** (numbers used to measure continuous quantities, like 5.739 inches) and **integers** (discrete numbers used for counting, like 5 apples). In the above example, `7.` indicates the real number seven, whereas `7` indicates the integer seven. For some computer languages (e.g., Fortran), the distinction between real numbers and integers is extremely important. In Fortran, the expression `1/3` indicates, "divide the integer one by the integer three," which is evaluated as the integer *zero*. Likely, this is *not* what you wanted. If you used this, your formula would lead to results that are unexpected and possibly catastrophic. To get the proper result, use the expression `1./3.` which indicates, "divide the real number 1 by the real number 3" and is evaluated as 0.333333.

There are other important distinctions between formulas used in algebra textbooks and formulas used in computers. For example, suppose the following two formulas are entered into a computer:

`y = x + 3` and `x + 3 = y`

Algebraically, they are identical. However, if entered into a computer, these two formulas are *not* identical; only the first formula is valid. The first formula is evaluated by a computer as "take the number stored in memory location `x`, add 3 to it, and store the result in memory location `y`." The first formula is successfully evaluated; therefore, in computers, there must be a single variable located left of the equal sign, indicating where the final result is to be stored. The second formula violates this rule, so it cannot be evaluated by a computer.

Another distinction is shown by the following formula, which could appear in computers:

```
y = y + 3
```

In algebra, there is no possible value for `y` that could satisfy this equation; yet this formula works perfectly well in computers. This formula says "take the current number stored in memory location `y`, add 3 to it, and store this new number back in memory location `y`." Such a formula might be encountered in a loop and would allow `y` to be incremented in steps of 3.

4.4 CREATING COMPUTER PROGRAMS

For the novice programmer (and sometimes the experienced programmer) the hardest part is getting started. This section presents several of the tools used by novice and professional alike, to help you marshal your thoughts and complete a program.

4.4.1 Flowcharts

A **flowchart** is a schematic of the logic needed in an algorithm; it describes the sequential order in which the steps are done. A standard set of symbols is used to denote particular operations, making flowcharts a universal medium of communication among programmers. Figure 4.8 shows the list of symbols and their use.

4.4.2 Structured Programming

Figure 4.9 illustrates two extremes of computer code: **spaghetti code** and **structured code**. Both codes will actually run on a computer and will give correct results. However, we think you will agree that the structured code is much easier to understand. As some programmers work on large programs for many months (or years), it is extremely important that the logic be clear so they can understand what they wrote many months before. Further, because large computer programs are often written by teams of programmers, it is important that the code be easy to follow by other team members. Therefore, even if spaghetti code is faster and more compact, structured code is preferred.

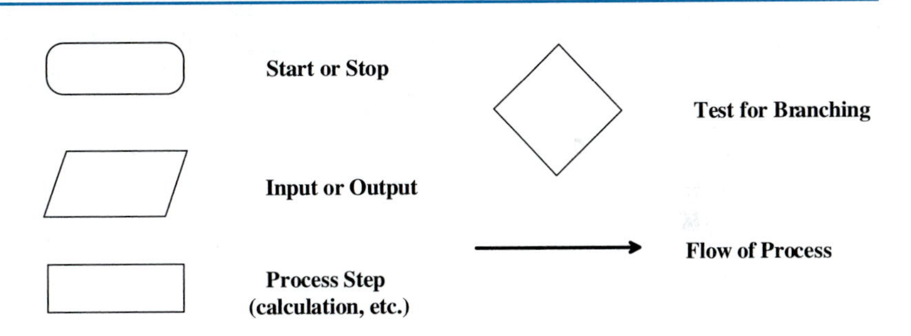

FIGURE 4.8
Symbols used in flowcharts.

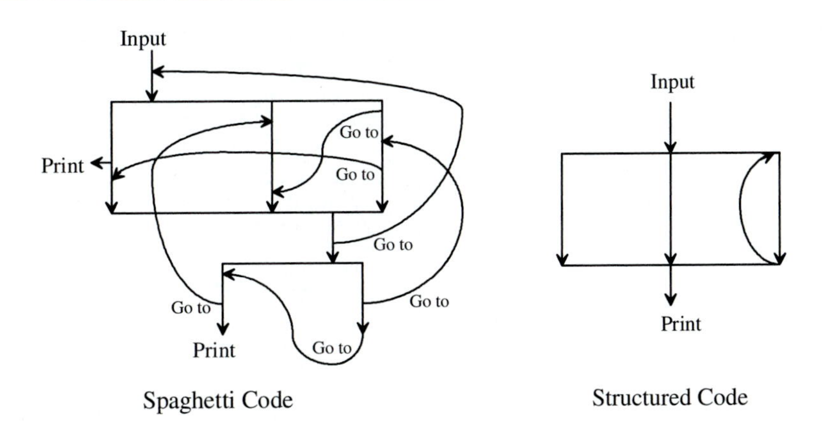

FIGURE 4.9
Illustration of spaghetti code and structured code. (The spaghetti code is characterized by many "go to" statements that direct the flow to various parts of the program. "Go to" statements are an essential part of the Basic computer language used by many high school students; therefore, it is difficult to write structured code in Basic.)

Because all computers work in similar manners, the basic structures in all computer languages are similar. **Sequential structures** (Figure 4.10) are performed one after the other in sequence. **Selection structures** (Figure 4.11) have a decision point in which a condition is evaluated (e.g., "is $a > 1000$?"). If the condition is true, the program follows one path, and if false, it follows another path. **Repetition structures** (Figure 4.12) can loop as long as a condition is true, but break out of the loop once it is false ("while loop") or loop a specified number of repetitions ("do loop").

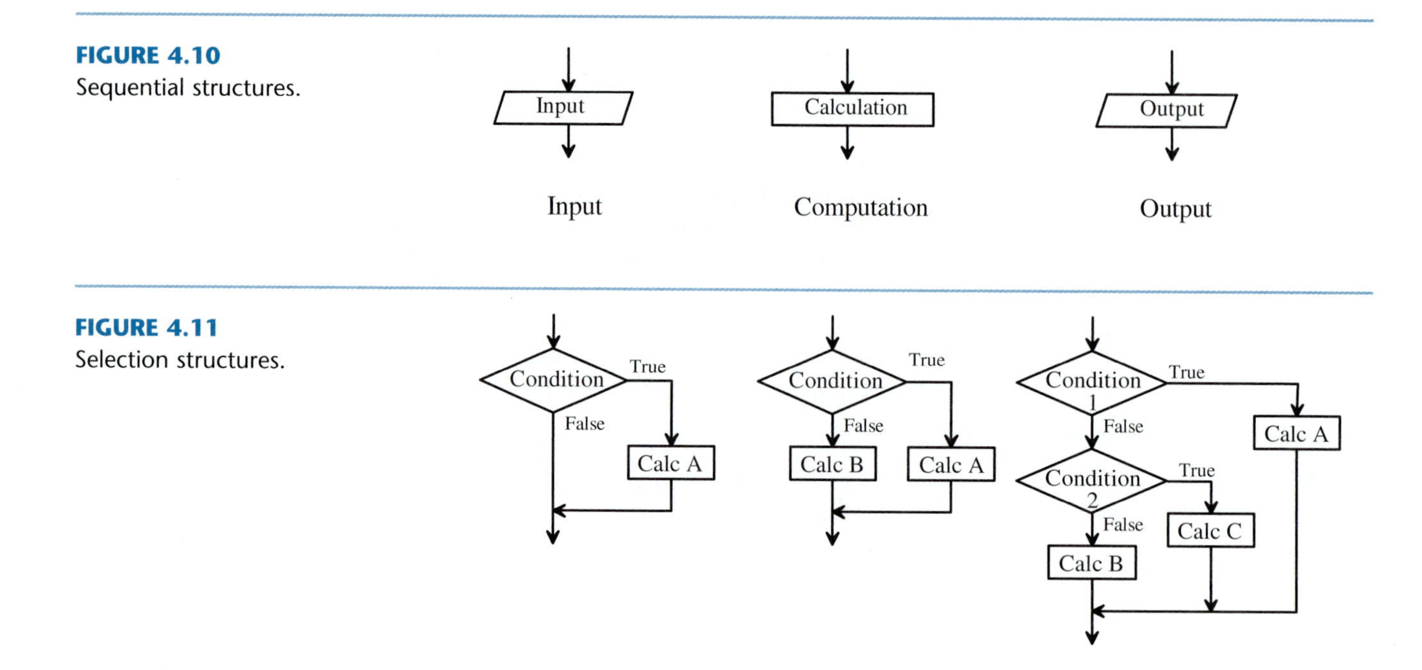

FIGURE 4.10
Sequential structures.

FIGURE 4.11
Selection structures.

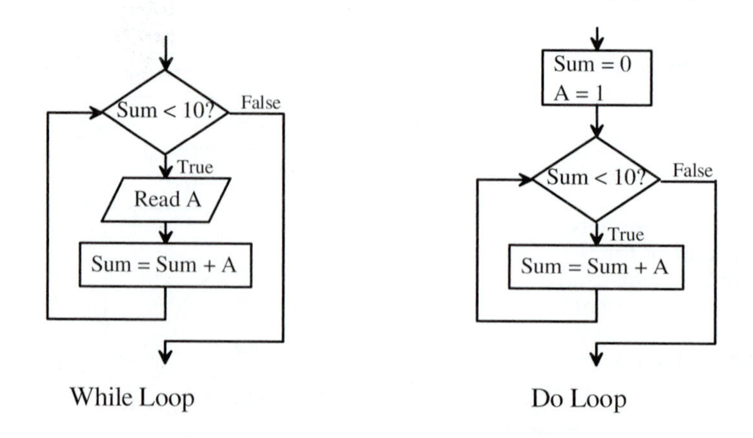

FIGURE 4.12
Repetition structures.

4.4.3 Top-Down Design

The concept of *top-down* design for producing computer code follows closely the basic engineering problem-solving techniques used in all of engineering:

1. State the problem clearly.
2. Describe the input and output information.
3. Work the problem by hand (or with a calculator) for a specific set of data.
4. Develop a general algorithm for the problem.
   a. Decompose
   b. Stepwise refinement
5. Test the solution with a variety of data sets.

The underlying theme is to have clearly in mind where you want to go before sitting down and encoding those parts of the problem you already understand. The bits of code you develop may in themselves be correct—but without the big picture, when put together to form the final code, they lead to subtle sequencing errors and produce wrong answers.

To illustrate the top-down approach, we will use a simple example program that adds two numbers together and reports their sum. The steps are:

1. **State the problem clearly.**

   The purpose of the program is to add two numbers together.

2. **Describe the input and output information.**

   Input = `A` and `B` (the two numbers to be added)
   Output = `C` (the sum of the two numbers)

3. **Work the problem by hand (or with a calculator) for a specific set of data.**

   $2 + 3 = 5$

4. **Develop a general algorithm for the problem.**

   This step is the real heart of the problem. Flowcharts are useful tools for decomposing the problem. The flowchart is translated into a high-level language (e.g., C^{++}). Generally, the code does not run properly when first written, so it must be refined in a stepwise manner. Figure 4.13 shows the flowchart and generic computer code for our simple example. Note that this simple program has all the elements of sophisticated programs: title, introduction, initialization of the variables, data input, computation of results, printing results, and finally ending the program.

5. **Test the solution with a variety of data sets.**

   The program shown in Figure 4.13 works fine for integer inputs (e.g., 4, 10, −8), but when it was tested with real inputs (57.6, 0.89, −11.6), it did not work properly. The program would need to be modified to calculate the sum of real numbers.

It is difficult to fully test whether a program functions with all possible data inputs (very large numbers, very small numbers, combinations of small numbers and large numbers, negative numbers, integers, real numbers, etc.). If we accomplish nothing else in this book, we hope to teach you to distrust everything that comes from a computer. For some reason, if a computer program provides something reasonably close to a right

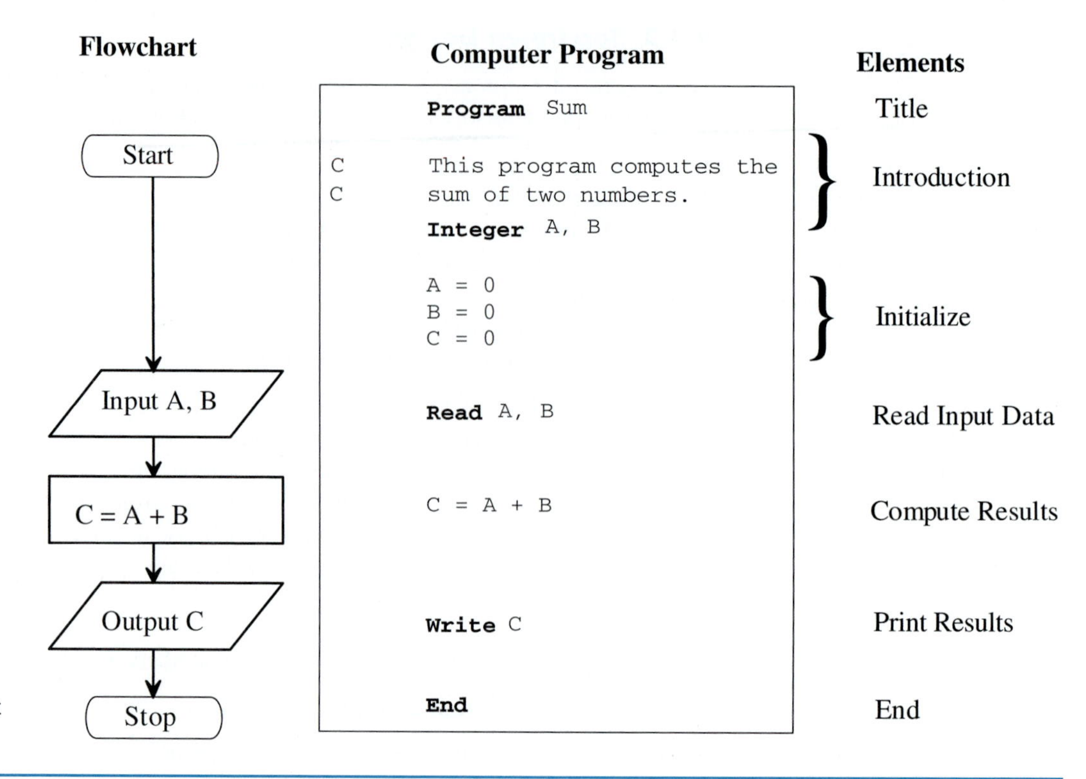

FIGURE 4.13
Flowchart and generic computer program.

answer twice in a row, our faith in its correctness is unbounded. As you will find when you program, there are many ways to write incorrect code but the path to fully correct coding is narrow indeed. Practice skepticism on computer output and you will save yourself much grief.

4.4.4 Algorithms and Flowchart Examples

The following pages show examples of how flowcharts are used.

Sample Program 1: addsub
The user enters two real numbers and the program calculates their sum and difference (Figure 4.14).

Variables:

A, B	user-supplied real numbers
SUM	sum of a and b
DIFF	absolute difference of a and b

Sample run:

```
        Enter A,B
(user)  3.0, 5.0
        8.0     2.0
```

Sample Program 2: factorial

This program returns the factorial of a positive integer selected by the user (Figure 4.15).

Variables:

N user-supplied positive integer
C counter
FACT factorial of N

Sample run:

Enter N
(user) 7
5040

FIGURE 4.14
Flowchart for Sample Program 1.

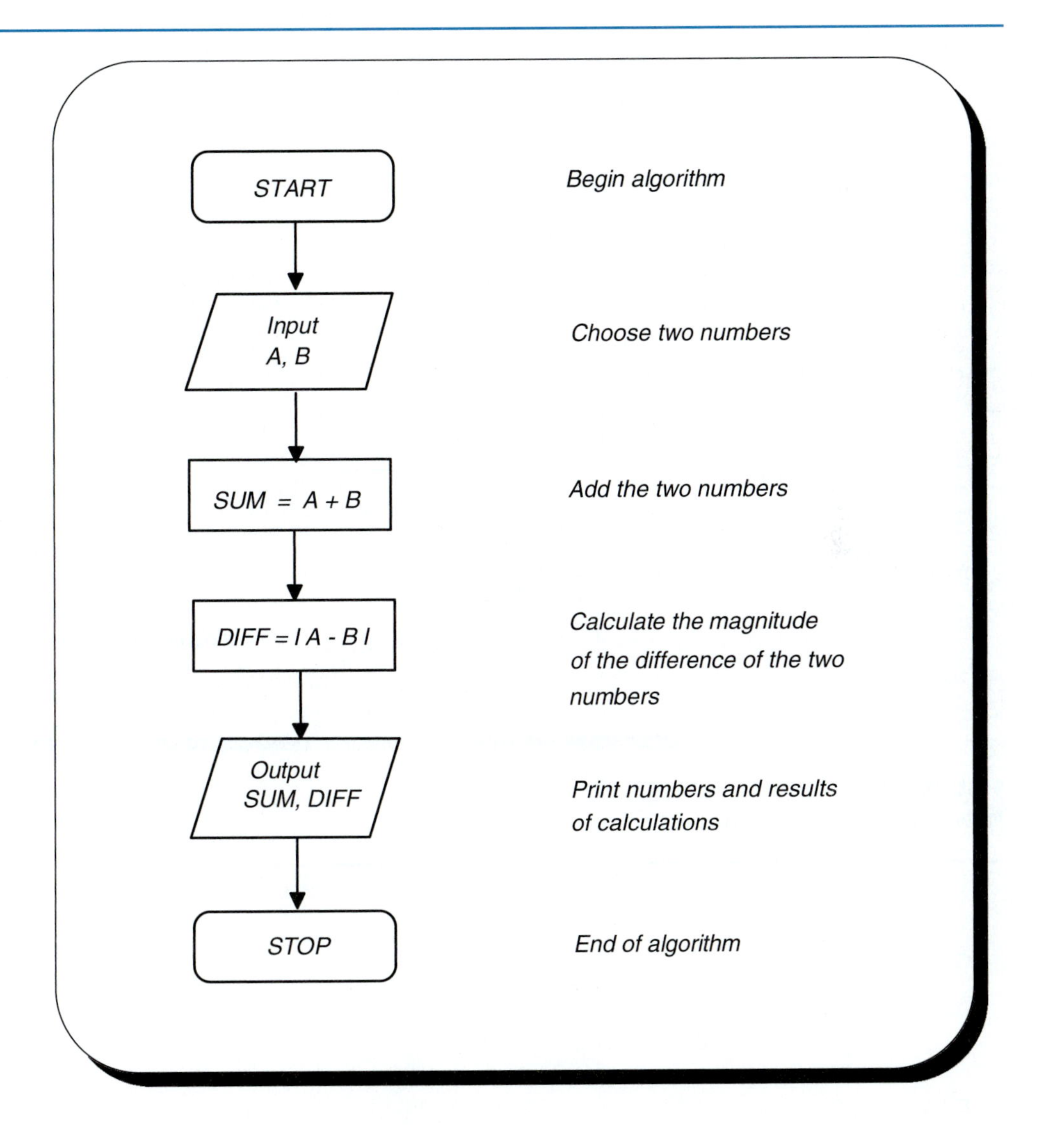

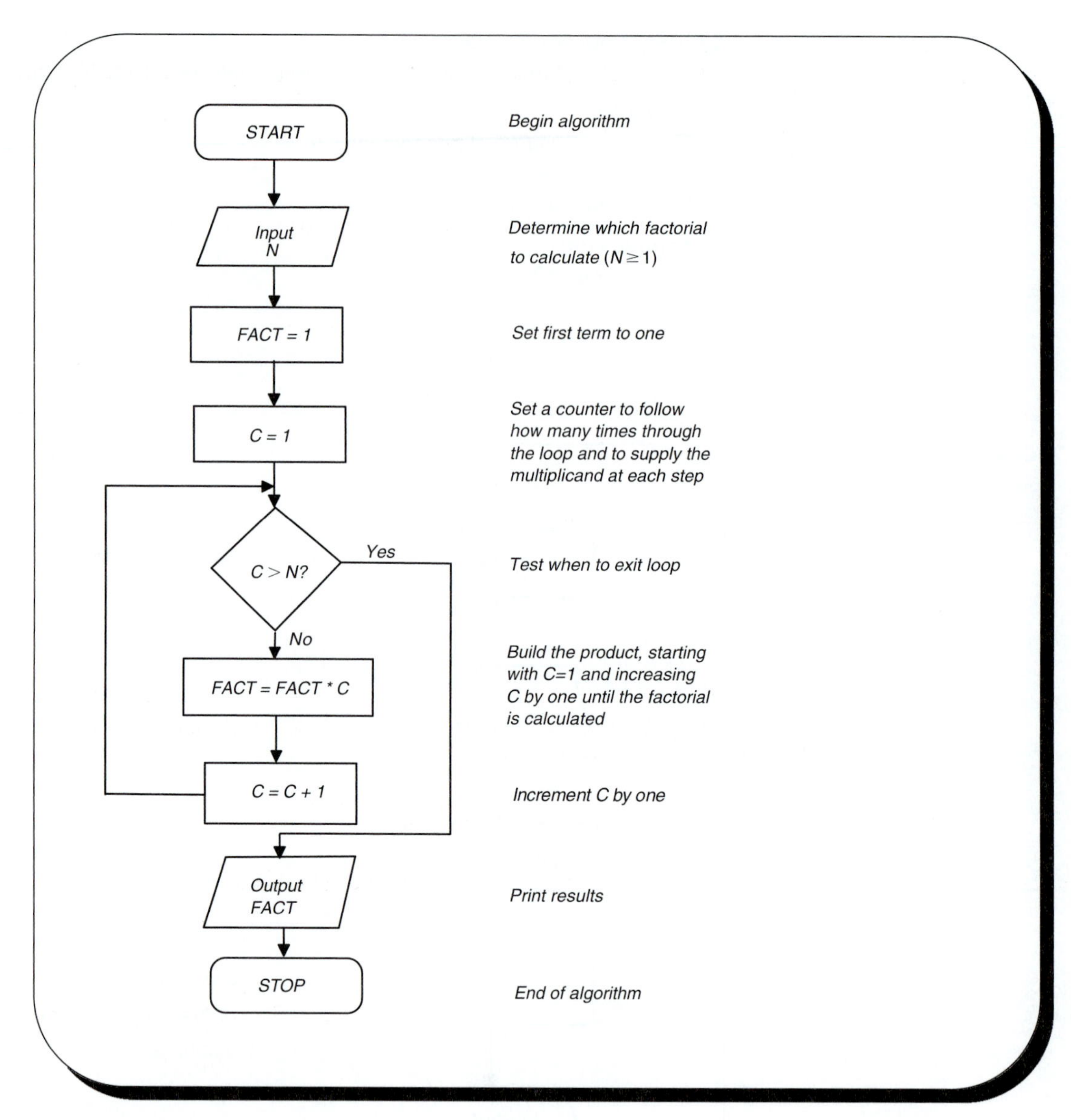

FIGURE 4.15
Flowchart for Sample Program 2.

Sample Program 3: Taylor

This program (Figure 4.16) calculates a value for e^x using the Taylor series approximation,

$$e^x = 1 + \frac{x}{1!} + \frac{x^2}{2!} + \frac{x^3}{3!} + \cdots$$

Addition of subsequent terms ceases when the value of $x^n/n!$ is below some user-specified criterion.

Variables:

X	user-supplied real number
MIN	user-supplied criterion for minimum useful value of a term in series
SUM	sum of all previous terms
FACT	factorial of i
TERM	individual term
I	counter

Sample run:

```
        Enter X and the minimum term value
(user)  3.0, 0.01
        20.08410
```

Sample Program 4: Eratosthenes Sieve

This program (Figure 4.17) finds all the prime numbers less than or equal to a user-specified integer. This algorithm, called **Eratosthenes' sieve,** is named after an ancient Greek mathematician.

Variables:

N	user-supplied integer
I, J	counters
X	array containing only 1s and 0s; if I is prime, X (I) = 0, otherwise, X (I) = 1

Sample run:

```
          Enter N
(user)    23
          1
          2
          3
          5
          7
          11
          13
          17
          19
          23
```

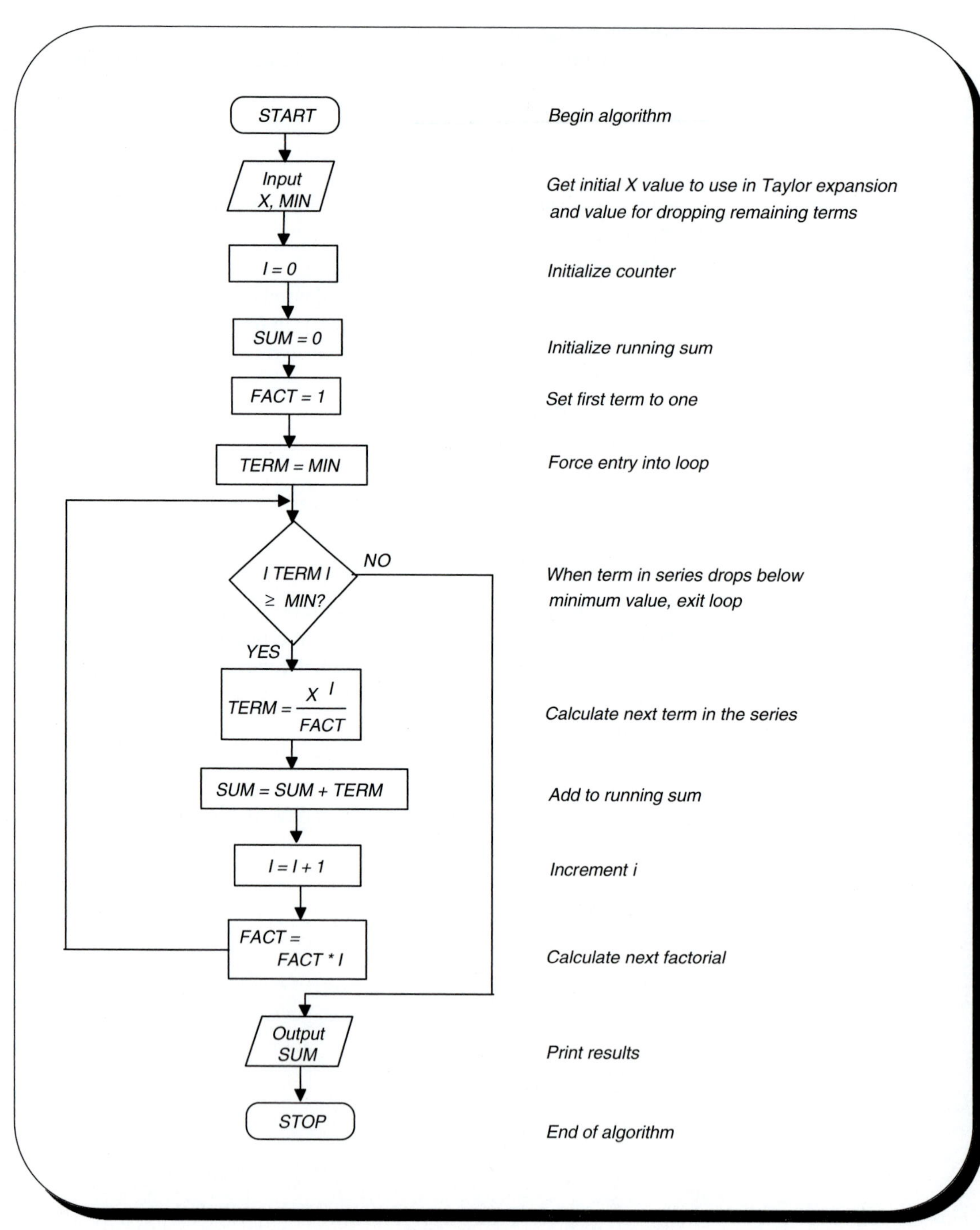

FIGURE 4.16
Flowchart for Sample Program 3.

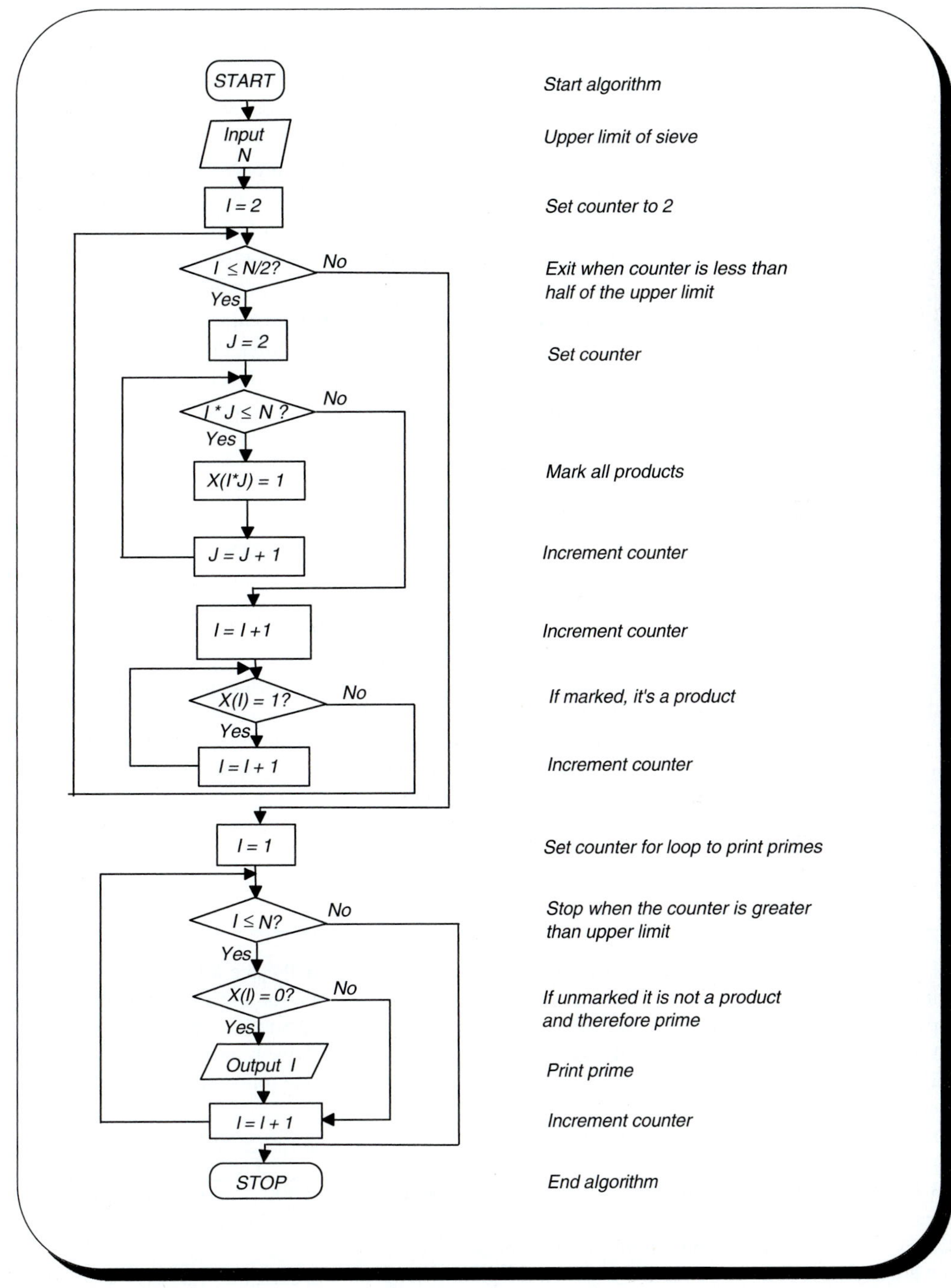

FIGURE 4.17
Flowchart for Sample Program 4.

4.5 SUMMARY

Computers are general-purpose electronic machines that can solve a wide variety of engineering problems. Various hardware devices in the computer allow storage of numbers in the computer memory, routing of numbers, simple addition, and nonsequential execution of stored instructions. It is the central processing unit (CPU) that actually performs these functions, based on instructions encoded in machine language using binary (base-2) numbers. In general, the binary instructions given to the CPU are created and controlled by software—instructions encoded to allow the computer to perform specific tasks. Software is usually created by using a high-level language, such as C, and operating system instructions. Many peripherals are available to expand the capability of the computer, including printers, modems, local area networks (LANs), and storage disks (hard drives and floppy disks).

In creating software using a high-level language, the top-down design often works best. The logic of the main and individual pieces of an algorithm (a set of instructions to effect the solution to a given problem) can be visualized by using flowcharts.

Further Readings

Alt, F. L. *Electronic Digital Computers.* New York: Academic Press, 1958.

Aspray, W. *John von Neuman and the Origins of Modern Computing.* Cambridge, MA: MIT Press, 1990.

Burks, A. R., and A. W. Burks. *The First Electronic Computer: The Atanasoff Story.* Ann Arbor: University of Michigan Press, 1988.

Cole, R. W. *Introduction to Computing.* New York: McGraw-Hill, 1969.

Davis, G. B. *Introduction to Electronic Computers.* 2nd ed. New York: McGraw-Hill, 1971.

Dorf, R. C. *Computers and Man.* 2nd ed. San Francisco: Boyd & Fraser, 1977.

Etter, D. M. *Structured FORTRAN 77 for Scientists and Engineers.* 3rd ed. Redwood City, CA: Benjamin/Cummings, 1990.

Hopper, G. M., and S. L. Mandell. *Understanding Computers.* 3rd ed. St. Paul, MN: West Publishing, 1990.

PROBLEMS

4.1 Convert the following numbers to binary, octal, and hexadecimal notation.

(a) 5 **(c)** 38

(b) 16 **(d)** 89

4.2 Convert the following numbers to decimal, octal, and hexadecimal notation.

(a) 10010010 **(c)** 11111111

(b) 00111111 **(d)** 10101010

4.3 Prepare a flowchart that finds the largest number in a series of numbers.

4.4 Prepare a flowchart that sorts 10 numbers from highest to lowest.

4.5 Prepare a flowchart that takes a list of grade point averages (GPAs) for 100 students and determines the number of students that made the President's List (GPA $\geq$ 3.80), Dean's List (3.50 $\leq$ GPA $\leq$ 3.79), and Honor Roll (3.00 $\leq$ GPA $\leq$ 3.49).

4.6 Prepare a flowchart that determines the surface-area-to-volume ratio for a sphere in which the radius increases from 0.5 to 20.0 m in 0.5-m increments.

4.7 Each month, the electric company bills consumers according to the following schedule:

Electricity Usage (kWh)	Marginal Rate ($/kWh)
Electricity $\leq$ 1000	0.10
1000 $<$ Electricity $\leq$ 10,000	0.08
Electricity $>$ 10,000	0.06

For example, if a customer used 5000 kWh of electricity in a given month, she would be charged $0.10/kWh for the first 1000 kWh and $0.08/kWh for the next 4000 kWh. Prepare a flowchart that calculates the bill for each customer.

4.8 The Fibonacci series is 1, 1, 2, 3, 5, 8, 13, 21, 34, ... Prepare a flowchart to calculate the first 50 terms of the Fibonacci series.

4.9 Prepare a flowchart that calculates the first 50 terms of the following summations.

(a) $\sum_{i=1}^{n} \frac{1}{i}$ (b) $\sum_{i=1}^{n} i^2$ (c) $\sum_{i=5}^{n} (5-i)^2$ (d) $\sum_{i=0}^{n} \frac{i}{1+i}$

(e) $\sum_{i=1}^{n} \frac{1}{i^2}$ (f) $\sum_{i=2}^{n} \frac{i}{i-1}$ (g) $\sum_{i=-50}^{n} i$ (h) $\sum_{i=0}^{n} i(i-1)$

4.10 Prepare a flowchart that calculates the summations described in Problem 4.9 (a), (d), (e), and (f). Stop the program when an additional term changes the value of the summation by less than one part in 1000. In no case should more than 1000 terms be evaluated.

4.11 Prepare truthz tables for the logic circuits shown in Figure 4.18.

FIGURE 4.18
Logic circuits for Problem 4.11.

a.
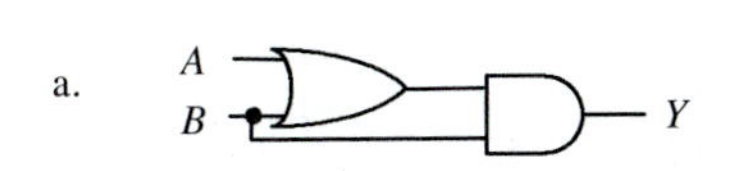

c.
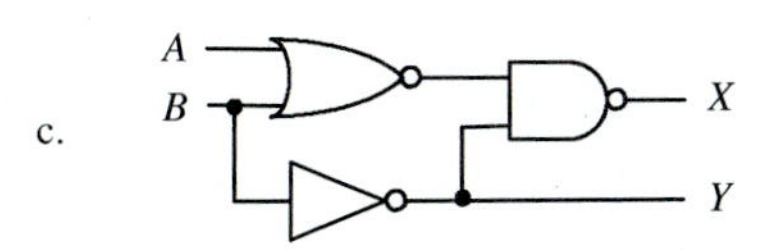

b.
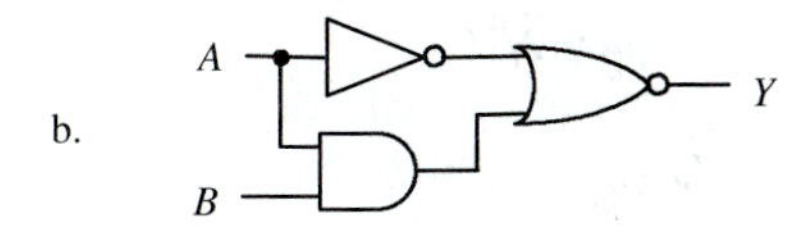

d.
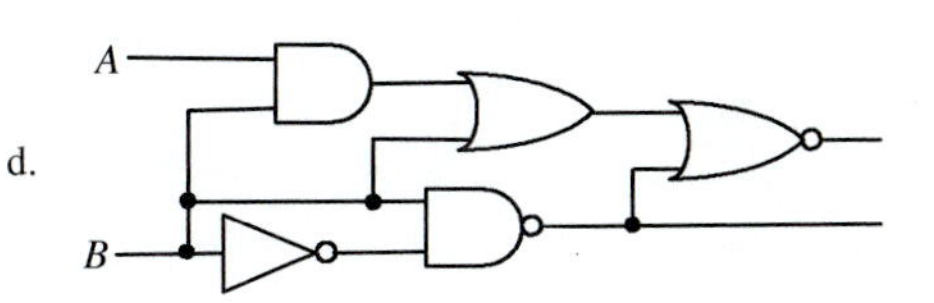

Glossary

address A numerical designation for a memory location.
algorithm A sequence of actions that will solve a particular problem.
assembly language The use of Englishlike commands to generate machine language instructions.
binary Base-2 number system.
bit A unit of information consisting of either 0 or 1.
byte Eight bits bundled together.
central processing unit The part of the computer that interprets and executes instructions.
clock cycle The speed at which the computer performs a single operation.
denary Base-10 number system.
Eratosthenes' sieve An algorithm that finds all the prime numbers less than or equal to a user-specified integer.
flip-flop An electronic circuit that can store either of two stable states.
flowchart A schematic that describes the sequential order in which the steps of a logarithm are done.
hardware A term for the computer and the associated peripherals, such as the printer, disk drives, and networks.
Hero's odometer A 2000-year-old aid for computation that used rotating pegged wheels.
hexadecimal Base-16 number system.
integer A discrete number used for counting.
interpolating To estimate a value of a function or series between two known values.
local area network A system that links computers together and allows them to share files, programs, and use common input/output devices.

logarithm The power to which a base must be raised to produce a given number.
machine language A series of binary numbers that directs the CPU.
memory A specific part of a computer used solely to store numbers.
multiplicand The number that is to be multiplied by another.
Napier's bones A set of ivory rods used to perform multiplication quickly.
operating system The fundamental interface through which the user loads instructions on the CPU, routes output to a printer, or saves work for later use.
octal Base-8 number system.
parallel processing The simultaneous work of multiple computers on a single problem.
program counter The storage of the memory location of the next instruction.
real number A number that is used to measure continuous quantities.
repetition structures Program structures that can loop as long as a condition is true, but break out of the loop once it is false, or loop a specified number of repetitions.
sequential structures Program structures that perform sequentially.
selection structures Program structures that have a decision point where a condition is evaluated.
server A computer that is used for routing communication between multiple computers and is a centralized repository for files and programs.
slide rule A simple device that uses logarithms to perform multiplication and division.
software A collection of instructions that directs computer hardware to perform specific tasks.
spaghetti code Computer code characterized by many "go to" statements that direct the flow to various parts of the program.
structured code Computer code characterized by statements that are logically organized.
virtual storage A programming technique in which programs and data are swapped out of active memory into peripheral storage devices.

Mathematical Prerequisites

- Algebra (Appendix E, Mathematics Supplement)
- Geometry (Appendix I, Mathematics Supplement)
- Calculus (Appendix L, Mathematics Supplement)

CHAPTER 5

Introduction to Design

The ability to create something out of nothing makes design one of the most exciting aspects of engineering. To be successful, design engineers require a broad set of talents, including knowledge, creativity, people skills, and planning ability. As we briefly described in the introductory chapter, design engineers follow the *engineering design method.*

In this chapter, we examine the engineering design method in detail. Although there are many variations, we will use the steps illustrated in Figure 5.1. The first four steps of the engineering design method proceed straightforwardly. The next four steps are repeated three times; the first pass is a **feasibility study** where ideas are roughed out, the second pass is a **preliminary design** where some of the more promising ideas are explored in more detail, and the third pass is a **detailed design** where highly detailed drawings and specifications are prepared for the best design option. Finally, the last two steps proceed straightforwardly. The end result of the engineering design method is a product, service, or process that serves the needs of humankind. Often, after the method is complete, the end result can still be improved, so the process is repeated by returning to the first step.

The engineering design method contains the following elements:

- *Synthesis*—combining various elements into an integrated whole.
- *Analysis*—using mathematics, science, engineering techniques, and economics to quantify the performance of various options.
- *Communication*—writing and oral presentations.
- *Implementation*—executing the plan.

The engineering design method is often an *iterative* procedure, meaning some of the steps must be repeated because information needed at the beginning is not known until later steps are completed. Further, the engineering design method is not a rigid procedure to be slavishly followed; rather, it is a general guide.

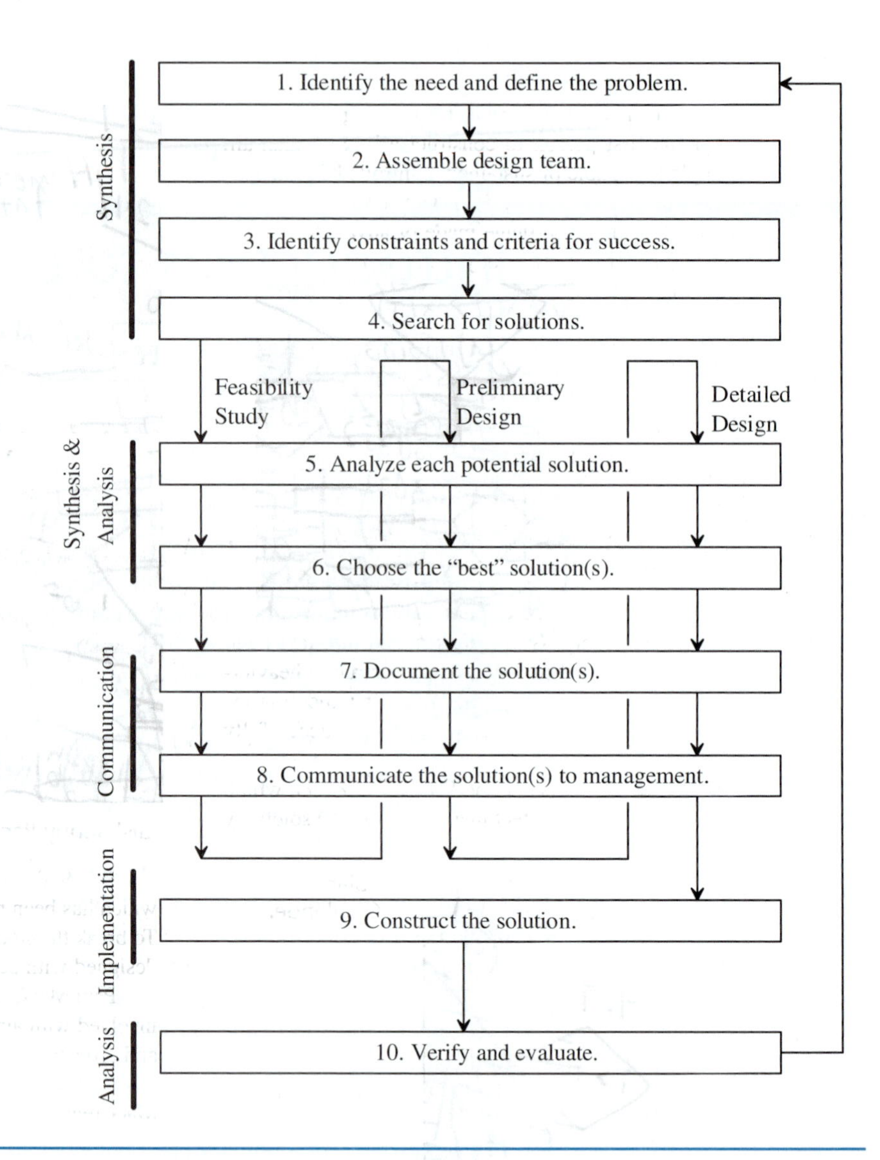

FIGURE 5.1
The engineering design method.

5.1 THE ENGINEERING DESIGN METHOD

Let us examine each step in the engineering design method in detail.

5.1.1 Step 1: Identify the Need and Define the Problem

An engineer is a person who applies science, mathematics, and economics to meet the needs of humankind. Therefore, the job of an engineer starts when a need is identified.

Needs may be identified by a creative engineer who goes through life saying, "There must be a better way." The military may identify a need when its intelligence sources reveal that the enemy has a new capability, and a countermeasure must be developed. (Of

Paul MacCready, Engineer of the Century

In 1977, Paul MacCready became famous for winning the $95,000 Kremer Prize for the first person to construct a heavier-than-air, human-powered plane capable of sustained, controlled flight. This prize stood unawarded for 18 years. Paul won it by building the *Gossamer Condor,* an innovative plane made of advanced materials and employing a sophisticated aerodynamic design.

Gossomer Condor.

British industrialist Henry Kremer then upped the stakes and offered a $213,000 prize to the first person to construct a heavier-than-air, human-powered plane capable of crossing the English Channel. In 1979, MacCready's *Gossamer Albatross* successfully met this challenge.

In 1981, MacCready constructed the *Solar Challenger,* which carried a pilot 163 miles at 11,000 feet in a craft powered solely by sunlight.

Solar Challenger.

In 1984, his human-powered *Bionic Bat* won two Kremer Prizes for speed. Later, he developed a radio-controlled, wing-flapping replica of a giant pterodactyl, which was featured in the Imax film *On the Wing.*

In 1987, he built the *Sunny Racer,* a solar-powered car that won an Australian race by going 50% faster than his nearest competitor.

Pterodactyl in flight.

GM Sunny Racer.

In 1990, he introduced the *Impact* electric automobile, which has been mass produced by General Motors as the EV1. To break the stodgy image of electric cars, this automobile was designed with performance in mind.

Paul MacCready founded AeroVironment, Inc., a company involved with air quality, hazardous wastes, alternative energy, and efficient vehicles for land, sea, and air.

In recognition of his outstanding achievements, Dr. MacCready has received many awards including the *Engineer of the Century Gold Medal* offered by the American Society of Mechanical Engineers.

Impact electric car, prototype of the General Motors EV1.

Source: Information provided by AeroVironment, Inc.

course, the enemy will soon learn of our countermeasure, so they will develop a counter-countermeasure. We will counter with a counter-counter-countermeasure, and so on.) Some needs are identified by management or sales personnel who are very familiar with the market and can spot a need for a new product. Government regulations may create needs by setting new standards for safety or pollution control. Politicians may create needs by promising their constituents new roads or buildings. The growing world population creates stresses on our environment and creates new needs to reduce the stress.

Notice that often the engineer does not identify the need; rather, she is there to serve humankind when the need has been identified by others.

Once the need is identified, the problem must be defined. Without this step, we might solve the wrong problem. For example, suppose that a highway is congested and causing troublesome delays for commuters. Once identified, we might *define* the problem as, "How do we widen the road to accommodate more traffic?" However, experience has shown that widening the road often results in larger traffic jams, because commuters soon learn of the extra capacity and swarm to the wider road. Perhaps the problem is not with the road. Rather, perhaps there is a need for a commuter railway, or a high-occupancy lane that allows cars with multiple passengers to travel more quickly. Perhaps a better statement of the problem is, "How do we create a transportation system to move more people quickly and efficiently?"

5.1.2 Step 2: Assemble Design Team

Because of the complexity of modern engineering projects, rarely is a design tackled by a single individual; rather, design is done with teams of individuals who have complementary skills. In the past, a design problem was tackled by specialists who worked in a compartmentalized manner via **sequential engineering**. For example, suppose an automobile was being designed. First, the stylists would decide on the body shape, then the mechanical engineers would determine how to form the body panels and fit the engine under the hood, then the electrical engineers would design the electrical system, then the production engineers would design the production line, and finally the marketers would develop an advertising campaign. Although sequential engineering worked, it was not optimal. Each specialist could find a **local optimum** for each step, but only within the constraints of what they were given by the previous specialist. Such an approach missed the **global optimum.**

To find the global optimum requires the specialists to work together right from the beginning using an approach called **concurrent engineering.** To illustrate the benefits of concurrent engineering, suppose that while the automobile is being conceptualized, the marketer and the stylist worked together to establish a highly salable design. Further, the mechanical engineer would be involved so that the body style can accommodate the engine. Also, suppose the design objectives can be achieved only by using new materials, such as aluminum space frames or polymer body panels. Obviously, the production engineer must be involved because these new materials will make a big impact on the manufacturing methods. Further, suppose the design objectives can be met only by using a hybrid engine, in which a gasoline engine provides baseline power but an electric motor provides peak power during rapid accelerations. Clearly, with this design, the electrical system is an integral part of the automobile and cannot be designed as an afterthought.

5.1.3 Step 3: Identify Constraints and Criteria for Success

Every project has constraints or limitations, because resources are never infinite. The constraints must be identified early, as they affect the project planning. Typical constraints are listed here:

- *Budget.* Before a project is initiated, the engineers must know the proposed budget, because it affects the resources they can amass for the project.
- *Time.* Some projects must be completed quickly, because the need is urgent. The engineers must know the time allocated to the project, because it determines the number and type of options they can consider.
- *Personnel.* As the project team is assembled, the engineer must know the number of people assigned to the team and their skills. A large budget with ample time does not forecast success unless skilled individuals are working on the project.
- *Legal.* Legal constraints can be restrictive in today's world. Before a large project is undertaken, it must be coordinated with numerous government bureaucracies, each with a different responsibility (water pollution, air pollution, sewage, traffic control, etc.). Legal constraints can cause severe delays and cost overruns if not entered in the planning process.
- *Material properties and availability.* Engineers are always constrained by material properties. For example, it is well known that engine efficiency improves as higher temperatures are employed. However, we are constrained by materials capable of withstanding the high temperatures. Perhaps laboratories have developed new materials (e.g., ceramics) with the needed properties, but until they are available on a commercial scale, they are of no use to the project.
- *Off-the-shelf construction.* Early in the project, the engineers must understand whether they are restricted to assembling off-the-shelf components, or whether they are permitted to custom-design the required equipment. Off-the-shelf items are available quickly and are well tested; however, they may compromise ultimate success if they are not customized to the project requirements.
- *Competition.* The engineers must understand whether the ultimate product is a unique item or whether it will compete against other similar products.
- *Manufacturability.* Many items can be made in small quantities in a laboratory or machine shop, but may be unsuitable for mass production. For example, fighter jets can use high-performance, exotic, lightweight materials (e.g., graphite fiber composites) because they are hand-built in small numbers (maybe 50 per year). These materials are not suitable for automobiles, which are produced in higher quantities (maybe 100,000 of a given model per year).

Once the constraints of the project have been identified, it is necessary to determine the criteria for success; that is, what are the goals for the project? Engineering projects have many goals, some of which are listed here:

- *Aesthetics.* With consumer products, aesthetics play a large role in success. It does not matter how rugged, reliable, or functional the product is; if it is ugly, it will not sell. Aesthetics are difficult to define and are highly subjective, but a product that is well balanced and in proportion and has coordinated colors will usually have aesthetic appeal. An important aesthetic principle is that form follows function, meaning that each component of the product serves a useful purpose. Products that violate this principle are often short-lived. For example, automobiles in the 1950s and 1960s sometimes had large fins added for styling purposes. Because these fins had no function, they were a passing fad that thankfully has not reappeared.

- *Performance.* The performance of a product is generally determined by the producer, unless the project responds to a specific customer request. For example, the producer may specify the following automobile performance: 0 to 60 mph acceleration = 10 s, 60 to 0 mph braking = 150 ft, and fuel economy = 25 miles per gallon.
- *Quality.* The quality of a product is determined by the consumer. Quality is often defined as "fitness for use." For example, a consumer may expect an automobile to have the following qualities: 0 to 60 mph acceleration = 6 s, 60 to 0 mph braking = 110 ft, and fuel economy = 40 miles per gallon. An automobile that does not meet these criteria will be deemed of low quality even if it is reliable and stylish.
- *Human factors.* Because most products are used by humans, successful products must be designed with the human user in mind. In automotive design, human factors would include easy-to-read gages, controls at the fingertips, pedals that are well spaced and do not require excessive force to push, steering wheels that are easily turned and at the right height, padded seats that do not cause backaches, and so on.
- *Cost.* A product that is aesthetically pleasing, high quality, and user friendly may still fail in the marketplace if it is too costly. There are two types of costs to consider: **initial capital cost** and **life-cycle cost.** The initial capital cost is just the purchase price of the product; the life-cycle cost includes the purchase price but also other costs such as labor, operation, insurance, and maintenance. If it has high maintenance, fuel, and insurance costs, an automobile with a low purchase price actually might be uneconomical compared with a moderately priced automobile. Unfortunately, many consumers only look at the initial capital cost and neglect the life-cycle cost.
- *Safety.* An engineer must always design products that are safe for the end user and the artisans who construct the product. It is impossible to design completely safe products because they would be too costly; therefore, the engineer often must design to industry standards for similar products. Automotive safety standards have been increasing; for example, automobiles now are equipped with air bags.
- *Operating environment.* The engineer must design the product with the operating environment in mind. What temperature and pressure range will the product experience during storage and use? Is it a corrosive environment? What vibration levels will the product experience? Automobiles must be designed to operate in the Arctic to the Tropics, at sea level to the mountains, in the presence of corrosive road de-icing salt, and on roads filled with potholes.
- *Interface with other systems.* Many products must interface with others: Computers must be compatible with software and printers, televisions must be compatible with broadcast signals. An automobile must be compatible with fuels in common use and must have a turning radius and width compatible with roads.
- *Effect on surroundings.* The creation and use of a product may affect its surroundings adversely. With increasing environmental constraints, products must be designed with lower chemical, noise, and electromagnetic emissions. In the United States, automobiles are equipped with catalytic converters that reduce the air pollution emitted from vehicles. In Germany, automobiles are designed so that when the vehicle reaches the end of its service life, it can be disassembled and the components recycled.

- *Logistics.* Many products require support systems such as electricity, cooling, steam, fuel, and spare parts. Depending on where the product is used, these support systems may or may not be readily available. For example, a product used in outer space has little logistical support, so it must be designed to operate independently from earth. For automobiles, there is a tremendous infrastructure with refueling stations and repair centers widely available, so logistics are less of a problem.
- *Reliability.* A reliable product will always perform its intended function for the required time period and in the environment specified by the user. No product is 100% reliable, although some are close. NASA requires spacecraft components to be highly reliable. This is often accomplished through **redundancy,** that is, using multiple components each with the same capabilities. For example, the space shuttle is controlled by three computers. This provides backup in case two computers completely fail. Also, if the computers disagree, a "vote" can be taken to settle the dispute. Normally, automobiles are not designed with redundancy, because the consequences of failure are generally not catastrophic.
- *Maintainability.* A product that is maintainable can have the required maintenance performed upon it at the necessary frequency. A satellite is an example of a product that is not easily maintained, because of the difficulty in reaching it. On the other hand, automobiles are highly maintainable, because there is always ready access to a repair shop. *Preventive maintenance* is performed at regular time intervals or when components are nearing failure. Rotating automobile tires at regular mileage intervals and replacing them when they are worn are both examples of preventive maintenance. *Corrective maintenance* is employed after a part fails. Replacing a flat tire with a new tire is an example of corrective maintenance.
- *Serviceability.* If a product can be maintained easily, it is serviceable. For example, if a special tool is required to change an automobile oil filter because it is inaccessible with ordinary tools, it is not serviceable.
- *Availability.* If a product is always ready for use, it is available. If an automobile is frequently in the repair shop and does not operate at temperatures above 80°F or temperatures below 40°F, it is not available a high percentage of the time.

Once the desired properties of the product are identified, it is necessary to weight them, that is, specify their relative importance.

5.1.4 Step 4: Search for Solutions

The search for solutions requires engineers to generate ideas that solve the design problem. Inherently, idea generation is a creative process that cannot be accomplished simply by following a prescribed algorithm. Instead, we offer the following heuristics:

- *Can I eliminate the need?* Suppose you are an automobile designer and your boss defines the need as follows: "The suspension springs are rusting, so I want you to design a coating that protects them." Perhaps you can eliminate the need altogether by specifying nonrusting polymer springs rather than metal springs.
- *Challenge basic assumptions.* The human heart pumps blood through the body in a pulsatile manner, rather than smoothly and continuously. A designer might assume that an artificial mechanical heart must operate in a pulsatile manner also. However, recent

medical studies show that after about 5 days, the human body can adapt to continuous blood flow. A designer who challenges the pulsatile assumption has many more design options available.

- *Be knowledgeable.* In engineering, useful ideas don't spring from nothing; we work from a knowledge base. While searching for solutions, the design engineer must become as knowledgeable about the problem as possible. Information may be obtained from libraries, the Internet, government documents, professional organizations, trade journals, vendor catalogs, and other individuals.
- *Employ analogies.* By using analogies, a design engineer can exploit information from other fields and bring this information to the design problem. Nature is rich with design solutions to problems. For example, in their search for methods to fly a heavier-than-air craft, the Wright brothers employed bird wings as an analogy. Recently, engineers at MIT have developed an extremely efficient method to propel boats by making an analogy to fish swimming. Nature is not the only place to find solutions. Many engineers have come before you, so you can make analogies to their solutions to design problems. There are books that catalog engineering solutions. For example, *Pictorial Handbook of Technical Devices* (Schwarz and Grafstein, 1971) has over 5000 illustrations of such technical devices as pumps, gears, bearings, fasteners, machine tools, electronic circuits, knots, bridges, and many others. Thumbing through such a book can be very stimulating when you confront a design problem.
- *Personalize the problem.* For some problems, you can gain insight by imagining yourself shrunken down and literally entering the device being designed. Suppose you are tasked with designing a low-shear pump for transporting fragile polymer solutions. By shrinking yourself down and entering the pump, you can identify the high-shear zones and create a design that eliminates them.
- *Identify the critical parameters.* Many engineering designs have a critical feature that must be overcome for the design to be functional. Perhaps a component must operate at high speeds, or be very reliable, or be made to close tolerances. After these *critical parameters* are identified and addressed, the design will more quickly converge to a workable solution.
- *Switch functions.* Normally, a pump housing is stationary and the internals move. Perhaps you can find an elegant solution to a design problem by switching functions, using stationary internals and a rotating housing.
- *Alter sequence of steps.* Processes involve a prescribed sequence of steps. Perhaps an elegant solution can be developed by altering the sequence. Suppose you are trying to improve the process for making coffee. In the traditional process, first the coffee beans are roasted and then they are ground. This approach works best because traditional grinders function only with roasted beans, which are more friable. But suppose a new grinding technology allows fresh, wet beans to be pulverized. Using this new grinder would allow the traditional steps to be reversed: beans could be pulverized and then roasted. Perhaps the high surface area of the ground beans would allow more flavors to be generated in the roaster, or perhaps the capacity of the roaster would increase when ground beans rather than whole beans are used.
- *Reverse the problem.* Suppose your goal is to develop a lightweight wrench for NASA astronauts. By reversing the problem and thinking of ways you could make the wrench heavier, you will become sensitive to where the weight is and then can think of ways to engineer the weight out.

- *Repeat components or process steps.* Sometimes if one is good, two are better, and three are better yet. Suppose you wish to compress a gas that decomposes if it gets too hot. If the compression is done in a single step, so much compression energy must be added to the gas that it will become too hot and decompose. However, if you break the compression into multiple steps and cool between the steps (intercooling), then you can compress the temperature-sensitive gas to high pressures.
- *Separate functions.* Sometimes an elegant design results when functions are separated. For example, the following functions occur in the piston/cylinder of an automobile engine: air input, fuel input, air/fuel mixing, compression, combustion, expansion, exhaust, lubrication, and wall cooling. Fuel combustion and wall cooling are incompatible functions; perhaps the engine could be improved by performing these functions in separate pieces of equipment.
- *Combine functions.* Sometimes an elegant design results when functions are combined, rather than separated. For example, many industrial processes need both electricity and heat. To supply its energy needs, a plant could convert fuel into electricity in one facility and convert fuel into steam in another. However, these functions may be combined so that fuel is converted into both electricity and steam in a single facility. Such an approach is called *combined heat and power* and is more efficient; the heat that inevitably results from electricity production is used to generate steam rather than being wasted.
- *Use vision.* While you are working on the design, imagine that you have accomplished the goals and are inspecting the final product. What features would this product have if it were built?
- *Employ basic engineering principles.* Throughout your engineering studies, you will learn basic principles that may be applied to your design problem. For example, a basic thermodynamic principle is that unused "driving forces" lead to inefficiencies. Stated another way, when designing a machine or process, every pressure difference, temperature difference, voltage difference, or concentration difference that is not harnessed to make energy will increase total energy expenditures. By reviewing the design and taking steps to minimize these differences, you make the machine or process more efficient.

To help your mind find design solutions, you must put yourself in a physical environment that stimulates creativity. The proper environment will differ from person to person. For some, a busy room filled with people stimulates creativity, whereas others find it distracting. Some people find that everyday activities (driving, gardening) stimulate creativity, for these activities do not require focused attention but allow the mind to wander freely and explore creative solutions. Sometimes while trying to find a solution, you may hit a "brick wall" and make no progress. In this situation, it pays to take a break and do something completely different, like play music or sports, or even sleep. When you return to the problem, you may find that your subconscious solved the problem while you were engaged elsewhere. Sometimes the biggest aid to creativity is simply dogged perseverance; by focusing on the problem for an extended period of time, you completely immerse yourself in it and increase your chances of finding a solution.

Although an individual can certainly generate creative ideas, it helps to work with others. The interaction stimulates ideas that no lone individual could find. The following techniques have been developed as formats to help people work together to solve problems:

- *Brainstorming.* The group (generally 6 to 12) has a leader that sets the tone for free expression. All members reveal their ideas to the group as soon as they form them. The most important rule is not to belittle any ideas or rate them. Often, seemingly silly ideas lead to valuable solutions. A person is designated to record the ideas as they are generated.
- **Nominal Group Technique.** In this technique, the leader assembles the group and poses the problem. Each group member independently works on the problem and writes ideas on paper. Then, when the group members have stopped generating ideas, each group member sequentially explains one idea to the entire group. Each idea is recorded on a blackboard or flipchart for all members to see. Although discussion is permitted for clarification purposes, no critiques are allowed. Later, when the ideas are all identified, the group members may be asked to rate the ideas. If a few personalities in the group are dominant and tend to sway discussions, this technique reduces the potential for "groupthink," which might occur with brainstorming.
- **Delphi Technique.** This technique is similar to the previous technique except that the members of the group are intentionally separated. The leader mails each group member the problem statement. The group members then return their solutions by mail. The leader pools the responses and asks the group members to rate the ideas. Those ideas receiving low ratings can be clarified by the originator, if desired.

The Color TV War

The foundations for television were laid in 1883 when Germany's Paul Nipkow developed a spinning disk with spiral apertures that broke an image into a series of scanned lines.

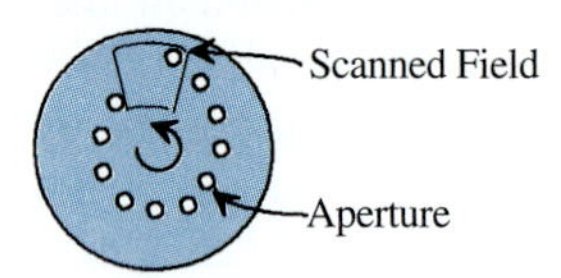

The image is scanned once per revolution.

In 1889, Russia's Polumordvinov conceptualized a system in which a Nipkow disk was combined with three color filters that would break the image into the three primary colors (red, green, blue). His efforts, and those of other inventors (Adamian, von Jaworski, and Frankenstein), never resulted in a workable system.

In 1928, a workable spinning-disk color television was demonstrated in England. A year later, Bell Labs demonstrated a similar system in the United States. However, during the 1930s, most development efforts focused on black-and-white television.

In 1940, Peter Goldmark of CBS developed a color television that combined a spinning multicolor disk with a black-and-white television tube in which an electron gun "painted" the image on a

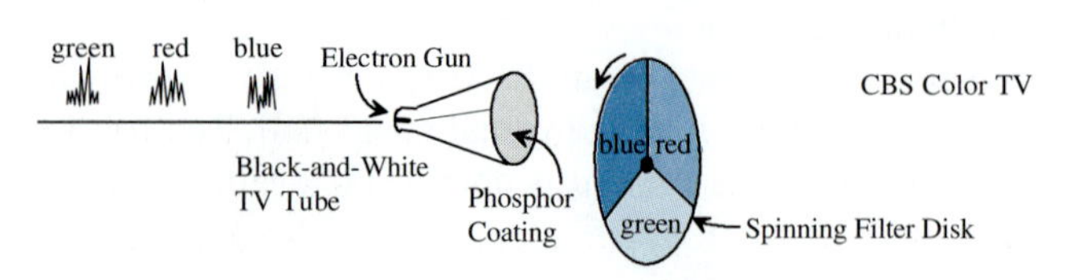

CBS Color TV

phosphor coating. In Goldmark's system, the three primary colors were broadcast *sequentially.* When the blue filter aligned with the TV tube, the blue signal reached the TV tube and the eye would see only the blue information. The other primary colors were received by the eye in a similar manner. Because the disk was spinning rapidly, the eye integrated the three primary colors into a single image.

In July 1941, commercial black-and-white broadcasting began to an audience of a few thousand, but in December, World War II intervened and halted progress on both black-and-white and color television. After the war, in 1946, RCA sold 10,000 black-and-white sets at a cost of about $385 each. In the same year, CBS demonstrated Goldmark's television to the Federal Communications Commission (FCC), seeking approval for this to be the standard color television. RCA waged a campaign against adopting the CBS television because its signal would be incompatible with black-and-white sets already being sold. To make the black-and-white sets compatible would require the consumer to spend about $100 for a special converter.

In 1946, RCA was developing its own all-electronic color television in which the three primary colors were broadcast *simultaneously.* Their television tube had three electron guns, one for each color. The viewing surface of the tube had three phosphors; each would glow a separate primary color when struck by the electron beam. Further, through clever engineering, their signal would be completely compatible with black-and-white television.

In 1950, the CBS and RCA systems were tested side by side. The *Variety* headline announced "RCA Lays Colored Egg" and proclaimed the failure of RCA compared to CBS. The FCC authorized CBS to begin producing its sets.

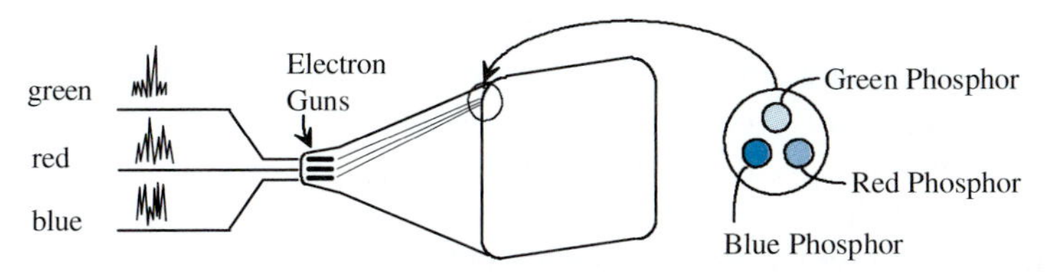

RCA Color TV

RCA countered by stepping up production of black-and-white sets, which were incompatible with the CBS color broadcast signal. According to RCA's David Sarnoff, "Every set we get out there makes it that much tougher on CBS." Sarnoff increased RCA's efforts to perfect its system by requiring employees to work 18 hours per day, including weekends. Rewards of thousands of dollars were offered for key developments. Sarnoff tried, unsuccessfully, to get the courts to block the FCC ruling.

In 1951, CBS broadcast the official color premier, a one-hour Ed Sullivan show, to a few dozen color sets. In contrast, there were now 12 million black-and-white sets in the United States.

RCA's development efforts paid off. In 1951, a few days after the CBS premier broadcast, RCA demonstrated its new color television with a 20-minute program. The press responded enthusiastically to the demonstration. Clearly, the RCA system was the way of the future. In 1953, CBS abandoned its efforts in color television, and the FCC officially adopted the RCA system.

In 1954, the first RCA color sets were offered for $1000 each (a quarter of the average annual salary). Only 5000 sets were sold, out of a projected 75,000. Further, the color sets were failure prone; twice as many service calls were required for the few thousand color sets than all the millions of black-and-white sets put together. Over the years, prices were cut but sales were sluggish. *Time* magazine declared color television "the most resounding industrial flop of 1956."

By 1959, RCA had spent $130 million developing and marketing color television and still had not made a profit. However, in 1960, they acquired the Walt Disney Show from ABC. NBC announced it would broadcast the show in color as "Walt Disney's Wonderful World of Color." By 1960, RCA made its first profit in color television.

Interestingly, although the CBS spinning-disk system was not successful in the consumer market, it was successful with NASA. During the Apollo moon mission, NASA needed a color TV camera that was compact, lightweight, robust, energy-efficient, simple to use, and sensitive to low light levels. At the time, the RCA system could not meet the objectives, but the CBS system could. From the moon, the CBS sequential signal was broadcast to earth where it was converted to the RCA simultaneous signal so it could be viewed by millions of people all over the world.

Adapted from: D. E. Fisher and M. J. Fisher, "The Color War," *American Heritage of Invention & Technology* 12, no. 3 (1997), pp. 8–18; and S. Lebar, "The Color War Goes to the Moon," *American Heritage of Invention & Technology* 13, no. 1 (1997), pp. 52–54.

5.1.5 Feasibility Study

A feasibility study is meant to quickly eliminate ideas without consuming much engineering time. The goal is to see the big picture and address the most important features of the problem.

Step 5: Analyze Each Potential Solution.

During the feasibility study, when first analyzing each potential solution, an engineering team may use simple calculations to characterize each design. Alternatively, the team may use simple heuristics (e.g., number of parts, number of process steps, number of high-precision components, simplicity of components, logistical complexity, or need for exotic materials) to analyze different design options.

Step 6: Choose the Best Solution(s).

Although engineers would like to investigate each candidate technology in great detail, this is rarely possible. Because we do not have infinite time to work on a problem, it is necessary to eliminate some technologies based upon the results of the feasibility study. Depending upon the methods used to analyze each design option, engineers develop a rating scheme to choose the best option(s). For example, designs with fewer numbers of parts tend to be less expensive and more reliable. So the team may select the designs that have the fewer number of parts.

Step 7: Document the Solution(s).

After the engineers have searched for a solution and made some choices, they should document their results in written form. The report should define the problem, identify criteria for success, describe and analyze the various options, describe the rating scheme used to evaluate the various options, and recommend the best option(s). Graphics and good writing are essential to the report. This report is distributed to team members so everyone is working from the "same sheet of music." Before the report is released, all members of the team must concur with it.

Step 8: Communicate the Solution(s) to Management.

The engineers must communicate their results to management by sending them a copy of the written report developed in Step 7. In addition, they may wish to discuss the results in person, by phone, or in a formal oral presentation.

5.1.6 Preliminary Design

The purpose of the feasibility study was to do a "quick and dirty" survey to determine if more effort is warranted. Provided the outcome was positive, the engineers now engage in a preliminary design that is more detailed than the feasibility study.

Step 5: Analyze Each Potential Solution.

During the preliminary design, the engineers will use detailed calculations to analyze the promising designs that emerged from the feasibility study. Analysis of the design options may rely heavily on engineering courses such as thermodynamics, statics, strength of materials, circuit analysis, heat transfer, fluid flow, and others.

Step 6: Choose the Best Solution(s).

The preliminary design analysis provides us additional information with which to identify the best solution(s). One approach to identifying the best solutions is to list the advantages and disadvantages of each option. Another approach uses an **evaluation matrix** (Table 5.1),

TABLE 5.1
Evaluation matrix

		Option A		Option B		Option C	
Property	Weighting Factor	Score	Weighted Score	Score	Weighted Score	Score	Weighted Score
Property 1	α	A1	$\alpha \times$ A1	B1	$\alpha \times$ B1	C1	$\alpha \times$ C1
Property 2	β	A2	$\beta \times$ A2	B2	$\beta \times$ B2	C2	$\beta \times$ C2
Property 3	γ	A3	$\gamma \times$ A3	B3	$\gamma \times$ B3	C3	$\gamma \times$ C3
Total			Σ above		Σ above		Σ above

in which the desired properties and their relative importance (i.e., weighting factors) are listed on the left. Each option has a separate column in which a score is given for each property. The scores multiplied by the weighting factors are added for each option. The option with the highest total weighted score is selected.

Steps 7 and 8: Document the Solution(s) and Communicate with Management.

The results of the preliminary design should be documented in a report. First the report should be circulated among the design team to get everyone's concurrence; then, it is sent to management. This report contains the same topics as the feasibility report, but with more details. In addition to the written report, the results of the preliminary design are presented to management in a formal oral report that allows them to ask questions to determine if the project is worth continuing. Normally, a single solution will emerge from the preliminary design phase.

5.1.7 Detailed Design

When a project successfully passes through the preliminary design phase, it enters the detailed design phase. The detailed design involves a very large team that works on *the* solution that emerged from the preliminary design phase.

Steps 5 through 8 of the Detailed Design.

Each and every component of the solution must now be specified in great detail. Issues such as materials, dimensions, tolerances, and processing steps all must be documented in detailed drawings and reports that allow artisans to construct a prototype. For some complex projects, such as airplanes, the detailed drawings and reports literally weigh hundreds or thousands of pounds and require a truck to transport them.

5.1.8 Step 9: Construct Solution.

Typically, a prototype will be constructed from the documents produced in the detailed design. If the prototype testing goes well, the design will be finalized. A production facility must be built to construct the product to be marketed to the consumer. Many people are involved in this phase. Material suppliers must be identified and contracts written. Personnel to operate the production facility must be hired. Large-scale processes must be refined, because the prototype was constructed in a small shop. Marketing strategies must be finalized and financing arranged. Owner's manuals must be written, to go with the product. Spare parts must be manufactured and stocked to provide future customer support.

5.1.9 Step 10: Verify and Evaluate.

The engineer's job is not complete once the first product is made in a large-scale production facility. Product samples should now be taken from the production facility to determine that the product is meeting the design specifications. If there are problems in the production facility, they must be corrected.

The engineer now should be thinking about the next generation of products. Because technology changes rapidly, the engineer should be thinking about improved manufacturing techniques, substituting new materials with superior properties, or improving the design to meet the latest market requirements. If management approves the next generation of products, the engineer returns to Step 1, where the design problem is defined.

5.2 FIRST DESIGN EXAMPLE: IMPROVED PAPER CLIP

To illustrate the steps of the engineering design method, we consider the design of an ordinary object: the paper clip.

Step 1: Identify the Need and Define the Problem. A salesman working for Office Supply, Inc. (OSI) returned from a recent sales call to a hospital. The nurses informed him that a patient received improper treatment and nearly died because a critical document was lost from a stack of papers fastened with ordinary paper clips. Apparently, a nurse was carrying a stack of 20 bound medical records and dropped them. The paper clips failed, thus scattering the papers. In the resulting scuffle, a critical document was placed in the wrong file.

The salesman approaches OSI management, requesting that a new paper clip be developed for holding critical documents, such as medical records. Management likes the idea; high-end products often command a greater profit margin.

The salesman has identified the need for a highly reliable paper clip. Management has defined the problem as, "Develop a highly reliable paper clip for use in high-end applications, such as binding medical documents."

Step 2: Assemble Design Team. To design an improved paper clip, management assembles a design team composed of the following individuals: mechanical engineer for specifying the shape of the clip, manufacturing engineer for specifying the equipment to manufacture the clip, and marketing expert to determine how to market the clip to high-end users.

Step 3: Identify Constraints and Criteria for Success. The marketing expert determines that to be successful, the maximum acceptable price is $1.00 per clip. The clip system must be capable of binding anywhere from 2 to 200 pages. The pages cannot separate if the papers are dropped from a height of 10 feet, or lower.

The design team decides that the paper clip should have the properties shown in Table 5.2. The relative importance of each property is indicated with a weight; justifications are given for each assigned weight.

Step 4: Search for Solutions. The design team did some research on paper clips and found an article on the subject.[1] The article states that before there were paper clips, papers

TABLE 5.2
Properties of improved paper clip

Property	Weight	Justification
Reliability	5	The paper clip is intended for critical documents, so reliability is essential.
Compact design	3	Although a compact design is desirable, it is not critical.
Weight	3	Although light weight is desirable, it is not critical.
Convenience	5	The marketing expert indicated that no matter how good the new paper clip is, if it is inconvenient to use, it will not sell.
Low cost	1	This paper clip will be marketed to high-end customers, so cost is not as much of a factor.

1. H. Petroski, "The Evolution of Artifacts," *American Scientist* 80 (1992), pp. 416–420.

were joined with a straight pin. This technology had obvious problems, such as being limited to only a few sheets of paper, sticking the reader with the sharp point, and grabbing extraneous papers. In the mid-19th century, clothespins and elaborate wooden clips were used to bind papers. By the late-19th century, machines for bending wire into paper clips became available, and a myriad of shapes were developed (Figure 5.2). Near the turn of the century, the standard "Gem" paper clip became available from the British company Gem, Ltd.

Commonly available paper clips all rely on the springiness of the wire to hold the papers together. Other options should be identified for holding the paper more securely. Analogous situations should be explored.

From the Petroski article, the design team learned that clothespins have been used to hold papers. The design team discussed this possibility and developed Options 1 and 2 (Figure 5.3). However, rather than rely on a spring to clamp the papers, they proposed two more secure possibilities. Option 1 employs a screw that, when extended, tightly clamps the papers. Option 2 uses a wedge to clamp the papers.

Options 1 and 2 are bulky, so the design team did some brainstorming and came up with Options 3 and 4. Both these design options are analogous to a vise or clamp. The sandwiched metal strip pivots as the screw is rotated or the wedge is inserted. Options 5 and 6 are very similar to Options 3 and 4, except that the sandwiched metal strip slides along a guide instead of pivoting.

Options 1, 3, and 5 require a screwdriver to adjust the screw; Options 2, 4, and 6 require a wedge that could be lost. Desiring a self-contained paper clip, the team devised Option 7 (Figure 5.4). The paper is inserted into the paper clamp, which has a square hole on the top. The retractable pin is threaded at the top and square at the bottom, where it fits

FIGURE 5.2
Examples of wire paper clips.

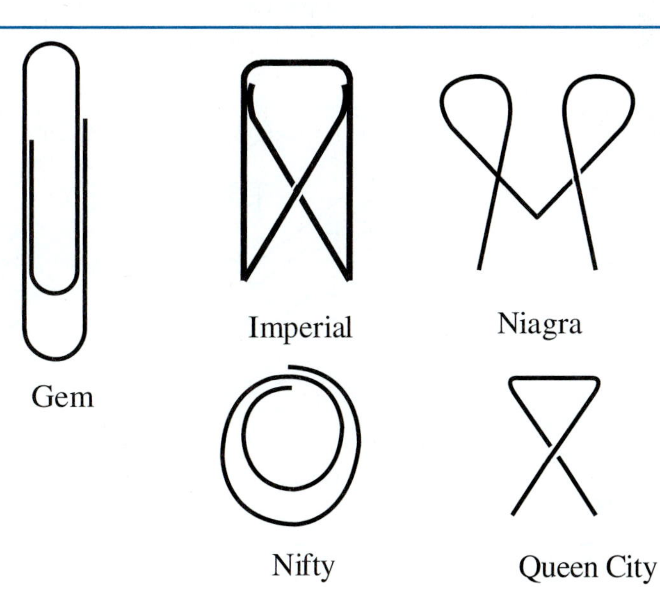

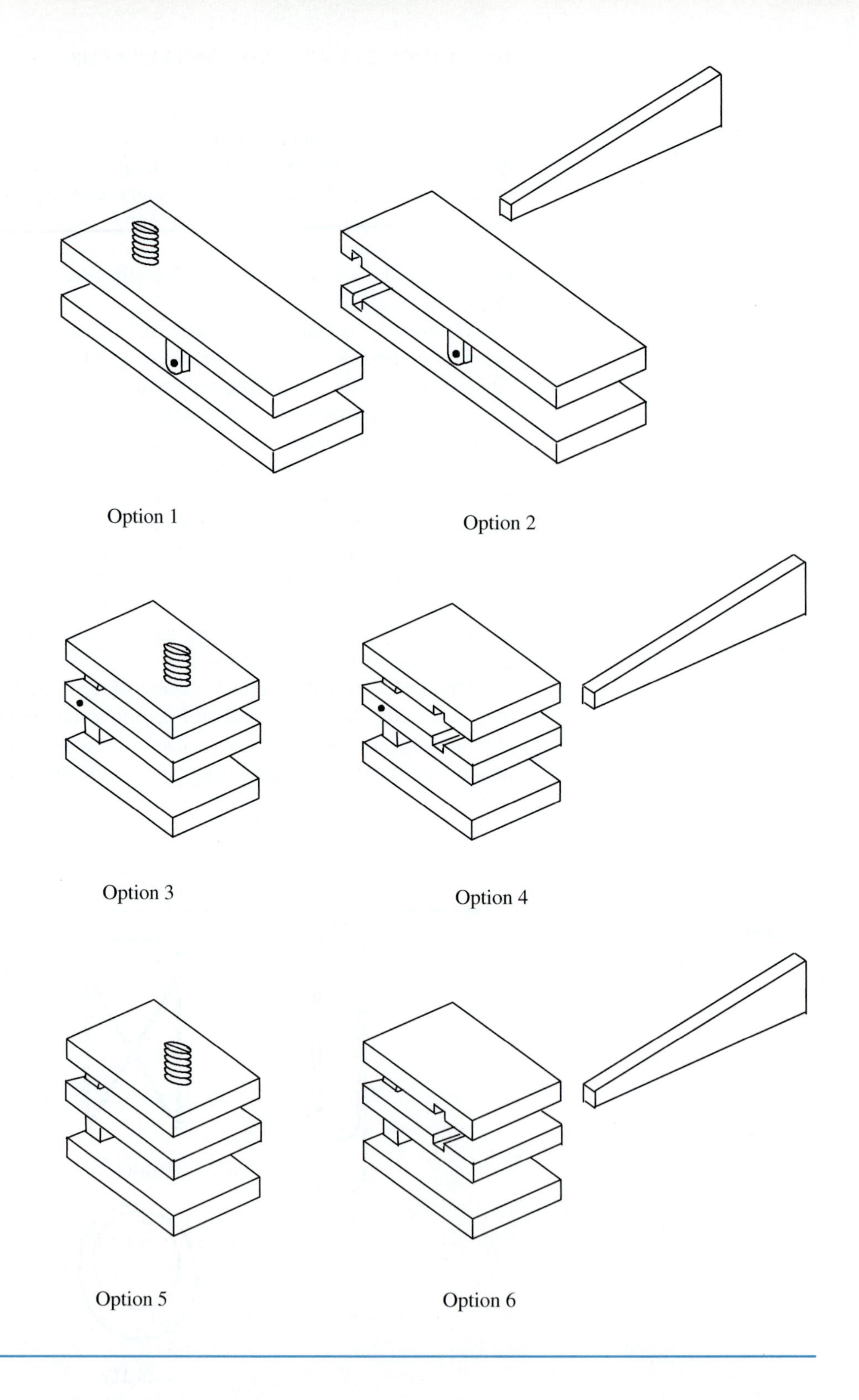

FIGURE 5.3
Six options for an improved paper clip.

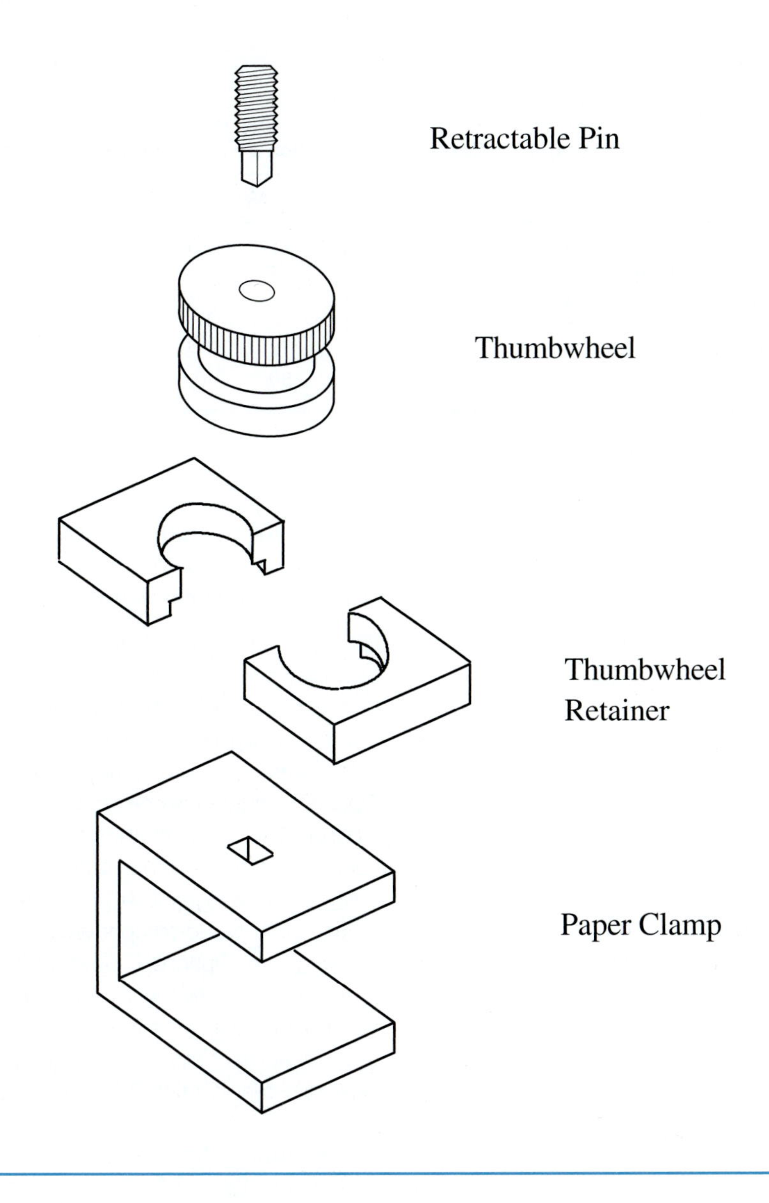

FIGURE 5.4
Option 7 for an improved paper clip.

into the square hole of the paper clamp; the square hole prevents the retractable pin from rotating. The threaded portion of the retractable pin fits into the threaded hole in the thumbwheel. The thumbwheel is captured by the thumbwheel retainers, which allow the thumbwheel to rotate, but prevent movement in the axial direction. As the thumbwheel rotates, it causes the retractable pin to move downward and clamp the papers. A variety of sizes could be manufactured to bind different numbers of papers.

TABLE 5.3
Evaluation matrix for improved paper clip

	Weight	Option 1		Option 2		Option 3		Option 4		Option 5		Option 6		Option 7	
Property	**W**	**S**	**S × W**	**S**	**S × W**	**S**	**S × W**	**S**	**S × W**	**S**	**S × W**	**S**	**S × W**	**S**	**S × W**
Reliability	5	5	25	3	15	5	25	3	15	5	25	3	15	5	25
Compact design	3	1	3	1	3	3	9	2	6	3	9	2	6	2	6
Weight	3	1	3	1	3	3	9	3	9	3	9	3	9	3	9
Convenience	5	1	5	1	5	1	5	1	5	1	5	1	5	5	25
Low cost	1	4	4	4	4	3	3	3	3	3	3	3	3	5	5
Total			40		30		51		38		51		38		70

1 = Poor, 5 = Excellent

Step 5: (Feasibility Study) Analyze Each Potential Solution. The design team decided to use an evaluation matrix (Table 5.3) to help select the best option. They rated the five properties of each option as follows:

- Reliability. Options 1, 3, 5, and 7 use a screw, which is unlikely to fail; therefore, these options scored the highest points. Options 2, 4, and 6 use a wedge that might fall out; hence, these options scored fewer points.
- Compact design. Options 1 and 2 are very large, so they scored poorly. Also, the options with a wedge scored poorly because the wedge takes up space. The thumbwheel on Option 7 requires some space, so this option scored poorly also. Of all the options, Options 3 and 5 were the most compact; even so, they were not as compact as traditional paper clips, so they scored only 3 points.
- Weight. The weight scores are similar to the compactness scores, except for Option 7, where some of the components could be made from plastic to reduce weight.
- Convenience. Options 1, 3, and 5 require a screwdriver, which is inconvenient. Options 2, 4, and 6 require the wedge, which could easily be lost. Only Option 7 was deemed convenient, so it got maximum points.
- Cost. Option 7 is the most complex, so it has the highest cost. The other options are less expensive than Option 7 but still costly compared to traditional paper clips.

Step 6: (Feasibility Study) Choose the Best Solution(s). Based on the total weighted scores shown in Table 5.3, Option 7 is the best by a wide margin.

Step 7: (Feasibility Study) Document the Solution(s). The design team prepares a written report documenting the various options and their evaluation.

Step 8: (Feasibility Study) Communicate the Solution(s) to Management. The design team makes an oral presentation to management to explain the options and why they think Option 7 is the best. They submitted their written report to management one week prior to the oral presentation so management had time to review the options and think of good questions.

After the oral presentation, management approves the project for continued support.

Steps 5 through 8 of the Preliminary Design. The engineering team performs some detailed calculations to determine the required thickness of the metal in the paper clamp. They decide it will be constructed from ordinary carbon steel that is electroplated to resist corrosion. The retractable pin will also be constructed of electroplated metal. To save weight and cost, the thumbwheel will be plastic, with a metal insert for the threads. The thumbwheel retainers will also be plastic.

The machine shop makes some prototype clips and tests them to work satisfactorily when papers are dropped from a height of 10 feet.

To document their design, the team prepares engineering drawings and written reports. After a design review with management, the project receives approval to continue.

Steps 5 through 8 of the Detailed Design. Although the prototype performed successfully, still there are many issues to address, primarily related to manufacturability. To meet the cost objective of $1 per paper clip, it is critical to identify manufacturing methods that can produce and assemble the components cost-effectively. Once the manufacturing machines and methods are identified, the final engineering drawings of the paper clip can be prepared. The drawings address key issues such as material suppliers and manufacturing tolerances for each component.

Step 9: Construct the Solution. The machinery to construct the paper clips is installed and paper clips are sold to high-end customers.

Step 10: Verify and Evaluate. The paper clips must be constantly sampled and inspected to ensure that the manufacturing equipment continues to make high-quality products. Even though the process is operating successfully, the engineers should always be looking for improved manufacturing methods to reduce costs and prevent loss of market share to competitors.

5.3 SECOND DESIGN EXAMPLE: ROBOTIC HAND FOR SPACE SHUTTLE

To continue our familiarization with the engineering design method, we explore the design of a robotic hand used to build the International Space Station.

Step 1: Identify the Need and Define the Problem. The U.S. Congress has authorized funding for NASA to construct the International Space Station, an inhabited satellite providing permanent manned presence in near-earth orbit. Much of the hardware is being provided by other nations, so the project is truly international in scope. The purpose of the International Space Station is to provide a microgravity environment for scientific and engineering investigations. Also, it allows physicians to study the long-term effects of weightlessness on human physiology.

Although the hardware is manufactured on earth, it must be assembled in space. Some components are very large and cannot be manipulated directly by free-floating astronauts wearing protective space suits; therefore, large items will be manipulated by a robotic hand.

The robotic hand will be attached to the end of a robotic arm mounted on the space shuttle. To assemble a large component, the operator manipulates the object using the robotic arm and hand, which are governed by remote controls located inside the space shuttle.

Although the space shuttle already has a robotic arm and hand, the present hand is too small to manipulate the large components of the space station; therefore, your company has been hired to design and construct a large robotic hand to replace the small hand.

Step 2: Assemble Design Team. To satisfy the NASA contract, your management has formed a design team composed of individuals with the following expertise: project management, mechanical engineering, electrical engineering, control engineering, and manufacturing.

Step 3: Identify Constraints and Criteria for Success. To meet the contract, the robotic hand must be constructed within 1 year at a cost of $10 million. Because it costs $10,000 to lift 1 pound into space, the contract specifies that the hand can have a mass no larger than 100 lb_m. Because the robotic hand will be used in either full sun or complete darkness, it must operate in a temperature range from −150 to +100°C. It must function in a complete vacuum and will be subjected to micrometeor showers, which are frequent in space. Because it is difficult to provide service in space, the robotic hand must be very reliable. If it fails, it cannot damage the space shuttle or space station. The robotic hand must have a "soft grip," meaning it cannot crush the space station components while assembling them. Because energy is precious in space, the robotic hand must be energy efficient.

The design team discusses the relative importance of each of these features. They establish weighting factors for each feature, with 10 indicating the most weight and 1 the least weight. Table 5.4 shows the weighting factors and an explanation for each.

The space shuttle can manipulate components of the International Space Station using its robotic arm.

TABLE 5.4
Evaluation matrix for robotic hand actuators

Property	Weighting Factor	Explanation
Soft grip	10	Essential so that space station components are not crushed.
Failure does not damage space station or shuttle	10	Failure of the robotic hand should not lead to other failures, risking lives.
Energy efficient	5	The robotic hand will not be utilized frequently, so energy efficiency is not critical.
Mass	5	If extra mass is needed for functionality, the extra cost is affordable.
Reliability	9	A failure could cause a mission to be "scrubbed," which is very expensive.
Operates over full temperature range	5	The temperature changes frequently, so it can be used when the temperature is acceptable.
Compatible with vacuum	7	Incompatibilities are undesirable, but acceptable if they do not damage space station or shuttle.
Cost	6	Cost overruns are undesirable, but are preferable to developing inferior equipment.
Time schedule	6	Time overruns are undesirable, but are preferable to developing inferior equipment.

Step 4: Search for Solutions. The design team meets in a brainstorming session and discusses the robotic hand. Because it will be used to manipulate large objects, fine dexterity is not required. Team members decide that it can be more like a lobster claw than a human hand. After discussing a number of ideas, they sketch the two concepts in Figure 5.5. Option 1 has an **actuator** that pulls on cables that manipulate two spring-loaded fingers. The finger tips are rubber, thus providing a soft grip. Option 2 has two stationary fingers and one actuated opposable thumb.

FIGURE 5.5
Sketches of two robotic hands.
(Drawings kindly provided by Gerald Vinson.)

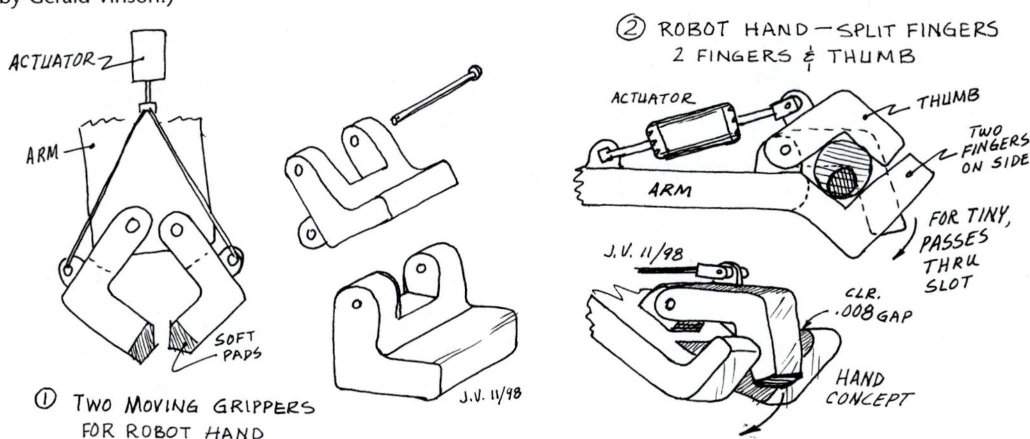

Each of the robotic hands requires an actuator. The design team develops the three options illustrated in Figure 5.6. Notice that numbers are used to indicate component parts, rather than words. This numbering technique is often used in patents so that the drawings do not become cluttered. Also, it allows for completely unambiguous labeling of components. For the same reasons, we use this numbering technique here.

Option A uses pneumatic pressure as the actuator. Low-pressure gas is stored in tank 1. Compressor 2 pressurizes the low-pressure gas so it can flow into high-pressure tank 3. Pressure regulator 4 adjusts the downstream gas pressure; a high pressure gives a firm grip and a low pressure gives a soft grip. Double-acting piston 12 actuates push rod 11. By opening valves 5 and 8 and closing valves 6 and 7, the pressure in chamber 13 is greater than in chamber 14, so push rod 11 extends. Similarly, by closing valves 5 and 8 and opening valves 6 and 7, the pressure in chamber 13 is less than chamber 14, so push rod 11 retracts. Relief valves 9 and 10 protect chambers 13 and 14 in case of overpressure.

FIGURE 5.6
Schematic of actuators. Options: A = pneumatic, B = hydraulic, and C = electric motor.

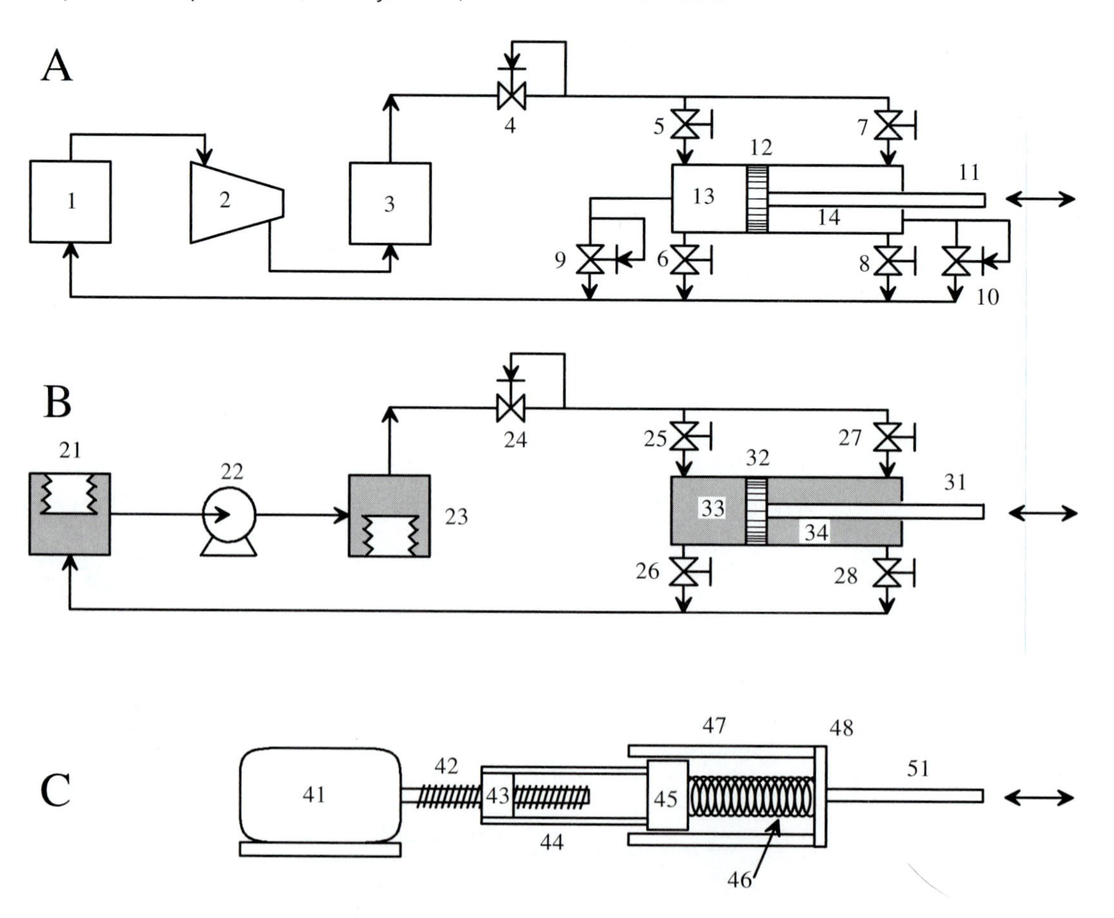

Option B uses hydraulic pressure as the actuator. Low-pressure hydraulic fluid is stored in accumulator 21, a tank with a gas-filled bladder. Pump 22 pressurizes the low-pressure hydraulic fluid so it can flow into high-pressure accumulator 23. Pressure regulator 24 adjusts the downstream hydraulic pressure; a high pressure gives a firm grip and a low pressure gives a soft grip. Double-acting piston 32 actuates push rod 31. By opening valves 25 and 28 and closing valves 26 and 27, the pressure in chamber 33 is made greater than in chamber 34, so push rod 31 extends. Similarly, by closing valves 25 and 28 and opening valves 26 and 27, the pressure in chamber 33 is made less than in chamber 34, so push rod 31 retracts.

Option C uses an electric servomotor 41 as the actuator. In a servomotor, the number of shaft rotations can be controlled. The threaded shaft 42 has a nut 43 that can travel along the length of shaft. Extensions 44 connect to piston 45 located inside of cylinder 47. When nut 43 travels rightward, spring 46 pushes against baseplate 48 to extend push rod 51. Similarly, when nut 43 travels leftward, spring 46 pulls against baseplate 48 to retract push rod 51. When grasping an object, the more spring 46 is compressed, the tighter the grip.

Step 5: (Feasibility Study) Analyze Each Potential Solution. To analyze which of the two robotic hands is better, the engineers list the advantages and disadvantages of each.

	Disadvantages	**Advantages**
Option 1	Difficult to grasp both large- and small-diameter objects.	Spring load on the fingers sets the maximum squeezing pressure. It can be adjusted to prevent damage to space station components.
Option 2	Actuator could squeeze too hard and damage space station components.	Can grasp both large- and small-diameter objects.

To analyze which of the actuators is best, the design team creates an evaluation matrix (Table 5.5). Here are explanations of the scores:

TABLE 5.5
Evaluation matrix for robotic hand actuators

		Option A		Option B		Option C	
Property	**Weighting Factor**	**Score**	**Weighted Score**	**Score**	**Weighted Score**	**Score**	**Weighted Score**
Soft grip	10	10	100	10	100	10	100
Failure does not damage space station or shuttle	10	8	80	1	10	10	100
Energy efficient	5	1	5	5	25	10	50
Mass	5	5	25	1	5	10	50
Reliability	9	3	27	7	63	10	90
Operates over full temperature range	5	8	40	1	5	10	50
Compatible with vacuum	7	4	28	6	42	10	70
Cost	6	10	60	10	60	10	60
Time schedule	6	10	60	10	60	10	60
Total			425		370		630

1 = Poor, 10 = Excellent

- Soft grip. All designs have the soft grip feature.
- Failure does not damage space station or shuttle. If a micrometeor ruptures the pneumatic lines, the robotic hand could flail around because the exhausting gas behaves like a miniature rocket motor. However, the robotic arm is fairly rigid, so movement should be limited. If a micrometeor ruptures the hydraulic lines, hydraulic fluid could be sprayed onto the space station or shuttle and may damage critical components. None of these failures can occur with the electric motor.
- Energy efficient. Because the gas is compressible, large gas volumes must be compressed to achieve a given pressure, so much energy is needed. The pressure regulator in the hydraulic line causes energy losses. No such losses occur with the electric motor.
- Mass. The gas tanks and compressor are fairly heavy. Also, the hydraulic fluid is heavy. The electric motor eliminates these masses.
- Reliability. The pneumatic and hydraulic options are complex because of the many valves and pressure regulators, and therefore violate the KISS principle. In contrast, the electric motor is the simplest design.
- Operates over full temperature range. The gas pressure in the pneumatic system will fluctuate as the temperature changes, making control difficult. The viscosity of the hydraulic fluid will vary greatly with temperature. None of these problems occur with the electric motor.
- Compatible with vacuum. Both the pneumatic and hydraulic systems have a critical seal around the push rod. An imperfect seal means fluid will leak and must be replaced. This problem does not exist with the electric motor.
- Cost. All designs have a comparable cost.
- Time schedule. All designs can meet the design schedule.

Step 6: (Feasibility Study) Choose the Best Solution(s). The design team discusses the advantages and disadvantages of each robotic hand. Because space station components come in many sizes, they feel Option 2 is the better design.

For the actuator, according to the evaluation matrix in Table 5.5, Option C is clearly the best design.

Step 7: (Feasibility Study) Document the Solution(s). The design team prepares a written report documenting the various options and their evaluation.

Step 8: (Feasibility Study) Communicate the Solution(s) to Management. To allow time to review it, the design team gives the written report to management 1 week prior to the design review meeting. At the meeting, the team verbally presents their work and answers questions raised by management. Management agrees with the design team's conclusions and authorizes them to proceed with the preliminary design.

Steps 5 through 8 of the Preliminary Design. The engineering team prepares a preliminary design of robotic hand Option 2 and actuator Option C. They address such issues as identifying potential vendors for the servomotor 41, materials of construction for various components, lubrication for the threaded rod 42 and nut 43, methods for joining the spring 46 to the piston 45 and baseplate 48, methods to control the servomotor 41, sensors to determine the amount of compression in the spring 46, and methods of machining or casting the fingers. When addressing each issue, multiple options are available, so they must choose the best option using an evaluation matrix, or a table listing the advantages and disadvantages of each option. Again, they must document their decisions in a written report and defend their choices to management. Once management approves the preliminary design, the design team can proceed with the final design.

Steps 5 through 8 of the Detailed Design. During the detailed design, team members draw each component, indicating dimensions, tolerances, and materials of construction. They also prepare assembly drawings and instructions. Figure 5.7 shows an example of a drawing of the assembled components. These drawings must be approved by management before construction can begin.

Step 9: Construct the Solution. While the robotic hand is being constructed, the design team answers questions from the artisans (e.g., machinists, electronic technicians). Because the artisans have more hands-on experience than the engineers, they provide many useful suggestions during construction. During this phase, the engineers must be sensitive to problems that may arise and modify the design if necessary.

Step 10: Verify and Evaluate. Once the robotic hand construction is complete, it must be tested to ensure that it meets the design specifications and requirements. If testing shows there are deficiencies, the team must modify the design and the artisans must construct replacement components. Modifications at this stage are extremely expensive because they cause delays; however, if the hardware is deficient, these delays may be unavoidable.

5.4 THIRD DESIGN EXAMPLE: ZERO-EMISSION VEHICLE (ADVANCED TOPIC)

To illustrate how scientific and mathematical knowledge are applied to solve a practical, meaningful problem, we present the design of a zero-emission vehicle. To fully understand this case study, you will need a good command of Newton's laws, energy, and calculus. If you are shaky on these topics, just skim this case study to get a flavor of the role science and mathematics play in the engineering design method.

Step 1: Identify the Need and Define the Problem. For many years, your company has produced conventional automobiles powered by internal combustion engines. Recent California air quality regulations require that your company develop a zero-emission vehicle. Sales of conventional automobiles will be restricted; thus, a need has been created by government mandate. The zero-emission vehicle will be a "commuter car" not intended for long outings with the family. Management assumes the zero-emission vehicle will be an electric automobile.

Step 2: Assemble Design Team. After prolonged discussions, management has decided to commit a team of 10 engineers to the zero-emission vehicle project for 1 year. The team consists of mechanical engineers expert in structural strength and drive trains, aerospace engineers expert in aerodynamic body design, electrical engineers expert in batteries and electric motors, and chemical engineers expert in advanced polymers.

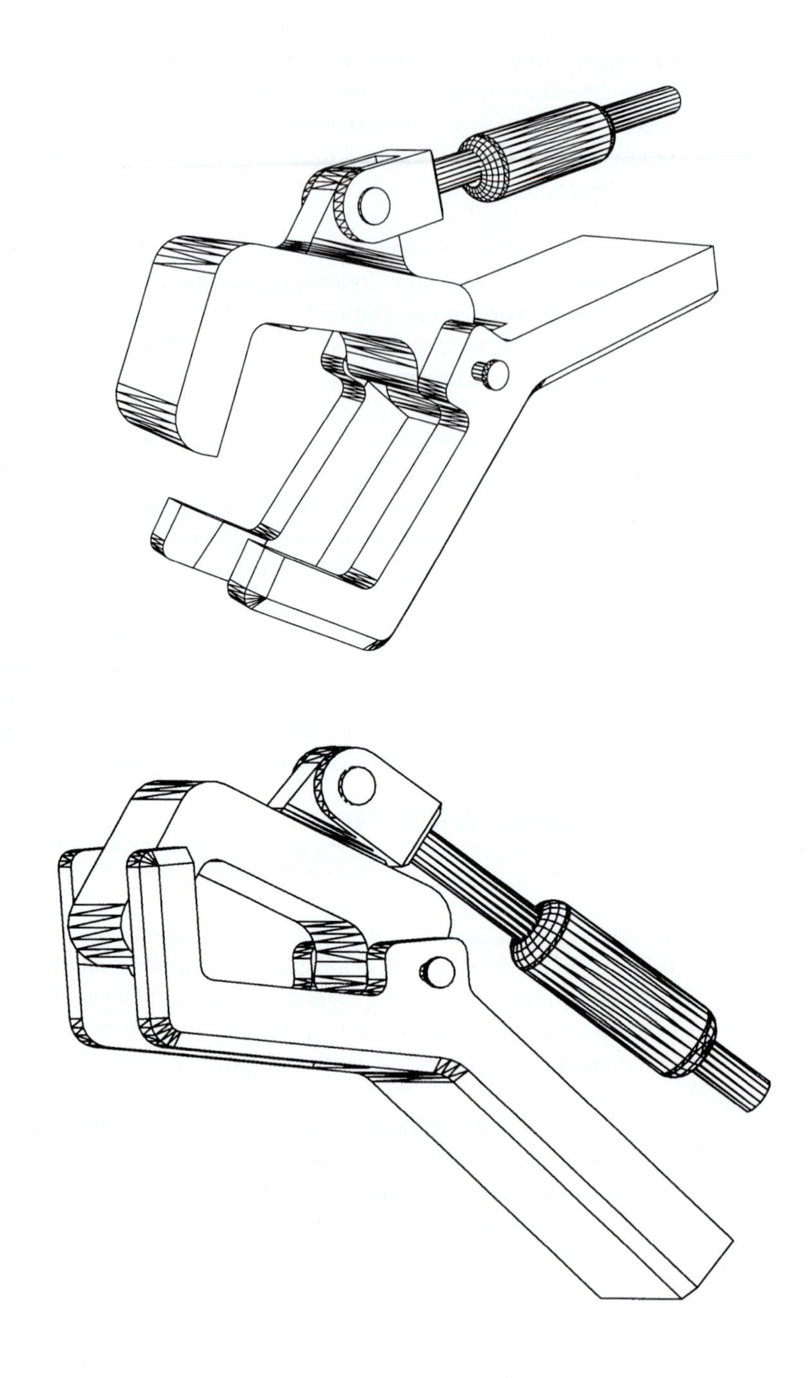

FIGURE 5.7
Example of a detailed drawing. (Drawings kindly provided by Gerald Vinson.)

Step 3: Identify Constraints and Criteria for Success. The design team will have $1 million for the first year. During that time, they are to explore different options and plan a course of action. The plan should recommend not only a vehicle but also a budget, and should specify the engineering support required to construct a prototype in future years.

The engineers are permitted to specify technology that will be commercially available within 5 years, but not laboratory "curiosities." The team is aware that there is intense competition from other automakers both in the United States and abroad. The team will attempt to use off-the-shelf components as much as possible, but they realize that some customized components will probably be necessary. Because the zero-emission vehicle must be mass produced, initially at a rate of 1000 units per year, only technologies that can be mass produced may be considered.

The engineers will keep an open mind, but management assumes the zero-emission vehicle will be electric. Although electric automobiles have a long history—they were produced at the end of the 19th century—they have a reputation for being slow with very limited range. The engineers decide to design an automobile with performance comparable to those powered by internal combustion gasoline engines. They decide that the zero-emission vehicle should have the properties shown in Table 5.6.

TABLE 5.6
Desired properties of the zero-emission vehicle

Property	Weight*
Aesthetics: It should look sleek and sexy. Nonconventional futuristic designs are allowed, provided they are attractive.	8
Performance: The performance should be comparable to a conventional automobile.	
Specification 1: Range, 120 miles @ 60 mph on level ground	4
Specification 2: Acceleration, 0 to 60 mph on flat road in 10 s	6
Specification 3: Climbing ability, maintain 60 mph on a 5% grade	3
Human factors: Although gages should be easy to use and controls readily accessible, the automobile will be designed to be comfortable for customers under 5′ 10″. Taller customers will have to continue using gasoline engines.	2
Cost: The initial purchase cost will be about $25,000. The operating cost should be comparable to conventional gasoline automobiles (about $0.25/mile).	5
Safety: The vehicle will be equipped with air bags and crush zones to absorb impact energy.	6
Operating environment: The automobile must be fully usable at temperatures from −20 to 100°F and altitudes from sea level to 10,000 ft.	3
Interface with other systems: The automobile must be charged from household 20-A, 110-V alternating current.	10
Effect on environment: Spent batteries must be recyclable and must not release dangerous toxins in the event of a crash.	6
Logistics: Replacement parts, especially batteries, must be carried by all dealers in the California area.	5
Reliability: The weakest component should have a 30,000-mile life before failure.	4
Maintainability: Sensors should indicate when service is required.	2
Serviceability: All components should be readily accessible and serviceable using standard tools.	1
Availability: The vehicle should be working 99.5% of the time.	4

* Ten-point scale: 10 = important, 0 = not important

Step 4: Search for Solutions. The engineers spend one month reviewing the literature and thinking of ideas. They then assemble for a brainstorming session. During the session, all agree that electric motors and controllers are well-developed technologies. The main technology limitation is the method for storing the energy. They generate the following ideas for storing the energy:

- Battery
- Compressed air
- Ultracapacitor
- Flywheel
- Spring

Step 5: (Feasibility Study) Analyze Each Potential Solution. Table 5.7 summarizes some properties of the energy storage ideas generated by the engineers. The **gravimetric energy density** (Wh/kg) describes the total ***amount*** of energy that can be stored in a given mass, the **volumetric energy density** (Wh/L) describes the total ***amount*** of energy that

TABLE 5.7
Energy and power densities of energy storage systems

Energy Storage System	Gravimetric Energy Density (Wh/kg)	Volumetric Energy Density (Wh/L)	Gravimetric Power Density (W/kg)	Storage Cost ($/kWh)
Gasoline	2500†	1670†	500‡	1 to 2
Battery				
Nickel–metal-hydride battery[a]	100 (80 to 150)[c] (54 to 80)[a]	152 to 215	200	200
Sodium-sulfur battery[a]	100 (80 to 140)	76 to 120	90 to 130	100+
Nickel-cadmium battery[a]	35 (33 to 70)	60 to 115	100 to 200	300
Lead-acid battery[a]	35 (18 to 42)	50 to 82	67 to 138	70 to 100
Compressed air				
10,000 psig (aluminum core, graphite-wound tank)*	57	84	500‡	200 to 400
10,000 psig (polymer core, graphite-wound tank)*	77	84	500‡	—
Ultracapacitor (goals)[e]	5	10	500	1000
Flywheel[d]	56.5	—	278	—
Spring				
Polyurethane[b]	5.7	—	—	—
Steel	0.020	0.048	—	—

* Safety factor ≈3.5 (walls are about 3.5 times thicker than the minimum requirement). Assume that 80% of tank is discharged when it is refilled. Gravimetric energy density includes average mass of air assuming tank is refilled at 20% discharge.
† Delivered shaft energy per unit of fuel at 20% efficiency.
‡ Includes engine and transmission.

Adapted from: (a) D. L. Illman, *C&E News,* August 1, 1994, p. 8.
(b) R. U. Ayres and R. P. McKenna, *Alternatives to the Internal Combustion Engine* (Baltimore: Johns Hopkins University Press, 1972).
(c) L. O'Connor, *Mechanical Engineering,* July 1993, p. 73.
(d) Oak Ridge National Laboratory, FESS#2.
(e) S. Ashley, *Mechanical Engineering,* February 1995, p. 76.

can be stored in a given volume, and the **gravimetric power density** (W/kg) describes the ***rate*** at which energy is released. The *storage cost* ($/kWh) is the initial capital cost of storing 1 kilowatt-hour of energy.

Step 6: (Feasibility Study) Choose the Best Solution(s). The results of the feasibility study are shown in Table 5.7. The engineers note that even though the nickel–metal-hydride battery is a new laboratory technology, it warrants further investigation because it has a higher gravimetric energy density than most of the other options. (Even so, the engineers notice that this option has about 15 to 30 times less gravimetric energy density than gasoline.) The sodium-sulfur battery also has a high gravimetric energy density, but it must be maintained at about 350°C, which causes many operational problems. The compressed air gravimetric energy density is better than lead-acid and nickel-cadmium batteries.

The engineers rule out flywheels as being potentially unreliable because of their high rotation rate. The ultracapacitors are eliminated because they are not yet available with acceptable performance. Springs are no longer considered because of the low gravimetric energy density.

Step 7: (Feasibility Study) Document the Solution(s). The results of the feasibility study (Table 5.7) are documented in a short report that includes detailed calculations and references from which information was drawn.

Step 8: (Feasibility Study) Communicate the Solution(s) to Management. The engineers make a short 30-minute presentation to management. Management is pleased with the results and authorizes the engineers to continue with the project.

Step 5: (Preliminary Design) Analyze Each Potential Solution. Table 5.6 presented three performance specifications for the zero-emission vehicle. We now examine each of these specifications in some detail. The symbols used are listed again at the end of the chapter.

Specification 1: 120-mile range

Figure 5.8 shows an automobile driving up a hill with angle ϕ. The motor allows a force F_{motive} to be applied to the automobile, allowing it to accelerate, overcome gravity to climb the hill, overcome the rolling resistance of the tires, and overcome air drag.

$$F_{motive} = F_{acceleration} + F_{gravity} + F_{rolling} + F_{air\ drag} \tag{5-1}$$

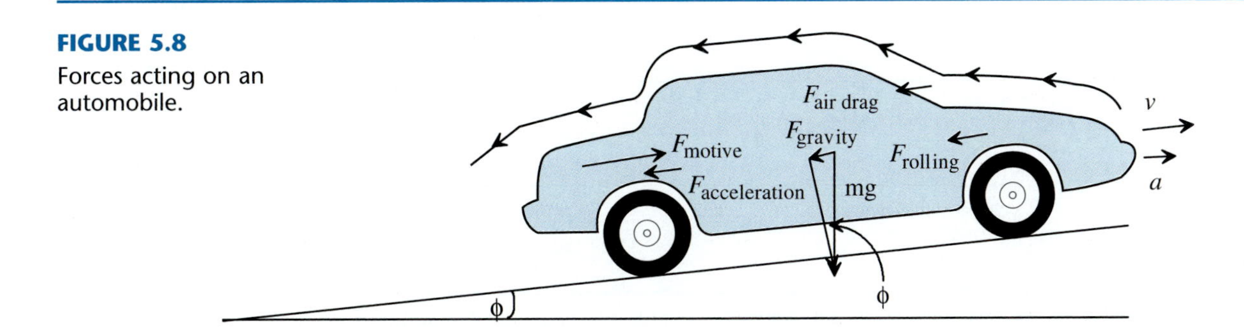

FIGURE 5.8
Forces acting on an automobile.

TABLE 5.8
Rolling resistance k

	Crossply	Radial
Automobiles	0.018	0.013
Truck tires	0.009	0.006
Earth-moving tires	0.014	0.008

Source: G. G. Lucas, *Road Vehicle Performance* (New York: Gordon and Breach, 1986).

We can formulate these forces into the following equation:

$$F_{\text{motive}} = ma + mg \sin \phi + kmg + \tfrac{1}{2}C_d\rho A v^2 \tag{5-2}$$

where m is the total mass including the vehicle and passengers (kg); a is the vehicle acceleration (m/s^2); g is the acceleration resulting from gravity (m/s^2), which also may be viewed as the specific force due to gravity (N/kg); ρ is the air density (1.23 kg/m^3); A is the projected frontal area of the automobile (m^2); v is the automobile velocity (m/s); k describes the rolling resistance of the tires (dimensionless); and C_d is the drag coefficient (dimensionless).

The hill angle ϕ is generally given as "percent grade," that is, rise over run expressed in percent. The hill angle can be calculated from the percent grade G by using

$$\phi = \tan^{-1} \frac{G}{100} \tag{5-3}$$

Rolling resistance results from tire deformation. Truck tires are inflated to higher pressures than automobile tires so they deform less and have a lower rolling resistance. Crossply tires have about 40% more rolling resistance than radial tires. Table 5.8 gives typical values for k. These constants are valid for speeds up to about 60 mph (100 km/h). Above this speed, the rolling resistance increases.

The projected frontal area A can be determined by looking directly at the front of the automobile and measuring the cross-sectional area (Figure 5.9). For most automobiles, the projected frontal area is about 81% of the width times the height. Table 5.9 shows some typical frontal areas for automobiles.

The drag coefficient C_d is determined by the aerodynamic shape of the vehicle. Table 5.10 gives some drag coefficients for simple shapes. Notice that a teardrop shape has an extremely low drag coefficient. Table 5.11 lists some drag coefficients for vehicles. The General Motors EV1 electric car was designed to have an extremely low drag coefficient.

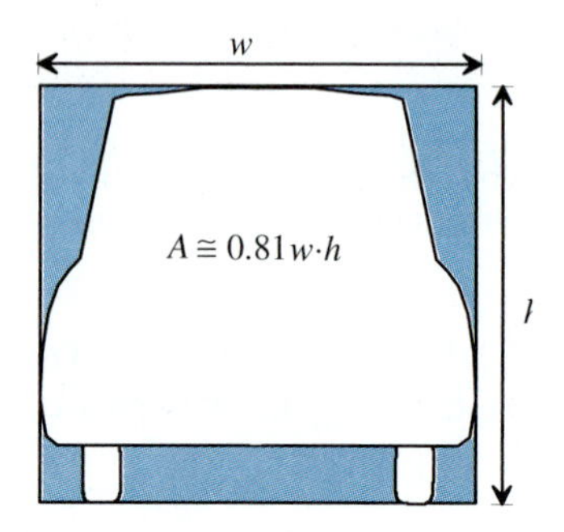

FIGURE 5.9
Projected front area of an automobile.

TABLE 5.9
Projected frontal area A of automobiles

Automobile Class	Projected Frontal Area (m^2)
Mini	1.8
Medium size	1.9
Upper medium size	2.0
Full size	2.1

Source: W. H. Hucho, *Aerodynamics of Road Vehicles* (Boston: Butterworth, 1987).

TABLE 5.10
Drag coefficient of some simple shapes

Body	Flow Situation	C_d (dimensionless)
Circular plate		1.17
Sphere		0.47
Cube		1.05
Streamlined body $L/D = 2.5$		0.04

Source: W. H. Hucho, *Aerodynamics of Road Vehicles* (Boston: Butterworth, 1987).

General Motors designed this automobile body to have a drag coefficient of 0.163. This extremely low drag coefficient was achieved by eliminating rear-view mirrors, locating the engine and its cooling system in the rear, shielding wheels from airflow, and smoothing the underbody.

In part, this was achieved by making the underside smooth, unlike conventional automobiles, which have very rough undersides.

The mass of the automobile is the sum of the body, motor, batteries (or other energy storage system), and passengers

$$m = m_{\text{body}} + m_{\text{motor}} + m_{\text{passengers}} + m_{\text{batteries}} \tag{5-4}$$

The commuter car will carry two passengers at most. Assuming each has a mass of 80 kg (175 lb_m), the total passenger mass is 160 kg (350 lb_m). A modern electric motor with controller has a mass of 73 kg (160 lb_m) and an output of 85 kW (114 hp). A small automobile with a polymer body has a mass of about 545 kg (1200 lb_m).

TABLE 5.11

Drag coefficients C_d for vehicles

Vehicle	C_d (dimensionless)
General Motors EV1	0.19
1982 Audi 100 III	0.30
1982 Mercedes Benz 190 ("Baby Benz")	0.33
1981 Porsche 944	0.35
1974 Volkswagen Scirocco	0.41
Typical passenger car	0.3 to 0.5
Typical van	0.3 to 0.6
Typical truck	0.4 to 1.0

Source: W. H. Hucho, *Aerodynamics of Road Vehicles* (Boston: Butterworth, 1987) (except EV1).

The battery mass depends upon the energy expended during the trip, the gravimetric energy density, and the motor and transmission efficiency, as

$$m_{\text{batteries}} = \frac{E}{E_d \eta_{\text{motor}} \eta_{\text{transmission}}} \tag{5-5}$$

where E is the energy expended on the trip (J), E_d is the gravimetric energy density of the battery (J/kg), η_{motor} is the motor efficiency (dimensionless fraction), and $\eta_{\text{transmission}}$ is the transmission and axle efficiency (dimensionless fraction). The energy expended on the trip is the motive force exerted over the length of the trip L. Because Specification 1 calls for a steady speed on level ground, Equation 5-2 simplifies to

$$\begin{aligned} E &= F_{\text{motive}} L = \left[kmg + \tfrac{1}{2} C_d \rho A v^2\right] L \\ &= \left[k(m_{\text{body}} + m_{\text{motor}} + m_{\text{passengers}} + m_{\text{batteries}})g + \tfrac{1}{2} C_d \rho A v^2\right] L \end{aligned} \tag{5-6}$$

We can substitute this equation into Equation 5-5, which gives

$$m_{\text{batteries}} = \frac{\left[k(m_{\text{body}} + m_{\text{motor}} + m_{\text{passengers}} + m_{\text{batteries}})g + \tfrac{1}{2} C_d \rho A v^2\right] L}{E_d \eta_{\text{motor}}} \tag{5-7}$$

This equation may be solved explicitly for the mass of the batteries,

$$m_{\text{batteries}} = \frac{\left[k(m_{\text{body}} + m_{\text{motor}} + m_{\text{passengers}})g + \tfrac{1}{2} C_d \rho A v^2\right] L}{E_d \eta_{\text{motor}} \eta_{\text{transmission}} - kgL} \tag{5-8}$$

Specification 1 calls for a range of 120 miles (193,000 m) at a speed of 60 mph (26.8 m/s). Modern electric automobile motors are about 90% efficient and the transmission/axle efficiency is about 95%. High-pressure tires (such as those used on trucks) have a rolling resistance of 0.006. Assume the body design has a drag coefficient of 0.19 and a frontal area of 1.7 m^2. The nickel–metal-hydride battery has a gravimetric energy density of about 100 Wh/kg (360,000 J/kg). From this information, we calculate the mass of the batteries as

$$m_{\text{batteries}} = \frac{\left[0.006(545\text{ kg} + 73\text{ kg} + 160\text{ kg})(9.8\,\tfrac{\text{m}}{\text{s}^2}) + \tfrac{1}{2}0.19(1.23\,\tfrac{\text{kg}}{\text{m}^3})(1.7\text{ m}^2)(26.8\,\tfrac{\text{m}}{\text{s}})^2\right] 193{,}000\text{ m}}{(360{,}000\,\tfrac{\text{J}}{\text{kg}})(0.90)(0.95) - 0.006(9.8\,\tfrac{\text{m}}{\text{s}^2})(193{,}000\text{ m})} = 122\text{ kg}$$

The mass of the nickel–metal-hydride batteries is 122 kg—less than the mass of the passengers. Table 5.12 shows the required mass of the other batteries and also compressed air tank with stored air (assuming a 70% air-motor efficiency).

Specification 2: Accelerate from 0 to 60 mph in 10 seconds

We can determine the force required to accelerate the automobile from 0 to 60 mph on level ground from Equation 5-2 as

$$F_{\text{motive}} = ma + kmg + \tfrac{1}{2} C_d \rho A v^2 \tag{5-9}$$

This equation can be solved explicitly for the acceleration. Also, recall that acceleration is the time derivative of velocity:

TABLE 5.12
Required energy storage mass*

Energy Storage	Mass (kg)
Nickel–metal-hydride battery	122
Sodium-sulfur battery	122
Nickel-cadmium battery	377
Lead-acid battery	377
Compressed air (10,000 psig aluminum core, graphite-wound)	294

* Assumptions: 120 miles between recharge, 60 mph, level ground, $A = 1.7\ m^2$, $C_d = 0.19$, $k = 0.006$, $\eta_{transmission} = 0.95$, $\eta = 0.9$ (electric) or 0.70 (compressed air).

$$\frac{dv}{dt} = a = \frac{F_{\text{motive}} - kmg - \frac{1}{2}C_d\rho A v^2}{m} \tag{5-10}$$

This equation can be rearranged, giving

$$\frac{m}{F_{\text{motive}} - kmg - \frac{1}{2}C_d\rho A v^2}\, dv = dt \tag{5-11}$$

Multiplying the numerator and denominator by $\dfrac{1}{\frac{1}{2}C_d\rho A}$ gives

$$\frac{\dfrac{1}{\frac{1}{2}C_d\rho A}}{\dfrac{1}{\frac{1}{2}C_d\rho A}} \frac{m}{(F_{\text{motive}} - kmg - \frac{1}{2}C_d\rho A v^2)}\, dv = \frac{\dfrac{m}{\frac{1}{2}C_d\rho A}}{\dfrac{F_{\text{motive}} - kmg}{\frac{1}{2}C_d\rho A} - v^2}\, dv$$

$$= \left(\frac{m}{\frac{1}{2}C_d\rho A}\right) \frac{1}{\dfrac{F_{\text{motive}} - kmg}{\frac{1}{2}C_d\rho A} - v^2}\, dv = dt \tag{5-12}$$

If we define

$$c = \sqrt{\frac{F_{\text{motive}} - kmg}{\frac{1}{2}C_d\rho A}} \tag{5-13}$$

we can substitute this into Equation 5-12, giving

$$\left(\frac{m}{\frac{1}{2}C_d\rho A}\right) \frac{1}{c^2 - v^2}\, dv = dt \tag{5-14}$$

This expression may now be integrated

$$\left(\frac{m}{\frac{1}{2}C_d\rho A}\right) \int_0^{v_f} \frac{1}{c^2 - v^2}\, dv = \int_0^{t_f} dt \tag{5-15}$$

The solution to the integral is provided by the *CRC Handbook of Chemistry and Physics.*[2]

$$\left(\frac{m}{\frac{1}{2}C_d\rho A}\right)\left[\frac{1}{c}\tanh^{-1}\frac{v}{c}\right]_0^{v_f} = [t]_0^{t_f}$$

$$\left(\frac{m}{\frac{1}{2}C_d\rho A}\right)\left[\frac{1}{c}\left(\tanh^{-1}\frac{v_f}{c} - \tanh^{-1}\frac{0}{c}\right)\right] = [t_f - 0]$$

$$\left(\frac{m}{\frac{1}{2}C_d\rho A}\right)\left[\frac{1}{c}\tanh^{-1}\frac{v_f}{c}\right] = t_f \tag{5-16}$$

Substituting the definition of c (Equation 5-13) gives

$$\left(\frac{m}{\frac{1}{2}C_d\rho A}\right)\sqrt{\frac{\frac{1}{2}C_d\rho A}{F_{\text{motive}} - kmg}}\tanh^{-1}\left(v_f\sqrt{\frac{\frac{1}{2}C_d\rho A}{F_{\text{motive}} - kmg}}\right) = t_f \tag{5-17}$$

$$\frac{m}{\sqrt{\frac{1}{2}C_d\rho A(F_{\text{motive}} - kmg)}}\tanh^{-1}\left(v_f\sqrt{\frac{\frac{1}{2}C_d\rho A}{F_{\text{motive}} - kmg}}\right) = t_f \tag{5-18}$$

This equation cannot be solved explicitly for F_{motive}. Therefore, it is best to independently specify F_{motive} and final velocity v_f and then calculate the required time t_f to achieve the final velocity. By trial and error, we determine that a motive force of 2511 N will accelerate the nickel–metal-hydride-powered automobile to 60 mph in 10 s:

$$\frac{900\text{ kg}}{\sqrt{\frac{1}{2}(0.19)\left(1.23\frac{\text{kg}}{\text{m}^3}\right)(1.7\text{ m}^2)\left(2511\text{ N} - 0.006(900\text{ kg})\left(9.8\frac{\text{m}}{\text{s}^2}\right)\right)}}$$

$$\cdot\tanh^{-1}\left(26.9\frac{\text{m}}{\text{s}}\sqrt{\frac{\frac{1}{2}0.19\left(1.23\frac{\text{kg}}{\text{m}^3}\right)(1.7\text{ m}^2)}{2511\text{ N} - 0.006(900\text{ kg})\left(9.8\frac{\text{m}}{\text{s}^2}\right)}}\right) = 10.0\text{ s}$$

The required motor power to attain 60 mph (26.9 m/s) is

$$P_{\text{motor}} = \frac{F_{\text{motive}}v_f}{\eta_{\text{transmission}}} = \frac{(2511\text{ N})\left(26.9\frac{\text{m}}{\text{s}}\right)}{0.95}$$

2. R. C. Weast (ed.), *CRC Handbook of Chemistry and Physics,* 58th ed. (West Palm Beach, FL: CRC Press, 1997).

$$= 71{,}100 \frac{\text{N} \cdot \text{m}}{\text{s}} = 71{,}100 \text{ W} = 71.1 \text{ kW} \tag{5-19}$$

Table 5.13 lists the required motor power for the other energy storage options.

Specification 3: Maintain 60 mph on a 5% Grade

An automobile that maintains a constant 60 mph (26.8 m/s) on a 5% grade is not accelerating. From Equation 5-2, the required motive force F_{motive} is

$$F_{\text{motive}} = mg \sin \phi + kmg + \tfrac{1}{2} C_d \rho A v^2 \tag{5-20}$$

For a 5% grade, we can determine the angle from Equation 5-3,

$$\phi = \tan^{-1} \frac{G}{100} = \tan^{-1} \frac{5}{100} = 2.862°$$

The motive force required to move the nickel–metal-hydride-powered automobile is

$$F_{\text{motive}} = (900 \text{ kg})\left(9.8 \frac{\text{m}}{\text{s}^2}\right) \sin 2.862° + 0.006(900 \text{ kg})\left(9.8 \frac{\text{m}}{\text{s}^2}\right)$$

$$+ \tfrac{1}{2}(0.19)\left(1.23 \frac{\text{kg}}{\text{m}^3}\right)(1.7 \text{ m}^2)\left(26.8 \frac{\text{m}}{\text{s}}\right)^2$$

$$= 636 \frac{\text{kg} \cdot \text{m}}{\text{s}^2} \times \frac{1 \text{ N}}{\frac{\text{kg} \cdot \text{m}}{\text{s}^2}} = 636 \text{ N}$$

This motive force is much less than that required to accelerate the automobile from 0 to 60 mph in 10 seconds, so there is more than enough power to meet this requirement.

Specifications Summary

Table 5.14 summarizes the properties of each option.

TABLE 5.13

Required motor power

Energy Storage Option	Total Mass m (kg)	Motive Force F_{motive} (N)	Motor Power (kW)	Motor Power (hp)
Nickel–metal-hydride battery	900	2511	71.1	95.3
Sodium-sulfur battery	900	2511	71.1	95.3
Nickel-cadmium battery	1155	3200	90.6	121
Lead-acid battery	1155	3200	90.6	121
Compressed air	1072	2993	84.7	114

Assumptions: 120 miles between recharge, 0 to 60 mph in 10 seconds, level surface, drag coefficient = 0.19, rolling resistance = 0.006, frontal area = 1.7 m^2, transmission efficiency = 95%, electric-motor efficiency = 90%, air-motor efficiency = 70%.

TABLE 5.14
Summary of each zero-emission vehicle option

Option	Energy Storage Mass (kg)	Total Mass* (kg)	Motor Power (kW)	Recharge Distance (miles)	Acceleration From 0 to 60 mph (s)	Climb 5% Grade
Nickel–metal-hydride battery	122	900	71.1	120	10	Yes
Sodium-sulfur battery	122	900	71.1	120	10	Yes
Nickel-cadmium battery	377	1155	90.6	120	10	Yes
Lead-acid battery	377	1155	90.6	120	10	Yes
Compressed air	335	1072	84.7	120	10	Yes

* Includes two passengers.

Assumptions: Drag coefficient = 0.19, rolling resistance = 0.006, frontal area = 1.7 m^2, electric-motor efficiency = 90%, air-motor efficiency = 70%, transmission efficiency = 95%.

Step 6: (Preliminary Design) Choose the Best Solution(s). Table 5.15 lists the advantages and disadvantages of each automotive energy storage option. Table 5.16 shows an evaluation matrix in which the indicated weighting factors and scores reveal compressed air energy storage to be superior. Thus, the initial assumption that the zero-emission vehicle would be a battery-powered electric vehicle was ill founded. Through creativity and analysis, an apparently superior option has been identified. This unexpected result emphasizes the importance of keeping an open mind, the hallmark of a good design engineer.

Steps 7 and 8: (Preliminary Design) Document the Solution(s) and Communicate with Management. The design team documents the results of their analysis in a written report and presents their results to management in an oral presentation.

TABLE 5.15
Advantages and disadvantages of automotive energy storage options

Option	Advantage	Disadvantage
Nickel–metal-hydride battery	Very high gravimetric energy density Rapid recharge (about 1 h)	Not a commercial product Expensive
Sodium-sulfur battery	Very high gravimetric energy density	High temperature (350°C) Potential recycling problems Low lifespan in use
Nickel-cadmium battery	Commercially available High lifespan in use Maintains charge on storage	Low gravimetric energy density Cadmium is toxic
Lead-acid battery	Commercially available Relatively low cost Operates in wide temperature range	Low gravimetric energy density Medium lifespan in use Poor charge retention
Compressed air	Infinitely rechargeable Rapidly recharged Air conditioning provided by chilled exhaust air	Moderate gravimetric energy density Possible explosion in an accident*

* Explosion hazard is greatly reduced by designing tanks with ~3.5 safety factor.

TABLE 5.16
Evaluation matrix for the zero-emission vehicle

Property	Weighting Factor*	Nickel–Metal-Hydride S†	Nickel–Metal-Hydride S × W	Sodium Sulfur S	Sodium Sulfur S × W	Nickel-Cadmium S	Nickel-Cadmium S × W	Lead-Acid S	Lead-Acid S × W	Compressed Air S	Compressed Air S × W
Aesthetics	8	10	80	10	80	8	64	8	64	8	64
Performance											
Spec 1: Range	4	10	40	10	40	10	40	10	40	10	40
Spec 2: Acceleration	6	10	60	10	60	10	60	10	60	10	60
Spec 3: Climbing	3	10	30	10	30	10	30	10	30	10	30
Human factors	2	10	20	10	20	8	16	8	16	8	16
Cost	5	5	25	5	25	5	25	7	35	10‡	50
Safety	6	7	42	5	30	8	48	8	48	5	30
Operating environment	3	10	30	10	30	10	30	10	30	10	30
Interface with other systems	10	10	100	10	100	10	100	10	100	10	100
Effect on environment	6	6	36	6	36	3	18	8	48	10	60
Logistics	5	2	10	6	30	6	30	10	50	10	50
Reliability	4	10	40	3	12	10	40	10	40	10	40
Maintainability	2	10	20	10	20	10	20	10	20	10	20
Serviceability	1	10	10	8	8	10	10	10	10	10	10
Availability	4	10	40	3	12	10	40	10	40	10	40
Total			583		533		571		631		640

* Ten-point scale: 10 = important, 0 = not important.

† Score: 10 = excellent, 0 = poor.

‡ Excellent cost because air tanks last the life of the car, unlike batteries.

Steps 5 through 8 of the Detailed Design. Management approves the report and authorizes expenses for the detailed design. At this point, the team is greatly expanded, because each and every component must be specified in detail. The drawings that document the design must specify materials, dimensions, tolerances, and processing steps. A prototype zero-emission vehicle is constructed and tested. Engineers document deficiencies so they can make revisions in the final design.

Step 9: Construct Solution. Full-scale production of the zero-emission vehicle requires an enormous capital investment. In addition, many people must be hired, including production engineers to supervise the manufacturing process, laborers to construct the vehicles, lawyers to write contracts, advertisers to sell the product, and accountants to keep track of finances. Contracts are written with primary material suppliers. Secondary suppliers must also be identified in case primary supplies are disrupted by strikes or natural disasters. Spare parts must be manufactured and stocked to provide future customer support.

Step 10: Verify and Evaluate. For quality control purposes, the test engineer must periodically test vehicles from the production line. As problems are identified in the production line, improvements can be made. Information gained during production can be used to design future generations of zero-emission vehicles.

5.5 SUMMARY

An engineer is a person who applies science, mathematics, and economics to meet the needs of humankind; they meet these needs by implementing the engineering design method to design new products, services, or processes.

The engineering design method involves synthesizing new technologies, analyzing their performance, communicating within the engineering team and to management, and implementing the new technology. This design method consists of 10 steps, many of which are iterative.

Design involves conceptualizing new technologies to meet human needs and bringing the product to market. In many ways, design is the heart of engineering. To be successful, design engineers must have excellent creative, technical, organizational, and people skills. As a design engineer, you will work with engineers from many disciplines as well as many other professions (lawyers, marketers, advertisers, financiers, distributors, politicians, etc.). It is unlikely that your formal schooling will teach you everything you need to know, so you must continue to learn after graduation.

Nomenclature

A projected frontal area (m^2)
a acceleration (m/s^2)
C_d drag coefficient (dimensionless)
E energy (J)
E_d energy density (J/kg)
F force (N)
G grade (percent)
g acceleration due to gravity (m/s) or specific force due to gravity (N/kg)
k rolling resistance (dimensionless)
L length of trip (m)
m mass (kg)
P power (W)
t time (s)
v velocity (m/s)
η efficiency (dimensionless)
ρ air density (kg/m^3)
ϕ hill angle (rad)

Further Readings

Beakley, G. C., and H. W. Leach. *Engineering: An Introduction to a Creative Profession.* 4th ed. New York: Macmillan, 1983.

Norman, D. A. *The Design of Everyday Things.* New York: Doubleday, 1988.

Sandler, B. Z. *Creative Machine Design.* New York: Solomon Press, 1985.

Schwarz, O. B., and P. Grafstein. *Pictorial Handbook of Technical Devices.* New York: Chemical Publishing, 1971.

Glossary

actuator A mechanism that uses pneumatic, hydraulic, or electrical signals to activate equipment.

concurrent engineering A design approach in which specialists work together right at the beginning of a project.

Delphi technique A technique in which the members of a group are intentionally separated and given the same problem. They return their solutions to the leader, the solutions are pooled, and the group members rate the ideas.

detailed design The step where highly detailed drawings and specifications are prepared for the best design option.

evaluation matrix A mathematical evaluation that identifies the best solution by weighting desired properties on the basis of their relative importance.

feasibility study The step of the engineering design method where ideas are roughed out.

global optimum The best condition found without constraints.

gravimetric energy density The total amount of energy that can be stored in a given mass.

gravimetric power density The rate at which energy is released from a given mass.

initial capital cost The purchase price of a product.

life-cycle cost The purchase price and additional costs such as labor, operation, insurance, and maintenance.

local optimum The best condition found with constraints.

nomenclature A system of names used in arts and science.

nominal group technique A technique in which a leader poses a problem to a group and each group member then works on the problem and comes up with ideas. The ideas are discussed and then ranked.

preliminary design The step of the engineering design method where some of the more promising ideas are explored in more detail.

redundancy The use of multiple components in which each has the same capabilities.

sequential engineering An approach in which specialists work in a compartmentalized manner.

volumetric energy density The total amount of energy that can be stored in a given volume.

CHAPTER 6

Engineering Communications

In the curriculum for your chosen engineering discipline, you will find many "hard" courses (mathematics, science, and engineering) with just a smattering of "soft" courses (English, history, and other humanities). The hard courses emphasize computations, whereas the soft courses emphasize communication, primarily written. Given that the engineering curriculum overwhelmingly emphasizes hard courses, a student might logically conclude that communications are not important in engineering.

Nothing could be farther from the truth. The emphasis on hard courses merely reflects the necessary trade-offs that faculty must make when designing a curriculum to fit within severe time constraints. In fact, writing and oral communications are an integral part of a practicing engineer's job; some engineers report that they spend 80% of their time in these activities. Likely, these soft skills will affect the promotions of an engineer more than the hard skills, particularly if the ultimate objective is to become a manager.

A recent survey of corporations posed the following question: "What skills are lacking in recent engineering graduates?" The number one response was that engineers lack communication skills. Because corporations are collections of individuals working toward a common goal, good communication skills are essential and highly prized.

Engineers communicate both orally and in writing. Regardless of which method, engineers use both words and graphics to present their ideas. Until now, most of your education has focused on communication via words. As a budding engineer, you also must learn to communicate graphically. Many engineering ideas are simply too complex to be described by words and can be communicated only through drawings. (Undoubtedly, you have heard the expression, One picture is worth a thousand words.) Engineering graphics is a huge topic well beyond the scope of this text. There are many fine engineering graphics texts available; we trust that you will have the opportunity to study one.

In school, your goal should be to develop a set of skills that will enable you to become a successful practicing engineer. Honing your communication skills is essential to reaching that goal. Take every essay, report, and oral presentation seriously. This chapter will help you get started.

6.1 PREPARATION

Whether writing or giving an oral presentation, you must prepare by using the following three steps: topic selection, research, and organization.

6.1.1 Topic Selection

Your topic may be given, or you may be able to select it. If you are selecting your own topic, you may wish to choose something you are already familiar with or perhaps something you wish to know more about.

6.1.2 Research

There are many sources for obtaining information, as described below:

- ***Technical journals*** are generally devoted to a single topic (e.g., heat transfer). Authors submit their papers to journal editors who then have the papers reviewed by experts in the field. This peer review process can take a year or longer, so the results reported in technical journals are often a few years old; however, they tend to be high quality because of the review process. Technical journals are the primary means by which new information is introduced into the engineering community.
- *Books* are written by authors who are familiar with a field and wish to describe it in a consistent, coherent manner. The primary source of their information is knowledge that was first reported in technical journals, so the information in books tends to be even older than that in technical journals. The great advantage of a book is that the information is in a single source rather than in multiple articles spread over many years in a plethora of journals.
- ***Conference proceedings*** are a collection of papers written by authors who speak at a meeting devoted to a particular topic. Sometimes the proceedings are made available at the meeting, so the information can be extremely recent—literally, data taken a few days or weeks prior to the meeting. However, in this case, the information has not been peer reviewed, so some of the information may be of lesser quality. To overcome this problem, some conferences peer review the proceedings, but this takes time and delays publication of the information.
- *Encyclopedia articles* are very short descriptions of a particular topic. They are peer reviewed, so the information is of high quality. Like books, the information in an encyclopedia is at least a few years old.
- *Government reports* are collections of research data taken by government-sponsored researchers. The reports are required by the funding agency and are maintained at the agency. In some cases, the reports are copied onto microfilm by the National Technical Information Service, so they are more widely available. The final reports are written immediately after each project is completed, so the information is very recent; however, it usually is not peer reviewed. Generally, if the information is important to a wide audience, it will be translated into a journal article where it will be subjected to peer review.
- *Patents* describe technology that is novel, useful, and nonobvious. To be valid, a patent must fully disclose the technology so that a person "skilled in the art" can translate the patent into a working device or process. A patent protects the intellectual property of the inventor for a fixed time period, typically 20 years from the time the patent was filed.

- *Popular press articles* appear in widely circulated magazines and newspapers. Usually, they are written by people with a journalism degree who have little technical background. Further, the information often must be printed quickly to meet a publication deadline, so the information may not be scrutinized. As a consequence, popular press articles dealing with technical subjects often report erroneous information. However, such articles may be useful for getting to the human side of a technical issue and may indicate how the lay public or politicians feel about controversial technical topics, such as nuclear energy, landfills, or dams. Also, they may indicate how a technical issue affects certain people, groups, or institutions.
- *Course notes* can be a good source of information; however, the information is not peer reviewed.
- *Internet sites* can be established by anyone with a computer and enough money to pay monthly connection fees. The Internet is an extremely democratic means for disseminating information. Because it lacks the scrutiny of peer review, however, erroneous information or extreme viewpoints can easily make their way into the information marketplace. Further, information on the Internet is extremely volatile and is available only as long as the computer and its connection are maintained.

Finding information is a technical skill; in fact, libraries have experts trained to find information in their vast archives. When doing your research, freely consult these experts. But be aware that they are not experts in your field, so there are benefits to becoming more self-sufficient. The following resources will help you find information:

- *Abstracts* are brief, one-paragraph descriptions of the contents in a journal or popular press article. The abstracts are accessible through keywords or author names. There are abstracts for chemistry, biology, physics, engineering, and popular press articles (*Reader's Guide to Periodical Literature*). Abstracts are now accessible through computer searches using either CD-ROMs or the Internet.
- *Citations* or *references* are listed at the end of a publication, detailing where the information was obtained. Suppose you are a civil engineer interested in improving asphalt roads and you find a particularly good paper on asphalt chemistry by Charles Glover published in 1995. Glover's citations are an excellent way to connect to other related literature *before* 1995.
- *Citations indices* list authors and publications that have cited them. For example, knowing that Charles Glover wrote an excellent asphalt article in 1995, you can look up his name in the citations index. If another author cited his work in 1998, it is likely that this author also is writing about asphalt chemistry, so you may be interested in obtaining her paper as well. Using the citations index, it is possible to find related articles *after* 1995.
- *Library catalogs* list the holdings of the library, often by subject and author, so this is a rapid way to find relevant literature in your library.

6.1.3 Organization

When organizing your presentation or writing, the most important rule to follow is

Know your audience

Again, imagine you are a civil engineer who seeks to improve our highways. If you were invited to an elementary school to speak to eight-year-olds about your work, you would certainly give a different talk than if you were invited to speak at a technical symposium on our highway infrastructure.

Once you know your audience, the next step is to determine the most important points you wish to make. Whether writing or speaking, it is usually impossible to communicate everything you know about a subject because of space or time constraints. Instead, you must carefully choose the key points and determine the most logical sequence for the ideas to flow together smoothly. An outline is very helpful for achieving this goal.

When structuring your writing or speech, you may want to employ the following strategies:

- A **chronological strategy** gives a historical account of the topic. Again, imagining you are a civil engineer, you could present a history of roads by sequentially describing simple dirt paths, gravel roads, cobblestone streets, asphalt roads, and concrete multilane highways.
- A **spatial strategy** describes the component parts of an object. In the case of a road, you could describe its various features (gravel substructure, asphalt surface, drainage system).
- A **debate strategy** would describe the pros and cons of a particular approach. For example, you could describe the advantages and disadvantages of asphalt and concrete roads, with the goal of choosing which is best for a given situation.
- A **general-to-specific strategy** presents general information first, and then gives increasingly detailed information and specific examples. For example, as a civil engineer describing methods for connecting one road with another, you could first describe general considerations (e.g., number of lanes, vehicle speed, amount of traffic) and then describe specific types of connections (four-way stop signs, traffic circles, stoplights, highway cloverleafs, etc.). In some cases, a *specific-to-general* strategy is a more effective way to communicate.
- A **problem-to-solution strategy** is very effective for communicating with engineers because they are trained problem solvers. For example, a road for which you are responsible may have too many potholes. In a presentation to your boss, you could first describe this problem and then offer a variety of solutions (use a high-grade asphalt, deepen the roadbed, improve the drainage system, ban heavy trucks).
- A **motivational strategy** is often employed by sales engineers. For example, imagine you sell a high-quality asphalt that will improve road life. Your presentation could have the following components:

  1. You could get your client's *attention* by showing a picture of a pothole swallowing an automobile.
  2. You could create a *need* for your product by showing how much money your client spends fixing potholes.
  3. You could *satisfy* their need by showing how your product reduces the formation of potholes.
  4. You could help them *visualize* a better, pothole-free world.
  5. You could get them to *act* by signing a purchase contract for your asphalt product.

No matter which strategy you employ, you should have an introduction, body, and conclusion.

6.2 ORAL PRESENTATIONS

When you become an engineer, you will give oral presentations in a variety of situations—you might need to make proposals to prospective clients, explain why your company should be allowed to build a new facility in a community, explain the results of a recent analysis to your boss, or present research results at a conference. Mastering oral presentations will increase your chances of being promoted to high-visibility positions within your company.

6.2.1 Introduction

During the introduction of your oral presentation, your goal is to win your audience over. If you cannot win them over within the first few minutes, you never will. Jokes are a classic way to win the audience; if you are skilled at telling jokes, use them. However, if you are unskilled, win your audience in other ways; there is no quicker way to lose them than by telling a joke badly. Instead, you may win your audience by using anecdotes, particularly if they are personal.

During the introduction, you must connect the audience with your world. They may not have thought about your topic before, so you must grab their attention. Find an aspect of your topic that everyone can relate to. For example, everyone can relate to potholes, so this is a good way to introduce the topic of high-quality asphalt.

Commonly, the first slide of an oral presentation is the title. There is nothing wrong with this approach; however, it may not help win your audience. There is little you can do with a title but read it to the audience, which insults their intelligence. Further, many engineering presentations have titles with technical phrases that are unintelligible to most members of the audience. A more effective opening is to find an aspect of your topic that everyone can relate to so you win your audience immediately. Then, present them with enough information that they can understand every word in the title slide. Using this approach, the title slide may appear a few minutes into the presentation. This may seem awkward—but how many television shows start with the title? It is much more common to start with an opening skit that grabs the audience's attention and dissuades them from changing channels. After the skit, they show the title. This approach is very successful in television, and it can work effectively for you, too.

Often, a speaker includes an outline of the talk immediately following the title slide. Technically, this is not wrong; however, again it does little to win your audience. It is difficult to make an outline interesting, particularly if it contains technical words that few audience members understand. Instead, use "chapter" designators, which we describe next.

6.2.2 Body

The body is the heart of the presentation where you will spend the majority of your time. About 80% of the presentation is the body, with about 15% devoted to the introduction and about 5% devoted to the conclusion.

In the body, use "chapter" designators to let the audience know when you have changed topics. Suppose you were dividing your talk on road construction into the following topics: surveying, grading, roadbed preparation, and surfacing. Rather than listing these topics at the beginning of the presentation, it is more effective to have a chapter designator with the single word "Surveying" in large letters. This lets the audience know that the next few slides relate to this topic. Then, have a chapter designator with the word "Grading" in large letters to let the audience know that you have switched topics. In this way, you can step through the various topics in your presentation. An alternative approach is to indicate all the topics in a list, but highlight the particular topic being considered during the particular chapter; thus, the same list appears multiple times throughout your presentation, but each topic in the list is highlighted only once. If you organize your presentation according to a spatial strategy, a very effective chapter designator is a graphic image of the object being described. Each chapter in the presentation would begin with a portion of the graphic image highlighted. For example, you could show a cutaway view of a road indicating the soil, gravel bed, and asphalt surface. You could begin each chapter by highlighting the particular feature of the road you will discuss next.

6.2.3 Conclusion

In the conclusion, you wrap up your presentation and summarize your key points. When preparing your conclusion, think hard about what you want to be the take-home message; that is, if the audience remembers only one or two things from your presentation, what do you want those to be?

6.2.4 Visual Aids

Table 6.1 shows the results of a study in which participants were given a presentation and then later tested to determine their retention of information. This study clearly shows that combining visual aids with oral comments is the best approach.

When preparing your visual aids, employ the KISS (Keep It Simple, Stupid) principle. Each visual aid must communicate your idea rapidly; otherwise, the audience will spend too much time interpreting your visual aid and not listening to your oral comments. Obviously, you would like to make the visual aid as interesting as possible to maintain their attention. Sometimes you can accomplish this using exaggeration. Also, as much as possible, use graphic images rather than words. As the speaker, you can easily create slides containing only words; however, the audience must read the words and form a mental image all while listening to your oral comments. Most likely you will lose them.

Among the many types of visual aids are the following:

- **Word charts** convey information using short phrases, or even single words. Use bullets when the order is of no particular importance; use numbered lists when the order is

TABLE 6.1
Message retention

Presentation	Testing 3 Hours Later (%)	Testing 3 Days Later (%)
Oral only	70	10
Visual only	72	20
Oral and visual	85	65

Source: Cassagrande, D. O. and R. D. Cassagrande, *Oral Communication in Technical Professions and Businesses* (Belmont, CA: Wadsworth Publishing, 1986).

TABLE 6.2
Example of a word chart

Types of Heat Transfer
• Conduction • Radiation • Convection —Natural convection —Forced convection

TABLE 6.3
Example of a table

Typical Thermal Conductivity

Material	Thermal Conductivity (W/(m·K))
Liquid	0.1–0.6
Nonmetallic solid	0.04–1.2
Metallic solid	15–420

important. Table 6.2 shows an example of a word chart with bullets. Note that there are no complete sentences; the key is to use few words so the audience can read the chart in just a few seconds and then turn their attention to your oral comments.

- *Tables* present numerical data. Table 6.3 is an example of a table; note that both a title and units are included. Because it is difficult to visualize information in tabular form, it is better to convert the data into a chart or graph, if possible.
- *Charts* and *graphs* are images that rapidly convey numerical information. Figure 6.1 shows an example of a bar chart, a pie chart, and a graph. Note that each has a title, and units are indicated for the reported quantities.
- *Photographs* are very effective ways to communicate still images, whereas *videos* or *movies* are needed to convey moving images. All of these can add real force to a presentation.
- *Schematics* (Figu re 6.2) are line drawings of objects or processes, and *illustrations* (Figure 6.3) are graphic images used to explain concepts.
- *Maps* (Figure 6.4) rapidly communicate geographic information.
- *Physical objects* can be very effective in presentations. People are interested in examining physical objects (perhaps because it brings back childhood memories of show-and-tell) and it can make the abstract become concrete. However, passing the objects can be very disruptive, so this aid works best with smaller audiences, or if done during the question-and-answer period.

The visual media you select can make a big impact on your oral presentation. Here are some options:

- *Transparencies* have the advantage of being rapidly and inexpensively reproduced by a copier or computer printer. Also, the information is available in a random-access manner, making it possible to select easily any desired transparency at any point in your presentation. Generally, color transparencies are not as vibrant as slides.
- *Slides* have stunning colors and resolution, so they are the visual medium of choice. However, they require a longer lead-time to prepare, and the information is available only in a sequential-access manner, making it difficult to select any desired slide at any point in your presentation.

FIGURE 6.1

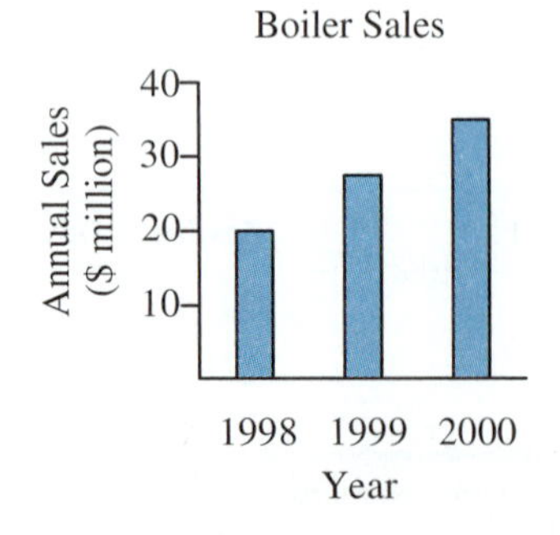

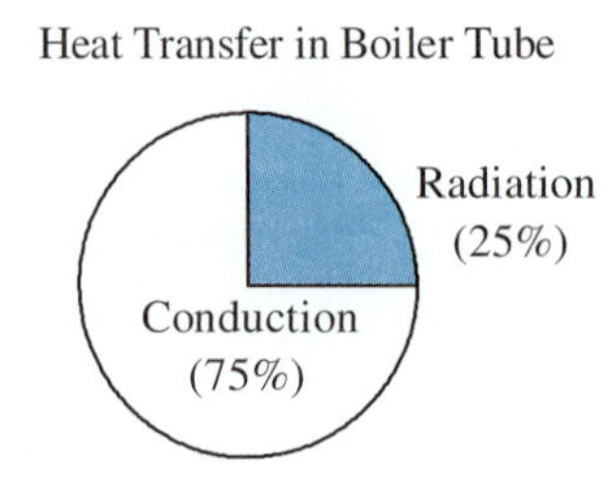

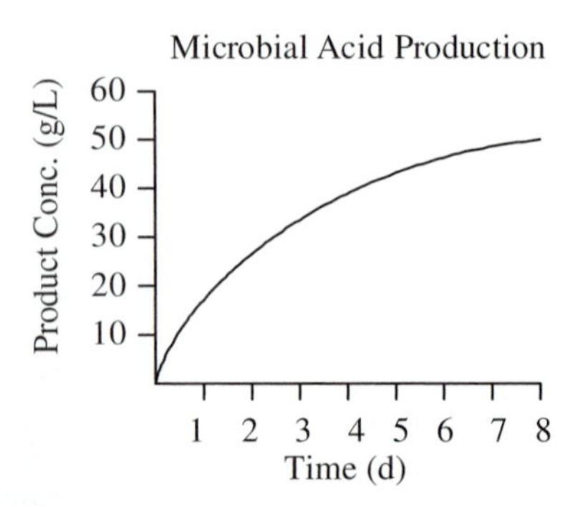

FIGURE 6.2
Example of a schematic.

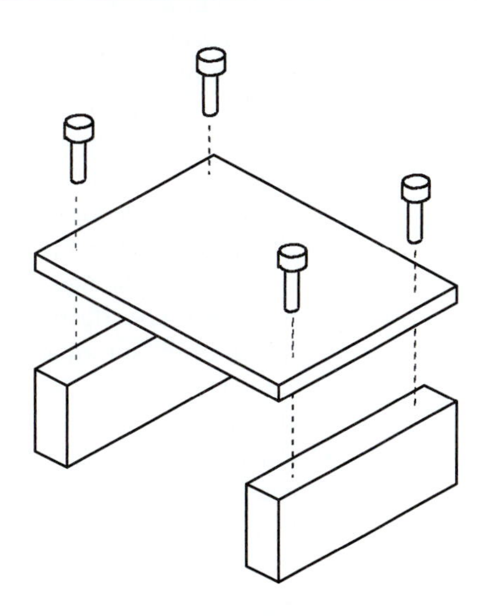

FIGURE 6.3
Example of an illustration showing how acid rain is produced from sulfur in coal.

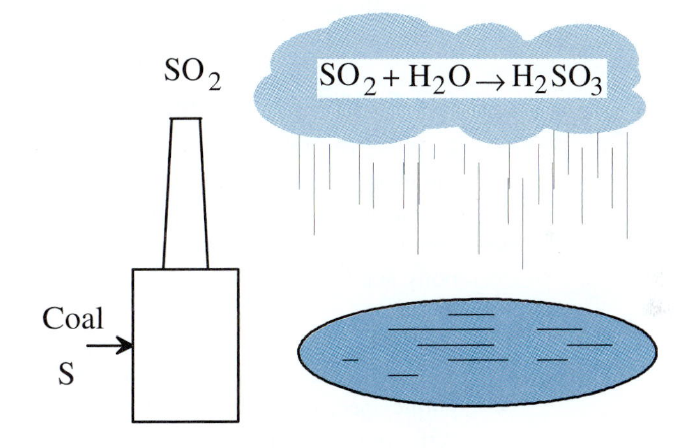

FIGURE 6.4
Map showing locations of manufacturing facilities.

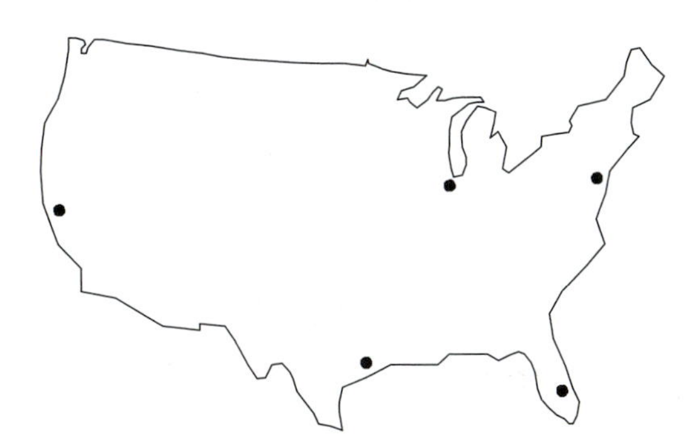

- *Computer projections*, using programs such as PowerPoint, are very flexible; you can change a presentation minutes before giving the talk. Also, you can embed eye-catching visual effects, sound effects, and even moving images in the presentation. However, the resolution of the projected images is not as high as slides. Also, given the complexity of computer technology, unexpected glitches can happen in the middle of your presentations, which can detract from your message. Often, people who give presentations using computer projections also have back-up slides in case there is a computer glitch.
- *Blackboards*, even though they are old-fashioned, are effective for interacting spontaneously with the audience. However, because of the time it takes to draw on a blackboard, it is not effective for rapidly conveying information. Also, there is no permanent record of the information placed on the blackboard. And, because your back is to the audience when you write, using a blackboard to excess can impede interaction with the audience.
- *Butcher paper* is large sheets of paper that are typically mounted on an easel. Butcher paper allows for spontaneous interaction with the audience, and has the added advantage of maintaining a permanent record. Pre-prepared presentations on butcher paper can allow for rapid information transfer. Butcher paper is a favorite visual medium in brainstorming sessions.
- *Handouts* of the presentation are often used with a small, important audience. For example, if making a sales presentation, provide copies of your presentation to the clients so they can pay attention to you rather than getting distracted by taking copious notes.

6.2.5 Speech Anxiety

In 1974, *The Bruskin Report*[1] revealed that some adults are more fearful of public speaking than financial problems, loneliness, and even death. Why would people rather die than give a speech? Most likely, the reason is that at least once in our lives, each of us has been embarrassed in front of a group. The experience may have been so humiliating that the body does not want to repeat it. Uncontrollable physiological responses to speech anxiety include sweating, shakiness, stomach distress, and an increased heart and breathing rate.

How should you respond to speech anxiety? Well, you could surrender yourself to it and become a basket case every time you must speak publicly, or you could refuse to give public speeches. Neither of these is a viable option for an aspiring engineer such as yourself; your job will require you to give oral presentations. You might try to ignore the physiological symptoms, but sometimes they are simply too powerful to ignore. Instead, you should learn to harness the energy that comes from speech anxiety and direct it into your presentation.

There are a number of tricks you can use to overcome speech anxiety:

- The most powerful trick is to be well-prepared; those who are ill-prepared have good reason to be nervous.
- The first few minutes of a presentation are the most important. Recall that this is the critical time when you are trying to win the audience. It is also the time you will be most nervous, as the transition from being an anonymous audience member to becoming the

1. "What are Americans Afraid of?" *The Bruskin Report* (New Brunswick, N.J.: R. H. Bruskin Associates, July 1973), p. 1.

center of attention puts a strain on your body. The best way to survive this initial period is to memorize the first few sentences so you can deliver them in "auto-pilot" mode. Also, take a deep breath before you utter your first words to calm your nerves.

- If you were to meet each audience member individually, you could become friends, so think of them collectively as your friends. You can reinforce this notion by picking out a few friendly faces in various parts of the room and speaking to them. Don't look at the people who are obviously bored or disinterested; they will suck the energy out of you.
- Allow yourself to make mistakes. We all trip over words or drop things. Most people in the audience will filter out your mistakes and not even notice them; however, they will notice if you respond to your mistake by becoming agitated or confused.
- The physiological manifestations of speech anxiety are based upon the "fight-or-flight" response. When your body is in a stressful situation, it is poised to do battle or to flee. Adrenaline flushes through the body, putting the nerves on edge and tightening the muscles. You can counteract these effects with endorphins that are released with intense physical exercise, such as running. By exercising one or two hours before your presentation, you will find yourself to be less nervous, which will help your presentation go well. After having many positive public speaking experiences, you will find that you get less speech anxiety each time, and the need for endorphins will eventually subside.

6.2.6 Style

According to the *Mehrabian Study*,[2] only 7% of what you communicate is verbal. The remainder is nonverbal communication, such as body language (55%) and voice strain (38%). Thus, you can be well-prepared and still give a bad presentation if your nonverbal communication is poor. Some tips on nonverbal communication follow.

- Look the audience members in the eye. It is said that the eyes are the gateway to the soul. If you will not look audience members in the eye, you will be perceived as being either a liar or a coward, neither of which is desirable. If you are nervous, you can look at their foreheads; they will never know the difference.
- As you speak, scan the audience so you are not talking to a lone friendly face. You want everyone to be involved with your presentation.
- Speak forcefully and confidently. Use your singing voice, which is supported by the diaphragm. Do not speak from the throat; doing so will make you hoarse. If you speak softly, the audience may not hear you, and may even see you as insecure.
- Use a pointer to direct the audience's attention to a particular part of your visual aid. If you want the audience to stop looking at the visual aid and direct their attention to you, step away from the visual aid.
- Do not stand where you block the view to the visual aids. If you use a transparency, do not point directly to the transparency, as this usually blocks somebody's view. Instead, point directly to the screen.
- Avoid distracting habits such as jiggling change in your pocket. Do not overuse distracting phrases, such as "you know" or "ummm."

2. Albert Mehrabian, *Nonverbal Communication* (Chicago, IL: Aldine Publishing Company, 1972), p. 182; Albert Mehrabian, *Silent Messages* (Belmont, CA: Wadsworth Publishing Co., 1971), pp. 43–44.

- Watch your body language. Do not be stiff, meaning you are scared, or overly relaxed, meaning you do not care. To show respect for the audience, wear nice clothing and be well groomed.
- Be enthusiastic. If you do not care about your presentation, why should the audience?

6.3 WRITING

Engineers employ technical writing, which is very different from the literary writing you learn in most English classes. Consider this passage:

> *Helen, thy beauty is to me*
> *Like those Nicèan barks of yore*
> *That gently, o'er a perfumed sea,*
> *The weary way-worn wanderer bore*
> *To his own native shore.*
>
> *Edgar Allan Poe*

To express these ideas in technical writing, we would simply say

He thinks Helen is beautiful.

Although this technical writing does not elicit the emotions of Poe's passage, that is not the goal. Instead, the goals are that technical writing be

- *Accurate.* In engineering, it is essential that the information be correct.
- *Brief.* Readers of technical writing are busy and do not have time to sift through a lot of words.
- *Clear.* Be sure that your technical writing can be interpreted only one way.
- *Easy to understand.* Your goal is to express, not impress.

6.3.1 Organization

In engineering, the typical types of written communications are business letters (which are sent outside the organization), memoranda (which are sent inside the organization), proposals (which are solicitations for funding), technical reports (which are used internally), and technical papers (which are published in the open literature). Table 6.4 summarizes the content of each document. The specific order of the content and the format depend upon the organization for which you are writing.

6.3.2 Structural Aids

In your written communication, be sure to employ headings and subheadings; these break the document into digestible chunks and let the reader know when you have changed topics. Paragraphs should begin with a topic sentence to inform the reader what the paragraph is about. In general, paragraphs should contain more than one sentence, unless it is presenting a particularly emphatic idea. When connecting ideas in a paragraph, be sure to use transition words, such as *but, however, in addition,* and so forth.

TABLE 6.4
Contents of typical engineering documents

Business Letter	Memorandum	Proposal	Technical Report	Technical Paper
Date	To:	Title page	Title page	Title page
Recipient address	Through:	Contents	Contents	Abstract
Salutation (Dear . . .)	From:	Background	List of figures	Text
Text	Date:	Scope of work	List of tables	Introduction
Introduction	Subject:	Methods	Executive summary	Math derivations
Body	Text	Time table	Text	Methods
Conclusion	Introduction	References	Introduction	Results
Closing (Sincerely)	Body	Appendixes	Math derivations	Conclusions
Signature	Conclusion	Facilities	Methods	Acknowledgments
Stenographic reference	Enclosures	Budget	Results	Nomenclature
Enclosures		Personnel	Conclusions	References
		Letters of support	Recommendations	Appendixes
			Acknowledgments	
			Nomenclature	
			References	
			Appendixes	
1–5 pages without enclosures	1–5 pages without enclosures	1–1000+ pages	5–1000+ pages	1–40 pages

6.3.3 Becoming a Good Writer

Everyone has problems with writing; the problems differ only by degree. Unlike some engineering problems that you can solve by applying an algorithm that results in the correct answer every time, there is no such algorithm that guarantees good writing. Instead, you learn to write by trial and error and by reading examples of good writing. Learning to write properly requires a lifelong commitment. Slowly, over time and with practice, this skill sinks in.

Good writing requires editing; rarely does a well-written document emerge extemporaneously. The author actually must wear two hats: those of the writer and the reader. After writing a passage, you must clear your mind and read it from the viewpoint of your readers, considering their backgrounds, biases, and knowledge. (Remember: Know your audience.) Can you understand what was written, from the reader's viewpoint? This is not easy to do; after all, you just wrote it, and you know what you were *trying* to communicate. If you have the luxury of time, put the writing away for a while so you can forget what you were trying to say, and later see what you actually wrote. Alternatively, you can have someone else read your work.

Unlike natural laws, which are valid for all time and all locations, language "laws" constantly evolve. Although some grammatical rules are fairly fixed (e.g., end a sentence with a period), other rules change with time or location (e.g., "color" in the United States, "colour" in Britain). The French language has *L'Académie Française* to define proper French, but no such governing body exists for the English language. This lack is both a blessing and a curse. English freely borrows words from all over the world, allowing for many subtle shades of meaning; but then we are stuck with inconsistent spelling. Without a governing authority, English is a matter of convention. Some of the conventions are arbitrary—and some are absurd—but many allow the brain to process words rapidly and unambiguously into understanding. To impose some order on the chaotic English language, many organizations employ a style manual. Here, we describe some of the most common conventions; however, you will certainly find exceptions as you go out into the world.

6.3.4 Building Better Sentences

As you construct sentences, consider the following issues:

1. ***Use parallel construction.*** **When comparing related ideas, use similar sentence construction.**

   Incorrect Scientists acquire knowledge and engineers are concerned with applying knowledge.

   Correct Scientists are concerned with acquiring knowledge, whereas engineers are concerned with applying knowledge.

Also employ parallel construction with lists.

   Incorrect Civil engineers build roads, building construction, and waterworks planning.

   Correct Civil engineers build roads, construct buildings, and plan waterworks.

2. ***Avoid sentence fragments.*** **Use complete sentences in your writing.**

   Incorrect Joining a professional society, important to furthering your career.

   Correct Joining a professional society is important to furthering your career.

3. ***Use clear pronoun references.*** **Be sure that the noun referenced by the pronoun is clear.**

   Incorrect Procedure A is used for a high-concentration sample. *This* results from the benefits of advanced technology.

   Correct Procedure A is used for a high-concentration sample. *This procedure* results from the benefits of advanced technology.

   Correct Procedure A is used for a high-concentration sample. *This sample* results from the benefits of advanced technology.

4. ***Avoid long sentences.*** **Break overly long sentences into multiple short sentences.**

   Incorrect The procedure for operating the chemical reactor starts by first opening Valve A by turning the handle counterclockwise when viewed from the top, then turning on Pump A and waiting 15 min while simultaneously watching the temperature gauge to ensure that the reactor does not overheat, in which case, open Valve B, which introduces cooling water to cool the reactor.

   Correct The procedure for operating the chemical reactor follows: First, open Valve A by turning the handle counterclockwise when viewed from the top. Then, turn on Pump A and wait 15 min while simultaneously watching the temperature gauge. If the reactor overheats, open Valve B, which introduces cooling water to cool the reactor.

5. ***Avoid short sentences.*** **Combine overly short sentences into longer sentences that flow more fluidly.**

   Incorrect The procedure for operating the chemical reactor follows: First, open Valve A. Open Valve A by turning the handle counterclockwise. The proper viewing position is from the top. Then, turn on Pump A. Wait 15 min. Simultaneously, watch the temperature gauge. If the reactor overheats, open Valve B. Opening Valve B introduces cooling water. Cooling water cools the reactor.

   Correct (See previous item.)

(*Note:* Occasionally using short sentences can make writing more interesting and varied.)

6. ***Use active voice.*** **Active sentences require fewer words and are more interesting to read.**

   Incorrect Temperature is dependent upon the heat input.
   Correct Temperature depends on the heat input.

Other examples are shown in the table below:

Incorrect	Correct	Incorrect	Correct
place emphasis on	emphasize	is an indication of	indicates
is compliant with	complies with	is a representation of	represents

7. ***Avoid vague words.*** **Use precise words to replace general words.**

   Incorrect The sensor read 150°F.
   Correct The thermometer read 150°F.
   Incorrect The community suffered through a period of economic troubles.
   Correct For five years, the community had over 10% unemployment.

8. ***Reduce prepositions.*** **Overusing prepositions (*of, in, by, on, out, to, under,* etc.) makes understanding difficult.**

   Incorrect The establishment of a panel of experts in safety was needed for investigation of accidents by miners in Pennsylvania.
   Correct A panel of safety experts was established to investigate accidents by Pennsylvania miners.

9. ***Eliminate redundancies.*** **Excess words take time to process and may lead to confusion.**

   Incorrect The pH value was 7.2.
   Correct The pH was 7.2.

10. ***Avoid bureaucratic language.*** **The following table shows that bureaucratic phrases can be replaced by fewer words:**

Incorrect	Correct
by means of	by
in the event of	if
for the reason that	because
with regard to	about

11. ***Avoid informal language.*** **Using informal language is analogous to wearing jeans and a T-shirt while giving an important sales presentation.**

    Incorrect We plugged numbers into the equation.
    Correct We substituted numbers into the equation.
    Incorrect The shaft can't rotate. (Contractions are considered casual language.)
    Correct The shaft cannot rotate.

12. ***Avoid pompous language.*** **Do not use a 50-cent word when a nickel word will do the job.**

Incorrect	Correct	Incorrect	Correct
prior to	before	personnel	people
utilize	use	subsequent	next
initiate	start	terminate	end

13. ***Avoid sexist language.*** **In the past, if the sex of a person was indeterminate, the default sex was "he." In modern usage, "he or she" can be used, although this can lead to clunky sentences. An alternative approach is to mix "he" and "she" throughout your writing, or you can use plurals by saying "they."**

14. ***Avoid dangling modifiers. Dangling modifiers*** **are words or phrases that describe something that has been left out of the sentence.**

    Incorrect Determining the experiment to be a failure, the entire project was canceled.
    Correct Determining the experiment to be a failure, the project manager canceled the project.
    Correct Because the experiment was a failure, the entire project was canceled.

15. ***Avoid split infinitives. Infinitives*** **are verbs preceded by the word *to*. In many places, it is best to keep these two words together.**

    Undesirable Eddie decided to quickly drive down the road.
    Preferred Eddie decided to drive quickly down the road.

However, to provide emphasis, an infinitive may be split.

Example To pass the course, Edna needs to thoroughly study the class notes.

Increasingly, split infinitives are accepted unless they are awkward or ambiguous, so this is an example of a grammatical rule that is changing.

6.3.5 Punctuation

Although punctuation may seem insignificant, improper punctuation can lead to gross misunderstandings.

1. ***Hyphens.* Use hyphens to avoid ambiguity.**

Incorrect The specifications call for eight foot long pipes.
Correct The specifications call for eight foot-long pipes.
Correct The specifications call for eight-foot-long pipes.

Use a hyphen when spelling out fractions or numbers less than 100.

Examples two-thirds
seventy-eight

A hyphen can be used to combine nouns of equal things.

Example Because he does both fundamental and applied work, John is a scientist-engineer.

Use a hyphen to create compound units.

Example The accident rate is reported per person-mile.

Use a hyphen to create compound adjectives.

Examples We need a face-to-face meeting to resolve this dispute.
The process requires high-pressure pipe.
Use 5-in-diameter pipe.

Incorrect The pipe diameter is 5-in.
Correct The pipe diameter is 5 in.
Example This car can be powered by a six- or eight-cylinder engine.

Do not use a hyphen with adverbs ending in "ly."

Incorrect This is a highly-explosive process.
Correct This is a highly explosive process.

2. ***Colons.* Use colons to introduce a list.**

Example The following skills are used by engineers: analysis, creativity, and communication.

Colons can be used to introduce equations, provided a complete sentence precedes the colon.

Example The following equation results from Newton's second law:

$$F = ma$$

where

F = force
m = mass
a = acceleration

Note that a colon does not appear after the word *where.* Similarly, do not put a colon after the following words: *when, if, therefore, is, by, are, such as, especially,* and *including.*

3. ***Commas.* Use commas to separate items in a list of three or more items.**

   Example The primary tools of an engineer are a pencil, a calculator, and a computer.

Use commas to separate nonessential or nonrestrictive clauses, that is, clauses that are parenthetical or that could be deleted without changing the meaning of the sentence.

   Example The shaft seal, which is supposed to work at high speeds, failed.

Use commas to separate long independent clauses joined by *and, or, but*, or *nor.*

   Example In the United States we use an English measurement system, but slowly we are adopting the SI system.

Use commas to set off introductory clauses.

   Example Before turning on the amplifier, be sure it is grounded.

Use commas to separate multiple adjectives that could be joined by *and.*

   Example The automobile has shiny, red paint.

4. ***Parentheses.* Use to set off parenthetical lists, clarifications, acronyms, abbreviations, or asides.**

   Example Ellen's technical courses (heat transfer, fluids, and thermodynamics) are canceled.

   Example Mike suggested that we use high-pressure (schedule 80, not schedule 40) pipe.

   Example Units of measure are regulated by the National Institute of Standards and Technology (NIST).

(*Note:* Always spell out an abbreviation upon first use.)

   Example The primary salt in seawater is sodium chloride (NaCl).

   Example The boss announced that because of his fine work, Fred will be promoted (although everyone knows it's because he married the boss's daughter).

5. ***Dashes.* Use — to emphasize parenthetical statements. (The proper symbol is —, but -- will do in a pinch.)**

   Example Open the steam valve—the one with the red handle, not the blue handle—by turning it counterclockwise.

   Example Open the steam valve--the one with the red handle, not the blue handle--by turning it counterclockwise.

6. ***Semicolon.* Do not use a comma to separate two phrases that could stand alone as independent sentences; instead, use a semicolon.**

   Incorrect The engineering student worked hard in school, therefore he landed a good job.

   Correct The engineering student worked hard in school; therefore, he landed a good job.

Use a semicolon to separate phrases that have commas.

Example The following individuals attended the meeting: Martin Fields, vice president, Ford Motor Company; Alfred Reno, chief executive officer, General Motors; and Jennifer Anderson, president, Chrysler.

7. ***Apostrophe.* To show possession for a single individual, use *'s*.**

Example The engineer's book was published.

To show possession for a plural noun ending in *s* or *es*, simply add the apostrophe.

Example When the fraternity house burned, the engineers' books were lost.

Generally, if the noun ends in an *s* sound, add *'s*.

Example The waitress's order

However, with some names that end in the *s* sound, simply add the apostrophe.

Example Gauss' law

It is best to use the possessive form only for animate creatures; however, it is sometimes employed with inanimate nouns.

Example The earth's orbit

8. ***Quotation marks.* Use to identify quotations.**

Example The astronaut's exact words were, "Houston, we have a problem."

Also, quotation marks identify a word or phrase that is used in an unconventional way.

Incorrect With this word processor, press enter to start a new line.
Correct With this word processor, press "enter" to start a new line.

In proper American usage, a comma or period appears inside the quotation.

Incorrect To use this computer program, memorize the following commands: "start", "stop", and "help".
Correct To use this computer program, memorize the following commands: "start," "stop," and "help."

What idiot made this rule?

6.3.6 Word Demons

"Word demons" have similar meanings or similar spellings, and are often confused with one another.

1. ***Effect* (n.): result**
   ***Effect* (v.): to cause; to bring about**
   ***Affect* (v.): to act upon**

Example The effect of high-intensity noise is ear damage.
Example High-intensity noise effects ear damage.
Example High-intensity noise affects the cochlea.

2. ***Compliment:* to praise**
   ***Complement:* to complete or balance**

Example Mary's boss complimented her fine presentation.
Example Humanities courses complement technical courses, leading to a well-balanced education.

3. ***Continuous:* uninterrupted**
   ***Continual:* recurring**

Example Because it must operate 24 hours per day, the electric motor was designed for continuous duty.
Example Although it operated much of the time, the motor was plagued by continual overheating.

4. ***Datum:* single piece of information**
   ***Data:* multiple pieces of information**

Example Last night, we added another datum to our temperature log book.
Example These data indicate that over the past century, the average nighttime temperature has increased by 1°F.

Note: Some style manuals treat *data* as a collective noun. In this case, the following sentence would be accepted:

Example The data indicates that over the past century, the average nighttime temperature has increased by 1°F.

5. ***Fewer:* used with integers**
   ***Less:* used with real numbers or things that cannot be counted (e.g., love)**

Example Because we sold fewer automobiles, our company had less income last month.

6. ***Farther:* greater physical distance**
   ***Further:* to a greater degree or extent**

Example The sun is farther away from the earth than the moon.
Example Further investigation showed that the engineer was unqualified to sign the construction documents.

7. ***i.e.:* abbreviation of Latin *id est* (that is to say)**
   ***e.g.:* abbreviation of Latin *exempli gratia* (for example)**

Example Engineers are better lovers; i.e., they have fewer divorces than many other professionals.
Example A mechanical engineer must take technical courses (e.g., heat transfer, fluid flow, design).

Note: A comma always follows i.e. or e.g.

8. ***Principle* (n.): a rule, law, or code of conduct**
   ***Principal* (adj.): most important**
   ***Principal* (n.): leader, head, person in authority**

Example KISS (Keep It Simple, Stupid) is a basic engineering principle.
Example John is the principal investigator for the project.
Example Because of her misbehavior, Joyce was sent to see the principal.

9. ***It's:*** **contraction of** ***it is***
Its: **possessive of it**

Example It's raining outside.
Example Its engine overheated.

10. ***There:*** **that place**
Their: **relating to them**
They're: **contraction of** ***they are***

Example There is the wrench.
Example Their wrench is on the ground.
Example They're going outside and taking the wrench.

11. ***That:*** **First word of a phrase that is essential for meaning.**
Which: **First word of a phrase that is not essential for meaning. (If adding "by the way" makes sense, use** ***which.*****)**

Example Of the tools that are carried on a sailboat, the screwdriver is needed most.
Example The screwdriver, which can also open cans of paint, is needed to tighten loose screws.

12. ***Lose:*** **to misplace**
Loose: **not tight**

Example Do not lose that screw or we will have to find another.
Example That screw is loose.

13. ***Since:*** **from a time in the past until now**
Because: **for the reason that**

Example Since the first hominid made a tool, humankind has had engineers.
Incorrect Since the heating element failed, the oven could no longer maintain the desired temperature.
Correct Because the heating element failed, the oven could no longer maintain the desired temperature.

(*Note:* This rule is not always rigidly enforced as some style manuals allow *since* to be used in place of *because.*)

14. ***While:*** **during the time that**
Whereas: **although**

Correct While studying for the exam, Fred suddenly became hungry.
Incorrect Cargo planes are used to carry things, while passenger planes are used to carry people.
Correct Cargo planes are used to carry things, whereas passenger planes are used to carry people.

(*Note*: This rule is not always rigidly enforced as some style manuals allow *while* to be substituted for *whereas.*)

15. ***Respectively:*** **in the order given**
Respectfully: **showing respect**

Example Experiments 1, 2, and 3 were performed on Tuesday, Thursday, and Friday, respectively.

Example Respectfully, I disagree with your conclusion.

6.3.7 Equations and Numbers

To aid understanding, separate equations from the text line, as follows:

$$F = \frac{Q\rho}{A}$$

Notice that the numerator is written above the denominator. Although this equation could have been written $F = Q\rho/\mathrm{A}$, this should be done only if the equation is incorporated into the text line.

Properly used italics can help clarify mathematical symbols. The following table shows when italics should, and should not, be used.

Italics	No Italics
Latin letters (*a, A, b, B,* etc.)	Greek letters (α, β, γ, etc.) Abbreviations (e.g., LMTD) Units (lb, gal, mL) Numbers Words Mathematical functions (e.g., sin, cos) Chemical formulas (e.g., NaCl) Parentheses ( ) and brackets [ ]

Consult Chapter 13 on the SI system for more rules on units.

For decimal numbers less than one, use a leading zero.

Incorrect .756

Correct 0.756

In technical writing, a common dilemma is whether to report the Arabic number, or spell it out. To address this dilemma, consult the following table:

Arabic Numbers	Examples
With units of measure	The reaction takes 4 min. The mass is 5 g.
In mathematical or technical contexts	The voltage is 3 orders of magnitude greater. The velocity increased by 2 fold.
For items and sections	Use Wrench 3 to tighten the pipe. Section 7 shows a cross-sectional view.
For numbered objects (e.g., tables, figures, experiments)	Use the data from Experiment 3. Refer to Figure 4 to see the correlation. Table 2 shows the chemical formulas.
For all numbers in a series, even if some numbers normally would be spelled out	The tests involved 3, 8, or 15 subjects.
Ordinal numbers 10 and above	That is the 11th explosion this year.

Spell Out	Examples
Integers below 10 with no associated units	We sold eight valves today.
Ordinal numbers below 10	That is the fifth time he crashed his car.
Integers below 10, with units, if not in a mathematical or technical context	It took me seven years to write this book.
Common fractions	In today's lecture, half the students were asleep.
Numbers that begin sentences*	Fifteen valves were sold today.
Consecutive numerical expressions*	It took fifteen 2-lb weights to balance the load.

*Sometimes, a sentence can be reformulated to avoid using these rules.

6.3.8 Subject/Verb Agreement

A common error in technical writing occurs when subjects and verbs do not agree.

Incorrect The telescope and associated hardware is the major tool of astronomers.
Correct The telescope and associated hardware are the major tools of astronomers.

The following table shows the conventions for subject/verb agreement:

Plural	Example
Compound subjects (joined by *and*)	Cattle and goats are ruminant animals.
Either/or and *neither/nor* constructions with plural nouns	Either scientists or engineers are going to fly on the space shuttle.
Singular	
Compound subjects that appear as a single unit	Research and development is the main activity.
Subjects modified by *each* or *every*	Each nut and screw is made of titanium.
When the subject is one of the following pronouns: *each, either, neither, one, anybody, somebody*	Neither of the animals is alive.
Collective nouns	The engineering staff is getting a raise.
Units of measure	To the beaker, 5 g of salt is added.
Either/or and *neither/nor* constructions with plural nouns	Either a scientist or an engineer is going to fly on the space shuttle.
Noun clauses that are subjects	What the schools need is donations.

6.3.9 Miscellaneous Issues

1. ***Verb tense.* In technical writing, the present tense is preferred. Use past tense only for those things that you did in the past. Consider the following examples:**

   Present The correlation in Figure 3 shows that machining cost increases with smaller tolerances.
   Past The flask was cleaned with chromic acid.

2. ***First person.* In formal writing, use of the first person is discouraged.**

   Informal We studied the combustion of methane.
   Formal The combustion of methane was studied.

However, avoiding first person often prevents use of the active voice, so some style manuals allow use of first person, even in formal writing.

3. ***Capitalization.* Only proper nouns are capitalized.**

Incorrect I want to become a Mechanical Engineer.
Correct I want to become a mechanical engineer.

Incorrect To learn more about this field, visit the mechanical engineering department.
Correct To learn more about this field, visit the Mechanical Engineering Department.

Incorrect As a new hire, you are required to visit the President of the company.
Correct As a new hire, you are required to visit the president of the company.

Incorrect Our new leader is president James Garland.
Correct Our new leader is President James Garland.

Naming a common noun transforms it into a proper noun. Study the following examples:

Example This procedure has five steps.
Example In this new procedure, Step 1 should be altered.

Example John is so efficient, he completed four experiments in a single day.
Example The most interesting results were reported in Experiment 3.

Example The reactor has 10 valves.
Example Cooling water is introduced by opening Valve C.

4. ***Italics.* Italics are used for foreign words employed in English sentences, scientific names of organisms, defined words, and names of long publications. (Books and journals are italicized; articles have quotation marks.)**

Example Well, as they say in France, *vive la différence.*
Example Her intestinal problems were caused by a new strain of *E. coli.*

Example A *heptagon* is a seven-sided figure.
Example Ed was elated when his article was accepted in the prestigious journal *Science.*

5. ***Articles. The* is used for a specific noun.**

Example The plane was overloaded when it took off, so it crashed.

A is used for an unspecified noun starting with a consonant sound.

Example If a plane is overloaded when it takes off, it will crash.

An is used for an unspecified noun starting with a vowel sound.

Examples an ulcer a unicorn
an RSVP a rebel
an hour a human

6. ***Figures and Tables.* Figures and tables must be described and referenced in the text. Place the figure or table just after the first reference to it.**

Example On the next page, Figure 1 shows that costs increase exponentially with tighter tolerances.

7. ***Lists.* Start lists with simple items and proceed to the more complex items.**

   Example Rosa owns the following vehicles: a bicycle, a motor scooter, and a red Mercedes with a sunroof.

8. ***Spelling.* Use a dictionary to check your spelling. Word processors can check spelling, but they are not perfect. Consider the following sentence from a student's report:**

   Example All computer engineering students take curses in electrical circuits.

9. ***References.* Each publication has its own standards for citing references. Below are some examples:**

   Example Mifflin, W. B., and R. L. Jones (1978) *Engineering Design*, New York: McGraw-Hill, pp. 32–78.

   Example Mifflin, W. B. and R. L. Jones (1978) *Journal of Engineering Design* 33, 45–64.

10. ***Consistency.* Sometimes, grammatical rules are unclear or arbitrary. In these cases, choose what you think is best and be consistent throughout your document.**

6.4 SUMMARY

Mastering engineering communications is essential. Not only will it help your career, but it could also avert disaster. Imagine the possible consequences of a poorly written operating manual for a nuclear power plant.

Before you start writing or preparing your speech, you must select a topic, conduct research, and organize your thoughts. When organizing, remember that the overwhelming consideration is, "Know your audience." Regardless of whether you are communicating orally or in writing, the communication will have an introduction, a body, and a conclusion.

For a speech, you must provide visual aids to help convey your ideas. The key is for the visual aids to communicate rapidly so the audience can focus on what you are saying; they cannot decipher a complex slide and listen to you talk at the same time. Because graphical images are more rapidly processed by the audience, convey your thoughts graphically, if possible, rather than with written text. In oral presentations, words convey only a small part of your message; body language and voice strain convey the majority of the information.

In technical writing, the goals are accuracy, brevity, clarity, and ease of understanding. With these goals, you are fortunate to be able to communicate in English. The English language has more words than any other language in the world, allowing those who master it to convey subtle shades of meaning. As with any language, there are numerous rules and conventions that mark good writing. Those who ignore these rules send out the message that they are sloppy thinkers. If an engineering report is written poorly, the reader could logically conclude that an author who cannot master the rules of English probably cannot master technology either, a conclusion that invalidates the whole report.

SECTION TWO

MATHEMATICS

In this section, we look at some selected topics in mathematics. Numbers are so fundamental to engineering, we will start with them. Next, we look at tables and graphs, because engineers use these to communicate—with fellow engineers and with others outside the profession. Finally, we take a brief look at statistics, a topic of growing importance in engineering.

Mathematical Prerequisite

- Algebra (Appendix E, Mathematics Supplement)

CHAPTER 7

Numbers

As you pursue your engineering studies, you will surely notice that numbers are everywhere. Engineers are obsessed with numbers and want to quantify everything. If an average person were to describe his new car, he would perhaps describe its color, upholstery, and stereo system. In contrast, an engineer would most likely describe its horsepower, acceleration, and weight, all of which can be quantified with numbers.

Just as a writer is concerned with accurately communicating with words, engineers are concerned with accurately communicating with numbers. Here, we describe how to use numbers properly.

7.1 NUMBER NOTATION

The U.S. standard decimal notation for numbers is

4,378.1 (U.S. standard decimal notation)

where the comma indicates three orders of magnitude and the period indicates decimals. However, in Europe, the comma replaces the period to indicate decimals and the period replaces the comma to indicate three orders of magnitude:

4.378,1 (European decimal notation)

To avoid confusion, an accepted convention is to use a space rather than a comma to indicate three orders of magnitude:

4 378.1 (Accepted convention)

Numbers written in these ways are suitable for most of the quantities we encounter in our everyday lives. However, many numbers in science and engineering are much too large or small to be recorded in decimal notation. For example, Avogadro's number (the number of molecules in a mole) would be

602,213,670,000,000,000,000,000

Because this is obviously cumbersome, **scientific notation** is generally used to represent Avogadro's number:

$$6.0221367 \times 10^{23}$$

In computers, scientific notation is often represented with a leading zero:

$$0.60221367 \times 10^{24}$$

Leading zeros are often wrongly dropped. In the popular press (e.g., *Time* magazine), positive numbers less than one are indicated as follows:

.593 (Popular press notation)

THIS CONVENTION MUST NEVER BE FOLLOWED BY ENGINEERS. A tiny dot separates this number from a number that is three orders of magnitude larger (593). In a world with dirty copy machines, a stray dot could easily change the order of magnitude of a number. To solve this problem, engineers must ALWAYS use leading zeros for decimal numbers less than one:

0.593 (**Engineering notation**)

7.2 SIMPLE ERROR ANALYSIS

One use of numbers is for counting things. For example, if one were to ask, "How many marbles are in the following picture?"

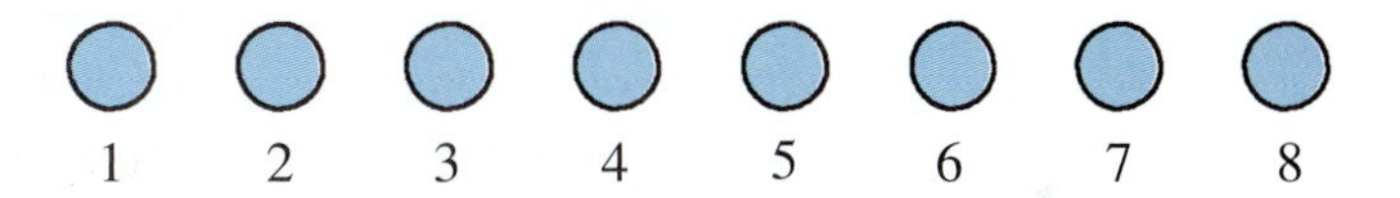

the answer is obviously the *integer* 8. Presuming there is no mistake made in the counting, the answer can be known exactly, without error.

Another use of numbers is to measure continuous properties. Assume that one were to ask the question, "What is the length of the following rod?"

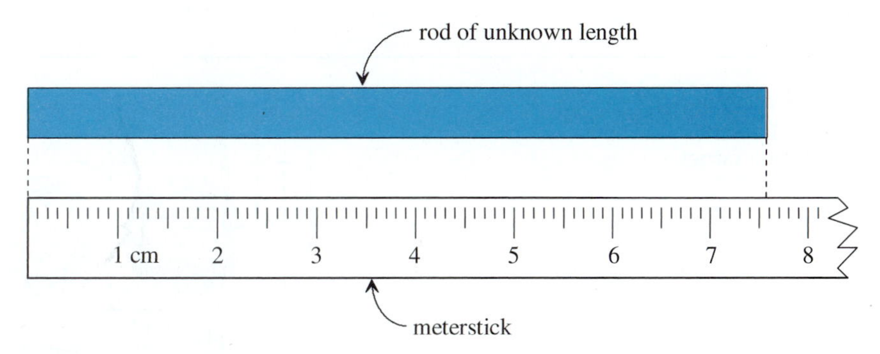

The approach to answering this question is to compare the unknown length of the rod to the known length of a meterstick. Depending on the care with which you measure the rod, you could answer using the following *real numbers:*

The rod is between 7 and 8 cm, so the length is 7.5 ± 0.5 cm.
The rod is between 7.5 and 7.6 cm, so the length is 7.55 ± 0.05 cm.
The rod is between 7.57 and 7.59 cm, so the length is 7.58 ± 0.01 cm.

If you really needed to know the length precisely, you could employ more sophisticated measuring methods, such as micrometers or even lasers. The essential point here is that no one can know the exact length of the rod because that would require an infinite number of digits. There will always be some error in the reported *real number.*

The distinction between *integers* and *real numbers* is very important in computers. Computers can represent integers exactly, provided the number does not exceed the limits of the machine. However, when computers manipulate real numbers (in computer jargon, the computer is performing "floating-point operations"), there are inherent errors, because an infinite number of digits are required to represent real numbers exactly.

Whenever measurements are made, important distinctions arise such as **accuracy** versus **precision, systematic errors** versus **random errors,** and **uncertainty** versus **error.** The differences between these concepts are a source of confusion.

- *Accuracy* is the extent to which the reported value approaches the "true" value and is free from error (Figure 7.1).
- *Precision* is the extent to which the measurement may be repeated and the same answer obtained (Figure 7.1).
- *Random errors* result from many sources, such as random noise in electronic circuits and the inability to reproducibly read instruments. For example, it is very difficult to read the meterstick the same way every time. Even though you may close one eye and try to read the number scale from a perpendicular position, you may report a slightly different measurement each time you read the scale.
- *Systematic errors* result from a measurement method that is inherently wrong. Consider an engineer who finds an old rusty scale made of steel and iron. He has a service technician clean it, calibrate it, and certify that it correctly weighs a set of standard weights. The engineer uses the scale for many months and comes to trust its measurements. One day, while measuring the weight of a very powerful magnet, he notices the scale is giving an unusually high reading. He then realizes there is a systematic error; the magnet

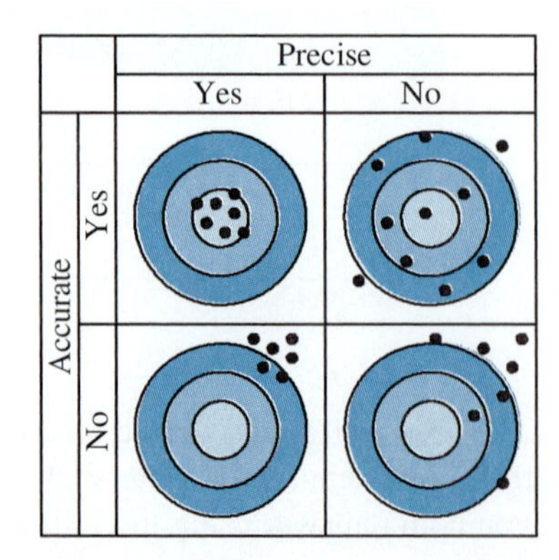

FIGURE 7.1
Accuracy and precision of bullets hitting a target.

is attracted to the iron and steel components of the scale, which causes a high reading. To eliminate this systematic error, he must use a scale constructed from nonmagnetic materials such as plastic or stainless steel.

- *Uncertainty* results from random errors and describes the lack of precision. The uncertainty in the rod measurement may be expressed on a fractional or percentage basis:

$$\text{Fractional uncertainty} = \frac{\text{uncertainty}}{\text{best value}} = \frac{0.01 \text{ cm}}{7.58 \text{ cm}} = 0.0013 \quad (7\text{-}1)$$

$$\text{Percentage uncertainty} = \frac{\text{uncertainty}}{\text{best value}} \times 100\% = \frac{0.01 \text{ cm}}{7.58 \text{ cm}} \times 100 = 0.13\% \quad (7\text{-}2)$$

Note that the rod length was precise to about 1 part in 1000, which is typical of most measurements we make.

- *Error* may be defined as the difference between the reported value and the true value:

$$\text{Error} = \text{reported value} - \text{true value} \quad (7\text{-}3)$$

Error results from systematic errors and describes the lack of accuracy. To determine the true value, it is necessary to correct the systematic error. The error may be reported as the fractional error or the percentage error:

$$\text{Fractional error} = \frac{\text{error}}{\text{true value}} \quad (7\text{-}4)$$

$$\text{Percentage error} = \frac{\text{error}}{\text{true value}} \times 100\% \quad (7\text{-}5)$$

EXAMPLE 7.1

Problem Statement: An Arctic researcher measures the length of a rod at −60°C. She records the length to be 7.58 cm. Her assistant notes a systematic error, in that the aluminum meterstick was calibrated at room temperature (20°C), not −60°C. Because the meterstick shortened at the cold temperature, the true length of the rod is actually shorter than 7.58 cm. The assistant goes to a handbook and learns that aluminum shortens by 23.6 $\times 10^{-6}$ cm/cm for each C° that it is cooled.

He calculates that the rod shortened by 0.0143 cm, so the actual length of the rod is 7.57 cm. What is the fractional error and the percentage error?

Solution:

$$\text{Fractional error} = \frac{0.0143 \text{ cm}}{7.57 \text{ cm}} = 0.0019$$

$$\text{Percentage error} = \frac{0.0143 \text{ cm}}{7.57 \text{ cm}} \times 100\% = 0.19\%$$

Trouble with Hubble

Hubble is a space-based telescope that is designed to observe distant galaxies without interference from earth's atmosphere. Soon after it went into operation, astronomers were disappointed by its blurry images. They had great expectations for this telescope; the 2.4-m (94.5-inch) diameter mirror was **precisely** smooth to within 9.7 nm (0.00000038 inch). To put this smoothness into perspective, if the Hubble mirror were the diameter of the Gulf of Mexico, ripples on its surface would be only 0.5 cm (0.2 inch) high.

The main problem with Hubble originated from a **systematic error.** During the manufacture of the mirror, the null corrector (an instrument that measures mirror geometry) was improperly positioned. The mirror was supposed to be ground to a perfect hyperbolic shape. However, because it was not **accurately** made, the shape deviates slightly from a perfect hyperbola; the edges are about 2 μm (0.00008 inch) too low, making it too flat. If it were a perfect hyperbola, all the light would focus onto the light collector (see figure), but because of the defect, the light did not focus properly, thus greatly reducing the light collected and giving star images a "halo." Thus, the Hubble mirror was **precisely inaccurate.** Fortunately, because it was precisely made, the distortion was later corrected by modifying the light collector.

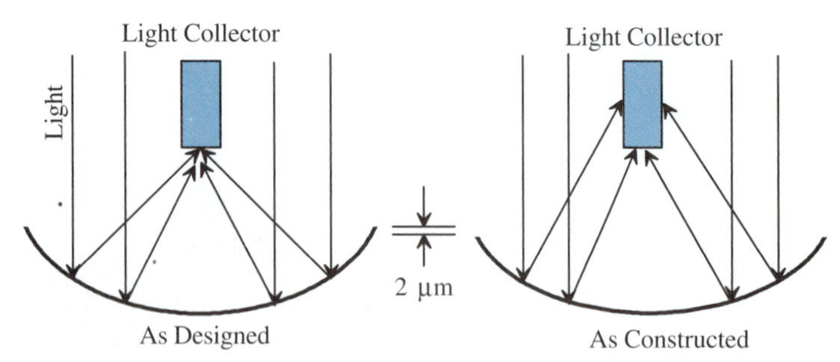

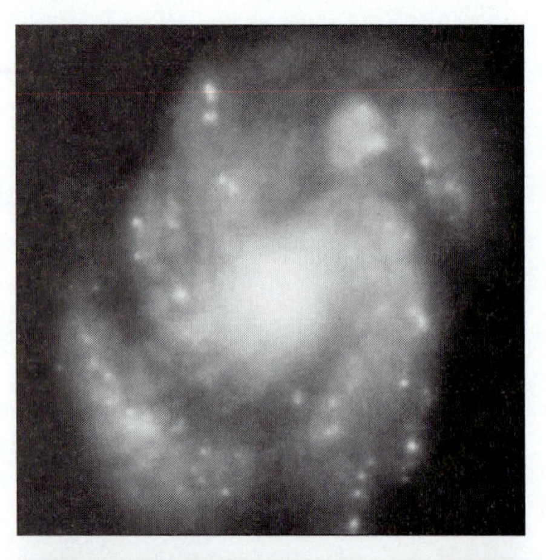

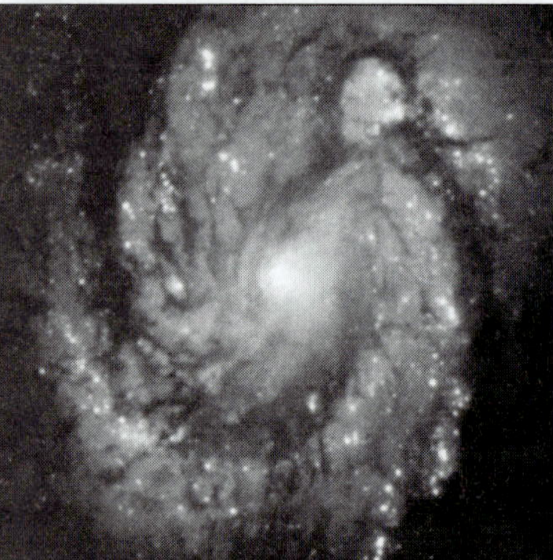

Hubble images of the M100 galactic nucleus. Top image before correction, bottom image after correction.

Suppose we wished to express the length of the rod in meters, rather than centimeters. Recalling that one meter is exactly 100 centimeters (by definition), we can convert the measured length into meters:

$$7.58 \text{ cm} \times \frac{1 \text{ meter}}{100 \text{ cm}} = 0.0758 \text{ m}$$

Note that the original measurement had two decimal places, whereas the new reported measurement has four decimal places. Clearly, the new measurement is no more accurate than the original measurement, so the number of decimal places is completely irrelevant as a measure of accuracy. Rather than describing accuracy according to the number of decimal places (as is often erroneously done), the concept of **significant figures** is required. Both the above numbers have three significant figures, so they both convey the same degree of accuracy.

7.3 SIGNIFICANT FIGURES

The issue of *significant figures* is central to how we use numbers and how much we believe them. A common question is, How many significant figures should I use when reporting a number? The answer to the question depends on how well you know the number. The following table should help you decide.

If you know the number to:	Then report this many significant figures:
1 part per 10	1
1 part per 100	2
1 part per 1000	3
1 part per 10,000	4
1 part per 100,000	5
1 part per 1,000,000	6
etc.	etc.

The uncertainty in the rod length measurement was about 1 part per 1000, so three significant figures are appropriate. The accepted convention is that the last reported digit has some error, whereas the first digits are known exactly.

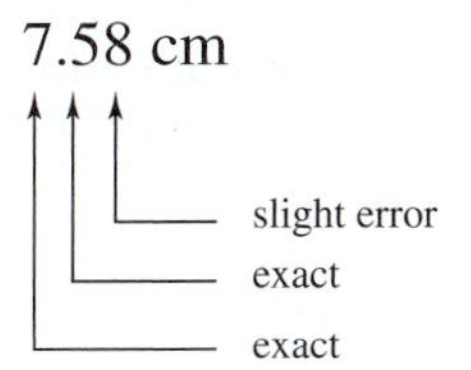

The following rule of thumb should help you decide how many significant figures to report:

> Many engineering measurements are accurate to 1 part in 1000, so three significant figures are appropriate. For estimates, only one or two significant figures should be reported. If a measurement is made using an extremely accurate instrument, then four or more significant figures are warranted.

To determine the number of significant figures in a number, use the following rules:

- A *significant figure* is an accurate digit, although the last digit is accepted to have some error.
- The number of significant figures does not include the zeros required to place the decimal point.

The latter rule causes some confusion. By looking at a number, it is not always possible to determine if the zeros are truly significant, or whether they are simply locating the decimal point. Consider these examples:

Number	Number of Significant Figures
0.00342	3
342	3
340	2 or 3

The first two numbers clearly have three significant figures, whereas the last number is ambiguous. It is impossible to determine whether the last zero is truly significant, or whether it is needed to locate the decimal place. To avoid ambiguity, scientific notation should be used.

Number	Number of Significant Figures
3.42×10^{-3}	3
3.42×10^{2}	3
3.40×10^{2}	3
3.4×10^{2}	2

Two important points must be made about significant figures:

1. Exact definitions have an infinite number of significant figures. For example, one inch is **defined** to be **exactly** 2.54 centimeters,

   $$1.00000000000^{+} \text{ inch} \equiv 2.5400000000000^{+} \text{ centimeters}$$

   where the superscript "+" indicates there are an infinite number of zeros. Generally, the zeros would not be written; it is understood that there are an infinite number of them.

2. Numbers resulting from exact mathematical relationships have an infinite number of significant figures. For example, the area A of a circle can be calculated from the radius r as

   $$A = \pi r^2$$

   The radius is exactly half the diameter; therefore,

   $$A = \pi\left(\frac{D}{2}\right)^2 = \frac{\pi}{4}D^2$$

   In this formula, the 4 is equivalent to 4.0000000^{+} and the 2 is equivalent to 2.0000000^{+}.

Numbers are *rounded* when there are more digits than are appropriate. For example, calculators generally display many more digits than are significant, so the final answer must be rounded. The following rules determine how to properly round:

- Round up if the number following the cut is between 5 and 9.
- Leave alone if the number following the cut is between 0 and 4.

This is better understood by considering the following examples:

Calculator Display	Desired Number of Significant Figures	Reported Number
5,937,458	3	5,940,000
0.23946	3	0.239
0.23956	3	0.240

It should be emphasized that the rounding should be performed only when the final answer is reported. Do not round during the intermediate calculations. Some students diligently round at each calculation step, thus introducing so much error that the final answer is wrong. This is called **rounding error.**

Rounding error is often encountered in computer calculations, because computers represent real numbers with a finite number of digits. The adverse effects of rounding error can be reduced by declaring variables as "double precision," meaning that twice the normal number of digits are used to represent real numbers. Some programming languages even support "variable precision," which allows the user to represent real numbers with as many digits as they require.

When real numbers are used in calculations, the uncertainty in the final answer is dictated by the most uncertain real number. There are sophisticated techniques for determining how the uncertainties (or errors) propagate in the calculation. Here, we present some simple rules to help you determine how many significant figures can be reported in the final answer.

Significant Figures: Multiplication/Division.

Consider the following example for $A \times B = C$:

		Calculator Display	Reported Value
Lower bound	$4.9 \times 10.623 =$	52.0527	
"Best" value	$5.0 \times 10.624 =$	53.1200	53 ± 1
Upper bound	$5.1 \times 10.625 =$	54.1875	

From this example, it is clear that the two-digit number limits the final reported number to two significant figures. The appropriate procedure for multiplying/dividing numbers follows:

1. Indicate the number of significant figures for each number.
2. Calculate the answer.
3. Round the answer to have the same number of significant figures as the least precise number.

$$\underset{(2)}{5.0} \times \underset{(5)}{10.624} = 53.120 \Rightarrow \underset{(2)}{53}$$

Significant Figures: Addition/Subtraction.

Consider the following calculation for $A + B = C$:

		Calculator Display	Reported Value
Lower bound	$4.9 + 14.696 =$	19.596	
"Best" value	$5.0 + 14.697 =$	19.697	19.7 ± 0.1
Upper bound	$5.1 + 14.698 =$	19.798	

From this example, it is clear that the two-digit number limits the accuracy of the reported value. But notice that the reported value actually has *three* significant figures. The appropriate procedure for adding/subtracting numbers follows:

1. Align the decimal points.
2. Mark the last significant figure of each number with an arrow.
3. Calculate the answer.
4. The arrow farthest to the left dictates the last significant figure of the answer.

$$\begin{array}{r} \downarrow \\ 5.0\downarrow \\ +14.697 \\ \hline 19.697 \rightarrow 19.7 \\ \uparrow \end{array}$$

7.4 SUMMARY

Numbers are indicated according to a variety of conventions. In the United States, a period indicates the decimal and commas mark three orders of magnitude. This convention is switched in Europe. Very large and very small numbers are often written in scientific notation. Decimal numbers between −1.0 and +1.0 are indicated with a leading zero.

Numbers may be classified as integers (used for counting) or reals (used to measure continuous properties). Integers have no error if the counting is done correctly. In contrast, real numbers do have errors, because an infinite number of significant figures are required to represent them exactly. An accurate real number means it is very close to the true value, whereas a precise real number means that the same number is obtained after repeated measurements. Random errors make a measured number less precise. Even though a measurement may be precise, systematic errors may cause it to be inaccurate. Error (i.e., the difference between the measured value and the true value) results from systematic errors. Uncertainty results from random errors.

The better a number is known, the more significant figures are reported. Numbers resulting from exact definitions, or exact mathematical relationships, have an infinite number of significant figures. When performing mathematical operations on real numbers, it is important to report the final answer with an appropriate number of significant figures.

Further Readings

Eide, A. R.; R. D. Jenison; L. H. Mashaw; and L. L. Northup. *Engineering Fundamentals and Problem Solving.* 3rd ed. New York: McGraw-Hill, 1997.

PROBLEMS

7.1 How many significant figures are reported in each of the following numbers?

(a) 385.35
(b) 0.385×10^3
(c) 40 001
(d) 40 000
(e) 4.00×10^4
(f) 0.400×10^4
(g) 389,592
(h) 0.0000053
(i) 345
(j) 3.45

7.2 Round the following numbers to three significant figures.

(a) 356,309
(b) 0.05738949
(c) 0.05999999
(d) 583,689
(e) 3 556
(f) 0.004555
(g) 400,001
(h) 730 999

7.3 Perform the following computations and report the correct number of significant figures. Ensure that the reported answer is rounded properly.

(a) 39.4×3.4
(b) $39.4 \div 3.4$
(c) $39.4 + 3.4$
(d) $39.4 - 3.4$
(e) $(0.0134)(5.58 \times 10^2)$
(f) $(248{,}287 \text{ in}^2)(\text{ft}^2/144 \text{ in}^2)$
(g) $(452 \text{ cm})(\text{m}/100 \text{ cm})$
(h) $(34.7 - 49.0456)/7$
(i) $(0.00034)(48{,}579) - 345.984$
(j) $x^2 + 3.4x + 3.982$ where $x = 9.4$
(k) $4.0568 \times 10^{-3} - 0.492 \times 10^{-2}$
(l) $8.9245 \times 10^4/6.832 \times 10^{-5}$

7.4 An experienced engineer is designing a bridge to cross a river. He needs to know the distance between two fixed points on opposite banks of the river. He assigns a young engineer the task of measuring the distance between the two points. The young engineer purchases a very long tape measure and nails the end to one fixed point. He then gets in a boat and rows to the other bank while taking the tape measure with him. When he finds the other fixed point on this bank, he stretches the tape measure as tightly as he can and measures a distance of 163 meters. When the experienced engineer hears how the measurement was made, he laughs and says that the measurement is completely inaccurate because the tape measure sags in the middle because of gravity. No matter how tightly it is stretched, the sag will still introduce too much error. The experienced engineer recommends using a laser range finder, which bounces a laser pulse off a mirror placed on the opposite bank and measures the time it takes to return. Because gravity has a negligible effect on light, this method can measure the straight-line distance between the two points. The laser range finder reports the distance as 138 meters. What is the fractional error and percentage error of the original measurement?

7.5 A laser range finder consists of a laser and a light receiver. It operates by bouncing a laser pulse off a mirror and measuring the "time of flight," the time it takes for the pulse to leave the laser, bounce off the mirror, and return to the receiver. The speed of light in a vacuum is exactly 299,792,458 m/s, so the accuracy of the distance measurement depends strictly on the accuracy of the time measurement. Using a laser range finder, an engineer measures the time of flight to be 3.45 ± 0.03 μs. [*Note*: A microsecond (μs) is 10^{-6} s.]

(a) Derive a formula for the distance between the mirror and the laser/receiver, and calculate the distance using the measured time of flight.
(b) What is the fractional uncertainty and percentage uncertainty in the time of flight?
(c) What is the fractional uncertainty and percentage uncertainty in the distance measurement?
(d) What is the uncertainty (in meters) in the distance between the mirror and the laser/receiver?
(e) The speed of light in air is 0.02925% slower than in a vacuum. If a correction were applied to account for this, is it correcting for random error or systematic error?
(f) Do you think it is necessary to correct the reported distance measurement because the speed of light in air is slightly slower than in a vacuum?

7.6 A European engineer works for an American automobile manufacturer. She measures the diameter of an automobile drive shaft using a micrometer, an accurate measuring device. The micrometer reads 2.0573. When she records the reading, she instinctively writes "2.0573 cm." An American engineer is checking the European's work and determines there is a mistake; the micrometer measures inches, not centimeters. What is the fractional error and percentage error of the diameter recorded by the European engineer?

7.7 Avogadro's number is measured by determining the number of atoms in exactly 0.012 kg of carbon 12. As you might imagine, it is very difficult to count individual atoms, so there is necessarily some error in the reported value. Other sources of error are contaminants in the carbon 12 and the difficulty of measuring exactly 0.012 kg. The reported value is 6.0221367×10^{23}. The "true" value probably lies between 6.0221295×10^{23} and 6.0221439×10^{23}. (There is only one chance in twenty that the true value would be outside this range.) What is the fractional uncertainty and percentage uncertainty in the reported value of Avogadro's number?

7.8 Write a computer program that allows the user to input the "true" value and the "reported" value. The program calculates and reports the fractional error and the percentage error. Use the data presented in Problem 7.4.

7.9 Write a computer program that multiplies two real numbers A and B and calculates the answer C. The two inputs A and B may have an arbitrary number of significant figures (up to 8). The reported answer C must have the correct number of significant figures as determined by the rules for multiplication/division. The program may ask the user how many significant figures are in A and B, but it must calculate the number of significant figures in C.

7.10 Write a computer program that accomplishes the same task as Problem 7.9, but the program may not ask the user how many significant figures are in A and B. The program must determine the number of significant figures from the user input. For example, if the user says A is 89.43, the program must determine that there are four significant figures.

7.11 Write a computer program that adds two real numbers A and B and calculates the answer C. The two inputs A and B may have an arbitrary number of significant figures (up to 8). The reported answer C must have the correct number of significant figures as determined by the rules for addition/subtraction. The program may ask the user how many significant figures are in A and B, but it must calculate the number of significant figures in C.

7.12 Write a computer program that accomplishes the same task as Problem 7.11, but the program may not ask the user how many significant figures are in A and B. The program must determine the number of significant figures from the user input. For example, if the user says A is 63, the program must determine that there are two significant figures.

Glossary

accuracy The extent to which the reported value approaches the "true" value and is free from error.
engineering notation The use of leading zeros for decimal numbers less than 1.
error The difference between the reported value and the true value.
precision The extent to which the measurement may be repeated and the same answer obtained.
random error An error that does not result from a measurement method that is inherently wrong.
rounding error Rounding in intermediate calculations which results in an incorrect final answer.
scientific notation Numbers expressed in terms of a decimal number between 1 and 10 multiplied by a power of 10.
significant figure An accurate digit, excluding the zero required to place the decimal point.
systematic error An error that results from a measurement method that is inherently wrong.
uncertainty The result from random errors and describes the lack of precision.

Mathematical Prerequisite

- Algebra (Appendix E, Mathematics Supplement)
- Logarithms and Exponents (Appendix H, Mathematics Supplement)

CHAPTER 8

Tables and Graphs

Many professions, such as law, rely almost exclusively on the written and oral word. Although engineers also must write and speak well, this alone is insufficient to convey complex engineering information. For this, graphical or visual communication is required. In just a few seconds, a well-prepared graph can accurately communicate information that would require many pages of written text. In addition, a graph can provide readers with insight they can obtain through no other means. Graphs are prepared from tabulated data, so understanding tables goes hand-in-hand with understanding graphs.

Presenting graphical information in a coherent, visually appealing manner is one of the engineering arts. Mastering it will help your career tremendously. Although personal style and taste affect graphical presentations, we will present the rules and guidelines that are almost universally agreed upon.

8.1 DEPENDENT AND INDEPENDENT VARIABLES

We generally understand nature to work in a cause/effect manner. When an engineer studies a system, he often classifies some variables as *independent* (i.e., cause) and other variables as *dependent* (i.e., effect). For example, let's assume an engineer is studying an automobile (the system) and is interested in the factors that affect its speed. The **dependent variable** in this case is speed s. Some **independent variables** that affect the speed are the rate of fuel entering the engine f; tire pressure p; air temperature T; air pressure P; road grade r; car mass m; frontal area A; and drag coefficient C_d. This can be mathematically stated as

$$s = s(f, p, T, P, r, m, A, C_d) \tag{8-1}$$

which says the automobile speed depends on variables such as fuel rate, tire pressure, and so forth. Obviously, because an automobile is a complex system, a simple algebraic formula will not describe the functional relationship between the dependent variable and independent variables. Sophisticated computer modeling or experimentation will be required.

8.2 TABLES

A table is a convenient way to list dependent and independent variables. The independent variable(s) are usually listed in the left column(s) and the dependent variable(s) are usually listed in right columns. The values in a given row correspond to each other.

For example, let's assume the engineer actually did an experiment using a car. He measured the car speed at different fuel rates and road grades. All the other parameters were kept constant. A properly constructed table is shown in Table 8.1. The independent variables (road grade and fuel rate) are listed in the left two columns, and the dependent variable (speed) is listed in the right column. The data were taken by driving the car on three roads, one flat (0% grade) and the other two on hills (5 and 10% grades). On each road, the gas pedal was adjusted to give various fuel rates from 1 to 5 gal/h, as measured by an electronic flow meter installed in the fuel line. The speed was measured by an electronic instrument that reports the speed to an accuracy of 0.1 mi/h. The objective is to produce a table that is self-contained, so the reader does not have to read accompanying text to fully understand the experiment. Therefore, other relevant parameters are specified in the table. For example, the car model is identified, which specifies the frontal area, drag coefficient, and mass. The tire pressure, and air temperature and pressure, are also indicated.

Notice that the numbers presented in the table have an appropriate number of significant figures. The instruments were capable of reporting three significant figures, which is typical of most engineering instruments. The decimal points are vertically aligned, mak-

TABLE 8.1
Properly constructed table

Effect of Fuel Rate and Road Grade on Car Speed ← Title

	Car Model = XLR	
	Tire Pressure = 30 psig	
	Air Temperature = 70°F	
	Air Pressure = 0.985 atm	
Road Grade (%)	**Fuel Rate (gal/h)**	**Speed (mi/h)**
0.00	1.00	38.2
	2.00	64.3
	3.00	81.0
	4.00	93.4
	5.00	99.2
5.00	1.00	32.2
	2.00	54.5
	3.00	68.4
	4.00	78.1
	5.00	84.8
10.00	1.00	25.8
	2.00	43.6
	3.00	54.7
	4.00	62.5
	5.00	67.8

Annotations: Title; Additional Information; Column Heads; Units; Numerical Values ing the numbers easy to read. Another important feature is that all the numbers are reported **with their units.** If the units were omitted, the numbers would have absolutely no meaning and the engineer's effort would be completely wasted.

8.3 GRAPHS

Tables are useful for presenting technical information because the numbers can be easily entered into computer programs or calculators. However, it is very difficult to interpret tabulated data. For this, graphs are much better suited. A wide variety of graphs are available to help visualize data. Graphically presenting data is actually an art form; those who are good at it are often able to see things in data that are missed by others.

A well-constructed graph is self-contained, just like a well-constructed table. Further, a well-constructed graph must communicate information accurately and rapidly. These goals are accomplished using titles, axis labels (including units), readable fonts, and legible symbols. This will become more clear as you read further.

Figure 8.1 shows the car speed data from Table 8.1 presented in a number of formats: (a) line graph with rectilinear axes, (b) line graph with semilog axes, (c) line graph with log-log axes, (d) three-dimensional surface, and (e) bar graph. For data that sum to 100%, pie charts are also very popular. Figure 8.1(f) shows a pie chart indicating the percentage of emitted energies from the fission of uranium-235. Note that a legend is employed in which the shading indicates each type of energy emitted.

Figures 8.1(a) to 8.1(c) label the three road-grade lines directly. Alternatively, a legend could have been used indicating that closed boxes are for 0% grade, open diamonds are for 5% grade, and closed circles are for 10% grade. In situations where a wide variety of cases are interpreted by a single line going through all the data, a legend is needed. However, when the data for the individual cases are clearly separated (such as Figures 8.1(a) to 8.1(c)), it is preferable to label the lines directly. This reduces the amount of mental gymnastics required to match legend symbols to the symbols appearing on the line.

Graphs (and tables) should have a descriptive title. It does not add any information to title the graph "Car Speed versus Fuel Consumption" because this is readily apparent just by looking at the axes. A complete title must be more specific and provide additional information. A proper title would be as follows:

FIGURE 8.1. Effect of fuel rate and road grade on car speed.
(Car model = XLR, Tire pressure = 30 psig,
Air temperature = 70°F, Air pressure = 0.985 atm)

This title is appropriate for describing Figures 8.1(a) through 8.1(e). (*Note:* To reduce redundancy and to save space, this title is not shown in Figure 8.1.)

The dependent variable is traditionally plotted on the **ordinate** (y-axis) and the independent variable is plotted on the **abscissa** (x-axis). (*Note:* This convention is usually followed in science and engineering, but is often reversed in economics.) When describing the graph, we say the dependent variable is plotted versus the independent variable. For example, we would say, "Figure 8.1 shows the car speed versus the fuel consumption." The ordinate and abscissa must have labels **with the units.** The units are enclosed in parentheses or separated from the label by a comma.

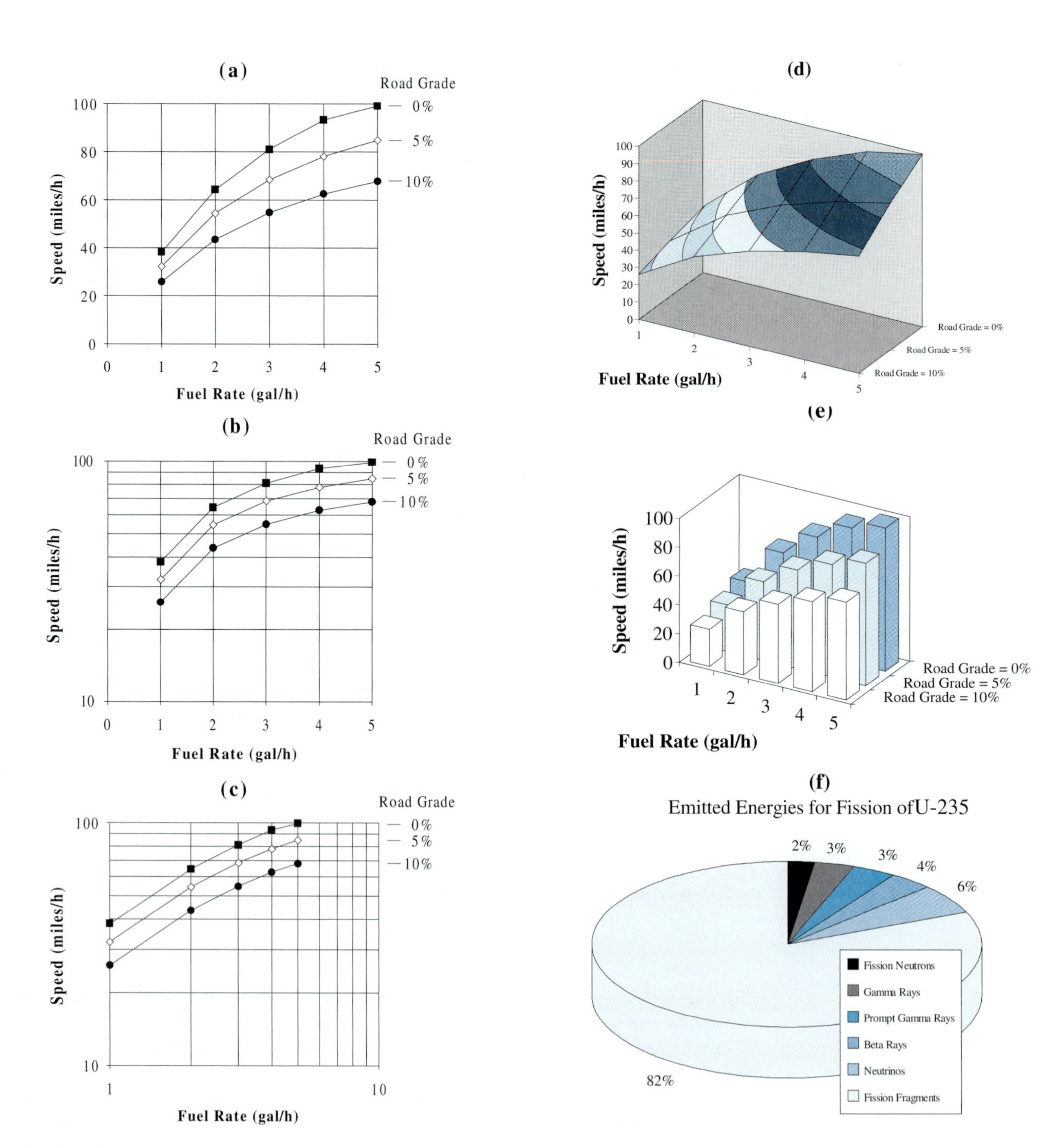

FIGURE 8.1
Examples of graphs. [*Note:* See text for titles of Figures (a) to (e).]

Each axis is graduated with *tick marks.* It is preferred that the tick marks appear outside the graph field so that they do not interfere with the data.

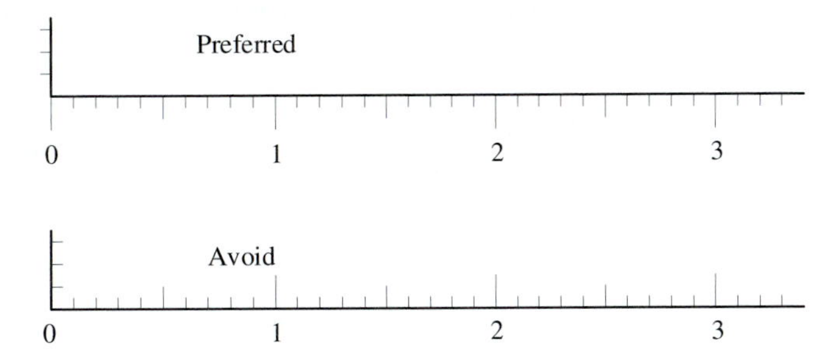

The numbers on the axes should be spaced so they can be easily read. Compare the following two axes. Obviously the first one is easier to read.

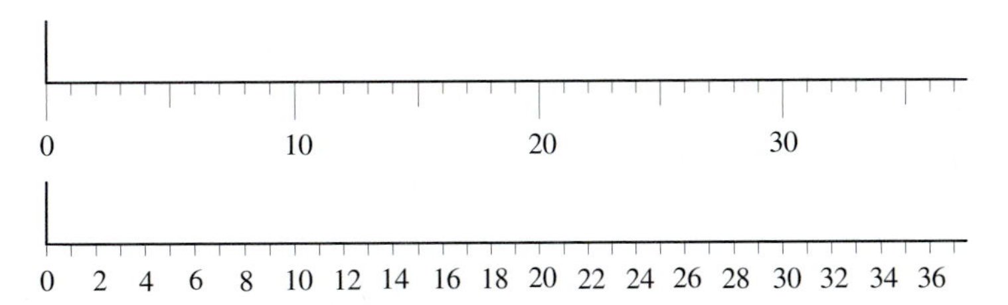

The smallest graduations on the scale are selected to follow the *1, 2, 5 rule,* meaning that if the number were written in scientific notation, the mantissa would be a 1, 2, or 5. The following axes show some examples of acceptable and nonacceptable graduations:

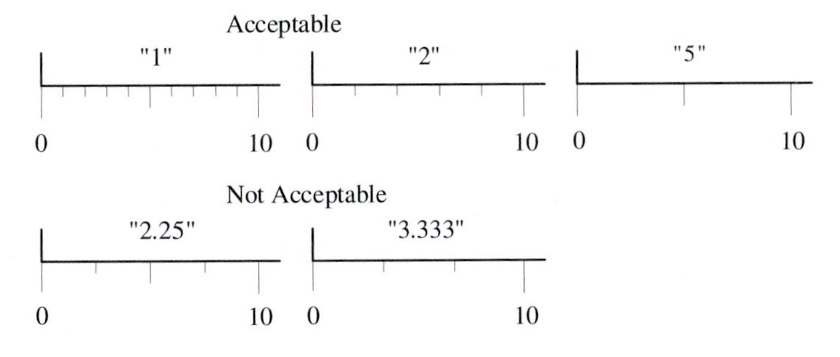

The allowable exceptions to the 1, 2, 5 rule include units of time (days, weeks, years, etc.) because these are not decimal numbers.

Problems can result if the numbers on the axis are extremely large or small, because they will crowd.

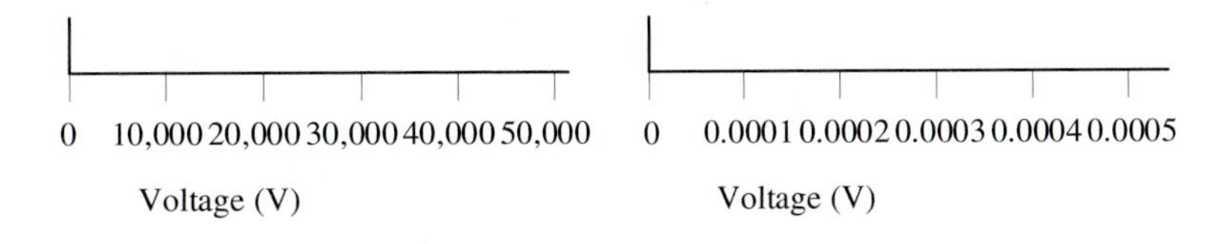

This problem can be solved in a number of ways. The best method is to use the SI system multipliers (see Chapter 13). In the SI system, "k" means "1000 ×" and "m" means "0.001 ×." Therefore, these axes can be cleaned up as follows:

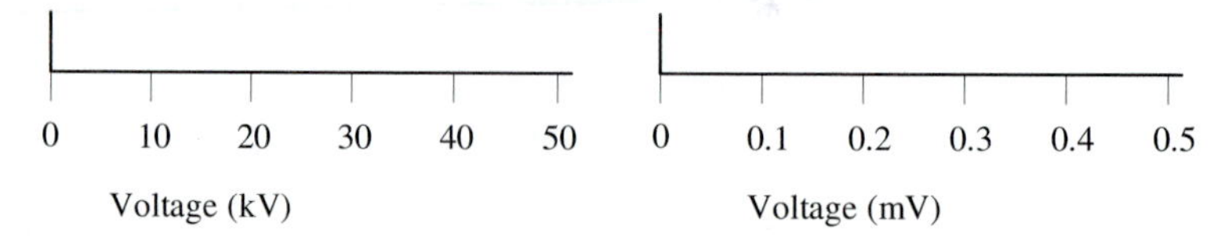

It is **extremely** important to observe the case of the units and multipliers. For example, if an uppercase "M" is used instead of the lowercase "m," the meaning is completely changed. "M" means "1,000,000 ×," so this would change the meaning by 9 orders of magnitude!

The use of the SI multipliers is convenient for solving the problem of numbers that crowd together. However, many engineering units do not have multipliers; and, by tradition, some SI units do not have multipliers (e.g., °C). There are two conventions to solving this problem. These conventions are not universally followed or understood; therefore, it is incumbent upon the reader to know the meaning from the context.

The first convention uses numbers or words as multipliers, rather than SI symbols.

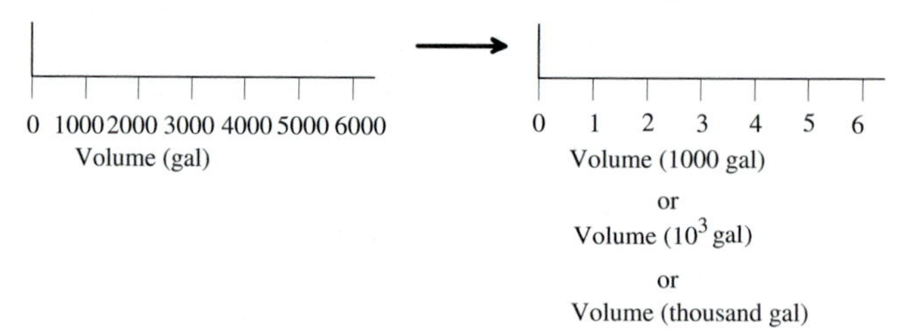

The second convention is to take the actual value and multiply it by an appropriate number so that the reported value has fewer digits.

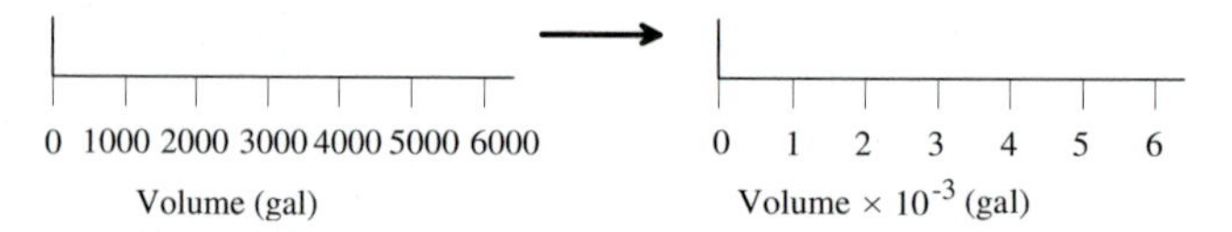

Note that here we are multiplying the actual volume by ten to the **minus** three so that the number reported on the axis is three orders of magnitude smaller. The natural tendency is for the reader to take the reported value and multiply by 10^{-3} giving a "milligallon" (if such a unit existed). The reader would be wrong by 6 orders of magnitude! Because these two conventions are so similar, it is easy for the reader (and author) to be confused. It is best to check the number from the context and determine if it makes sense. If an error were made, it would be huge; errors of this magnitude can usually be caught using good thinking.

Another problem often encountered in graphs is how to plot numbers that span many orders of magnitude. For example, what if you wanted to have the following numbers on the abscissa (x-axis):

2, 23, 467, 3876, and 48,967

This problem is solved by using a logarithmic scale rather than a linear scale.

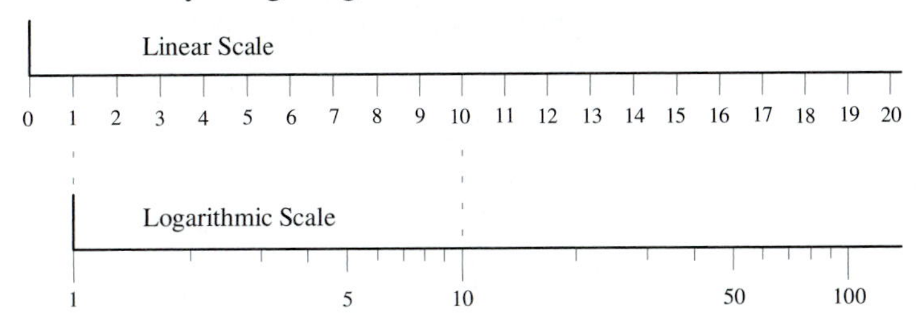

This particular logarithmic scale has 2 orders of magnitude, so it is called a *two-cycle* log scale. If it had 3 orders of magnitude, it would be called a *three-cycle* log scale. Notice that the log scale has no zero; it occurs at minus infinity on the linear scale.

Data points are plotted with symbols. The following symbols are commonly used:

Complex symbols such as these should be avoided:

No

The reader must look at each complex symbol very carefully to distinguish between them. This is difficult because graphs are often printed in a small format. Also, many photocopiers blur the symbols. An open circle with a dot in it could easily become a closed circle when blurred, thus leading to confusion. The size of the symbols is also very important. They must be large enough for the reader to easily distinguish them, but not so large that they run into each other.

A different symbol is used for each data set. For example, the car speed data were taken using three different road grades, so it would be appropriate to use three different symbols, one for each road grade.

The data points are often connected together with lines. Different line styles are commonly encountered:

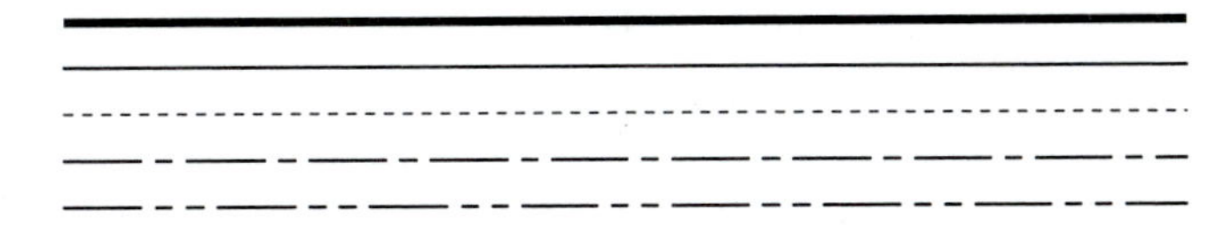

Although the line style may also be used to differentiate data sets, the preferred procedure is to differentiate data sets with different data point symbols and connect the symbols with solid lines of uniform width. It should be noted that lines must not penetrate into open symbols, because they could easily be mistaken for closed symbols.

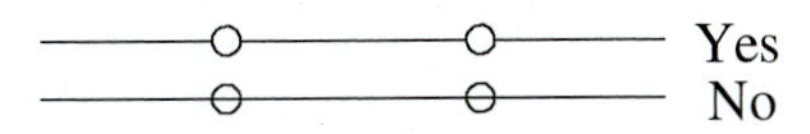

The meanings of the symbols or lines must be identified on the graph. This is generally done in one of three ways: (1) in the figure title, (2) in a legend, or (3) adjacent to the lines. The third method is preferred, because the reader can instantly identify the meaning of the symbols/lines without having to make a connection between the symbols listed in the legend or figure title and those presented in the graph.

Data may be categorized as: observed, empirical, or theoretical. *Observed* data are often simply presented without an attempt to smooth them or correlate them with a mathematical model [Figure 8.2(a)]. *Empirical* data are presented with a smooth line, which may be determined by a mathematical model, or perhaps it is just the author's best judgment of where the data points would have fallen had there been no error in the experiment [Figure 8.2(b)]. *Theoretical* data are generated by mathematical models [Figure 8.2(c)]. Note that data points are shown with both observed and empirical data but **not** with theoretical data. No data points are indicated with theoretical data because the calculated points are completely arbitrary and of no interest to the reader.

FIGURE 8.2
Data for a chemical reaction: (a) observed, (b) empirical, and (c) theoretical.

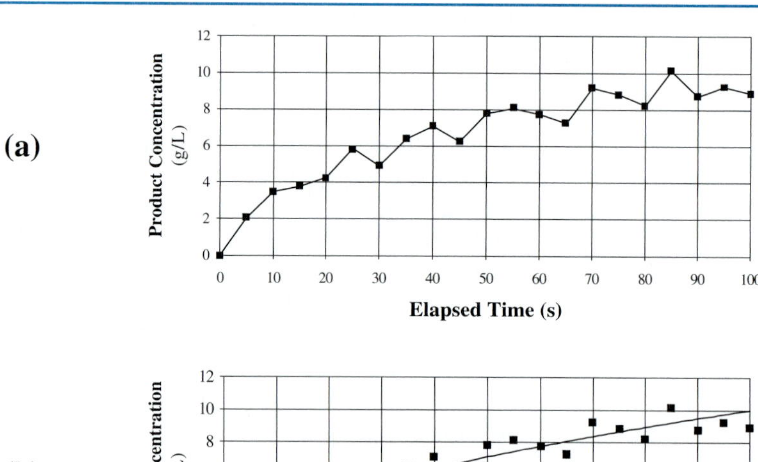

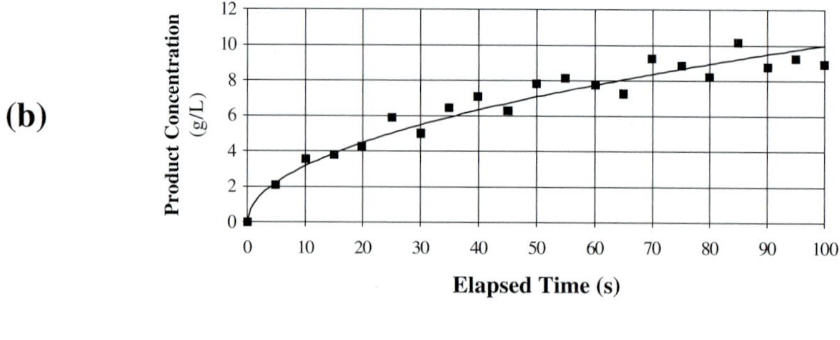

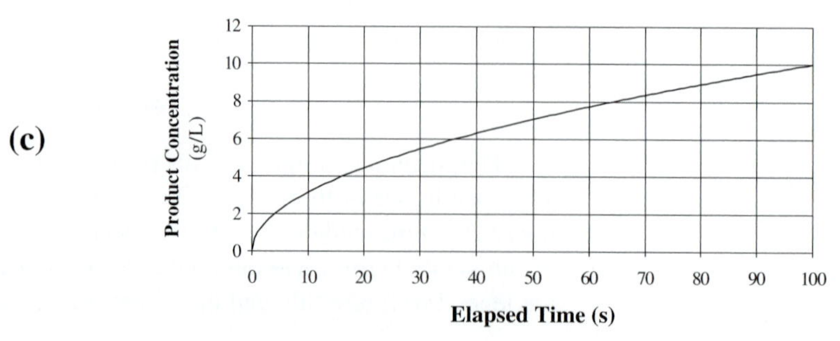

8.4 LINEAR EQUATIONS

Figure 8.3 shows that the two distinct points (x_1, y_1) and (x_2, y_2) establish a straight line. The point (x, y) is an arbitrary point on the line.

The **slope,** m, of this line is defined as "rise over run," or

$$m = \frac{y_2 - y_1}{x_2 - x_1} \tag{8-2}$$

Because all the quantities in this equation are known, the slope can be calculated. (*Note*: This equation is valid only for $x_2 \neq x_1$ to avoid division by zero. If $x_2 = x_1$, then the line is vertical, with the equation $x = x_1 = x_2$.)

The equation for the slope may also be written by using the arbitrary point (x, y):

$$m = \frac{y - y_1}{x - x_1} \tag{8-3}$$

Both sides of this equation may be multiplied by $(x - x_1)$, yielding

$$y - y_1 = m(x - x_1) = mx - mx_1 \tag{8-4}$$

$$y = mx + y_1 - mx_1 \tag{8-5}$$

If the constant b is defined as $(y_1 - mx_1)$, this equation becomes

$$y = mx + b \tag{8-6}$$

The constant b is interpreted as the y-**intercept** because $x = 0$ when $y = b$. The *x-intercept, a,* is where $y = 0$. From Equation 8-6, it is easy to show that

$$a = -\frac{b}{m} \tag{8-7}$$

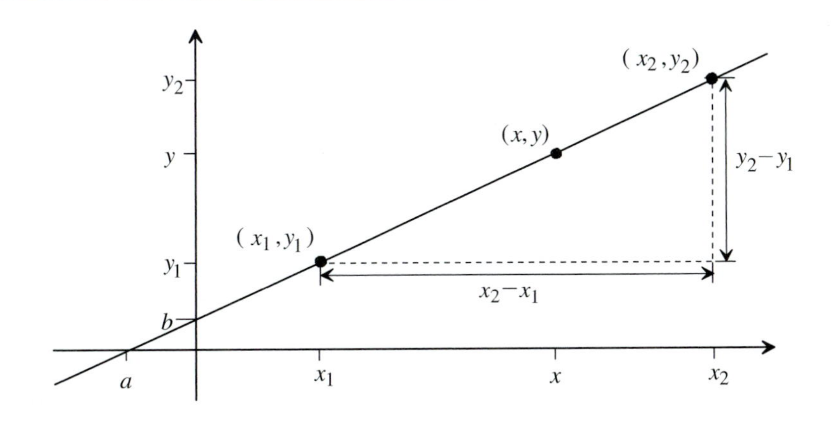

FIGURE 8.3
The straight line established by points (x_1, y_1) and (x_2, y_2).

8.5 POWER EQUATIONS

A **power equation** has the form

$$y = kx^m \tag{8-8}$$

If a logarithm is taken of both sides, then the power equation becomes linear:

$$\log y = \log (kx^m) = \log (x^m k)$$

$$\log y = \log x^m + \log k$$

$$\log y = m \log x + \log k \tag{8-9}$$

Thus a plot of $\log y$ versus $\log x$ gives a straight line with a slope m and y-intercept $\log k$, which is analogous to b in Equation 8-6. One can derive this linear equation using any desired base (2, e, or 10).

If the exponent m is positive, then the power equation plots as a *parabola.* Figure 8.4(a) shows a plot of

$$y = 2x^{0.5} \tag{8-10}$$

using linear axes. This equation becomes linear when logarithms are taken of both sides:

$$\log y = 0.5 \log x + \log 2 \tag{8-11}$$

The following table shows some selected values of x and y, along with the corresponding logarithms.

x	y	$\log x$	$\log y$
1	2.000	0.000	0.301
2	2.828	0.301	0.452
3	3.464	0.477	0.540
5	4.472	0.699	0.651
10	6.325	1.000	0.801
25	10.000	1.398	1.000

When plotting these data, we have a choice. We may plot y versus x directly on a **log-log graph** [Figure 8.4(b)], or we may plot $\log y$ versus $\log x$ on a **rectilinear graph** [Figure 8.4(c)]. The advantage of a log-log graph is that x and y may be read directly from the axes. Also, it eliminates the need to calculate the logarithms; in effect, the logarithmic axis is doing the calculation for you. **Unfortunately, the slope of the log-log graph is not meaningful.** (See this for yourself. Determine the slope at two places on the line, and you will see that they differ.) The advantage of the rectilinear graph is that the slope is meaningful. A rectilinear graph is particularly useful for plotting experimental data where the exponent m will be determined from the measured slope and the constant k will be determined from the measured y-intercept.

Astute readers will notice that in Figure 8.4, "data" points are plotted from the **calculated** numbers in the table. Properly, as stated in Section 8.3, data points should not

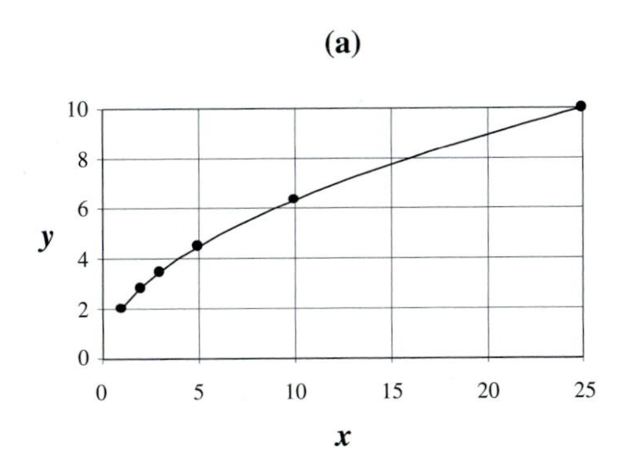

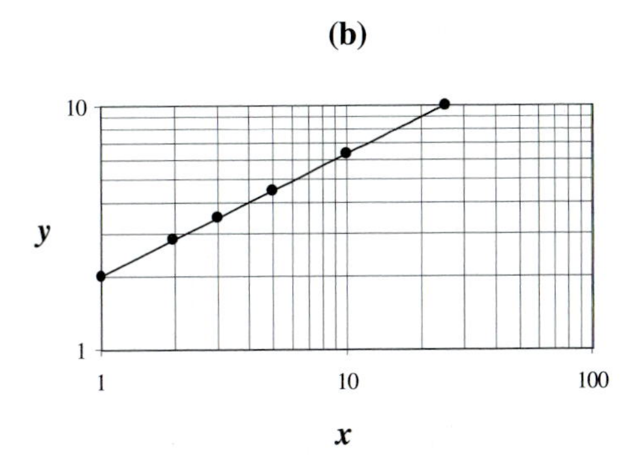

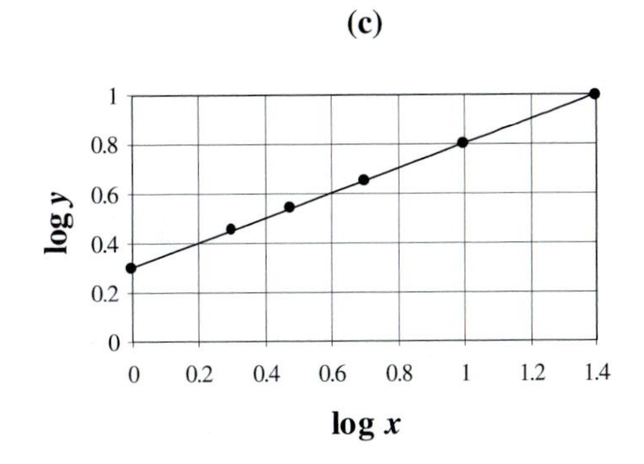

FIGURE 8.4
Plots of the parabolic equation $y = 2x^{0.5}$.

be indicated because calculated numbers are not really data. Here, for teaching purposes, we violate the rule to show you how the numbers in the table plot on the figure.

If the exponent m is negative, then the power equation plots as a *hyperbola.* Figure 8.5(a) shows a plot of

$$y = 10x^{-0.8} \tag{8-12}$$

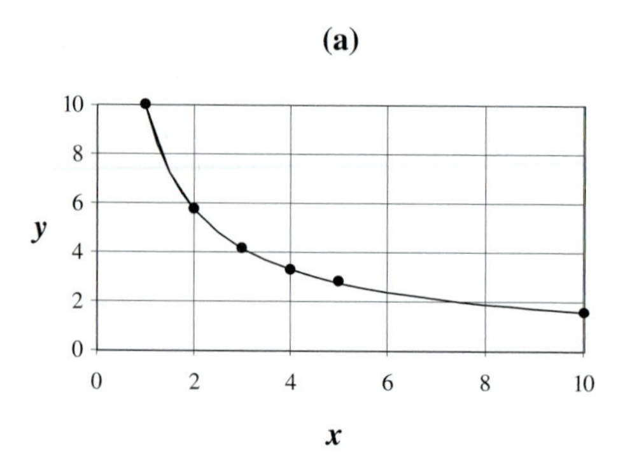

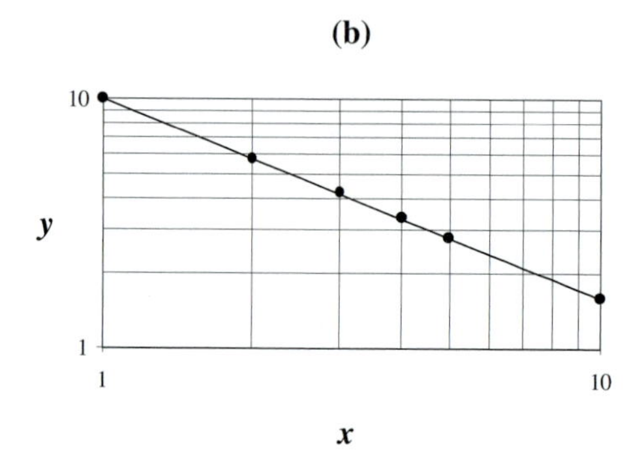

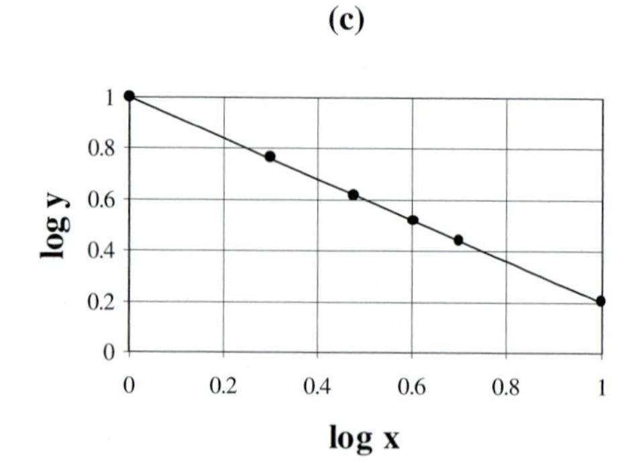

FIGURE 8.5
Plots of the hyperbolic equation $y = 10x^{-0.8}$.

using linear axes. This equation becomes linear when logarithms are taken of both sides:

$$\log y = -0.8 \log x + \log 10 \tag{8-13}$$

The following table shows some selected values of x and y, and the corresponding logarithms.

x	y	$\log x$	$\log y$
1	10.000	0.000	1.000
2	5.743	0.301	0.759
3	4.152	0.477	0.618
4	3.299	0.602	0.518
5	2.759	0.699	0.441
10	1.585	1.000	0.200

Figure 8.5(b) shows y versus x on a log-log graph, and Figure 8.5(c) shows $\log y$ versus $\log x$ on a rectilinear graph.

8.6 EXPONENTIAL EQUATIONS

An **exponential equation** has the form

$$y = kB^{mx} \tag{8-14}$$

where B is the desired base (e.g., 2, e, or 10). Assuming base 10 is used, this equation becomes

$$y = k10^{mx} \tag{8-15}$$

Logarithms (base 10) are taken of both sides to give a linear equation:

$$\log y = \log (k10^{mx}) = \log (10^{mx}k)$$

$$\log y = \log 10^{mx} + \log k$$

$$\log y = mx + \log k \tag{8-16}$$

Thus, a plot of $\log y$ versus x gives a straight line with slope m and intercept $\log k$, which is analogous to b in Equation 8-6.

Figure 8.6(a) shows a plot of

$$y = 6 \times 10^{-0.5x} \tag{8-17}$$

using linear axes. It becomes linear when logarithms are taken of both sides:

$$\log y = -0.5x + \log 6 \tag{8-18}$$

The following table shows some selected values of x and y, and the corresponding value of $\log y$.

x	y	$\log y$
0	6.000	0.778
1	1.897	0.278
2	0.600	−0.222
3	0.190	−0.722
4	0.060	−1.222
5	0.019	−1.722

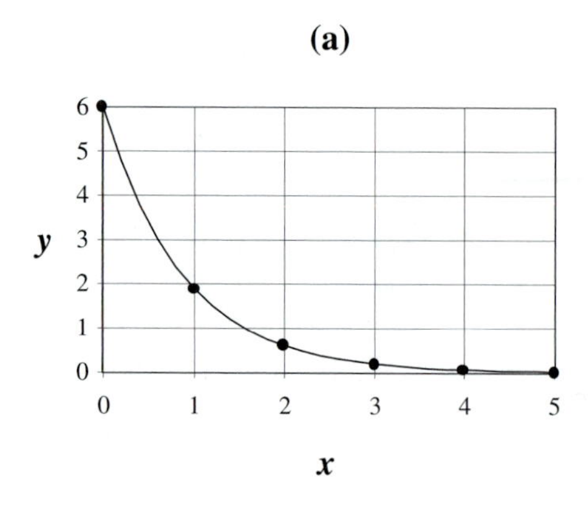

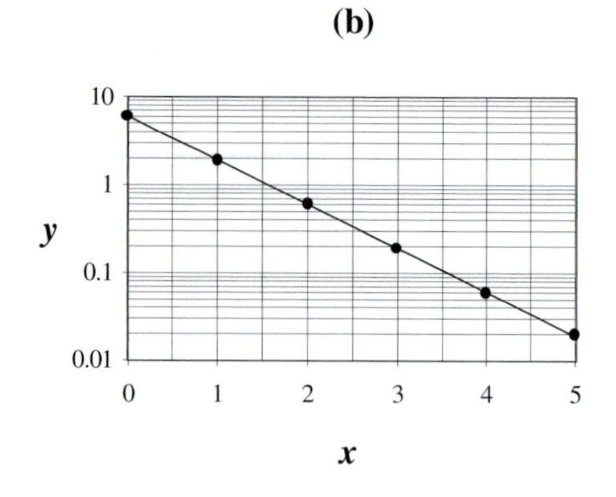

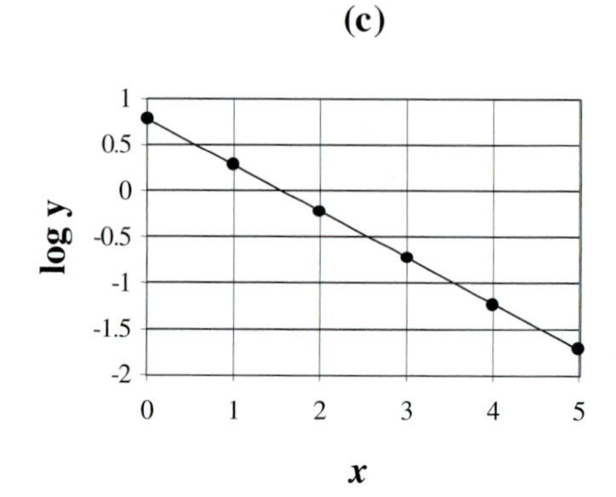

FIGURE 8.6
Plots of the exponential equation $y = 6 \times 10^{-0.5x}$.

Figure 8.6(b) shows y versus x on a **semilog graph,** and Figure 8.6(c) shows log y versus x on a rectilinear graph. The semilog graph has the advantages that y can be read directly and that there is no need to calculate log y because the axis does the calculation. However, the slope is meaningless (as with log-log graphs). If experimental data are being plotted to determine the slope and intercept, it is necessary to plot the data on a rectilinear graph.

8.7 TRANSFORMING NONLINEAR EQUATIONS INTO LINEAR EQUATIONS

To an engineer, there is something marvelous about a straight line. One can always argue about a curve; does it fit this equation or that? But if the data plot is a straight line, there is no question; a straight line is a straight line. Therefore, engineers often manipulate nonlinear equations into a linear form, as you have already seen with the power and exponential equations.

Table 8.2 shows some examples of nonlinear equations that may be transformed into linear equations through appropriate manipulations.

TABLE 8.2
Transforming nonlinear equations into linear equations

	Redefined Equation				
Original Equation	**y**	**m**	**x**	**b**	**Graph**
$c = a \sin q + d$	c	a	$\sin q$	d	
$c = aq^2 + f/g$	c	a	q^2	f/g	
$\frac{1}{c} = \frac{q^2 - 3}{a} + f/g$	$\frac{1}{c}$	$\frac{1}{a}$	$q^2 - 3$	f/g	
$c = \frac{1}{aq^2 + f}$	$\frac{1}{c}$	a	q^2	f	

EXAMPLE 8.1

Problem Statement: If a liquid and vapor coexist in the same vessel and come to equilibrium (i.e., the temperature and pressure are the same everywhere in the vessel and do not change with time), the pressure in the vessel is called the **vapor pressure** (Figure 8.7).

A nonlinear equation predicts the vapor pressure,

$$P = 10^{A - B/T} \tag{8-19}$$

where P is the vapor pressure, T is the absolute temperature, and A and B are constants. Figure 8.8(a) shows vapor pressure data for water (steam) as a function of temperature. Manipulate this equation so that it plots linearly.

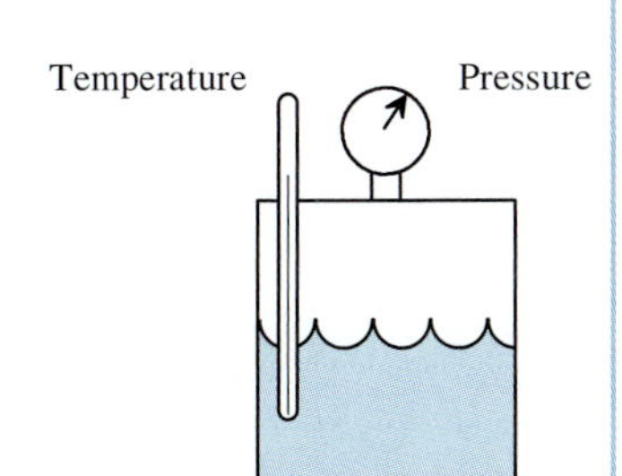

FIGURE 8.7
Vapor pressure exerted by a liquid.

Solution: We can transform this equation into a linear equation by taking the logarithm of both sides, writing

$$\log P = A - \frac{B}{T} = A - B\left(\frac{1}{T}\right) \tag{8-20}$$

Thus, a plot of $\log P$ versus $1/T$ will be linear [Figure 8.8(b) and (c)] with slope $-B$ and intercept A.

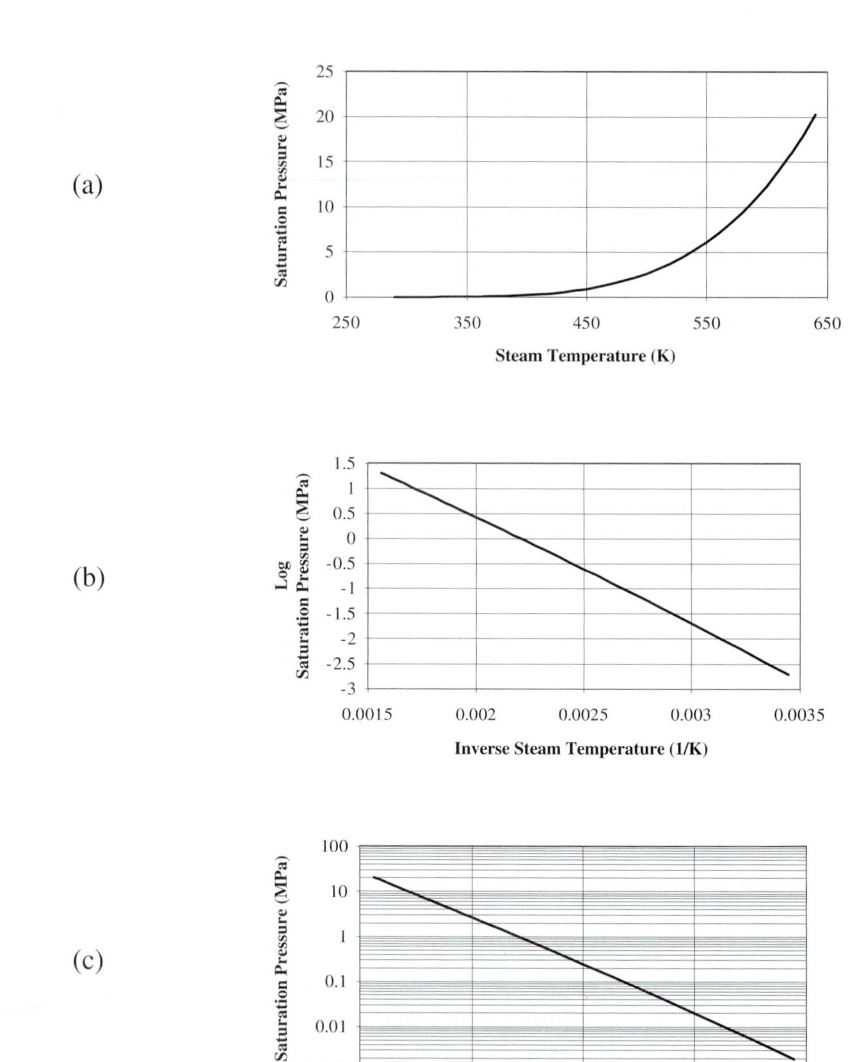

FIGURE 8.8
Vapor pressure of water (steam).

8.8 INTERPOLATION AND EXTRAPOLATION

Interpolation is extending between the data points and **extrapolation** is extending beyond the data points (Figure 8.9). The smooth curve drawn between the data points is actually an interpolation, because there are no data between the points. Provided there are a sufficiently large number of closely spaced data points, interpolation is safe. Extrapolation, on the other hand, can be quite risky, particularly if the extrapolation extends far beyond the data.

Linear interpolation approximates a curve with a straight line (Figure 8.10). A straight line passes through the points (x_1, y_1) and (x_2, y_2) which are on the curve. Provided

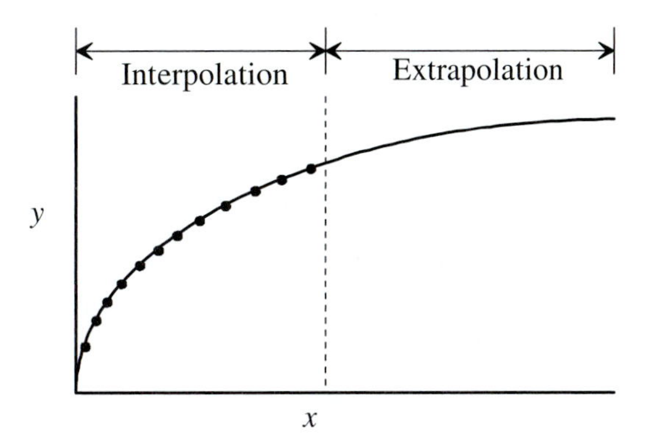

FIGURE 8.9
Interpolation and extrapolation.

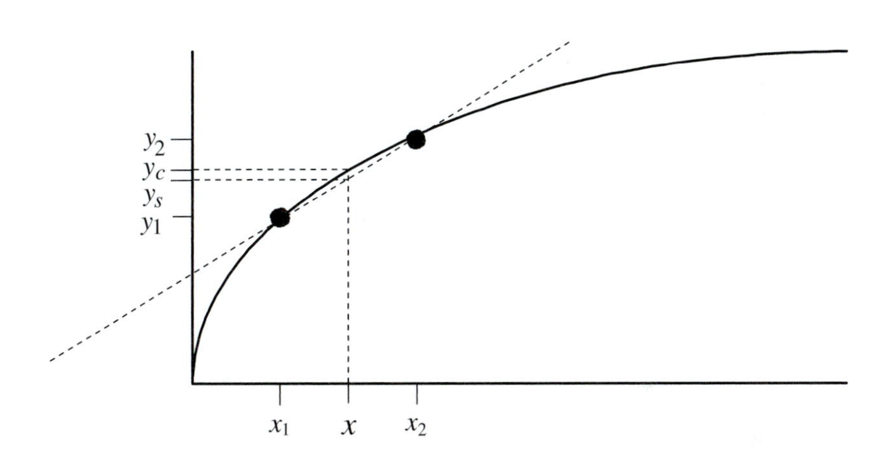

FIGURE 8.10
Linear interpolation.

these points are close to each other and the curve is continuous, the straight line approximates the curve well. We can be more explicit and say that for an arbitrary value of x that lies between x_1 and x_2, the corresponding value of y that lies on the curve (y_c) is very close to the corresponding value of y that lies on the straight line (y_s). The values for y that lie on the straight line are easily calculated from the equation for a straight line,

$$y_s = mx + b \tag{8-21}$$

The slope m is easily determined from the two points that lie on the curve, by

$$m = \frac{y_2 - y_1}{x_2 - x_1} \tag{8-22}$$

The y-intercept is determined by substituting this expression for the slope and solving for b at the known point (x_1, y_1).

$$y_1 = \left(\frac{y_2 - y_1}{x_2 - x_1}\right)x_1 + b \tag{8-23}$$

$$b = y_1 - \left(\frac{y_2 - y_1}{x_2 - x_1}\right)x_1 \tag{8-24}$$

Because we now have equations for the slope and intercept in terms of known quantities, we can substitute these into the equation for a line and derive a formula for y_s in terms of an arbitrary x.

$$y_s = \left(\frac{y_2 - y_1}{x_2 - x_1}\right)x + \left(y_1 - \frac{y_2 - y_1}{x_2 - x_1}x_1\right) \tag{8-25}$$

$$y_s = y_1 + \left(\frac{y_2 - y_1}{x_2 - x_1}\right)(x - x_1) \tag{8-26}$$

This final expression allows calculation of y_s, which is approximately equal to the desired value y_c.

Linear interpolation may also be performed on tabulated data. Suppose there is a table that has data in the range of interest, but the specific value you are seeking is not listed. You know the independent variable x and are seeking the dependent variable y.

Independent Variable — Dependent Variable

x_1 — y_1

x (fractional difference) — y (same fractional difference)

x_2 — y_2

x_3 — y_3

⋮ — ⋮

Using linear interpolation, we can say the fractional difference between the dependent variables is the same as the fractional difference between the independent variables. Mathematically, this is stated as

$$\text{Fractional difference} = \frac{x - x_1}{x_2 - x_1} = \frac{y - y_1}{y_2 - y_1} \tag{8-27}$$

This may be solved explicitly for y:

$$y = y_1 + \left(\frac{y_2 - y_1}{x_2 - x_1}\right)(x - x_1) \tag{8-28}$$

Note that this equation is identical to the one developed from the graphical approach to linear interpolation (Equation 8-26).

EXAMPLE 8.2

Problem Statement: The *steam tables* list properties of water (steam) at different temperatures and pressures. Because water is one of the most common materials on earth, these tables are widely used by a variety of engineers. Among the properties listed in the steam tables is the vapor pressure of water at various temperatures. Suppose you wish to know the vapor pressure of water at 114.7°F. Although this temperature is not explicitly listed in the table, estimate it by linear interpolation.

Temperature (°F)	Pressure (psia)
112	1.350
114	1.429
116	1.512
118	1.600
120	1.692

Solution: $$P = 1.429 + \frac{1.512 - 1.429}{116 - 114}(114.7 - 114) = 1.458 \text{ psia}$$

8.9 LINEAR REGRESSION

In mathematics, we are normally given a formula from which we calculate numbers. If we reverse this (i.e., determine the formula from the numbers), the process is called **regression,** meaning "going backward." If the formula we seek is the equation of a straight line, then the process is called **linear regression.** (*Nonlinear regression* seeks an equation other than that for a straight line. This is an advanced topic beyond the scope of this book. A number of commercially available computer programs are able to perform nonlinear regression.)

A linear regression problem may be stated as follows: Given a data set, what are the slope and y-intercept that best describe these data? Two approaches are generally taken, the *method of selected points* and **least-squares linear regression.**

One starts the *method of selected points* by plotting the data (Figure 8.11) and then manually drawing a line that best describes the data as judged by the person analyzing the data. In principal, the person analyzing the data is attempting to draw a line that is close to **all** the data points, not just a few. Two arbitrary points at opposite ends of the line are selected. (*Note*: These points are **not** necessarily data points; they are merely two convenient points that fall on the line.) These two selected points (x_1, y_1) and (x_2, y_2) are substituted into Equations 8-22 and 8-24 so that the slope and y-intercept may be calculated.

The problem with the method of selected points is that it relies on the judgment of the person analyzing the data. If 100 people were to analyze the data, there likely would be 100 different slopes and 100 different y-intercepts. The next method is not affected by personal bias; all 100 people would produce the same "best line" and the same slope and y-intercept.

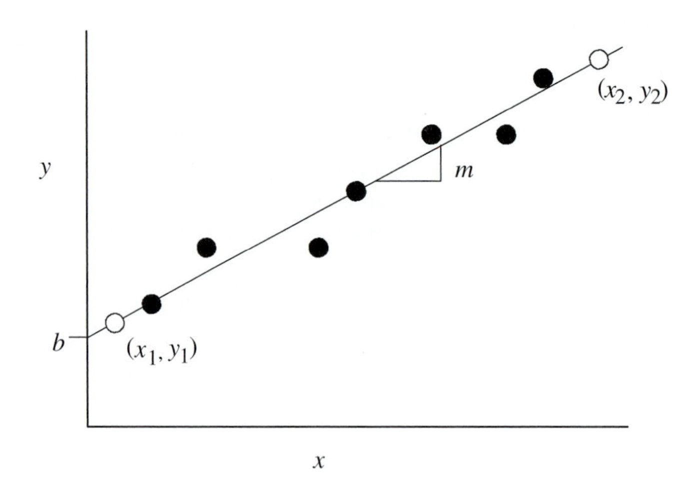

FIGURE 8.11
Method of selected points.

Least-squares linear regression uses a rigorous mathematical procedure to find a line that is close to all the data points (Figure 8.12). The difference between the actual data point y_i and the point predicted by the straight line y_s is the residual d_i:

$$d_i = y_i - y_s \tag{8-29}$$

$$d_i = y_i - (mx_i + b) \tag{8-30}$$

The residual will assume both positive and negative values, depending on whether the data point is above or below the line; however, the square of the residual d_i^2 is always a positive number. The "best" straight line that describes the data would have the smallest possible d_i^2. This is rigorously stated as "find m and b such that the sum d_i^2 is a minimum,"

$$\text{Sum} = \sum_{i=1}^{n} d_i^2 = \sum_{i=1}^{n} [y_i - (mx_i + b)]^2 \tag{8-31}$$

where n is the number of data points. Performing this minimization is the subject of differential calculus and is beyond the scope of this book. Here the results are simply presented for two cases:

1. Best line: $y = mx + b$

$$m = \frac{n(\Sigma x_i y_i) - (\Sigma x_i)(\Sigma y_i)}{n(\Sigma x_i^2) - (\Sigma x_i)^2} \tag{8-32}$$

$$b = \frac{\Sigma y_i - m(\Sigma x_i)}{n} \tag{8-33}$$

2. Best line through the origin: $y = mx$

$$m = \frac{\Sigma x_i y_i}{\Sigma x_i^2} \tag{8-34}$$

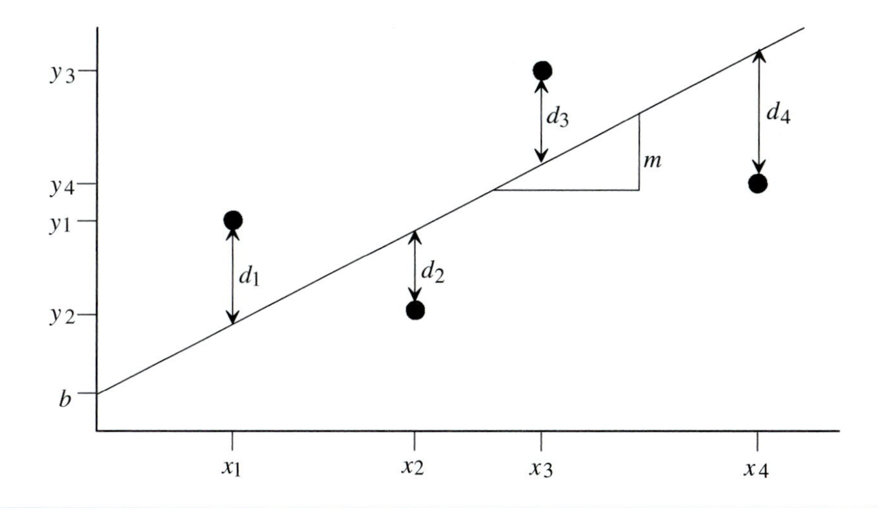

FIGURE 8.12
Least-squares linear regression.

The *correlation coefficient* r is used to determine how well the data fit a straight line. It is defined as

$$r \equiv \pm\sqrt{1 - \frac{\Sigma(y_i - y_s)^2}{\Sigma(y_i - \bar{y})^2}} \tag{8-35}$$

where $\bar{y}$ is the mean value of y, defined as

$$\bar{y} \equiv \frac{\Sigma y_i}{n} \tag{8-36}$$

In Equation 8-35, the positive sign is used for a positively sloped line and the negative sign is used for a negatively sloped line. If all the data lie precisely on the line, then $\Sigma(y_i - y_s)^2 = 0$ and $r = 1$ (positively sloped line) or $r = -1$ (negatively sloped line). If the data are scattered randomly and do not fit a straight line, then $\Sigma(y_i - y_s)^2 = \Sigma(y_i - \bar{y})^2$ and $r = 0$.

Although Equation 8-35 is useful for understanding the meaning of r, it is inconvenient to use. The following equation is an alternative version of r that is more convenient:

$$r \equiv \frac{n(\Sigma x_i y_i) - (\Sigma x_i)(\Sigma y_i)}{\sqrt{n(\Sigma x_i^2) - (\Sigma x_i)^2}\sqrt{n(\Sigma y_i^2) - (\Sigma y_i)^2}} \tag{8-37}$$

EXAMPLE 8.3

Problem Statement: A group of engineering students who study together decide they must use their time more efficiently. Because they have many classes, they must allocate their study time to each class in an optimal manner. They decide that an equation that predicts their exam grade on the basis of number of hours studied will help them allocate their time more efficiently. They prepare Table 8.3 on the basis of their performance on the last exam. What linear equation correlates their performance? What is the correlation coefficient?

TABLE 8.3
Effect of studying on grades

Student	Hours Studied (x_i)	Grade (y_i)	(y_i^2)	(x_iy_i)	(x_i^2)
1	5	63	3969	315	25
2	10	91	8281	910	100
3	2	41	1681	82	4
4	8	75	5625	600	64
5	6	69	4761	414	36
6	12	95	9025	1140	144
7	0	32	1024	0	0
8	4	50	2500	200	16
9	8	80	6400	640	64
$n = 9$	$\sum x_i = 55$	$\sum y_i = 596$	$\sum y_i^2 = 43{,}266$	$\sum x_i y_i = 4301$	$\sum x_i^2 = 453$

Solution:

$$m = \frac{9(4301) - (55)(596)}{9(453) - (55)^2} = 5.64 \qquad \text{(Equation 8-32)}$$

$$b = \frac{(596) - 5.64(55)}{9} = 31.8 \qquad \text{(Equation 8-33)}$$

$$r = \frac{9(4301) - (55)(596)}{\sqrt{9(453) - (55)^2}\sqrt{9(43{,}266) - (596)^2}} = 0.98878 \qquad \text{(Equation 8-37)}$$

As you can see, this calculation is rather tedious. Fortunately, many calculators with statistical packages can perform linear regression with just a few keystrokes.

EXAMPLE 8.4

Problem Statement: An engineer determines the spring constant k of a spring by applying a force F, causing it to compress (shorten) by a displacement d. The engineer correlates the data with the equation

$$F = kd \qquad \text{(8-38)}$$

The engineer will therefore determine the best line that goes through the origin. She collected the data in Table 8.4. What is the best value for the spring constant? What is the correlation coefficient?

Solution:

$$k = m = \frac{2150.2}{204.0} = 10.5 \text{ kN/mm} \qquad \text{(Equation 8-34)}$$

TABLE 8.4
Spring displacement due to applied force

Measurement	Displacement† (mm) (x_i)	Force† (kN) (y_i)	(y_i^2)	(x_iy_i)	(y_i^2)
1	1.00	10.3	106.09	10.3	1.00
2	2.00	20.8	432.64	41.6	4.00
3	3.00	31.3	979.69	93.9	9.00
4	4.00	42.1	1772.41	168.4	16.0
5	5.00	52.7	2777.29	263.5	25.0
6	6.00	63.2	3994.24	379.2	36.0
7	7.00	73.9	5461.21	517.3	49.0
8	8.00	84.5	7140.25	676.0	64.0
$n = 9$	$\sum x_i = 36.00$	$\sum y_i = 378.8$	$\sum y_i^2 = 22633.82$	$\sum x_i y_i = 2150.2$	$\sum x_i^2 = 204.0$

† Experimentally, force is the independent variable and displacement is the dependent variable. However, Equation 8-38 treats displacement as the independent variable and force as the dependent variable; therefore, displacement is shown in the left column and force in the right column.

$$r = \frac{8(2150.2) - (36.00)(378.8)}{\sqrt{8(204.0) - (36.00)^2}\sqrt{8(22{,}663.82) - (378.8)^2}} = 0.999996 \qquad \text{(Equation 8-37)}$$

8.10 SUMMARY

Tables list corresponding independent and dependent variables on the same row. A properly constructed table has a descriptive title and columns that are identified with both heads and units. Although tables are very useful for putting data into a computer or calculator, they are difficult to comprehend and interpret; for this, graphs are required.

A properly constructed graph has a descriptive title and each axis is identified with a label and units. Engineers typically construct three types of graphs: rectilinear, semilog, and log-log. A linear equation plots as a straight line on rectilinear graphs, an exponential equation plots as a straight line on semilog graphs, and a power equation plots as a straight line on log-log graphs. When regressing data (i.e., determining an equation from data), it is best to use rectilinear graphs. If the data fit a linear equation, a plot of y versus x is straight; if the data fit an exponential equation, a plot of log y versus x is straight; and if the data fit a power equation, a plot of log y versus log x is straight. (*Note:* Alternatively, the natural logarithms may be used.) Many other equations will plot linearly on rectilinear graphs by properly manipulating the variables. For example, log P versus $1/T$ will plot linearly if P is the absolute vapor pressure and T is the absolute temperature of a liquid in equilibrium with vapor.

Interpolation is the process of extending between data points and extrapolation is the process of extending beyond the data points. Although extrapolation is fairly risky, interpolation is generally safe. Most commonly, linear interpolation is used.

Linear regression is a process of finding the best straight line that fits the data. The method of selected points may be used to "eyeball" the data and find the best straight line. More rigorous is the method of least-squares linear regression.

Further Readings

Eide, A. R.; R. D. Jenison; L. H. Mashaw; and L. L. Northup. *Engineering Fundamentals and Problem Solving.* 3rd ed. New York: McGraw-Hill, 1997.

Felder, R. M., and R. W. Rousseau. *Elementary Principles of Chemical Processes.* New York: Wiley, 1986.

PROBLEMS

8.1 A *resistor* is a device that converts electrical energy into heat by passing electric current through a poor electrical conductor (e.g., carbon) rather than a good electrical conductor (e.g., copper). Electrons (i.e., electric current) will flow through the resistor when a voltage, such as that from a battery, is applied (Figure 8.13).

The current I through a resistor is proportional to the amount of applied voltage V:

$$I = \left(\frac{1}{R}\right)V$$

where the inverse resistance, $1/R$, is the proportionality constant.

A single battery cell has a 1.5-volt output. By placing the battery cells in series (by stacking them end to end), it is possible to obtain voltages that are multiples of 1.5 volts. The voltages and currents in Table 8.5 were measured for a resistor with an unknown resistance.

Plot the data using rectilinear axes with current on the y-axis and voltage on the x-axis. The slope of this line is the inverse resistance, $1/R$. Determine the slope by two methods:

(a) Method of selected points.

(b) Least-squares linear regression.

What is the value of the resistance? (*Note:* The unit of electrical resistance is the ohm (Ω), which is identical to volt/ampere.) What is the correlation coefficient? You may use a spreadsheet if you wish. (*Hint:* Current flows **only** when voltage is applied.)

8.2 As a car gets heavier, fuel consumption increases because of greater friction losses in wheels, and because more kinetic energy is converted into heat each time the car brakes to a stop. In response to the energy crisis in the 1970s, automakers started a downsizing program in which cars were made lighter by decreasing their size and substituting plastic and aluminum for steel. Of course, other factors affect fuel consumption, such as aerodynamic shape and engine design.

The data in Table 8.6 were obtained from a random sampling of automobiles. Using rectilinear axes, plot fuel consumption versus car weight. Determine the slope and y-intercept of a straight line that correlates the data, using:

(a) Method of selected points.

(b) Least-squares linear regression.

FIGURE 8.13

Electric circuit in which current flows through a resistor.

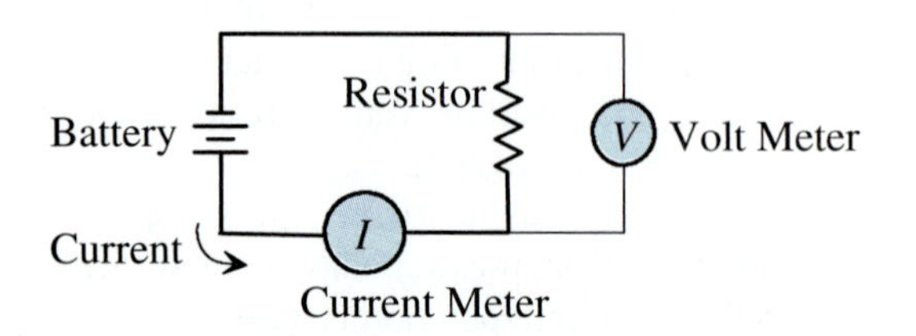

TABLE 8.5

Voltage drop through a resistor

Applied Voltage (volt)	Current (ampere)
1.5	0.11
3.0	0.26
4.5	0.35
6.0	0.50
7.5	0.61
9.0	0.68
10.5	0.81
12.0	0.92
13.5	1.02

TABLE 8.6
Effect of automobile mass on fuel consumption

Automobile Mass (lb_m)	Fuel Consumption (miles per gallon)
2534	24.3
3023	15.9
2294	30.7
3797	12.5
2876	20.4
2382	35.8
3498	22.8
2475	40.3
2103	45.5

What is the correlation coefficient? You may use a spreadsheet if you wish.

8.3 The decay of a radioactive element is described by the equation

$$A = A_0 e^{-kt}$$

where A is the amount at time t, A_0 is the amount at time zero, and k is the decay constant. The data in Table 8.7 were taken for the highly radioactive element balonium-245. Plot the following:

(a) Balonium versus time using rectilinear axes.
(b) Napierian logarithm of balonium versus time using rectilinear axes (calculate correlation coefficient).
(c) Balonium versus time on semilog paper.

Determine the decay constant, k. (*Note:* The units of k are days^{-1}.) You may use a spreadsheet if you wish.

The *half-life* is defined as the time it takes for half of the balonium-245 to decay. Alternatively, we could say it is the time at which half the balonium-245 still remains, or $A/A_0 = 0.5$. What is the half-life of balonium-245?

TABLE 8.7
Decay of balonium-245

Time (day)	Balonium-245 (grams)
0.00	45.3
0.05	30.4
0.10	20.9
0.15	14.1
0.20	9.4
0.25	6.1
0.30	4.1
0.35	3.0
0.40	2.0

8.4 John, a skeptical engineering student, reads in a physics book that "any object, regardless of its mass, has the same acceleration a_0 when it falls under the influence of earth's gravity in the absence of other forces. Provided the object starts from rest, the position x as a function of time t is given by the equation $x = \frac{1}{2} a_0 t^2$." The engineer doubts this statement because he knows that if you drop a rock and a leaf from the same height, the rock will hit the ground first.

To test the claim made in the physics book, he decides to do an experiment. He enlists the help of six friends. After much discussion, they decide that the most practical way to make the measurements is to drop a baseball and a cannonball down the center shaft of the stairwell in their dormitory. Each friend will stand on a different floor while holding a stopwatch. When John releases the ball from the top floor, he yells "NOW," and each friend starts his stopwatch. Each friend agrees to stop his watch when the ball passes by his feet. (They were going to stop the watch when the ball passed at eye level, but because they are all different heights, they realized this would introduce systematic error into their experiment.) With a tape measure, they determine that the floors are separated by twelve feet. After some practice, they take the data in Table 8.8.

TABLE 8.8
Distance traveled by dropping a baseball and cannonball

Distance (ft)	Baseball time (s)	Cannonball time (s)
12	0.85	0.87
24	1.24	1.22
36	1.43	1.55
48	1.73	1.75
60	1.93	1.93
72	2.20	2.15

Using John's data, prove to yourself that the physics equation is correct by doing the following:

(a) Plot distance versus time using rectilinear axes.
(b) Plot distance versus time on log-log paper.
(c) Plot $\log_{10}$ distance versus $\log_{10}$ time using rectilinear axes (calculate the correlation coefficient for the baseball data and cannonball data).
(d) Plot ln distance versus ln time using rectilinear axes.
(e) Determine the acceleration and the exponent for time using the method of selected points.
(f) Determine the acceleration and the exponent for time using least-squares linear regression.

You may use a spreadsheet to make your plots, if you wish.

On the basis of the results of the experiment, do you agree with the physics book that all objects have the same acceleration under the pull of earth's gravity regardless of their mass? If John had performed the experiment with a rock and a leaf, what results would you expect and why?

8.5 After performing the experiment described in Problem 8.4, John believes the equation given in his physics book is correct. He would like to revise his estimate of the acceleration by interpreting the data with the time exponent exactly equal to 2. He does not know how to do this, so he comes to you for help. How would you plot the data? Using this new approach, what is your revised estimate of the acceleration due to gravity?

8.6 In the following equations, y is the dependent variable, x is the independent variable, and a and b are constants. Indicate what type of graph you would use to obtain a straight line (i.e., rectilinear, semilog, log-log). Also, indicate what will be plotted on the ordinate and the abscissa.

(a) $y = ax$
(b) $1/y = ax + b$
(c) $y = (ax + b)^{-1}$
(d) $y = (ax + b)^{-2}$
(e) $y = (ax + b)^2 + 5$
(f) $y = a10^{bx}$
(g) $y = ae^{bx}$
(h) $y = a2^{bx}$
(i) $y^2 = 3 + a10^{b(x-4)}$
(j) $y = \left[3 + \left(ae^{\frac{b}{x-4}}\right)\right]^{-2}$
(k) $y = ax^b$
(l) $y = [a(x - 5)^b]^{-1}$
(m) $\sin y = ax^b$
(n) $y = a(\cos x)^{-b}$
(o) $y = [ax^b]^{-1/2}$
(p) $y = 4 - \left(\frac{1}{ax^b}\right)$

8.7 The following variables plot linearly on a log-log graph. Develop an equation that relates the variables.

(a) y versus x (Answer: ax^b where a and b are constants.)
(b) $1/y$ versus $1/x$
(c) $1/(y - 3)$ versus x
(d) $\sin y$ versus $1/x$
(e) $1/y^2$ versus x

8.8 The following variables plot linearly on a semilog graph. Develop an equation that relates the variables.

(a) y versus x (Answer: $y = ae^{bx}$ or $y = a10^{bx}$ where a and b are constants.)
(b) $1/y$ versus $1/x$
(c) $1/(y - 3)$ versus x
(d) $\sin y$ versus $1/x$
(e) $1/y^2$ versus x

8.9 Determine the equation $y = f(x)$ for each of the following cases. Report the equation in its simplest form. All of the plots are straight lines. All coordinates are indicated with the abscissa first and the ordinate second, i.e., (x, y).

(a) y versus x on a rectilinear graph passes through (5, 7) and (2, 3).
(b) $\ln y$ versus $\ln x$ on a rectilinear graph passes through (5, 7) and (2, 3).
(c) $\log y$ versus $\log x$ on a rectilinear graph passes through (5, 7) and (2, 3).
(d) y versus x on a log-log graph passes through (5, 7) and (2, 3).
(e) $\ln y$ versus x on a rectilinear graph passes through (2, 3) and (4, 6).
(f) $\log y$ versus x on a rectilinear graph passes through (2, 3) and (4, 6).
(g) y versus x on a semilog graph passes through (2, 3) and (4, 6).
(h) y^2 versus x on a rectilinear graph passes through (3, 2) and (6, 4).
(i) $\sqrt{y}$ versus x on a rectilinear graph passes through (3, 2) and (6, 4).
(j) y^2 versus x on a log-log graph passes through (3, 2) and (6, 4).
(k) $\sqrt{y}$ versus x on a log-log graph passes through (3, 2) and (6, 4).
(l) y^2 versus x on a semilog graph passes through (3, 2) and (6, 4).
(m) $\sqrt{y}$ versus x on a semilog graph passes thorugh (3, 2) and (6, 4).
(n) $(y - 2)^2$ versus x on a rectilinear graph passes through (1, 2) and (3, 4).
(o) $(y - 2)^2$ versus x on a log-log graph passes through (1, 2) and (3, 4).
(p) $(y - 2)^2$ versus x on a semilog graph passes through (1, 2) and (3, 4).

8.10 Table 8.9 shows some thermodynamic properties of water. Find the vapor pressure P, liquid and vapor specific volume $\hat{V}$, liquid and vapor specific internal energy $\hat{U}$, and liquid and vapor specific enthalpy $\hat{H}$ at 31.3°C.

8.11 By hand, or using a spreadsheet, plot the power equation $y = 5x^4$ on a rectilinear graph in the range from $x = 0$ to $x = 4.5$.

(a) On the plot with the power equation, show a straight line that passes through (2, 80) and (4, 1280). Through linear interpolation, use this straight line to estimate the value of the power equation at $x = 3$. What is the fractional error and percentage error?
(b) On another graph, plot the power equation in the range from $x = 2.5$ to $x = 3.5$. Show a straight line that passes through (2.9, 353.64) and (3.1, 461.76). Through linear interpolation, use this straight line to estimate the value of the power equation at $x = 3$. What is the fractional error and percentage error?
(c) By comparing the errors in Parts (a) and (b), what do you conclude?

8.12 Write a computer program that allows the user to input an arbitrary number of x and y values that may be correlated by the equations $y = mx + b$ or $y = mx$. The program must ask the user

TABLE 8.9
Thermodynamic properties of water

		$\hat{V}$ (m^3/kg)		$\hat{U}$ (kJ/kg)		$\hat{H}$ (kJ/kg)	
T (°C)	*P* (bar)	Liquid	Vapor	Liquid	Vapor	Liquid	Vapor
28	0.0378	0.001004	36.7	117.3	2414.0	117.3	2552.7
30	0.0424	0.001004	32.9	125.7	2416.7	125.7	2556.4
32	0.0475	0.001005	29.6	134.0	2419.4	134.0	2560.0
34	0.0532	0.001006	26.6	142.4	2422.1	142.4	2563.6

which equation she wants to use to correlate the data. The program will calculate the best slope and *y*-intercept (if there is one) according to least-squares linear regression. The program will also calculate the correlation coefficient. Test the program with the data from Problems 8.1 and 8.2.

8.13 Write a computer program that accomplishes the same task as Problem 8.12, but reads the data from a file rather than having it input directly by the user.

8.14 Write a computer program that linearly interpolates between two data pairs.

Glossary

abscissa The *x*-axis.
dependent variable A variable that cannot be arbitrarily selected and is determined by the independent variable.
exponential equation $y = kB^{mx}$
extrapolation Extending beyond the data points.
independent variable A variable that can be arbitrarily selected.
intercept The value where a line intersects a coordinate axis.
interpolation Extending between the data points.
least-squares linear regression A rigorous mathematical procedure that is used to find a line that best fits all the data points.
linear interpolation The approximation of a curve with a straight line.
linear regression The process of finding the equation of a straight line that best fits the data.
log-log graph A graph in which both axes are logarithmic.
ordinate The *y*-axis.
power equation $y = kx^m$
rectilinear graph A graph in which both axes are linear.
regression Going backward from the data to the equation.
semilog graph A graph in which one axis is logarithmic and the other is linear.
slope Rise over run.
vapor pressure The pressure of the vapor in equilibrium with the liquid or the solid.

Mathematical Prerequisite

- Probability (Appendix G, Mathematical Supplement)

CHAPTER 9

Statistics

Statistics encompasses scientific methods for collecting, organizing, summarizing, presenting, and analyzing data as well as making conclusions based on the data. Statistics plays a very important role in our lives. For example, without statistics, our health would be in jeopardy, as there would be no methods for identifying ineffective medicines. When a pharmaceutical company spends millions of dollars to develop a new drug, it has an obvious incentive to market this new drug to recoup its investment. Before the drug can be sold commercially, the government requires the company to demonstrate it has minimal side effects and helps cure the disease for which it is intended. An obvious way to prove drug effectiveness is to divide the people with the disease into two groups. One group receives the drug and the other receives a **placebo** (a pill with no medication). When the data are collected, there will always be a difference between the two groups of people, but are the differences statistically significant? That is, if the group receiving the drug improves relative to the placebo group, is the improvement large enough to be attributed to the drug, or is it attributable to random fluctuations? Statistics provides unbiased tools to answer this question.

Statistics is widely used in engineering. Civil engineers use statistics to determine road use. You have undoubtedly driven on a road with a rubber hose stretched across it that connects to a box at the roadside. Every vehicle that runs over the rubber hose sends a signal to the box, which accumulates the vehicle count. Armed with these data, civil engineers can project the road life and plan new roads to relieve traffic congestion.

Electrical engineers use statistics to remove noise from signals. For example, the signals sent by the *Voyager I* space probe are extremely weak (only 28.3 watts, the equivalent of a refrigerator lightbulb) and must travel long distances (about 4.5 billion miles). When it arrives at earth, the signal is only 10^{-18} watts and is cluttered with noise. (To put this power level in perspective, the 70 m diameter receiver would have to operate for about 600 years to collect the amount of heat released when a single snowflake hits the earth.) The signals are processed using statistical techniques to reveal beautiful images of distant planets and moons.

The reliability of complex systems can be analyzed by using statistical techniques. For example, nuclear engineers are very concerned with the reliability of nuclear cooling systems, because failure can cause a meltdown. By determining the reliability of each valve, pump, pipe, computer interface, and so on, they can calculate the overall reliability of the complete cooling system.

The above examples are but a few engineering applications of statistics. In fact, statistics is widely used by nearly all engineers for a variety of applications such as quality control, forecasting, and experimental design.

9.1 STATISTICAL QUALITY CONTROL

Statistical quality control uses statistical methods to assess the quality of a product or process. It is an extremely important manufacturing tool that is largely responsible for the success of the Japanese automotive industry. Although the methods were originally developed in the United States and implemented here in the early 20th century, our industry abandoned them after World War II. Because the rest of the world's industrial base was essentially destroyed in the war, there was little international competition and hence little incentive to produce quality products. Now, of course, the situation has changed. There is plenty of international competition, and to meet it, U.S. industry is now readopting these techniques. Virtually every major U.S. industry now has a quality program that incorporates aspects of statistical quality control.

Broadly, *quality* is defined as "fitness for use." However, statistical quality control defines a high-quality product as one that closely matches the design specifications. (*Note:* There is an implicit assumption that the design is fit for use.) As an example, consider a rotating engine shaft and the bearing that guides it (Figure 9.1). The bearing is slightly larger than the shaft and the gap is filled with lubricating oil. If the system is properly designed, the shaft never touches the bearing because the shaft rotation pumps oil into the gap, creating a high-pressure protective film. For this to work properly, the gap must neither be too small (causing binding) nor too large (creating a sloppy fit). In Figure 9.1, the design gap on the radius is only 0.002 in. If the shaft diameter were manufactured to be 1.497 in and the sleeve diameter were 1.499 in (both off by only 1 part in 1500), the radial gap reduces to 0.001 in, or 50% smaller than the design gap! You can see from this example that it is extremely important to maintain close manufacturing tolerances. Statistical quality control provides the methods by which measurements are taken and analyzed in order to assess quality. To manufacture a quality automobile requires that all the components (steel, paint, plastic, glass, electronics, etc.) also be of high quality (i.e., closely approach design specifications), so many industries and engineering disciplines require statistical methods.

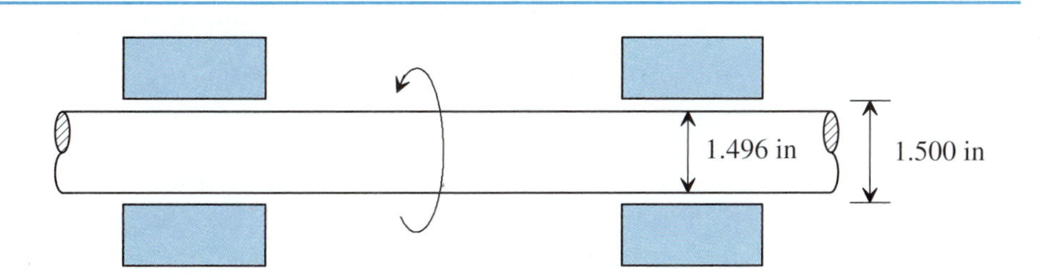

FIGURE 9.1
Rotating shaft in bearing.

9.2 SAMPLING

Statistical measurements are used to characterize a **population.** For example, a Gallup poll may be taken to determine how the U.S. population will vote in a presidential election. Because it is very costly to ask each and every American how he or she will vote, the Gallup company will select a **sample** from the general population. It is important to select a random sample that is representative of the population as a whole.

Classic sampling errors were made by the *Literary Digest* in its "straw polls" taken between 1916 and 1936. In an attempt to predict presidential election results, they mailed as many as 20 million ballots and received about 3 million returns. Because their mailing addresses were taken from sources that did not represent a true cross-section of the United States (e.g., automobile owners, telephone subscribers), and because those from the upper income levels tended to return the ballots, their predictions were wrong by as much as 39%. In the 1936 election between Roosevelt and Landon, they predicted that Landon would win; of course, Roosevelt actually won the election.

9.3 DESCRIPTIVE STATISTICS

There are two basic branches of statistics: **descriptive statistics,** which seeks only to describe data, and **statistical inference,** which seeks to make conclusions from the data. To illustrate the differences between these two, consider Table 9.1, which reports the shaft diameters made by two machines from different manufacturers.

Descriptive statistics were used to calculate the average shaft diameter produced by each machine. Statistical inference would be required to determine if Machine A is better than Machine B. Here, we will cover only descriptive statistics, which characterizes the data **range,** its **central tendency,** and its *variation.* Generally, the data first must be *sorted* before descriptive statistics can be determined.

9.3.1 Sorting Data

Raw data are data that have not been processed. Usually, raw data are in a random order, making them difficult to understand. Therefore, we typically *sort* the data from the lowest to highest value (or vice versa). Once the data are sorted, they may be categorized into *classes* on an *entry table.* Generally, the data are categorized into 5 to 20 evenly spaced

TABLE 9.1
Comparison of shaft diameters manufactured by two different machines

Shaft Diameters from Machine A (in)	Shaft Diameters from Machine B (in)
1.49650	1.49580
1.49590	1.49630
1.49620	1.49660
1.49550	1.49560
1.49570	1.49590
Arithmetic average = 1.49596	Arithmetic average = 1.49604

classes that accommodate all the data. The classes are defined such that each data point fits into only one class. Figure 9.2 shows an example in which student scores on an engineering exam are sorted into classes.

9.3.2 Range

The *range* of the data is determined by subtracting the smallest value (the *minimum*) from the largest value (the *maximum*):

$$\text{Range} = \text{maximum} - \text{minimum} \tag{9-1}$$

In Figure 9.2, the maximum value is 99 and the minimum value is 55; therefore, the range is 44.

9.3.3 Central Tendency

The *central tendency* of a data set is measured by the **mean, median,** and **mode.** The *mean* is defined as the arithmetic mean. If the entire population is measured, then the mean μ is

FIGURE 9.2
Engineering exams from 24 students sorted into letter-grade classes.

Raw Data		Entry Table: Data Number	Entry Table: Sorted Data	
81		1	99	← Maximum Value
87		2	96	
91		3	94	
70		4	91	A = 90 to 100
85		5	89	B = 80 to 89
71		6	87	
65		7	87	
85		8	86	
76		9	85	Mode = 85
87		10	85	
68		11	85	
89		12	85	
99	Sort →	13	83	← Median = $\frac{83+85}{2} = 84$
85		14	81	
94		15	79	C = 70 to 79
55		16	76	
83		17	75	
61		18	71	
86		19	70	
96		20	68	D = 60 to 69
58		21	65	
75		22	61	
85		23	58	F = less than 60
79		24	55	← Minimum Value
			Total = 1911	Mean = $\frac{1911}{24} = 79.6$

Range = 99 − 55 = 44

$$\mu = \frac{1}{n_T} \sum_{i=1}^{n_T} x_i \tag{9-2}$$

where n_T is the total number of data points in the entire population. If a sample of the population is measured, the mean $\bar{x}$ is

$$\bar{x} = \frac{1}{n} \sum_{i=1}^{n} x_i \tag{9-3}$$

where n is the number of data points in the sample. (Obviously $n < n_T$.) In Figure 9.2, 24 students took the exam, so the scores were collected from the entire population. Therefore, Equation 9-2 is used to calculate the mean. If a sample of students, say 13, were randomly selected from the entire population of 24 students, then Equation 9-3 would have been appropriate to use.

The *median* is the middle value of sorted data. If the data have an even number of points, then there is no one middle value. In this case, the median is reported as the arithmetic average of the two middle values. For the data shown in Figure 9.2, there are an even number of data points (24), so the median calculation is based upon the average of the 12th and 13th values in the entry table. If there were only 23 data points, then the 12th value in the entry table would be the median.

The *mode* is the value that appears most frequently in a data set. In Figure 9.2, a score of 85 occurs most often, so it is the mode.

9.3.4 Variation

The *variation* in a data set is measured using the **deviation, mean absolute deviation, standard deviation,** and **variance.**

The *deviation* of a particular data point from the mean is

$$d_i = x_i - \mu \tag{9-4}$$

or

$$d_i = x_i - \bar{x} \tag{9-5}$$

depending on whether the entire population is measured, or just a sample. For any data set, the sum of all the deviations is zero, as the negative deviations cancel the positive deviations. Therefore, the sum of the deviations provides no information about how individual data points vary from the mean.

To avoid the problem of the deviations summing to zero, we may define the *mean absolute deviation* as

$$\text{Mean absolute deviation} = \frac{1}{n_T} \sum_{i=1}^{n_T} |x_i - \mu| \tag{9-6}$$

for the entire population, or

$$\text{Mean absolute deviation} = \frac{1}{n} \sum_{i=1}^{n} |x_i - \bar{x}| \tag{9-7}$$

for a sampled population.

A more common solution to the problem of deviations summing to zero is to square the deviation, because the square is always positive. This is used in the definition of the *standard deviation* as

$$\sigma = \sqrt{\frac{1}{n_T} \sum_{i=1}^{n_T} (x_i - \mu)^2} \tag{9-8}$$

for the entire population. If the population is sampled, then the *standard deviation* is

$$s = \sqrt{\frac{1}{n-1} \sum_{i=1}^{n} (x_i - \bar{x})^2} \tag{9-9}$$

Statisticians have determined that the $n - 1$ form (Equation 9-9) better approximates the true standard deviation for the entire population when samples are taken. For large samples (e.g., $n > 30$), the two versions (Equations 9-8 and 9-9) are essentially identical.

The *variance* is defined as the standard deviation squared:

$$\text{Variance} = \sigma^2 \quad \text{or} \quad \text{Variance} = s^2 \tag{9-10}$$

EXAMPLE 9.1

Problem Statement: Suppose Machine A produces 10,000 shafts each day. In statistical terms, we would say that the population is 10,000. Because it may be impractical to measure the diameter of each shaft, we may decide to *sample* the population and randomly select 36 shafts. Their diameters are reported in Table 9.2. These data may be considered *raw data* because they are listed as they were collected with no attempt to sort them or classify them. What are the (1) maximum, (2) minimum, (3) range, (4) mode, (5) median, (6) mean, (7) mean absolute deviation, (8) standard deviation, and (9) variance?

Solution: The first four questions are easily answered by *sorting* the data in order from the lowest to the highest value. Tables 9.3 and 9.4 provide the necessary information.

1. Maximum = 1.4974 in
2. Minimum = 1.4948 in
3. Range = 0.0026 in

TABLE 9.2
Raw data

Shaft Diameters from Machine A (in)					
1.4961	1.4963	1.4958	1.4957	1.4963	1.4962
1.4955	1.4948	1.4952	1.4963	1.4968	1.4962
1.4958	1.4952	1.4966	1.4962	1.4965	1.4960
1.4953	1.4963	1.4973	1.4969	1.4956	1.4949
1.4971	1.4964	1.4958	1.4964	1.4974	1.4952
1.4963	1.4967	1.4963	1.4962	1.4959	1.4964

TABLE 9.3
Entry table

Class	Shaft Diameter (in)							Frequency
1.4945–49	1.4948	1.4949						2
1.4950–54	1.4952	1.4952	1.4952	1.4953				4
1.4955–59	1.4955	1.4956	1.4957	1.4958	1.4958	1.4958	1.4959	7
1.4960–64	1.4960	1.4961	1.4962	1.4962	1.4962	1.4962	1.4963	15
	1.4963	1.4963	1.4963	1.4963	1.4963	1.4964	1.4964	
	1.4964							
1.4965–69	1.4965	1.4966	1.4967	1.4968	1.4969			5
1.4970–74	1.4971	1.4973	1.4974					3

TABLE 9.4
Deviation, absolute deviation, and square deviation

x_i	$x_i - \bar{x}$	$\lvert x_i - \bar{x}\rvert$	$(x_i - \bar{x})^2$
1.4948	−0.0013	0.0013	0.00000169
1.4949	−0.0012	0.0012	0.00000144
1.4952	−0.0009	0.0009	0.00000081
⇓	⇓	⇓	⇓
1.4971	0.0010	0.0010	0.00000100
1.4973	0.0012	0.0012	0.00000144
1.4974	0.0013	0.0013	0.00000169
$\Sigma x_i = 53.8599$		$\Sigma\lvert x_i - \bar{x}\rvert = 0.0177$	$\Sigma(x_i - \bar{x})^2 = 0.00001426$

4. Mode = 1.4963 in
5. Median = 1.4962 in
6. The mean is

$$\bar{x} = \frac{1}{n}\sum_{i=1}^{n} x_i = \frac{53.8599}{36} = 1.4961 \text{ in}$$

7. The mean absolute deviation is

$$\text{Mean absolute deviation} = \frac{1}{n}\sum_{i=1}^{n} |x_i - \bar{x}| = \frac{0.0177}{36} = 0.000492 \text{ in}$$

8. The standard deviation is

$$s = \sqrt{\frac{1}{36-1}\Sigma(x_i - \bar{x})^2} = \sqrt{\frac{0.00001426}{36-1}} = 0.0006383 \text{ in}$$

9. The variance is

$$\text{Variance} = s^2 = (0.0006383)^2 = 0.0000004074 \text{ in}^2$$

Notice that the three measures of central tendency (mean, median, and mode) are nearly identical. This is often true, but not always. For example, if one were to calculate the mean and median salaries of 10 randomly selected Americans, one of whom happened to be Bill Gates, the mean and median would be vastly different. In this case, the median would be a better measure of central tendency.

9.4 HISTOGRAMS

The entry table (Table 9.3) reported the number of data points in each class, that is, the **frequency.** The frequency may be plotted versus each class in a **histogram** (Figure 9.3) in which the length of the bar indicates the frequency of each class. A **frequency polygon** is plotted by connecting the midpoints of the bars together. The right axis of Figure 9.3 indicates the **relative frequency,** which is defined as

$$\text{Relative frequency} = \frac{\text{frequency}}{n} \tag{9-11}$$

We determine the **cumulative frequency** by adding the frequency of a given class to the sum total of all the lower classes. For example, the first class has a frequency of 2 and the second class has a frequency of 4, therefore the cumulative frequency is 6. The third class has a frequency of 7, so the cumulative frequency is 13. Figure 9.4 shows a plot of the cumulative frequency for all the classes. The midpoints of the rectangles may be connected together to give a **cumulative frequency polygon.**

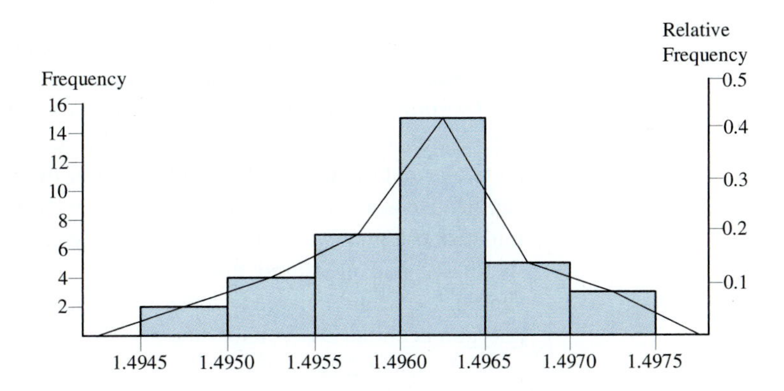

FIGURE 9.3
Histogram.

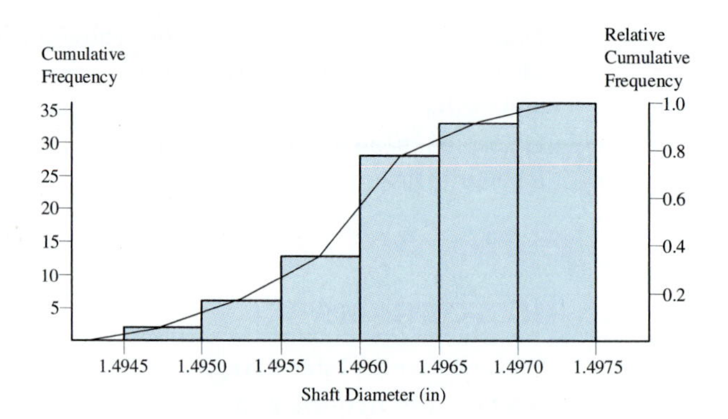

FIGURE 9.4
Cumulative frequency.

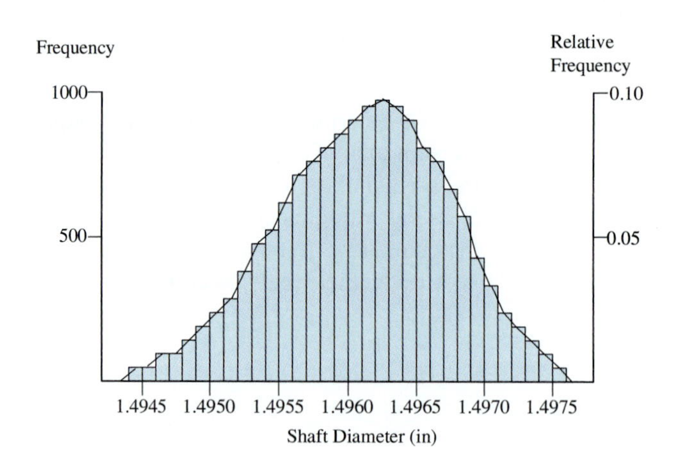

FIGURE 9.5
Histogram for a large population.

The right axis of Figure 9.4 shows the **relative cumulative frequency,** which is simply the accumulated sum of the relative frequencies. This axis allows **percentiles** to be easily determined. (You were classified into percentiles when you took the SAT or ACT college entrance exam.) A *percentile* indicates what percentage of the data are below a particular data point. For example, a shaft diameter of 1.4965 inches is in the 80th percentile, because 80% of the data are smaller. Note that when all the classes have been accumulated, the relative cumulative frequency is 1.0.

If we had measured the shaft diameters of an entire day's production (all 10,000 shafts), we would be able to have many more classes. As Figure 9.5 shows, the frequency polygon becomes essentially smooth.

9.5 NORMAL DISTRIBUTION AND STANDARD NORMAL DISTRIBUTION

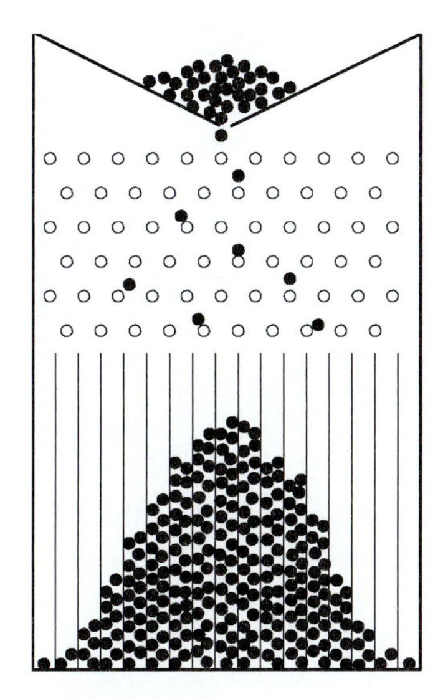

FIGURE 9.6
Quincunx.

Figure 9.6 shows a **quincunx,** a device into which marbles are dropped at the top center. As the marbles travel through the quincunx, they encounter evenly spaced pegs. When a marble encounters a peg, it falls to the left as often as it falls to the right; that is, the probability of falling to the left is the same as the probability of falling to the right. After traveling through many banks of pegs, the marbles enter chutes. After putting hundreds of marbles into the quincunx, they will sort as shown in Figure 9.6. Very few of the marbles will fill the chutes at the extreme ends, whereas most of the marbles tend to stay near the center.

The number of marbles in each quincunx chute describes a frequency polygon. In an idealized situation with an infinitely large number of marbles with infinitesimally small diameters and chutes (classes) that are infinitesimally narrow, the frequency polygon becomes completely smooth. The shape of this frequency polygon is the **normal distribution** or **bell-shaped curve.** It has the mathematical form

$$\text{Relative frequency} = \frac{1}{\sigma\sqrt{2\pi}} e^{-\frac{1}{2}(x-\mu)^2/\sigma^2} \tag{9-12}$$

This bell-shaped curve is centered about the mean μ using the variable x (Figure 9.7). The normal distribution appears often in nature. For example, the distribution of grades on an exam generally is a normal distribution.

If we define $z \equiv (x - \mu)/\sigma$, then the normal distribution becomes the **standard normal distribution**

$$\text{Relative frequency} = \frac{1}{\sqrt{2\pi}} e^{-\frac{1}{2}z^2} \tag{9-13}$$

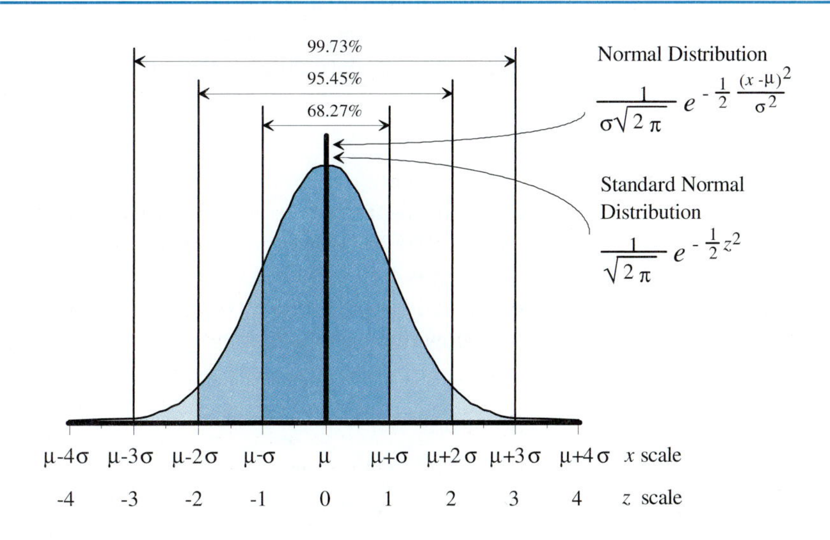

FIGURE 9.7
Normal distribution and standard normal distribution.

This bell-shaped curve is centered about zero using the variable z (Figure 9.7). When the standard normal distribution is integrated from $z = -1$ to $z = +1$, the area under the curve (i.e., the relative cumulative frequency) is 0.6827.

$$\text{Relative cumulative frequency} = 0.6827 = \frac{1}{\sqrt{2\pi}} \int_{-1}^{+1} e^{-\frac{1}{2}z^2} dz \tag{9-14}$$

This means that ± 1 standard deviation about the mean encompasses 68.27% of the data. When the standard normal distribution is integrated from $z = -2$ to $z = +2$, the area under the curve (i.e., the relative cumulative frequency) is 0.9545.

$$\text{Relative cumulative frequency} = 0.9545 = \frac{1}{\sqrt{2\pi}} \int_{-2}^{+2} e^{-\frac{1}{2}z^2} dz \tag{9-15}$$

This means that ± 2 standard deviations about the mean encompasses 95.45% of the data. When the standard normal distribution is integrated from $z = -3$ to $z = +3$, the area under the curve (i.e., the relative cumulative frequency) is 0.9973.

$$\text{Relative cumulative frequency} = 0.9973 = \frac{1}{\sqrt{2\pi}} \int_{-3}^{+3} e^{-\frac{1}{2}z^2} dz \tag{9-16}$$

This means that ± 3 standard deviations about the mean encompasses 99.73% of the data. To encompass all of the data (i.e., the relative cumulative frequency = 1) requires that the standard normal distribution be integrated from minus infinity to plus infinity. (*Note:* Observant readers noticed that when z was substituted for $(x - \mu)/\rho$, the constant preceding the exponent changed. This change was required to ensure that the integral from minus infinity to plus infinity remained 1. For most practical purposes, "infinity" is 4, because integrating from $z = -4$ to $z = +4$ is essentially 1.)

Appendix C shows the relative cumulative frequencies determined by integrating the standard normal distribution from 0 to z. Because the standard normal distribution is symmetrical, this same table also gives the relative cumulative frequency determined by integrating the standard normal distribution from $-z$ to 0.

EXAMPLE 9.2

Problem Statement: "High-quality" shafts will be defined as those having a diameter from 1.4955 to 1.4965 inches. From the sample of 36 shafts, we determined that the mean diameter is 1.4961 inches with a standard deviation of 0.0006383 inch. Assuming that the entire daily population of 10,000 shafts can be described by the standard normal distribution, how many shafts may be considered high quality?

Solution: $$z_1 = \frac{x_1 - \mu}{\sigma} \approx \frac{x_1 - \bar{x}}{s} = \frac{1.4955 - 1.4961}{0.0006383} = -0.94$$

Relative cumulative frequency = 0.3264 (from Appendix C: see Figure 9.8)

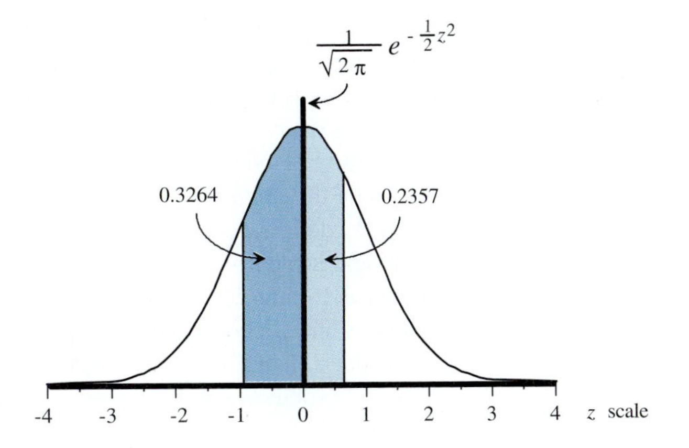

FIGURE 9.8
Standard normal distribution for determining the number of high-quality shafts.

$$z_2 = \frac{x_2 - \mu}{\sigma} \approx \frac{x_2 - \bar{x}}{s} = \frac{1.4965 - 1.4961}{0.0006383} = 0.63$$

Relative cumulative frequency = 0.2357 (from Appendix C: see Figure 9.8)

Number of high-quality shafts

$$= (0.3264 + 0.2357) \times 10{,}000 = 0.5621 \times 10{,}000 = 5621$$

Thus, slightly more than half the shafts may be considered high quality.

9.6 SUMMARY

Statistics provides scientific methods for collecting, organizing, summarizing, presenting, and analyzing data as well as making conclusions based on the data. There are two basic branches: descriptive statistics and statistical inference. Here, we consider descriptive statistics.

Three types of descriptors used to characterize data are the range, central tendency, and variation. Range is easily calculated as the maximum value minus the minimum value. Three measures of central tendency are typically used: mean, median, and mode. Measures of variation include deviation, mean absolute deviation, standard deviation, and variance.

When data are described, they are often sorted from high to low (or vice versa). Then, the sorted data may be grouped into classes. The number of data points within a particular class is called the frequency. A histogram plots the frequency versus the class using bars; if the bar centers are connected by straight lines, then the data are presented as a frequency polygon. The cumulative frequency is the sum of the frequency in a given class plus the frequencies in all lower classes.

As the number of data points increase, the range of each class can be narrowed and the polygons become smooth curves. In the limit where each class is only differentially wide, then the data may be described by the normal distribution, or bell-shaped curve.

(*Note:* There are many other possible distributions, but the normal distribution is the most common.) In the normal distribution, we plot the frequency versus x, the raw data. The peak of the curve is the mean. Plus or minus one standard deviation encompasses 68.27% of the data, plus or minus two standard deviations encompasses 95.45% of the data, and plus or minus three standard deviations encompasses 99.73% of the data. All normally distributed data have the same bell-shaped curve; however, different sets of data may be scaled differently. By defining a new variable, called z, which rescales the original x data relative to the mean and standard deviation, we can plot all normally distributed sets of data on a single *standard* normal distribution. When plotted in this manner, the peak occurs at $z = 0$. Plus or minus one z encompasses 68.27% of the data, plus or minus two z's encompasses 95.45% of the data, and plus or minus three z's encompasses 99.73% of the data. A wide variety of practical problems can be solved using the standard normal distribution.

Further Readings

Eide, A. R.; R. D. Jenison; L. H. Mashaw; and L. L. Northup. *Engineering Fundamentals and Problem Solving.* 3rd ed. New York: McGraw-Hill, 1997.

Spiegel, M. R. *Statistics.* Schaum's Outline Series in Mathematics. New York: McGraw-Hill, 1961.

PROBLEMS

9.1 Table 9.5 shows the maximum temperature each day in the month of June. Sort the data into an entry table consisting of classes that span 5°F (70 to 74.9, 75 to 79.9, etc.). By hand or using a spreadsheet, prepare a histogram (which indicates the frequency and relative frequency) and a cumulative frequency plot (which indicates the cumulative frequency and relative cumulative frequency). For this set of data, determine the following: maximum, minimum, range, mode, mean, median, mean absolute deviation, standard deviation, and variance.

9.2 A random sample of bolts was taken from a manufacturing line. Table 9.6 shows the measured length of each bolt. Sort the data into an entry table consisting of classes that span 0.010 in (2.4700 to 2.4799, 2.4800 to 2.4899, etc.). By hand or using a spreadsheet, prepare a histogram (which indicates the frequency

TABLE 9.5
Maximum air temperature for each day in June

Day	Temp. (°F)	Day	Temp. (°F)	Day	Temp. (°F)	Day	Temp. (°F)	Day	Temp. (°F)
1	89	7	88	13	85	19	93	25	92
2	73	8	83	14	77	20	96	26	86
3	71	9	84	15	72	21	98	27	83
4	76	10	95	16	76	22	101	28	81
5	83	11	97	17	84	23	102	29	94
6	92	12	83	18	89	24	94	30	92

TABLE 9.6
Length of sampled bolts (in inches)

2.489	2.511	2.521	2.502	2.516	2.483	2.471	2.512	2.502	2.505
2.503	2.507	2.492	2.519	2.499	2.479	2.491	2.522	2.516	2.496
2.498	2.508	2.476	2.482	2.496	2.504	2.523	2.487	2.495	2.519
2.478	2.515	2.517	2.492	2.509	2.513	2.509	2.497	2.517	2.507

and relative frequency) and a cumulative frequency plot (which indicates the cumulative frequency and relative cumulative frequency). For this set of data, determine the following: maximum, minimum, range, mode, mean, median, mean absolute deviation, standard deviation, and variance.

9.3 The bolt manufacturing line described in Problem 9.2 makes 50,000 bolts each day. A "quality" bolt is defined as one that is 2.500 ± 0.010 inches long. Assuming the bolt lengths are normally distributed, how many quality bolts are produced each day?

9.4 Using the standard normal distribution, prepare a cumulative frequency plot. (*Hint:* Appendix C will be useful.)

9.5 Monochlorobenzene (MCB) is an industrial chemical made by reacting chlorine gas with benzene in the presence of an iron chloride catalyst. In addition to the desired MCB, undesired products (e.g., dichlorobenzene and trichlorobenzene) are also made. The MCB is separated from the undesired products by a distillation column.

The customer requires that the MCB be at least 99.0% pure. Whenever the MCB purity drops below 99.0%, the impure product must be "reworked;" that is, it is returned to the distillation column to remove impurities. An on-line sensor continuously monitors the purity of the MCB. One day, the sensor reports that the mean purity was 99.2% with a standard deviation of 0.1%. What percentage of the product must be reworked?

9.6 A civil engineer is designing a road for a posted speed limit of 55 mph. The road has a sharp turn that must be "banked," or sloped downward toward the pivot of the turn. The slope is required so cars do not slide off the road. (You have undoubtedly noticed that racing cars require a very steep bank to prevent them from sliding off the racetrack in a sharp curve.) He is considering specifying a slope with the safety record reported in Table 9.7. This table indicates, for example, that 15% of cars going 76 mph slide off the curve and fatally crash.

TABLE 9.7
Safety record for banked curve

Automobile Speed Entering the Turn (mph)	Fatal Crashes (%)
70–74.9	2
75–79.9	15
80–84.9	45
85–89.9	75
90–94.9	95
95–100	100

Using a police radar gun, the civil engineer decides to survey the automobile speed on a highway with a posted speed of 55 mph. He determines that the mean speed is 62 mph with a standard deviation of 7 mph. The highway being designed by the engineer has an anticipated usage of 50 million automobiles each year. How many fatal crashes are expected on this curve each year? In your view, is this an acceptable number of crashes? Do you think redesign is warranted?

9.7 A *resistor* is a device that converts electrical energy into heat, causing the voltage to reduce as electrical current flows through the resistor. Resistors are extremely useful devices as they are found in nearly all electronic products. One way to make electrical resistors is to pack a plastic or ceramic tube with carbon powder and a suitable binder. The resistance is determined by the tube geometry and the composition of the carbon/binder. Because it is very difficult to control the variables to get a precise resistance, manufacturers make a batch of resistors and measure the resistance of each one. Then, they are sorted and sold.

At Resistors-R-Us Corporation (where the slogan is, "You can't resist our resistors"), a large order was received for resistors that are 55 ± 5 Ω. A production run will be considered successful provided 80% of the resistors meet the specifications for the order. You are assigned the task of determining if the machine that makes resistors can successfully fill the order. To test this machine, you prepare a small batch of resistors. Table 9.8 shows the resistances, in ohms (Ω), measured for each resistor in the batch.

TABLE 9.8
Measured resistances

45.3	67.8	34.3	51.2	48.5	61.3	59.3	65.1
49.3	42.4	63.5	69.8	71.2	39.8	55.5	53.2
56.7	48.8	61.5	51.2	58.9	63.1	67.5	62.4
52.4	50.2	49.8	56.8	59.7	60.4	45.8	43.8
51.3	54.8	55.1	52.3	56.2	59.7	63.0	46.7
63.1	58.2	41.9	59.2	57.2	67.3	68.2	38.9
51.3	63.8	53.4	58.9	56.3	58.9	53.2	56.8

Prepare an entry table for the above data. Separate the data into classes that are 5 Ω wide (the first class is 30.0 to 34.9 Ω). Prepare a histogram by hand or using a spreadsheet. Determine the following:

(a) Maximum
(b) Minimum
(c) Range
(d) Mode (use only two significant figures)
(e) Median
(f) Mean
(g) Mean absolute deviation
(h) Standard deviation
(i) Variance

Assume the above data can be described by the normal distribution and that a large production run will follow the same normal distribution. If 100,000 resistors are produced by this machine, how many will fall within the specification range of 55 ± 5 Ω? Is this good enough to be considered successful?

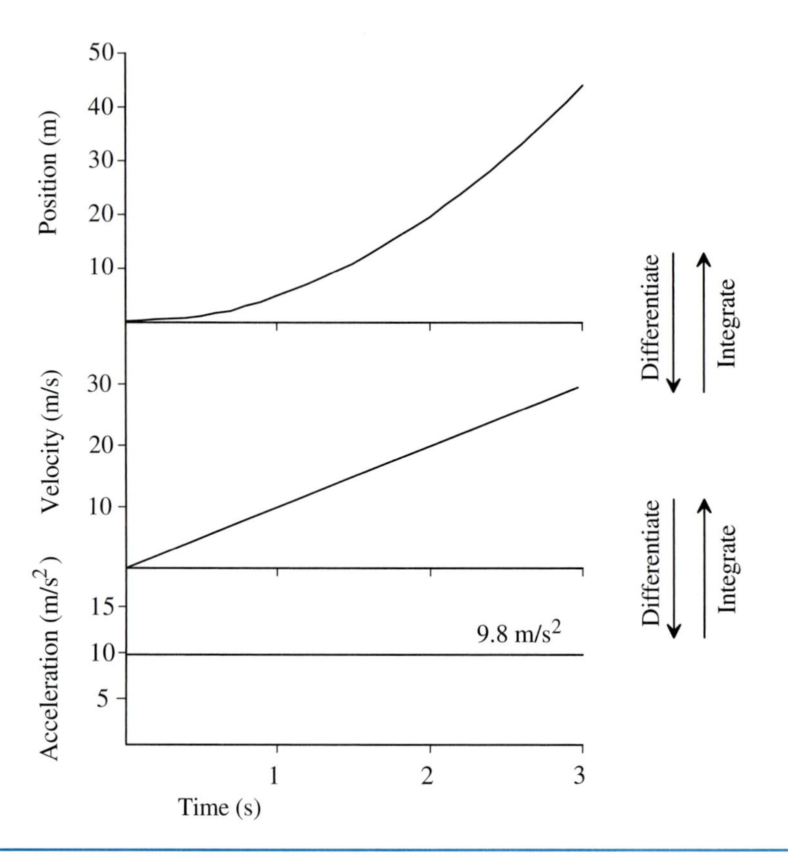

FIGURE 10.5
Position, velocity, and acceleration of a dense object dropped under the influence of earth's gravity.

where a_0 is assumed to be a constant acceleration as the object falls. This equation may be integrated as an indefinite integral (i.e., one in which the limits are not specified),

$$\int dv = \int a_0 dt = a_0 \int dt$$

$$v = v_0 + a_0 t \tag{10-15}$$

where v_0 is the integration constant.

Velocity is the derivative of position:

$$v = \frac{dx}{dt} \tag{10-16}$$

We can rearrange this equation as

$$dx = v\, dt \tag{10-17}$$

Equation 10-15 may be substituted for v, thus

$$dx = (v_0 + a_0 t)dt \tag{10-18}$$

This equation may also be evaluated as an indefinite integral,

$$\int dx = \int (v_0 + a_0 t)\, dt = \int v_0\, dt + \int a_0 t\, dt = v_0 \int dt + a_0 \int t\, dt$$

$$x = x_0 + v_0 t + \tfrac{1}{2} a_0 t^2 \tag{10-19}$$

where x_0 is the integration constant.

If a dense object is dropped with an initial position defined as zero (i.e., $x_0 = 0$) and an initial velocity of zero (i.e., $v_0 = 0$), then Equation 10-19 simplifies to

$$x = \tfrac{1}{2} a_0 t^2 \tag{10-20}$$

EXAMPLE 10.2

Problem Statement: An automobile starts from rest and moves in a straight line with a constant acceleration of 2 m/s^2. In 8 s, what is the velocity and displacement of the automobile?

Solution: The velocity is given by Equation 10-15,

$$\text{Velocity} = v = v_0 + a_0 t = 0\frac{\text{m}}{\text{s}} + \left(2\frac{\text{m}}{\text{s}^2}\right)8\text{ s} = 16\text{ m/s}$$

and the displacement is given by Equation 10-19,

$$\text{Displacement} = x - x_0 = v_0 t + \tfrac{1}{2} a_0 t^2$$

$$= \left(0\frac{\text{m}}{\text{s}}\right)(8\text{ s}) + \frac{1}{2}\left(2\frac{\text{m}}{\text{s}^2}\right)(8\text{ s})^2 = 64\text{ m}$$

10.1.3 Multidimensional Motion

So far, we have confined motion to a single dimension. Here, we describe motion in all three dimensions using a coordinate system with x-, y-, and z-axes.

The motion in each dimension is completely independent of the motion in the other dimensions. To illustrate this point, consider Figure 10.6, which shows Fred and Jane standing at the edge of a cliff, each with a ball. Fred decides to throw the ball horizontally outward and Jane decides to drop the ball straight down. If the balls leave their hands at the same time, the balls will strike the ground **at the same time,** regardless of how hard Fred throws the ball horizontally outward. This is because the motion in the y-direction is completely independent of motion in the x-direction. In the y-direction, gravity acts equally on each ball, so they each strike the ground at the same time.

Figure 10.7 shows the position vector **r**, the velocity vector **v**, and the acceleration vector **a** in a three-dimensional coordinate system. Table 10.2 shows the relationships between these vectors.

TABLE 10.2
Relationship between position, velocity, and acceleration in three dimensions for the special case of constant acceleration

	Vector Equations		Scalar Equations	
Position	$\mathbf{r} = \mathbf{r}_0 + \mathbf{v}_0 t + \frac{1}{2}\mathbf{a}_0 t^2$	(10-21a)	$x = x_0 + v_{x,0}t + \frac{1}{2}a_{x,0}t^2$ $y = y_0 + v_{y,0}t + \frac{1}{2}a_{y,0}t^2$ $z = z_0 + v_{z,0}t + \frac{1}{2}a_{z,0}t^2$	(10-21b)
	Integrate ↑ ↓ Differentiate			
Velocity	$\frac{d\mathbf{r}}{dt} = \mathbf{v} = \mathbf{v}_0 + \mathbf{a}_0 t$	(10-22a)	$\frac{dx}{dt} = v_x = v_{x,0} + a_{x,0}t$ $\frac{dy}{dt} = v_y = v_{y,0} + a_{y,0}t$ $\frac{dz}{dt} = v_z = v_{z,0} + a_{z,0}t$	(10-22b)
	Integrate ↑ ↓ Differentiate			
Acceleration	$\frac{d\mathbf{v}}{dt} = \mathbf{a} = \mathbf{a}_0$	(10-23a)	$\frac{dv_x}{dt} = a_x = a_{x,0}$ $\frac{dv_y}{dt} = a_y = a_{y,0}$ $\frac{dv_z}{dt} = a_z = a_{z,0}$	(10-23b)

The position vector $\mathbf{r}$ is

$$\mathbf{r} = x\mathbf{i} + y\mathbf{j} + z\mathbf{k} \tag{10-24}$$

The magnitude of the position vector $\mathbf{r}$ is

$$|\mathbf{r}| = \sqrt{x^2 + y^2 + z^2} \tag{10-25}$$

The velocity vector $\mathbf{v}$ is

$$\mathbf{v} = v_x\mathbf{i} + v_y\mathbf{j} + v_z\mathbf{k} \tag{10-26}$$

where

$$v_x = \frac{dx}{dt} \qquad v_y = \frac{dy}{dt} \qquad v_z = \frac{dz}{dt} \tag{10-27}$$

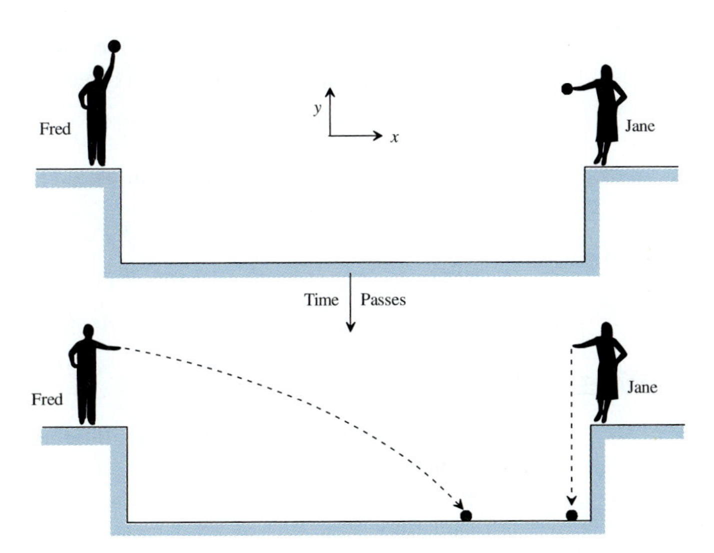

FIGURE 10.6
Illustration showing that motions in each direction are independent.

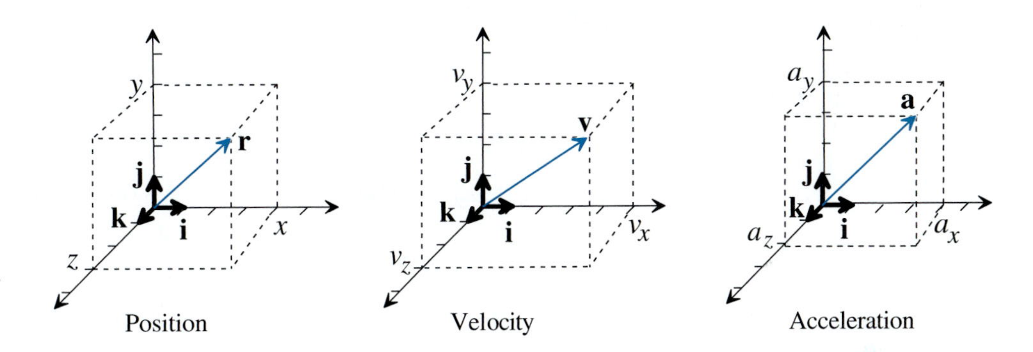

FIGURE 10.7
Three-dimensional position, velocity, and acceleration vectors.

The magnitude of the velocity vector **v** is

$$|\mathbf{v}| = \sqrt{v_x^2 + v_y^2 + v_z^2} \tag{10-28}$$

The acceleration vector **a** is

$$\mathbf{a} = a_x\mathbf{i} + a_y\mathbf{j} + a_z\mathbf{k} \tag{10-29}$$

where

$$a_x = \frac{dv_x}{dt} = \frac{d^2x}{dt^2} \qquad a_y = \frac{dv_y}{dt} = \frac{d^2y}{dt^2} \qquad a_z = \frac{dv_z}{dt} = \frac{d^2z}{dt^2} \tag{10-30}$$

The magnitude of the acceleration vector **a** is

$$|\mathbf{a}| = \sqrt{a_x^2 + a_y^2 + a_z^2} \tag{10-31}$$

If the three components of **a** are constants, i.e., $a_x = a_{x,0}$, $a_y = a_{y,0}$, and $a_z = a_{z,0}$, then at any time t, the velocity in each direction is given by

$$v_x = v_{x,0} + a_{x,0}t \qquad v_y = v_{y,0} + a_{y,0}t \qquad v_z = v_{z,0} + a_{z,0}t$$

where $v_{x,0}$, $v_{y,0}$, and $v_{z,0}$ are the initial velocities in each direction. The position is given by

$$x = x_0 + v_{x,0}t + \tfrac{1}{2}a_{x,0}t^2 \qquad y = y_0 + v_{y,0}t + \tfrac{1}{2}a_{y,0}t^2 \qquad z = z_0 + v_{z,0}t + \tfrac{1}{2}a_{z,0}t^2$$

where x_0, y_0, and z_0 are the initial positions in each direction. Table 10.2 summarizes the above relationships.

EXAMPLE 10.3

Problem Statement: At a given time in its flight (which we will designate as $t = 0$), an airplane is at the position $x_0 = 500$ m, $y_0 = 1000$ m, and $z_0 = 3000$ m; the velocity in each dimension is $v_{x,0} = 100$ m/s, $v_{y,0} = 150$ m/s, and $v_{z,0} = 5.00$ m/s; and the acceleration in each dimension is $a_{x,0} = 15.0$ m/s^2, $a_{y,0} = 3.00$ m/s^2, and $a_{z,0} = 1.00$ m/s^2. After 8.00 s, what are its speed, new position, and distance traveled?

Solution:

$$v_x = v_{x,0} + a_{x,0}t = 100 \text{ m/s} + 15.0 \text{ m/s}^2(8.00 \text{ s}) = 220 \text{ m/s}$$

$$v_y = v_{y,0} + a_{y,0}t = 150 \text{ m/s} + 3.00 \text{ m/s}^2(8.00 \text{ s}) = 174 \text{ m/s}$$

$$v_z = v_{z,0} + a_{z,0}t = 5.00 \text{ m/s} + 1.00 \text{ m/s}^2(8.00 \text{ s}) = 13 \text{ m/s}$$

$$\text{Speed} = |\mathbf{v}| = \sqrt{v_x^2 + v_y^2 + v_z^2}$$

$$= \sqrt{(220 \text{ m/s})^2 + (174 \text{ m/s})^2 + (13 \text{ m/s})^2} = 280 \text{ m/s}$$

$$x = x_0 + v_{x,0}t + \tfrac{1}{2}a_{x,0}t^2$$

$$= 500 \text{ m} + (100 \text{ m/s})(8.00 \text{ s}) + \tfrac{1}{2}(15.0 \text{ m/s}^2)(8.00 \text{ s})^2 = 1780 \text{ m}$$

$$y = y_0 + v_{y,0}t + \tfrac{1}{2}a_{y,0}t^2$$

$$= 1000 \text{ m} + (150 \text{ m/s})(8.00 \text{ s}) + \tfrac{1}{2}(3.00 \text{ m/s}^2)(8.00 \text{ s})^2 = 2296 \text{ m}$$

$$z = z_0 + v_{z,0}t + \tfrac{1}{2}a_{z,0}t^2$$

$$= 3000 \text{ m} + (5.00 \text{ m/s})(8.00 \text{ s}) + \tfrac{1}{2}(1.00 \text{ m/s}^2)(8.00 \text{ s})^2 = 3072 \text{ m}$$

$$\text{Distance} = \sqrt{(x - x_0)^2 + (y - y_0)^2 + (z - z_0)^2}$$

$$= \sqrt{(1789 \text{ m} - 500 \text{ m})^2 + (2296 \text{ m} - 1000 \text{ m})^2 + (3072 \text{ m} - 3000 \text{ m})^2}$$

$$= 1829 \text{ m}$$

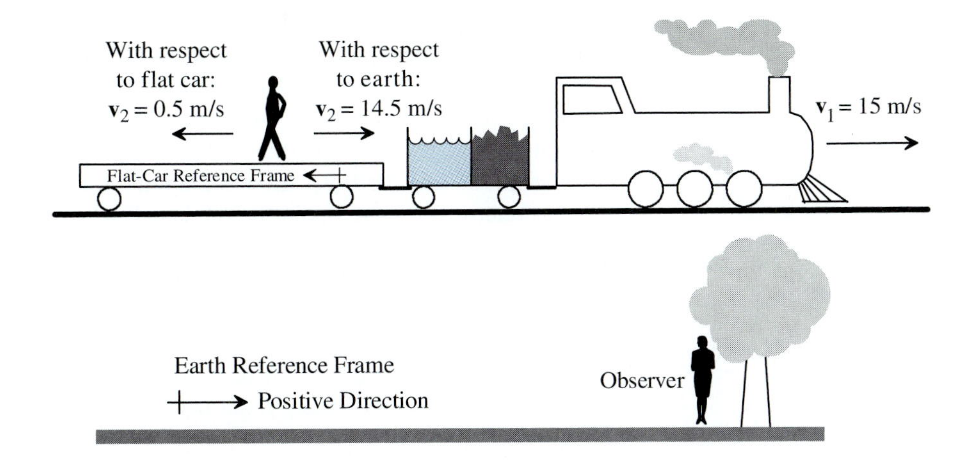

FIGURE 10.8
A person walking on a flat car. This illustration shows the importance of specifying a reference frame; there are two possible values for $\mathbf{v}_2$ depending upon the reference frame.

10.1.4 Relative Motion

Figure 10.8 shows a train moving at constant velocity $\mathbf{v}_1$ with a person walking on the flat car with constant velocity $\mathbf{v}_2$. An observer is standing under a tree and watching as the train moves by. A natural question to ask about this situation is, How fast is the person walking? The answer to this question depends upon the reference frame. In the reference frame of the flat car, the person has a velocity of 0.5 m/s; but in the earth reference frame, the person is moving with a velocity of 14.5 m/s. Either answer is correct; the only restriction is that the reference frame be an **inertial** reference frame (i.e., it is not accelerating). However, to avoid ambiguity, in this book we will always use the earth reference frame unless otherwise stated.

10.2 THEORY OF RELATIVITY (ADVANCED TOPIC)

Although classical physics seems to describe our ordinary world almost perfectly, at the atomic level strange things are observed, so physicists invented quantum mechanics to describe what they saw. Likewise subtle experiments by physicists Albert Michelson (1852–1931) and Edward Morley (1838–1923) produced results that could not be explained by classical Newtonian physics and, indeed, defied common sense. In 1905, the young Albert Einstein (1879–1955) proposed the special theory of **relativity** to explain these observations. Let's look at the Michelson-Morley experiment, then take a quick tour of Einstein's special theory.

One bedrock concept in classical physics is that velocities sum between frames of reference. To understand what these words mean, consider the following thought experiment. Eager to set the record for throwing the world's fastest fastball, you equip a friend with a radar gun and practice until you can throw a baseball pretty well. Alas, your top speed is about 70 mi/h, well below the record. Hmmm, how can you make it go faster? Easy. Conscript a friend to drive a car, lean out the window, and heave the ball at the radar gun while driving past. Neglecting air resistance, and any thought of personal safety, you find, as expected, that the velocity of the car is added to the velocity of your throw. If the car is going 60 mi/h, you now clock speeds of 130 mi/h, and you're in the big time! (Don't really try this; besides, *The Guinness Book of Records* thinks this is cheating.)

Now another, more subtle, thought experiment and we're ready to go. Tired of throwing baseballs, you decide to become a world-class swimmer. You train by swimming when the ocean's tide is coming in to help build your endurance (Figure 10.9). Jumping into the ocean, you have two choices: swim straight out against the tide and then back to the beach, swimming with the tide on the return trip (Path A), or swim parallel to the beach at right angles to the tide (Path B). After examining your times, you notice that you always swim faster relative to a fixed point on the beach when swimming Path A. After some thought you deduce why this is so.

Along Path A, your velocities relative to the point on the beach are easy to calculate. When swimming with the current, your velocity relative to the beach is $c + v$, where v is the velocity of the incoming tide, and c is your constant swimming velocity in still water. When you swim against the current, your velocity relative to the beach is $c - v$. The time it takes you to swim distance L out from the beach and back is

$$t_A = \frac{L}{c + v} + \frac{L}{c - v} = \frac{2cL}{c^2 - v^2} \tag{10-32}$$

If you take Path B, you must swim at an angle to maintain a constant position parallel to the beach. By looking at the vectors, your velocity relative to the beach is $\sqrt{c^2 - v^2}$; therefore the time to swim distance L and back is

$$t_B = \frac{2L}{\sqrt{c^2 - v^2}} \tag{10-33}$$

Now we're getting somewhere. If we divide the Path A race time by the Path B race time, we get

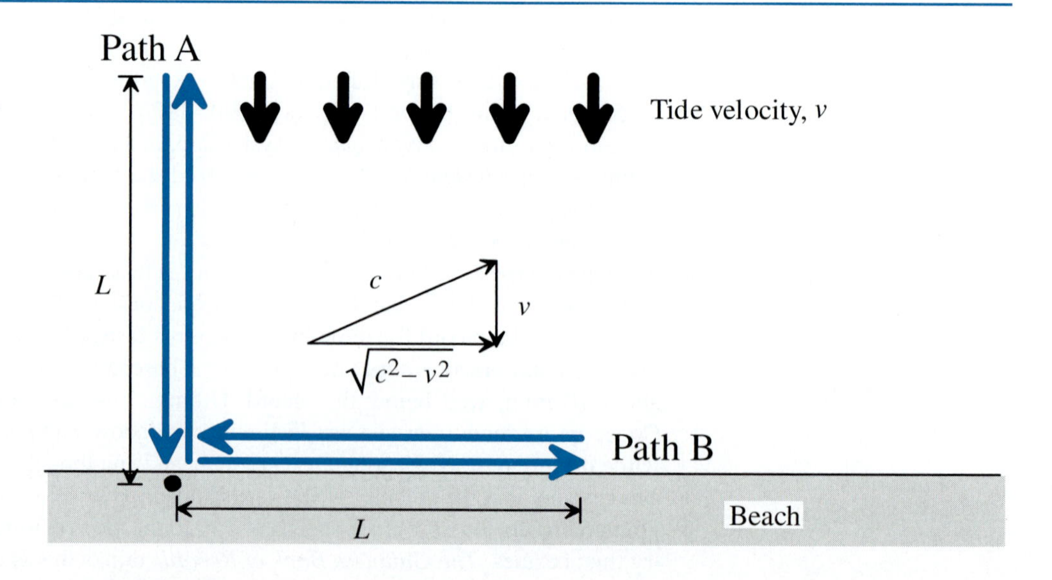

FIGURE 10.9
Two swimming paths from a point on the beach.

$$\frac{t_A}{t_B} = \frac{1}{\sqrt{1 - (v^2/c^2)}} \tag{10-34}$$

Looking at this we see that with no current, $t_A = t_B$ as it should. However, any current at all will make the ratio less than 1 and $t_A > t_B$. Although this may not seem very profound, notice that the summation of velocities is inherent in the derivation.

What do you think would happen if we ran the race using light instead of a swimmer? This is exactly what Michelson and Morley did by building an apparatus based on the schematic shown in Figure 10.10.

In principle, the time it takes the light to travel each of the two paths could be measured and differences in the speed of light along the paths could be found. However, because light travels so quickly, Michelson and Morley had the challenge of measuring extremely short travel times. To meet this challenge, they decided to use the wave properties of the light itself to "clock" the time it took to travel each path. For a fixed frequency, the rise and fall of the light's wave peaks are a function of the distance traveled. By using the **interference patterns** that result from the alignment or misalignment of the light's wave peaks, the relative velocity of the two light beams could be determined rather than the absolute velocities. The light wave peaks are aligned entering the *splitter,* a lightly silvered mirror. If the arms are of equal length (or differ by an integral number of wavelengths), and the earth reference frame is stationary relative to a distant light source, then the light wave peaks

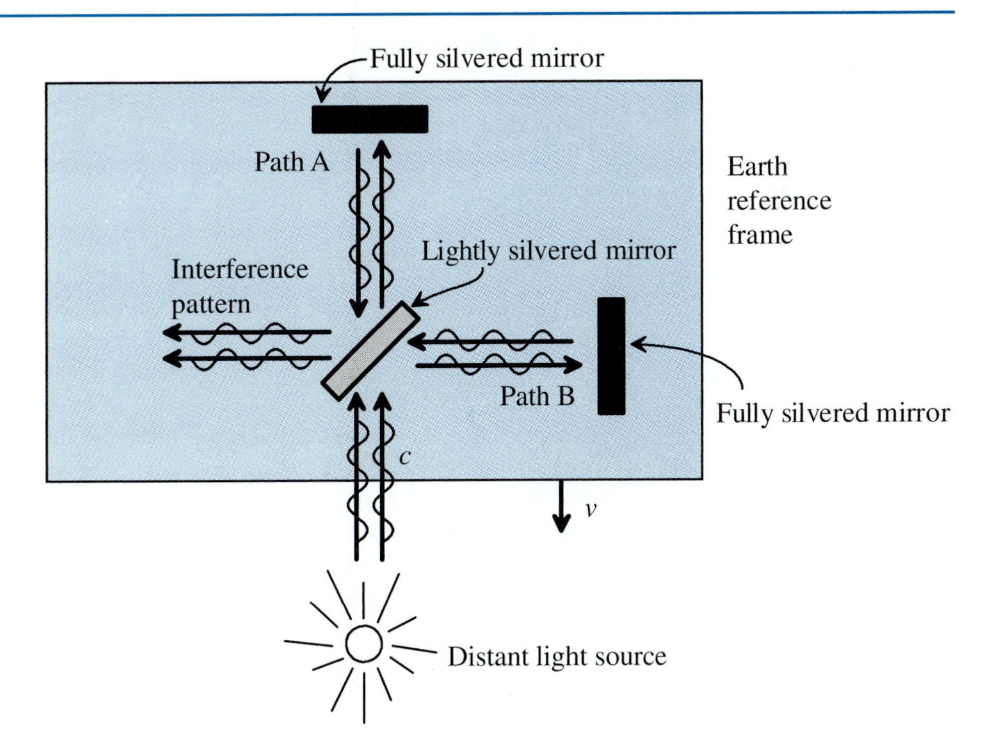

FIGURE 10.10
Schematic of the Michelson-Morley interferometer.

will align again when reformed. If one arm is parallel with the motion of the earth (Path A), and the other at right angles to the motion (Path B), then, just as in the swimming example, Path A should take longer than Path B and the light wave peaks should no longer align. Michelson and Morley floated their whole apparatus on liquid mercury so that it could be freely turned. By rotating the apparatus, no matter how the experiment was flying through space because of the earth's motion, one arm could be more or less pointed in the direction of travel and the other would be more or less perpendicular. They calculated that their **interferometer** was sensitive enough to see the effects of velocities 10 times lower than the orbital velocity of the earth around the sun.

After observing for over a year, no shift in light wave peaks was ever observed, no matter what direction the arms pointed! This meant that the velocity of the reference frame (earth) was somehow not summing with the speed of the light beams. How could this be explained? Several theories were proposed, but Einstein proposed two postulates that were the foundation of his special theory of relativity that led not only to explaining the Michelson-Morley experiment, but also to many strange and wonderful things. Einstein's postulates were

1. There is no preferred reference frame. The laws of physics are the same for any two observers moving with constant velocity relative to each other.
2. The speed of light in free space is the same for all observers regardless of their relative motion.

These two ideas seem reasonable enough, but they contradict Newtonian physics particularly as the velocity between two reference frames approaches the speed of light.

10.2.1 Relativistic Transformations

A physicist named Hendrik Lorentz (1853–1928) tried to understand the Michelson-Morley experiment at the same time as Dr. Einstein. He suggested that if the Path A arm of the interferometer shrank by γ, defined as

$$\gamma \equiv \frac{1}{\sqrt{1 - (v^2/c^2)}} \tag{10-35}$$

then the speed of light would measure the same. Looking at it another way, if the distances did not change, but if time slowed by the same factor given above, then the frequencies would match along the two arms. Through elegant arguments, Lorentz and Einstein showed that for two reference frames moving at velocity v relative to each other, the speed of light is constant for both reference frames and physical events are described the same way in both frames, even if v is approaching the speed of light, provided the transforms in Table 10.3 are used. Note that if $v \ll c$, then γ is very close to 1 and the **relativistic transforms** reduce to classical relations.

As a concrete example, let's look at an exotic subatomic particle called a **muon**. The half-life of the muon is known to be 1.52×10^{-6} s, measured by an observer moving the same speed as the muons (i.e., the muons are at rest relative to the observer). By **half-life** we mean that after 1.52×10^{-6} s, half of the original muons will be gone, having decayed into some other subatomic particle. After another 1.52×10^{-6} s, only one-fourth will remain—half of the remaining half have decayed, and so on. When cosmic rays strike the

TABLE 10.3
Classical and relativistic transformations

Galileo-Newton (Classical)	Lorentz-Einstein (Relativistic)
$x' = x - vt$	$x' = \gamma(x - vt)$
$y' = y$	$y' = y$
$z' = z$	$z' = z$
$t' = t$	$t' = \gamma\left(t - \frac{vx}{c^2}\right)$

Note: $\gamma = \frac{1}{\sqrt{1 - v^2/c^2}}$ and motion is along x only.

upper atmosphere, they create high-energy muons, many of which move toward the earth at speeds of $0.99c$ or greater (c is the speed of light, about 3×10^8 m/s). If we assume that the interactions occur 10 km above the earth, then it takes the muons

$$t = \frac{10^4 \text{ m}}{0.99(3 \times 10^8 \text{ m/s})} = 3.3 \times 10^{-5}\text{s}$$

to reach the surface of the earth. This elapsed time is equivalent to approximately 20 half-lives, so $(1/2)^{20}$ or only one out of a million muons produced would reach the earth's surface according to "classical" physics.

However, muons are readily measured at the earth's surface. In fact, many more than one out of a million make it. What's going on? The relativistic transformations given earlier tell the story. From earth, the "clocks" (i.e., half-lives) of the approaching muons run slow by a factor of γ, an effect called **time dilation.** For $v = 0.99c$, $\Delta t' = 7.08\ \Delta t$; therefore, their apparent half-life (as seen from earth) is $7.08 \times 1.52 \times 10^{-6}$ s, or 1.08×10^{-5} s. This means that the muons decay by only roughly three half-lives ($3.3 \times 10^{-5}/1.08 \times 10^{-5}$), so that one-eighth of the muons make it to the surface!

From the muon's viewpoint, which Einstein states is just as valid as ours, the half-life is certainly 1.52×10^{-6} s, but the distance to the surface of the earth is reduced to 1.41 km (i.e., 10 km/7.08), an effect called **length contraction.** The time to travel this distance is $1.41 \times 10^3\text{m}/(0.99 \times 3 \times 10^8\text{m/s}) = 4.75 \times 10^{-6}$ s, or roughly three half-lives, the same as before. So from the muon's viewpoint also, one-eighth make it to the surface! Although an observer on earth and one moving with the muons would disagree on the times and distances involved, their calculations produce a common result.

10.3 FORCES

Force is a concept with which you are certainly familiar. If you have been pushed in a crowd or played tug-of-war, you have had firsthand experience with force. **Force** is the influence on a body that will cause it to accelerate in the absence of any other counteracting forces. For example, if you were pushed in a crowd, the force caused you to accelerate in a direction you had not intended. If forces exactly counteract each other,

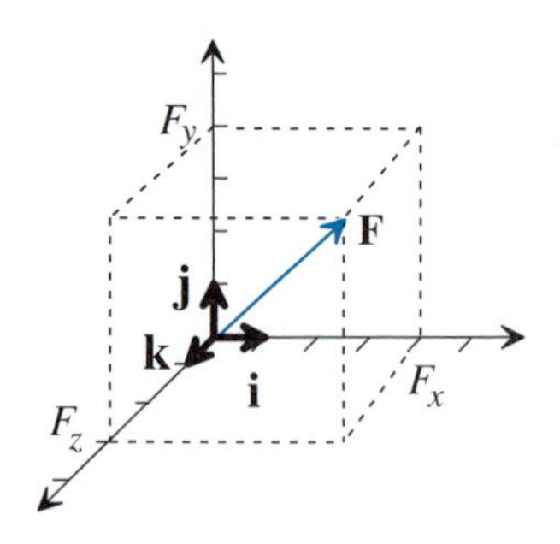

FIGURE 10.11
The force vector is resolved into its components.

there is no acceleration of the object. For example, in the case of tug-of-war with two teams perfectly matched, the forces on the rope are perfectly balanced so the rope does not accelerate.

Force is a vector quantity, just like position, velocity, and acceleration. Figure 10.11 shows a force vector **F** resolved into its three components F_x, F_y, and F_z.

$$\mathbf{F} = F_x\mathbf{i} + F_y\mathbf{j} + F_z\mathbf{k} \tag{10-36}$$

10.3.1 Fundamental Forces

There are only four known fundamental forces in nature: gravity force, electromagnetic force, strong force, and weak force. The **strong force** acts only at very short distances and is responsible for holding atomic nuclei together. The **weak force** is involved in radioactive decay. Both these forces are beyond the scope of this text; most engineers (except nuclear engineers) rarely encounter them. However, engineers commonly encounter the gravity and electromagnetic forces. The **electromagnetic force** has two manifestations, electrostatic force and magnetic force. Although superficially the electrostatic and magnetic forces may appear to be different phenomena, each results from electromagnetism, so they have been unified into a single force.

Gravity is an attractive force between two objects that have mass (Figure 10.12). The gravity force magnitude between two "point masses" (two objects that have mass but no volume) is given by

$$F = G\frac{m_1 m_2}{r} \tag{10-37}$$

where m_1 and m_2 are the masses (kg) of each object, r is the distance (m) that separates them, and G is the proportionality constant ($G = 6.67259 \times 10^{-11}$ N·m²/kg²). (*Note:* The units of G are easily determined. They must cancel the mass units of the numerator and the distance units of the denominator while providing the force units needed by the lefthand side of the equation.) It would seem that this equation is not very useful because it applies only to point masses; after all, how many practical objects have mass but no volume? However, it can be shown that Equation 10-37 exactly represents the force of attraction between two spheres, each with a uniform density, provided the distance r is measured from the centers of the spheres. Generally, as long as the two objects can be modeled as point masses (i.e., the distance separating them is large compared to the size of the object), the object shape is not important and this equation may be used.

The **electrostatic force** is an attractive force (Figure 10.12) when the two objects are charged differently (one is positive and one is negative) and is a repulsive force when the two objects are charged alike (both positive or both negative). Provided the size of the

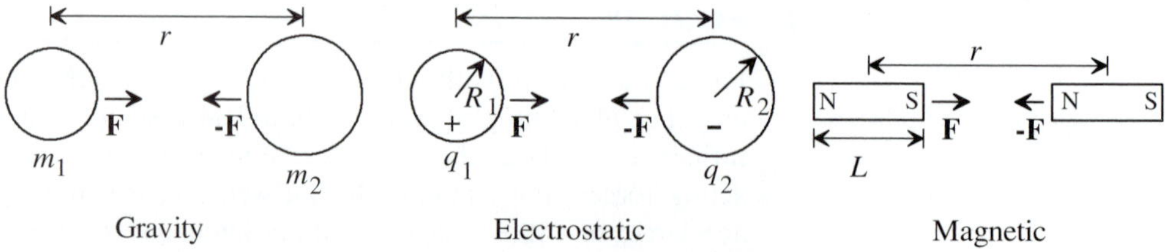

FIGURE 10.12
Some fundamental forces of nature.

r F   -F m_1 m_2 Gravity

r R_1 +   F   -F R_2 – q_1 q_2 Electrostatic

r N S   F   -F   N S L Magnetic

charged objects is small compared to the distance separating them (i.e., they are "point charges" with $R_1 << r$ and $R_2 << r$), the magnitude of the force is

$$F = \left(\frac{1}{4\pi\varepsilon_0}\right)\frac{q_1 q_2}{r^2} \tag{10-38}$$

where q_1 and q_2 are the charges (C) on each body, r is the distance (m) separating them as measured from the center of each body, and ε_0 is part of the proportionality constant [e.g., ε_0 = permittivity of vacuum = $8.854187817 \times 10^{-12}$ $C^2/(N \cdot m^2)$]. The direction of the force vector depends upon whether the force is attractive or repulsive.

An example of a **magnetic force** is shown in Figure 10.12 in which two bar magnets are brought together. If like poles (two norths or two souths) are brought together, the force is **repulsive,** and if unlike poles (one north and one south) are brought together, the force is **attractive.** Provided the length of the bar magnet is small compared to the distance separating them (i.e., they are "point magnets" with $L << r$), the magnitude of the force is

$$F = \left(\frac{3\mu_0}{2\pi}\right)\frac{\mu_1 \mu_2}{r^4} \tag{10-39}$$

where μ_1 and μ_2 are the "magnetic dipole moments" of each magnet (a measure of magnetic strength with units $A \cdot m^2$), r is the distance (m) separating the centers of each magnet, and μ_0 is part of the proportionality constant (e.g., μ_0 = permeability of vacuum = $4\pi \times 10^{-7}$ N/A^2).

10.3.2 "Other" Forces

In addition to the fundamental forces described above, engineers frequently encounter "other" forces, such as friction, drag, and spring forces. These other forces invariably result from the more fundamental forces.

The following description of other forces is not exhaustive; however, it does describe forces commonly encountered in engineering.

Friction.

Friction results when two solid surfaces contact each other. Because solid surfaces are not completely smooth, they literally weld at the points of contact (Figure 10.13). When small forces $\mathbf{F}_1$ are applied to Body 1 in Figure 10.13, there is no motion, because the two bodies are fused together ($\mathbf{v}_1$ = const = 0). The **static frictional** force $\mathbf{F}_{fs}$ is exactly equal and opposite to the applied force $\mathbf{F}_1$, so there is no motion. Similarly, force $\mathbf{F}_2$ is equal

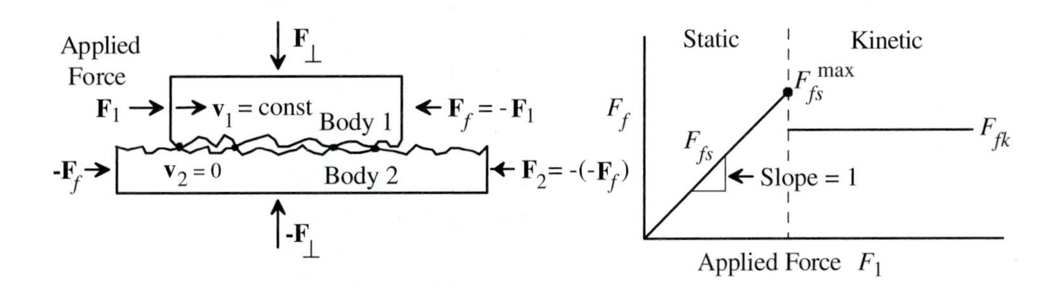

FIGURE 10.13
Frictional forces acting on two solid bodies.

TABLE 10.4
Coefficients of **static** and **kinetic friction**

	Static		Kinetic	
	Dry	Greasy	Dry	Greasy
Hard steel on hard steel	0.78	0.0052–0.23	0.42	0.029–0.12
Hard steel on graphite	0.21	0.09		
Hard steel on babbitt (ASTM No. 8)	0.42	0.08–0.17	0.35	0.065–0.14
Mild steel on lead	0.95	0.50	0.95	0.30
Aluminum on mild steel	0.61		0.47	
Teflon on Teflon	0.04			0.04
Teflon on steel	0.04			0.04
Nickel on nickel	1.10		0.53	0.12
Aluminum on aluminum	1.05		1.40	
Glass on glass	0.94	0.005–0.01	0.40	0.09–0.116
Oak on oak (parallel to grain)	0.62		0.48	0.067–0.164
Oak on oak (perpendicular to grain)	0.54		0.32	0.07

Adapted from: T. Baumeister, E. A. Avallone, and T. Baumeister III, *Mark's Standard Handbook for Mechanical Engineers,* 8th ed. (New York: McGraw-Hill, 1978), pp. 3–26.

and opposite to the frictional force acting on Body 2, $-\mathbf{F}_{fs}$; otherwise it would move. (We are assuming here that Body 2 is always stationary, i.e., $\mathbf{v}_2 = 0$.) When the magnitude of the applied force F_1 reaches a critical value $F_{fs}^{\max}$, the static welds break, allowing Body 1 to move at constant velocity $\mathbf{v}_1 \neq 0$. This critical value is related to the perpendicular force $\mathbf{F}_\perp$, the force exerted by gravity or by a bolt that holds the two bodies together. The critical force where the static welds break is

$$F_{fs}^{\max} = \mu_s F_\perp \tag{10-40}$$

where μ_s is the **coefficient of static friction.** Once Body 1 is in motion, the frictional force is determined by

$$F_{fk} = \mu_k F_\perp \tag{10-41}$$

where μ_k is the **coefficient of kinetic friction.** Both coefficients of friction are dimensionless. Table 10.4 lists some coefficients of friction for various materials.

Drag.

Drag is another type of frictional force exerted on a body as it moves through a fluid (e.g., gas or liquid). Figure 10.14 shows an automobile moving through the air, which puts a drag force on the automobile. The magnitude of the drag force is

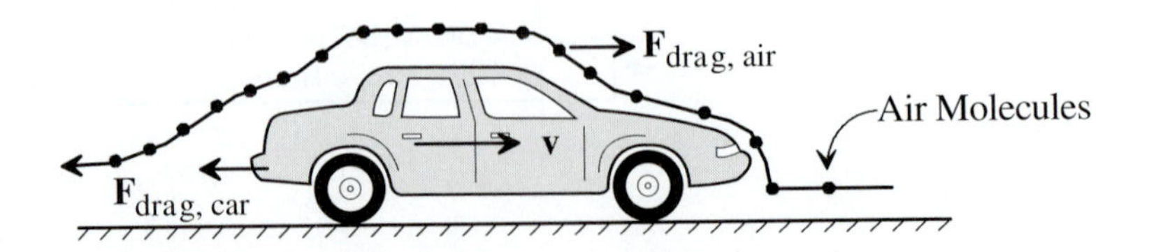

FIGURE 10.14
Drag on an automobile.

The drag on a bicyclist is measured in a wind tunnel.

TABLE 10.5

Drag coefficient of some simple shapes

Body	Flow Situation	C_d (dimensionless)
Circular plate		1.17
Sphere		0.47
Cube		1.05
Streamlined body $L/D = 2.5$		0.04

Source: W.-H. Hucho, *Aerodynamics of Road Vehicles* (Boston: Butterworth, 1987).

$$F = \tfrac{1}{2} C_d \rho A v^2 \tag{10-42}$$

where F is the drag force (N) on the body, ρ is the density (kg/m^3) of the fluid, v is the speed (m/s), and C_d is the drag coefficient, a dimensionless number; A is the "projected" frontal area (m^2) of the body, that is, the area of the shadow if the automobile were uniformly illuminated from the front. Table 10.5 shows the drag coefficient for a number of simple shapes.

Spring Force.

Spring force results when a spring is compressed or stretched (Figure 10.15). Hook's law describes the spring force

$$\mathbf{F} = -k\mathbf{x} \tag{10-43}$$

where $\mathbf{F}$ is the spring force, $\mathbf{x}$ is the position vector at the end of the spring using the unloaded spring position as the origin, and k is the spring constant (a scalar). The minus sign indicates that the spring force direction is opposite the position vector.

FIGURE 10.15

Spring forces due to compression and extension.

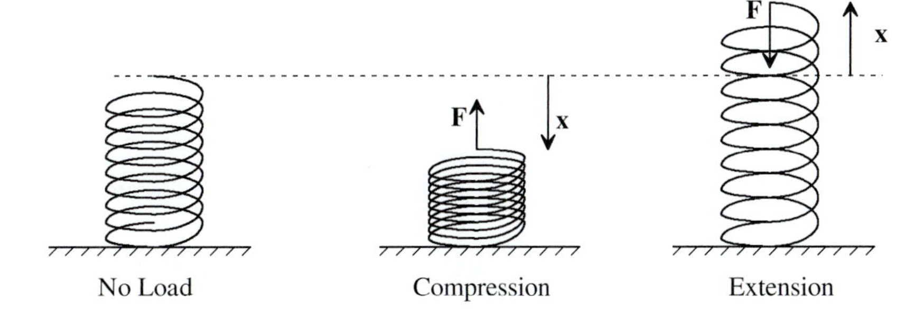

10.4 NEWTON'S FIRST LAW

Galileo (1564–1642) and Newton (1642–1727) thought about the question, What is the "natural state" of an object? The common sense answer is, The natural state of an object is when it comes to rest. Clearly, this answer agrees with our everyday experience. When you slide a hockey puck on a smooth ice surface, you know that eventually it will come to rest.

The genius of Galileo and Newton was to see beyond everyday experience. They both concluded that in the absence of applied forces, an object is in its natural state when it maintains its velocity. Thus, if an object is at rest, it will stay at rest in the absence of any applied forces, which is consistent with common sense. Less consistent with common sense is their assertion that an object in motion will maintain its velocity in the absence of any applied forces. This is known as the **law of inertia.**

In 1686, Newton described the law of inertia in his *Philosophiae Naturalis Principia Mathematica*. In his own words, "Every body persists in its state of rest or of uniform motion in a straight line unless it is compelled to change that state by forces impressed on it." This statement is often called **Newton's first law,** although it more properly should be attributed to Galileo, who preceded Newton by a generation. It can be stated mathematically as

$$\mathbf{v} = \text{constant} \qquad \text{or} \qquad \frac{d\mathbf{v}}{dt} = 0 \qquad \text{(no forces)} \tag{10-44}$$

Newton's first law can be written in an alternative way by defining a vector quantity called **momentum p,** which is just the mass m of the object (a scalar) multiplied by its velocity $\mathbf{v}$ (a vector):

$$\mathbf{p} \equiv m\mathbf{v} \tag{10-45}$$

Because the mass of the object is not changing, we can also write

$$\mathbf{p} = m\mathbf{v} = \text{constant} \qquad \text{or} \qquad \frac{d\mathbf{p}}{dt} = \frac{d(m\mathbf{v})}{dt} = 0 \quad \text{(no forces)} \tag{10-46}$$

How is it these great thinkers could propose that an object maintains its velocity, a conclusion so contrary to everyday experience? Figure 10.16 shows a "thought experiment" in which a ball is placed on an inclined plane and rolls downward. When it reaches the bottom of the incline, it travels in a horizontal direction until it rolls up the opposing

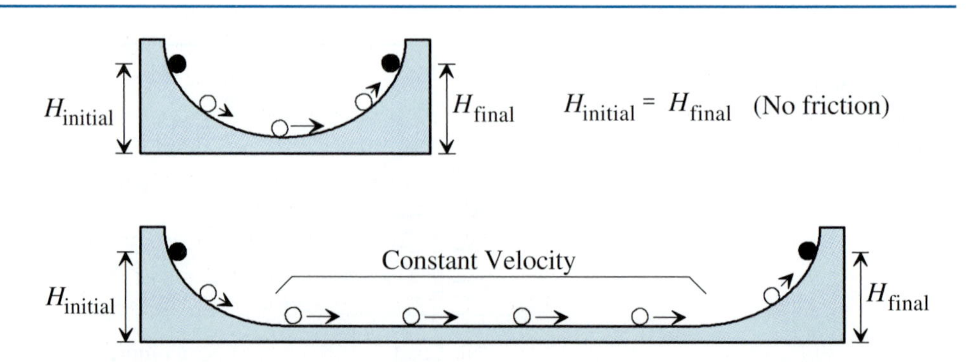

FIGURE 10.16
Thought experiment to illustrate Newton's first law.

incline. If we were to actually conduct this experiment, we would notice that the ball would rise to almost the same height at which it initially began. If it were a light ball and a rough-surfaced incline, the final height would be substantially less than the initial height. However, by using a heavy ball with smooth surfaces and lubrication, we would find that the final height is nearly indistinguishable from the initial height. Using such a perfected system, if the inclines were separated by farther and farther distances, we would expect the ball velocity on the horizontal surface to be maintained until it reaches the opposing incline. If we were to extrapolate this process and imagine the inclines to be separated by an infinite distance, then the ball velocity on the horizontal surface would be maintained forever.

The key idea in the above thought experiment is the ability to imagine the effects of frictionless motion. On the face of it, this is complete nonsense. How many things have you observed going on their way forever? Even on ice, friction will bring things to a halt pretty quickly. Ah, there's the rub (all puns intended)—friction is an outside force acting to slow the object down. In fact, friction is ubiquitous in our everyday life. It is an example of the genius of Galileo and Newton that long before the days of ball bearings or any low-friction devices, they were able to imagine what would happen if there were no friction.

On the other hand, you might ask: What good is it imagining friction doesn't exist, if friction is really inescapable? It is this ability to remove complications to arrive at a system simple enough to understand that is the hallmark of good science or engineering. The examples given here are, of course, flights of genius. But the ability to simplify and analyze complex systems is an art that can be mastered with practice. After the simple system is understood, complications can be added back one at a time until the original system is understood in sufficient detail for the purpose at hand.

As predicted by Newton's first law, we can actually observe real systems that maintain a given velocity in the absence of applied forces. For example, a spacecraft needs rocket power to pull away from earth's gravity and obtain a desired velocity. But, once it reaches this desired velocity, the rocket can be turned off and the spacecraft will continue moving at that same velocity. In fact, to stop the spacecraft (e.g., to land on Mars), a force is required to counter the spacecraft inertia.

10.5 NEWTON'S SECOND LAW

Newton's first law states that in the absence of a force, an object will maintain its velocity. By implication, to change the velocity of an object, a force must be applied. **Newton's second law** deals with the manner in which forces change the velocity of an object.

You are familiar with many different kinds of forces in your everyday life. Gravity, friction, muscle action, and wind resistance all generate forces that you experience every day. What do they have in common? These forces all have the **potential** to alter the motion of objects they act on. Muscle action **can** act on a stone, causing it to accelerate to some desired velocity. Wind resistance **can** act on an automobile, causing it to slow down. However, if there is a counterbalancing force, then the object will not change its motion. For example, if two people push on a stone with equal force in opposite directions, the stone will not move. Similarly, if the action of the automobile engine exactly counteracts wind resistance, the velocity of an automobile will not change. The essential point here is that to change an object's motion, there must be a **net** force acting on it.

Galileo Galilei: The First Physicist

About 400 years ago, an Italian astronomer and scientist named Galileo Galilei began a series of experiments to discover the principles of motion. Many people take this effort as the beginning of physics. Galileo challenged many common ideas of the time and was eventually threatened with being burned at the stake for his crazy idea that the earth is not at the center of the solar system. Along the way, one of the most important things that he did was to suggest that all theories are not proved true by logical argument only, but are also subject to confirming tests. He disproved many ideas that were common sense during his time. For example, everyone "knew" that heavy objects fell faster than light objects. "Just look at a feather and a cannonball," they said, "it's obvious what's going to happen." Legend has it that Galileo, in one of his many famous experiments, dropped the same-size wood and iron cannonballs from the Leaning Tower of Pisa and demonstrated that the two balls hit the ground at the same time even though the iron ball was seven times heavier than the wooden one. (Actually, Galileo never did this experiment; however, an experiment of this type was performed by Simon Stevinus in 1586.)

When trying to follow falling objects, Galileo realized that human perception is quickly outstripped by the speed of the objects after they fall a few seconds and positional observations are impossible. He ingeniously "diluted" gravity by rolling balls down inclined planes, and, using his heart pulse for timing, noted that for uniform units of time 1, 2, 3, 4, . . . , the distances

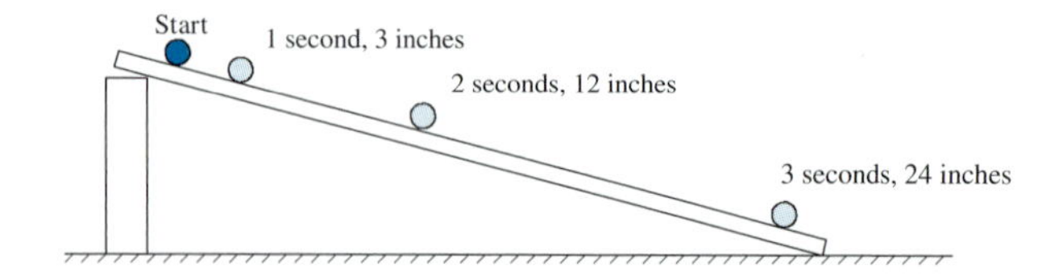

covered by the rolling ball were proportional to squares of integers, 1, 4, 9, 16. As we would say today, the distance is proportional to the square of time,

$$D \propto t^2 = kt^2$$

where D is the distance, t is the time since release, and k is a proportionality constant. By changing the slant of the inclined track, Galileo and an assistant could slow the motion of the rolling balls enough so that the position of a ball could be marked as time was called out.

Galileo in prison after being tried by the Inquisition for advocating that the sun is the center of the solar system.

Does this equation look familiar? This is the form of Equation 10-7a, the equation of motion for systems with constant acceleration and $\mathbf{x}_0$ and $\mathbf{v}_0$ both being zero. Galileo was the first to demonstrate by experiment that gravity acts by supplying a constant acceleration to all objects in its gravitation field.

Another law that Galileo discovered is essentially the same as Newton's first law, namely the *law of inertia.* Galileo proposed that, if free from external interference, an object will continue whatever motion it has forever.

Adapted from: I. Asimov, *Asimov's Biographical Encyclopedia of Science and Technology* (Garden City, NY: Doubleday, 1982).

Newton, in formulating the second law from these general concepts, asked himself what were the quantitative relationships between force and change in direction. He concluded that a proportionally larger force is needed to move a heavy object than to move a light object. This makes sense because it takes twice as much force to pick up a 100-pound rock as to pick up a 50-pound rock. His other conclusion was that force caused objects to accelerate in direct proportion to the magnitude of the force. Being the genius that he was, he cast his law more generally in terms of momentum rather than in terms of mass and acceleration. In words, his second law may be stated as: the time rate of change of momentum is proportional to the **net** force. In symbols,

$$\mathbf{F} = \frac{d\mathbf{p}}{dt} = \frac{d(m\mathbf{v})}{dt} \tag{10-47}$$

where $\mathbf{F}$ is the **net** force applied to the object, m is the mass, and $\mathbf{v}$ is the velocity. If the mass is not changing with time, then the second law reduces to its more familiar form,

$$\mathbf{F} = m\frac{d\mathbf{v}}{dt} = m\mathbf{a} \tag{10-48}$$

where $\mathbf{a}$ is the acceleration. For the case of one-dimensional motion, we can write the above equation using the components of force and acceleration as

$$F\mathbf{i} = m(a\mathbf{i}) = ma\mathbf{i} \tag{10-49}$$

Thus, the components of the vector are related as

$$F = ma \tag{10-50}$$

To emphasize the vector nature of the force, let's take the simplest case: a one-dimensional example with the acceleration acting opposite the initial velocity.

EXAMPLE 10.4

Problem Statement: Suppose that because of the action of brakes, a constant force of 10,000 N is applied to a car. If the brakes are applied in a 1000-kg automobile moving at 100 km/h, what is the acceleration, how long will it take to stop the car, and how far will the car go?

Solution: So many questions. First, the acceleration is pretty easy. Because the mass of the car is not changing, we can use Equation 10-50. Dividing 10,000 N by 1000 kg, we get an acceleration of 10 m/s^2:

$$a = \frac{F}{m} = \frac{10{,}000\ \text{N}}{1000\ \text{kg}} \times \frac{\frac{\text{kg}\cdot\text{m}}{\text{s}^2}}{\text{N}} = 10\ \text{m/s}^2$$

(Note the conversion factor from N to kg·m/s^2.)

Using Equation 10-8b, we can calculate when the velocity will be zero:

$$0 = v = v_0 + a_0 t$$

$$0 = 100\frac{\text{km}}{\text{h}} \times \frac{1000\text{ m}}{\text{km}} \times \frac{\text{h}}{3600\text{ s}} + 10\frac{\text{m}}{\text{s}^2} \times t$$

or

$$t = -2.77\text{ s}$$

Huh? This answer says that about 3 seconds ago, the car was standing still? What went wrong? Is Equation 10-8b wrong? Nope. Because the acceleration and the velocity are vector quantities, they have direction, and in this case, they oppose each other. The correct formulation of Equation 10-8b is then

$$0 = 100\frac{\text{km}}{\text{h}} \times \frac{1000\text{ m}}{\text{km}} \times \frac{\text{h}}{3600\text{ s}} - 10\frac{\text{m}}{\text{s}^2} \times t = 0$$

or

$$t = 2.77\text{ s}$$

Likewise, when using Equation 10-7a to calculate distance, we have to keep the vectors straight:

$$x = x_0 + v_0 t + \tfrac{1}{2}a_0 t^2$$

$$x = 0 + 100\frac{\text{km}}{\text{h}} \times \frac{1000\text{ m}}{\text{km}} \times \frac{\text{h}}{3600\text{ s}} \times 2.77\text{ s} - \frac{1}{2} \times 10\frac{\text{m}}{\text{s}^2} \times (2.77\text{ s})^2$$

$$= 38.6\text{ m}$$

For multidimensional problems, Newton's second law can be specified for each direction; thus,

$$\mathbf{F}_x = \frac{d\mathbf{p}_x}{dt} = \frac{d(m\mathbf{v}_x)}{dt} = m\frac{d\mathbf{v}_x}{dt} = m\mathbf{a}_x$$

$$\mathbf{F}_y = \frac{d\mathbf{p}_y}{dt} = \frac{d(m\mathbf{v}_y)}{dt} = m\frac{d\mathbf{v}_y}{dt} = m\mathbf{a}_y$$

$$\mathbf{F}_z = \frac{d\mathbf{p}_z}{dt} = \frac{d(m\mathbf{v}_z)}{dt} = m\frac{d\mathbf{v}_z}{dt} = m\mathbf{a}_z \quad (10\text{-}51)$$

where it is assumed that the mass m is constant. We can express the above relationships using the components of the vectors

$$F_x = ma_x \qquad F_y = ma_y \qquad F_z = ma_z \quad (10\text{-}52)$$

10.5.1 An Alternative Interpretation of *a*

Conventionally, a is the acceleration, the time derivative of velocity, a fact that we do not dispute. However, we can gain more insight into the significance of a if we were to rearrange Newton's second law and solve explicitly for it.

$$a = \frac{F}{m} \tag{10-53}$$

Here, we see an alternative interpretation of a; a also is the amount of force applied per unit mass.

We can think of a as a **specific quantity,** a topic discussed in more detail in Chapter 17. A specific quantity is the amount of something per unit mass. Specific quantities are indicated by placing a "hat" over the variable. For example, the specific volume $\hat{V}$ of a rock would be determined by dividing its volume V by its mass m.

$$\hat{V} = \frac{V}{m} \tag{10-54}$$

Because a is the amount of force per unit mass, it could be written as a specific quantity

$$a = \hat{F} = \frac{F}{m} \tag{10-55}$$

EXAMPLE 10.5

Problem Statement: A 70.0-kg skater is pushed with a force of 10.0 N for 1.00 s (Figure 10.17). Assuming the skates have negligible friction on the ice and the skater is initially at rest, what is the final velocity?

Solution: $v = v_0 + a_0 t$

$$a_0 = \frac{F}{m} = \frac{10.0 \text{ N}}{70.0 \text{ kg}} \times \frac{\frac{\text{kg}\cdot\text{m}}{\text{s}^2}}{\text{N}} = 0.143 \text{ m/s}^2$$

(Notice a_0 is the specific force as well as the acceleration.)

$$v = 0 + \left(0.143 \frac{\text{m}}{\text{s}^2}\right)(1.00 \text{ s}) = 0.143 \text{ m/s}$$

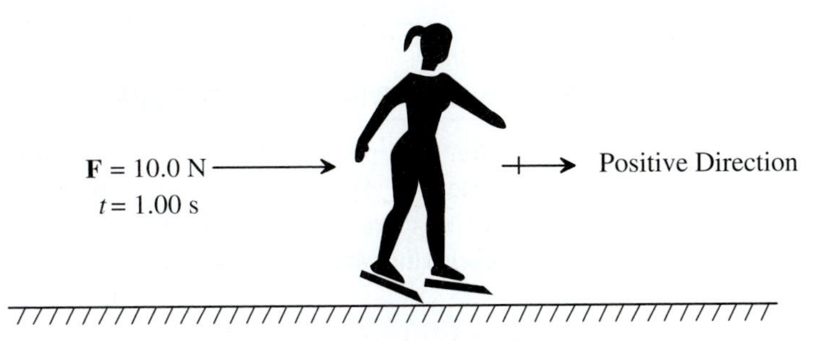

FIGURE 10.17
Skater being pushed with a force of 10.0 N for 1.00 s.

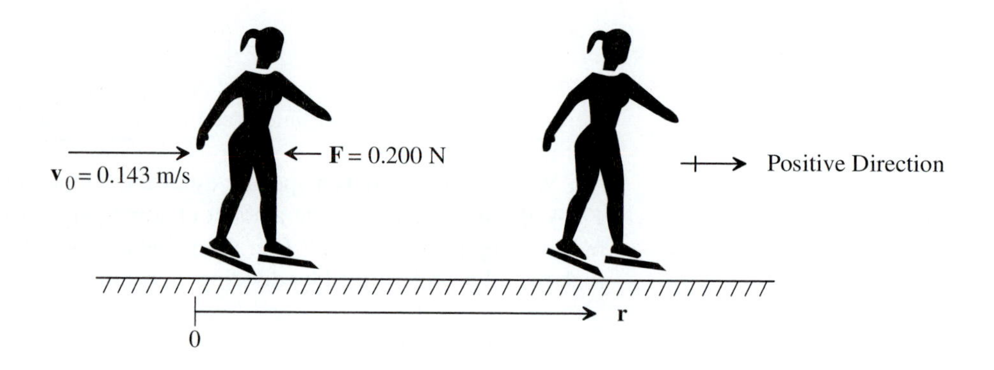

FIGURE 10.18
A skater's motion is opposed by friction, causing her to stop.

EXAMPLE 10.6

Problem Statement: The friction on the ice causes a constant 0.200 N force opposing the 70.0-kg skater's motion (Figure 10.18). If the skater is initially traveling at 0.143 m/s, how long before she comes to a complete stop? How far will she travel?

Solution: $v = v_0 + a_0 t$

$$t = \frac{v - v_0}{a_0}$$

$$a_0 = \frac{F}{m} = \frac{-0.200\text{ N}}{70.0\text{ kg}} \times \frac{\frac{\text{kg}\cdot\text{m}}{\text{s}^2}}{\text{N}} = -0.00286\text{ m/s}^2$$

(Again, notice a_0 is the specific force as well as the acceleration.)

$$t = \frac{0 - 0.143\text{ m/s}}{-0.00286\text{ m/s}^2} = 50\text{ s}$$

$$x = x_0 + v_0 t + \tfrac{1}{2} a_0 t^2 = 0 + \left(0.143\,\frac{\text{m}}{\text{s}}\right)(50\text{ s}) + \frac{1}{2}\left(-0.00286\,\frac{\text{m}}{\text{s}^2}\right)(50\text{ s})^2$$

$$= 3.57\text{ m}$$

10.6 NEWTON'S THIRD LAW

You have probably heard the expression "For every action, there is an equal and opposite reaction." This is a statement of **Newton's third law.** In Newton's own words, "To every action there is always opposed an equal reaction; or, the mutual actions of two bodies upon each other are always equal, and directed in contrary parts."

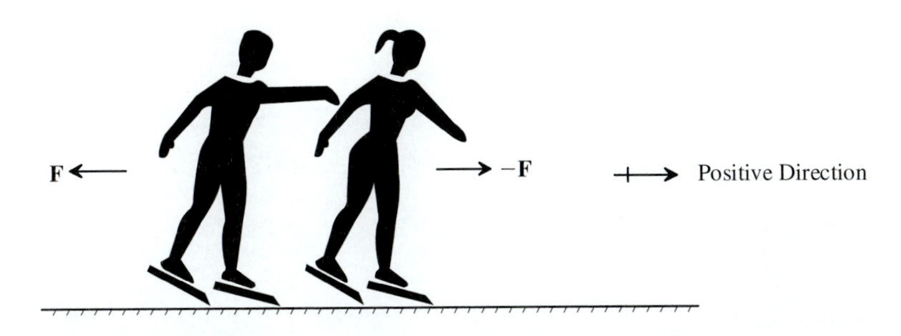

FIGURE 10.19
Two skaters pushing away from each other to illustrate Newton's third law.

The implications of Newton's third law are profound. It means:

- Forces **always** exist by the interaction of two (or more) bodies.
- The force on one body is equal and opposite to the force on the other body.
- It is impossible to have a single isolated force.
- The designation of an "action force" and a "reaction force" is arbitrary, because there is a mutual interaction between the two bodies.

EXAMPLE 10.7

Problem Statement: Henry and Mary are two ice skaters. Henry's mass is 110 kg and Mary's mass is 50.0 kg. They push against each other with a force of 20.0 N for 2.00 s (Figure 10.19). What is their final velocity?

Solution: *Henry*

$$v = v_0 + a_0 t$$

$$a_0 = \frac{F}{m} = \frac{-20.0\text{ N}}{110\text{ kg}} \times \frac{\frac{\text{kg·m}}{\text{s}^2}}{\text{N}} = -0.182\text{ m/s}^2$$

$$v = 0 + (-0.182\text{ m/s}^2)(2.00\text{ s}) = -0.364\text{ m/s}$$

Mary

$$v = v_0 + a_0 t$$

$$a_0 = \frac{F}{m} = \frac{20.0\text{ N}}{50\text{ kg}} \times \frac{\frac{\text{kg·m}}{\text{s}^2}}{\text{N}} = -0.400\text{ m/s}^2$$

$$v = 0 + (0.400\text{ m/s}^2)(2.00\text{ s}) = 0.800\text{ m/s}$$

Notice that the lighter person travels much faster.

Historical Perspective on the Laws of Motion

Nicolas Copernicus.

Johann Kepler.

There were few measuring instruments available to the ancient Greeks, so they tended to rely on observation and logic rather than careful measurement. Aristotle (B.C. 384–322) observed that light objects (e.g., feathers) inherently fall more slowly than heavy objects (e.g., rocks). Because this is such a common observation, this view of motion was widely held to be true, although it incorrectly neglects the effects of air friction. Aristotle also believed that an object in motion required a continuous force to maintain its motion. When analyzing how a catapulted object stays in motion, he neglected air friction and reasoned that the air continuously supplied the force necessary to keep the object moving.

About B.C. 260, Aristarchus (ca. B.C. 310–230) observed that the motion of all the planets and earth could be easily explained by assuming they move around the sun. A competing perspective was offered by Claudius Ptolemy (ca. A.D. 100–170), who propagated the views of Hipparchus (ca. B.C. 190–120) concerning an earth-centered universe with the planets and sun revolving around us. This Ptolemaic system was able to predict the motion of planets with reasonable accuracy given the lack of precise instruments available.

The late Middle Ages philosopher Jean Buridan (1295–1358) challenged Aristotle's views on motion. He proposed that although a force was required to put an object in motion, once in motion, the force was no longer required and the object would continue moving indefinitely. This notion challenged the belief that angels were needed to constantly push planets to keep them moving.

In 1507, Nicolas Copernicus (1473–1543) expanded on the ideas of Aristarchus. In 1530, he prepared a manuscript summarizing mathematical details about the proposed sun-centered universe. The full book was published in 1543, just before his death. His ideas were opposed by Martin Luther and the pope; the Roman Catholic Church finally removed his book from the Roman Catholic Index of prohibited literature in 1835.

Tycho Brahe (1546–1601), a Danish astronomer, was born a nobleman. At age 19, being rather arrogant and quarrelsome, he had a midnight duel with a rival over a fine point in mathematics. It cost him his nose; he wore a false metal nose the rest of his life. Being rather conservative, he rejected the Copernicus model of the universe. Instead, he proposed a compromise between the Copernican and Ptolemaic models in which the earth was the center of the universe, the sun revolved around the earth, and the planets revolved around the sun. Tycho is most famous for his very careful measurements of planetary motion.

In 1601, Johann Kepler (1571–1630) inherited Tycho's careful measurements of planetary motion. Tycho asked Kepler to interpret the data according to his rather complex model of the universe, but of course this failed. The best fit to the data had all the planets, including the earth, orbiting the sun. Further, rather than the expected circular orbits, they were elliptical. From these data, Kepler derived three empirical laws. *Kepler's first law* states that the sun is located at one of the **foci** of the elliptical orbit (if you do not know what "foci" are, see Appendix I, *Mathematics Supplement*). *Kepler's second law* states that a line connecting a planet and the sun will sweep over equal areas in equal times as the planet moves about its orbit. *Kepler's third law* states that the square of the period of planetary revolution is proportional to the cube of its distance from the sun.

Galileo Galilei (1564–1642) used measurement to quantify his ideas. He challenged Aristotle's views and argued that the reason light objects fall more slowly than heavy objects is due to air friction. Further, he reasoned that if they were placed in a vacuum, all objects would fall at the same rate. (This was later demonstrated by astronauts on the moon.) He made measurements showing that objects accelerate as they fall, a notion understood by Leondardo da Vinci (1452–1519). Galileo accepted the heliocentric model of the universe proposed by Copernicus. In 1632, he published *Dialogue on the Two Chief World Systems* in which the characters debated the relative merits of the Ptolemaic and Copernican universes. For supporting the Copernican universe, he was brought before the Inquisition, which forced him to renounce his views under threat of death. The Roman Catholic Church finally removed his book from the Roman Catholic Index of prohibited literature in 1835.

After graduating from Cambridge University, Sir Isaac Newton (1642–1727) returned home to the farm to avoid the plague that was sweeping London. He observed an apple fall from a tree and wondered if the same force acting on the apple also acted on the moon, causing it to constantly "fall" toward earth during its orbit. (This is a true story, not legend.) In 1687, he published his ideas related to gravity and motion in *Philosophiae Naturalis Principia Mathematica*, which some consider to be the greatest scientific work ever written. Here, he expanded on the ideas originally presented by Galileo and others and formalized them into Newton's three laws.

Adapted from: I. Asimov, *Asimov's Biographical Encyclopedia of Science and Technology* (Garden City, NY: Doubleday, 1982).

10.7 RELATIVISTIC MOMENTUM AND MASS AND ENERGY CHANGES (ADVANCED TOPIC)

Considering the relativistic transformation equations in Table 10.3, we see that it makes no sense for something to go faster than the speed of light. For velocities above the speed of light, the γ factor relating time and distance between frames of reference becomes imaginary, suggesting that the speed of light is the upper limit of velocity. This upper limit imposes further puzzles in relating classical and relativistic physics. Newton's laws tell us that **momentum** (mass times velocity) is constant in the universe; i.e., it is a *conserved* quantity. Ordinarily when kinetic energy is added to a particle, its velocity and therefore its momentum increases. If c is the upper limit, how can energy and momentum be related for relativistic particles? Specifically, if a particle is moving at $0.99c$, and we add enough energy to double its kinetic energy, then its velocity cannot increase proportionally because of the imposed upper limit. Because velocity is bounded by the relativistic transformations, then the only way to ensure that momentum is the same for each reference frame is for the moving mass m to increase as shown:

$$m = \frac{m_0}{\sqrt{1 - (v^2/c^2)}} \tag{10-56}$$

where m_0 is the **rest mass.** Does this really happen? That is, does this formula really imply that when energy is added to a relativistic particle that its mass increases? Indeed, as odd as it sounds, this is exactly what happens. Electrons are fairly easy to bring to relativistic speeds. When Joseph Thomson (1856–1940) measured the ratio of charge to mass (q/m) for electrons using the first primitive cathode-ray tubes in the late 1800s, he found that he could bring electrons to 90% of the speed of light, and that, for this and higher energies, the ratio of q/m would fall. We know now that the mass of the electrons was increasing so that q/m decreased.

10.7.1 Mass/Energy Equivalence

From fundamental definitions, we know that the differential work done on a particle is the force F acting over a small distance ds. Therefore the increased kinetic energy is

$$dE_k = F\,ds \tag{10-57}$$

Newton defined force in terms of changes to momentum, not acceleration. (How did he know?) Then,

$$F = \frac{d\,(mv)}{dt} \tag{10-58}$$

Substituting into Equation 10-57:

$$dE_k = \frac{d\,(mv)}{dt}\,ds \tag{10-59}$$

But $ds/dt = v$, the velocity between reference frames. Making this substitution

$$dE_k = v\,d\,(mv) = v^2dm + mv\,dv \tag{10-60}$$

Squaring Equation 10-56 and simplifying, we see

$$m^2c^2 = m^2v^2 + m_0^2c^2 \tag{10-61}$$

Differentiating, and then dividing both sides by $2m$, then

$$c^2dm = v^2dm + mv\,dv \tag{10-62}$$

Noting that the righthand sides of Equation 10-60 and Equation 10-62 are identical, we see

$$dE_k = c^2\,dm \tag{10-63}$$

that is, a change in kinetic energy can be related to a change in mass! It becomes even better. One can easily integrate Equation 10-63 by reasoning that when $v = 0$, then $E_k = 0$ and $m = m_0$, so

$$E_k = \int_0^{E_k} dE_k = c^2\int_{m_0}^{m} dm = c^2(m - m_0) \tag{10-64}$$

or

$$E_k = mc^2 - m_0c^2 \tag{10-65}$$

This implies that the classical equation for kinetic energy ($E_k = \frac{1}{2}mv^2$) is wrong! But we "know" this classical expression is right. What's going on? By taking the binomial expansion (Appendix J, *Mathematics Supplement*) of γ (Equation 10-35), and combining Equation 10-56 with Equation 10-65, we get

$$\begin{aligned} E_k &= \frac{m_0c^2}{\sqrt{1 - (v^2/c^2)}} - m_0c^2 \\ &= m_0c^2\left(1 + \frac{v^2}{2c^2} + \frac{3}{8}\frac{v^4}{c^4} + \cdots - 1\right) \\ &= \frac{1}{2}m_0v^2 + \frac{3}{8}m_0\frac{v^4}{c^2} + \cdots \end{aligned} \tag{10-66}$$

So we see that the first term is the classical expression for kinetic energy. For velocities smaller than about $0.1c$, this first term dominates.

More interesting yet is to take Equation 10-65 and turn it around so that

$$E = mc^2 = m_0c^2 + E_k \tag{10-67}$$

where E is now the total energy of a particle and the rest mass and kinetic energy are separate "energy" terms feeding the total energy. On the basis of this definition, we take Equation 10-62, multiply through by c^2, and, using the definition of $E = mc^2$ and $p = mv$, we get

$$E^2 = p^2c^2 + m_0^2c^4 \tag{10-68}$$

Several magical things spring from this equation. First, Equation 10-68 shows us that light, which has no rest mass, still has momentum! Measurements confirm this; in fact, there are serious proposals to propel objects out of the solar system using large "sails" to capture the momentum carried by light from the sun.

If the momentum is zero in Equation 10-68, then energy and rest mass are directly related, by $E = mc^2$, the most famous equation in science. In fact, nuclear physicists usually refer to the masses of subatomic particles by using their "rest mass energy." For example, the rest mass energy of an electron is 511 keV. (An *electron volt,* eV, is the energy gained by an electron moving through a potential of 1 volt.) This strange idea that mass and energy convert is readily confirmed. When an antielectron meets with a normal electron, two 511-keV photons are created from the loss of mass when the electron and antielectron annihilate each other.

This is but a quick look into relativity. We hope to have piqued your interest in one of the great discoveries of the 20th century.

10.8 EXAMPLE APPLICATIONS OF NEWTON'S LAWS

The following examples show Newton's laws applied in a variety of ways to show how powerful they are.

EXAMPLE 10.8

Problem Statement: On top of a 300-ft building, a man leans over the building edge and throws a ball vertically upward with an initial velocity of 64.4 ft/s. Find the position and velocity of the ball at successive 1-s intervals after it leaves the man's hand, and find the time it takes to reach the sidewalk on the ground at the side of the building.

Solution: We can use Equations 10-7b and 10-8b to analyze the motion of a freely falling body, that is, a body that experiences no force other than that of the earth's gravity, by letting $a_0 = g = -32.2$ ft/s^2. In this example, our equations are

$$v = v_0 + a_0t = 64.4 \text{ ft/s} + (-32.2 \text{ ft/s}^2)t$$

$$x = x_0 + v_0t + \tfrac{1}{2}a_0t^2 = 0 + (64.4 \text{ ft/s})t + \tfrac{1}{2}(-32.2 \text{ ft/s}^2)t^2$$

Figure 10.20 shows the solution to these equations in graphical and tabular form. Obviously, the time it takes to reach the ground is between 6 and 7 seconds. To find the exact time, solve the quadratic equation

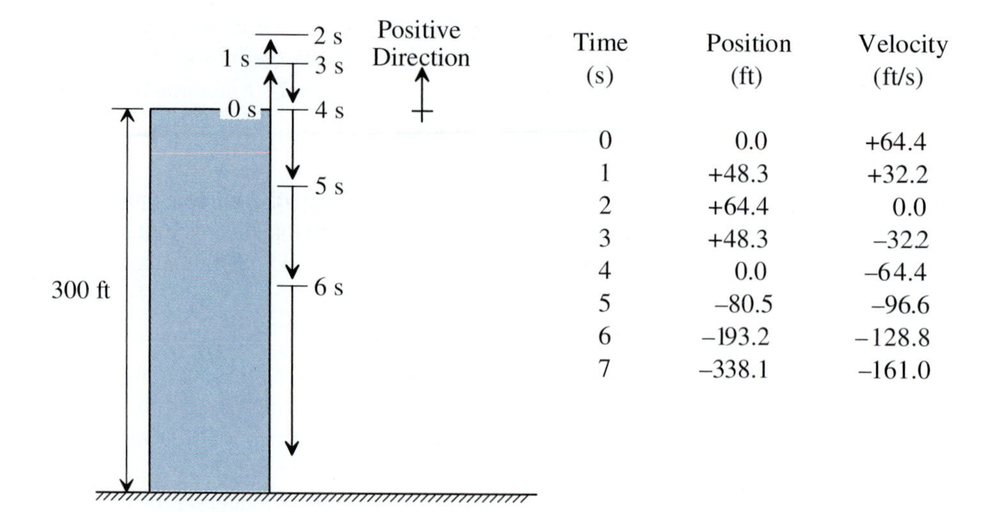

Time (s)	Position (ft)	Velocity (ft/s)
0	0.0	+64.4
1	+48.3	+32.2
2	+64.4	0.0
3	+48.3	–32.2
4	0.0	–64.4
5	–80.5	–96.6
6	–193.2	–128.8
7	–338.1	–161.0

FIGURE 10.20
Position and velocity of a ball thrown vertically into the air from a building.

$$x = -300 = 64.4t - 16.1t^2$$

or

$$0 = 16.1t^2 - 64.4t - 300$$

The two solutions are $t = 6.76$ s and $t = -2.76$ s, but, of course, the second solution is not physically possible.

EXAMPLE 10.9

Problem Statement: A cannonball is fired horizontally from a cannon located on a 64.4-ft cliff (Figure 10.21). The muzzle velocity is 1000 ft/s. How long does the cannonball remain in the air? What is its range? (*Note:* Assume air resistance is negligible.)

Solution: This problem can be solved by independently considering motion in the x- and y-directions:

$$y = -64.4 \text{ ft} = y_0 + v_{y,0}t + \tfrac{1}{2}a_{y,0}t^2 = 0 + 0t + \tfrac{1}{2}(-32.2 \text{ ft/s}^2)t^2$$

$$-64.4 \text{ ft} = (-16.1 \text{ ft/s}^2)t^2$$

$$t = \sqrt{\frac{-64.4 \text{ ft}}{-16.1 \text{ ft/s}^2}} = 2.00 \text{ s}$$

x-direction

$$x = x_0 + v_{x,0}t + \tfrac{1}{2}a_{x,0}t^2 = 0 + (1000 \text{ ft/s})(2.00 \text{ s}) + \tfrac{1}{2}0(2.00 \text{ s})^2 = 2000 \text{ ft}$$

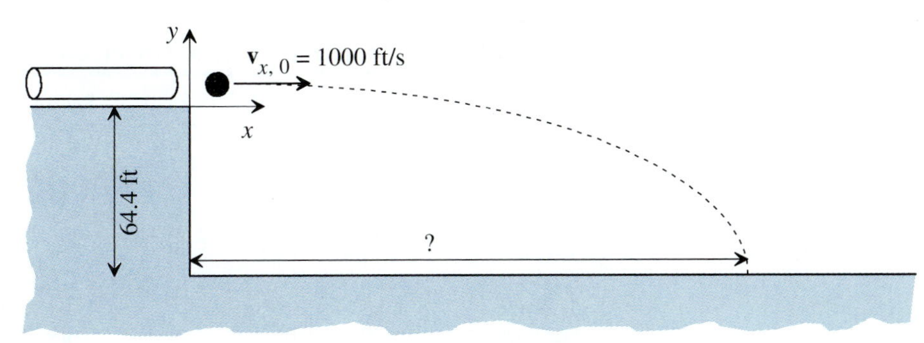

FIGURE 10.21
Cannonball fired horizontally from a cliff.

EXAMPLE 10.10

Problem Statement: The magnitude of the attractive gravity force between two spherical objects is given by the equation

$$F = G\frac{Mm}{r^2}$$

where M and m are the mass of each spherical object, r is the distance separating them as measured from the center of each spherical object, and G is the proportionality constant. Allow M to be the mass of the earth and m to be the mass of an arbitrary spherical object. What is the acceleration g of this arbitrary object toward the earth?

Solution: We can visualize the magnitude of the acceleration g as the magnitude of the force per unit mass $\hat{F}$,

$$g = \hat{F} = \frac{F}{m} = \frac{G\frac{Mm}{r^2}}{m} = G\frac{M}{r^2} = G\frac{M}{(R+y)^2}$$

FIGURE 10.22
Forces at the earth's surface.

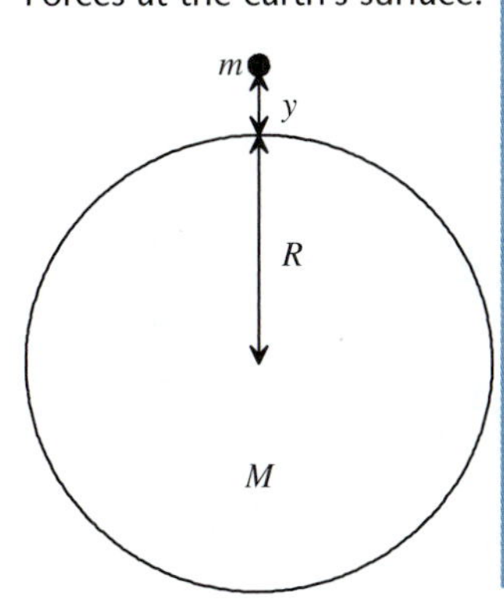

where R is the earth's radius and y is the height of the object above the earth's surface (Figure 10.22). Near the earth's surface, $R >> y$, therefore

$$g = G\frac{M}{R^2} \tag{10-69}$$

$$= \left(6.673 \times 10^{-11}\frac{\text{N·m}^2}{\text{kg}^2}\right)\frac{5.9785 \times 10^{24}\,\text{kg}}{(6.378 \times 10^6\,\text{m})^2} \times \frac{\frac{\text{kg·m}}{\text{s}^2}}{\text{N}}$$

$$= 9.806\ \text{m/s}^2$$

TABLE 10.6
Accelerations due to gravity on the surface of some heavenly bodies

Body	Mass, M (kg)	Radius, R (m)	Acceleration, g (m/s^2)
Moon	7.342×10^{22}	1.738×10^{6}	1.622
Mercury	3.28×10^{23}	2.57×10^{6}	3.31
Mars	6.34×10^{23}	3.43×10^{6}	3.60
Uranus	8.61×10^{25}	2.67×10^{7}	8.06
Venus	4.82×10^{24}	6.310×10^{6}	8.08
Earth	5.9785×10^{24}	6.378×10^{6}	9.806
Saturn	5.63×10^{26}	6.03×10^{7}	10.3
Neptune	9.98×10^{25}	2.49×10^{7}	10.7
Jupiter	1.880×10^{27}	7.180×10^{7}	24.33
Sun	1.9693×10^{30}	6.96×10^{8}	271

EXAMPLE 10.11

Problem Statement: Using Equation 10-69, prepare a table listing the accelerations due to gravity on the surface of a number of heavenly bodies.

Solution: The results are shown in Table 10.6.

EXAMPLE 10.12

Problem Statement: A 0.300-m spring ($k = 1.00 \times 10^5$ N/m) supports a 500-kg mass (Figure 10.23). How much does the spring compress? What is the spacing between the mass and the ground?

Solution: $L_{\text{final}} = L_{\text{initial}} - x$

$$F_{\text{spring}} = -kx \quad \text{(Hook's law)}$$

Once the mass comes to equilibrium (i.e., it stops moving), the force of gravity acting on the mass is opposite the force of the spring acting on the mass.

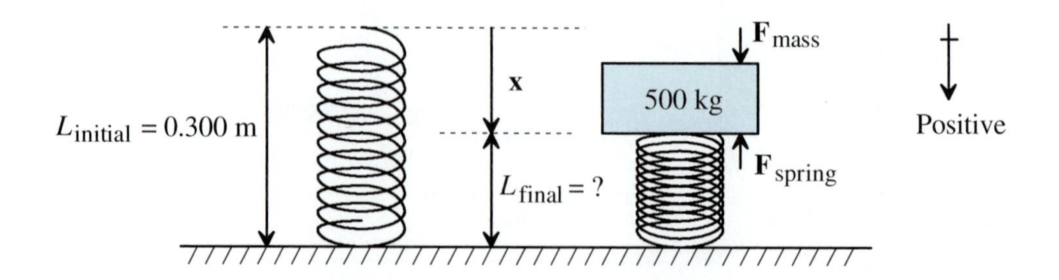

FIGURE 10.23
A spring supporting a 500-kg mass.

$$x = \frac{F_{\text{spring}}}{k} = \frac{F_{\text{mass}}}{k} = \frac{mg}{k} = \frac{(500\text{ kg})(9.8\text{ m/s}^2)}{1.00 \times 10^5\text{ N/m}} \times \frac{\text{N}}{\frac{\text{kg}\cdot\text{m}}{\text{s}^2}} = 0.049\text{ m}$$

$$L_{\text{final}} = 0.300\text{ m} - 0.049\text{ m} = 0.251\text{ m}$$

EXAMPLE 10.13

Problem Statement: (a) A 50-kg cannonball drops from the roof of a 200-m building (Figure 10.24). How long does it take for the cannonball to hit the ground? (b) According to Newton's third law, the earth is attracted upward toward the cannonball. How far does the earth move?

Solution: (a) *Cannonball*

$$x = 200\text{ m} = x_0 + v_0 t + \tfrac{1}{2}a_0 t^2 = 0 + 0t + \tfrac{1}{2}(9.8\text{ m/s}^2)t^2 = (4.9\text{ m/s}^2)t^2$$

$$t = \sqrt{\frac{200\text{ m}}{4.9\text{ m/s}^2}} = 6.4\text{ s}$$

(b) *Earth*

$$x = x_0 + v_0 t + \tfrac{1}{2}a_0 t^2$$

$$a_0 = \frac{F}{M} = \frac{G\left(\frac{Mm}{R^2}\right)}{M} = \frac{\left(\frac{Gm}{R^2}\right)m}{M} = \frac{gm}{M}$$

$$= \frac{(-9.8\text{ m/s}^2)(50\text{ kg})}{5.9785 \times 10^{24}\text{ kg}} = -8.2 \times 10^{-23}\text{ m/s}^2$$

$$x = 0 + 0t + \tfrac{1}{2}(-8.2 \times 10^{-23}\text{ m/s}^2)(6.4\text{ s})^2 = -1.7 \times 10^{-21}\text{ m}$$

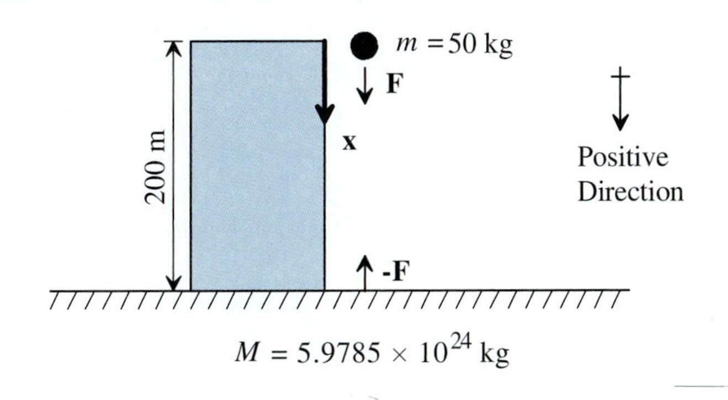

FIGURE 10.24
Cannonball dropping from a 200-m roof.

The earth moves only 1.7×10^{-21} m. As a point of reference, a helium atom is 1.9×10^{-10} m in diameter; therefore, for all practical purposes, the earth is stationary. It is worth noting that in the preceding calculations, the acceleration due to gravity was assumed to be constant because the distance between the center of the cannonball and the center of the earth is approximately constant.

EXAMPLE 10.14

Problem Statement: A "magnetic separator" allows ferromagnetic materials (e.g., steel) to be separated from nonferromagnetic materials. For recycling purposes, this separator may be used to separate iron and steel from municipal solid waste. The mixed waste is first shredded to an average particle diameter of 0.1 cm. Then, the mixed waste is dropped adjacent to a "magnetic wall" that applies a force of 3.00 μN to each steel particle, but no force is applied to the rest of the municipal solid waste (Figure 10.25). Assuming air resistance is negligible, how long does it take for the pulverized waste to fall 30 m? How far must the two chutes be separated? (*Note:* $\rho_{steel} = 7.8$ g/cm^3.)

Solution: y-direction:

$$y = -30 \text{ m} = y_0 + v_{y,0}t + \tfrac{1}{2}a_{y,0}t^2 = 0 + 0t + \tfrac{1}{2}(-9.8 \text{ m/s}^2)t^2 = (-4.9 \text{ m/s}^2)t^2$$

$$t = \sqrt{\frac{-30 \text{ m}}{-4.9 \text{ m/s}^2}} = 2.5 \text{ s}$$

x-direction:

$$x = x_0 + v_{x,0}t + \tfrac{1}{2}a_{x,0}t^2$$

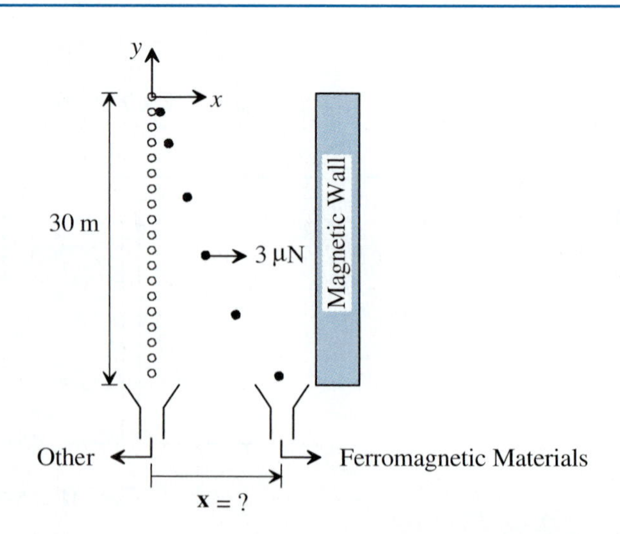

FIGURE 10.25
Magnetic separator to remove iron and steel from mixed municipal solid waste.

$$a_{x,0} = \frac{F}{M} = \frac{F}{\rho_{\text{steel}} V} = \frac{F}{\rho_{\text{steel}} \frac{4}{3}\pi\left(\frac{D}{2}\right)^3}$$

$$= \frac{3 \times 10^6\,\text{N}}{\left(7.8\frac{\text{g}}{\text{cm}^3}\right)\left(\frac{\text{kg}}{1000\,\text{g}}\right)\frac{4}{3}\pi\left(\frac{0.1\,\text{cm}}{2}\right)^3} \times \frac{\frac{\text{kg}\cdot\text{m}}{\text{s}^2}}{\text{N}} = 0.73\,\frac{\text{m}}{\text{s}^2}$$

$$x = 0 + 0(2.5\,\text{s}) + \tfrac{1}{2}(0.73\,\text{m/s}^2)(2.5\,\text{s})^2 = 2.3\,\text{m}$$

Having mastered some simple cases, let's tackle a more complicated example: planetary motion in two dimensions. We will also slip in the concept of solving equations numerically, just for fun.

10.9 PLANETARY MOTION

Sometimes you will need to calculate some motion that is difficult, or even impossible, to solve analytically. One example of this is when calculating the orbits of planets. The gravitational force F between two point masses can be calculated by

$$F = G\frac{Mm}{r^2} \tag{10-70}$$

where G is the universal gravitation constant, r is the distance between the centers of the two point masses, M is the mass of the sun (in this case), and m is the mass of the earth (in this case). At the **perihelion** (the point where the earth is nearest the sun), the distance r is 1.4708×10^{11} m and the earth's velocity is 30,280 m/s (Figure 10.26). Given this information, it is possible to calculate the motion of the earth around the sun. This is not an extremely difficult system to solve analytically, but for fun, let's do it numerically.

One numerical method that allows us to solve such problems is to divide the time interval of interest into many small slices. With the information given, we cannot directly calculate where the earth will be 180 days from now; however, we can make a reasonable guess where it will be in a day or two based on its current direction and velocity. After calculating the position 2 days from now, we could then calculate the new gravitational force and velocity, and then estimate where it will be 2 days from then. We then continue time step by time step until we reach the position after 180 days. Of course, the larger you make the time step, the more error slips into the calculation.

The topic of numerical methods is a huge one that could easily take a whole semester, but for now we will blindly charge ahead. Given an x, y coordinate position of the earth relative to the sun, we can calculate the distance between them as

$$r = \sqrt{x^2 + y^2} \tag{10-71}$$

From this and Equation 10-70, we can determine the force applied to the earth.

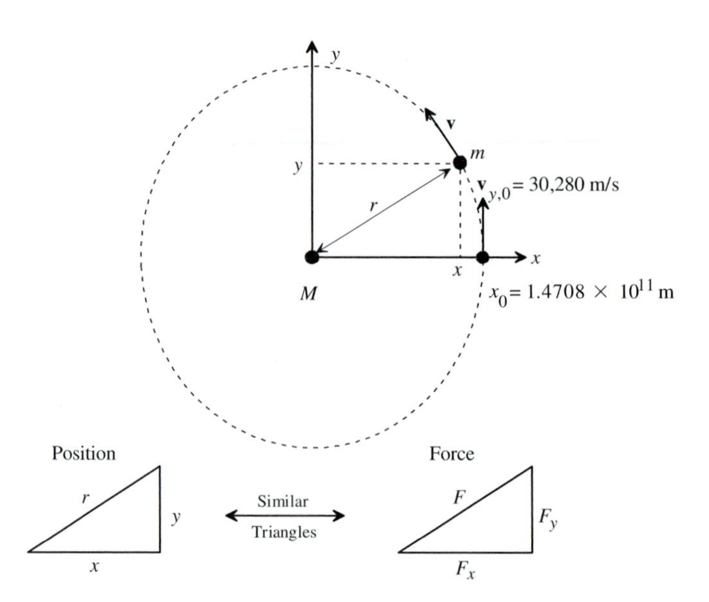

FIGURE 10.26
Analysis of earth orbit.

Next, we resolve the force vector **F** into its components F_x and F_y. From Figure 10.26, we see by similar triangles that

$$\frac{x}{r} = \frac{F_x}{F} \qquad \frac{y}{r} = \frac{F_y}{F}$$

$$F_x = F\frac{x}{r} \qquad F_y = F\frac{y}{r} \tag{10-72}$$

These are scalar equations; all indicated values are the magnitudes of the vectors. These equations can be formulated as vector equations

$$\mathbf{F}_x = -F\frac{\mathbf{x}}{r} \qquad \mathbf{F}_y = -F\frac{\mathbf{y}}{r} \tag{10-73}$$

where the minus sign indicates that the force acting on the earth is always opposite the position vector.

The acceleration in each direction is determined by the force

$$\mathbf{a}_{x,0} = \frac{\mathbf{F}_x}{m} \qquad \mathbf{a}_{y,0} = \frac{\mathbf{F}_y}{m} \tag{10-74}$$

both of which are constant at a given instant in time. Knowing the acceleration, we can calculate the velocity by manipulating Equation 10-5 and noting that $\mathbf{a}_{ave}$ becomes **a** at small Δt.

$$\mathbf{v}_x = \mathbf{v}_{x,0} + \mathbf{a}_{x,0}\Delta t \qquad \mathbf{v}_y = \mathbf{v}_{y,0} + \mathbf{a}_{y,0}\Delta t \tag{10-75}$$

Knowing the velocity, we can calculate the position by manipulating Equation 10-2 and noting that $\mathbf{v}_{ave}$ becomes **v** at small Δt.

TABLE 10.7
Computer printout giving information about the earth as it orbits the sun

Time (d)	x	y	r	F	F_x	F_y	V_x	V_y
0.0	1.47E+11	0.00E-01	1.47E+11	3.66E+22	−3.66E+22	0.00E−01	−2.65E+02	3.03E+04
0.5	1.47E+11	1.31E+09	1.47E+11	3.67E+22	−3.67E+22	−3.26E+20	−5.30E+02	3.03E+04
1.0	1.47E+11	2.62E+09	1.47E+11	3.67E+22	−3.66E+22	−6.52E+20	−7.95E+02	3.03E+04
1.5	1.47E+11	3.92E+09	1.47E+11	3.67E+22	−3.66E+22	−9.78E+20	−1.06E+03	3.03E+04
2.0	1.47E+11	5.23E+09	1.47E+11	3.67E+22	−3.66E+22	−1.30E+21	−1.32E+03	3.03E+04
2.5	1.47E+11	6.54E+09	1.47E+11	3.67E+22	−3.66E+22	−1.63E+21	−1.59E+03	3.02E+04
15.0	1.42E+11	3.88E+10	1.47E+11	3.67E+22	−3.54E+22	−9.68E+21	−8.12E+03	2.92E+04
30.0	1.27E+11	7.49E+10	1.47E+11	3.66E+22	−3.15E+22	−1.87E+22	−1.54E+04	2.61E+04
45.0	1.03E+11	1.06E+11	1.47E+11	3.65E+22	−2.54E+22	−2.62E+22	−2.16E+04	2.11E+04
60.0	7.15E+10	1.29E+11	1.48E+11	3.63E+22	−1.76E+22	−3.18E+22	−2.63E+04	1.48E+04
75.0	3.55E+10	1.44E+11	1.48E+11	3.61E+22	−8.63E+21	−3.50E+22	−2.91E+04	7.49E+03
90.0	−2.94E+09	1.49E+11	1.49E+11	3.58E+22	7.07E+20	−3.58E+22	−2.99E+04	−2.28E+02
105.0	−4.12E+10	1.44E+11	1.50E+11	3.55E+22	9.76E+21	−3.41E+22	−2.87E+04	−7.84E+03
120.0	−7.67E+10	1.29E+11	1.50E+11	3.51E+22	1.80E+22	−3.02E+22	−2.57E+04	−1.48E+04
135.0	−1.07E+11	1.06E+11	1.51E+11	3.49E+22	2.48E+22	−2.45E+22	−2.10E+04	−2.08E+04
150.0	−1.31E+11	7.61E+10	1.51E+11	3.46E+22	2.99E+22	−1.74E+22	−1.50E+04	−2.53E+04
165.0	−1.46E+11	4.13E+10	1.52E+11	3.44E+22	3.31E+22	−9.37E+21	−8.12E+03	−2.82E+04
180.0	−1.52E+11	3.86E+09	1.52E+11	3.43E+22	3.43E+22	−8.71E+20	−7.65E+02	−2.93E+04
195.0	−1.48E+11	−3.38E+10	1.52E+11	3.43E+22	3.34E+22	7.62E+21	6.61E+03	−2.85E+04
210.0	−1.35E+11	−6.94E+10	1.52E+11	3.43E+22	3.05E+22	1.57E+22	1.36E+04	−2.60E+04
225.0	−1.14E+11	−1.01E+11	1.52E+11	3.44E+22	2.58E+22	2.28E+22	1.97E+04	−2.17E+04
240.0	−8.50E+10	−1.25E+11	1.51E+11	3.46E+22	1.94E+22	2.86E+22	2.46E+04	−1.61E+04
255.0	−5.09E+10	−1.42E+11	1.51E+11	3.48E+22	1.17E+22	3.28E+22	2.80E+04	−9.43E+03
270.0	−1.34E+10	−1.50E+11	1.50E+11	3.51E+22	3.13E+21	3.49E+22	2.95E+04	−2.04E+03
285.0	2.49E+10	−1.48E+11	1.50E+11	3.54E+22	−5.87E+21	3.49E+22	2.92E+04	5.57E+03
300.0	6.15E+10	−1.36E+11	1.49E+11	3.57E+22	−1.47E+22	3.25E+22	2.69E+04	1.29E+04
315.0	9.41E+10	−1.15E+11	1.48E+11	3.60E+22	−2.28E+22	2.78E+22	2.28E+04	1.95E+04
330.0	1.20E+11	−8.61E+10	1.48E+11	3.62E+22	−2.95E+22	2.11E+22	1.71E+04	2.48E+04
345.0	1.38E+11	−5.15E+10	1.47E+11	3.65E+22	−3.42E+22	1.27E+22	1.02E+04	2.84E+04
360.0	1.47E+11	−1.34E+10	1.47E+11	3.66E+22	−3.65E+22	3.33E+21	2.44E+03	3.02E+04

$$\mathbf{x} = \mathbf{x}_0 + \mathbf{v}_{x,0}\Delta t \qquad \mathbf{y} = \mathbf{y}_0 + \mathbf{v}_{y,0}\Delta t \tag{10-76}$$

In this case, Δt is the time step and the subscript 0 indicates the position, velocity, and acceleration calculated from the previous time step. As shown in Program 10.1, we use half a day, or 43,200 seconds, for the time step (called `epsi` in the program). Computers make this problem easier to work, but it can be done by hand if necessary. The basic procedure is to set up a table and work through it step by step. We have calculated the motion through 360 days. In Table 10.7, the first 5 steps are printed, then only every 30th step is printed.

10.10 CARE AND FEEDING OF FORMULAS

The most important principle in the use of formulas is that the user must understand the formula. Formulas range from always-and-forever True (with a capital T) to a loose collection of data points gathered with little understanding of what's going on by the formulator. As shown by the examples, the user must understand what the variables stand for. If a formula

```
program orbits
      implicit none
      real*8 mass_sun, mass_earth, grav, gm12, epsi
      real*8 x, y, r, f, fx, fy, vx, vy
      integer time
c real*8 is required to hold gm12, below. It's a big number
      parameter (mass_sun = 1.9891E+30)
      parameter (mass_earth = 5.9758E+24)
      parameter (grav = 6.67E 11)
      parameter (epsi = 432.0E+2)
      open(15, file = 'orbit.dat')
c setting initial conditions
      x = 1.4708E+11
      y = 0
      vx = 0
      vy = 3.028E+04
      gm12 = grav*mass_sun*mass_earth
c write headers
      write(15,25)
25    format(1x, 'Time', t12,' X', t22,' Y',t32,' R',t42,
     & ' F',t52,' FX', t62,' FY', t72,' VX', t82,' VY')
      write(15,*)' (days)'
c starting main loop advancing 1/2 day at a time
      do time = 0, 730
         r = sqrt(x*x + y*y)
         f = gm12 / (r*r)
         fx =  abs(f)*x/r
         fy =  abs(f)*y/r
         vx = vx + fx*epsi/mass_earth
         vy = vy + fy*epsi/mass_earth
c print out 1st 5 time steps, then each 30 days
      if (time .lt. 5 .or. mod( time, 30) .eq. 0)then
         write(15, ' (f7.1,1p8E10.2) ' )
     &   0.5*real(time),x,y,r,f,fx,fy,vx,vy
      endif
c advance travel
      x = x + vx*epsi
      y = y + vy*epsi
      endo
      close (15)
      stop
      end
```

PROGRAM 10.1

Computer code to analyze earth orbit.

involves time, for example, does the t stand for time since the beginning of the universe, time since you last looked at your watch, or what? Avoid the "plug-and-chug" approach with formulas and you will be on your way to being a successful engineer.

10.11 SUMMARY

To understand Newton's laws, it is necessary to understand how motion is analyzed. The following definitions are important:

- *Position* is the location within a reference frame and can be described as a vector.
- *Displacement* is a vector describing the change in position.
- *Distance* is a nonnegative scalar that describes the length of a path.
- *Velocity* is a vector describing the rate that the position vector changes with respect to time.
- *Speed* is a scalar describing the magnitude of the velocity vector.
- *Acceleration* is a vector describing the rate that the velocity vector changes with respect to time.

Many of these words have similar meanings that differ in subtle ways. It is important to use them properly.

When specifying a position, velocity, or acceleration, it is necessary to establish a reference frame against which the motion is measured. In this text, we will use the earth reference frame unless otherwise specified. The reference frame can exist in one, two, or three dimensions, depending upon the problem being solved.

A force is the influence on a body that can cause it to accelerate. There are four fundamental forces in nature: strong force, weak force, gravity, and electromagnetism. Engineers often deal with "other" forces (e.g., friction, drag, and spring force) that are really manifestations of the fundamental forces.

Newton's first law states that in the absence of a net force, the momentum of an object does not change. Newton's second law states that a net force changes the momentum of an object. Newton's third law states that isolated forces cannot exist; the force on one body is equal and opposite the force on the second body.

We have all had daily experience with motion all our lives. We have had to understand motion using intuition, without the formal guidance of Newton's laws. For most of us, our intuition is actually wrong. It is necessary to unlearn your intuitive understanding of motion and then learn the Newtonian view of the world. Surprisingly, many students who have been repeatedly exposed to Newton's laws through their high school and college physics courses have not really "bought into" Newton's laws. Actually, only a small fraction of people are "Newtonian thinkers" when it comes to understanding motion. You will be well advised to carefully study Newton's laws and incorporate them into your understanding of the world.

Nomenclature

$\mathbf{a}$ acceleration vector (m/s^2)
a acceleration scalar (m/s^2)
c speed of light (m/s)
E energy (J)
$\mathbf{F}$ force vector (N)

F force scalar (N)
$\hat{F}$ specific force (N/kg)
G gravitational constant ($N \cdot m^2/kg^2$)
g acceleration due to gravity (m/s^2) or specific force due to gravity (N/kg)
H height (m)
$\mathbf{i}$ unit vector in x-direction (dimensionless)
$\mathbf{j}$ unit vector in y-direction (dimensionless)
$\mathbf{k}$ unit vector in z-direction (dimensionless)
k spring constant (N/m)
L position (m)
M mass (kg)
m mass (kg)
$\mathbf{p}$ momentum vector ($kg \cdot m/s^2$)
p momentum scalar ($kg \cdot m/s^2$)
q charge (C)
$\mathbf{r}$ displacement vector (m)
s distance (m)
t time (s)
V volume (m^3)
$\hat{V}$ specific volume (m^3/kg)
$\mathbf{v}$ velocity vector (m/s)
v velocity scalar (m/s)
$\mathbf{x}$ position vector in x-direction (m)
x position scalar in x-direction (m)
$\mathbf{y}$ position vector in y-direction (m)
y position scalar in y-direction (m)
$\mathbf{z}$ position vector in z-direction (m)
z position scalar in z-direction (m)
γ $1/\sqrt{1 - v^2/c^2}$ (dimensionless)
ε permittivity ($C^2/N \cdot m^2$)
μ magnetic dipole moment ($A \cdot m^2$)
μ_0 permeability (N/A^2)
μ_k coefficient of kinetic friction (dimensionless)
μ_s coefficient of static friction (dimensionless)

Further Readings

Halliday, D., and R. Resnick, *Fundamentals of Physics*. New York: Wiley, 1974.
Serway, R. A. *Principles of Physics.* Fort Worth: Saunders, 1994.

PROBLEMS

10.1 A car starts from rest and travels northward. It accelerates at a constant rate for 30 s until it reaches a velocity of 55 mph. Plot the position, velocity, and acceleration as a function of time.

10.2 A motorcycle moves with an initial velocity of 30 m/s. When its brakes are applied, it decelerates at 5.0 m/s^2 until it stops. Plot the position, velocity, and acceleration as a function of time.

10.3 A girl shoots an arrow upward. It strikes the ground 10.0 s later. What was its initial velocity and what was its maximum height? (Assume air resistance is negligible.)

10.4 A bullet is fired vertically into the air and reaches a maximum height of 15,000 ft. What was its initial velocity? (Assume air resistance is negligible.)

10.5 A man standing on a 200-ft tower throws a ball upward at 40 ft/s. How long does it take to reach the ground?

10.6 A cannon is pointed at a 35° angle relative to a flat, horizontal field. It fires a shell with a muzzle speed of 1000 ft/s. Plot the trajectory of the shell at 1-second intervals. The y-axis represents the height above the field and the x-axis represents the horizontal distance away from the gun. (Assume air resistance is negligible.)

10.7 In Problem 10.6, the height y was calculated as a function of time. Derive an equation in which height y is a function of x, the distance from the cannon. This equation should be able to reproduce the figure prepared for Problem 10.6. (Assume air resistance is negligible.)

10.8 A cannon is pointed at a 60° angle relative to a flat, horizontal field. It fires a projectile with a muzzle speed of 1500 ft/s. Then, the gunners change the angle to 30°, reload the cannon, and 5.0 seconds later, fire. What is the required muzzle speed for the second projectile to hit the ground at the same time as the first? (Assume air resistance is negligible.)

10.9 A baseball is thrown, leaving the baseball player's hand 6.0 ft above the ground and at a 45° angle with respect to the horizon. The initial speed is 80 ft/s. Find the maximum height attained, the length of time the ball stays in the air, and the horizontal distance where the ball hits the ground. (Assume air resistance is negligible.)

10.10 A room is uniformly illuminated from the ceiling. A 6.00-ft pole is erected in the room. A string is inserted through the center hole of a metal ring. One end of a string is attached to the top of the pole and the other end is attached to the floor 12.0 ft from the base of the pole. The metal ring is slid up to the top of the pole and released with an initial speed of zero. How long does it take for the ring to reach the floor? What is the magnitude of the acceleration of the shadow on the floor? (Assume friction of the metal ring on the string is negligible.)

10.11 A room is uniformly illuminated from the ceiling. The room measures 8 ft high, 12 ft wide, and 20 ft long. A string is inserted through the center hole of a metal ring. One end of the string is attached to the corner of the ceiling and the other end is attached to the corner of the floor at the complete opposite end of the room so the string passes through the middle of the room. The metal ring is slid to the ceiling and released with an initial speed of zero. How long does it take for the ring to reach the floor? What is the magnitude of the shadow's acceleration on the floor? (Assume friction of the metal ring on the string is negligible.)

10.12 With an initial speed of 5.00 m/s, a 1.00-kg Teflon disk is slid on a horizontal steel plate. How long will it take for the Teflon disk to stop, and how far does it travel? (Do not neglect friction.)

10.13 A 1.00-kg Teflon disk is slid down a 5.00-m steel plate that is mounted at a 30° angle with respect to the horizon. Starting from rest, the disk is released from the top of the steel plate. How long does it take for the disk to reach the bottom, and what is its final velocity? (Do not neglect friction.)

10.14 A 15-cm, vertical spring with a 500 N/cm spring constant is compressed by 2.00 cm. A 5.00-kg ball is placed on the compressed spring. The spring is released, thrusting the ball into the air. What is the maximum height attained by the ball? How long does it take for the ball to land on the ground, not on the spring?

10.15 A 1500-kg automobile has a projected frontal area of 1.9 m^2 and a drag coefficient of 0.35. It is traveling at 100 km/h on a flat road when suddenly both the engine and brakes fail. Write a computer program that calculates the position, velocity, and acceleration of the automobile until it reaches a velocity of 1 km/h. Print this information every 2 s. You may wish to use a smaller time step for your calculations, however. (Neglect wheel friction, but do not neglect air drag.)

10.16 In stagnant water, you are able to swim at a speed of 2 mi/h. You swim in a 1-mile-wide river with a current that is flowing at 1 mi/h. How long will it take to swim parallel to the shore to a point 1 mile upstream, and then return to your starting point? How long will it take to swim to a perpendicular point located on the opposite shore, and return to your starting point?

10.17 You have the opportunity to ride on a Klingon hyperdrive spaceship that cruises at 0.9 the speed of light. Its engines are so powerful that it accelerates from rest to cruising speed in just 5 s. Using advanced spacewarp technology, the occupants of the spaceship easily tolerate the resulting acelerations. Your 25-year-old brother watches you leave and wishes you well on your trip. When you return, the chronometer on the spaceship indicates that your trip lasted exactly 1 year. How old is your brother?

10.18 Prepare a graph in which the kinetic energy of a 1-kg object is plotted on the y-axis and its velocity is plotted on the x-axis. The velocity should range from 0 to 0.99 the speed of light. The single graph should consider the following two approaches: classical physics and relativistic physics.

Glossary

acceleration The rate that velocity changes with respect to time.

attractive force A force that causes two objects to move toward each other.

average acceleration The rate that velocity changes during a finite time period.

average velocity The rate that position changes during a finite time period.

coefficient of kinetic friction The ratio of the friction force between two bodies and the perpendicular force between the bodies when there is relative sliding motion between the two bodies.

coefficient of static friction The ratio of the friction force between two bodies and the perpendicular force between the bodies at the instant when the bodies transition from being stationary to being set in relative sliding motion.
displacement A change in position.
distance A nonnegative scalar that describes the length of a path.
drag Resistance exerted on a body as it moves through a fluid.
electromagnetic force A force that attracts two objects of opposite charge (or opposite magnetic poles) and repels two objects of like charge (or like magnetic poles).
electrostatic force A force of attraction or repulsion between electric charges.
foci The two points symmetrically located on the major axis of an ellipse either side of the center.
force The influence on a body that can cause it to accelerate in the absence of any other counteracting forces.
friction Bonding between two surfaces that contact each other.
gravity An attractive force between two objects that have mass.
half-life The time it takes for half the radioactive particles to decay.
inertial Not accelerating.
instantaneous acceleration The rate that velocity changes during a differential time period.
instantaneous velocity The rate that position changes during a differential time period.
interference pattern Light patterns resulting from aligned or misaligned light wave peaks.
interferometer A laboratory device used to observe interference patterns.
kinetic friction The frictional force tending to slow a body in motion.
law of inertia An object in motion will maintain its velocity in the absence of any applied forces.
length contraction The observation that length contracts in the direction of travel.
magnetic force The force that exists between two magnetic poles.
momentum The mass of an object multiplied by its velocity.
muon Subatomic particle.
Newton's first law In the absence of a net force, the momentum of an object does not change.
Newton's second law A net force changes the momentum of an object.
Newton's third law The force on one body is equal and opposite to the force on the second body; isolated forces cannot exist.
perihelion The point where the Earth is nearest the sun.
position A place or location within a reference frame.
quantum mechanics Atomic-scale physics.
reference frame A portion of the universe which the observer defines as stationary.
relativistic transformations A set of equations that relates the position and time in one reference frame to another reference frame.
relativity Einstein's theory stating that there is no preferred reference frame and that the speed of light in free space is constant regardless of the reference frame.
repulsive force A force that pushes two objects apart.
rest mass The mass of a stationary object.
scalar A quantity completely described by its magnitude; it has no direction.
specific quantity The amount of something per unit mass.
speed A scalar describing the magnitude of the velocity vector.
spring force The force that results when a spring is compressed or stretched.
static friction The bonding between two bodies at rest.
strong force A force that acts at very short distances and is responsible for holding atomic nuclei together.
time dilation The observation that time slows as speed increases.
vector A quantity with a magnitude and a direction.
velocity The time rate of change of a position of a body.
weak force A force involved in radioactive decay.

Mathematical Prerequisites

- Algebra (Appendix E, Mathematics Supplement)
- Calculus (Appendix L, Mathematics Supplement)

CHAPTER 11
Introduction to Thermodynamics

Thermodynamics is a study that spans all engineering branches. **Thermodynamics** is from the Greek for "heat" (*therme*) and "power" (*dynamis*). This science developed during the 1800s to explain how steam engines converted heat into work. Since then, it has been expanded to describe systems as diverse as automobile engines, jet engines, refrigeration and air conditioning, rockets, electricity production, electric motors, chemical plants, and living organisms.

Thermodynamics is the science that describes what is possible and what is impossible during energy conversion processes. Through the ages, inventors have attempted to build perpetual motion machines that create more energy than they consume. If they had studied thermodynamics, they would have realized the folly of their pursuit.

As powerful as thermodynamics is, it has one major limitation: it cannot describe the amount of time required for a process to occur. (That is the topic of the next chapter.) For example, thermodynamics tells us that diamonds at atmospheric pressure will spontaneously convert back to graphite, yet it says nothing about how long this process takes. Fortunately, this conversion is very slow, taking millions of years; otherwise, diamond engagement rings would soon lose their sparkle.

11.1 FORCES OF NATURE

All physical phenomena can be traced to four **forces** of nature: gravity force, electromagnetic force, strong force, and weak force. **Gravity** is an attractive force between two objects that have mass. The *electromagnetic force* attracts two objects of opposite charge (or opposite magnetic poles) and repels two objects of like charge (or like magnetic poles). The **strong force** acts only at very short distances and is responsible for holding atomic nuclei together. The **weak force** is involved in radioactive decay.

11.2 STRUCTURE OF MATTER

Matter is composed of **protons** (positive charge), **neutrons** (no charge), and **electrons** (negative charge). Protons and neutrons are attracted together by the strong force, which causes them to cluster together in a nucleus (Figure 11.1). The strong force is powerful enough to overcome the electromagnetic force that would cause the protons to repel each

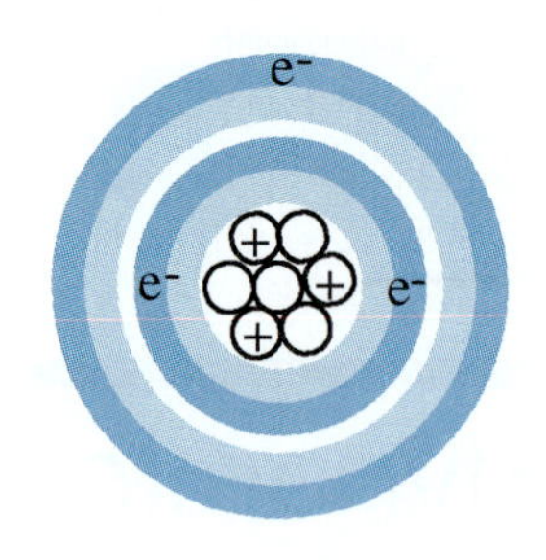

FIGURE 11.1
Atomic structure of lithium.

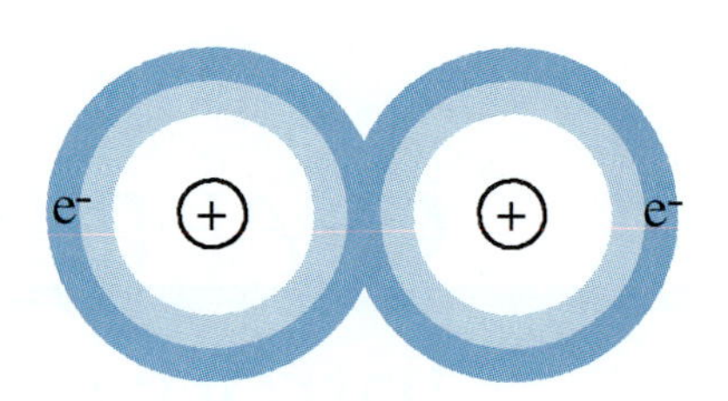

FIGURE 11.2
Hydrogen molecule (H_2).

other. Electrons are attracted to the protons by the electromagnetic force and form a diffuse cloud surrounding the nucleus. The electrons closest to the nucleus are strongly held, whereas the electrons farthest from the nucleus (**valence electrons**) are loosely held. The number of protons in an atom is called the **atomic number.** In a nonionized atom, the number of electrons equals the number of protons so that the charges in the atom balance.

A single nucleus with its accompanying electrons is an **atom.** Atoms often join together into **molecules** in which electrons are shared between nuclei (Figure 11.2). Chemistry is basically the study of how atoms interact and bond together.

11.3 TEMPERATURE

We have an intuitive sense of what is meant by temperature when we choose what to wear every day. In its simplest sense, **temperature** is a quantitative measure of the "degree of hotness." On a deeper level, temperature is related to the motion of individual atoms or molecules. At high temperatures, gaseous atoms (or molecules) are moving rapidly in random patterns, whereas at low temperatures they move more slowly (Figure 11.3).

At **absolute zero,** temperature is reduced as low as possible and motion, while not zero, is at its minimum. A temperature scale that designates absolute zero as "zero degrees" is called an **absolute temperature scale.** For example, the Kelvin temperature scale designates absolute zero as 0 K.

11.4 PRESSURE

Pressure is defined as the force exerted on a unit area (Figure 11.4). For example, pressure arises from the impact of high-velocity atoms (or molecules) against a surface [Figure 11.5(a)] or because an object's weight (gravitational force) is exerted over an area [Figure 11.5(b)]. A pressure scale that designates the pressure exerted by a perfect vacuum as zero is termed an **absolute pressure scale.**

FIGURE 11.3
Visualization of temperature in a gas.

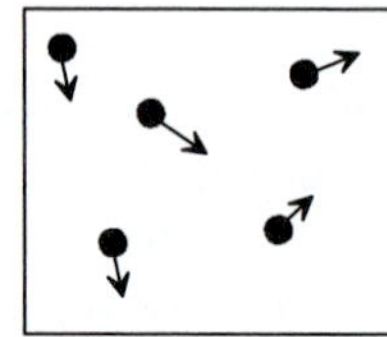

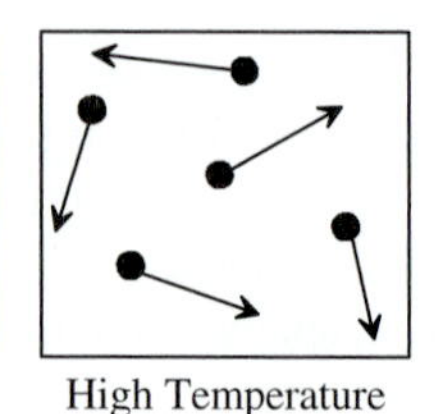

FIGURE 11.4
Visualization of pressure.

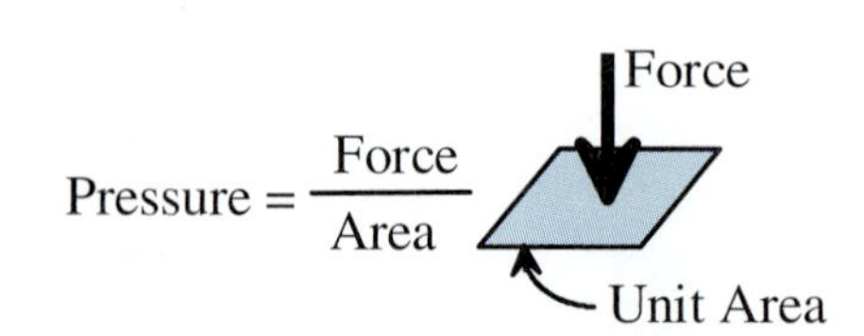

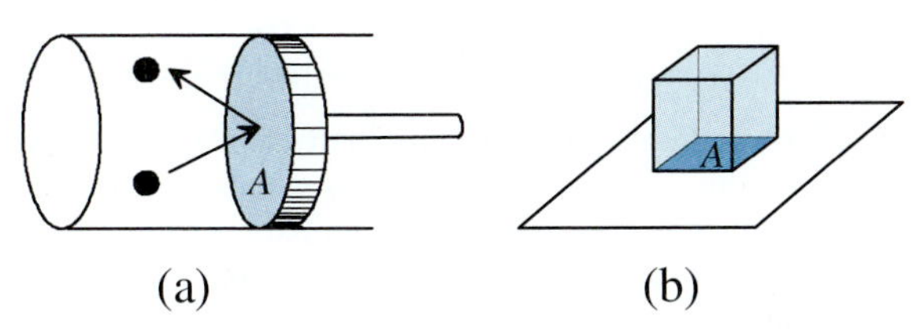

FIGURE 11.5
Pressure resulting from (a) impact or (b) weight.

$$\text{Density} = \frac{\text{Mass}}{\text{Volume}}$$

Low Density High Density

FIGURE 11.6
Visualization of density.

11.5 DENSITY

Density is defined as the amount of mass in a unit volume (Figure 11.6). At normal temperatures and pressures, the density of ordinary matter ranges from 0 kg/m^3 for a pure vacuum to 22,610 kg/m^3 for osmium (Table 11.1). The density of a neutron star is 7×10^{17} kg/m^3.

TABLE 11.1
Density of some common substances (Reference)

Gases†	kg/m^3
Air	1.284
Ammonia	0.771
Carbon dioxide	1.977
Carbon monoxide	1.251
Ethane	1.357
Helium	0.178
Hydrogen	0.090
Methane	0.718
Nitrogen	1.251
Oxygen	1.429
Sulfur dioxide	2.927

Liquids†	kg/m^3
Benzene	880
Ethanol	800
Gasoline	670
Glycerin	1,250
Kerosene	800
Oil	880
Turpentine	870
Water, fresh	1,000
Water, salt	1,030

Nonmetallic Solids	kg/m^3
Asbestos	2,500
Brick	1,800
Cedar	350
Cement, portland	1,440
Coal, anthracite	1,600
Coal, bituminous	1,350
Concrete	2,300
Douglas fir	500
Glass, common	2,650
Granite, solid	2,700
Gravel	2,550
Gypsum	2,310
Leather	940
Limestone, solid	2,450
Mahogany	540
Marble	2,750
Oak, white	770
Paper	930
Pine, white	430
Redwood	420
Rubber	940
Salt	1,000
Sand, loose	1,550
Sugar	1,610
Sulfur	2,100

Metals	kg/m^3
Aluminum	2,640
Brass, cast	8,570
Bronze	8,170
Copper, cast	8,900
Gold	19,300
Iridium	22,500
Iron, cast	7,050
Iron, ore	5,210
Iron, wrought	7,770
Lead	11,300
Manganese	7,400
Mercury	13,600
Nickel	8,900
Osmium	22,610
Platinum	21,450
Silver	10,500
Steel, cold drawn	7,830
Steel, structural	7,900
Steel, tool	7,700
Tin, cast	7,300
Titanium	4,500
Uranium	18,700
Zinc, cast	7,050

†$T = 0°C$, $P = 1$ atm.

Temperature Scales

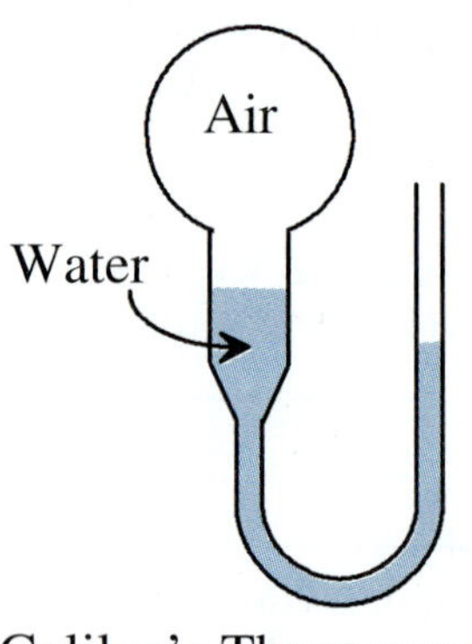

Galileo's Thermoscope

In 1593, Galileo Galilei (1564–1642) devised a "**thermoscope**" to measure temperature. It consisted of air trapped in a bulb that was connected to an adjacent water tube (see figure). When the air temperature increased, its volume expanded, causing the water level to rise in the adjacent water tube. Similarly, when the air temperature decreased, its volume decreased, causing the water level to drop in the adjacent water tube. It was a crude and inaccurate device, because the readings depended on atmospheric pressure, which changes daily.

Guillaume Amontons (1663–1705) improved on Galileo's thermoscope by replacing the water with mercury. He adjusted the mercury height in the adjacent tube until the gas volume was constant; thus he inferred temperature from changes in pressure.

Eventually, liquid-filled thermometers, much like those used today, became popular. Typically, the glass capillary contained alcohol or alcohol/water mixtures, but these thermometers worked over a limited temperature range. In 1714, Gabriel Fahrenheit (1686–1736) perfected the thermometer by replacing the alcohol with mercury; his critical contribution was to purify the mercury so it would not stick to the walls of the capillary tube.

With his new thermometer, Fahrenheit had to decide how to mark the divisions. Previously, in 1701, Newton had suggested that the freezing point of water be designated zero and human body temperature be designated twelve. Fahrenheit rejected Newton's zero reference point because readings would be negative in the winter. Instead, he designated zero as the lowest temperature then attainable in the laboratory using a slurry of salt, water, and ice. (Perhaps you have used this slurry when freezing homemade ice cream.) Initially, he designated human body temperature as 96, a number easily divided into halves, thirds, quarters, and so forth. Later, he made slight adjustments so the boiling point of water was 212 and the freezing point was 32, thus making human body temperature 98.6 on this revised scale. The Fahrenheit scale was adopted in Great Britain and the Netherlands.

In 1742, astronomer Anders Celsius (1701–1744) designated the boiling point of water as zero and the freezing point as 100. One year later, he reversed this, making the freezing point zero and the boiling point 100. His centigrade scale (i.e., "one hundred steps") is used the world over. In 1948, it was agreed to refer to it as the Celsius scale.

Adapted from: Isaac Asimov, *Asimov's Biographical Encyclopedia of Science and Technology* (Garden City, NY: Doubleday, 1982).

Absolute Zero

The discovery of absolute zero culminated from a series of observations spanning almost 2 centuries. In 1699, Guillaume Amontons (1663–1705) published his observations that the volume of constant-pressure gases increased the same percentage for the same rise in temperature. In 1787, Jacques Charles (1746–1823) repeated these experiments and showed that for each Celsius degree increase in temperature, a constant-pressure gas increased its volume by 1/273 of its 0°C volume. Similarly, for each Celsius degree decrease in temperature, a constant-pressure gas decreased its volume by 1/273 of its 0°C volume. It was not until 1848 that Lord Kelvin (1824–1907) discovered the consequences of Charles's work. Kelvin surmised that it would be impossible for a gas to go below −273°C and proposed a new temperature scale in which −273°C was assigned a value of zero. This temperature scale proposed by Kelvin is an absolute temperature scale and has SI units of kelvin (abbreviated K). Absolute zero is assigned a value of 0 K. The divisions on the Kelvin scale have the same width as the Celsius scale. Another absolute scale, called the Rankine temperature scale, has divisions with the same width as the Fahrenheit scale.

It is impossible to achieve a temperature of absolute zero, although it is possible to get arbitrarily close. According to *The Guinness Book of Records 1995,* the coldest temperature ever achieved was 2.8×10^{-9} K in a nuclear demagnetization device at the Low Temperature Laboratory, Helsinki University of Technology, Finland, in April 1993.

It is a common erroneous belief that all motion ceases at absolute zero. In fact, there is some motion at absolute zero, albeit very small. (*Note:* If there were no motion, both the position and momentum of a particle would be exactly known, a violation of the Heisenberg Uncertainty Principle, which you will learn about in your physics class.)

Unusual behavior is exhibited near absolute zero. Many metals become superconductors in which electrons flow without resistance. Helium becomes a superfluid in which it flows through a pipe without resistance.

In 1995, working near absolute zero, Carl Wieman and Eric Cornell observed a new state of matter, called Bose-Einstein condensate, in which about 2000 atoms fuse together into a single "super atom." They captured rubidium-87 atoms in a laser trap that cooled the atoms to 20×10^{-6} K. Then, they further cooled it in a magnetic trap to 170×10^{-9} K, where they observed this new state of matter, which had been predicted 70 years previously by Satyendra Bose and Albert Einstein. Ultimately, they cooled it to 20×10^{-9} K, which some people consider to be a record low temperature.

Adapted from: Isaac Asimov, *Asimov's Biographical Encyclopedia of Science and Technology* (Garden City, NY: Doubleday, 1982); *Science* 269 (July 14, 1995), p. 198.

11.6 STATES OF MATTER

Normally, matter exists in four states: solid, liquid, gas/vapor, and plasma (Figure 11.7). At low temperatures, matter is very well ordered (solids). Solid atoms or molecules can only vibrate, because they are held in place in the ordered structure. At slightly higher temperatures, matter becomes more disordered, but has a density similar to solids (liquid). Liquid atoms or molecules can vibrate and rotate around each other. As temperature increases further, matter separates into individual atoms (or molecules) that are widely separated (gas/vapor). Gas/vapor atoms or molecules can vibrate, rotate, and move from one location to another. At extremely high temperatures—such as those in a star or welding torch—electrons are stripped from their nuclei (plasma).

Figure 11.8 shows a typical **phase diagram,** which indicates the stable phase (or state) at a given temperature and pressure. Any temperature/pressure combination above Line C and left of Line B is solid. Any temperature/pressure combination above Line A and right of Line B is liquid. Any temperature/pressure combination below Lines A and C is a vapor. When a temperature/pressure combination falls exactly on Line A, both liquid and vapor exist simultaneously. When a temperature/pressure combination falls exactly on Line B, both liquid and solid exist simultaneously. When a temperature/pressure combination falls exactly on Line C, both solid and vapor exist simultaneously. The intersection of Lines A, B, and C is the **triple point,** where all three phases exist simultaneously.

FIGURE 11.7
Four states of matter.

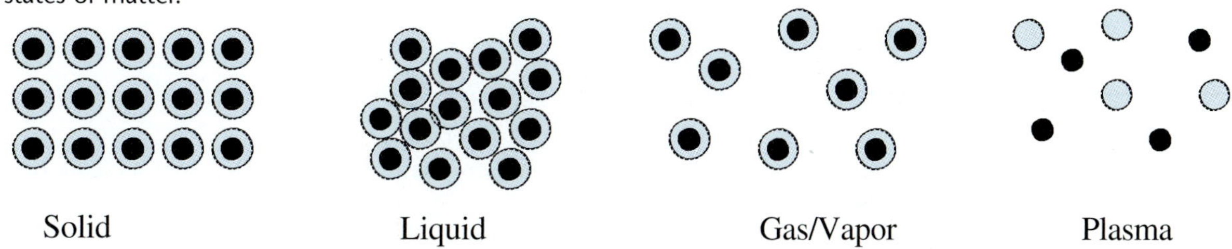

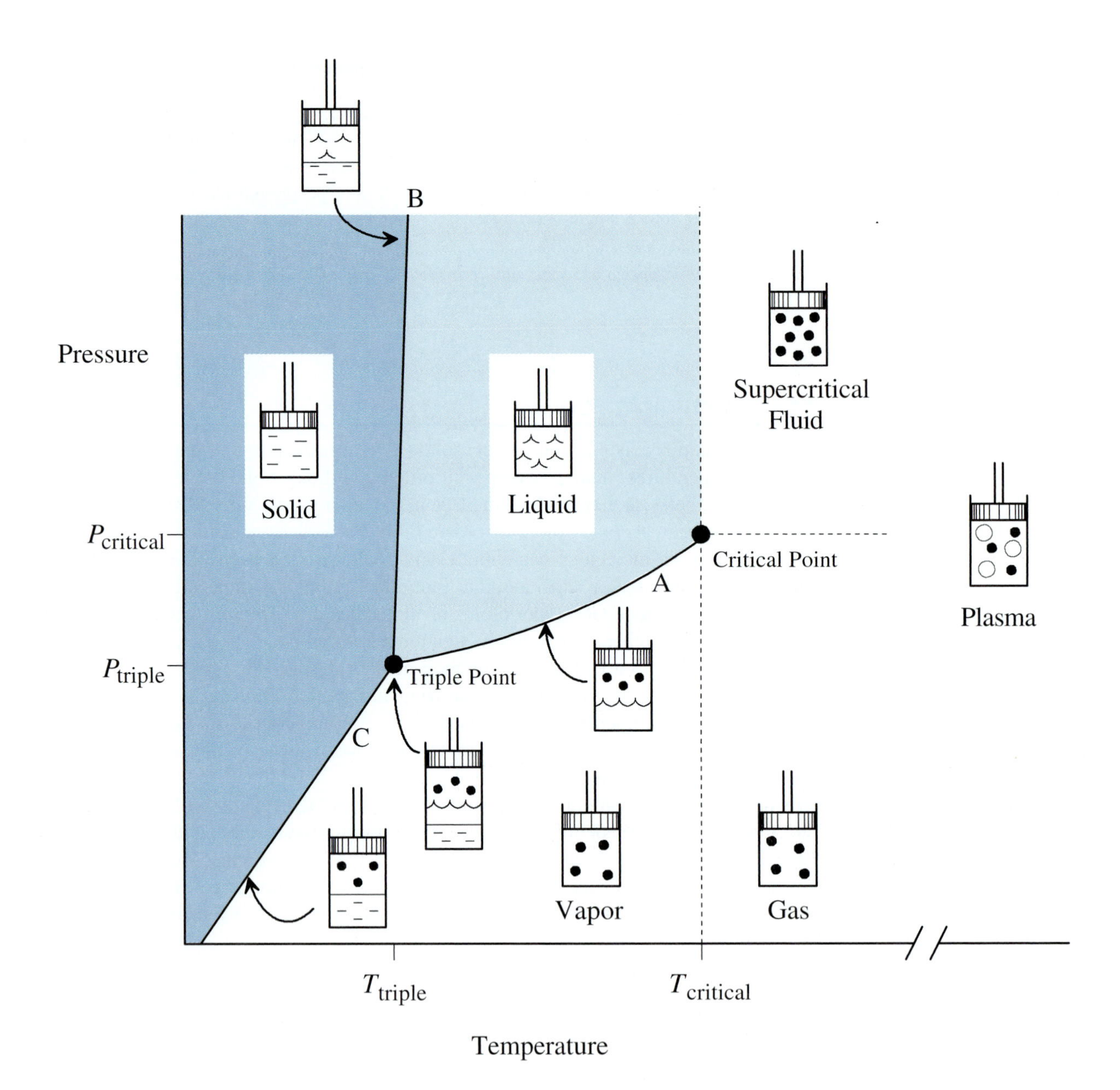

FIGURE 11.8

Phase diagram for a pure substance. (Solid lines show conditions where two phases coexist; dotted lines indicate divisions between vapor, gas, and supercritical fluid.)

Pressure Scales

Pressure is a force exerted on an area. The figure shows a large vessel divided into two chambers. On the right is the pressure of interest and on the left is the reference pressure.

Three types of pressure scales are commonly used: absolute, gage, and differential. Each scale uses a different reference pressure as shown in the following chart:

Pressure Scale	Reference Pressure
Absolute	Perfect vacuum
Gage	Atmospheric pressure
Differential	Arbitrary reference

There are a variety of units used to describe pressure. In the SI system, a newton of force exerted per square meter is called a pascal (abbreviated Pa). In engineering, we commonly use a pound-force exerted per square inch, abbreviated "psi." Depending upon the type of pressure measured, a descriptor may be added. Absolute pressure is designated "psia," gage pressure is designated "psig," and a pressure difference is designated "psid." Pressure can also be described in "atmospheres," the approximate ambient air pressure at sea level. In addition, it is common to describe pressure in terms of the height of a fluid column, such as centimeters of mercury or inches of water.

According to *The Guinness Book of Records 1995,* the lowest pressure ever attained in a laboratory is about 10^{-17} atmosphere at the IBM Thomas Watson Research Center in Yorktown Heights, New York, in 1976. The highest sustained pressure ever achieved is 1.7×10^6 atmospheres in a diamond-faced hydraulic press at the Carnegie Institution's Geophysical Laboratory, Washington, DC, in 1978.

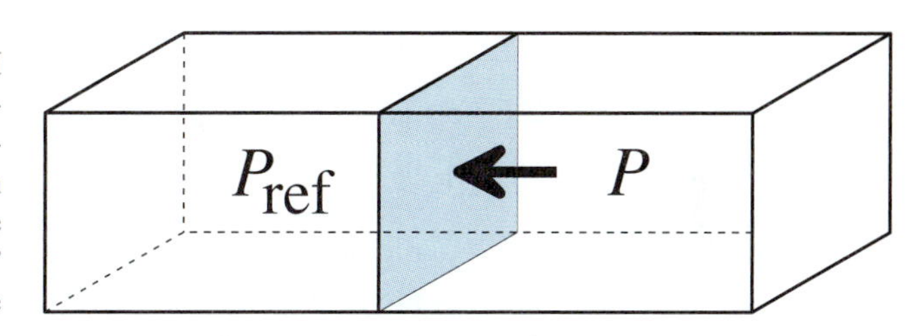

Classical Greece: Philosophical Engineering

Although Ancient Greece is probably best remembered for its multitude of great thinkers and remarkable temple architecture, it also contributed to the engineering world. Many of Greece's greatest minds, however, deplored application-oriented science; consider Plato's admonition in *The Republic:*

> . . . whether a man gapes at the heavens or blinks at the ground, seeking to learn some particular of sense, I would deny that he can learn, for nothing of that sort is science.

Despite such sentiments, Greek engineering and architecture flourished in the works of the *architekton,* the title for the Greek master builders. Accomplishments include urban planning, water-supply systems for agricultural and urban use, harbors, advanced weaponry, and the temples and theaters that remain to this day.

One Greek notable was Archimedes of Syracuse (287–212 B.C.), a mathematician who was challenged by the King of Sicily to verify the authenticity of his newly made gold crown. Though Archimedes could easily measure the crown's mass, calculation of its density required measurement of the volume as well, a much more difficult problem. In a well-known tale, Archimedes

The Acropolis in Athens, Greece.

stepped into his bathtub and, noting the rise of the water level, discovered a way to measure the crown's volume. He then proceeded to run naked through the streets of Syracuse shouting, "Eureka!" meaning "I have found it!" Upon measuring the density of the crown, he determined it was not dense enough to be pure gold; the goldsmith was thereby executed.

Every modern pupil of fluid mechanics learns about Archimedes' principle—that a solid body floating in a liquid displaces a mass of liquid equal to its own mass. Archimedes also invented the screw pump, a device for carrying water up an incline, and devoted much study to geometry, levers, and the existence of gravity.

Courtesy of: Seth Adelson, graduate student.

At high pressures along Line A, the density of the vapor phase is nearly identical to that of the liquid phase; at the **critical point,** the density of the liquid and vapor phases becomes identical. At the critical point, matter exhibits unusual behavior. For example, it is impossible to transmit sound through a critical material. Also, **opalescent** "ghosts" appear and disappear at the critical condition.

At temperatures above the critical point, it is impossible for two phases to coexist. Unlike vapors that condense into a liquid phase at high pressures, gases simply become dense gases at high pressures. When the gas pressure is above the critical pressure, it is designated a **supercritical fluid,** which has liquid-like densities, but is not actually a liquid. New uses are being found for supercritical fluids; for example, supercritical carbon dioxide is used to extract caffeine from coffee beans to make decaffeinated coffee. Vapors, gases, and supercritical fluids are all very similar; the different names are used to designate their relationship to the critical point.

11.7 EQUILIBRIUM CONDITIONS

Equilibrium conditions are those that do not change as time passes. Thermodynamics is used to determine where the final equilibrium lies, but it cannot be used to determine how quickly a process will come to equilibrium.

The easiest way to conceptualize an equilibrium condition is to visualize a two-pan scale like the one in Figure 11.9(a). Masses A and B are the same, so the gravity force acting on each pan is the same. Because the two forces balance, the scale does not move as time passes; hence the system is in equilibrium. Figure 11.9(b) shows a system in which liquid and vapor are in equilibrium. As an atom (molecule) transfers from the vapor phase to the liquid phase, it is replaced by an atom (molecule) that transfers from the liquid to the vapor phase. Similar equilibria exist between solid and liquids on Line B of Figure 11.8 and between solids and vapors on Line C of Figure 11.8.

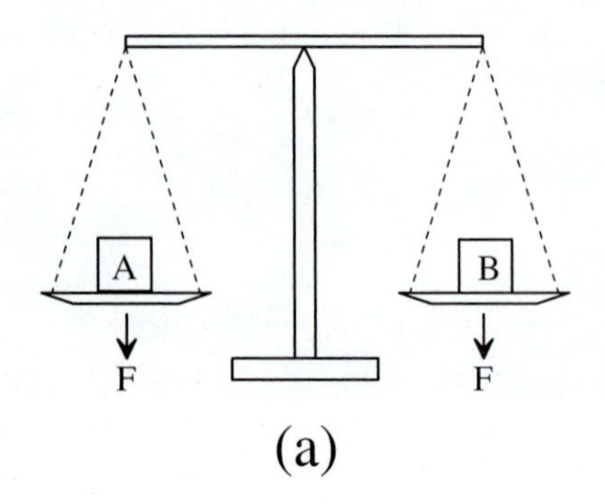

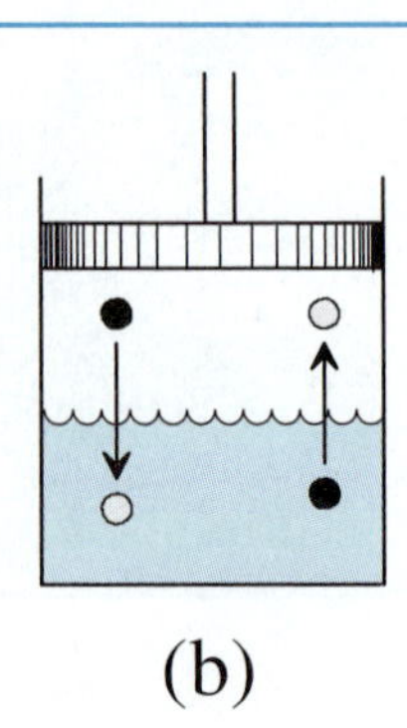

FIGURE 11.9
Visualization of equilibrium conditions: (a) two-pan scale and (b) liquid/vapor equilibrium.

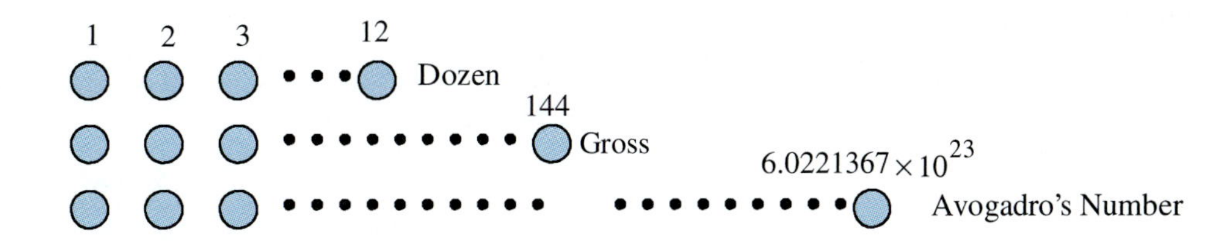

FIGURE 11.10
Visualization of Avogadro's number.

11.8 AMOUNT OF SUBSTANCE

In common business affairs, we attach names to special numbers. For example, 12 is designated as a dozen and 144 is a gross. Donuts are often sold by the dozen and pencils by the gross. When counting atoms (molecules), a dozen or a gross is simply too small to be of use. Therefore, a much larger special number, called **Avogadro's number N_A** is used (Figure 11.10).

$$N_A = 6.0221367 \times 10^{23} \tag{11-1}$$

This number is so large, it approximately equals the number of drops in the world's oceans. Avogadro's number is the number of atoms or molecules in one *gram-mole,* or simply *mole,* of substance. In SI units, the mole is abbreviated "mol."

Just as we would expect a dozen Ping-Pong balls to weigh less than a dozen bowling balls, so too would we expect a mole of hydrogen to weigh less than a mole of uranium. The mass of one mole of atoms is the **atomic mass** (commonly called **atomic weight**). Table 11.2 shows the atomic mass of each element.

The mass of one mole of molecules is the **molecular mass** (commonly called **molecular weight**). (For historical reasons, atomic mass is commonly given in grams rather than kilograms. Be aware of this convention, as it can lead to errors.) Because molecules are composed of atoms, it is easy to calculate the molecular mass by summing the mass of each atom in the molecule.

EXAMPLE 11.1

Problem Statement: Calculate the molecular mass of sodium chloride (NaCl) and carbon dioxide (CO_2) using Table 11.2.

Solution: $NaCl = 1(22.9898) + 1(35.453) = 58.443$ g/mol

$CO_2 = 1(12.01115) + 2(15.9994) = 44.0100$ g/mol

11.9 GAS LAWS

The gas laws have resulted from observations made centuries ago. Strictly, these laws apply only to **perfect gases** (sometimes called *ideal gases*). A perfect gas is a hypothetical gas in which the volume of the individual atoms (molecules) is negligible compared to the total volume of the container. Also, because the distance between atoms (molecules) is

TABLE 11.2
Atomic number and atomic mass of the elements (Reference)

Element	Symbol	Atomic Number	Atomic Mass (g/mol)	Element	Symbol	Atomic Number	Atomic Mass (g/mol)
Actinium	Ac	89	(227)	Mercury	Hg	80	200.59
Aluminum	Al	13	26.98154	Molybdenum	Mo	42	95.94
Americium	Am	95	(243)	Neodymium	Nd	60	144.24
Antimony	Sb	51	121.75	Neon	Ne	10	20.179
Argon	Ar	18	39.948	Neptunium	Np	93	237.0482
Arsenic	As	33	74.9216	Nickel	Ni	28	58.70
Astatine	At	85	(210)	Niobium	Nb	41	92.9064
Barium	Ba	56	137.33	Nitrogen	N	7	14.0067
Berkelium	Bk	97	(247)	Nobelium	No	102	(255)
Beryllium	Be	4	9.01218	Osmium	Os	75	190.2
Bismuth	Bi	83	208.9804	Oxygen	O	8	15.9994
Boron	B	5	10.81	Palladium	Pd	46	106.4
Bromine	Br	35	79.904	Phosphorous	P	15	30.97376
Cadmium	Cd	48	112.41	Platinum	Pt	78	195.09
Calcium	Ca	20	40.08	Plutonium	Pu	94	(244)
Californium	Cf	98	(251)	Polonium	Po	84	(209)
Carbon	C	6	12.011	Potassium	K	19	39.0983
Cerium	Ce	58	140.12	Praseodymium	Pr	59	140.9077
Cesium	Cs	55	132.9054	Promethium	Pm	61	(145)
Chlorine	Cl	17	35.453	Protactinium	Pa	91	231.0359
Chromium	Cr	24	51.996	Radium	Ra	88	226.0254
Cobalt	Co	27	58.9332	Radon	Rn	86	(222)
Copper	Cu	29	63.546	Rhenium	Re	75	186.207
Curium	Cm	96	(247)	Rhodium	Rh	45	102.9055
Dysprosium	Dy	66	162.50	Rubidium	Rb	37	84.4678
Einsteinium	Es	99	(254)	Ruthenium	Ru	44	101.07
Erbium	Er	68	167.26	Samarium	Sm	62	150.4
Europium	Eu	63	151.96	Scandium	Sc	21	44.9559
Fermium	Fm	100	(257)	Selenium	Se	34	78.96
Fluorine	F	9	18.998403	Silicon	Si	14	28.0855
Francium	Fr	87	(223)	Silver	Ag	47	107.868
Gadolinium	Gd	64	157.25	Sodium	Na	11	22.98977
Gallium	Ga	31	69.72	Strontium	Sr	38	87.62
Germanium	Ge	32	72.59	Sulfur	S	16	32.06
Gold	Au	79	196.9665	Tantalum	Ta	73	180.9479
Hafnium	Hf	72	178.49	Technetium	Tc	43	(97)
Helium	He	2	4.00260	Tellurium	Te	52	127.60
Holmium	Ho	67	164.9304	Terbium	Tb	65	158.9254
Hydrogen	H	1	1.0079	Thallium	Tl	81	204.37
Indium	In	49	114.82	Thorium	Th	90	232.0381
Iodine	I	53	126.9045	Thulium	Tm	69	168.9342
Iridium	Ir	77	192.22	Tin	Sn	50	118.69
Iron	Fe	26	55.847	Titanium	Ti	22	47.90
Krypton	Kr	36	83.80	Tungsten	W	74	183.85
Lanthanum	La	57	138.9055	Uranium	U	92	238.029
Lawrencium	Lr	103	(260)	Vanadium	V	23	50.9414
Lead	Pb	82	207.2	Xenon	Xe	54	131.30
Lithium	Li	3	6.941	Ytterbium	Yb	70	173.04
Lutetium	Lu	71	174.97	Yttrium	Y	39	88.9059
Magnesium	Mg	12	24.305	Zinc	Zn	30	65.38
Manganese	Mn	25	54.9380	Zirconium	Zr	40	91.22
Mendelevium	Md	101	(258)				

so great, there are no forces of attraction or repulsion; thus, one atom (molecule) is not affected by other atoms (molecules). A real gas (vapor) behaves like a perfect gas at low densities resulting from low pressures and/or high temperatures.

Boyle's law states that the absolute pressure P and the volume V of a perfect gas are inversely proportional, provided the temperature T and number of moles n are constant [Figure 11.11(a)],

$$\frac{P_1}{P_2} = \frac{V_2}{V_1} = k_1 \quad (T = \text{const}, n = \text{const}) \tag{11-2}$$

where k_1 is a constant.

Charles' law states that the volume V and absolute temperature T are directly proportional, provided the pressure P and number of moles n are constant [Figure 11.11(b)].

$$\frac{V_2}{V_1} = \frac{T_2}{T_1} = k_2 \quad (P = \text{const}, n = \text{const}) \tag{11-3}$$

Gay-Lussac's law states that the absolute pressure P and absolute temperature T are directly proportional, provided the volume and number of moles are constant [Figure 11.11(c)].

$$\frac{P_2}{P_1} = \frac{T_2}{T_1} = k_3 \quad (V = \text{const}, n = \text{const}) \tag{11-4}$$

FIGURE 11.11
Gas laws.

(a) Boyle's law

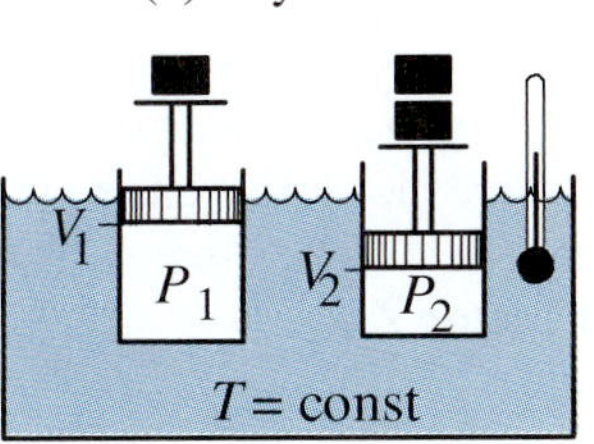

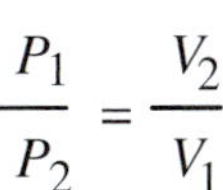

$n = \text{const}$

(b) Charles' law

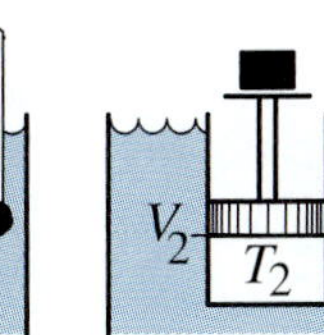

$\frac{V_2}{V_1} = \frac{T_2}{T_1}$

$n = \text{const}$

(c) Gay-Lussac's law

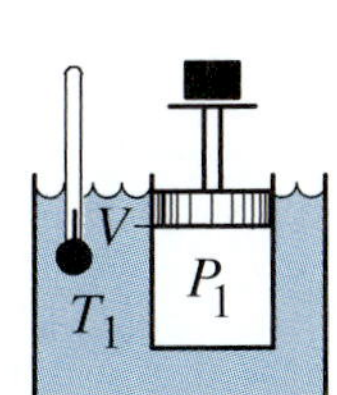

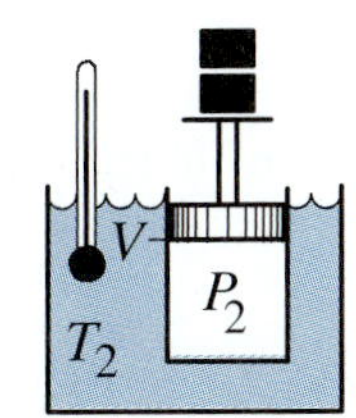

$\frac{P_2}{P_1} = \frac{T_2}{T_1}$

$n = \text{const}$

(d) Mole proportionality law

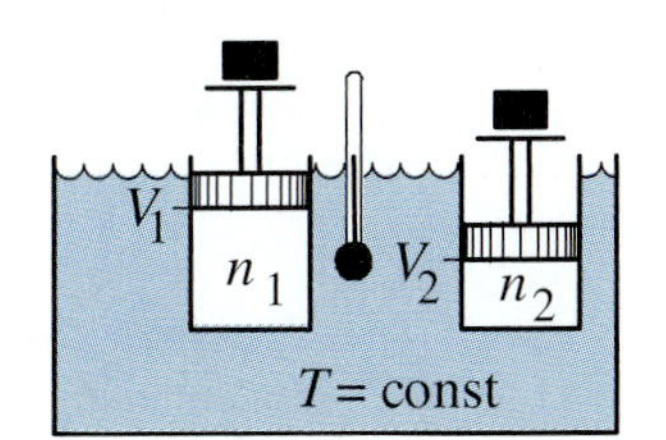

$\frac{V_2}{V_1} = \frac{n_2}{n_1}$

TABLE 11.3
Values of the universal gas constant R

$8.314 \frac{\text{J}}{\text{mol} \cdot \text{K}}$	$8.314 \frac{\text{Pa} \cdot \text{m}^3}{\text{mol} \cdot \text{K}}$	$1.987 \frac{\text{cal}}{\text{mol} \cdot \text{K}}$	$83.14 \frac{\text{bar} \cdot \text{cm}^3}{\text{mol} \cdot \text{K}}$	$0.08205 \frac{\text{atm} \cdot \text{L}}{\text{mol} \cdot \text{K}}$
$82.05 \frac{\text{atm} \cdot \text{cm}^3}{\text{mol} \cdot \text{K}}$	$0.7302 \frac{\text{atm} \cdot \text{ft}^3}{\text{lbmol} \cdot \text{R}}$	$10.73 \frac{\text{psia} \cdot \text{ft}^3}{\text{lbmol} \cdot \text{R}}$	$1545 \frac{\text{ft} \cdot \text{lb}_\text{f}}{\text{lbmol} \cdot \text{R}}$	$1.986 \frac{\text{Btu}}{\text{lbmol} \cdot \text{R}}$

The **mole proportionality law** states that volume V and number of moles n are directly proportional provided the pressure and temperature are constant [Figure 11.11(d)].

$$\frac{V_2}{V_1} = \frac{n_2}{n_1} = k_4 \qquad (P = \text{const}, T = \text{const}) \tag{11-5}$$

The four gas laws described above can be combined into a single gas law,

$$\frac{P_1 V_1}{n_1 T_1} = \frac{P_2 V_2}{n_2 T_2} = R \tag{11-6}$$

where R is the **universal gas constant** (Table 11.3). This relationship may be rewritten in the more familiar form of

$$PV = nRT \tag{11-7}$$

which is known as the **perfect gas equation** (also commonly called the *ideal gas equation*).

11.10 REAL GASES (ADVANCED TOPIC)

The perfect gas equation applies only to low-pressure gases and vapors. However, engineers often must use high-pressure gases where the perfect gas equation no longer applies. The simple form of the perfect gas equation is so appealing, we keep it by introducing a "fudge factor" called the **compressibility factor Z:**

$$PV = ZnRT \tag{11-8}$$

which typically ranges from 0.2 to about 4. There are many methods to find Z, the simplest of which is the **van der Waals equation**

$$Z = \frac{V}{V - nb} - \frac{na}{RTV} \tag{11-9}$$

where the constants a and b are shown in Table 11.4.

EXAMPLE 11.2

Problem Statement: At 500 K, 2.00 mol of CO_2 is stored in a 0.000800-m^3 container. Calculate the pressure using (a) the perfect gas equation and (b) the van der Waals equation.

TABLE 11.4
Van der Waals constants

Gas	$a\left(\frac{(m^3)^2 Pa}{mol^2}\right)$	$b\left(\frac{m^3}{mol}\right)$
He	0.00345	2.34×10^{-5}
H_2	0.0247	2.66×10^{-5}
O_2	0.138	3.18×10^{-5}
CO_2	0.366	4.29×10^{-5}
H_2O	0.580	3.19×10^{-5}
Hg	0.292	0.55×10^{-5}

Solution: (a) Perfect gas equation

$$P = \frac{nRT}{V} = \frac{(2.00 \text{ mol})\left(8.314 \frac{\text{Pa}\cdot\text{m}^3}{\text{mol}\cdot\text{K}}\right)(500 \text{ K})}{0.000800 \text{ m}^3} = 1.04 \times 10^7 \text{ Pa}$$

(b) Van der Waals equation

$$Z = \frac{V}{V - nb} - \frac{na}{RTV}$$

$$= \frac{0.000800 \text{ m}^3}{0.0008 \text{ m}^3 - (2.00 \text{ mol})\left(4.29 \times 10^{-5} \frac{\text{m}^3}{\text{mol}}\right)} - \frac{(2.00 \text{ mol})\left(0.366 \frac{\text{m}^3\cdot\text{Pa}}{\text{mol}^2}\right)}{\left(8.314 \frac{\text{Pa}\cdot\text{m}^3}{\text{mol}\cdot\text{K}}\right)(500 \text{ K})(0.0008 \text{ m}^3)}$$

$$= 0.900$$

$$P = \frac{ZnRT}{V} = \frac{0.900\,(2.00 \text{ mol})\left(8.314 \frac{\text{Pa}\cdot\text{m}^3}{\text{mol}\cdot\text{K}}\right)(500 \text{ K})}{0.000800 \text{ m}^3} = 0.935 \times 10^7 \text{ Pa}$$

11.11 ENERGY

Energy is one of the central ideas that thermodynamics brings to science and engineering. Thermodynamics provides a precise definition of energy; however, we use the term in a vague way in our everyday language. For example, if you are tired or sick, you might say, "I have no energy today." The dictionary defines energy as "the capacity to do work," but this definition is rather circular, because **work** is defined as a type of energy.

Energy is best understood as a unit of exchange, much like money is a unit of exchange. For example, dollars are used as units of exchange to equate the value of a car relative to a house:

1 car = $20,000
1 house = $100,000
5 cars = 1 house

The dollars are used to relate the value of one object relative to another; the dollars themselves have no intrinsic value. (You could prove this to yourself by being stranded on a deserted island with a million dollars. The money would be useful to start fires for cooking your dinner, but not much else.) In an analogous manner, energy is used to relate physical phenomena. For example, electric power plants can be fueled by coal or nuclear energy. The energy unit of exchange (in this case, joules) can be used to relate the relative amounts of coal and uranium needed to produce the same amount of electricity.

Combust 1 kg of coal = 42,000,000 joule
Fission 1 kg of uranium = 82,000,000,000,000 joule
Fission 1 kg of uranium = Combust 2,000,000 kg of coal

This analogy between money and energy is deeper than you might think (see box). Money is often budgeted for specific purposes. For example, you budget money for food, books, rent, and entertainment. Although it's all money you have spent for each of these purposes, when accounting for this money, each dollar has been assigned to a different category. Similarly, energy can be classified by using a variety of accounting schemes. Different accounting schemes distinguish between potential and kinetic energies, macroscopic and microscopic energies, and heat and work.

11.11.1 Potential and Kinetic Energies

Figure 11.12 shows a "thought experiment" in which a mass is suspended from a spring attached to a ceiling. To remove the effects of gravity, imagine the ceiling is on a spacecraft heading for Mars. Initially, the spring is compressed and the mass is at rest. When the mass is released, the spring pushes downward on the mass, causing it to accelerate to its maximum speed s_{max}. Then the spring stretches, slowing down the mass until it stops. The stretched spring pulls upward on the mass, causing it to accelerate again to its maximum speed. As the mass moves upward, the spring compresses, pushing downward until the mass stops and the cycle begins again.

Dollars and Energy

The figure shows the correlation of total U.S. energy consumption with gross national product (i.e., the total value of all goods and services) from 1929 to 1968. This very tight correlation shows that energy is absolutely essential to our economy, for without it, we cannot transform raw materials into finished goods.

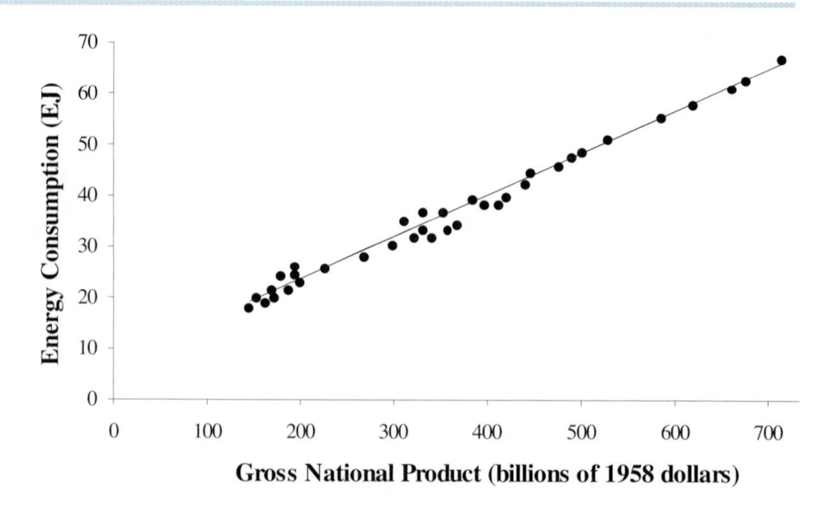

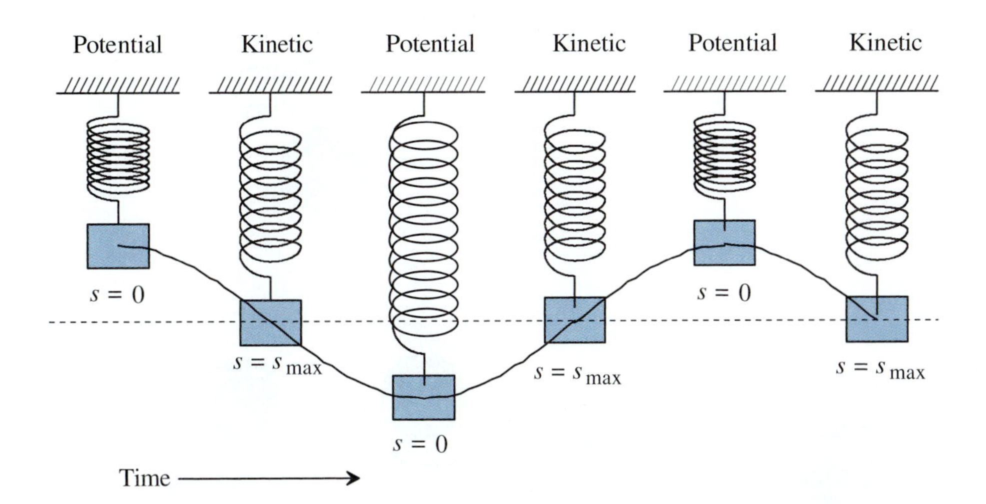

FIGURE 11.12
A spring/mass system suspended from a ceiling.

The spring/mass system has both **kinetic energy** (the energy associated with a moving mass) and **potential energy** (the energy associated with position). When the mass is moving at its maximum speed, the spring/mass system has the maximum kinetic energy, and when the spring is completely compressed or stretched, it has the maximum potential energy.

In this example, the mass was acted upon by a spring force; the amount of force depended upon its position. Other position-dependent forces (gravity, electrostatic or magnetic force) could have acted upon the mass, giving rise to other forms of potential energy.

11.11.2 Macroscopic and Microscopic Energies

Figure 11.13 shows a "thought experiment" in which a spring launches a cannonball, which then falls into a pool of water. When the spring is compressed, the system has potential energy. As soon as the cannonball disengages from the spring, it has its maximum speed, so the system has maximum kinetic energy. As the cannonball rises above the earth, gravity slows it down until it reaches zero speed, where it has maximum potential energy. When it falls back to the earth, it regains maximum kinetic energy. Finally, the cannonball falls into the pool of water. If a highly accurate thermometer were placed in the pool of water, you could observe an increase in temperature from the cannonball falling in the water.

The initial stages of this sequence involve the **interconversion** of potential and kinetic energy at the macroscopic scale. In the final step, macroscopic kinetic energy is converted to kinetic and potential energies at the microscopic scale. Let us examine this conversion a little further.

Figure 11.14 schematically represents the water in the pool as dumbbell-shaped molecules. The water molecules have microscopic kinetic and potential energies. There are three types of microscopic kinetic energy (translation, rotation, and vibration) and two types of microscopic potential energy (vibration and molecular interactions). Notice that the vibrational energy involves the interconversion of kinetic and potential energy, just like the spring/mass system of Figure 11.12. The potential energy in the molecular interactions

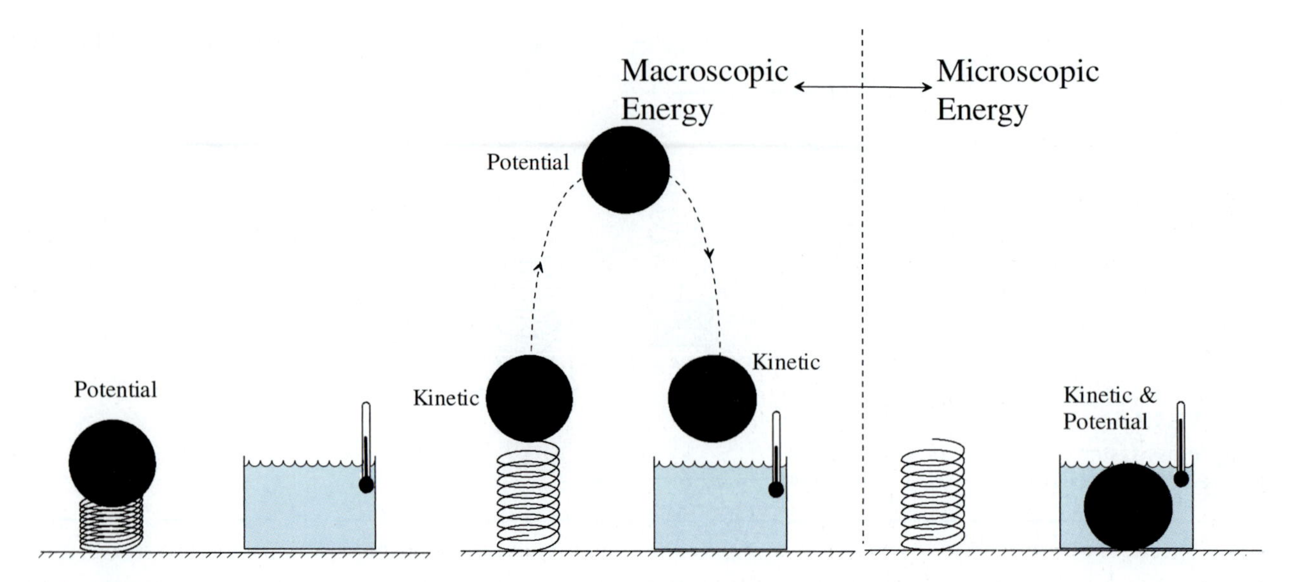

FIGURE 11.13
Thought experiment showing macroscopic and microscopic energies.

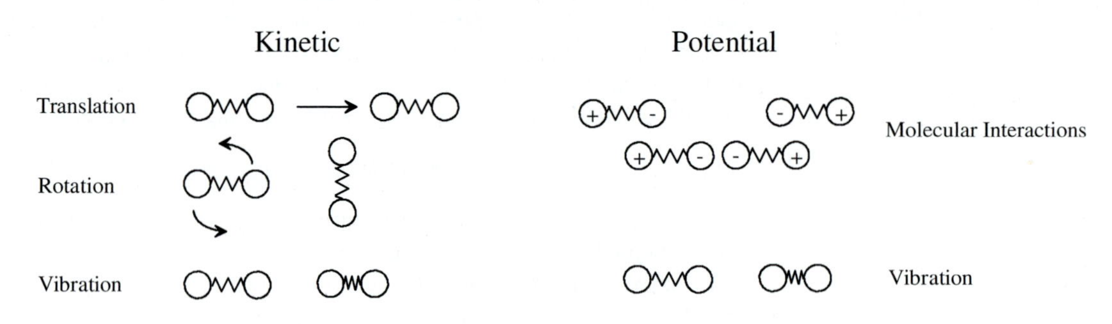

FIGURE 11.14
Visualization of internal energy.

result from the repulsion of like charges or the attraction of opposite charges. The collection of microscopic kinetic and potential energies is called **internal energy** and is given the symbol U.

11.11.3 Energy Flow

Energy that is flowing may be classified as either heat or work.

Heat.

Heat is visualized as energy flow resulting from atoms or molecules of different temperatures contacting each other. Most people confuse "heat" with "temperature" even though **heat and temperature are not the same.**

As these paint cans show, there is confusion regarding proper use of words *heat* and *temperature.*

The distinction between temperature and heat can be understood by considering Figure 11.15, which shows a copper bar bridging a bath of boiling water and an ice bath. As the water boils, the liquid stays at 100°C, and as the ice melts, the liquid stays at 0°C; thus, each bath temperature stays constant. The copper atoms in direct contact with the boiling water bath oscillate vigorously (indicated by the long double arrow) and the atoms in contact with the ice bath oscillate much less (indicated by the short double arrow). The graph above the copper bar indicates the temperature profile along its length. Thus, the atoms at the left are oscillating with a degree of vigor corresponding to 100°C and the atoms on the right are oscillating at 0°C.

The atoms in the copper bar are closely packed, so they touch each other. An atom oscillating at 100°C is adjacent to an atom with a slightly cooler temperature (say 99°C). When they touch, they share their vibrational energy and become 99.5°C. When the atoms separate, the leftmost atom will contact the boiling water and again return to 100°C. The net effect of this interaction is that a packet of energy is transferred from the 100°C atom to the 99°C atom. The 99°C atom will hand off this packet of energy to the next adjacent atom (say 98°C), and so on. In a process analogous to a "fire bucket brigade" in which people stand in line and hand buckets of water from one to the next, energy is transferred from atom to atom along the copper bar. This flow of energy from the boiling water bath to the ice bath is termed *heat* and will eventually cause the ice to melt.

Heat is a term that is often used in a sloppy manner. Just like the words *weight* and *mass* or *speed* and *velocity* are wrongly used interchangeably, *heat* is often used where *internal energy* would be more accurate. Some of the confusion over the proper use of the term results from the caloric theory of heat discussed in the box.

Work.

Heat is energy flow resulting from a temperature difference. The temperature difference may be visualized as a **driving force** that causes heat to flow. *Work* is energy flow resulting from **any** other driving force, as summarized in Table 11.5. The driving force concept will be explored in more detail in the next chapter.

The easiest form of work to understand is mechanical work. Imagine a large stone block destined for a pharaoh's tomb being dragged by a gang of slaves (Figure 11.16). The force the slaves apply to the rope causes the stone block to slowly move. The more force they apply and the farther they drag it, the more work they do. This can be stated mathematically as

$$W = \int_{x_1}^{x_2} F\,dx \tag{11-10}$$

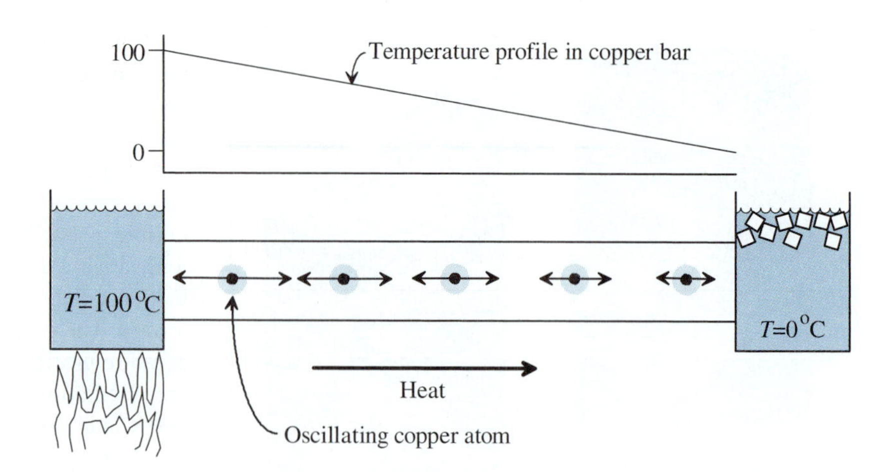

FIGURE 11.15
A copper bar bridging a boiling water bath and an ice bath illustrates the difference between "heat" and "temperature."

TABLE 11.5
Summary of heat and work

Heat	Driving Force	Work	Driving Force
Heat conduction	Temperature	Mechanical	Force
Blackbody radiation	Temperature	Shaft work	Torque
		Hydraulic	Pressure
		Electrical	Voltage
		Chemical	Concentration†
		Monochromatic radiation	Electromagnetic fields

†"Chemical potential" is the actual driving force, but it is related to concentration.

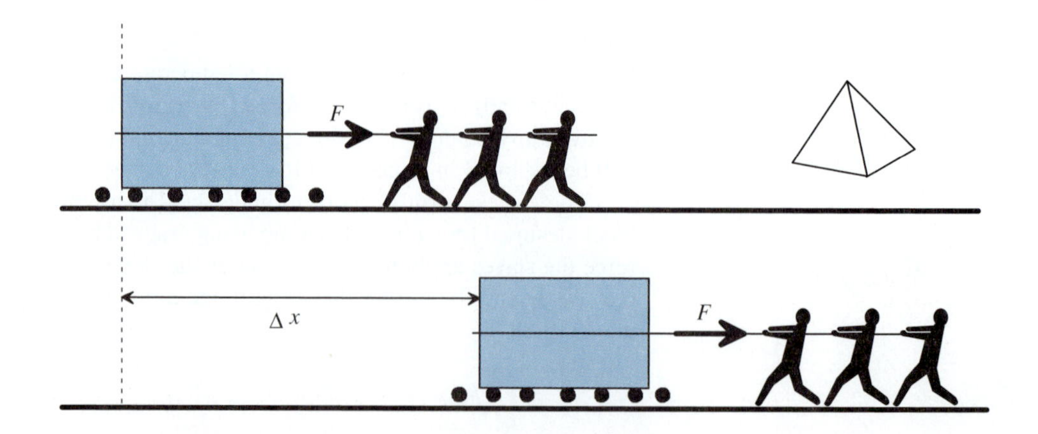

FIGURE 11.16
Visualization of mechanical work. With a constant force, slaves are pulling a large stone block to a pyramid. Logs placed under the block roll, reducing friction.

where W is the work, F is the force, and x is the distance. If the force is constant over the distance (e.g., the slaves are pulling with a constant force as they drag the block), Equation 11-10 becomes

$$W = F \int_{x_1}^{x_2} dx = F|x|_{x_1}^{x_2} = F(x_2 - x_1) = F\Delta x \quad (11\text{-}11)$$

Because this equation is very simple and understandable, it provides a convenient basis for defining an energy unit. In the SI system (see Chapter 13 for more detail), the units of F are $kg \cdot m/s^2$ and Δx is m, so the units of work are $kg \cdot m^2/s^2$, which is also known as a joule (J). In the SI system, all forms of energy are expressed in joules.

From this definition of work as a force exerted over a distance, we can extend the concept of work to include a pressure exerted over a change in volume:

$$W = F\Delta x = \left(\frac{F}{A}\right)(A\Delta x) = P\Delta V \quad (11\text{-}12)$$

Notice in Table 11.3 that the numerator of the universal gas constant is sometimes expressed in pressure × volume units and sometimes in energy units. By solving Equation 11-12 explicitly for pressure, we see that pressure has units of energy per unit volume, in addition to the more familiar units of force per unit area.

The Caloric Theory of Heat

In the 1700s, heat was considered an "imponderable fluid" that could be poured from one substance to another. In 1789, Count Rumford (1753–1814) observed a cannon being bored in Munich and noticed that the metal became extremely hot during the process. According to conventional wisdom, the metal contained "caloric," which was loosened from the metal as the shavings were released during the boring process. However, he observed that when the drill was dull, the heating was more severe even though there were no metal shavings produced. He concluded that heat must have been produced from the mechanical motion of the borer and that heat must be a form of motion, a view held by Francis Bacon (1561–1626), Robert Boyle (1627–1691), and Robert Hooke (1635–1703).

According to the caloric theory, liquid water has more caloric than frozen water. In 1799, Count Rumford carefully weighed a quantity of ice and observed no weight change as it melted; he concluded that if caloric existed, it must be weightless.

In the same time period, Sir Humphrey Davy (1778–1829) mechanically rubbed ice, causing it to melt, even though the surrounding temperature was maintained one degree below the melting point. He also concluded that mechanical motion was converted to heat.

These experiments and ideas proved unconvincing to the scientific community; great scientists, such as Antoine Lavoisier (1743–1794), maintained their belief in caloric.

About 1860, James Maxwell (1831–1879) studied gases and concluded that the atoms (molecules) moved more vigorously at higher temperatures, thus giving the caloric theory the final blow. Nonetheless, the caloric theory influences our modern language. A unit of heat is the **calorie,** and heat effects are measured in a **calorimeter.**

According to modern thermodynamic theory, caloric is actually analogous to internal energy, the energy associated with the random motion of atoms (molecules). We reserve the word *heat* for the *flow* of internal energy from a body at a higher temperature to one at a lower temperature.

Adapted from: Isaac Asimov, *Asimov's Biographical Encyclopedia of Science and Technology* (Garden City, NY: Doubleday, 1982).

Concepts of Work

Some of the earliest notions of work were connected with the trebuchet, an ancient engine of war discussed in Chapter 1. The trebuchet has a large beam with a fulcrum point. One end of the beam has a heavy counterweight and the other end contains the projectile. When the counterweight drops, it throws the projectile great distances. In medieval times, Jordanus of Nemore was studying ways to improve the trebuchet and determined that the performance critically depended upon the weight of the counterweight and the distance that it dropped.

Nicolas Carnot (1796–1832) was a French military engineer who fought against invading armies in 1814. In 1824, he published his only work, entitled *On the Motive Power of Fire,* his investigation into the limits of steam engines and their ability to convert heat into useful work. In this book, he defined work as a "weight lifted through a height." In our modern view, this definition is somewhat restrictive because it implies that only vertically moving objects can have work done on them.

In 1829, Gustave Coriolis (1792–1843) published a textbook in which he defined work in more general terms. Specifically, the work done on an object equals the force exerted on it times the distance through which it is moved against a resistance. This definition agrees with our modern view of work.

Adapted from: Isaac Asimov, *Asimov's Biographical Encyclopedia of Science and Technology* (Garden City, NY: Doubleday, 1982); P. E. Chevedden, L. Eigenbrod, V. Foley, and W. Soedel, "The Trebuchet," *Scientific American,* July 1995, p. 66.

11.12 REVERSIBILITY

We can understand the concept of *reversibility* by returning to our money analogy. Imagine that we visit England and go to the bank to exchange dollars for pounds. The official exchange rate is $2.50 per pound. If the bank charges no service fee for the transaction, it is a **reversible** process. We can change dollars into pounds and vice versa as many times as we wish, and lose no value (Table 11.6). If, however, the bank charges a 5% service fee for each transaction, it is an **irreversible** process. Eventually, after many transactions, we would lose the value of our money.

Just as dollars can be transformed into pounds, energy can be transformed from one form to another. The spring/mass system in Figure 11.12 transforms potential energy into kinetic energy, and vice versa. If this process were performed in a perfect vacuum with a perfect spring, it would be a reversible process and oscillate forever. In reality, there is always a little friction, which irreversibly converts the macroscopic energy into micro-

TABLE 11.6
Illustration of reversibility using a money analogy†

	Reversible Transaction (0% service fee)		Irreversible Transaction (5% service fee)	
Day	**Dollars**	**Pounds**	**Dollars**	**Pounds**
Monday	100.00	40.00	100.00	38.00
Tuesday	100.00	40.00	90.25	34.30
Wednesday	100.00	40.00	81.45	30.95
Thursday	100.00	40.00	73.51	27.93
Friday	100.00	40.00	66.34	25.20

†Each morning, we convert dollars to pounds; each evening, we convert pounds to dollars.

scopic energy. As a consequence, the spring/mass will eventually stop oscillating and warm the surroundings.

Examples of devices that transform one form of energy into another are: electric motors (electrical energy into shaft energy), generators (shaft energy into electrical energy), turbines (pressure energy into shaft energy), and pumps (shaft energy into pressure energy). Figure 11.17(a) shows an electric generator coupled to an electric motor and Figure 11.17(b) shows a turbine coupled to a pump. If these processes could be performed reversibly, they would run forever; the generator would produce the electricity needed by the motor, which would produce the shaft energy needed by the generator. Similarly, the pump would produce the pressure needed to operate the turbine, which would produce the shaft energy needed by the pump. We know from experience that this is impossible. Friction in bearings and flow losses (voltage drop in the wire and pressure drop in the pipe) cause some work to degenerate into heat. Eventually, the shaft spins slower and slower until it stops. Through good engineering, we can reduce the irreversibilities by improving the bearings and minimize flow losses by installing larger diameter wire or pipe. If we do a good job, the system could run much longer. But eventually, the outcome is the same; the shaft stops.

By using smaller driving forces (e.g., voltage, pressure, temperature), physical systems become more reversible; that is, less energy is degraded to low-temperature heat. This always presents interesting trade-offs for engineers; smaller driving forces require that equipment must operate more slowly or be larger to perform the same task. Thus, engineers must make compromises between energy efficiency and equipment costs.

11.13 THERMODYNAMIC LAWS

The thermodynamic laws cannot be proved by appealing to more fundamental axioms; they have been developed through years of observing the physical world. These laws have been shown to be valid for atomic interactions, for everyday processes, and for interactions on a galactic scale.

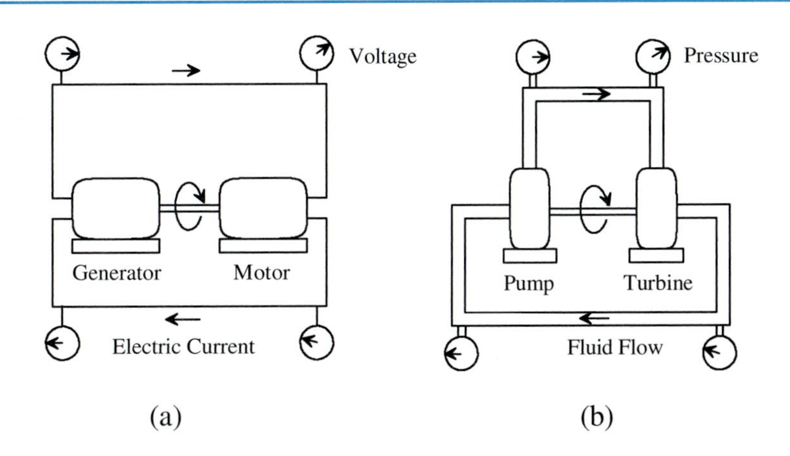

FIGURE 11.17
Demonstration of irreversibility: (a) motor powering generator powering motor and (b) turbine powering pump powering turbine.

11.13.1 First Law of Thermodynamics

The **first law of thermodynamics** states that energy can neither be created nor destroyed; therefore, it must be conserved. This law states that one form of work may be converted to another form of work, or work may be converted to heat, or heat may be converted to work. However, after the transformation is accomplished, the final amount of energy is the same as the initial amount.

If this law did not exist, the system shown in Figure 11.17(a) would not only run forever, but it could generate extra electricity with which we could run our homes. Although inventors have tried for years to build such systems, none have been successful. The first law of thermodynamics was developed by inductive reasoning.

11.13.2 Second Law of Thermodynamics

The **second law of thermodynamics** states that naturally occurring processes are directional. For example, if you saw a movie in which balloon fragments reassembled themselves into an intact balloon, you would know that the movie was run backward. You know from experience that this does not happen. Similarly, you know that your bedroom tends to become messy with time; clothes are left on the floor, dust settles on furniture. You would not expect your room to spontaneously become cleaner as time progresses. Although it is certainly possible for a room to become cleaner, it requires an expenditure of energy. Similarly, the balloon fragments could be gathered up and reshaped into a balloon, but this also requires an energy expenditure.

The second law of thermodynamics is closely tied to the idea of reversibility presented earlier. Referring to Figure 11.12, if you saw a movie of a perfect spring/mass oscillating in a perfect vacuum, the movie could be run forward or backward, and you could not tell the difference. Reversible processes have no directionality. Unfortunately, it is very difficult to build a reversible process, so most real processes have directionality.

If you saw a movie of Figure 11.15 in which ice cubes spontaneously formed as the water boiled at the other end of the copper bar, you would know this movie was running backwards. From experience, we know that heat always flows from a higher temperature to a lower temperature; it does not flow from lower temperature to higher temperature. (We can reverse the flow by installing a refrigerator, but this requires energy input.) Whereas processes that employ work can approach reversibility, processes that employ heat are inherently irreversible. Therefore, the second law imposes limits on the amount of heat that can be converted to work.

Heat can be converted to work by using a heat engine (Figure 11.18). Examples of heat engines are steam engines, internal combustion engines, and jet engines. Given an amount of heat Q_{hot} delivered to the engine at T_{hot} and waste heat Q_{cold} rejected at T_{cold}, **Carnot** derived an equation for the maximum amount of work W_{max} that can be produced:

$$\frac{W_{max}}{Q_{hot}} = 1 - \frac{T_{cold}}{T_{hot}} \tag{11-13}$$

where the temperatures must be absolute temperatures.

Although the second law imposes restrictions on the conversion of heat to work, there are no restrictions on converting work to heat. Work may always be converted to heat with 100% efficiency. A very famous demonstration of this principle was performed by Joule in

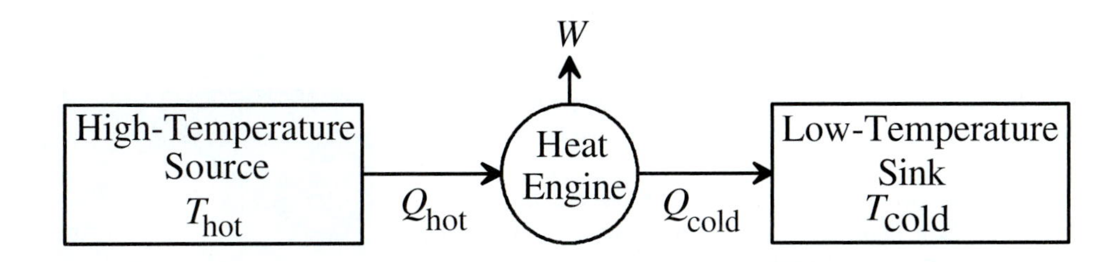

FIGURE 11.18
Schematic of heat engine.

which he showed the *work equivalent to heat.* Figure 11.19 shows his experiment, in which a stirrer churns the water, causing its temperature to rise. The work input is known, because gravity exerts a force F on the mass as it travels a distance Δx. A temperature rise could have also been obtained by putting heat into the water. For example, 1 kilocalorie of heat would cause the temperature of 1 kg water to rise by 1 kelvin. This experiment shows that 1 kilocalorie of heat equals 4184 joules of work.

In Joule's experiment, a natural question is, Where does the energy input go? We know that we can put energy into the water in the form of work or heat, but according to the first law, it must go somewhere, because energy cannot be created or destroyed. The answer to the question is that the added macroscopic energy (heat or work) increases the microscopic energy (i.e., internal energy) of the water.

11.14 HEAT CAPACITY

When heat is added to an object of mass m, its temperature rises, provided there is no phase change. If you were to add 10,000 joules of heat to a kilogram of liquid water, the temperature would increase by 2.4°C. However, if you added the same amount of heat to a kilogram of concrete, the temperature would increase by 15.4°C. The temperature change of each material is governed by its heat capacity.

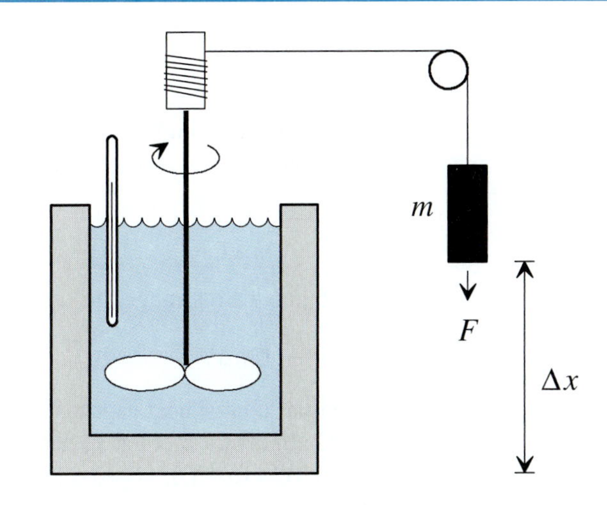

FIGURE 11.19
Joule's experiment showing the work equivalent to heat.

Origins of the First Law

The first law of thermodynamics is generally credited to three individuals who discovered it independently: Mayer, Helmholtz, and Joule.

The first person to discover the first law was Julius Mayer (1814–1878), a ship's physician. About 1840, on a long trip to Java, he thought about the generation of heat by animals. In 1842, he conducted an experiment in which he had horses stir paper pulp in a cauldron and measured the temperature rise. From this experiment, he determined a crude value for the mechanical equivalent of heat. Unfortunately, he received little credit for this work.

Hermann Helmholtz (1821–1894) was a German physician, physiologist, and physicist with many interests. He studied the eye, nerves, ears, solar energy, electrolysis, and the physics of music (he was also an accomplished musician). His studies on animal muscle contractions eventually led him to the first law. In 1847, he published his studies in such detail that he is often given credit for discovering the first law.

James Joule (1818–1889) was a brewer who had a passion for measurement. As a teenager, he measured the heat output from electric motors. In 1840, he determined that the heat released from an electric circuit equals the current squared times the resistance. He converted work into heat by churning water and mercury with paddles, passing water through small holes, and expanding and contracting gases. He was such a fanatic that on his honeymoon, he devised a special thermometer to measure the temperature rise resulting from water falling over a waterfall that he and his new bride were going to visit. In 1847, he published his careful measurements. Because he measured minor effects (some temperature increases were only 0.005°F) and because he was not a member of academia, his work was largely unnoticed until Lord Kelvin popularized his results.

James Prescott Joule. The device in the background was used to measure the work equivalent to heat.

Adapted from: Isaac Asimov, *Asimov's Biographical Encyclopedia of Science and Technology* (Garden City, NY: Doubleday, 1982).

The **heat capacity** C is defined as the amount of heat Q added per unit mass divided by the temperature change ΔT.

$$C \equiv \frac{Q}{m\Delta T} \tag{11-14}$$

which may be rearranged to

$$Q = Cm\Delta T \tag{11-15}$$

Water has a large heat capacity, so a given Q (e.g., 10,000 joules) causes a small ΔT (e.g., 2.4°C). In contrast, concrete has a small heat capacity, so a given Q (10,000 joules) causes a large ΔT (15.4°C).

Origins of the Second Law

Nicolas Carnot.

Rudolph Clausius.

In 1824, Nicolas Carnot (1796–1832) had the earliest glimpse of the second law when he studied the limits of converting heat into work. His work was largely neglected until Lord Kelvin (1824–1907) took notice. In 1851, Lord Kelvin deduced from Carnot's studies that all energy tends to degrade into heat. Further, the heat eventually becomes unusable as the temperature at which it is delivered becomes lower and lower. He concluded that the whole universe will eventually run down: the so-called "heat-death of the universe."

Rudolf Clausius (1822–1888), aware of Carnot's work, discussed the degradation of energy into heat with Kelvin. In 1850, Clausius was studying isolated systems in which nothing (e.g., no energy, no mass) entered or left. He devised a quantity equal to the ratio of heat to temperature that proved very useful. No matter what process he placed in his isolated system, he found that this ratio always increased or, at best, stayed the same. In 1865, he termed this ratio *entropy* and used the symbol S. In his own words, "I propose . . . to call S the entropy of a body, after the Greek word 'transformation.' I have designedly coined the word 'entropy' to be similar to energy, for these two quantities are so analogous in their physical significance, that an analogy of denominations seems to be helpful."

Adapted from: Isaac Asimov, *Asimov's Biographical Encyclopedia of Science and Technology* (Garden City, NY: Doubleday, 1982); R. A. Serway, *Principles of Physics* (Fort Worth: Saunders, 1994).

Figure 11.20(a) shows heat being added to a material (gas, liquid, or solid) that is maintained at constant volume. In this case, the measured heat capacity is C_v (where the subscript v stands for "constant volume"). The added heat causes the internal energy U of the material to increase; therefore, we can write the equation

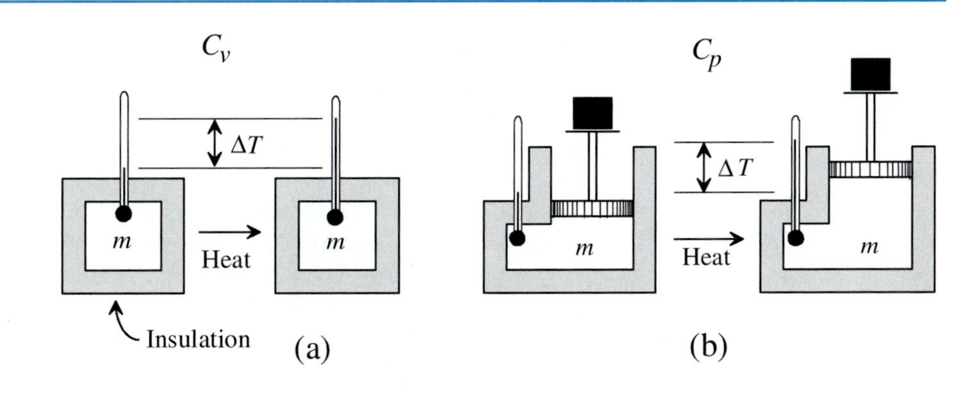

FIGURE 11.20
Heat capacity:
(a) constant volume and
(b) constant pressure.

James Watt: Master Mechanic

James Watt.

It may be impossible to give any individual credit for inventing the steam engine. The idea first occurred to Hero of Alexandria, a Hellenistic thinker from the first century A.D. Hero designed an impractical turbine, consisting of a rotating sphere with two tangential outlets, set over a fire (see figure). Seventeenth-century British prototypes, including those developed by Thomas Savery, Denys Papin, and Thomas Newcomen, were based on more practical piston-and-cylinder arrangements. But credit for improving the steam engine to the point of usefulness must be given to James Watt (1736–1819).

At the suggestion of a philosophy professor at the University of Glasgow in 1765, Watt set out to repair and improve the university's model of a Newcomen engine. He soon found that this beast was tremendously inefficient, and developed a design that would increase the power and decrease energy consumption. Newcomen's engine used steam to fill the cylinder, and then quickly condensed the steam with a shot of cold water. This created a vacuum inside the cylinder, and atmospheric pressure would force the piston into the evacuated cylinder. Watt recognized that much of the steam's energy was wasted in reheating the cylinder walls after each evacuation. He concluded that efficiency could be greatly improved by insulating the cylinder and evacuating it with an external condenser. Watt's second improvement was to use the pressure of the steam, which can be much higher than atmospheric pressure, to push the piston into the evacuated cylinder, thus generating much more power. With his partner, Birmingham industrialist Matthew Boulton, Watt succeeded in building an efficient, single-stroke engine. His later inventions included the centrifugal governor, which controlled the engine speed, and the double-stroke engine, in which steam pushed the piston in both directions.

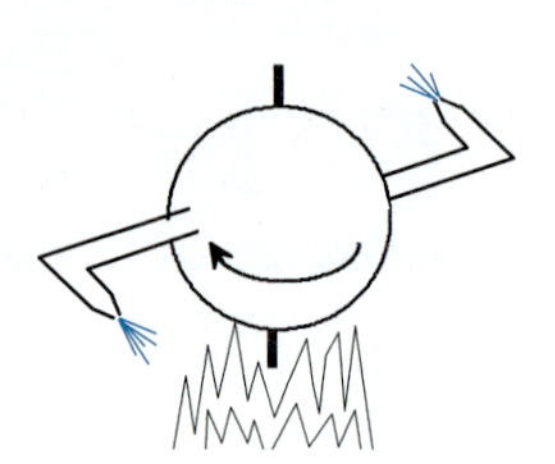

Courtesy of: Seth Adelson, graduate student.

$$Q = \Delta U = mC_v\Delta T \tag{11-16}$$

Figure 11.20(b) shows heat being added to a material (gas, liquid, or solid) that is maintained at constant pressure. In this case, the measured heat capacity is C_p (where the subscript p stands for "constant pressure"). The added heat causes the internal energy of the material to increase, **plus** it causes a weight to be lifted; thus, some PV work (see Equation 11-12) is also done. In this case, the added heat contributes to both a change in internal energy and PV work, so we can write the equation

$$Q = \Delta U + P\Delta V = mC_p\Delta T \tag{11-17}$$

Experimentally, it is much easier to add heat at constant pressure than constant volume. For example, if a liquid is placed in a constant-volume container, the liquid wants to expand at higher temperatures. But, because the liquid is confined to a constant-volume container, it has nowhere to go. As a consequence, the pressure goes extraordinarily high, making it difficult to contain the high-pressure liquid. In contrast, in a constant-pressure vessel, the liquid can easily expand. Because constant-pressure measurements are more easily made, the term $\Delta U + P\Delta V$ is encountered frequently; therefore, a new variable called **enthalpy H** is defined as

$$\Delta H \equiv \Delta U + P\Delta V \tag{11-18}$$

This equation is valid provided the pressure is constant, which is true in this case.[1] Using this newly defined variable, Equation 11-17 may be rewritten as

$$Q = \Delta H = mC_p\Delta T \tag{11-19}$$

11.15 SUMMARY

Thermodynamics is the science that describes what is possible and what is impossible during energy conversion processes. However, it does not address the length of time required to perform the energy conversion process (this is the subject of the next chapter).

Temperature measures the degree of random atomic (molecular) motion, pressure measures the force per unit area, and density measures the mass of material per unit volume. Depending upon the temperature and pressure, matter will assume a particular state, either solid, liquid, gas/vapor, or plasma. The phase diagram describes the state of matter at any given temperature/pressure combination. Along phase lines, matter can have two phases in equilibrium. At the triple point, three phases can be in equilibrium.

The gas laws describe the relationship between temperature, pressure, volume, and the number of moles of gas. The perfect gas equation (also called the ideal gas equation) describes the relationship between these variables for low-density gases. At high densities, more complex equations (e.g., van der Waals) are required.

Energy is an abstract concept used as a "unit of exchange" to relate physical phenomena. Energy flow is classified into two types: heat and work. Heat is energy flow resulting from a temperature driving force, whereas work is energy flow resulting from any other driving force.

In simple terms, reversible thermodynamic processes are infinitely slow and frictionless. Reversible processes are generally idealizations, except in a few rare cases (e.g., electrical superconductors and superfluids).

The first law of thermodynamics states that energy can neither be created nor destroyed; rather, it is conserved. The second law of thermodynamics states that naturally occurring processes are directional. An important implication of the second law is that there are inherent limits on the ability to convert heat into work, as dictated by the Carnot equation.

Internal energy is the energy associated with the random motion of atoms (molecules) and their position relative to one another. Changes in internal energy are determined using the constant-volume heat capacity. Enthalpy is a variable defined as the sum of internal energy and *PV* work. Changes in enthalpy are determined using the constant-pressure heat capacity.

Nomenclature

A area (m^2)
C heat capacity (J/(kg·K))
C_p constant-pressure heat capacity (J/(kg·K))

1. In general, enthalpy is defined as $H \equiv U + PV$. From the rules of calculus, a differential change in enthalpy is $dH = d(U + PV) = dU + d(PV) = dU + PdV + VdP$. At constant pressure, $dP = 0$ and this becomes $dH = dU + PdV$. At constant pressure, a nondifferential change in enthalpy is $\Delta H = \Delta U + P\Delta V$.

C_v constant-volume heat capacity (J/(kg·K))
F force (N)
H enthalpy (J)
k constant (dimensionless)
N_A Avogadro's number (dimensionless)
n moles (mol)
P pressure (Pa)
Q heat (J)
R universal gas constant [Pa·m^3/(mol·K) or J/(mol·K)]
S entropy (J/K)
T temperature (K)
U internal energy (J)
V volume (m^3)
W work (J)
x distance (m)
Z compressibility factor (dimensionless)

Further Readings

Eide, A. R.; R. D. Jenison; L. H. Mashaw; and L. L. Northup. *Engineering Fundamentals and Problem Solving.* 3rd ed. New York: McGraw-Hill, 1997.

Felder, R. M., and R. W. Rousseau. *Elementary Principles of Chemical Processes.* New York: Wiley, 1986.

Smith, J. M., and H. C. Van Ness. *Introduction to Chemical Engineering Thermodynamics.* 3rd ed. New York: McGraw-Hill, 1975.

Whitten, K. W., and K. D. Gailey. *General Chemistry.* 2nd ed. Philadelphia: Saunders, 1984.

PROBLEMS

11.1 Show that the perfect gas equation is consistent with each of the following laws:

(a) Boyle's
(b) Charles'
(c) Gay-Lussac's
(d) mole proportionality

11.2 The *standard state* for a gas is defined as $T = 273.15$ K (0°C) and $P = 1.0133 \times 10^5$ Pa (1 atm). Using the perfect gas equation, find the volume of 1 mol of perfect gas.

11.3 Do the following:

(a) Using the perfect gas equation, find the number of moles in one cubic meter (m^3) at standard conditions, i.e., $T = 273.15$ K (0°C) and $P = 1.0133 \times 10^5$ Pa (1 atm).
(b) Calculate the molecular weight of the following gases:
 i. CO_2 **iii.** NH_3
 ii. N_2 **iv.** H_2
(c) Calculate the density (kg/m^3) of each gas listed in Part (b).
(d) Compare the densities you calculated with those listed in Table 11.1. What do you conclude?

11.4 A gas cylinder is stored on a loading dock. In the morning, the temperature is 290 K and the pressure is 1.00×10^6 Pa. In the afternoon, the temperature of the gas cylinder rises to 350 K because it is exposed to the hot sun. Assuming there were no leaks in the gas cylinder and that the gas behaves as a perfect gas, what is the pressure in the afternoon?

11.5 Natural gas is stored in very large vessels consisting of a floating tank and a liquid seal (Figure 11.21). The pressure in the tank is constant because the weight of the floating tank is constant. In the morning, the temperature of the natural gas is 300 K but in the afternoon, the temperature rises to 330 K. In the morning, the top of the tank is 50 m above ground level. What is the height above ground level in the afternoon? Assume that the natural gas behaves as a perfect gas and that no gas has been removed or added during the day.

11.6 A truck uses gas springs (Figure 11.22) instead of coil or leaf springs to support its weight and the weight of its cargo. There is one gas spring at each of the four wheels. The truck's mass is 1500 kg. When the truck is empty, it rides 0.5 m above the ground,

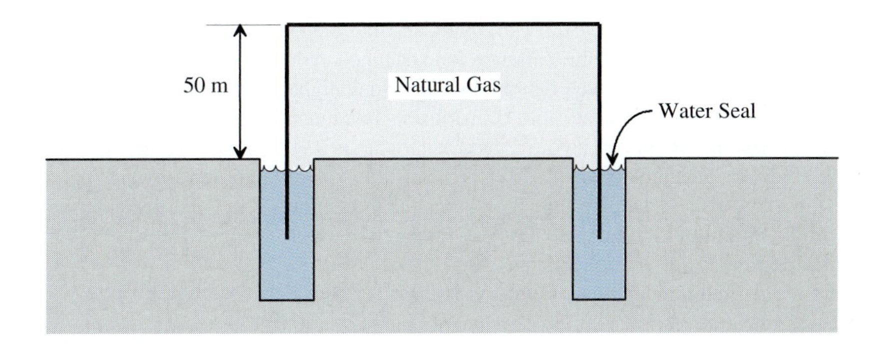

FIGURE 11.21
Natural gas storage tank.

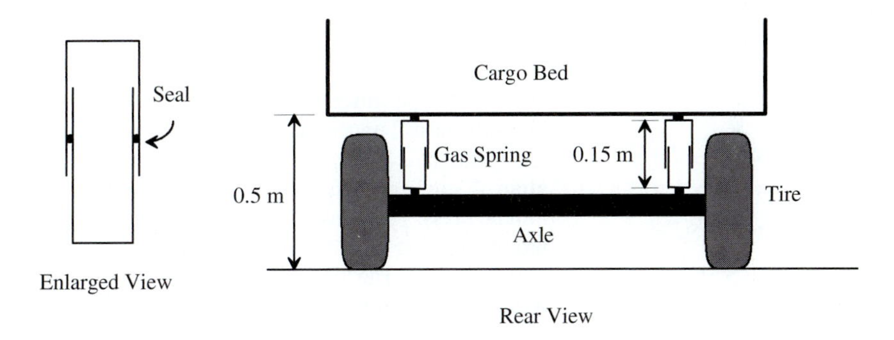

FIGURE 11.22
Gas springs used to support the weight of a truck and its cargo.

and the length of the gas spring is 0.15 m. With a 500-kg cargo, what is the height of the truck bed above the ground? (Assume the load distributes evenly on each wheel.) Assume the gas behaves as a perfect gas and that no gas leaks out of the gas springs. Also, assume that the gas spring is always at the temperature of the surroundings.

11.7 The Carnot equation (Equation 11-13) is used to calculate the maximum amount of work that can be produced from a given high-temperature heat input. The first law says that energy is neither created nor destroyed. Therefore, any heat input not converted to work must be rejected as waste low-temperature heat. Derive an equation that gives the minimum amount of waste heat that must be rejected for a given high-temperature heat input.

11.8 At 300 K, 1.000 mol of carbon dioxide (CO_2) is stored in a 0.000400-m^3 container. Calculate the pressure using the

(a) perfect gas equation

(b) van der Waals equation

11.9 The heat capacity C_p of water is 1.00 kcal/(kg·K). Joule's experiment (Figure 11.19) is conducted using 5.00 kg water. Gravity exerts a force of 10,000 N on the mass as it drops a 1-m distance. Before the experiment began, the water temperature was 300.00 K. What is its temperature when the experiment is completed?

11.10 Ocean Thermal Energy Conversion (OTEC) is a method for producing energy from temperature differences in the ocean. Calculate the maximum possible work produced per unit of high-temperature heat, assuming heat is taken into the OTEC system at the water surface (300 K) and waste heat is rejected below the surface (290 K).

11.11 Pollution-free transportation has been an engineering goal for many years. One suggestion is to have cars carry a tank of liquid nitrogen and operate a heat engine based upon the temperature difference between the air (300 K) and boiling nitrogen (77.36 K). Calculate the maximum possible work produced per unit of high-temperature heat.

11.12 You are a sales engineer for a company located on the U.S./Canada border. In the morning you service U.S. customers, and in the afternoon you service Canadian customers. Your car is rather old, so you decide to keep cash on you at all times in case the car breaks down and needs servicing. On Monday morning, you put $100 (U.S.) into the car-repair fund. Every time you cross the border, you go to the bank and convert U.S. dollars into

Canadian dollars, and vice versa. Both U.S. and Canadian banks charge a 3% service fee every time currency is exchanged. You work every day that week. When you return home on Friday night, you convert your car-repair fund back into U.S. dollars. How much money do you have in the fund? The official exchange rate is 0.90 U.S. dollar = 1.00 Canadian dollar.

11.13 Using a spreadsheet, prepare a graph showing the maximum possible work per unit of high-temperature heat. Prepare the graph using high temperatures ranging from 300 K to 3000 K. Regardless of the high temperature, assume waste heat is rejected at ambient temperature (300 K).

11.14 Write a computer program that allows the gas volume of carbon dioxide to be calculated. The user specifies the number of moles, the absolute temperature, and absolute pressure. The program will calculate the volume using both the perfect gas equation and the van der Waals equation. The perfect gas equation is straightforward, but the van der Waals equation requires that you find the zero of the equation. (See Appendix K, *Mathematics Supplement*.) Use a method of your choice.

Glossary

absolute pressure scale A pressure scale that designates the pressure exerted by a perfect vacuum as zero.

absolute temperature scale A temperature scale that designates absolute zero as "zero degrees."

absolute zero Temperature is reduced as low as possible and motion, although not zero, is at its minimum.

atom A single nucleus with its accompanying electrons.

atomic mass The mass of 1 mole of molecules; commonly called *atomic weight.*

atomic number The number of protons in an atom.

atomic weight The weight of 1 mole of molecules; also called *atomic mass.*

Avogadro's number N_A The number of atoms or molecules in 1 gram-mole.

Boyle's law The absolute pressure P and the volume V of a perfect gas are inversely proportional, provided the temperature T and number of moles n are constant.

calorie The amount of heat needed to increase the temperature of 1 gram of water by 1 degree Celsius.

calorimeter A device that measures heat effects.

Carnot equation The equation that describes the maximum amount of work W_{max} that can be produced from a unit of high-temperature heat (Equation 11-13).

Charles' law Volume V and absolute temperature T are directly proportional, provided the pressure P and number of moles n are constant.

compressibility factor Z A factor used to correct the perfect gas equation so it can describe real gases.

critical point The temperature and pressure where the density of the liquid and vapor phases become identical.

density The amount of mass in a unit volume.

driving force An influence that causes change.

electron A negatively charged subatomic particle.

enthalpy H Internal energy plus the product of pressure and volume.

equilibrium conditions Conditions that do not change as time passes.

first law of thermodynamics Energy can neither be created nor destroyed.

force An influence that can cause a body to accelerate.

Gay-Lussac's Law The absolute pressure P and absolute temperature T are directly proportional, provided the volume and number of moles are constant.

gravity An attractive force between two objects that have mass.

heat The energy flow resulting from atoms or molecules of different temperatures.

heat capacity C The amount of heat per unit mass divided by the temperature change.

interconversion Mutual conversion.

internal energy The collection of microscopic kinetic and potential energies.

irreversible The inability of a system to return to its initial state without inputs from the external environment.

kinetic energy The energy associated with a moving mass.

molecular mass The mass of 1 mole of molecules; commonly called *molecular weight.*

molecular weight The weight of 1 mole of molecules; also called *molecular mass.*

molecules A group of atoms held together by chemical bonds.

mole proportionality law The volume V and number of moles n are directly proportional provided the pressure and volume are constant.

neutron A subatomic particle with no charge.

opalescent A milky iridescence like that of an opal.

perfect gas A hypothetical gas in which the volume of the individual atoms (molecules) is negligible compared to the total volume of the container and there are no forces between the atoms (molecules); also called an *ideal gas.*

perfect gas equation The relationship between temperature, pressure, volume, and the number of moles of low-density gas; also called *the ideal gas equation.*

phase diagram A diagram that indicates the stable phase or state at a given temperature and pressure.

potential energy Energy associated with position.

pressure The force exerted on a unit area.

proton A positively charged subatomic particle.

reversible The ability of a system to return to its initial state without inputs from the external environment.

second law of thermodynamics Naturally occurring processes are directional.

strong force A force that acts at very short distances and is responsible for holding atomic nuclei together.

supercritical fluid A gas with liquid-like densities in which the pressure and temperature are above the critical point.

temperature A measure of the degree of random atomic motion.

thermodynamics The science that describes what is possible and what is impossible during energy conversion processes.

thermoscope A device used by Galileo to measure temperature.

triple point The three phases (or states) in equilibrium.

universal gas constant The proportionality constant in the perfect gas equation.

valence electrons The electrons farthest from the nucleus.

van der Waals equation An equation used to calculate the compressibility factor Z.

weak force A force involved in radioactive decay.

work Energy flow resulting from any driving force other than temperature.

Mathematical Prerequisite

- Algebra (Appendix E, Mathematics Supplement)

CHAPTER 12

Introduction to Rate Processes

Previously, we learned that thermodynamics tells us where a process is going. However, it cannot tell us how long the process takes to get there. This is the topic of **rate processes,** which we now study.

12.1 RATE

Rate is the amount a quantity N changes per unit time t.

$$\text{Rate} = r = \frac{\Delta N}{\Delta t} \tag{12-1}$$

For example, if water flows into a tank, the water level increases at a particular rate (Figure 12.1). If the liquid volume in the tank is plotted versus time, the slope of the curve is the rate r;

$$r = \frac{V_t - V_0}{t - 0} = \frac{\Delta V}{\Delta t}$$

A **flow rate** is the amount of quantity N that flows by a point in a given length of time t.

$$r = \frac{N}{t} \tag{12-2}$$

For example, if we were to place a screen at the water outlet in Figure 12.1, the volume of water V that passes through the screen per unit time t would be the flow rate r:

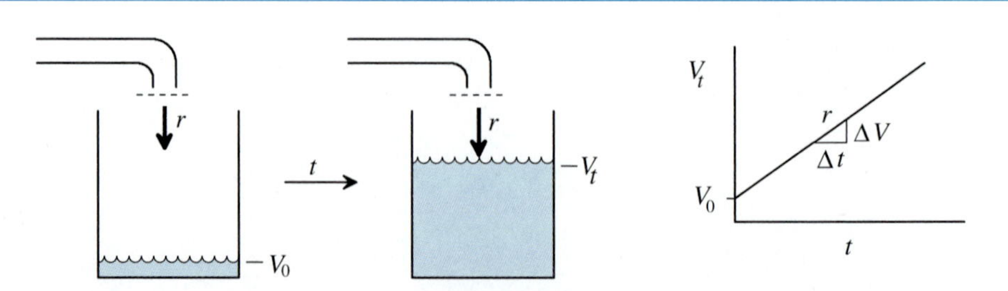

FIGURE 12.1
Water filling a tank to illustrate the rate concept.

$$r = \frac{V}{t} \tag{12-3}$$

In this example, the tank has no outlet, so the flow rate of water through the pipe is identical to the rate of water accumulation in the tank.

12.2 FLUX

Flux is the flow rate r per unit of cross-sectional area A:

$$J = \text{flux} = \frac{\text{flow rate}}{\text{cross-sectional area}} = \frac{r}{A} \tag{12-4}$$

Because flow rate is the amount of quantity N that flows per unit time t, we can substitute Equation 12-2 into Equation 12-4 to get

$$J = \text{flux} = \frac{\text{amount}}{(\text{cross-sectional area})(\text{time})} = \frac{N}{At} \tag{12-5}$$

A simple illustration of flux is a machine gun firing bullets at a target that is 1 m on a side; that is, it has a 1-m^2 cross-sectional area (Figure 12.2). If the machine gun fires five bullets per second, then the flux of bullets through the target is

$$\text{Flux} = \frac{5 \text{ bullets}}{(1 \text{ m}^2)(1 \text{ s})} = 5 \frac{\text{bullets}}{\text{m}^2\text{·s}}$$

12.3 DRIVING FORCE

We have already briefly considered the concept of a **driving force** in our discussion of thermodynamics. Recall that energy (either heat or work) flows as a result of a driving force.

Flux J is directly proportional to the driving force ΔD but inversely proportional to the distance x over which the driving force is applied (Figure 12.3),

$$J = \frac{r}{A} = \frac{N}{At} \propto \frac{\Delta D}{x} = K\frac{\Delta D}{x} \tag{12-6}$$

where K is the proportionality constant and $\Delta D = D_{in} - D_{out}$. Table 12.1 shows the types of fluxes we will consider in this chapter.

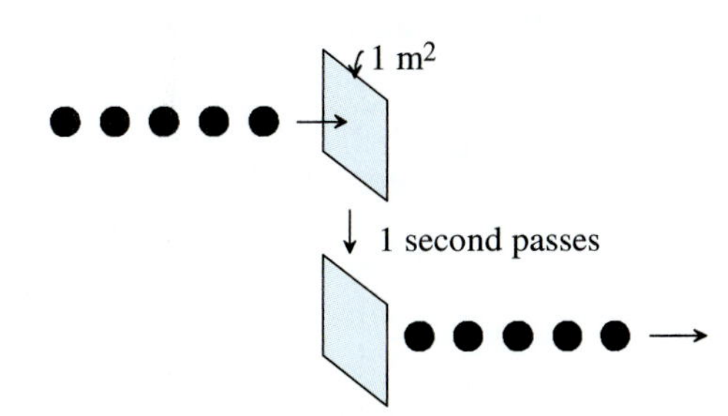

FIGURE 12.2
Visualization of flux using bullets passing through a target.

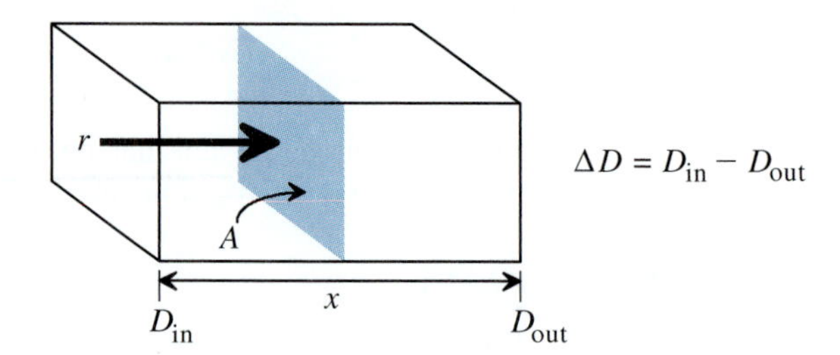

FIGURE 12.3
Flux resulting from a driving force.

TABLE 12.1
Fluxes discussed in this chapter

Quantity *N*	Driving Force *D*
Heat	Temperature
Fluids	Pressure
Electric current	Voltage
Chemical species	Concentration

12.4 HEAT

Heat is energy flow resulting from a temperature difference. Heat flux J_{heat} is the amount of heat Q that flows per unit cross-sectional area per unit time. It is proportional to the temperature difference ΔT and inversely proportional to the distance x separating the two temperatures:

$$J_{heat} = \frac{Q}{At} = k\frac{T_{in} - T_{out}}{x} = k\frac{\Delta T}{x} \tag{12-7}$$

The proportionality constant k is called the **thermal conductivity.** Table 12.2 lists the thermal conductivity of some common materials. When the thermal conductivity is large, the material is called a **conductor,** and when it is low, the material is called an **insulator.**

EXAMPLE 12.1

Problem Statement: A copper rod is 0.500 m long and 0.0250 m in diameter. One end is placed in a boiling water bath (373.15 K) and the other is placed in an ice bath (273.15 K). The walls of the rod are well insulated, so negligible heat leaks out the sides. What is the heat flow q (J/s) down the rod?

Solution:

$$q = \frac{Q}{t} = \frac{kA(T_{in} - T_{out})}{x} = \frac{k\frac{\pi}{4}d^2(T_{in} - T_{out})}{x}$$

$$= \frac{\left(386 \frac{\text{J}}{\text{s·m·K}}\right)\frac{\pi}{4}(0.0250\text{ m})^2\,(373.15\text{ K} - 273.15\text{ K})}{0.500\text{ m}} = 37.9\text{ J/s}$$

TABLE 12.2
Thermal conductivity (Reference)

Gases†	$k\left(\frac{J}{s \cdot m \cdot K}\right)$	Metallic solids‡	$k\left(\frac{J}{s \cdot m \cdot K}\right)$
Air	0.026240	Aluminum	229
Carbon dioxide	0.016572	Copper	386
Carbon monoxide	0.025240	Gold	293
Hydrogen	0.181660	Iron	73.2
Nitrogen	0.026052	Platinum	70.1
Oxygen	0.026760	Silver	415
		Stainless steel	16
Liquids*		Steel, mild	42.9
Ammonia	0.5512	**Nonmetallic Solids‡**	
Benzene	0.145		
n-Butanol	0.171	Asbestos	0.159
Glycerin	0.287	Brick, masonry	0.66
Kerosene	0.138	Concrete	1.21
Mercury	8.16	Corkboard	0.043
Water	0.610	Glass, window	0.78
		Rock wool	0.040
		Oak, perpend. to grain	0.21
		Oak, parallel to grain	0.40

†$T = 300$ K (26.85°C), $P = 1.0133 \times 10^5$ Pa (1 atm).
*$T = 300$ K.
‡$T = 293$ K.

Adapted from: J. R. Welty, C. E. Wicks, and R. E. Wilson, *Fundamentals of Momentum, Heat, and Mass Transfer,* 2nd ed. (New York: Wiley, 1976), pp. 730–758.

12.5 FLUID FLOW

Figure 12.4 shows a pressure difference ΔP forcing a fluid volume V through a circular pipe with cross-sectional area A and length x. The fluid flux is given by the **Poiseuille equation**

$$J_{\text{fluid}} = \frac{V}{At} = \left(\frac{A}{8\pi\mu}\right)\frac{P_{\text{in}} - P_{\text{out}}}{x} = \left(\frac{A}{8\pi\mu}\right)\frac{\Delta P}{x} \tag{12-8}$$

where μ is the viscosity (Table 12.3). The **viscosity** is a measure of the "thickness" of a fluid. For example, maple syrup has a higher viscosity than water because it is thicker. The Poiseuille equation is valid for laminar (i.e., nonturbulent) flow and for $x >> d$. Laminar flow is assured provided

$$\frac{d\rho}{\mu} J_{\text{fluid}} < 2300 \tag{12-9}$$

where d is the pipe diameter and ρ is the fluid density. According to Equation 12-8, if a fluid had no viscosity ($\mu = 0$), then fluid could flow without a pressure difference. Such fluids are called **superfluids,** an example of which is helium near absolute zero.

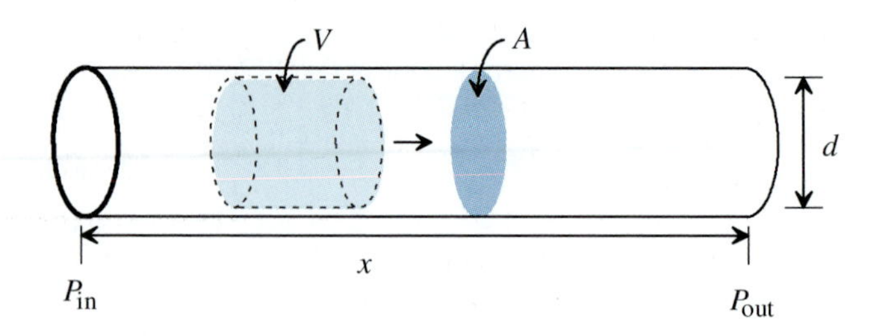

FIGURE 12.4
Flux of fluid volume through a circular pipe.

TABLE 12.3
Viscosity of fluids (Reference)

Gas†	μ (Pa·s)	Liquid‡	μ (Pa·s)
Air	1.8464×10^{-5}	Ammonia	21.1×10^{-5}
Carbon dioxide	1.4948×10^{-5}	Benzene	56.6×10^{-5}
Carbon monoxide	1.7854×10^{-5}	*n*-butanol	268×10^{-5}
Hydrogen	0.89594×10^{-5}	Glycerin	0.9
Nitrogen	1.7855×10^{-5}	Kerosene	729×10^{-5}
Oxygen	2.0633×10^{-5}	Mercury	149×10^{-5}
		Water	86.0×10^{-5}

†$T = 300$ K (26.85°C), $P = 1.0133 \times 10^5$ Pa (1 atm).
‡$T = 300$ K.

Adapted from: J. R. Welty, C. E. Wicks, and R. E. Wilson, *Fundamentals of Momentum, Heat, and Mass Transfer,* 2nd ed. (New York: Wiley, 1976), pp. 733–758.

EXAMPLE 12.2

Problem Statement: Calculate the flow rate of water ($T = 300$ K) through a tube that is 2.00 m long with an inner diameter of 0.00100 m. The inlet pressure is 1.0153×10^5 Pa and the outlet pressure is 1.0133×10^5 Pa.

Solution:

$$\frac{V}{t} = A\left(\frac{A}{8\pi\mu}\right)\frac{P_{in} - P_{out}}{x} = \left(\frac{A^2}{8\pi\mu}\right)\frac{P_{in} - P_{out}}{x}$$

$$= \left(\frac{(\frac{\pi}{4}d^2)^2}{8\pi\mu}\right)\frac{P_{in} - P_{out}}{x} = \left(\frac{\pi d^4}{128\mu}\right)\frac{P_{in} - P_{out}}{x}$$

$$= \left(\frac{\pi(0.00100\text{ m})^4}{128(86.0 \times 10^{-5}\text{ Pa·s})}\right)\frac{1.0153 \times 10^5\text{ Pa} - 1.0133 \times 10^5\text{ Pa}}{2.00\text{ m}}$$

$$= 2.85 \times 10^{-9}\,\frac{\text{m}^3}{\text{s}}$$

To verify that the flow is laminar,

$$\frac{d\rho}{\mu} J_{\text{fluid}} = \frac{d\rho}{\mu}\left(\frac{V}{t}\right)\frac{1}{A} = \frac{d\rho}{\mu}\left(\frac{V}{t}\right)\frac{1}{\frac{\pi}{4}d^2} = \frac{4\rho}{\pi\mu d}\left(\frac{V}{t}\right)$$

$$= \frac{4\left(1000\,\frac{\text{kg}}{\text{m}^2}\right)}{\pi\left(86.0 \times 10^{-5}\,\frac{\text{kg}}{\text{m}\cdot\text{s}}\right)(0.00100\text{ m})}\left(2.85 \times 10^{-9}\,\frac{\text{m}^3}{\text{s}}\right) = 4.22$$

Because this number is less than 2300, the flow is laminar and the Poiseuille equation is valid.

12.6 ELECTRICITY

Electricity is a mystery to most students. After all, you cannot see it—but it can kill you through carelessness. Many engineering students graduate with a poor understanding of electricity, which is unfortunate because it is one of our most powerful engineering tools. Computers, motors, sensors, and actuators all rely on electricity to function.

Electricity is flowing electrons. The best electrical conductors are metal. The atoms in metals have, in their outer shells, just a few electrons that are loosely bound to the nucleus. Although the outer electrons are negative and hence attracted to the positive nucleus, they are shielded by the inner electron shells. Therefore, these outer electrons are free to jump from one nucleus to another (Figure 12.5).

To help visualize electron flow, imagine a straw completely filled with peas. If a new pea is added to the left end of the straw, a pea at the right end of the straw must fall out. Similarly, Figure 12.5 shows that as an electron is added at the left, it causes a cascade effect, forcing an electron to exit to the right. Although the electrons move slowly through the wire (about 10^{-3} m/s), the effect of the added left electron is felt at the right electron almost instantly, causing it to immediately jump. [The "jump signal" actually travels down the wire at nearly the speed of light (3×10^8 m/s)].

Just as gas atoms (molecules) are counted by a special number called a gram-mole (1 mol = 6.0221367×10^{23} atoms or molecules), electrons are counted by a special number called the **coulomb** (1 C = $6.24150636 \times 10^{18}$ electrons). The flow rate of electrons is called **current.** It is measured in **amperes** (A), which is defined as one coulomb per second (1 C/s).

Electricity may be visualized by making an analogy to gas flowing through a pipe filled with sand [Figure 12.6(a)]. The gas is "energized" by putting work into a pump or compressor. The gas flows from the high-pressure pump outlet, through the sand-filled pipe, to the low-pressure pump inlet. As the gas flows, there is friction with the sand, causing heat to be released. Thus, this system converts work into heat, an irreversible process.

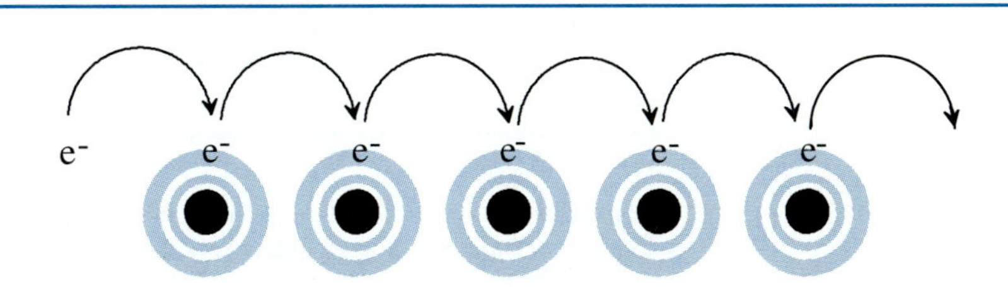

FIGURE 12.5
Electricity is electrons jumping from one atom to the next.

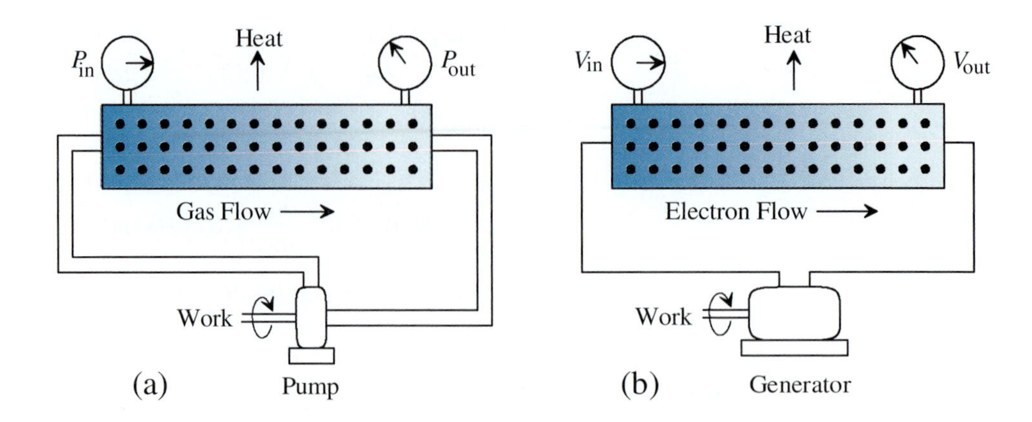

FIGURE 12.6
Analogy between (a) gas flow through a bed of sand and (b) electron flow through a wire.

Electron flow [Figure 12.6(b)] is analogous to this gas flow. The nuclei and inner electron shells of the atoms comprising the wire correspond to the sand in the analogy, and the mobile outer-shell electrons correspond to the gas. The generator "energizes" the flowing electrons by squeezing them together. (Recall that because electrons are negatively charged, they repel each other. Therefore, energy is required to squeeze them together. For example, a 1-volt generator adds 1 joule of energy to 1 coulomb of electrons, i.e., 1 *volt* = 1 J/C). Electrons flow from the high-voltage generator outlet, cascade from atom to atom in the wire, and then are collected at the low-voltage generator inlet. The flowing electrons bump into the nuclei/inner electron shells, causing friction and releasing heat. In good electrical conductors (e.g., copper wire), small amounts of heat are produced, whereas in poor electrical conductors (e.g., stove heating elements), large amounts of heat are produced. In superconductors, the electrons are able to flow without bumping into the nuclei/inner electron shells, and hence no heat is released.

TABLE 12.4
Electrical conductivity of various materials (Reference)

Material	κ ($C^2/(J{\cdot}s{\cdot}m)$)	Material	κ ($C^2/(J{\cdot}s{\cdot}m)$)
Aluminum	3.54×10^7	Nickel	1.17×10^7
Brass	1.61×10^7	Platinum	0.933×10^7
Carbon, amorphous	0.0025×10^7	Silver	6.14×10^7
Copper wire	5.80×10^7	Steel, transformer	0.902×10^7
Gold	4.10×10^7	Tin	0.860×10^7
Iron wire	0.102×10^7	Tungsten	1.81×10^7
Monel metal	0.230×10^7	Zinc	1.68×10^7
Mercury	0.103×10^7		

T = 293 K (20°C).

Adapted from: T. Baumeister, E. A. Avallone, and T. Baumeister III, *Mark's Standard Handbook for Mechanical Engineers,* 8th ed. (New York: McGraw-Hill, 1978), pp. 15-6.

Electron flux $J_{electron}$ is the number of electrons q that flow per unit cross-sectional area per unit time. It is proportional to the voltage difference ΔV and inversely proportional to the distance x separating the two voltages, or

$$J_{electron} = \frac{q}{At} = \kappa \frac{V_{in} - V_{out}}{x} = \kappa \frac{\Delta V}{x} \quad (12\text{-}10)$$

where the proportionality constant κ is the *electrical conductivity.* Table 12.4 shows the **electrical conductivity** of various materials. Silver has an extremely high conductivity followed closely by copper. Carbon has a low conductivity, which is why it is often used in electrical resistors. **Electrical superconductors** have essentially infinite electrical conductivity, meaning current can flow without a voltage difference.

EXAMPLE 12.3

Problem Statement: A voltage difference of 10.0 V is placed across a copper wire that is 1000 m long and 0.00100 m in diameter. What is the current i?

Solution: $$i = \frac{q}{t} = \kappa A \frac{\Delta V}{x} = \kappa \frac{\pi}{4} d^2 \frac{\Delta V}{x}$$

$$= \left(5.80 \times 10^7 \frac{C^2}{J{\cdot}s{\cdot}m}\right) \frac{\pi}{4} (0.00100\ m)^2 \frac{10.0\ J/C}{1000\ m} = 0.456 \frac{C}{s} = 0.456\ A$$

AC/DC

Direct current (DC) is the flow of electrons in a single direction through a wire, whereas **alternating current** (AC) is the oscillation of electrons back and forth through a wire.

DC e^- (→) AC e^- (↔)

Direct current is produced by a battery and the discharge of static electricity; it is the earliest form of current known. Early electric motors were powered by direct current.

In an early electric generator, called a **Gramme machine,** a coil rotated within a stationary horseshoe magnet (see figure on next page). As the coil rotated through the magnetic field, an alternating current was induced in the coil whereby the current would reach a maximum, reduce to zero, and then reach a maximum in the opposite direction. This alternating current was transformed into direct current using brushes and a *commutator,* a conducting cylinder split lengthwise that rotated between the stationary brushes. Each half of the commutator was connected to an opposite end of the coil. The brushes contacted the commutator and allowed electricity to flow through the "load" (in this case, a lightbulb). Just at the point where the alternating current in the coil reduced to zero and was about to reverse directions, the commutator would reverse contact with the brushes, allowing the electricity to continue flowing in the same direction through the load. Although the current from the commutator went in only one direction, its rate varied. The Gramme machine could also function as an electric motor; if supplied with direct current, it would rotate.

In 1876, Croatia's Nikola Tesla saw a Gramme machine in his sophomore year at college. He reasoned that the sparks generated by the commutator and brushes were very inefficient and commented to his professor that it should be possible to power a motor with alternating current. His professor spent the next lecture ridiculing young Tesla's idea, claiming that it would be a perpetual motion machine and therefore impossible.

Undaunted, in 1881, Tesla took up the problem of AC motors and figured out how to eliminate the commutator and brushes. By running two or more out-of-phase alternating currents through stationary coils on the periphery of the rotor, he could create a rotating magnetic field that would cause the rotor to rotate.

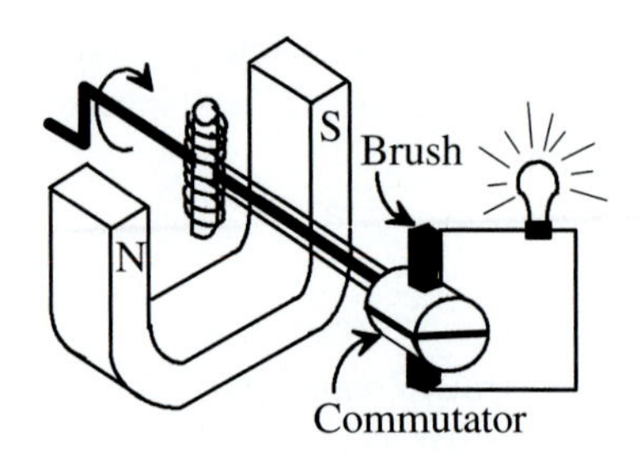

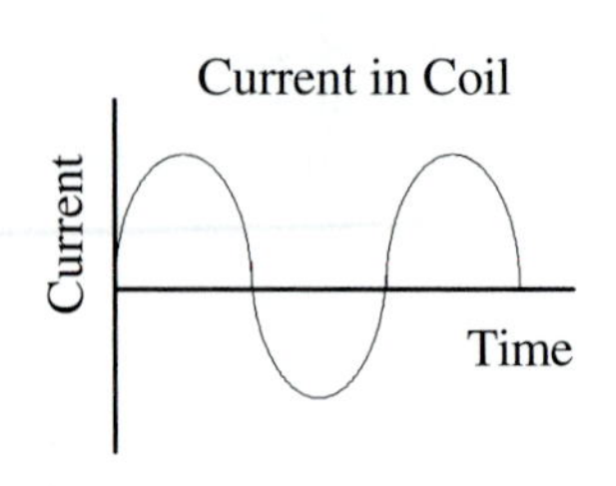

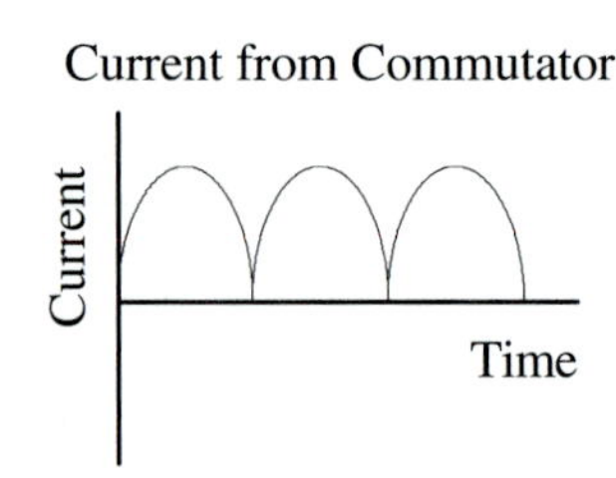

While working for the Continental Edison Company in Paris, Tesla tried to interest them in building his "polyphase" electric motor. He failed. Edison was already installing DC electrical systems around the world and had a great dislike for AC. Edison's systems generated low-voltage, high-current electricity that could be transported only a few miles.

In 1884, Tesla came to the United States and worked for Thomas Edison directly. They had a falling out, so Tesla began his own company, called the Tesla Electric Company, located only a few blocks from Edison. There, he perfected motors, dynamos, and transformers (devices that convert high-voltage AC to low-voltage AC).

In 1888, Tesla aligned with George Westinghouse (inventor of the train air brake), who was also convinced AC electricity was the future. High-voltage AC could be produced at the dynamos and transported long distances with little loss. At the final site, a transformer would step down the voltage to safer levels.

Edison, realizing the threat, waged a campaign against AC, citing the dangers of high-voltage electricity. To underscore the dangers, he developed an AC electrocution system that was tested on horses, calves, and stray dogs and cats collected by children for a 25-cent bounty. New York adopted electrocution as the method for carrying out the death penalty and executed murderer William Kemmler in Sing Sing prison's electric chair. Despite the unfavorable publicity this generated for AC, its many merits have made it the preferred means of transporting electricity the world over.

Adapted from: Curt Wohleber, "The Work of the World," *American Heritage of Invention & Technology* 7, no. 3 (1992), pp. 44–52.

12.7 DIFFUSION

Diffusion is a process in which chemical species A flows from a high concentration to a low concentration while dispersed in chemical species B. For example, if a bottle of perfume (chemical species A) is opened in a stagnant room filled with air (chemical species B), the perfume will diffuse out of the bottle and eventually fill the room.

Figure 12.7 shows two large tanks connected by a small-diameter pipe. Each tank is filled with chemical species A (e.g., salt) and chemical species B (e.g., water). The concentration of species A in each tank is different. For example, the left tank could contain seawater and the right tank could contain freshwater. The salt will diffuse from the high concentration to the low concentration, until eventually the concentration in each tank is the same.

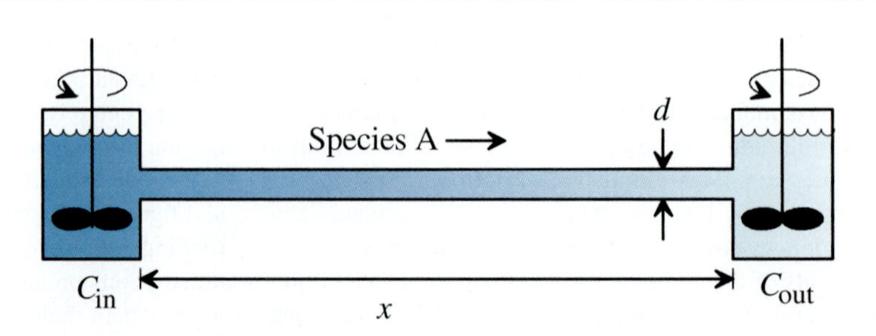

FIGURE 12.7
Diffusion of species A through species B.

TABLE 12.5
Diffusivity in liquids (Reference)

Diffusing Species A	Stationary Species B	Temperature (K)	δ (m^2/s)*
Chlorine	Water	289	1.26×10^{-9}
Carbon dioxide	Water	293	1.77×10^{-9}
Sodium chloride	Water	291	1.26×10^{-9}
Methanol	Water	288	1.28×10^{-9}
Acetic acid	Water	286	0.91×10^{-9}
Ethanol	Water	283	0.83×10^{-9}
n-butanol	Water	288	0.77×10^{-9}
Carbon dioxide	Ethanol	290	3.2×10^{-9}

*Valid only at dilute concentrations.

Adapted from: J. R. Welty, C. E. Wicks, and R. E. Wilson, *Fundamentals of Momentum, Heat, and Mass Transfer,* 2nd ed. (New York: Wiley, 1976), p. 761.

The flux J_A is the mass M_A of species A that flows per unit cross-sectional area per unit time. It is proportional to the concentration difference ΔC between the two tanks and inversely proportional to the length x of the pipe,

$$J_A = \frac{M_A}{At} = \delta \frac{C_{in} - C_{out}}{x} = \delta \frac{\Delta C}{x} \tag{12-11}$$

where δ is the **diffusivity** (Table 12.5).

EXAMPLE 12.4

Problem Statement: Two large tanks are connected together by a pipe that is 0.500 m long and 0.0100 m in diameter. One tank has sodium chloride dissolved in water (2.00 kg salt/m^3) and the other has distilled water (0.00 kg salt/m^3). At 291 K, what rate does salt diffuse from the saltwater tank to the distilled-water tank?

Solution:

$$\frac{M_A}{t} = A\delta \frac{C_{in} - C_{out}}{x} = \frac{\pi}{4} d^2 \delta \frac{C_{in} - C_{out}}{x}$$

$$= \frac{\pi}{4}(0.0100 \text{ m})^2\left(1.26 \times 10^{-9} \frac{\text{m}^2}{\text{s}}\right)\frac{2.00 \frac{\text{kg}}{\text{m}^3} - 0 \frac{\text{kg}}{\text{m}^3}}{0.5 \text{ m}}$$

$$= 3.96 \times 10^{-13} \frac{\text{kg}}{\text{s}}$$

12.8 RESISTANCE

Resistance is defined as the driving force ΔD required to produce a given flow rate r:

$$\bar{R} = \text{resistance} \equiv \frac{\text{driving force}}{\text{flow rate}} = \frac{\Delta D}{r} \tag{12-12}$$

TABLE 12.6
Resistance for each type of flux

Heat	$R_{heat} = \frac{x}{Ak}$
Fluid flow	$R_{fluid} = \frac{x 8 \pi \mu}{A^2}$
Electricity	$R_{electron} = \frac{x}{A\kappa}$
Diffusion	$R_A = \frac{x}{A\delta}$

We can manipulate Equation 12-6 to give

$$\Delta D = \frac{r}{A} \frac{x}{K} \tag{12-13}$$

We can substitute this into Equation 12-12, obtaining

$$\overline{R} = \frac{x}{AK} \tag{12-14}$$

Table 12.6 shows the resistance for each type of flux we have considered.

When dealing with electricity, resistance is so frequently considered that it is given a special unit called the *ohm* (Ω). A wire with 1 Ω of resistance passes 1 A of electric current when a voltage difference of 1 V is applied.

$$R_{\text{electron}} = 1\ \Omega = \frac{\Delta V}{i} = \frac{1\frac{\text{J}}{\text{C}}}{\frac{\text{C}}{\text{s}}} = \frac{\text{J}\cdot\text{s}}{\text{C}^2} \tag{12-15}$$

Table 12.6 shows the required properties of the wire (electrical conductivity, length, cross-sectional area) to produce a given resistance.

Generally, resistance is understood to mean electrical resistance, so the subscript in Equation 12-15 may be dropped, thus

$$R = \frac{\Delta V}{i} \tag{12-16}$$

12.8.1 Resistances in Series

Figure 12.8(a) shows two electrical resistors R_1 and R_2 arranged in series, and Figure 12.8(b) shows a single resistor with equivalent resistance R. The same current i flows through each circuit, that is,

$$i = \frac{V_0 - V_1}{R_1} = \frac{V_1 - V_2}{R_2} = \frac{V_0 - V_2}{R} \tag{12-17}$$

From this equation, we can write the following relationships:

$$iR_1 + iR_2 = (V_0 - V_1) + (V_1 - V_2) = V_0 - V_2 = iR \tag{12-18}$$

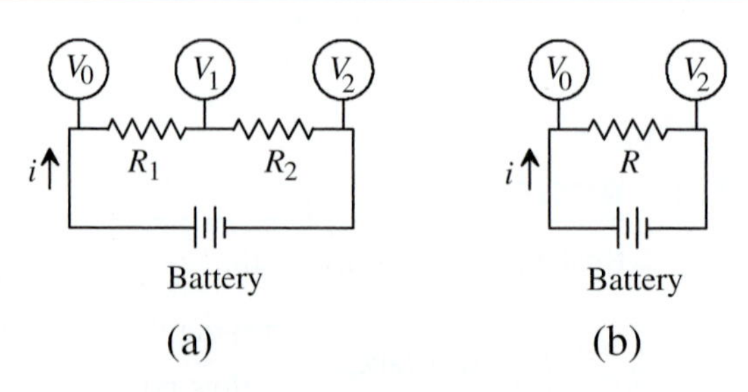

FIGURE 12.8
Electrical resistors: (a) two resistors in series and (b) a single resistor with equivalent resistance.

The current i can be factored out of the left side, giving

$$i(R_1 + R_2) = iR \tag{12-19}$$

The i's cancel, giving

$$R = R_1 + R_2 \tag{12-20}$$

Thus, when resistors R_1 and R_2 are placed in series, the equivalent resistance R is the sum of the individual resistances. This may be generalized to n resistors as

$$R = \sum_{i=1}^{n} R_i \tag{12-21}$$

12.8.2 Resistances in Parallel

Figure 12.9(a) shows two resistors R_1 and R_2 arranged in a parallel circuit. Figure 12.9(b) shows an equivalent circuit with a single resistor R. The total current i is the sum of the current through each resistor i_1 and i_2:

$$i = i_1 + i_2 \tag{12-22}$$

The current through each resistor is the voltage drop divided by the resistance:

$$\frac{V_0 - V_1}{R} = \frac{V_0 - V_1}{R_1} + \frac{V_0 - V_1}{R_2} \tag{12-23}$$

The voltage drop across each resistor is the same, therefore,

$$\frac{1}{R} = \frac{1}{R_1} + \frac{1}{R_2} \tag{12-24}$$

This equation may be generalized for n resistors as

$$\frac{1}{R} = \sum_{i=1}^{n} \frac{1}{R_i} \tag{12-25}$$

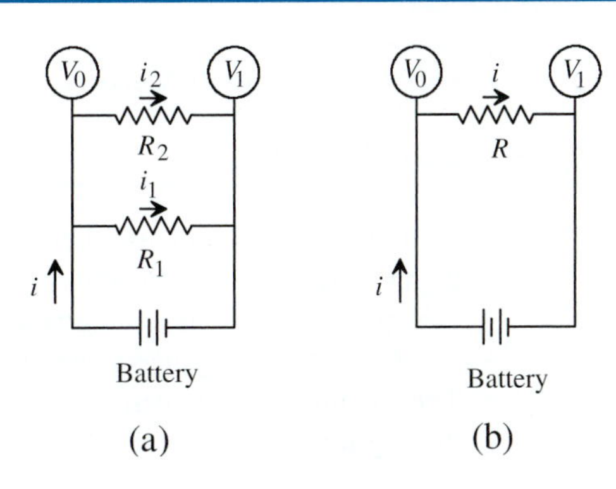

FIGURE 12.9
Electrical resistors: (a) two resistors in parallel and (b) a single resistor with equivalent resistance.

Electrical Superconductors

The electrical resistance of metals decreases as temperature decreases. In 1911, Dutch physicist Heike Kamerlingh Onnes was studying the resistance of mercury at very low temperatures. His data looked as follows:

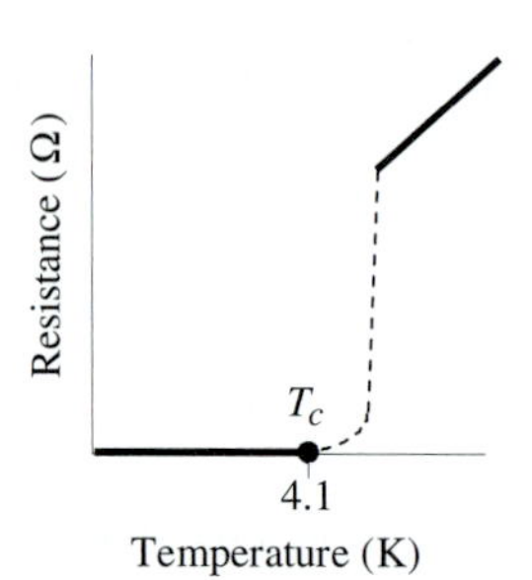

He measured no resistance at 4.1 K and below. Materials that have no resistance are called *electrical superconductors* and the temperature where the resistance becomes zero is called the **critical temperature,** T_c. Below this temperature, once the electrons are set in motion, they can continue to flow without stopping, even without an applied voltage. Superconducting loops have had electrons circulate for years without any observed decrease in current, true perpetual motion machines. The electrical conductivity of superconductors is known to be above 2.5×10^{24} $C^2/(J \cdot s \cdot m)$, about 10^{17} greater than the best ordinary metals.

Through the years, thousands of superconductors have been found. The following table provides a partial list.

Material	T_c (K)	Material	T_c (K)
Zn	0.88	Nb_3Sn	18.05
Al	1.19	Nb_3Ge	23.2
Sn	3.72	$YBa_2Cu_3O_7$	92
Hg	4.15	Bi-Sr-Ca-Cu-O	105
Pb	7.18	Tl-Ba-Ca-Cu-O	125
Nb	9.46		

Since 1987, new ceramics have been developed that are superconductors at temperatures slightly above the boiling point of liquid nitrogen (77 K). Given that liquid nitrogen is about the same cost as milk, this opens many possible new applications for superconductors. Some exciting applications include powerful electric motors, high-speed computers, long-distance electrical transmission lines, extremely sensitive magnetic field detectors, and strong magnets for medical imaging. In addition, the magnetic field surrounding the superconductor can store energy for electric-powered automobiles or peak demand in electric utilities.

Superconductor levitating above a magnet.

Adapted from: R. A. Serway, *Principles of Physics* (Fort Worth: Saunders, 1994), pp. 511–512.

12.9 SUMMARY

Thermodynamics tells us where we are going and rate processes tell us how long it takes to get there. A *rate* is the amount of change per unit time, and a *flow rate* is the amount of flow past a fixed point in a given time. Flux is a flow rate per cross-sectional area.

Commonly, engineers encounter fluxes of heat, fluids, electrons, or chemical species. The equations that govern these fluxes all have the same form; flux is proportional to the driving force and inversely proportional to the distance separating the driving force.

A resistance is the driving force divided by the flow rate. Generally, when people refer to resistance, they mean electrical resistance; however, resistance is a valid concept for the other fluxes as well. Resistances can be in series or parallel.

Nomenclature

A area (m^2)
C concentration (kg/m^3)
D driving force
d pipe diameter (m)
J flux [amount/($m^2 \cdot s$)]
K proportionality constant
k thermal conductivity [J/(s·m·K)]
i electrical current (C/s)
M_A mass of species A (kg)
N amount
P pressure (Pa)
Q heat (J)
q heat rate (J/s) or electrical charge (C)
R resistance, generally electrical resistance ($J \cdot s/C^2$)
r rate (amount/s)
T temperature (K)
t time (s)
V volume (m^3) or voltage (V or J/C)
x spacing (m)
δ diffusivity (m^2/s)
κ electrical conductivity [$C^2/(J \cdot s \cdot m)$]
μ viscosity (Pa·s)
ρ density (kg/m^3)

Further Readings

Johnson, D. E.; J. L. Hilburn; and J. R. Johnson. *Basic Electrical Circuit Analysis.* 4th ed. Englewood Cliffs, NJ: Prentice-Hall, 1990.
Serway, R. A. *Principles of Physics.* Fort Worth: Saunders, 1994.
Welty, J. R.; C. E. Wicks; and R. E. Wilson. *Fundamentals of Momentum, Heat, and Mass Transfer.* 2nd ed. New York: Wiley, 1976.

PROBLEMS

12.1 A child's toy consists of a clear tube that is 3.00 m long and 0.0250 m in diameter. The tube is flexible and can be bent into a variety of shapes. Children form the tube into a roller coaster shape and delight in dropping marbles into the tube and watching them twist through the various bends. Each marble has a mass of 0.005 kg. A child places six marbles into the tube each minute. Answer the following questions:

(a) What is the rate that marbles are placed in the tube (marbles/s)?
(b) What is the rate that mass is placed in the tube (kg/s)?
(c) What is the marble flux through the tube [marbles/($m^2 \cdot s$)]?
(d) What is the mass flux through the tube [kg/($m^2 \cdot s$)]?

12.2 A concrete building is designed to store nuclear waste. The building is constructed entirely of concrete with no windows and a very small opening through which the waste is placed into the building. The building is shaped like a shoe box; that is, it has a flat roof and vertical walls. The building is 10 m high, 50 m long, and 30 m wide. The walls and roof are 0.050 m thick. Because the waste is radioactive, it decays and produces heat at a rate of 10^5 J/s. The outdoor temperature is 300 K. What is the temperature inside the building? (Assume the air temperature in the building is uniform and that a strong wind is blowing outside. Assume negligible heat is transferred through the floor. Assume the major thermal resistance is the concrete wall.)

12.3 An engineer is performing some laboratory studies and wishes to put a constant water flow rate of 10^{-12} m^3/s into a reactor. This is a very low flow rate and she has trouble finding a pump that is accurate and steady. She devises the system shown in Figure 12.10, in which a reservoir is filled with water and a piston is placed on top. A weight is placed on top of the piston so the pressure in the reservoir is a constant 2.346 Pa. The liquid flows out of the reservoir through a 0.0005-m-diameter capillary tube into the reactor. The reactor pressure is 1.000 Pa. What is the required length of the capillary tube to achieve the desired flow?

FIGURE 12.10
Capillary tube used to regulate water flow to a reactor.

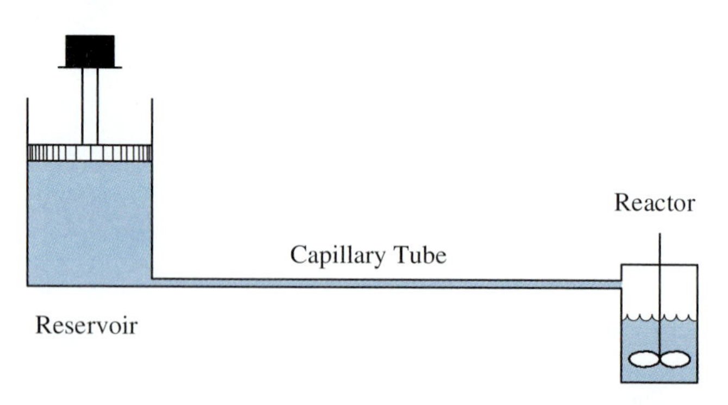

12.4 An engineer wishes to flow 1.00 m^3/s of water through a pipe. To minimize pump power requirements, it is important that the flow stay laminar. He specifies an initial pipe diameter, but finds that it is in the turbulent region. Should he increase or decrease the pipe diameter?

12.5 Copper wire will be used to carry 1.00×10^3 A from a power plant to a community located 10,000 m away. The allowable voltage drop through the wire is 200 V. What is the required wire diameter?

12.6 A stagnant puddle 1.0 m in diameter and 0.010 m deep has a very thin, uniform layer of algae growing on the bottom. In the presence of sunlight, algae are able to convert carbon dioxide into sugar and oxygen. On a sunny day, the algae are so efficient that the bottom of the puddle is free of carbon dioxide. The puddle surface gets carbon dioxide from the air and has a concentration of 0.00042 kg/m^3. At what rate (kg/s) does carbon dioxide diffuse to the algae?

12.7 Calculate the equivalent resistance for a 5-Ω and a 10-Ω resistor connected in (a) series and (b) parallel.

12.8 Calculate the equivalent resistance for the circuit shown in Figure 12.11 for $R_1 = 20\ \Omega$, $R_2 = 10\ \Omega$, and $R_3 = 5\ \Omega$.

FIGURE 12.11
Electrical circuit for Problem 12.8.

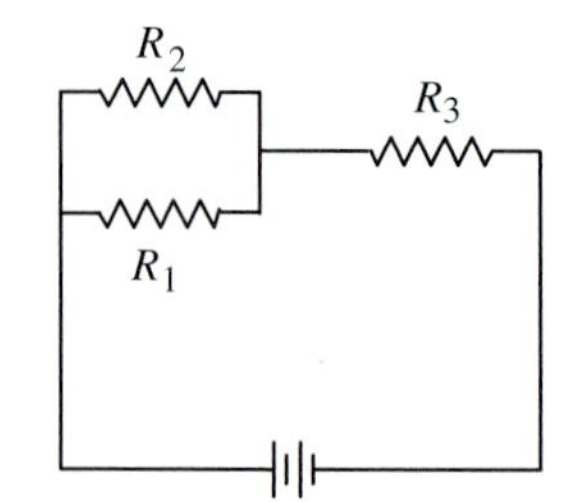

12.9 Sheets of 0.005-m-thick mild steel, 0.010-m-thick copper, and 0.040-m-thick stainless steel are sandwiched together. What thickness of aluminum would have a thermal resistance equivalent to these three sheets?

12.10 A strain gage is a device for measuring small movements. Figure 12.12 shows a strain gage attached to a block that is being pulled in tension. The strain gage consists of a small wire that is bonded to the block [Figure 12.12(a)]. As the block stretches, it lengthens the strain gage wire and narrows its cross-sectional area. (*Note:* The volume of the strain gage wire is the same regardless of how far it is stretched.)

(a) The strain gage wire is constructed from Monel metal. Calculate its resistance in the normal, unstretched geometry.

FIGURE 12.12
Strain gage: (a) single wire, (b) parallel wires, and (c) series wires.

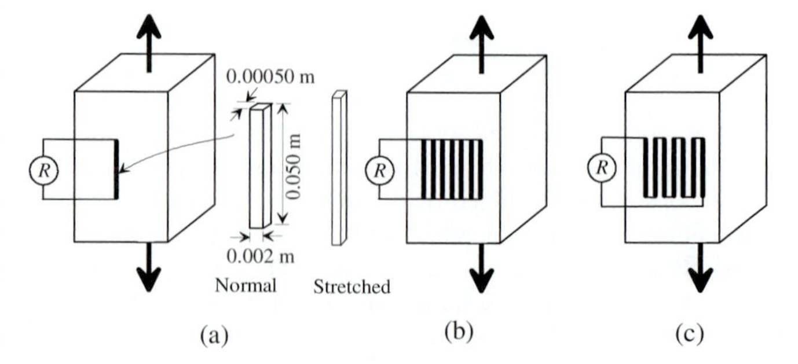

(b) The block is stretched so the strain gage wire increases its length by 0.50%. What is the new resistance of the strain gage wire?

(c) Seven strain gage wires can be mounted in parallel [Figure 12.12(b)] or series [Figure 12.12(c)] onto the block. In each case, calculate the resistance before being stretched and after being stretched to a 0.50% increase in length. To get the most precise measurement of the amount of stretching, which arrangement (parallel or series) would you recommend?

12.11 A civil engineer is designing a pipe to transport 178 m^3/s of water a distance of 4.35×10^4 m. To save capital costs, he would like the pipe diameter to be as small as possible, but to save pumping costs, he would like the flow to stay laminar and not become turbulent. What is the smallest diameter that allows the flow to stay laminar? The actual pipe diameter will be 20% larger than this minimum diameter in order to provide a margin of safety. What ΔP must be generated by the pump? Comment on the practicality of the engineer's desire for the flow to be laminar.

12.12 A high-efficiency solar collector is being designed. It has a mirror that reflects sunlight up onto the collector surface (Figure 12.13). The top of the collector surface is insulated with rock wool and the bottom is insulated by stagnant air. This arrangement takes advantage of the fact that stagnant, noncirculating air is a good insulator. The air does not circulate because the less-dense warm air is above the more-dense cold air, a stable configuration. (*Note:* If the air were heated from below, it would circulate in a process called *natural convection*; the less-dense warm air would be below the more-dense cold air, an unstable configuration.)

The collector surface consists of a thin black channel that has water flowing through it to collect the solar energy. The water temperature is 323 K and the ambient air temperature is 268 K. Estimate the amount of heat loss from the solar collector (J/s). You may neglect heat loss from the sides and consider only the heat loss from the top and bottom. The black channel has negligible thickness.

12.13 An ice skating rink decides it will save energy by insulating the rink each night when not in use. The ice temperature is 265 K and the air temperature is 290 K. The rink is rectangular with measurements of 20.0 m by 100.0 m. The insulation consists of corkboard panels that are 1.0 m × 2.0 m × 0.020 m. The rink bought a total of 3000 panels. They can lay the panels onto the ice in a variety of ways. For example, they can place a three-panel layer uniformly over the ice, or they can provide half the ice with a four-panel layer and the other half with a two-panel layer. Of these two configurations, which would you recommend? Justify your answer with a calculation.

12.14 Write a computer program that asks the user how many resistors he has, the resistance of each resistor, and whether they are connected in series or parallel. The program will report the value of a single resistor that has the same resistance as the multiple resistors.

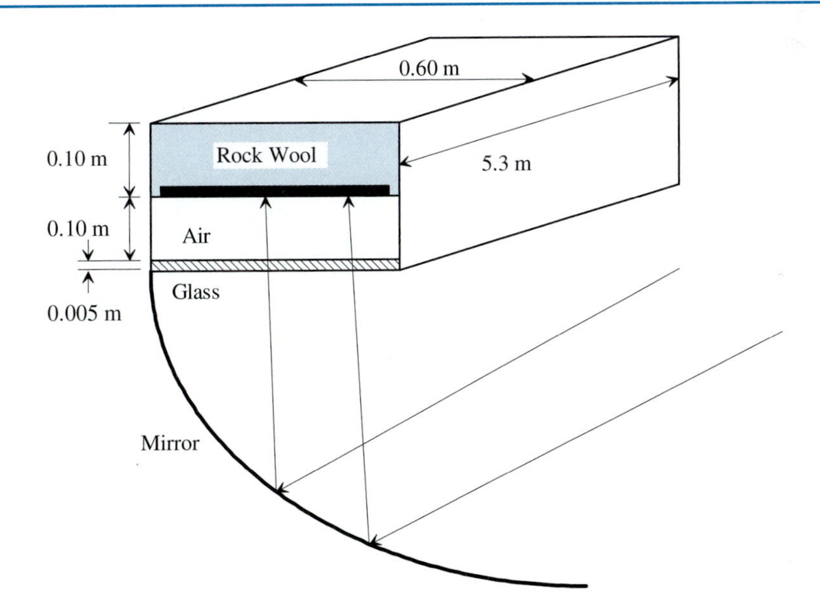

FIGURE 12.13
Solar collector.

Glossary

alternating current The oscillation of electrons back and forth through a wire.
amperes The measure of current, defined as 1 coulomb per second (1 C/s).
conductor A material with a large electrical or thermal conductivity.
coulomb A special number used to count electrons: 1 C = $6.24150636 \times 10^{18}$ electrons.
critical temperature T_c The temperature where electrical resistance becomes zero.
current The flow rate of electrons.
diffusion A process in which chemical species A flows from a high concentration to a low concentration while dispersed in chemical species B.
diffusivity The proportionality constant that describes the ability of chemical species A to diffuse through chemical species B.
direct current The flow of electrons in a single direction through a wire.
driving force An influence that causes change.
electrical conductivity The ability to conduct current.
electrical superconductors Materials that have no electrical resistance.
electricity The flow of electrons.
flow rate The amount of flow past a fixed point in a given time.
flux The flow rate per cross-sectional area.
Gramme machine An early electric generator.
heat Energy flow resulting from a temperature difference.
insulator A material with a low electrical or thermal conductivity.
Poiseuille equation The equation that describes the laminar flow rate of a viscous fluid through a pipe of a given length and diameter resulting from an applied pressure (Equation 12-8).
rate The amount of change per unit time.
rate processes Processes where fluxes occur as a result of driving forces.
resistance The driving force divided by the flow rate.
superfluid A fluid with no viscosity that can flow without a pressure difference.
thermal conductivity The proportionality constant that describes the ability of heat to flow through a material.
viscosity A measure of the "thickness" of a fluid.

Mathematical Prerequisites

- Algebra (Appendix E, Mathematics Supplement)
- Geometry (Appendix I, Mathematics Supplement)

CHAPTER 13

SI System of Units

The **SI System of Units** (*Le Système International d'Unités*) is probably familiar to you as the "metric system." This system is used worldwide except for the United States, Liberia, and Myanmar (Burma). It is the system favored by science, so you undoubtedly have been exposed to it in your science classes.

13.1 HISTORICAL BACKGROUND

The need for units of measure was evident as soon as human commerce began. If two farmers were to trade grain for a goat, they needed to quantify the amount of grain and the weight of the goat. In early commerce, the units of measure were based on commonly available items. For example, the bushel basket used to transport the grain became the unit of measure. (In Britain, the **bushel** was eventually standardized to be eight imperial gallons.) The weight of the goat could be measured by placing the animal on a scale and determining the number of stones required to counterbalance the animal. (In Britain, the **stone** was eventually defined to be 14 pounds.) The unit of length, based on a man's foot, has been used in Britain for over a thousand years. It quickly became evident that units of measure had to be subdivided. Many ancient measuring systems were based upon fractions of the base unit, such as halves, thirds, and quarters. Thus, the unit was subdivided into a number of segments that is easily divided into fractions. For example, the foot is divided into 12 inches, which may be evenly divided by 2, 3, 4, and 6 with no remainder.

For units of measure to be useful, they must be standardized so that business transactions are unambiguous. Thus, it fell upon governments to establish official units of measure. For example, the Egyptian Royal **Cubit** was equivalent to the length from the Pharaoh's elbow to the farthest fingertip of his extended hand (20.62 inches). A block of granite was fashioned to this length to become a standard. (After all, the Pharaoh was much too busy to help carpenters measure the lengths of boards.) This standard was further divided into finger widths, palms, hands, remens (20 finger widths), and a small cubit (18 inches) equal to six palms (3 inches). The small cubit was used widely in construction and was fashioned into wood or granite copies that were regularly checked against the standard. We continue to use the Pharaoh's system of measure—the height of horses is often measured in **hands,** which are now defined to be exactly four inches.

In the 16th century, decimal systems were conceived in which the units of measure were divided into 10 parts, 100 parts, 1000 parts, and so on, rather than fractional divisions. This allowed for more accurate and convenient subdivisions; however, as there were no standards, confusion abounded. In 1790, the French National Assembly requested that the French Academy of Sciences establish a system of units that could be adopted the world over. It used the meter as the unit of length and the gram as the unit of mass. This system was legalized in the United States in 1866.

In 1870, an international meeting was held in Paris in which 15 nations were represented. This led to the establishment of the International Bureau of Weights and Measures near Paris. They agreed to hold the General Conference on Weights and Measures at least every 6 years to decide upon issues relating to units of measure. The National Institute of Standards and Technology (NIST), formerly the National Bureau of Standards (NBS), represents the United States at these meetings.

Any measuring system must establish **base units** from which all other units are derived. (For example, volume is derived from the base unit of length.) In 1881, time was added as a third base unit to establish the centimeter-gram-second (CGS) system. Outside of the laboratory, this system is inconvenient, so about 1900, the meter-kilogram-second (MKS) was adopted. In 1935, electrical measurements based on the **ampere** were added. Thus there was a fourth base unit in the MKSA system. In 1954, base units for temperature (kelvin) and luminous intensity (**candela**) were adopted, bringing the total base units to six. In 1960, the measurement system was given the formal title *Le Système International d'Unités,* which we abbreviate as SI. In 1971, the amount of substance (mole) was added as a base unit, bringing the total to seven.

13.2 DIMENSIONS AND UNITS

The distinction between a **dimension** and a **unit** is best understood by example. The *dimension* of length may be described by *units* of meters, feet, inches, cubits, and so forth. Thus, *dimension* is an abstract idea whereas *unit* is more specific. Table 13.1 shows common dimensions and the associated **SI base units.**

TABLE 13.1
Dimensions and SI base units

Dimension	Symbol	Unit	Abbreviation
Length	[L]	meter	m
Mass	[M]	kilogram	kg
Time	[T]	second	s
Electric current	[A]	ampere	A
Thermodynamic temperature	[θ]	kelvin	K
Luminous intensity	[I]	candela	cd
Amount of substance	[N]	mole	mol

13.3 SI UNITS

SI includes three types of units: supplementary, base, and derived.

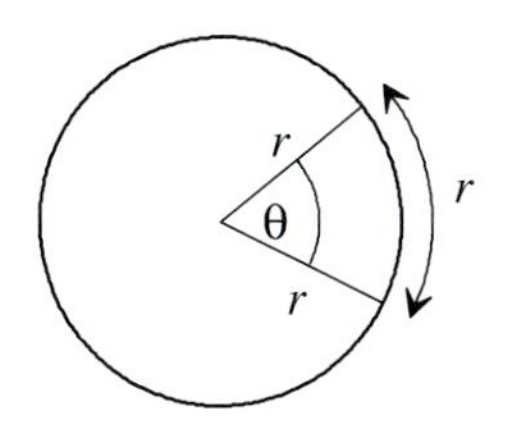

FIGURE 13.1
Plane angle.

13.3.1 SI Supplementary Units

The **SI supplementary units** were added in 1960. They are mathematical definitions that are needed to define both base and derived units.

1. Plane Angle (radian).
Figure 13.1 shows a circle in which two radii define a **plane angle** θ. If the length of the swept circumference is equal to the circle radius, then the plane angle θ is equal to one radian (1 **rad**).

Born in Revolution

The erratic behavior and bankruptcy of Louis XVI led to the disintegration of French social order and culminated in the storming of the Bastille on July 14, 1789. This event initiated the destruction of the feudal system as peasants were emboldened to torch chateaux, burn feudal contracts, take land, and drive out the gentry. Political power shifted from the king to the French National Assembly.

In 1790, the French National Assembly requested that the French Academy of Sciences revise the French system of weights and measures, which, under the monarchy, were chaotic, inconsistent, and complex. They appointed a "blue-ribbon" panel headed by Jean Charles Borda (1733–1799). Invitations to sit on the committee were sent to both Britain and the United States, but these were declined.

The committee decided the measuring system should be base 10, although base 11 and 12 were considered. The unit of length, the meter, was to be one 10-millionth of a "quadrant of meridian," that is, the distance from the north pole to the equator measured along a great circle passing through the poles. The unit of mass, the gram, was to be the mass of water, at its maximum density (4°C), occupying a volume of 10^{-6} m^3.

Two French astronomer-geodesists were given the task of surveying the distance from Dunkirk, France, to a site near Barcelona, Spain, from which the quadrant of meridian could be calculated. It took them seven years to complete the measurements; their work was impeded because they were arrested for spying while surveying foreign countries. They were remarkably accurate, with an error of only 2 parts in 10,000.

In 1792, the newly elected National Convention proclaimed France to be a republic. They severed all ties with the traditional Gregorian calendar and established this as Year 1 of the Republic of France. To create a rational calendar, they established a new commission, which devised a 12-month calendar with each month exactly 30 days. To complete the 365-1/4 days in a year, each year had a 5-day festival, except during leap year, which had a 6-day festival. Rather than the traditional 7-day week, the month was divided into three 10-day décades. Rather than naming each day after gods and goddesses, the days were numbered from one to ten. This calendar was employed for over 12 years, until Napoleon abandoned it in 1806.

The calendar commission also proposed a decimal system of time. Each day was divided into ten decidays (2.4 hours); smaller units were the milliday (86.4 seconds) and microday (0.0864 seconds). In 1793, the decimal time system was introduced, but was met with stiff resistance. Unlike the other weights and measures employed by the monarchy, the system for time measurement was rigorous, well structured, and universally followed. Changing all the clocks would be expensive. In addition, time is intimately connected with people's everyday lives, whereas the units for length and mass are less interwoven. There was little incentive for change, so in 1795, the proposed decimal time system was "tabled" and has remained so ever since.

In 1798, European scientists were invited to France to continue improving the new "metric system." Eventually, it became a measuring system that was adopted by nearly the entire world. Interestingly, the United States, which was invited to attend the very first meetings, has had the greatest difficulty adopting this measurement system.

Adapted from: H. A. Klein, *The World of Measurements* (New York: Simon and Schuster, 1974).

> The second is the duration of 9,192,631,770 periods of the radiation corresponding to the transition between two hyperfine levels of the ground state of the cesium-133 atom.

This definition is based on the atomic clock. One of the best atomic clocks (NIST-7) is precise to within about 1 second in 3 million years, or 1 part in 10^{14}. Commercially available atomic clocks are precise to within 3 parts in 10^{12}.

4. Unit of Electric Current (ampere).

When electric current flows through a wire, a magnetic field surrounds the wire. The ampere was defined in 1948 on the basis of the magnetic force of attraction between two parallel wires with electric current flowing.

> The ampere is that constant current which, if maintained in two straight parallel conductors of infinite length, of negligible circular cross section, and placed 1 meter apart in a vacuum, would produce between these conductors a force equal to 2×10^{-7} newton per meter of length.

This can be better understood by considering Figure 13.3.

5. Unit of Thermodynamic Temperature (kelvin).

Temperature is not to be confused with **heat.** Please review the chapter on thermodynamics if you do not know the difference between the two.

The definition of temperature is based on the **phase diagram** for water (Figure 13.4). The liquid/solid, liquid/vapor, and solid/vapor lines meet at the **triple point,** where all three phases coexist simultaneously. Although it would seem difficult to attain the triple point experimentally because the pressure and temperature combination must be **exact,** it is actually achieved rather easily. A glass vial is evacuated and then partially filled with liquid water, leaving a vapor-space above the liquid. The partially full vial is then frozen. As the ice melts, all three phases will coexist: ice, liquid, and vapor.

Because the triple point of water is rather easily obtained, it is ideal for defining a temperature scale. By definition, the triple point of water is assigned the value 273.16 K and **absolute zero** is assigned the value 0 K. The distance from absolute zero to the triple point of water is divided into 273.16 parts, which define the size of the **kelvin** unit.

FIGURE 13.3
The definition of the ampere.

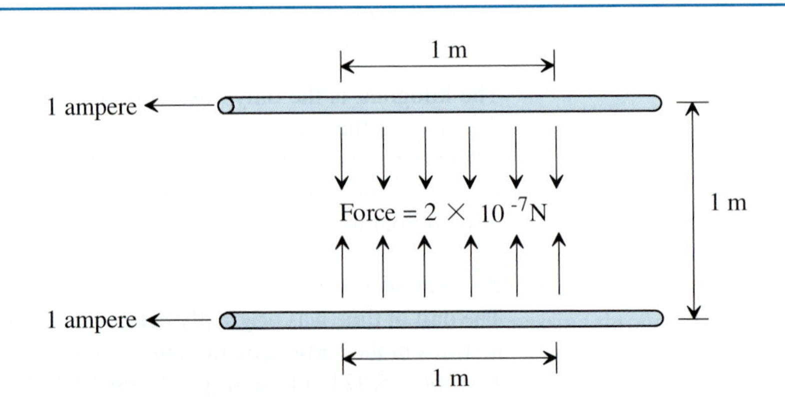

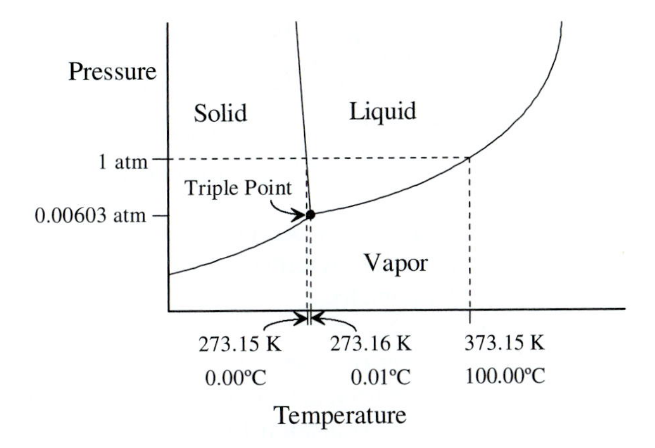

FIGURE 13.4
Phase diagram for water.

> The kelvin unit of thermodynamic temperature is the fraction 1/273.16 of the thermodynamic temperature of the triple point of water.

An auxiliary temperature scale is sanctioned in which Celsius temperature t (in °C) is related to Kelvin temperature T (in K) according to the relation

$$t = T - T_o \tag{13-3}$$

where $T_o = 273.15$ K. In the Celsius temperature scale, water freezes at 0°C and it boils at 100°C provided the pressure is 1 atm. For most engineering work, the Celsius temperature scale is more convenient than the Kelvin temperature scale.

An instrument is needed to divide the interval from absolute zero to the triple point of water, and to extend beyond. In practice, the interval is divided using many different types of instruments (e.g., constant-volume gas thermometers, acoustic gas thermometers, spectral and total radiation thermometers, and electronic noise thermometers). The easiest instrument to understand is the constant-volume gas thermometer. At very low pressures, real gases behave as perfect (ideal) gases. The perfect (ideal) gas equation defines the relationship between the pressure P, the volume V, the quantity of gas in moles n, and the temperature T,

$$PV = nRT \tag{13-4}$$

where R is the universal gas constant. By filling a fixed volume with a given amount of gas, then V and n are constant. The perfect gas equation is simplified to

$$P = \left(\frac{nR}{V}\right)T = kT \tag{13-5}$$

where k is the proportionality constant. Thus, pressure is directly proportional to temperature. To illustrate how this relationship could be used, imagine that we perform an experiment in which the pressure in the constant-volume gas thermometer is 0.010000 atm at the triple point of water. If we then reduce the temperature so the pressure in the thermometer becomes 0.0050000 atm, we can calculate the temperature as

$$k = \frac{P_1}{T_1} = \frac{P_2}{T_2} \tag{13-6}$$

$$T_2 = \frac{P_2}{P_1} T_1 = \frac{0.0050000 \text{ atm}}{0.010000 \text{ atm}} 273.16 \text{ K} = 136.58 \text{ K} \tag{13-7}$$

It is very difficult to make precise and accurate thermometer measurements. The convenient reference points presented in Table 13.2 were determined by very careful thermometry.

6. Unit of Amount of Substance (mole).

In chemistry, the number of molecules is extremely important. For example, the perfect gas equation (Equation 13-4) has the term n, which describes the number of gas molecules in terms of *moles.* The **mole** often gives students difficulty, perhaps because of its unusual name. It is a term that has been used since about 1902 and is short for "gram-**mole**cule."

> The mole is the amount of substance that contains as many elementary entities as there are atoms in 0.012 kg of carbon-12. When the mole is used, the elementary entities must be specified and may be atoms, molecules, electrons, other particles, or specified groups of such particles.

We can visualize what a mole means by imagining using some very tiny tweezers to count the number of atoms in 12 grams (0.012 kg) of carbon-12. The number we obtain is called *Avogadro's number* and is equal to 6.0221367×10^{23}. Just as we use the name *dozen* to describe the number 12, we give a name to this important number.

7. Unit of Luminous Intensity (candela).

A unit for **luminous intensity** is required to describe the brightness of light. Candle flames or incandescent lightbulbs were originally used as standards. The current standard uses a monochromatic (i.e., single-color) light source, typically produced by a laser, and an instrument called a *radiometer* to measure the amount of heat generated when light is absorbed.

> The candela is the luminous intensity, in a given direction, of a source that emits monochromatic radiation of a frequency 540×10^{12} cycles per second and that has a radiant intensity in that direction of (1/683) watt per steradian.

TABLE 13.2

International Temperature Scale (ITS-90) reference points (Reference)

^3He	Boiling point	3.2 K	Ga	Triple point	302.9146 K
^4He	Boiling point	4.2 K	In	Freezing point	429.7845 K
H_2	Triple point	13.8033 K	Sn	Freezing point	505.078 K
H_2	Boiling point	20.3 K	Zn	Freezing point	692.677 K
Ne	Triple point	24.5561 K	Al	Freezing point	933.473 K
O_2	Triple point	54.3584 K	Ag	Freezing point	1234.93 K
Ar	Triple point	83.8058 K	Au	Freezing point	1337.33 K
Hg	Triple point	234.3156 K	Cu	Freezing point	1357.77 K
H_2O	Triple point	273.16 K			

Note: Boiling and freezing points measured at P = 101.325 kPa

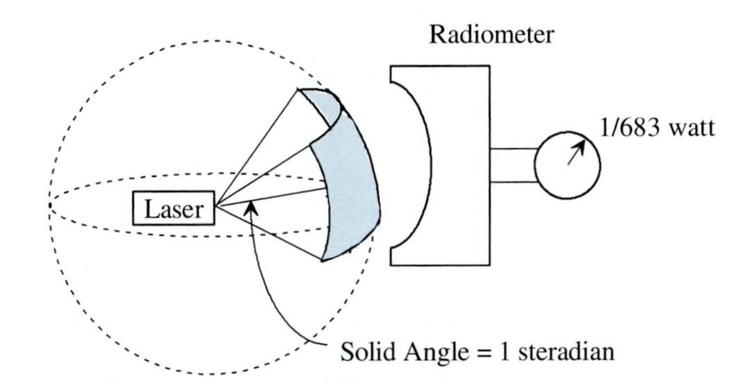

FIGURE 13.5
Schematic representation of the apparatus to measure light intensity.

Figure 13.5 is a schematic representation of the measuring system for the candela.

13.3.3 SI Derived Units

The base units may be combined into the derived units shown in Table 13.3. Some derived units have been assigned special names (Table 13.4). These derived units with special names may even be combined with other units to form new derived units (Table 13.5).

13.4 SI PREFIXES

Because scientists and engineers describe quantities that span many orders of magnitude (e.g., the size of the atomic nucleus to distances between galaxies), SI includes the multipliers listed in Table 13.6. It is generally desirable to use the appropriate multipliers so that the number falls between 0.1 and 1000. (For example, the length 1340 m is best written as 1.34 km.) Exceptions are:

TABLE 13.3
Examples of SI derived units (Reference)

	SI Unit	
Quantity	**Name**	**Symbol**
Area	square meter	m^2
Volume	cubic meter	m^3
Speed, velocity	meter per second	m/s
Acceleration	meter per second squared	m/s^2
Wave number	reciprocal meter	m^{-1}
Density, mass density	kilogram per cubic meter	kg/m^3
Specific volume	cubic meter per kilogram	m^3/kg
Current density	ampere per square meter	A/m^2
Magnetic field strength	ampere per meter	A/m
Angular velocity	radian per second	rad/s
Angular acceleration	radian per second squared	rad/s^2
Concentration (amount of substance)	mole per cubic meter	mol/m^3
Luminance	candela per square meter	cd/m^2

TABLE 13.4
SI derived units with special names (Reference)

	SI Unit			
Quantity	Name	Symbol	Expression in Terms of Other Units	Expression in Terms of SI Base Units
Frequency	hertz	Hz		s^{-1}
Force	newton	N		$m \cdot kg \cdot s^{-2}$
Pressure, stress	pascal	Pa	N/m^2	$m^{-1} \cdot kg \cdot s^{-2}$
Energy, work, heat	joule	J	$N \cdot m$	$m^2 \cdot kg \cdot s^{-2}$
Power, radiant flux	watt	W	J/s	$m^2 \cdot kg \cdot s^{-3}$
Electric charge	coulomb	C		$s \cdot A$
Electric potential	volt	V	W/A	$m^2 \cdot kg \cdot s^{-3} \cdot A^{-1}$
Capacitance	farad	F	C/V	$m^{-2} \cdot kg^{-1} \cdot s^4 \cdot A^2$
Electric resistance	ohm	Ω	V/A	$m^2 \cdot kg \cdot s^{-3} \cdot A^{-2}$
Electric conductance	siemens	S	A/V	$m^{-2} \cdot kg^{-1} \cdot s^3 \cdot A^2$
Magnetic flux	weber	Wb	$V \cdot s$	$m^2 \cdot kg \cdot s^{-2} \cdot A^{-1}$
Magnetic flux density	tesla	T	Wb/m^2	$kg \cdot s^{-2} \cdot A^{-1}$
Inductance	henry	H	Wb/A	$m^2 \cdot kg \cdot s^{-2} \cdot A^{-2}$
Celsius temperature	degree Celsius	°C		K
Luminous flux	lumen	lm		$cd \cdot sr$
Illuminance	lux	lx	lm/m^2	$m^{-2} \cdot cd \cdot sr$
Activity (of a radionuclide)	becquerel	Bq		s^{-1}
Absorbed dose, specific energy imparted	gray	Gy	J/kg	$m^2 \cdot s^{-2}$
Dose equivalent	sievert	Sv	J/kg	$m^2 \cdot s^{-2}$

1. In particular applications, a single unit may be customary. For example, engineering mechanical drawings often express all dimensions in millimeters (mm), regardless of how large or small the number is. Clothing dimensions often are expressed in centimeters (cm).
2. When numbers are being compared or listed (as in a table), all numbers should be given with a single prefix.

The use of prefixes eliminates ambiguities associated with significant figures. The number 1340 m could have three or four significant figures, depending on if the last zero is needed merely to place the decimal point. By using the prefix, it is clear that 1.34 km has three significant figures and 1.340 km has four significant figures.

Although prefixes clearly communicate the size of a number, their use in calculations can lead to disasters. **IT IS STRONGLY RECOMMENDED THAT ALL NUMBERS IN CALCULATIONS BE CONVERTED TO SCIENTIFIC NOTATION.** Thus, if we want to calculate the distance d that light travels in a given time t (say 1 millisecond) given that the speed of light c is 299.8 Mm/s, then we must put these numbers into scientific notation:

$$d = ct \tag{13-8}$$

$$d = \left(299.8 \times 10^6 \frac{\text{m}}{\text{s}}\right)(1 \times 10^{-3}\ \text{s}) = 299.8 \times 10^3\ \text{m}$$

TABLE 13.5
Examples of SI derived units expressed by means of several names (Reference)

Quantity	SI Unit: Name	SI Unit: Symbol	SI Unit: Expression in Terms of SI base units
Dynamic viscosity	pascal second	Pa·s	$m^{-1}\cdot kg\cdot s^{-1}$
Moment of force	newton meter	N·m	$m^2\cdot kg\cdot s^{-2}$
Surface tension	newton per meter	N/m	$kg\cdot s^{-2}$
Heat flux density, irradiance	watt per square meter	W/m^2	$kg\cdot s^{-3}$
Heat capacity, entropy	joule per kelvin	J/K	$m^2\cdot kg\cdot s^{-2}\cdot K^{-1}$
Specific heat capacity, specific entropy	joule per kilogram kelvin	J/(kg·K)	$m^2\cdot s^{-2}\cdot K^{-1}$
Specific energy	joule per kilogram	J/kg	$m^2\cdot s^{-2}$
Thermal conductivity	watt per meter kelvin	W/(m·K)	$m\cdot kg\cdot s^{-3}\cdot K^{-1}$
Energy density	joule per cubic meter	J/m^3	$m^{-1}\cdot kg\cdot s^{-2}$
Electric field strength	volt per meter	V/m	$m\cdot kg\cdot s^{-3}\cdot A^{-1}$
Electric charge density	coulomb per cubic meter	C/m^3	$m^3\cdot s\cdot A$
Electric flux density	coulomb per square meter	C/m^2	$m^{-2}\cdot s\cdot A$
Permittivity	farad per meter	F/m	$m^{-3}\cdot kg^{-1}\cdot s^4\cdot A^2$
Permeability	henry per meter	H/m	$m\cdot kg\cdot s^{-2}\cdot A^{-2}$
Molar energy	joule per mole	J/mol	$m^2\cdot kg\cdot s^{-2}\cdot mol^{-1}$
Molar entropy, molar heat capacity	joule per mole kelvin	J/(mol·K)	$m^2\cdot kg\cdot s^{-2}\cdot K^{-1}\cdot mol^{-1}$
Radiant intensity	watt per steradian	W/sr	$m^2\cdot kg\cdot s^{-3}$
Radiance	watt per square meter steradian	$W/(m^2\cdot sr)$	$kg\cdot s^{-3}$
Exposure (x and γ rays)	coulomb per kilogram	C/kg	$kg^{-1}\cdot s\cdot A$
Absorbed dose rate	gray per second	Gy/s	$m^2\cdot s^{-3}$

TABLE 13.6
SI prefixes (Reference)

Multiples			Submultiples		
Factor	Prefix	Symbol	Factor	Prefix	Symbol
10^{24}	yotta	Y	10^{-1}	deci	d†
10^{21}	zetta	Z	10^{-2}	centi	c†
10^{18}	exa	E	10^{-3}	milli	m
10^{15}	peta	P	10^{-6}	micro	μ
10^{12}	tera	T	10^{-9}	nano	n
10^9	giga	G	10^{-12}	pico	p
10^6	mega	M	10^{-15}	femto	f
10^3	kilo	k	10^{-18}	atto	a
10^2	hecto	h†	10^{-21}	zepto	z
10^1	deka*	da†	10^{-24}	yocto	y

* Outside the United States, "deca" is used extensively.
† Generally to be avoided.

Now that we have the answer, we may wish to communicate it with the appropriate prefix. In this case, the distance could be reported as 299.8 km.

The prefixes that represent 1000 raised to a power are recommended. Thus, the prefixes *hecto, deka, deci,* and *centi* are generally to be avoided. Some units are so commonly expressed in this way (e.g., centimeter) that their use is accepted.

Use of words such as *billion* and *trillion* are to be avoided in describing multiples of a unit, because the American meaning is different from elsewhere:

Number	United States	Britain, Germany, France
10^9	billion	milliard
10^{12}	trillion	billion
10^{15}	quadrillion	—
10^{18}	quintillion	trillion

The Celsius temperature scale was designed to describe temperatures within the range of normal use. Therefore, it is customary not to attach prefixes to the °C symbol. For example, the temperature 5240°C would not be written as 5.24 k°C. For very large (or small) temperatures, it is preferable to use the Kelvin temperature scale.

In the United States, it is customary to express each multiple of 10^3 with the symbol "M." (This notation is derived from the Roman numeral for 1000.) For example, a chemical plant that produces 1,000,000 pounds per year of benzene might be described as a 1 MM lb/year plant. Although this notation is customary, its use should be avoided because of obvious conflicts with SI, in which the prefix "M" means a multiple of 10^6.

13.5 CUSTOMARY UNITS RECOGNIZED BY SI

Table 13.7 shows some customary units that are not formally a part of SI, but are so commonly used that their meaning is regulated by the General Conference on Weights and Measures. Note that the symbol for hour is "h," not "hr" as is commonly used. Also, note that although the symbol for liter is either "l" or "L," the use of the lowercase "l" should be avoided because it is easily confused with the number "1."

Table 13.8 shows some commonly used units that must be experimentally measured. Table 13.9 shows some customary units that are widely used in particular disciplines, but

TABLE 13.7
Customary units recognized by SI (Reference)

Name	Symbol	Value in SI Units
minute of time	min	1 min = 60 s
hour	h	1 h = 60 min = 3600 s
day	d	1 d = 24 h = 86,400 s
degree	°	$1° = (\pi/180)$ rad
minute of arc	′	$1' (1/60°) = (\pi/10{,}800)$ rad
second of arc	″	$1'' = (1/60') = (\pi/648{,}000)$ rad
liter	l, L	$1\ L = 1\ dm^3 = 10^{-3}\ m^3$
tonne, metric ton	t	$1\ t = 10^3$ kg

TABLE 13.8
Experimentally determined units recognized by the SI (Reference)

Name	Symbol	Definition	Measured Value
electron volt	eV	kinetic energy acquired by an electron passing through a potential difference of 1 volt in a vacuum	$1.60217733(49) \times 10^{-19}$ J
unified atomic mass unit	u	(1/12) of the mass of a single atom of ^{12}C	$1.6605402(10) \times 10^{-27}$ kg

Note: The ( ) indicates the uncertainty of the last two significant digits at ±1 standard deviation.

TABLE 13.9
Units temporarily recognized by SI for use in particular disciplines (Reference)

Name	Discipline or Use	Symbol	Value in SI Units
nautical mile	marine and aerial navigation		1 nautical mile = 1852 m
knot	marine and aerial navigation		1 knot = 1 nautical mile per hour = (1852/3600) m/s
ångström	chemistry, physics	Å	1 Å = 0.1 nm = 10^{-10} m
are	agriculture	a	1 a = 1 dam^2 = 10^2 m^2
hectare	agriculture	ha	1 ha = 1 hm^2 = 10^4 m^2
barn	nuclear physics	b	1 b = 100 fm^2 = 10^{-28} m^2
bar	meteorology	bar	1 bar = 0.1 MPa = 100 kPa = 10^5 Pa
gal	geodesy and geophysics	Gal	1 Gal = 1 cm/s^2 = 10^{-2} m/s^2
curie	nuclear physics	Ci	1 Ci = 3.7×10^{10} Bq
roentgen	exposure to x or γ rays	R	1 R = 2.58×10^{-4} C/kg
rad	absorbed dose of ionizing radiation	rad, rd	1 rad = 1 cGy = 10^{-2} Gy
rem	dose equivalent of radiation protection	rem	1 rem = 1 cSv = 10^{-2} Sv

are only temporarily recognized by the General Conference on Weights and Measures. The Conference discourages their introduction into new disciplines.

13.6 RULES FOR WRITING SI UNITS (REFERENCE)

1. **Regular upright type (not italics) is used. The *symbol* is written in lowercase except if it was derived from a proper name. The first letter of a symbol derived from a proper name is capitalized.**

   Example: m *is the symbol for* meter *and is written in lowercase letters*
   N *is the symbol for* newton *and is written with an uppercase letter because it originates from the proper name* Newton

   The symbol for liter (L) is an exception, because it was not derived from a proper name. It is capitalized to avoid confusion with the number "1."

2. **The unit *names* are always written in lowercase letters, even if they are derived from a proper name.**

   Example: meter *is the name for the unit of length*
   newton *is the name for the unit, whereas* Newton *is the name of the person*

   An exception is when the unit starts a sentence.

   Example: Newton is the SI unit of force. *Correct*
   newton is the SI unit of force. *Incorrect*

3. **Unit *symbols* are unaltered in the plural (i.e., do not add an "s" to the end of a symbol).**

   Example: The rod length is 3 m. *Correct*
   The rod length is 3 ms. *Incorrect*

   (*Note:* The addition of the "s" to the symbol for meter completely changed the meaning to "millisecond.")

4. **Plurals of the unit *names* are made using the rules of English grammar.**

   Example: The rod length is three meters. *Correct*
   The rod length is three meter. *Incorrect*

   The following units are identical in the singular and plural:

Singular	Plural
lux	lux
hertz	hertz
siemens	siemens

5. **Do not use self-styled abbreviations.**

   Example: s, A *Correct*
   sec, amp *Incorrect*

6. **A space is placed between the symbol and the number.**

   Example: 5 m *Correct*
   5m *Incorrect*

   Exceptions are: degrees Celsius (°C), degree (°), minutes (′), and seconds (″), with which there is no space.

   Example: 10°C *Correct*
   10 °C *Incorrect*

7. **There is no period following the symbol except if it occurs at the end of the sentence.**

   Example: It took 5 s for the reaction to occur. *Correct*
   It took 5 s. for the reaction to occur. *Incorrect*

   The rod is 3 m. *Correct*
   The rod is 3 m.. *Incorrect*

8. When a quantity is expressed as a number and unit, and is used as an adjective, then a hyphen separates the number and unit.

Example:	The 3-m rod buckled.	*Correct*
	The 3 m rod buckled.	*Incorrect*
	The rod is 3 m.	*Correct*
	The rod is 3-m.	*Incorrect*

9. The product of two or more unit *symbols* may be indicated with a raised dot or a space.

Example: N·m *or* N m

The raised dot is preferred in the United States. Where a raised dot is impossible (e.g., computer printouts), a period may be used instead. An exception is the symbol for watt hour, in which the space or raised dot may be eliminated.

Wh	*Correct*

10. The product of two or more unit *names* is indicated by a space (preferred) or a hyphen.

Example: newton meter *or* newton-meter

In the case of watt hour, the space may be omitted.

watthour	*Correct*

11. A solidus (oblique stroke, /), a horizontal line, or negative exponents may be used to express a derived unit formed from others by division.

Example: m/s *or* $\frac{\text{m}}{\text{s}}$ *or* $\text{m}\cdot\text{s}^{-1}$

12. The solidus must not be repeated on the same line unless ambiguity can be avoided by parentheses. In complicated cases, negative exponents or parentheses should be used.

Example:	m/s^2 *or* $\text{m}\cdot\text{s}^{-2}$	*Correct*
	m/s/s	*Incorrect*
	$\text{m}\cdot\text{kg/(s}^3\cdot\text{A)}$ *or* $\text{m}\cdot\text{kg}\cdot\text{s}^{-3}\cdot\text{A}^{-1}$	*Correct*
	$\text{m}\cdot\text{kg/s}^3\text{/A}$	*Incorrect*

13. When using the solidus notation, multiple symbols in the denominator must be enclosed in parentheses.

Example:	$\text{m}\cdot\text{kg/(s}^3\cdot\text{A)}$	*Correct*
	$\text{m}\cdot\text{kg/s}^3\cdot\text{A}$	*Incorrect*

14. For SI unit names that contain a ratio or quotient, use the word *per* rather than the solidus.

Example:	meters per second	*Correct*
	meters/second	*Incorrect*

15. Powers of units use the modifier *squared* or *cubed* after the unit name.

Example:	meters per second squared	*Correct*
	meters per square second	*Incorrect*

Exceptions are when the unit describes area or volume.

Example:	kilograms per cubic meter	*Correct*
	kilograms per meter cubed	*Incorrect*

16. Symbols and unit names should not be mixed in the same expression.

Example:	joules per kilogram *or* J/kg *or* $J \cdot kg^{-1}$	*Correct*
	joules per kg *or* J/kilogram *or* $J \cdot kilogram^{-1}$	*Incorrect*

17. SI prefix symbols are written in regular upright type (no italics). There is no space or hyphen between the prefix and the unit symbol.

Example:	5 ms	*Correct*
	5 m s	*Incorrect*
	5 m-s	*Incorrect*

18. The entire name of the prefix is attached to the unit name. No space or hyphen separates them.

Example:	five milliseconds	*Correct*
	five milli seconds	*Incorrect*
	five milli-seconds	*Incorrect*

The final vowel is commonly dropped from the prefix in three cases:

	megohm, kilohm, hectare	*Correct*
	megaohm, kiloohm, hectoare	*Incorrect*

19. The grouping formed by the prefix symbol attached to the unit symbol constitutes a new inseparable symbol that can be raised to a positive or negative power and that can be combined with other unit symbols to form compound unit symbols.

Example:

$1\ cm^3 = (10^{-2}\ m)^3 = 10^{-6}\ m^3$

$1\ cm^{-1} = (10^{-2}\ m)^{-1} = 10^2\ m^{-1}$

$1\ \mu s^{-1} = (10^{-6}\ s)^{-1} = 10^6\ s^{-1}$

$1\ V/cm = (1\ V)/(10^{-2}\ m) = 10^2\ V/m$

20. Compound prefixes formed by combining two or more SI prefixes are not permitted.

Example:	1 mg	*Correct*
	1 μkg	*Incorrect*

Note that even though the kilogram is the base SI unit, multiples are still formed from the gram.

21. A prefix must have an attached unit and should never be used alone.

Example:	$10^6/m^3$	*Correct*
	M/m^3	*Incorrect*

22. Modifiers are not to be attached to the units.

Example:	MW of electricity	*Correct*
	MWe	*Incorrect*
	V of alternating current	*Correct*
	Vac	*Incorrect*
	Pa of gage pressure	*Correct*
	Pag	*Incorrect*

If space is limited, the modifier may be placed in parentheses. For example, "Pa (gage)" could be replaced for "Pa of gage pressure."

23. Use only one prefix in compound units. Normally, the modifier is attached to the numerator.

Example:	mV/m	*Correct*
	mV/mm	*Incorrect*

An exception is when the kilogram occurs in the denominator.

Example:	MJ/kg	*Correct*
	kJ/g	*Incorrect*

24. Dimensionless numbers are not required to have the units reported. For example, the refractive index n is the speed of light in a vacuum c_2 relative to its speed in another medium, c_1.

$$n = \frac{c_2}{c_1} \tag{13-9}$$

Water has a refractive index of 1.33. It is not necessary to report the units; the same units are used in the numerator (m/s) and denominator (m/s), so they cancel.

In some cases, it is desirable to report the units of dimensionless numbers to avoid confusion. For example, in a mixture containing species A, B, and C, the mass fraction x_A expresses the mass of species A m_A relative to the total mass m_T.

$$x_A = \frac{m_A}{m_T} \tag{13-10}$$

Both the numerator and denominator have units of kilograms, so x_A is a dimensionless number. However, to be absolutely clear, it is best to report the species with the mass.

Example:	The mass fraction was 0.1 kg benzene/kg total.	*Preferred*
	The mass fraction was 0.1.	*Avoid*

25. Units such as "parts per thousand" and "parts per million" may be used. However, it is absolutely necessary to explain what the "part" is.

Example:	The mass fraction of CO_2 was 3.1 parts per million.	*Correct*
	The mole fraction of CO_2 was 3.1 parts per million.	*Correct*
	The fraction of CO_2 was 3.1 parts per million.	*Incorrect*

The adjectives "mass" and "mole" are absolutely essential to clarify the meaning.

26. Unit symbols are preferred to unit names.

Example: 15 m — *Preferred*
15 meters — *Avoid*

Many writing conventions require that integers from one to ten be written using words rather than numbers. Therefore, if the number is written in words, then the unit name should be used rather than the symbol.

Example: three meters — *Correct*
three m — *Incorrect*

13.7 SUMMARY

The SI System of Units has three types of units: supplementary, base, and derived. The supplementary units relate to geometry and define the radian and the steradian. There are seven base units: meter, kilogram, second, ampere, kelvin, mole, and candela. Each unit is precisely defined using a transportable standard, except for the kilogram, which still has a prototype standard. The base units may be combined into a variety of derived units, some of which are abbreviated with special names (e.g., Pa for N/m^2).

Because scientists and engineers deal with wide ranges of magnitude, prefixes are employed as multipliers to increase or decrease the size of the units. Great care must be taken when writing units to prevent miscommunication.

Nomenclature

c speed of light (m/s)
d distance (m)
k proportionality constant (atm/K)
m mass (kg)
n moles (mol) or refractive index (dimensionless)
P pressure (atm)
R universal gas constant ($atm\cdot m^3/(mol\cdot K)$)
T temperature (K)
t temperature (°C) or time (s)
V volume (m^3)
x mass fraction (dimensionless)
β solid angle (dimensionless)
θ plane angle (dimensionless)

Further Readings

ASTM Standard for Metric Practice, E 380-79. Philadelphia: American Society for Testing and Materials, 1980.

The International System of Units (SI), NIST Special Publication 330. United States Department of Commerce, National Institute of Standards and Technology, 1991.

PROBLEMS

13.1 Correct the following units to reflect the proper SI rules:
(a) 18.3 Newton
(b) 45.6 n
(c) 29.0 meter
(d) 56.9 meter/sec
(e) five m
(f) 23 m/second
(g) 493°K
(h) 89.6μ m
(i) 68.5 Kg
(j) 98.4 m/s/s
(k) 10 m's
(l) Mm/ms

13.2 Use an appropriate prefix so the number ranges from 0.1 to 1000:
(a) 9.8×10^5 m
(b) 9.56×10^{10} J
(c) 0.000056 s
(d) 1,984,000 m^3
(e) 35.6×10^{-4} g
(f) 92.4×10^7 N

13.3 A constant-volume gas thermometer is used to measure thermodynamic temperature. At the triple point of water, its pressure is 1.00000×10^2 Pa. What is its pressure at:
(a) triple point of H_2
(b) triple point of Ne
(c) triple point of O_2
(d) triple point of Ar
(e) triple point of Hg

13.4 Two astronomer-geodesists, J. B. J. Delambre (1749–1822) and P. F. A. Méchain (1744–1804), were appointed to determine the length of a "quadrant of meridian," that is, the length from the north pole to the equator along a great circle passing through the poles. They worked from 1792 to 1799 to complete their task. This length was divided into 10^{-7} parts, the original definition of the meter. In 1799, based upon their measurements, two fine scratches were placed on a platinum bar separated by the distance of one meter. When it was later discovered that there were some small surveying errors, the original definition was abandoned in favor of the platinum bar. For many years, this prototype was the standard that defined the meter.

Later, as measurements improved, the length of the quadrant of meridian was determined to be 10,002,288.3 m. What was the fractional error and percentage error of the measurements made by Delambre and Méchain? If they had access to modern high-precision equipment, would the two fine scratches on the platinum bar have been placed farther apart or closer together? How much longer, or shorter, would the meter be (in millimeters)?

Glossary

absolute zero Temperature is reduced as low as possible and motion, although not at zero, is at its minimum.
ampere The base unit of electric current.
base units The units from which all other units in a measuring system are derived.
bushel A British unit used to measure dry goods; equals 8 imperial gallons.
candela A base unit of luminous intensity.
cubit An ancient form of measure, equivalent to the length from a Pharaoh's elbow to the farthest fingertip of his extended hand.
dimension An abstract idea described by units of measure.
gram-molecule The base unit of the amount of substance (usually referred to as mole); Avogadro's number.
hands A unit of length used to determine the height of a horse; equal to 4 inches.
heat The energy flow resulting from atoms or molecules of different temperatures contacting each other.
kelvin The base unit of thermodynamic temperature.
kilogram The base unit of mass.
luminous intensity The amount of light flux into a specified solid angle.
meter The base unit of length.

mole The base unit of the amount of substance; short for gram-molecule.
phase diagram A diagram that indicates the stable phase or state of a substance at a given temperature and pressure.
plane angle In a circle, the length of the swept circumference divided by the radius.
rad The abbreviation for radian.
SI base units The unit of length (meter), unit of mass (kilogram), unit of time (second), unit of electric current (ampere), unit of thermodynamic temperature (kelvin), unit of amount of substance (mole), and unit of luminous intensity (candela).
SI supplementary units The mathematical definitions needed to define both base and derived units.
SI system of units The metric system.
second The base unit of time.
solid angle In a sphere, the swept area divided by the radius squared.
solidus Oblique stroke: /.
stone A unit of weight in Great Britain; equal to 14 pounds.
temperature A measure of the degree of random atomic motion.
triple point Condition where liquid, solid, and vapor are in equilibrium.
unit A quantity used as a standard of measurement.

Mathematical Prerequisites
- Algebra (Appendix E, Mathematics Supplement)
- Logarithms (Appendix H, Mathematics Supplement)

CHAPTER 14
Unit Conversions

Mistakes in unit conversions are the most frequent cause of errors in engineering calculations. This is particularly true in the United States, where we employ a mixture of customary and scientific systems. Therefore, the engineer must be well versed in all systems and able to convert between them with ease.

14.1 WHAT DOES IT MEAN TO "MEASURE" SOMETHING?

Whenever we make a measurement, it is always made with respect to a standard. For example, suppose we wished to measure the length of the rod in Figure 14.1. We could approach this problem by obtaining three metersticks and laying them end to end. Because the length of the rod is equivalent to the three metersticks, we report, "the unknown rod is 3 meters in length."

In the United States, we would probably measure the length of the unknown rod using yardsticks. The length of the unknown rod is equivalent to 3 + 0.28 yardsticks, so we report, "the length of the unknown rod is 3.28 yards in length."

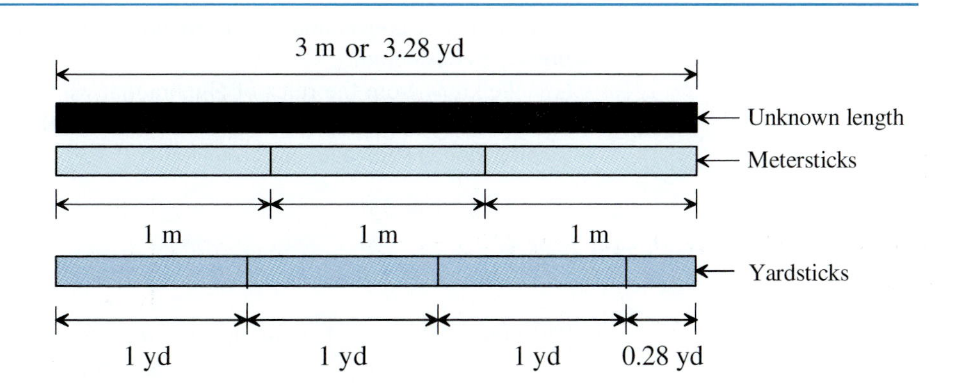

FIGURE 14.1
Measuring a rod of unknown length with calibrated sticks.

The Right and Lawful Rood

In his 16th-century surveying book, Master Koebel describes how to determine legal lengths. He instructs the surveyor to wait at the church door on Sunday and "bid sixteen men to stop, tall ones, and short ones, as they happen to come out." (Note the use of random selection to prevent bias from having all short, or all tall, men.) These selected men are to stand in line with "their left feet one behind the other"; the accumulated length of their feet defines the "right and lawful rood." The "right and lawful foot" was defined as the sixteenth part of the rood.

Adapted from: H. A. Klein, *The World of Measurements* (New York: Simon and Schuster, 1974), pp. 66–67.

Notice that the reported answer has two parts, the number "3" and the unit "meters" (or "3.28" and the unit "yards"). If we reported "the length of the rod is 3," the reader would not know whether the units were meters, yards, feet, inches, or whatever. **It is absolutely essential to report the unit with the number.** Students often perform elaborate calculations and forget to report the units. Even though the number may be correct, the answer is completely **wrong** without the units.

14.2 CONVERSION FACTORS

When switching between unit systems, it is necessary to use a **conversion factor.** You can find conversion factors in handbook tables and in Appendix A. Conversion factors are developed from identities that are determined experimentally or by definition. For example,

$$1 \text{ ft} \equiv 0.3048 \text{ m}$$

is an exact definition. Because we have this identity where $A = B$, then we know that $A/B = 1$.

$$\frac{1 \text{ ft}}{0.3048 \text{ m}} = 1 = \text{conversion factor} = F$$

$$\frac{0.3048 \text{ m}}{1 \text{ ft}} = 1 = \text{conversion factor} = F$$

Although these conversion factors are not numerically identical to 1, they equal 1 when the units are considered.

We know from the rules of algebra that we are permitted to multiply a quantity by 1 and the quantity remains unchanged. For example, we can multiply the quantity 5 ft by 1 and it remains unchanged.

$$5 \text{ ft} \times 1 = 5 \text{ ft}$$

$$5 \text{ ft} \times F = 5 \text{ ft} \times \frac{0.3048 \text{ m}}{1 \text{ ft}} = 1.524 \text{ m}$$

The units of ft cancel, leaving the unit of m. Because we have merely multiplied 5 ft by 1, we know that 1.524 m must be identical to 5 ft.

Some students employ an alternative technique:

$$\begin{array}{c|c} 5\ \text{ft} & 0.3048\ \text{m} \\ \hline & 1\ \text{ft} \end{array} = 1.524\ \text{m}$$

This method gives the same answer as the other method. It has the advantage of keeping the units neatly organized, but it is less apparent that we are simply multiplying by 1. You may use whichever method you find the most convenient.

A **VERY** common error occurs when converting a dimension that is raised to a power. Carefully review the following three examples:

$$5\ \text{ft}^2 \times \frac{(0.3048\ \text{m})^2}{(1\ \text{ft})^2} = 0.4676\ \text{m}^2 \qquad \textit{Correct}$$

$$5\ \text{ft}^2 \times \left(\frac{0.3048\ \text{m}}{1\ \text{ft}}\right)^2 = 0.4676\ \text{m}^2 \qquad \textit{Correct}$$

$$5\ \text{ft}^2 \times \frac{0.3048\ \text{m}^2}{1\ \text{ft}^2} = 1.524\ \text{m}^2 \qquad \textit{Incorrect}$$

The first example says that a square that is 1 ft on a side could also be described as a square that is 0.3048 m on a side. The second example says that when the conversion factor F is squared, it is still 1 (i.e., $1^2 = 1$). The third example is completely wrong because the conversion factor for linear dimensions has been confused with the conversion factor for area dimensions.

14.3 MATHEMATICAL RULES GOVERNING DIMENSIONS AND UNITS

In engineering, quantities continually appear in mathematical formulas. The following rules must be obeyed:

Addition/Subtraction: All the terms that are added (or subtracted) must have the same dimensions. For example, if

$$D = A + B - C \tag{14-1}$$

then the dimensions of A, B, C, and D must all be identical.

Multiplication/Division: The dimensions in multiplication/division are treated as though they were variables and cancel accordingly. For example, if A has dimensions of $[\text{M/T}^2]$, and B has dimensions of $[\text{T}^2/\text{L}]$, and C has dimensions of $[\text{M/L}^2]$, then D has dimensions of [L] as shown below:

$$D = \frac{AB}{C} = \frac{[\text{M/T}^2][\text{T}^2/\text{L}]}{[\text{M/L}^2]} = [L] \tag{14-2}$$

Transcendental Functions: A **transcendental function** cannot be given by algebraic expressions consisting only of the argument and constants. Examples of transcendental functions are

$$A = \sin x \qquad B = \ln x \qquad C = e^x$$

The argument of a transcendental function (x in these equations) cannot have any dimensions. Similarly, when the transcendental function is evaluated, the result has no dimensions. That is, A, B, and C in these equations have no dimensions.

One can understand the requirement for this rule by realizing that transcendental functions are generally represented by an infinite series. For example, e^x is evaluated with the infinite series:

$$e^x = 1 + x + \frac{x^2}{2!} + \frac{x^3}{3!} + \cdots \tag{14-3}$$

If the argument x had dimensions, then the rule for addition/subtraction would be violated. For example, if x had dimensions [L], then we would be adding [L], $[L]^2$, $[L]^3$, and so forth, which we cannot do. Because all the terms on the right must be dimensionless, then clearly e^x must also be dimensionless.

This rule requiring the argument of a transcendental function to be dimensionless applies only to scientific equations with fundamental significance. The rule may be violated in the case of empirical equations. For example, a mechanical engineer interested in the drag force F on an automobile could construct a scale model and test it in a wind tunnel. He would measure the force on the model resulting from different wind velocities v. The data could be correlated with the empirical equation

$$F = av^b \tag{14-4}$$

where a and b are **empirical constants.** This equation can be manipulated by taking the logarithm of both sides:

$$\log F = \log a + \log v^b \tag{14-5}$$

$$\log F = \log a + b \log v \tag{14-6}$$

Plotting $\log F$ versus $\log v$ results in a straight line with a slope b and a y-intercept of $\log a$. Force has dimensions of $[ML/T^2]$ and velocity has dimensions of $[L/T]$, so the argument of the logarithm has dimensions, a violation of the rule. The consequence of this violation is that the constant a will depend on the units chosen to measure force and velocity. For example, the constant a will have one value if force were measured in newtons and velocity in meters per second, and a would have another value if force were measured in poundals (discussed later) and velocity in feet per second.

Dimensional Homogeneity: For an equation to be valid, it must be **dimensionally homogeneous,** that is, the dimensions on the left-hand and right-hand sides of the equation must be the same. For example, Newton's second law

$$F = ma \tag{14-7}$$

$$\left[\frac{ML}{T^2}\right] = [M]\left[\frac{L}{T^2}\right]$$

is dimensionally homogeneous because the dimensions on both sides of the equal sign are identical. In contrast, the following equation is not homogeneous,

$$P = \rho T \qquad \textit{Incorrect} \qquad (14\text{-}8)$$

$$\left[\frac{\text{M}}{\text{LT}^2}\right] \neq \left[\frac{\text{M}}{\text{L}^3}\right][\theta]$$

because the dimensions on both sides of the equal sign are not identical.

14.4 SYSTEMS OF UNITS

The SI System of Units, which we discussed in the previous chapter, has evolved in modern times. Many other systems of units have preceded it. Table 14.1 summarizes the most important of these systems. Note that this table divides the unit systems into two major categories: coherent and noncoherent. Within the coherent systems, there are two further subcategories: absolute and gravitational. Coherent/noncoherent and absolute/gravitational systems will be discussed in the next two sections.

Table 14.1 shows some collections of units that are given a name; for example, a $lb_m \cdot ft/s^2$ is called a **poundal**. Be sure to become familiar with the other names given to collections of units.

14.4.1 Absolute and Gravitational Systems of Units

Absolute systems define mass [M], length [L], and time [T]. Force [F] is a derived quantity determined from Newton's second law:

$$\underbrace{[\text{F}]}_{\text{Derived}} = F = ma = \underbrace{[\text{M}]\frac{[\text{L}]}{[\text{T}^2]}}_{\text{Defined}} \qquad (14\text{-}9)$$

SI is an absolute system in which mass (kg), length (m), and time (s) are defined, and force (N) is derived.

TABLE 14.1
Systems of units

	Coherent						Noncoherent
	Absolute			Gravitational			
Fundamental Dimensions	**MKS (SI)**	**CGS**	**FPS**	**MKS**	**CGS**	**USCS**	**AES**
Length [L]	m	cm	ft	m	cm	ft	ft
Time [T]	s	s	s	s	s	s	s
Mass [M]	kg_m	g_m	lb_m	—	—	—	lb_m
Force [F]	—	—	—	kg_f	g_f	lb_f	lb_f
Derived Dimensions							
Mass [FT²/L]	—	—	—	$kg_f \cdot s^2/m$ (mug)	$g_f \cdot s^2/cm$ (glug)	$lb_f \cdot s^2/ft$ (slug)	—
Force [ML/T²]	$kg_m \cdot m/s^2$ (newton)	$g_m \cdot cm/s^2$ (dyne)	$lb_m \cdot ft/s^2$ (poundal)	—	—	—	—
Energy [LF] or [ML²/T²]	N·m (joule)	dyne·cm (erg)	ft·poundal	$m \cdot kg_f$	$cm \cdot g_f$	$ft \cdot lb_f$	$ft \cdot lb_f$
Power [LF/T] or [ML²/T³]	J/s (watt)	erg/s	ft·poundal/s	$m \cdot kg_f/s$	$cm \cdot g_f/s$	$ft \cdot lb_f/s$	$ft \cdot lb_f/s$

Gravitational systems define force [F], length [L], and time [T]. Mass [M] is a derived quantity, also determined from Newton's second law:

$$\underbrace{[\mathrm{M}]}_{\text{Derived}} = m = \frac{F}{a} = \underbrace{\frac{[\mathrm{F}][\mathrm{T}^2]}{[\mathrm{L}]}}_{\text{Defined}} \tag{14-10}$$

Historically, a number of gravitational systems evolved.

Gravitational systems define their units of [F] as pound-force (lb_f), kilogram-force (kg_f), and gram-force (g_f), which are the forces exerted by a pound-mass (lb_m), kilogram-mass (kg_m), and gram-mass (g_m), respectively, under the influence of earth's gravity field (Figure 14.2). Gravity's force can be calculated according to the formula

$$F = mg \tag{14-11}$$

where g is the acceleration of an object caused by earth's gravity. The value of g is not uniformly the same everywhere on earth. It varies by over 0.5%, depending on the height above sea level and whether extremely dense rock is nearby. Therefore, the **standard** acceleration due to gravity g^o has been defined as

$$g^o = 9.80665 \text{ m/s}^2 = 32.1740 \text{ ft/s}^2 \tag{14-12}$$

(*Note*: The *absolute systems* do not depend on the local strength of the gravity force, hence their name.)

The derived units for mass in the gravitational systems are the **slug, mug,** and **glug** (no joke), derived according to Figure 14.3. The values for the gravitational masses are 1 slug = 32.1740 lb_m, 1 mug = 9.80665 kg_m, and 1 glug = 980.665 g_m. Although the slug is widely used in engineering, the mug and glug are rarely used; they are presented here for the sake of completeness.

Table 14.2 shows some frequently encountered quantities and their associated dimensions in the absolute and gravitational systems. The symbol indicated for each quantity (e.g., "t" for time) is generally encountered in formulas in science and engineering literature, but can change depending on the discretion of the author.

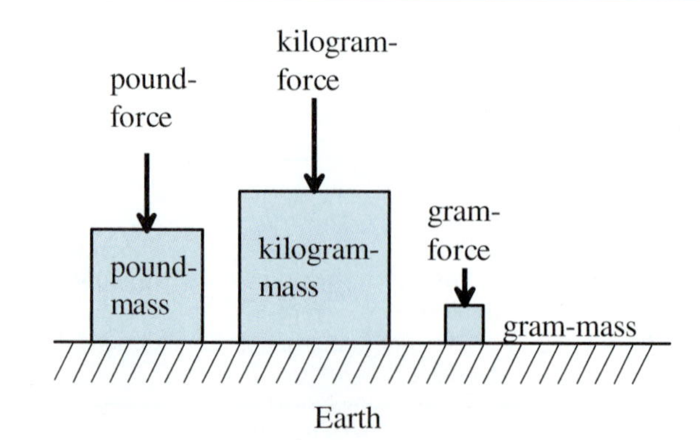

FIGURE 14.2
Gravitational systems. Definitions of pound-force, kilogram-force, and gram-force.

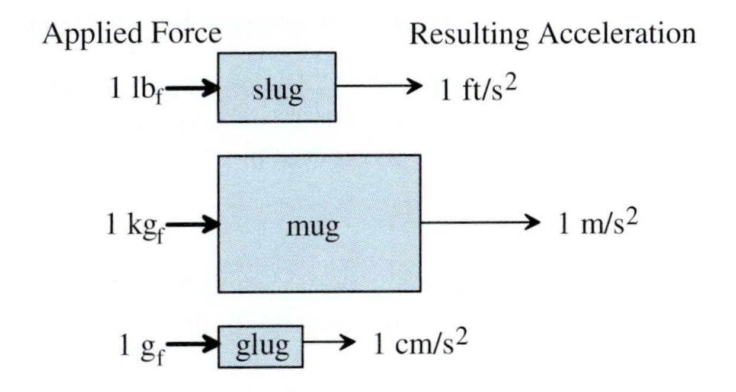

FIGURE 14.3
Gravitational systems. The slug, mug, and glug are derived units.

TABLE 14.2
Frequently encountered engineering quantities (Reference)

Quantity	Symbol	Absolute Dimensions	Gravitational Dimensions
Length	*L, l, d*	[L]	[L]
Time	*t*	[T]	[T]
Mass	*m*	[M]	[FT²/L]
Force	*F*	[ML/T²]	[F]
Electric current	*I, i*	[A]	[A]
Temperature	*T*	[θ]	[θ]
Amount of substance	*n*	[N]	[N]
Luminous intensity	*R*	[I]	[I]
Area	*A*	[L²]	[L²]
Volume	*V*	[L³]	[L³]
Velocity	*v*	[L/T]	[L/T]
Acceleration	*a*	[L/T²]	[L/T²]
Angular velocity	ω	[T⁻¹]	[T⁻¹]
Angular acceleration	α	[T⁻²]	[T⁻²]
Frequency	ω	[T⁻¹]	[T⁻¹]
Electric charge	*Q, q*	[AT]	[AT]
Heat capacity	C_p	[L²/T²θ]	[L²/T²θ]
Kinematic viscosity	ν	[L²/T]	[L²/T]
Momentum	*p*	[ML/T]	[FT]
Pressure	*P*	[M/LT²]	[F/L²]
Stress	σ	[M/LT²]	[F/L²]
Energy (work)	*W*	[ML²/T²]	[FL]
Energy (heat)	*Q, q*	[ML²/T²]	[FL]
Torque	τ	[ML²/T²]	[FL]
Power	*P*	[ML²/T³]	[FL/T]
Density	ρ	[M/L³]	[FT²/L⁴]
Dynamic viscosity	μ	[M/LT]	[FT/L²]
Thermal conductivity	*k*	[ML/T³θ]	[F/Tθ]
Voltage	*V, E*	[ML²/AT³]	[FL/AT]
Electrical resistance	*R*	[ML²/A²T³]	[FL/A²T]

14.4.2 Coherent and Noncoherent Systems of Units

Both the absolute and gravitational systems are **coherent,** meaning that no additional conversion factors are required if the units within that system are used exclusively. For example, the equation

$$F = ma \tag{14-13}$$

may be used in both the FPS (absolute) and USCS (gravitational) systems simply by putting numbers (with their corresponding units) directly into the formula:

$$F = (1\ \text{lb}_\text{m})(10\ \text{ft/s}^2) = 10\ \text{lb}_\text{m}\cdot\text{ft/s}^2 = 10\ \text{poundal} \qquad \text{FPS}$$

$$F = (1\ \text{slug})(10\ \text{ft/s}^2) = 10\ \text{slug}\cdot\text{ft/s}^2 = 10\ \text{lb}_\text{f} \qquad \text{USCS}$$

In both cases, the equations are correct; no additional conversion factors were required.

In contrast, the AES (American Engineering System) of units is **noncoherent.** (Some would say it is incoherent, but that's another story.) For example, if we put numbers with their corresponding units into Equation 14-13, **we do not get the correct answer:**

$$F = (1\ \text{lb}_\text{m})(10\ \text{ft/s}^2) \neq 10\ \text{lb}_\text{f} \qquad \text{AES}$$

This inconsistency in units results from historical reasons. The pound is a unit that was established in the year 1340. A pound was a pound; there was no distinction between pound-mass and pound-force. It was centuries before Newton taught us the difference between mass, force, and weight.

Mass is the property that causes a body to resist acceleration. A *force* causes a body to accelerate. (The known forces are: gravity, electromagnetism, strong force, and weak force; the latter two are involved with the atomic nucleus.) *Weight* is the force exerted on a body due to local gravity. An object in space can become "weightless" if there is no local gravity. However, it does not lose its mass, because that is a property of the object itself and does not depend on the local environment. In everyday language, the terms *mass* and *weight* are used interchangeably, which adds to the confusion.

Because our forefathers were also confused about the distinction between mass and weight, we have inherited the noncoherent AES system of units. Fortunately, we can overcome the noncoherence by using a conversion factor. This particular conversion factor is given the special name g_c, **but it is still a conversion factor just like the conversion factor F described earlier.** This conversion factor allows us to convert from mass to force or vice versa.

Table 14.3 shows values for g_c in the various systems of units. Notice that the number is unity for all the coherent systems and it differs from unity in the noncoherent system. Because the number is the same as that for the standard acceleration due to gravity, students confuse g^o and g_c. **THEY ARE NOT THE SAME.** The dimensions differ; the dimensions of g^o are [L/T^2] whereas the dimensions of g_c are [ML/T^{2}F]. Students also confuse g_c with g, the local acceleration due to gravity. The conversion factor g_c is a constant that is the same everywhere in the universe (think of it as "g sub constant"), whereas g changes with the local environment. For example, g on the moon is 1/6th that of the earth, but g_c on the moon is unchanged.

TABLE 14.3
The conversion factor g_c

System	Definition		g_c
MKS (absolute)	$1\ N \equiv 1\ kg_m \cdot m/s^2$	$\Rightarrow$	$g_c = 1\ kg_m \cdot m/(N \cdot s^2)$
CGS (absolute)	$1\ dyne \equiv 1\ g_m \cdot cm/s^2$	$\Rightarrow$	$g_c = 1\ g_m \cdot cm/(dyne \cdot s^2)$
FPS (absolute)	$1\ poundal \equiv 1\ lb_m \cdot ft/s^2$	$\Rightarrow$	$g_c = 1\ lb_m \cdot ft/(poundal \cdot s^2)$
MKS (gravitational)	$1\ mug \equiv 1\ kg_f \cdot s^2/m$	$\Rightarrow$	$g_c = 1\ mug \cdot m/(kg_f \cdot s^2)$
CGS (gravitational)	$1\ glug \equiv 1\ g_f \cdot s^2/cm$	$\Rightarrow$	$g_c = 1\ glug \cdot cm/(g_f \cdot s^2)$
USCS (gravitational)	$1\ slug \equiv 1\ lb_f \cdot s^2/ft$	$\Rightarrow$	$g_c = 1\ slug \cdot ft/(lb_f \cdot s^2)$
AES	$1\ lb_f \equiv 32.174\ lb_m \cdot s^2/ft$	$\Rightarrow$	$g_c = 32.174\ lb_m \cdot ft/(lb_f \cdot s^2)$

EXAMPLE 14.1

Problem Statement: What is the weight of 1 lb_m on earth where the local acceleration due to gravity is 32.174 ft/s^2? Express the answer in each system of units described in Table 14.3.

Solution: The weight is the force exerted by the object due to gravity. This force can be determined from the equation

$$F = mg \tag{14-14}$$

We place the original units in the equation and then apply appropriate conversion factors from Appendix A to convert the answer to the various systems of units. The factor $1/g_c$ is featured prominently in the conversion.

$$F = (1\ \text{lb}_\text{m})\left(32.174\ \frac{\text{ft}}{\text{s}^2}\right)\left(0.4536\ \frac{\text{kg}_\text{m}}{\text{lb}_\text{m}}\right)\left(0.3048\ \frac{\text{m}}{\text{ft}}\right)\left(1\ \frac{\text{N}\cdot\text{s}^2}{\text{kg}_\text{m}\cdot\text{m}}\right) = 4.448\ \text{N} \qquad \text{MKS-abs}$$

$$F = (1\ \text{lb}_\text{m})\left(32.174\ \frac{\text{ft}}{\text{s}^2}\right)\left(453.6\ \frac{\text{g}_\text{m}}{\text{lb}_\text{m}}\right)\left(30.48\ \frac{\text{cm}}{\text{ft}}\right)\left(1\ \frac{\text{dyne}\cdot\text{s}^2}{\text{g}_\text{m}\cdot\text{cm}}\right) = 444{,}800\ \text{dyne} \qquad \text{CGS-abs}$$

$$F = (1\ \text{lb}_\text{m})\left(32.174\ \frac{\text{ft}}{\text{s}^2}\right)\left(1\ \frac{\text{poundal}\cdot\text{s}^2}{\text{lb}_\text{m}\cdot\text{ft}}\right) = 32.174\ \text{poundal} \qquad \text{FPS-abs}$$

$$F = (1\ \text{lb}_\text{m})\left(32.174\ \frac{\text{ft}}{\text{s}^2}\right)\left(0.04625\ \frac{\text{mug}}{\text{lb}_\text{m}}\right)\left(0.3048\ \frac{\text{m}}{\text{ft}}\right)\left(1\ \frac{\text{kg}_\text{f}\cdot\text{s}^2}{\text{mug}\cdot\text{m}}\right)$$

$$= 0.4536\ \text{kg}_\text{f} \qquad \text{MKS-grav}$$

$$F = (1\ \text{lb}_\text{m})\left(32.174\ \frac{\text{ft}}{\text{s}^2}\right)\left(0.4625\ \frac{\text{glug}}{\text{lb}_\text{m}}\right)\left(30.48\ \frac{\text{cm}}{\text{ft}}\right)\left(1\ \frac{\text{g}_\text{f}\cdot\text{s}^2}{\text{glug}\cdot\text{cm}}\right)$$

$$= 453.6\ \text{g}_\text{f} \qquad \text{CGS-grav}$$

$$F = (1\ \text{lb}_\text{m})\left(32.174\ \frac{\text{ft}}{\text{s}^2}\right)\left(0.03108\ \frac{\text{slug}}{\text{lb}_\text{m}}\right)\left(1\ \frac{\text{lb}_\text{f}\cdot\text{s}^2}{\text{slug}\cdot\text{ft}}\right) = 1\ \text{lb}_\text{f} \qquad \text{USCS-grav}$$

$$F = (1\ \text{lb}_\text{m})\left(32.174\ \frac{\text{ft}}{\text{s}^2}\right)\left(\frac{\text{lb}_\text{f}\cdot\text{s}^2}{32.174\ \text{lb}_\text{m}\cdot\text{ft}}\right) = 1\ \text{lb}_\text{f} \qquad \text{AES}$$

Often in older engineering texts, g_c is incorporated directly into the equation. For example, Equation 14-14 would become

$$F = m\frac{g}{g_c} \tag{14-15}$$

This practice should be avoided because it gives more prominence to g_c than it deserves. After all, none of the other conversion factors were featured in the equation, and they were just as important as g_c. An engineer must always keep track of units; she needs no special reminder about this particular conversion factor.

Another potential source of noncoherence is in the units used to describe energy. For example, the equation for the kinetic energy of an object (i.e., the energy associated with motion) is

$$E = \tfrac{1}{2}mv^2 \tag{14-16}$$

where E is the kinetic energy, m is the mass, and v is the velocity. Energy may be expressed either as "work" (e.g., a force exerted over a distance) or "heat" (energy flow due to a temperature difference). This equation is coherent when energy is expressed as "work" (i.e., joules), but is noncoherent when energy is expressed as heat (e.g., calories).

$$E = \tfrac{1}{2}(1\ \text{kg})\left(2\,\frac{\text{m}}{\text{s}}\right)^2 = 2\ \text{kg·m}^2/\text{s}^2 = 2\ \text{J} \qquad \textit{Coherent}$$

$$E = \tfrac{1}{2}(1\ \text{kg})\left(2\,\frac{\text{m}}{\text{s}}\right)^2 = 2\ \text{kg·m}^2/\text{s}^2 \neq 2\ \text{cal} \qquad \textit{Noncoherent}$$

The noncoherent equation, which uses heat to measure energy, can be corrected by using the "mechanical equivalent of heat" (4.1868 J = 1 cal). Thus, by introducing a conversion factor, we can make the noncoherent equation correct:

$$E = \tfrac{1}{2}(1\ \text{kg})\left(2\,\frac{\text{m}}{\text{s}}\right)^2 = 2\ \text{kg·m}^2/\text{s}^2 = 2\ \text{J}\left(\frac{\text{cal}}{4.1868\ \text{J}}\right) = 0.4777\ \text{cal}$$

It is inconvenient to use this mechanical-equivalent-of-heat conversion factor. Therefore, in SI, we maintain coherence by always expressing heat in joules rather than calories.

14.5 THE DATUM

A **datum** is a reference point used when making a measurement. Suppose we wished to know the difference in height between the third floor and fifth floor of a building (Figure 14.4). We could choose the ground as our *datum.* Because the third floor is 30 ft above the ground and the fifth floor is 50 ft above the ground, we can calculate the difference in height between the third and fifth floors

$$\Delta h = h_5 - h_3 = 50\ \text{ft} - 30\ \text{ft} = 20\ \text{ft}$$

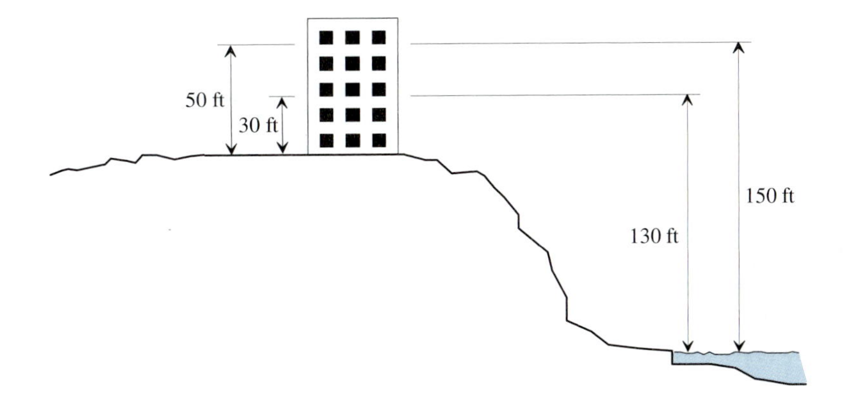

FIGURE 14.4
Illustration of a datum taken as local ground level and another datum taken as sea level.

If we had selected sea level as our datum, we can still calculate the difference in height between the third and fifth floors

$$\Delta H = H_5 - H_3 = 150 \text{ ft} - 130 \text{ ft} = 20 \text{ ft}$$

From this simple example, we see that, provided we are interested in the difference between two numbers, we may select any datum we wish. **The only requirement is that the datum not change in the middle of the calculation.** The use of an arbitrary datum is frequent in science and engineering. For example, the **Celsius temperature scale** uses the freezing point of water as the arbitrary datum.

14.6 PRESSURE

Pressure is a force exerted on an area. **Gas pressure** results from the impact of gas molecules on the container wall, thus exerting a force on a given area of surface (Figure 14.5). If the temperature is increased, the molecules move more vigorously and impact the walls with more force; hence, the pressure increases. For a perfect gas, the perfect (ideal) gas equation describes the relationship between pressure and temperature.

Hydrostatic pressure results from the weight of liquid or gas. If you have ever dived to the bottom of a swimming pool, you have experienced the effects of hydrostatic pressure on your ears. (Each 34 ft of water equals another atmosphere of pressure.) Atmospheric pressure results from the weight of the air above us.

The hydrostatic pressure at the bottom of a container with cross-sectional area A filled with liquid of density ρ can be calculated as follows:

$$P = \frac{F}{A} = \frac{mg}{A} = \frac{\rho V g}{A} = \frac{\rho (Ah) g}{A} = \rho g h \tag{14-17}$$

Thus, the hydrostatic pressure depends on the liquid density, the height of the liquid column h, and the local acceleration due to gravity g.

Three types of pressure are generally reported: *absolute, gage,* and *differential.* (A fourth type of pressure, *vacuum pressure,* is discussed later.) We can visualize these by

14.7 TEMPERATURE

Temperature is the measure of a body's thermal energy, that is, its molecular (atomic) motion. At high temperatures, molecules (atoms) move vigorously, whereas at low temperatures the motion is less vigorous.

The simplest way to measure temperature is with mercury-in-glass thermometers, which have existed for hundreds of years. At high temperatures, the mercury expands and rises in the thermometer. Once such a thermometer has been constructed, the markings on the glass are arbitrary.

A **temperature scale** is formed by placing two reference points on the mercury-in-glass thermometer and evenly subdividing them into **temperature intervals.** Because students often confuse "temperature scales" and "temperature intervals," here we will establish a nomenclature so the two are easily distinguished. A "temperature scale" will be indicated by placing the degree sign first (e.g., °C) whereas a "temperature interval" will be indicated by placing the degree sign last (e.g., C°). (*Note:* This convention is used in Halliday and Resnick, *Fundamentals of Physics,*[1] but it is rarely used in the general literature. In fact, SI removed the degree symbol from the Kelvin temperature scale in 1967, so it is simply "K," not "°K." However, here we will leave the degree symbol and manipulate it to suit our instructional purposes.)

Around 1714, G. D. Fahrenheit (1686–1736) developed the **Fahrenheit temperature scale.** The two reference points were the freezing point of water (32°F) and the boiling point of water (212°F). (The temperature 0°F corresponds to the coldest temperature achievable with water, salt, and ice.)

In 1743, Anders Celsius (1701–1744) devised the **Celsius temperature scale.** The two reference points were the melting point of ice (0°C) and the boiling point of water (100°C). It was originally known as the "centigrade scale" ("one hundred steps"), but this name has been abandoned in favor of the official SI name, Celsius.

The Fahrenheit scale is almost twice as sensitive as the Celsius scale; 180 F° span the melting point to the boiling point, whereas 100 C° span this same interval. Thus, we can find the relationship between the Fahrenheit interval (F°) and the Celsius interval (C°) as

$$100\ \text{C}^\circ = 180\ \text{F}^\circ \tag{14-21}$$

$$1\ \text{C}^\circ = \frac{180}{100}\ \text{F}^\circ = \frac{9}{5}\ \text{F}^\circ = 1.8\ \text{F}^\circ \tag{14-22}$$

Rather than using the arbitrary reference points of the Celsius and Fahrenheit scales, **thermodynamic temperature scales** (also called *absolute temperature scales*) specify that the temperature should be proportional to the thermal energy of an ideal gas. There are two thermodynamic temperature scales: Kelvin and Rankine. The **Kelvin temperature scale** has the same temperature interval as the Celsius scale:

$$1\ \text{K}^\circ = 1\ \text{C}^\circ \tag{14-23}$$

With this constraint, the triple point of water is defined as 273.16°K. The **Rankine temperature scale** has the same temperature interval as the Fahrenheit temperature scale

1. D. Halliday and R. Resnick, *Fundamentals of Physics* (New York: John Wiley & Sons, 1974) p. 351.

$$1\ R^\circ = 1\ F^\circ \tag{14-24}$$

With this constraint, the triple point of water is defined as 491.69°R. The thermodynamic temperature scales are named after J. K. Rankine (1820–1872) and Lord Kelvin (1824–1907), who helped establish them.

Figure 14.8 shows the four temperature scales: Celsius, Kelvin, Fahrenheit, and Rankine. The relationships between these various scales follow:

$$[^\circ R] = [^\circ F] + 459.67 \tag{14-25}$$

$$[^\circ K] = [^\circ C] + 273.15 \tag{14-26}$$

$$[^\circ C] = \frac{[^\circ F] - 32}{1.8} \tag{14-27}$$

$$[^\circ F] = 1.8\,[^\circ C] + 32 \tag{14-28}$$

$$[^\circ R] = 1.8\,[^\circ K] \tag{14-29}$$

$$[^\circ K] = \frac{[^\circ R]}{1.8} \tag{14-30}$$

where [°R] is a Rankine temperature, [°F] is a Fahrenheit temperature, [°K] is a Kelvin temperature, and [°C] is a Celsius temperature.

The conversion factors for the temperature intervals are

$$\text{Conversion factor} = \frac{1.8\ F^\circ}{C^\circ} = \frac{1.8\ R^\circ}{K^\circ} = \frac{1\ F^\circ}{R^\circ} = \frac{1\ C^\circ}{K^\circ} = \frac{1.8\ F^\circ}{K^\circ} = \frac{1.8\ R^\circ}{C^\circ} \tag{14-31}$$

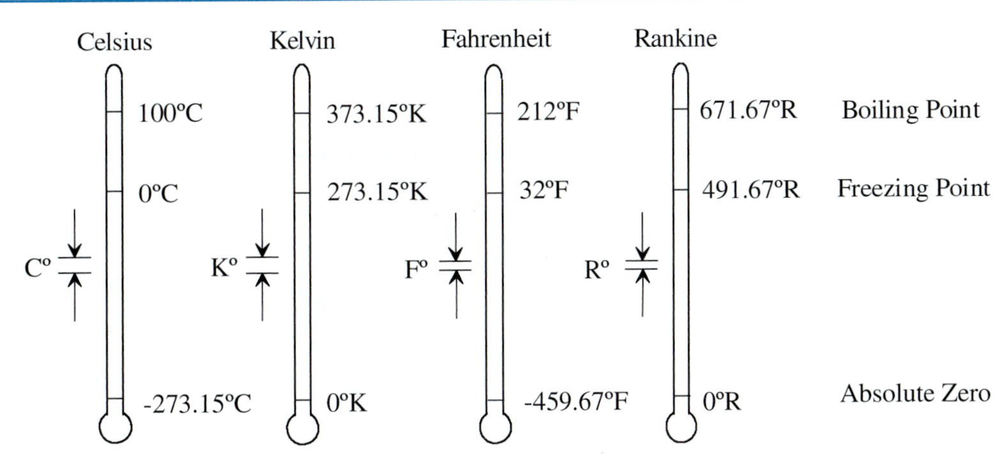

FIGURE 14.8
The four temperature scales. (*Note:* Temperature scales are indicated by placing the degree symbol first, whereas temperature intervals are indicated by placing the degree symbol last. According to SI conventions, the degree symbol has been eliminated from the Kelvin scale, but here it is included for instructional purposes.)

EXAMPLE 14.2

Problem Statement: Calculate the heat transfer Q/t through a block that has a cross-sectional area A of 1 ft^2, a length x of 2 ft, and a thermal conductivity k of 5 Btu/(h·ft·F°). One surface is maintained at 200°F and the other is maintained at 150°F (Figure 14.9). Do the calculation in both U.S. engineering units and SI units.

Solution: We obtain the equation for conductive heat transfer from Chapter 12 on rate processes:

$$\frac{Q}{t} = k\frac{A\Delta T}{x} \tag{14-32}$$

The temperature difference ΔT is a temperature interval calculated as follows:

$$\Delta T = 200°\text{F} - 150°\text{F} = 50\ \text{F}°$$

We may put this into Equation 14-32 to calculate the heat transfer Q/t,

$$\frac{Q}{t} = \left(5\ \frac{\text{Btu}}{\text{h·ft·F°}}\right)\frac{(1\ \text{ft}^2)(50\ \text{F°})}{2\ \text{ft}} = 125\ \text{Btu/h}$$

We can convert all the given information to SI:

$$\Delta T = 50\ \text{F°}\left(\frac{\text{K°}}{1.8\ \text{F°}}\right) = 27.8\ \text{K°}$$

$$k = 5\frac{\text{Btu}}{\text{h·ft·F°}}\left(\frac{\text{h}}{3600\ \text{s}}\right)\left(\frac{\text{ft}}{0.3048\ \text{m}}\right)\left(\frac{1055\ \text{J}}{\text{Btu}}\right)\left(\frac{1.8\ \text{F°}}{\text{K°}}\right) = 8.65\frac{\text{J}}{\text{s·m·K°}}$$

$$A = 1\ \text{ft}^2\left(\frac{0.3048\ \text{m}}{\text{ft}}\right)^2 = 0.0929\ \text{m}^2$$

$$x = 2\ \text{ft}\left(\frac{0.3048\ \text{m}}{\text{ft}}\right) = 0.610\ \text{m}$$

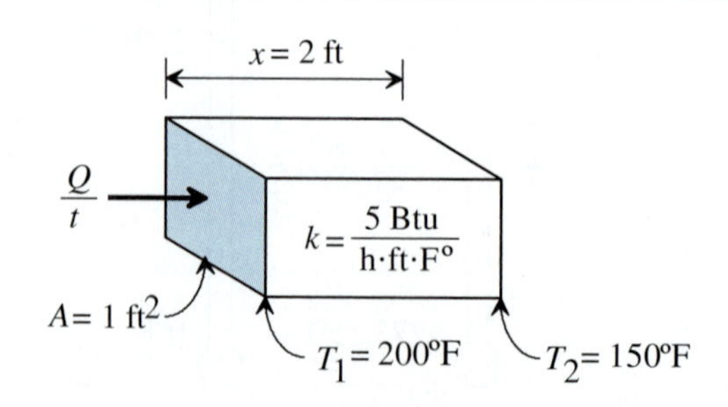

FIGURE 14.9
Heat transfer through a block.

These SI quantities may now be placed in Equation 14-32.

$$\frac{Q}{t} = \left(8.65 \frac{\text{J}}{\text{s·m·K°}}\right) \frac{(0.0929 \text{ m}^2)(27.8 \text{ K°})}{(0.610 \text{ m})} = 36.6 \text{ J/s} = 36.6 \text{ W}$$

The conversion factors in Appendix A show that these two answers (125 Btu/h and 36.6 W) are identical, as they must be.

14.8 CHANGING THE SYSTEM OF UNITS IN AN EQUATION

Engineers must often convert equations from one system of units to another. For example, the perfect gas equation

$$PV = nRT \tag{14-33}$$

is learned in chemistry classes but is widely used in engineering as well. The probable units used in each discipline are shown here:

Quantity	Chemistry (SI)	Engineering
P	Pa	psia
V	m^3	ft^3
n	mol	lbmol
T	K	°R
R	Pa·m³/(mol·K)	psia·ft³/(lbmol·°R)

The perfect gas constant R is empirically measured and contains the necessary units to make the equation dimensionally homogeneous (i.e., it makes the units to the left and right of the equal sign the same).

EXAMPLE 14.3

Problem Statement: Convert the SI perfect gas equation

$$PV = n\left(8.314 \frac{\text{Pa·m}^3}{\text{mol·K}}\right)T$$

into an equation in which engineering units may be used for the variables.

Solution: Because the perfect gas constant R contains the necessary units to make the equation dimensionally homogeneous, we merely have to change the units of R to engineering units:

$$R = \left(8.314 \frac{\text{Pa·m}^3}{\text{mol·K}}\right)\left(\frac{\text{psia}}{6895 \text{ Pa}}\right)\left(\frac{35.31 \text{ ft}^3}{\text{m}^3}\right)\left(\frac{453.6 \text{ mol}}{\text{lbmol}}\right)\left(\frac{\text{K}}{1.8°\text{R}}\right)$$

$$= 10.73 \frac{\text{psia·ft}^3}{\text{lbmol·°R}}$$

The perfect gas equation in engineering units now becomes

$$PV = n\left(10.73\,\frac{\text{psia·ft}^3}{\text{lbmol·°R}}\right)T$$

14.9 DIMENSIONAL ANALYSIS (ADVANCED TOPIC)

Dimensional analysis is widely used in engineering to solve problems about which there is little fundamental information. Simply by looking at the dimensions of the quantities involved, we can tell much about how the quantities are related.

As an example, consider the pendulum shown in Figure 14.10. In the design of a grandfather clock, it would be very useful to know the *period,* that is, the length of time it takes for the pendulum to swing back and forth. Before we perform a sophisticated analysis of the problem, we can use our intuition and experience to simply write down the relevant quantities and their dimensions:

p	period [T]
m	mass [M]
g	acceleration due to gravity [L/T^2]
L	length [L]

We propose that the period is proportional to these quantities according to the following equation:

$$p = km^a g^b L^c \tag{14-34}$$

Using dimensional analysis, we can determine the exponents required to make the equation dimensionally homogeneous. We consider each dimension one at a time.

	[T]	=	[M]^a		[L/T^2]^b		[L]^c		
[M]	0	=	a	+	0	+	0	⇒	$a = 0$
[T]	1	=	0	−	$2b$	+	0	⇒	$b = -1/2$
[L]	0	=	0	+	b	+	c	⇒	$c = +1/2$

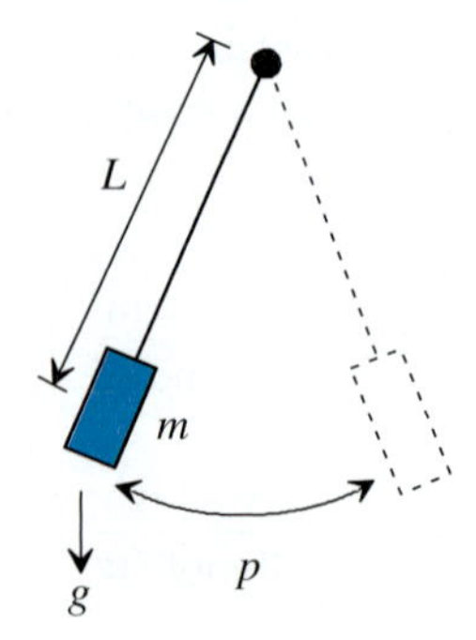

FIGURE 14.10
Pendulum.

In the first row, [M] appears only on the right side. Therefore, the exponent a must be zero. In the second row, [T] appears to the first power on the left and to the negative two power on the right. Therefore, the exponent b must be $-1/2$. In the third row, [L] does not appear on the left, so the lengths on the right must cancel. Because we have already determined that b is $-1/2$, then c must be $+1/2$.

Using dimensional analysis, we know that p must be proportional to $g^{-1/2}$ and $L^{+1/2}$:

$$p \propto g^{-1/2} L^{+1/2} = kg^{-1/2} L^{+1/2} = k\sqrt{\frac{L}{g}} \tag{14-35}$$

Although this simple analysis does not give the value of k, we have learned much about the system. Surprisingly, the period does not depend on the mass.

Dimensional analysis may be applied to fluid flow in a pipe (Figure 14.11). The pressure drop ΔP (i.e., $P_1 - P_2$) is needed to specify the pump size. The relevant quantities for this problem are

ΔP	pressure [M/LT²]
D	pipe diameter [L]
L	pipe length [L]
v_{ave}	average fluid velocity in pipe [L/T]
ρ	fluid density [M/L³]
μ	fluid dynamic viscosity [M/LT]

Dimensional analysis has shown that these quantities can be grouped into the following equation:

$$\left(\frac{\Delta P D}{2L\rho v_{ave}^2}\right) = k\left(\frac{D v_{ave} \rho}{\mu}\right)^a \tag{14-36}$$

Both terms in brackets are dimensionless. The left one is the *Fanning friction factor* f and the right one is the *Reynolds number* Re. Therefore, Equation 14-36 may be written as

$$f = k(\mathrm{Re})^a \tag{14-37}$$

Experimentation with many fluids (e.g., air, water, oil) has shown that $k = 16$ and $a = -1$ provided $\mathrm{Re} < 2300$.

We often encounter dimensionless groups in engineering because many systems are too complicated to be analyzed using mechanistic models. Instead, we write down the relevant quantities, make dimensionless groups using dimensional analysis, and then perform

FIGURE 14.11
Pressure drop of fluid flowing through a pipe.

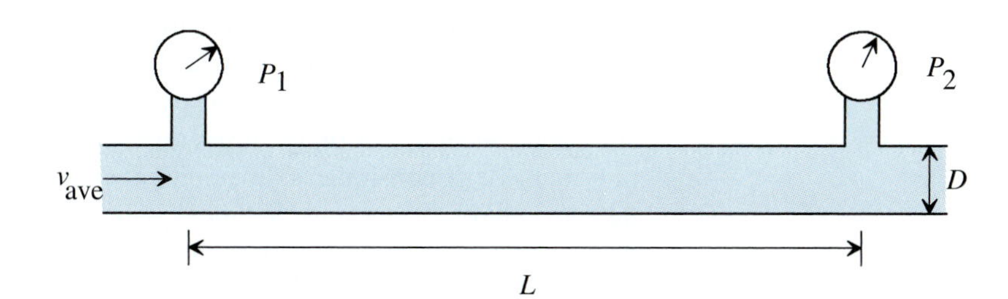

TABLE 14.4
Some common dimensionless groups (Reference)

Name	Symbol	Definition
Fanning friction factor	f	$\frac{\Delta PD}{2L\rho(v_{ave})^2}$
Reynolds number	Re	$\frac{Dv_{ave}\rho}{\mu}$
Drag coefficient	C_d	$\frac{2F}{\rho Av^2}$
Prandtl number	Pr	$\frac{C_p\mu}{k}$
Nusselt number	Nu	$\frac{hD}{k}$
Mach number	Ma	$\frac{v}{a}$

ΔP = pressure drop
ρ = density
C_p = heat capacity
v = speed
A = projected area
D = diameter
v_{ave} = average velocity
k = thermal conductivity
a = speed of sound
L = length
μ = viscosity
h = heat transfer coeff.
F = force

experiments to determine the value of the proportionality constant and the exponent(s). Table 14.4 shows some dimensionless groups that frequently occur in engineering. Many of these dimensionless groups have fundamental significance. For example, the Reynolds number gives the inertia force/viscous force.

14.10 SUMMARY

Because there are so many unit systems in use, engineers must be familiar with a wide variety of conversions. Mistakes with unit conversions are among the most common errors made by engineers.

When reporting a measurement or computed answer, it is absolutely essential to report the units along with the number. The units indicate the reference; without the reference, the number has no meaning.

When adding/subtracting quantities, the dimensions of each quantity must be identical. (Remember the old adage: You cannot add apples and oranges.) When multiplying/dividing quantities, the dimensions of each quantity are treated as though they were variables and cancel accordingly. In scientific equations, the arguments of transcendental functions must be dimensionless. In empirical equations, the arguments of transcendental functions may have dimensions, but the equation is valid only if you use the units employed to develop the equation. An equation is dimensionally homogeneous if the left-hand and right-hand sides have identical dimensions.

Why Do Golf Balls Have Dimples?

Experiments with smooth golf balls and dimpled golf balls reveal that those with dimples travel much faster and farther. As a smooth golf ball travels through the air, there is a vacuum created in its wake because the air is not able to quickly fill in the void behind the golf ball. The vacuum behind the golf ball slows it down and prevents it from traveling great distances. A dimpled golf ball also has a vacuum in its wake, but it is much less. The turbulent vortices caused by the dimples force some air behind the golf ball, thus lessening the vacuum and allowing the golf ball to travel faster and farther.

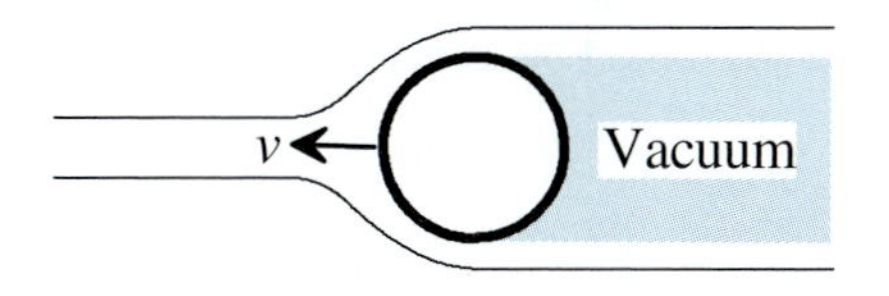

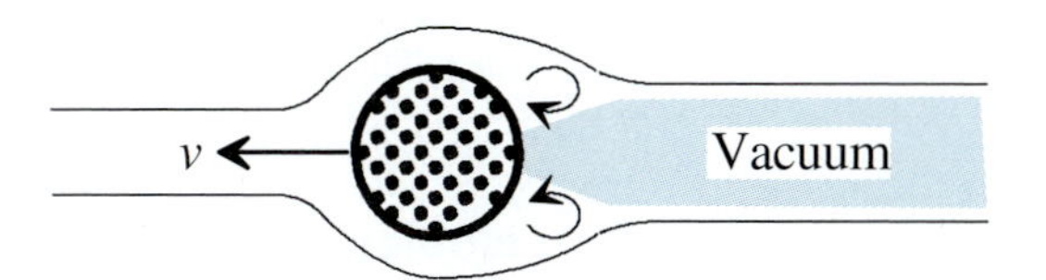

An engineer designing a golf ball is interested in the drag force as a function of velocity. Using dimensionless groups, this would be quantified by plotting the drag coefficient versus the Reynolds number. In the region of Reynolds numbers (i.e., velocities) achievable with a golf club, the drag coefficient is much less with a dimpled ball.

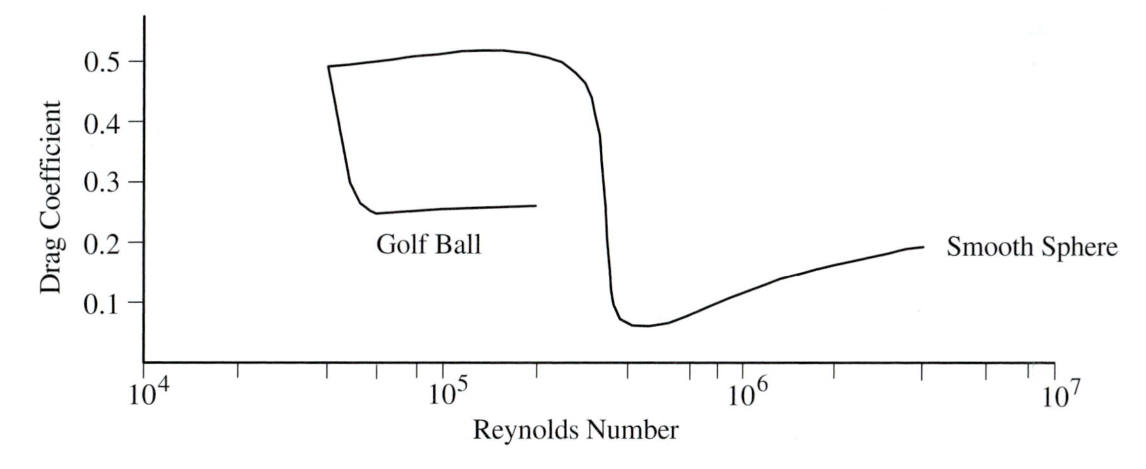

Adapted from: P. Moin and J. Kim, "Tackling Turbulence with Supercomputers," *Scientific American* 276, no. 1 (1997), pp. 62–68.

In coherent unit systems, the numerical part of any conversion factor is 1; however, in noncoherent unit systems, the numerical part of conversion factors differs from 1. For example, in coherent systems, the numerical part of g_c is 1 [e.g., 1 kg·m/(N·s^2)], but in noncoherent unit systems, the numerical part of g_c is not 1 [e.g., 32.174 lb$_m$·ft/(lb$_f$·s^2)]. There are two types of coherent unit systems: gravitational and absolute. Gravitational systems define force, length, and time; mass is derived. Absolute unit systems define mass, length, and time; force is derived. SI is an example of an absolute system.

A datum is a reference point used in measurement. When measuring pressure, the datum can be a perfect vacuum (absolute pressure), atmospheric pressure (gage pressure), or an internal reference (pressure difference). When measuring temperature, the datum can be absolute zero (Rankine and Kelvin temperature scales), the freezing point of water (Celsius temperature scale), or the coldest temperature achievable using water, ice, and salt (Fahrenheit temperature scale).

Dimensional analysis is a process of determining the functional relationship between quantities by using the dimensions of each quantity. It is commonly used in complex phenomena that defy mechanistic modeling, but are amenable to experiment. Typically, the quantities are arranged into dimensionless groups that are correlated using a power equation. The constants in the power equation are determined by experiment.

Nomenclature

A area (m^2)
a acceleration (m/s^2) or proportionality constant ($N \cdot s^c/m^c$) or speed of sound (m/s)
b exponent (dimensionless)
C_p constant-pressure heat capacity [J/(kg·K)]
D diameter (m)
E energy (J)
F force (N)
g acceleration due to gravity (m/s^2)
k heat transfer coefficient [J/(s·m·K)] or proportionality constant
H height (m)
h height (m) or heat transfer coefficient [$J/(s \cdot m^2 \cdot K)$]
L length (m)
m mass (kg)
n mole (mol)
P pressure (Pa)
p period (s)
Q heat (J)
R universal gas constant [$Pa \cdot m^3/(mol \cdot K)$]
T temperature (K)
t time (s)
V volume (m^3)
v velocity (m/s)
x spacing (m)
μ viscosity [kg/(m·s)]
ρ density (kg/m^3)

Further Readings

Eide, A. R.; R. D. Jenison; L. H. Mashaw; and L. L. Northup. *Engineering Fundamentals and Problem Solving.* 3rd ed. New York: McGraw-Hill, 1997.

Felder, R. M., and R. W. Rousseau. *Elementary Principles of Chemical Processes.* 2nd ed. New York: Wiley, 1986.

Jerrard, H. G., and D. B. McNeill. *A Dictionary of Scientific Units.* Englewood, NJ: Franklin Publishing, 1964.

PROBLEMS

Note: These problems are solved using the unit conversions in Appendix A.

14.1 Convert the following numbers to SI:

(a) 56.8 in
(b) 1 year
(c) 34.8 atm
(d) 8.3 Btu/min
(e) 38.96 ft/min
(f) 98.6 furlongs per fortnight
(g) 34.5 in·pdl
(h) 15.8 gal/min

14.2 Perform the following conversions:

(a) 34,589 Pa to psia
(b) 34.6 m/s to mph
(c) 89.6 L to m^3
(d) 68.4 in^3 to ft^3
(e) 964 slug to lb_m
(f) 569 pdl to lb_f
(g) 56.8 glug to g
(h) 78.5 mug to kg
(i) 358 atm·L to Btu
(j) 1.2 steradians to square degrees

14.3 Convert the following pressures to absolute pressure. Report the answer in the given units. You may assume that the atmospheric pressure at the time of the measurement was a standard atmosphere.

(a) 8.56 psig
(b) 18 inches of mercury vacuum
(c) 10 atm gage pressure
(d) 15.6 inches of water gage pressure

14.4 Convert the following temperatures to the Kelvin temperature scale:

(a) 259.6°R
(b) 145°C
(c) 98.6°F

14.5 Convert the following temperature intervals to Kelvin.

(a) 58.3 C°
(b) 89.7 F°
(c) 4.5 R°

14.6 A number of objects are taken to the moon, where the acceleration due to gravity is about 1/6 of earth's. Complete the following table:

Item	Mass on Earth	Mass on Moon	Weight on Moon	Weight on Moon
Paper clip	1.2 g	g	g_f	dyne
Can of Coke	0.56 lb_m	lb_m	lb_f	pdl
Hammer	1.3 kg	kg	kg_f	N

14.7 Are the following equations dimensionally homogeneous? (Show how you made your conclusion.)

(a) $F = PT$ — F = force, P = pressure, T = temperature
(b) $E = 0.5mv^2$ — E = energy, m = mass, v = velocity
(c) $p = mv$ — p = momentum, m = mass, v = velocity
(d) $E = C_p/P$ — E = energy, C_p = heat capacity, P = pressure
(e) $V = IR$ — V = voltage, I = current, R = resistance
(f) $P = I^2R$ — P = power, I = current, R = resistance

14.8 Compute the ideal gas constant R using the following units of measure:

P = inches of mercury
V = gal
n = short-ton moles
T = Rankine

14.9 A weir is a slot in an open channel with water flowing over it (Figure 14.12). For a weir with a rectangular shape, the following formula can be derived:

$$Q = 5.35Lh^{3/2}$$

where

Q = water flow rate, ft^3/s
L = length of weir, ft
h = height of liquid above weir gate, ft

Do the following:

(a) Determine the units associated with the constant that are needed to make the equation dimensionally homogeneous.
(b) Determine a new constant that allows the following units to be used:

Q = water flow rate, gal/min
L = length of weir, ft
h = height of liquid above weir gate, in

14.10 Compute the Reynolds number defined in Table 14.4.

(a) Use the following information:

v_{ave} = 0.3 ft/min

FIGURE 14.12
Water flowing over a weir.

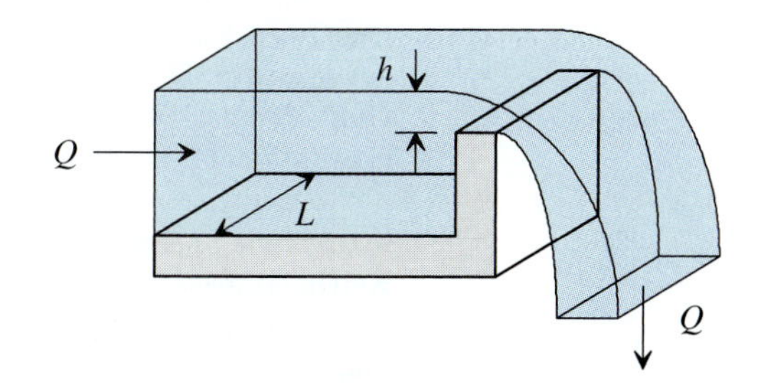

$D = 2$ in
$\rho = 62.3\ \text{lb}_m/\text{ft}^3$
$\mu = 2.42\ \text{lb}_m/(\text{ft}\cdot\text{h})$

(b) Use the equivalent SI units to the data given in Part (a).

(c) Use the following information:

$v_{ave} = 0.3$ ft/min
$D = 2$ in
$\rho = 62.3\ \text{lb}_m/\text{ft}^3$
$\mu = 2.09 \times 10^{-5}\ \text{lb}_f\cdot\text{s/ft}^2$

14.11 Determine the pressure difference between the following two pressures (use the units given). You may assume that the atmospheric pressure was one standard atmosphere when the pressure measurements were made.

(a) $P_1 = 1.5$ atm (absolute), $P_2 = 1.3$ atm (absolute)
(b) $P_1 = 1.5$ atm (absolute), $P_2 = 0.3$ atm (gage)
(c) $P_1 = 1.24$ psig, $P_2 = 1.00$ psig
(d) $P_1 = 10.3$ psig, $P_2 = 16.3$ psia

14.12 Decide whether you would use a mercury manometer (at 0°C) or a water manometer (at 4°C) to measure the following pressures. (*Note:* Each manometer will use a perfect vacuum as a reference. At 0°C, mercury has a negligible vapor pressure. At 4°C, water has a vapor pressure of 813 Pa.)

(a) 34 Pa (absolute)
(b) 1.45 psig
(c) 0.75 atm (absolute)
(d) 1.5 millibar (absolute)

14.13 Write a spreadsheet that converts lb_m to kg, ft to m, in to m, ft^2 to m^2, gal to m^3, and ft^3 to m^3.

14.14 Using Equation 14-37, write a computer program that calculates the pressure drop (in psi) of water flowing through a 1-mile pipe. Use the water properties and pipe diameter given in Problem 14.10. Use v_{ave} increments of 0.1 ft/min until the Reynolds number exceeds 2300, the upper limit where the equation is valid.

14.15 Perform the task described in Problem 14.14 using a spreadsheet. In addition, plot the pressure drop (in psi) versus the average velocity (in ft/min).

Glossary

absolute pressure The pressure relative to a perfect vacuum.

absolute systems A coherent system that defines mass, length, and time; force is a derived quantity.

Celsius temperature scale A temperature scale in which the two reference points are the melting point of ice (0°C) and the boiling point of water (100°C).

coherent unit system A unit system in which the numerical part of any conversion factor is 1.

conversion factor A numerical factor used to multiply or divide a quantity when converting one unit system to another.

datum A reference point used when making a measurement.

differential pressure The difference between two pressures.

dimensional analysis A process of determining the functional relationship between quantities by using the dimensions of each quantity.

dimensional homogeneity The dimensions on the left-hand and right-hand sides of the equation must be the same.

empirical constant The value of the constant is determined by experiment.

Fahrenheit temperature scale A temperature scale in which the two reference points are the freezing point of water (32°F) and the boiling point of water (212°F).

gage pressure The pressure relative to atmospheric air.

gas pressure The pressure resulting from the impact of gas molecules on the container wall.

glug A derived unit for mass in the CGS gravitational system; rarely used.

gravitational systems A coherent system that defines force, length, and time. Mass is a derived quantity.

hydrostatic pressure The pressure resulting from the weight of liquid or gas.

Kelvin temperature scale A thermodynamic temperature scale in which 0 K is defined as absolute zero and the temperature interval is the same as the Celsius scale.

mug A derived unit for mass in the MKS gravitational system; rarely used.

noncoherent unit system A unit system in which the numerical part of conversion factors differs from 1.
poundal A unit of force in the FPS absolute system.
pressure A force exerted on an area.
Rankine temperature scale A thermodynamic temperature scale in which 0°R is defined as absolute zero and the temperature interval is the same as the Fahrenheit scale.
slug A derived unit for mass in the USCS gravitational systems; widely used in engineering.
temperature interval The difference between two temperatures.
temperature scale Formed by placing two reference points on the mercury-in-glass thermometer and evenly subdividing the points into temperature intervals.
thermodynamic temperature scales Temperature scale that is linearly proportional to the thermal energy of an ideal gas; also called absolute temperature scales.
transcendental functions Mathematical functions represented by an infinite series.

Mathematical Prerequisites

- Algebra (Appendix E, Mathematics Supplement)
- Vectors (Appendix I, Mathematics Supplement)
- Trigonometry (Appendix I, Mathematics Supplement)
- Calculus (Appendix L , Mathematics Supplement)

CHAPTER 15

Introduction to Statics and Dynamics

Statics and dynamics are central to almost every engineering profession. **Statics** is the study of forces and moments on stationary objects (e.g., buildings and bridges), whereas **dynamics** is the study of forces and moments on mobile objects (e.g., automobiles and airplanes).[1] Dynamics can also be applied to objects that are normally stationary, but that move under unusual circumstances (buildings in an earthquake, bridges in high winds).

The principles of statics and dynamics can be applied to particles and to rigid bodies. A **particle** is an idealized object with mass but no dimensions (i.e., it has no volume). A **rigid body** is an idealized object with both mass and dimensions, with the restriction that the components of the object do not move with respect to each other. A long, slender rigid body may be modeled as having only the dimension of length, and a thin, planar rigid body may be modeled as having only two dimensions. Of course, real objects exist in three dimensions (i.e., they have volume), so a complete model of a rigid body would include all three dimensions.

Particles can have forces acting on them, but because particles have no volume, it is impossible to apply a moment. (A **moment** is a right-angle force exerted at a distance from a central pivot point. Because a particle has no distance for the force to be exerted upon, a moment is impossible.) Rigid bodies can have both forces and moments act on them.

With a static object, the sum of all forces **F** and moments **M** is zero. According to Newton's first law, such an object with no forces or moments maintains a constant velocity, in this case, a zero velocity. With a dynamic object, the sum of all forces **F** and/or moments **M** is not zero. According to Newton's second law, such an object with net forces or net moments applied to it will accelerate. The table summarizes.

1. A *moment* is a twisting force and is identical to *torque*. The word *moment* evolved within the engineering community, whereas the word *torque* evolved within the physics community. Here we deal primarily with engineering applications, so we adopt the engineering nomenclature.

	Statics	Dynamics
Particle	$\Sigma F = 0$	$\Sigma F \neq 0$
Rigid body	$\Sigma F = 0$	$\Sigma F \neq 0$
	$\Sigma M = 0$	$\Sigma M \neq 0$

15.1 STATICS OF PARTICLES

Because a particle is confined to a point with zero dimensions (no volume), it is impossible to apply a moment to it. Therefore, a static particle need only satisfy the condition

$$\Sigma \mathbf{F} = 0 \tag{15-1}$$

This simple equation states that the sum of the force vectors applied to the particle must be zero.

Figure 15.1 shows an arbitrary three-dimensional force vector. Force **F** can be described as the sum of the three unit vectors **i**, **j**, and **k**,

$$\mathbf{F} = F_x\mathbf{i} + F_y\mathbf{j} + F_z\mathbf{k} \tag{15-2}$$

where F_x is the x-component of force, F_y is the y-component of force, and F_z is the z-component of force. This particular force vector **F** is the sum of the three components 7**i**, 4**j**, and 5**k**,

$$\mathbf{F} = 7\mathbf{i} + 4\mathbf{j} + 5\mathbf{k}$$

(Prove this to yourself by summing 7**i** and 4**j**; i.e., place the tail of 4**j** to the head of 7**i** and draw the resultant vector. Then place the tail of 5**k** to the head of the resultant vector. The sum of these two vectors is **F**.)

Using component vectors, we can rewrite Equation 15-1 as

$$\Sigma \mathbf{F} = 0 = (\Sigma F_x)\mathbf{i} + (\Sigma F_y)\mathbf{j} + (\Sigma F_z)\mathbf{k} = 0\mathbf{i} + 0\mathbf{j} + 0\mathbf{k} \tag{15-3}$$

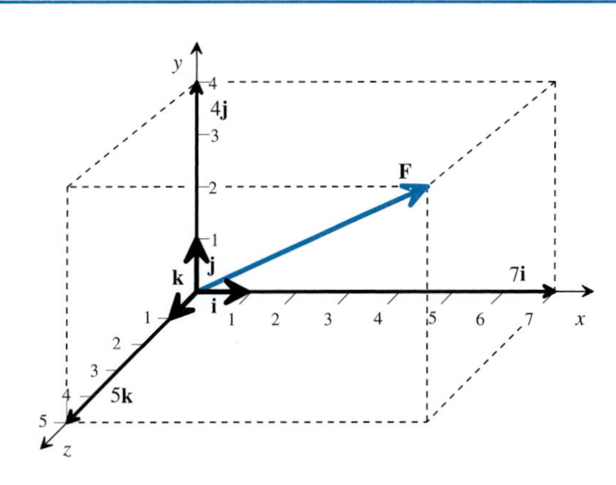

FIGURE 15.1
The force vector **F** can be described as the sum of three component vectors 7**i**, 4**j**, and 5**k**.

which states that the sum of the x-, y-, and z-components of all forces acting on the particle must be zero for the particle to be static.

In actuality, no real object is a particle—real objects have dimensions (i.e., volume). However, for many systems of interest, treating them as an idealized particle is sufficiently accurate to be useful. A drawing of the system with forces acting on it is called a **free body diagram.** If the object is treated as a particle, then the forces act on a point.

EXAMPLE 15.1

Problem Statement: To keep food away from bears, a camper suspends a food-filled, 32-lb_m knapsack in a tree with the ropes shown in Figure 15.2. Calculate the force on each rope.

Solution: The position vector for each rope is

$$\mathbf{R}_1 = (2 - 4)\mathbf{i} + (10 - 6)\mathbf{j} = -2\mathbf{i} + 4\mathbf{j}$$

$$\mathbf{R}_2 = (7 - 4)\mathbf{i} + (8 - 6)\mathbf{j} = 3\mathbf{i} + 2\mathbf{j}$$

The direction of each force vector must be the same as each rope direction; however, the magnitude differs. Mathematically, we can multiply each position vector by an appropriate scaling constant that converts the position vector (feet) into a force vector (lb_f).

$$\mathbf{F}_1 = a\mathbf{R}_1 = -2a\mathbf{i} + 4a\mathbf{j}$$

$$\mathbf{F}_2 = b\mathbf{R}_2 = 3b\mathbf{i} + 2b\mathbf{j}$$

The magnitude of the gravity force is

$$F_{gravity} = mg = (32\ lb_m)\left(32.174\ \frac{ft}{s^2}\right) \times \frac{lb_f \cdot s^2}{32.174\ lb_m \cdot ft} = 32\ lb_f$$

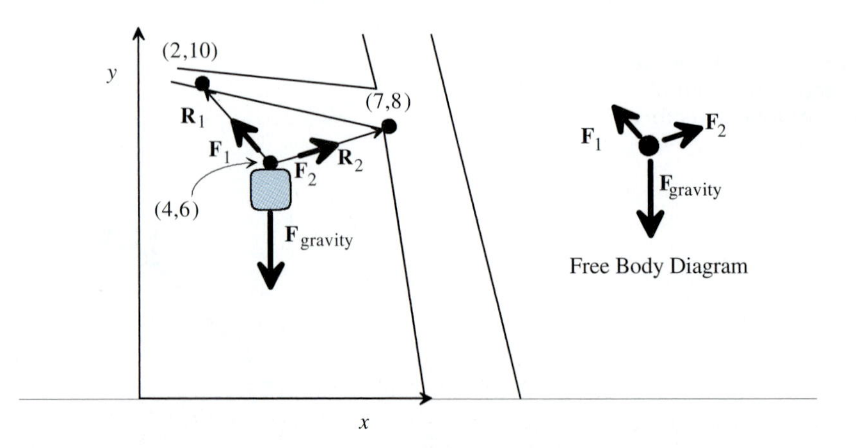

FIGURE 15.2
A food-filled, 32-lb_m knapsack suspended in a tree. (*Note:* All positions given in feet.)

The direction of the gravity force is directly downward, so the gravity force vector $\mathbf{F}_{\text{gravity}}$ is

$$\mathbf{F}_{\text{gravity}} = 0\mathbf{i} - 32\mathbf{j}$$

According to Equation 15-3, the sum of the vector components in each direction must be zero. Therefore,

$$\Sigma F_x = -2a + 3b + 0 = 0$$

$$\Sigma F_y = 4a + 2b - 32 = 0$$

These two equations must be solved simultaneously. We can solve the first equation for *a*,

$$a = \frac{3}{2}b$$

which we substitute into the second equation:

$$4\left(\frac{3}{2}b\right) + 2b - 32 = 0$$

$$6b + 2b - 32 = 0$$

$$8b = 32$$

$$b = \frac{32}{8} = 4$$

Knowing *b*, we can determine *a*:

$$a = \frac{3}{2}(4) = 6$$

Now we can calculate the force vectors as

$$\mathbf{F}_1 = -2(6)\mathbf{i} + 4(6)\mathbf{j} = -12\mathbf{i} + 24\mathbf{j}$$

$$\mathbf{F}_2 = 3(4)\mathbf{i} + 2(4)\mathbf{j} = 12\mathbf{i} + 8\mathbf{j}$$

The magnitude of each force vector is

$$F_1 = |\mathbf{F}_1| = \sqrt{(-12)^2 + (24)^2} = 26.8 \text{ lb}_f$$

$$F_2 = |\mathbf{F}_2| = \sqrt{(12)^2 + (8)^2} = 14.4 \text{ lb}_f$$

EXAMPLE 15.2

Problem Statement: A sports stadium has a 1000-kg scoreboard supported by four cables as shown in Figure 15.3. What is the force in each support cable? The cables will be designed with a safety factor of 3, meaning they will be 3 times stronger than necessary. What is the design specification for each cable?

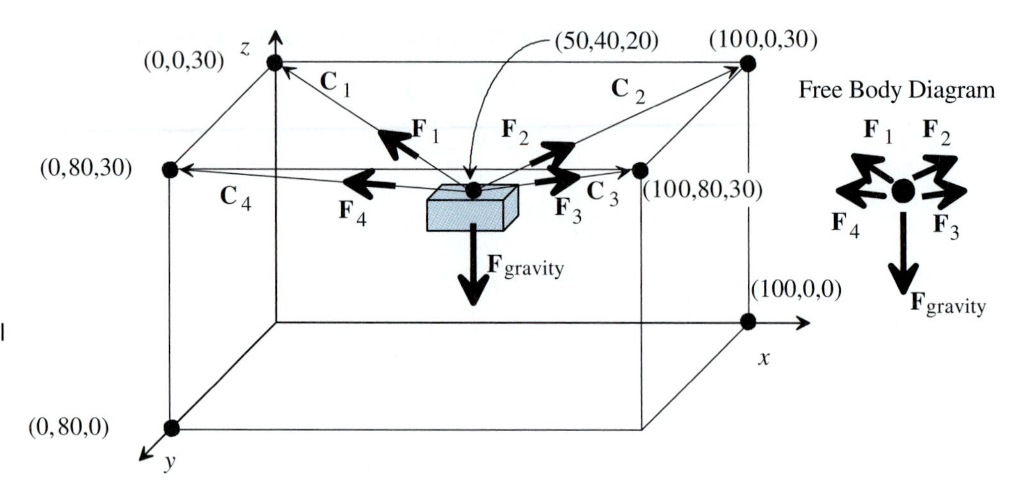

FIGURE 15.3
Forces acting on a 1000-kg sports scoreboard. (*Note:* All positions given in meters. The coordinate system chosen here uses the *x-y* plane for the floor and *z* to represent height.)

Solution: Each supporting cable can be described by a position vector

$$\mathbf{C}_1 = (0 - 50)\mathbf{i} + (0 - 40)\mathbf{j} + (30 - 20)\mathbf{k} = -50\mathbf{i} - 40\mathbf{j} + 10\mathbf{k}$$

$$\mathbf{C}_2 = (100 - 50)\mathbf{i} + (0 - 40)\mathbf{j} + (30 - 20)\mathbf{k} = 50\mathbf{i} - 40\mathbf{j} + 10\mathbf{k}$$

$$\mathbf{C}_3 = (100 - 50)\mathbf{i} + (80 - 40)\mathbf{j} + (30 - 20)\mathbf{k} = 50\mathbf{i} + 40\mathbf{j} + 10\mathbf{k}$$

$$\mathbf{C}_4 = (0 - 50)\mathbf{i} + (80 - 40)\mathbf{j} + (30 - 20)\mathbf{k} = -50\mathbf{i} + 40\mathbf{j} + 10\mathbf{k}$$

The direction of each force vector must be the same as each cable direction; however, the magnitude differs. Mathematically, we can multiply each position vector by an appropriate scaling constant that converts the position vector (meters) into a force vector (newtons). Because of symmetry, we know this scaling constant a (with units N/m) must be the same for each vector.

$$\mathbf{F}_1 = a\mathbf{C}_1 = -50a\mathbf{i} - 40a\mathbf{j} + 10a\mathbf{k}$$

$$\mathbf{F}_2 = a\mathbf{C}_2 = 50a\mathbf{i} - 40a\mathbf{j} + 10a\mathbf{k}$$

$$\mathbf{F}_3 = a\mathbf{C}_3 = 50a\mathbf{i} + 40a\mathbf{j} + 10a\mathbf{k}$$

$$\mathbf{F}_4 = a\mathbf{C}_4 = -50a\mathbf{i} + 40a\mathbf{j} + 10a\mathbf{k}$$

The magnitude of the gravity force acting on the scoreboard is

$$F_{\text{gravity}} = mg = (1000\text{ kg})\left(9.8\,\frac{\text{m}}{\text{s}^2}\right) \times \frac{1\text{ N}}{\dfrac{\text{kg}\cdot\text{m}}{\text{s}^2}} = 9800\text{ N}$$

This force operates only in the downward direction, so it can be described by the vector

$$F_{\text{gravity}} = 0\mathbf{i} + 0\mathbf{j} - 9800\mathbf{k}$$

According to Equation 15-3, the components of force in each direction must sum to zero:

$$\Sigma F_x = -50a + 50a + 50a - 50a + 0 = 0$$

$$\Sigma F_y = -40a - 40a + 40a + 40a + 0 = 0$$

$$\Sigma F_z = 10a + 10a + 10a + 10a - 9800 = 0$$

The first two equations give no information about the value of a. However, the third equation can be solved for a:

$$10a + 10a + 10a + 10a = 9800$$

$$40a = 9800$$

$$a = \frac{9800}{40} = 245\,\frac{\text{N}}{\text{m}}$$

Now that a is known, the force vectors are known:

$$\mathbf{F}_1 = -50(245)\mathbf{i} - 40(245)\mathbf{j} + 10(245)\mathbf{k} = -12{,}250\mathbf{i} - 9800\mathbf{j} + 2450\mathbf{k}$$

$$\mathbf{F}_2 = 50(245)\mathbf{i} - 40(245)\mathbf{j} + 10(245)\mathbf{k} = 12{,}250\mathbf{i} - 9800\mathbf{j} + 2450\mathbf{k}$$

$$\mathbf{F}_3 = 50(245)\mathbf{i} + 40(245)\mathbf{j} + 10(245)\mathbf{k} = 12{,}250\mathbf{i} + 9800\mathbf{j} + 2450\mathbf{k}$$

$$\mathbf{F}_4 = -50(245)\mathbf{i} + 40(245)\mathbf{j} + 10(245)\mathbf{k} = -12{,}250\mathbf{i} + 9800\mathbf{j} + 2450\mathbf{k}$$

The magnitude of each vector can now be determined as

$$F_1 = |\mathbf{F}_1| = \sqrt{(-12{,}250)^2 + (-9800)^2 + (2450)^2} = 15{,}900 \text{ N}$$

$$F_2 = |\mathbf{F}_2| = \sqrt{(12{,}250)^2 + (-9800)^2 + (2450)^2} = 15{,}900 \text{ N}$$

$$F_3 = |\mathbf{F}_3| = \sqrt{(12{,}250)^2 + (9800)^2 + (2450)^2} = 15{,}900 \text{ N}$$

$$F_4 = |\mathbf{F}_4| = \sqrt{(-12{,}250)^2 + (9800)^2 + (2450)^2} = 15{,}900 \text{ N}$$

(Notice that the force in each cable is **not** the gravity force divided by 4.) Because each cable will be designed with a safety factor of 3, the specification will call for each cable to support 3(15,900 N) = 47,700 N.

15.2 STATICS OF RIGID BODIES

Rigid bodies have mass and dimensions, unlike particles that have mass but no dimensions. The points in rigid bodies do not move with respect to each other. A rigid body is an engineering idealization used to analyze many common bodies such as roofs, chairs, tables, and bridges.

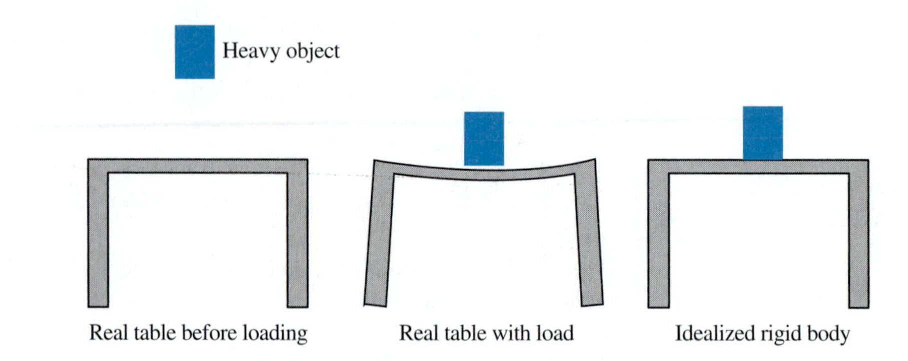

FIGURE 15.4
Illustration of a rigid body.

Figure 15.4 shows a table with a heavy object placed upon it. Originally, the table is perfectly flat, but once the object is placed upon it, it bends in the middle and the legs bow outward. The amount of bending and bowing is proportional to the mass of the object. We can visualize the table as having a spring-like surface in which the spring deformation is proportional to the mass of the object. The spring will deform until the force exerted by the spring equals the force exerted by the object. Finally, the table deformation stops, resulting in an equilibrium in which there is no further movement. In reality, any object, no matter how light, causes the table surface to deform. Even a pencil placed upon a table will cause it to bend. However, for practical purposes, the deformation is negligible so we can mathematically model the table as a "rigid body."

When modeling a real body as a rigid body, the engineer must first define the system to be studied; that is, he must specify which subset of the universe he will analyze. A drawing of the system with the forces and moments that act on it is called a free body diagram. Figure 15.5 shows a free body diagram of a child's seesaw.

A rigid body that is static has no unbalanced forces; therefore,

$$\Sigma \mathbf{F} = 0 \tag{15-4}$$

In terms of component vectors, Equation 15-4 can be written as

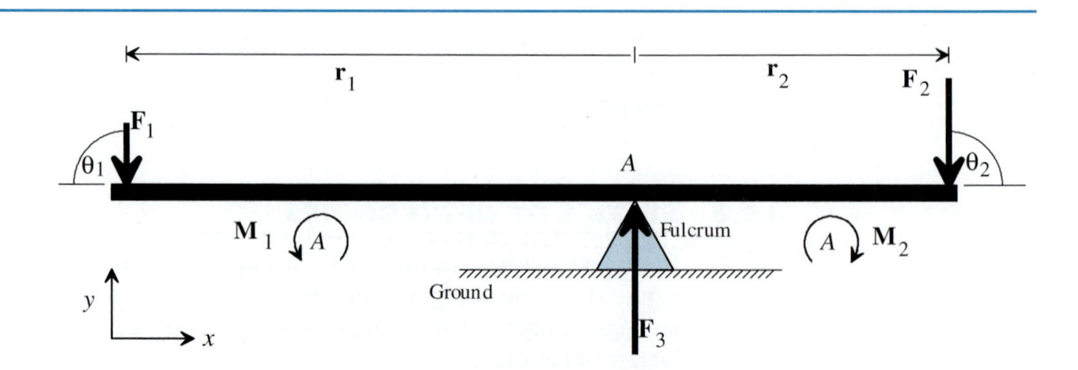

FIGURE 15.5
Free body diagram of a child's seesaw.

When a load is placed on an object, the object deforms. On a trampoline, the deformation is large and is readily apparent. On a rigid body, the deformation is small and is idealized as being zero.

$$\Sigma \mathbf{F} = (\Sigma F_x)\mathbf{i} + (\Sigma F_y)\mathbf{j} + (\Sigma F_z)\mathbf{k} = 0\mathbf{i} + 0\mathbf{j} + 0\mathbf{k} \tag{15-5}$$

A rigid body that is static also has no unbalanced moments; therefore

$$\Sigma \mathbf{M} = 0 \tag{15-6}$$

The magnitude of the moment is

$$M = rF \sin \theta \tag{15-7}$$

where r is the distance from the applied force to the pivot point A, and angle θ is defined in Figure 15.6. If the force vector $\mathbf{F}$ is at a right angle to the distance vector $\mathbf{r}$, then Equation 15-7 reduces to

$$M = rF \qquad \text{(right-angle force)} \tag{15-8}$$

According to the most popular convention, a counterclockwise moment is positive and a clockwise moment is negative.[2]

2. This convention is consistent with specifying the torque vector shown in Chapter 21, "Accounting for Angular Momentum," as positive when it points out of the page and negative when it points into the page.

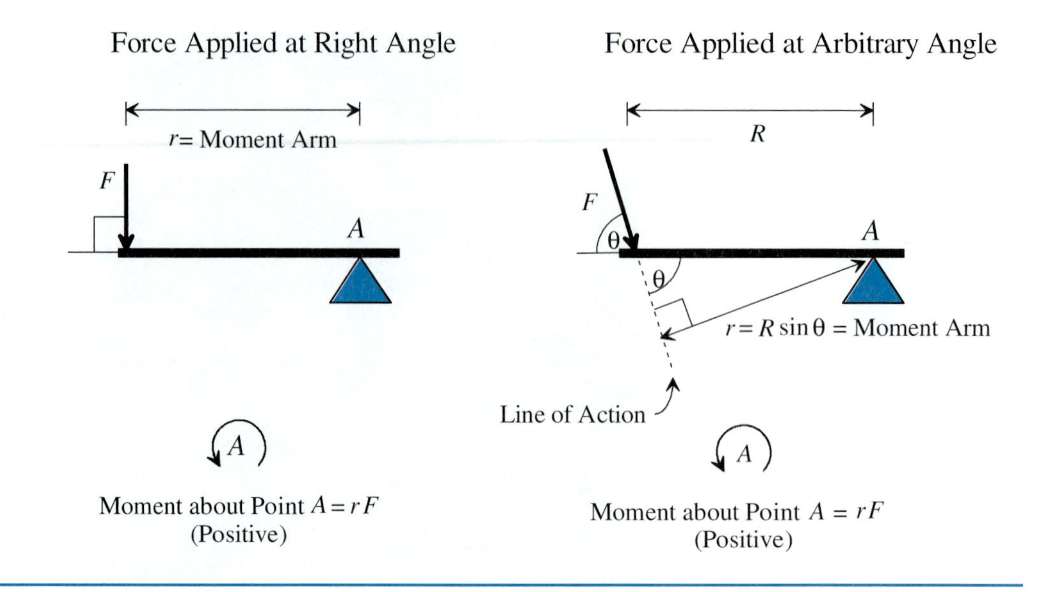

FIGURE 15.6
Moments and moment arms.

EXAMPLE 15.3

Problem Statement: A 50-lb_m child and a 75-lb_m child wish to play on a 10-ft-long seesaw (Figure 15.5). What upward force F_3 must be supplied by the **fulcrum**? What is the required length of r_1 and r_2 for the two children to balance each other if the light child sits on the left and the heavy child sits on the right? (Assume the seesaw has negligible mass.)

Solution:

$$F_1 = m_1 g = 50\ \text{lb}_\text{m}\left(32.174\,\frac{\text{ft}}{\text{s}^2}\right) \times \frac{\text{lb}_\text{f}\cdot\text{s}^2}{32.174\ \text{lb}_\text{m}\cdot\text{ft}} = 50\ \text{lb}_\text{f}$$

$$F_2 = m_2 g = 75\ \text{lb}_\text{m}\left(32.174\,\frac{\text{ft}}{\text{s}^2}\right) \times \frac{\text{lb}_\text{f}\cdot\text{s}^2}{32.174\ \text{lb}_\text{m}\cdot\text{ft}} = 75\ \text{lb}_\text{f}$$

$$\mathbf{F}_1 = F_{x,1}\mathbf{i} + F_{y,1}\mathbf{j} = 0\mathbf{i} - 50\mathbf{j}$$

$$\mathbf{F}_2 = F_{x,2}\mathbf{i} + F_{y,2}\mathbf{j} = 0\mathbf{i} - 75\mathbf{j}$$

$$\mathbf{F}_3 = F_{x,3}\mathbf{i} + F_{y,3}\mathbf{j} = F_{x,3}\mathbf{i} + F_3\mathbf{j}$$

$$\sum_{i=1}^{3} F_{x,i} = F_{x,1} + F_{x,2} + F_{x,3} = 0 + 0 + F_{x,3} = 0$$

$$F_{x,3} = 0$$

$$\sum_{i=1}^{3} F_{y,i} = F_{y,1} + F_{y,2} + F_{y,3} = (-50) + (-75) + F_3 = 0$$

$$F_3 = 50 + 75 = 125\ \text{lb}_\text{f}$$

The fulcrum must exert an upward force of 125 lb_f to resist the weight of the two children. The moments may be calculated relative to the fulcrum point A.

$$\mathbf{M}_1 = r_1 F_1 \qquad \text{(counterclockwise rotation)}$$

$$\mathbf{M}_2 = r_2 F_2 \qquad \text{(counterclockwise rotation)}$$

$$\mathbf{M}_1 + \mathbf{M}_2 = r_1 F_1 - r_2 F_2 = 0$$

$$L = r_1 + r_2 \Rightarrow r_2 = L - r_1 \qquad \text{(geometric constraint)}$$

$$r_1 F_1 = r_2 F_2 = (L - r_1) F_2 = L F_2 - r_1 F_2$$

$$r_1 F_1 + r_1 F_2 = L F_2$$

$$r_1 (F_1 + F_2) = L F_2$$

$$r_1 = \frac{L F_2}{F_1 + F_2} = \frac{(10 \text{ ft})(75 \text{ lb}_f)}{50 \text{ lb}_f + 75 \text{ lb}_f} = 6 \text{ ft}$$

$$r_2 = L - r_1 = 10 \text{ ft} - 6 \text{ ft} = 4 \text{ ft}$$

If you have been observant, you will have noticed that we have always been very careful to define the point about which the moment was calculated. In the case of the seesaw, we picked the fulcrum point A to be our reference point because this is the point about which the seesaw will rotate. However, there was no **requirement** that we choose the fulcrum point as our reference point when calculating the moments. We are allowed to choose **any** point we wish when we perform the calculations. The only requirement is that once we have selected a reference point, it cannot change in the middle of the calculations. You may wish to prove this to yourself by repeating the seesaw calculation using another reference point. Although the moment exerted by each force will depend upon the reference point, the net moment (the sum of all the individual moments) will equal zero in all cases, because the body is in static equilibrium.

15.2.1 Couples

A **couple** exists when two forces have the same magnitude and act parallel to each other, but in opposite directions (Figure 15.7). If you place your hands on opposite sides of an automobile steering wheel and apply pure rotation to the wheel, the twisting forces acting on the steering wheel would be described as a couple. Other examples of couples are a corkscrew twisting into the cork of a wine bottle and a screwdriver twisting a screw. In all these cases, a pure twisting force is applied, causing the body (steering wheel, corkscrew, screwdriver) to rotate, but not move from side to side. Within the engineering community, the word **torque** is used to mean a couple.

Unlike ordinary moments, such as those calculated in the seesaw example, **couples do not depend upon the reference point.** To see this, refer to Figure 15.7, which shows two forces $\mathbf{F}_1$ and $\mathbf{F}_2$ acting on the body, in this case, a ruler. Force $\mathbf{F}_1$ acts at the 4-in mark and

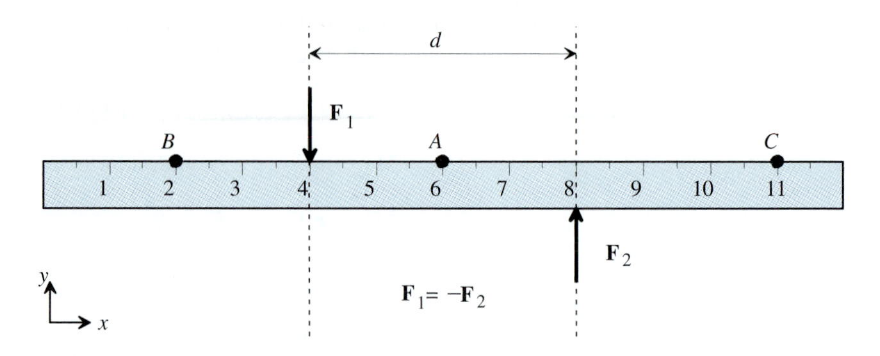

FIGURE 15.7
A couple.

force $\mathbf{F}_2$ acts at the 8-in mark. Because these forces have equal magnitudes, but act in opposite directions, the ruler will not linearly accelerate. However, there is a net moment on the ruler, which will cause it to rotate. When we calculate the net moment using Point *A* as a reference, we notice that each force will apply a counterclockwise (positive) rotation, and

$$\mathbf{M}_{\text{net}} = \mathbf{M}_1 + \mathbf{M}_2 = (6 - 4)F_1 + (8 - 6)F_2 = 2F_1 + 2F_2 = 2F_1 + 2(F_1) = 4F_1$$

Using Point *B* as a reference, we see that force $\mathbf{F}_1$ applies a clockwise (negative) rotation and force $\mathbf{F}_2$ applies a counterclockwise (positive) rotation.

$$\begin{aligned}\mathbf{M}_{\text{net}} &= \mathbf{M}_1 + \mathbf{M}_2 = -(4 - 2)F_1 + (8 - 2)F_2\\ &= -2F_1 + 6F_2 = 2F_1 + 6(F_1) = 4F_1\end{aligned}$$

Using Point *C* as a reference, we see that force $\mathbf{F}_1$ applies a counterclockwise (positive) rotation and force $\mathbf{F}_2$ applies a clockwise (negative) rotation.

$$\begin{aligned}\mathbf{M}_{\text{net}} &= \mathbf{M}_1 + \mathbf{M}_2 = (11 - 4)F_1 - (11 - 8)F_2\\ &= 7F_1 - 3F_2 = 7F_1 - 3(F_1) = 4F_1\end{aligned}$$

Notice that the net moment was the same regardless of the reference point. If the magnitudes of forces $\mathbf{F}_1$ and $\mathbf{F}_2$ had been different (as they are for most moments), we would have obtained a different answer for each reference point. The fact that couples are independent of the reference point is a particularly useful feature. In this book, we will use the following notations for couples and noncouple moments:

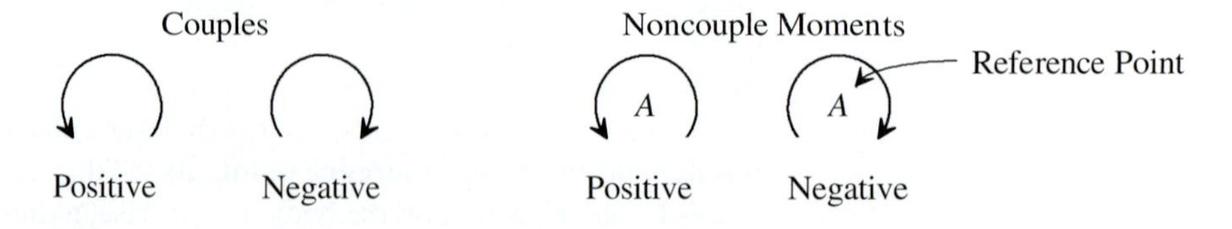

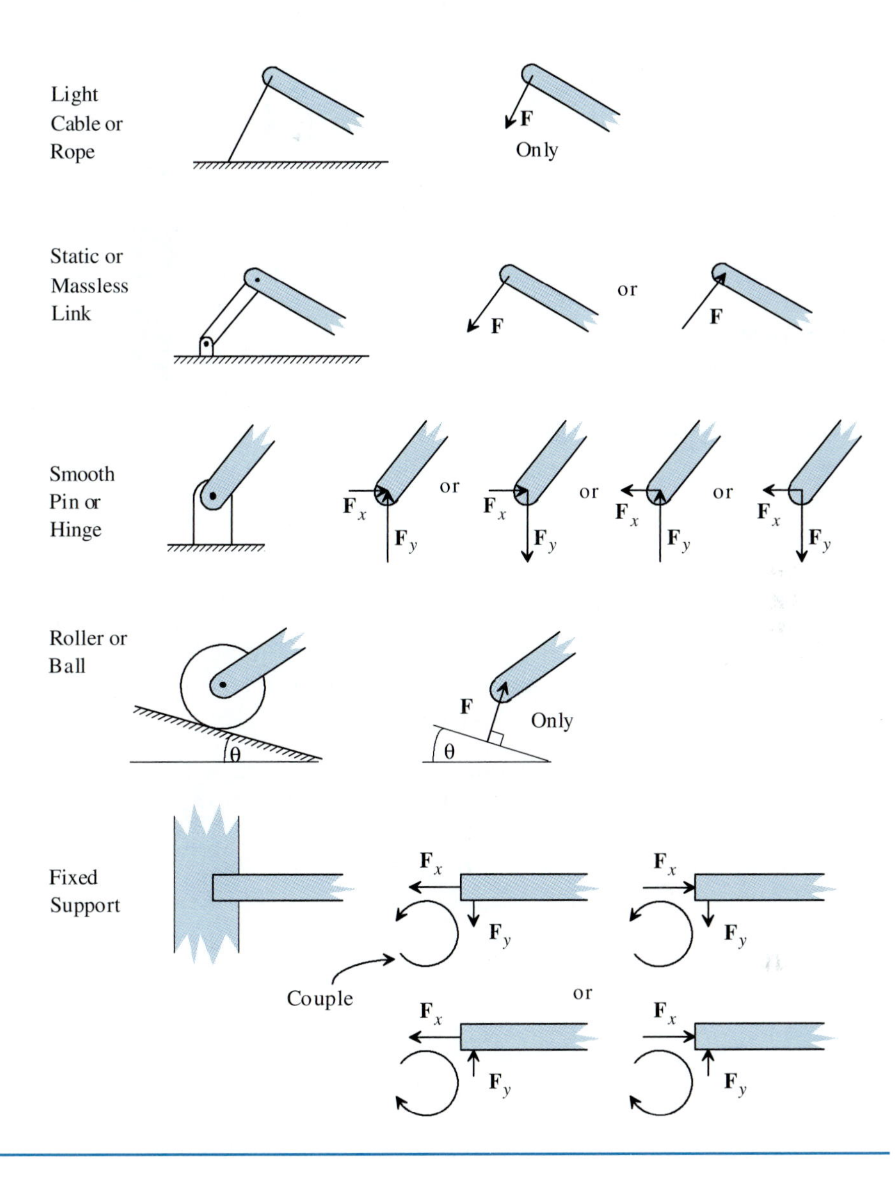

FIGURE 15.8
Free body connections.

Because the couple is independent of the reference point, no letter appears at the center of the circular arrow. In contrast, noncouple moments do depend on the reference point, so the reference point is indicated at the center of the moment.

15.2.2 Free Body Connections

In the seesaw example, it was clear how the forces interacted with the free body, because all forces were applied at right angles to the free body. However, in other situations, applied forces may interact in much more complex ways, depending on how the free body is mounted (Figure 15.8).

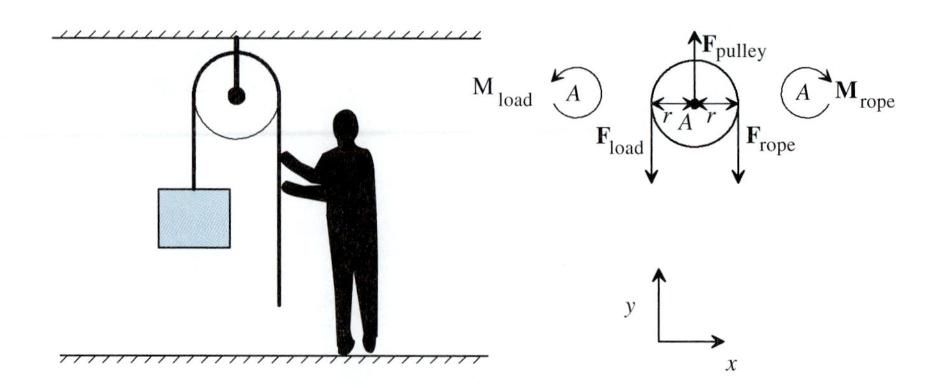

FIGURE 15.9
A single pulley used to lift a 100-lb_m load.

EXAMPLE 15.4

Problem Statement: The pulley shown in Figure 15.9 lifts a 100-lb_m mass. What is the required force on the rope to hold the mass stationary above the earth? What is the force on the pulley?

Solution:

$$\Sigma \mathbf{M} = \mathbf{M}_{\text{load}} + \mathbf{M}_{\text{rope}} = rF_{\text{load}} - rF_{\text{rope}} = 0$$

$$rF_{\text{rope}} = rF_{\text{load}}$$

$$F_{\text{rope}} = F_{\text{load}}$$

$$F_{\text{load}} = mg = 100 \text{ lb}_m \left(32.174 \frac{\text{ft}}{\text{s}^2}\right) \times \frac{\text{lb}_f\text{·s}^2}{32.174 \text{ lb}_m\text{·ft}} = 100 \text{ lb}_f$$

$$F_{\text{rope}} = 100 \text{ lb}_f$$

$$\Sigma \mathbf{F} = (\Sigma F_x)\mathbf{i} + (\Sigma F_y)\mathbf{j} = 0$$

$$\mathbf{F}_{\text{load}} = F_{x,\text{load}}\,\mathbf{i} + F_{y,\text{load}}\,\mathbf{j} = 0\mathbf{i} + (-100)\mathbf{j}$$

$$\mathbf{F}_{\text{rope}} = F_{x,\text{rope}}\,\mathbf{i} + F_{y,\text{load}}\,\mathbf{j} = 0\mathbf{i} + (-100)\mathbf{j}$$

$$\mathbf{F}_{\text{pulley}} = F_{x,\text{pulley}}\,\mathbf{i} + F_{y,\text{pulley}}\,\mathbf{j} = 0\mathbf{i} + F_{y,\text{pulley}}\,\mathbf{j}$$

$$\Sigma F_y = (-100) + (-100) + F_{y,\text{pulley}} = 0$$

$$F_{y,\text{pulley}} = 100 + 100 = 200 \text{ lb}_f$$

EXAMPLE 15.5

Problem Statement: The pulleys shown in Figure 15.10 lift a 100-lb_m mass. What is the required force on the rope to hold the mass stationary above the earth?

Solution: For Pulley B:

$$\Sigma \mathbf{M} = \mathbf{M}_2 + \mathbf{M}_3 = rF_2 - rF_3 = 0$$

$$rF_2 = rF_3$$

$$F_2 = F_3$$

$$\Sigma \mathbf{F} = (\Sigma F_x)\mathbf{i} + (\Sigma F_y)\mathbf{j} = 0$$

$$\mathbf{F}_2 = F_{x,2}\mathbf{i} + F_{y,2}\mathbf{j} = 0\mathbf{i} + F_2\mathbf{j}$$

$$\mathbf{F}_3 = F_{x,3}\mathbf{i} + F_{y,3}\mathbf{j} = 0\mathbf{i} + F_3\mathbf{j}$$

$$\mathbf{F}_{\text{load}} = F_{x,\text{load}}\mathbf{i} + F_{y,\text{load}}\mathbf{j} = 0\mathbf{i} - F_{\text{load}}\mathbf{j}$$

$$F_{\text{load}} = mg = 100\ \text{lb}_\text{m}\left(32.174\ \frac{\text{ft}}{\text{s}^2}\right) \times \frac{\text{lb}_\text{f}\cdot\text{s}^2}{32.174\ \text{lb}_\text{m}\cdot\text{ft}} = 100\ \text{lb}_\text{f}$$

$$\mathbf{F}_{\text{load}} = 0\mathbf{i} - 100\mathbf{j}$$

$$\Sigma F_y = F_2 + F_3 - 100 = 0$$

$$F_2 = -F_3 + 100 = -F_2 + 100$$

$$2F_2 = 100$$

$$F_2 = \frac{100}{2} = 50\ \text{lb}_\text{f}$$

Because the cable connecting Pulley B to Pulley A is not moving, we know that $F_1 = F_2 = 50\ \text{lb}_\text{f}$. In the previous example, we analyzed a pulley identical to Pulley A and learned that the force exerted by the two ropes is identical; therefore, $F_{\text{rope}} = F_1 = 50\ \text{lb}_\text{f}$.

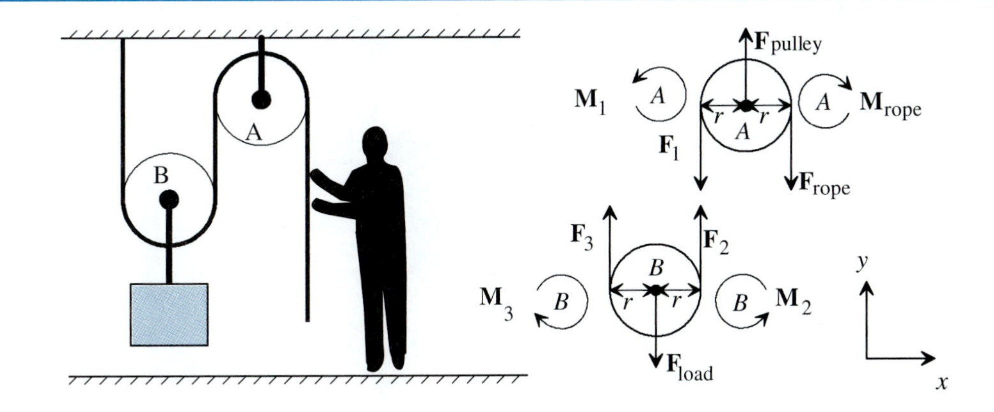

FIGURE 15.10
Lifting a load with two pulleys.

Here we have an interesting result. By using two pulleys connected as shown in Figure 15.10, the force exerted by the human is half that with the single pulley shown in Figure 15.9. Using two pulleys, we obtained a **mechanical advantage** in lifting the load. Similarly, the seesaw obtained a mechanical advantage for the 50-lb_m child, who was able to balance a 75-lb_m child.

EXAMPLE 15.6

Problem Statement: The crane shown in Figure 15.11 is rigidly attached to the earth. The pulley attached to the crane lifts a 100-lb_m load. What force does the pulley exert on the crane when holding the mass stationary above the earth? What couple does the earth have to produce on the buried end to resist the collapse of the crane?

Solution: In a previous example, we learned that the force on a pulley such as the one shown in Figure 15.9 is twice that of the load; therefore,

$$F_{\text{pulley}} = 2(100 \text{ lb}_\text{f}) = 200 \text{ lb}_\text{f}$$

$$\mathbf{M}_{\text{earth}} = -\mathbf{M}_{\text{pulley}} = -(-rF_{\text{pulley}}) = rF_{\text{pulley}} = (10 \text{ ft})(200 \text{ lb}_\text{f}) = 2000 \text{ ft}\cdot\text{lb}_\text{f}$$

EXAMPLE 15.7

Problem Statement: One day, the crane shown in Figure 15.11 lifts a heavy load and topples over. An engineer analyzes the situation and modifies the crane (Figure 15.12). What forces do the pulleys exert on the crane when holding a 100-lb_m load stationary above the earth? What couple does the earth have to produce on the buried end to resist the collapse of the crane?

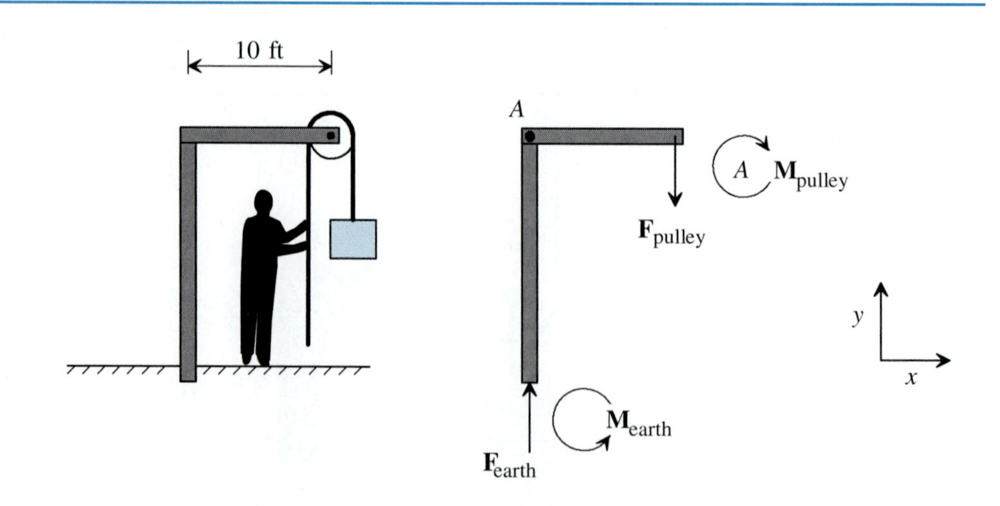

FIGURE 15.11
A crane used to lift a 100-lb_m load.

Solution: Because the tension in the rope is everywhere the same, the magnitude of the following force vectors is the same:

$$F = |\mathbf{F}_{\text{load}}| = |\mathbf{F}_{\text{rope}}| = |\mathbf{F}_{x,A}| = |\mathbf{F}_{x,B}| = 100 \text{ lb}_f$$

Static free body analysis of Pulley A shows that

$$|-\mathbf{F}_{y,A}| = |\mathbf{F}_{\text{load}}| = 100 \text{ lb}_f$$

$$|-\mathbf{F}_{x,A}| = |\mathbf{F}_{x,A}| = 100 \text{ lb}_f$$

and static free body analysis of Pulley B shows that

$$|-\mathbf{F}_{y,B}| = |\mathbf{F}_{\text{rope}}| = 100 \text{ lb}_f$$

$$|-\mathbf{F}_{x,B}| = |\mathbf{F}_{x,B}| = 100 \text{ lb}_f$$

The upward force on the pulleys is provided by the crane; therefore,

$$|\mathbf{F}_{y,A}| = |-\mathbf{F}_{y,A}| = 100 \text{ lb}_f$$

$$|\mathbf{F}_{y,B}| = |-\mathbf{F}_{y,B}| = 100 \text{ lb}_f$$

The total downward force of the pulleys on the crane is

$$F_{\text{pulleys}} = |\mathbf{F}_{y,A}| + |\mathbf{F}_{y,B}| = 100 \text{ lb}_f + 100 \text{ lb}_f = 200 \text{ lb}_f$$

which is identical to the crane shown in Figure 15.11.

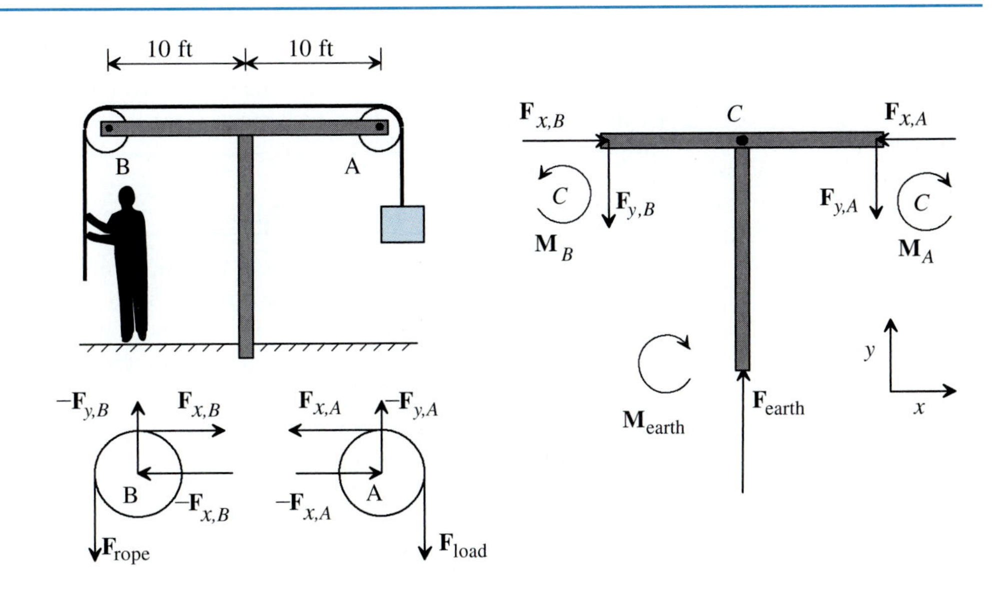

FIGURE 15.12
A counterbalanced crane used to lift a 100-lb_m load.

Leonardo da Vinci: Renaissance Visionary

Although Leonardo da Vinci crafted many wondrous ideas and fantastic designs in his extensive notebooks, he left this world in 1519 having actually constructed very little in his 67 years. An accomplished painter, sculptor, architect, and engineer, his life was marked by an extremely productive mind hindered by an inability to bring many projects to completion. One possible explanation for this may be that Leonardo's head was simply bursting with ideas, and he struggled to get them all down on paper before his death, unable to pursue anything beyond a drawing. Another is that he was too busy with the jobs that kept food on his table—most often, jobs commissioned for the entertainment of Italian and French nobles.

As an engineer, Leonardo is probably best known for his prototypical designs of modern mechanical devices such as the automobile, the airplane, and the submarine. His biggest contribution to the field of engineering is the establishment of the field of applied mechanics. He posed complex problems in the areas of statics and dynamics, playing with pulleys and weights, gears, and screws. He characterized the principles of solid-body motion nearly 200 years before Sir Isaac Newton did. Foreshadowing modern studies of structural deformation, he discussed the abilities of pillars in various conformations to withstand loading.

Some of the other machines that Leonardo designed are cannons; tanks; "machines for throwing fire"; lathes; oil presses; machines for grinding, polishing, filing, and planing; pumps; cranes; mechanical musical instruments; a parachute; a diving suit; and much more. Had he been paired with a production-oriented genius, the history of engineering may have been quite different.

Courtesy of: Seth Adelson, graduate student.

The following calculation finds the size of the couple the earth must produce to resist the two moments of the pulleys:

$$\Sigma \mathbf{M}_{\text{earth}} = \mathbf{M}_A + \mathbf{M}_B = -rF_{y,A} + rF_{y,B} = -r(100\ \text{lb}_f) + r(100\ \text{lb}_f) = 0$$

With this counterbalanced crane, the earth need not produce a couple to prevent the crane from toppling over. This makes it much more stable than the crane shown in Figure 15.11.

15.3 STRENGTH OF MATERIALS

Free body analysis allows the forces and moments to be calculated in a static structure. Generally, the motivation for these analyses is to determine if the structure is sufficiently strong to resist the forces and moments and therefore remain intact. If the forces and moments exceed the allowable limits of the structural materials, it can result in catastrophic failure.

Figure 15.13 shows a structural member with forces acting in *tension, compression,* or *shear. Tension stress* σ_{tension} is defined as the **tension force** F_{tension} applied at a right angle to cross-sectional area A:

$$\sigma_{\text{tension}} \equiv \frac{F_{\text{tension}}}{A} \tag{15-9}$$

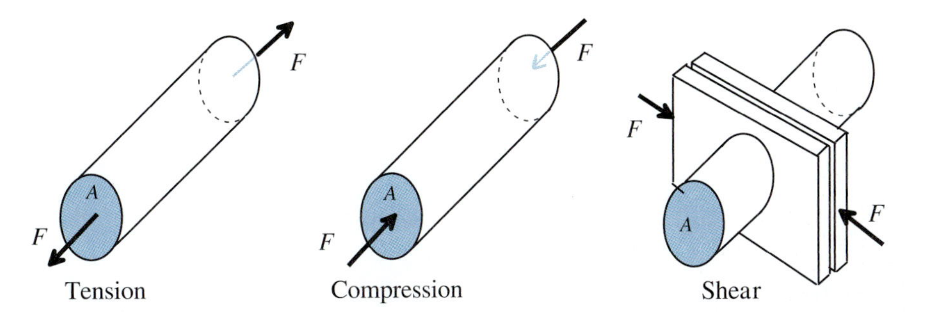

FIGURE 15.13
Tension, compression, and shear forces.

When a tension force is applied, the atoms (molecules) of the structure tend to separate.

Compression stress $\sigma_{\text{compression}}$ is defined as the **compression force** $F_{\text{compression}}$ applied at a right angle to cross-sectional area A:

$$\sigma_{\text{compression}} \equiv \frac{F_{\text{compression}}}{A} \tag{15-10}$$

When a compression force is applied, the atoms (molecules) of the structure tend to compact.

Shear stress τ is defined as the **shear force** F_{shear} applied in the same plane as cross-sectional area A:

$$\tau \equiv \frac{F_{\text{shear}}}{A} \tag{15-11}$$

When shear force is applied, the atoms (molecules) of the structure tend to slide past each other.

Regardless of the type of stress, it has units of force per area, the same as pressure (typically, lb_f/in^2 or Pa).

When a tension or compression force is applied to a structure, the structure changes dimensions (Figure 15.14). **Strain** ε is defined as the change in length δ per unit of original length L:

The cables of the Golden Gate Bridge are in tension, whereas the support towers are in compression.

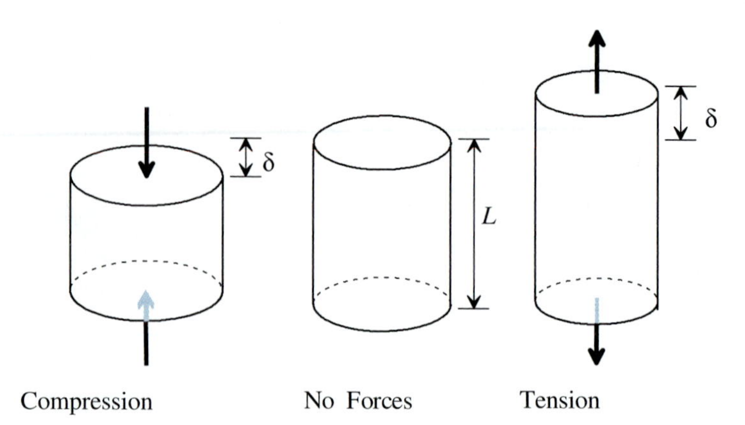

FIGURE 15.14
Dimensional changes when forces are applied to a structure. (*Note:* Dimensional changes are exaggerated.)

$$\varepsilon \equiv \frac{\delta}{L} \tag{15-12}$$

Because δ and L both have dimensions of length, ε is dimensionless.

When a structural material is under tension, it elongates. Figure 15.15 shows the tension stress for a typical metal as a function of strain. At low strains, the relationship is linear,

$$\sigma_{\text{tension}} = E\varepsilon \tag{15-13}$$

where the proportionality constant E is the **modulus of elasticity**. This linear relationship fails for strains above the **proportional limit.**

Above the proportional limit is the **elastic limit.** When a material is subjected to forces above the elastic limit, a permanent rearrangement of the structure occurs. It will not return to its original geometry even when the force is removed.

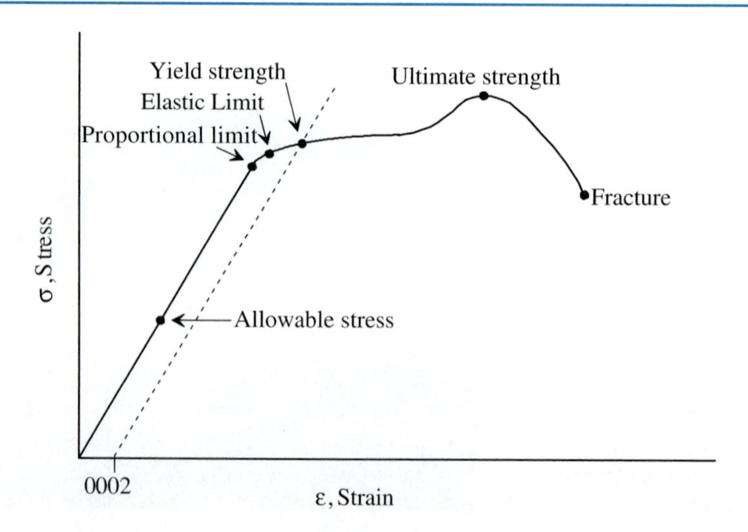

FIGURE 15.15
Typical stress-strain curve for a ductile metal.

The **yield strength** is generally defined as the intersection of the stress-strain curve and a line that passes through 0.002 strain and is parallel to the linear portion of the stress-strain curve. The yield strength is often considered the upper allowable stress for a structure.

The **ultimate strength** is the highest stress obtained prior to failure. Table 15.1 shows the modulus of elasticity, yield strength, and ultimate strength for a number of materials.

The **allowable stress** is used for design purposes. It includes a **safety factor** defined as

$$\text{Safety factor} = \frac{\text{yield strength}}{\text{allowable stress}} \tag{15-14}$$

TABLE 15.1
Mechanical properties of some materials (Reference) (Representative values; actual values may differ.)

Material	E (lb_f/in^2)	Yield Strength* (lb_f/in^2)	Ultimate Strength (lb_f/in^2)		Density (lb_m/in^3)
Structural steel (A-7)	30×10^6	35,000	Tension	60,000	0.283
High-carbon steel (SAE 1090)	30×10^6	67,000	Tension	122,000	0.283
Alloy steel (SAE 4130) heat treated	30×10^6	100,000	Tension	125,000	0.283
Stainless steel (18-8)	28×10^6	80,000	Tension	120,000	0.284
Gray cast iron (ASTM Class 30)	14.7×10^6		Tension	31,000	0.260
			Compression	124,000	
Cast iron (pearlitic malleable)	26.4×10^6	80,000	Tension	100,000	0.266
			Compression	300,000	
Aluminum 1100-0 (annealed)	10.0×10^6	3,500	Tension	11,000	0.098
Aluminum alloy 2024-T3	10.6×10^6	50,000	Tension	70,000	0.100
Aluminum alloy 7075-T6	10.4×10^6	70,000	Tension	80,000	0.101
Magnesium alloy (H K31A-H24)	6.5×10^6	23,000	Tension	34,000	0.0647
Titanium alloy (6Al-4V)	15.9×10^6	120,000	Tension	130,000	0.160
Brass, hard yellow	15×10^6	60,000	Tension	74,000	0.306
Douglas fir (air dry, parallel to grain)	1.7×10^6		Tension	8,100	0.020
			Compression	7,400	
Tungsten carbide (Carboloy, Grade 999)	100×10^6		Compression	600,000	
			Tension	1,300	
Glass (fused silica)	10.0×10^6		Compression	13,000	0.15
Concrete (low strength)	2×10^6		Compression	2,000	0.087
Concrete (high strength)	3×10^6		Compression	5,000	0.087
Polystyrene (average)	0.5×10^6		Compression	14,000	
Polyethylene (average)	1.8×10^6		Tension	2,000	0.033
Epoxy (cast, average)	0.65×10^6		Tension	7,000	
			Compression	30,000	
E-glass fiber	11×10^6		Tension	260,000	0.0918
Carbon fiber (high modulus)	49×10^6		Tension	360,000	0.0672
Carbon fiber (high strength)	33×10^6		Tension	460,000	0.0647
Kevlar fiber	18×10^6		Tension	400,000	0.0524

* Based upon intersection of the tensile stress-strain curve with a line that is parallel to the linear portion of the stress-strain curve and passes through 0.002 strain.

Adapted from: R. D. Snyder and E. F. Byars, *Engineering Mechanics: Statics and Strength of Materials* (New York: McGraw-Hill, 1973), p. 478. Values for last four materials were found in N. G. McCrum, C. P. Buckley, and C. B. Bucknall, *Principles of Polymer Engineering* (New York: Oxford University Press, 1988), p. 218.

Stress Is a Momentum Flux

Stress has units of pressure, that is, force per unit area. The dimensions of pressure may be algebraically manipulated as follows:

$$\text{Stress} = \frac{\text{Force}}{\text{Area}} = \frac{\frac{[M][L]}{[T]^2}}{[L]^2} = \frac{\left\{\frac{[M][L]}{[T]}\right\}}{[T][L]^2} = \frac{\text{Momentum}}{\text{Time·Area}}$$

The dimensions in the curly brackets are the same as those for momentum; thus, stress has dimensions of momentum per unit time per unit area, which is a momentum flux. (Recall that *flux* is a concept we discussed in Chapter 12, "Introduction to Rate Processes.") If the momentum flux through an object is too great, it breaks. This is analogous to trying to force too much water through a pipe; if the flow capacity is exceeded, the pipe breaks.

The safety factor provides a margin of safety to allow for material variability and imperfections. For example, if a safety factor of 3 were specified, a material with a yield strength of 45,000 lb_f/in^2 would be limited to a design stress of 15,000 lb_f/in^2.

EXAMPLE 15.8

Problem Statement: A 1000-lb_m load will be suspended by a single round rod made of structural steel. What is the minimum diameter of the rod using a safety factor of 3? Initially, the rod is 10.0000 ft long. After it is loaded, what is the length?

Solution: Figure 15.16 shows the load suspended by the rod.

$$\sigma_{\text{allowable}} = \frac{F}{A} = \frac{F}{\frac{\pi}{4}D^2}$$

$$D = \sqrt{\frac{F}{\frac{\pi}{4}\sigma_{\text{allowable}}}} = \sqrt{\frac{F}{\frac{\pi}{4}(\frac{1}{3}\sigma_{\text{yield}})}} = \sqrt{\frac{1000\ \text{lb}_f}{\frac{\pi}{4}(\frac{1}{3}\ 35{,}000\ \frac{\text{lb}_f}{\text{in}^2})}} = 0.33\ \text{in}$$

$$\sigma_{\text{allowable}} = E\varepsilon$$

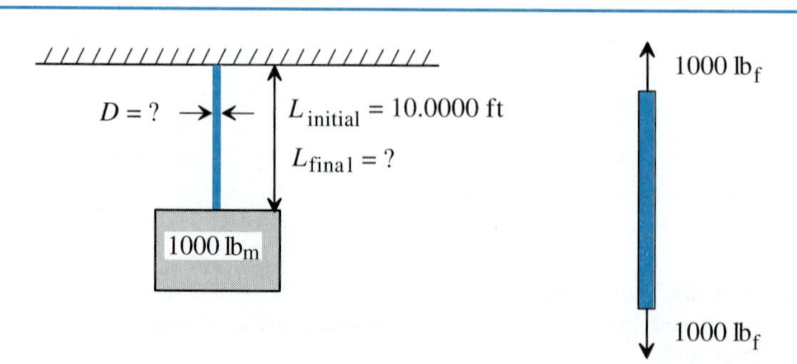

FIGURE 15.16
A 1000-lb_m load suspended by a rod.

$$\varepsilon = \frac{\sigma_{\text{allowable}}}{E} = \frac{\delta}{L_{\text{initial}}}$$

$$\delta = \frac{L_{\text{initial}}\sigma_{\text{allowable}}}{E} = \frac{L_{\text{initial}}\frac{1}{3}\sigma_{\text{yield}}}{E} = \frac{10\text{ ft}\frac{1}{3}\left(35{,}000\frac{\text{lb}_f}{\text{in}^2}\right)}{30 \times 10^6\frac{\text{lb}_f}{\text{in}^2}} = 0.0039\text{ ft}$$

$$L_{\text{final}} = L_{\text{initial}} + \delta = 10.0000\text{ ft} + 0.0039\text{ ft} = 10.0039\text{ ft}$$

15.4 DYNAMICS (ADVANCED TOPIC)

Until this point, all the systems we considered were static because the sum of all forces and moments equaled zero. Here, we consider dynamic systems, in which the forces and/or moments do not equal zero:

$$\Sigma \mathbf{F} \neq 0 \tag{15-15}$$

$$\Sigma \mathbf{M} \neq 0 \tag{15-16}$$

Figure 15.17 shows an example of a dynamic system. A mass of m is suspended by a spring with spring constant k. The mass oscillates about an equilibrium position. The force $\mathbf{F}$ on the mass is

$$\mathbf{F} = -k\mathbf{x} \tag{15-17}$$

where $\mathbf{x}$ is the position vector. The negative sign indicates that the force vector $\mathbf{F}$ and the position vector $\mathbf{x}$ always point in opposite directions.

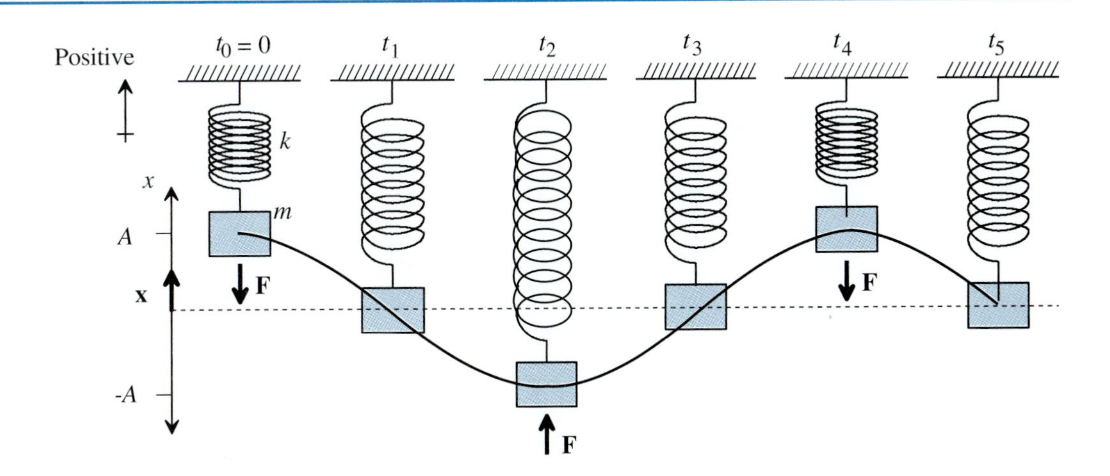

FIGURE 15.17 A dynamic spring-mass system.

Safety Factors in the Man-Made and Natural Worlds

A *safety factor* is the ratio of the minimum load that would cause failure divided by the actual peak load. The safety factor can vary widely depending upon the application. Issues that must be addressed when specifying a safety factor include: (1) variations in material strength, (2) variations in actual peak load, (3) the cost of overdesign both in money and performance, and (4) consequences of failure. Engineering safety factors are determined by government regulations, industry standards, and the marketplace. Nature too has safety factors that are determined by the pressures of natural selection. The following table shows some typical safety factors in both the man-made and natural worlds.

Man-Made Structures	Safety Factor
Cable of fast passenger elevator	11.9
Cable of elevator in shallow mine	8
Cable of slow passenger elevator	7.6
Cable of slow freight elevator	6.7
Cable of crane	6
Wooden building	6
Cable of elevator in deep mine	5
Cable of powered dumbwaiter	4.8
Steel building or bridge	2
Natural Structures	
Jawbone of biting monkey	7
Wing bones of flying goose	6
Leg bones of running turkey	6
Leg bones of galloping horse	4.8
Leg bones of running elephant	2.5 to 4
Leg bones of hopping kangaroo	3
Leg bones of running ostrich	2.5
Leg bones of jumping dog	2 to 3
Dragline of spider	1.5
Backbones of human lifting weights	1.0 to 1.7

Adapted from: J. Diamond, "Building to Code," Discover, May 1993, p. 93.

According to Newton's second law

$$\mathbf{F} = m\mathbf{a} = m\frac{d^2\mathbf{x}}{dt^2} \tag{15-18}$$

Both Equations 15-17 and 15-18 give expressions for the force acting on the mass; therefore, these two equations must be equal to each other:

$$m\frac{d^2\mathbf{x}}{dt^2} = -k\mathbf{x}$$

$$\frac{d^2\mathbf{x}}{dt^2} = -\frac{k}{m}\mathbf{x}$$

$$\frac{d^2|\mathbf{x}|}{dt^2} = -\frac{k}{m}|\mathbf{x}|$$

$$\frac{d^2x}{dt^2} = -\frac{k}{m}x \tag{15-19}$$

This is a **differential equation,** meaning it is an equation that contains a differential (in this case, d^2x/dt^2). Generally, differential equations are a topic reserved for the sophomore year in engineering curricula. However, this equation is relatively simple and can be solved with a general knowledge of freshman calculus.

Equation 15-19 says we are seeking a mathematical function that when differentiated twice gives us the original function times the constant $-k/m$. We propose the following three functions as possibilities:

Equation A: $$x = C\cos\sqrt{\frac{k}{m}}\,t$$

Equation B: $$x = C\sin\sqrt{\frac{k}{m}}\,t$$

Equation C: $$x = C\exp\left(\sqrt{\frac{k}{m}}\,it\right)$$

where C is an arbitrary constant and $i = \sqrt{-1}$.

We can test each of these proposed functions by differentiating them twice.

Equation A

$$\frac{dx}{dt} = -C\sqrt{\frac{k}{m}}\sin\sqrt{\frac{k}{m}}\,t$$

$$\frac{d^2x}{dt^2} = -C\left(\sqrt{\frac{k}{m}}\right)^2\cos\sqrt{\frac{k}{m}}\,t$$

$$= -C\frac{k}{m}\cos\sqrt{\frac{k}{m}}\,t$$

Equation B

$$\frac{dx}{dt} = C\sqrt{\frac{k}{m}}\cos\sqrt{\frac{k}{m}}\,t$$

$$\frac{d^2x}{dt^2} = -C\left(\sqrt{\frac{k}{m}}\right)^2\sin\sqrt{\frac{k}{m}}\,t$$

$$= -C\frac{k}{m}\sin\sqrt{\frac{k}{m}}\,t$$

Equation C

$$\frac{dx}{dt} = C\sqrt{\frac{k}{m}}\,i\exp\left(\sqrt{\frac{k}{m}}\,it\right)$$

$$\frac{d^2x}{dt^2} = C\left(\sqrt{\frac{k}{m}}\,i\right)^2\exp\sqrt{\frac{k}{m}}\,it$$

$$= -C\frac{k}{m}\exp\sqrt{\frac{k}{m}}\,it$$

To test each of these proposed solutions, substitute them into Equation 15-19.

$$-C\frac{k}{m}\cos\sqrt{\frac{k}{m}}\,t = -\frac{k}{m}\left(C\cos\sqrt{\frac{k}{m}}\,t\right) \qquad \text{(Equation A)}$$

$$-C\frac{k}{m}\sin\sqrt{\frac{k}{m}}\,t = -\frac{k}{m}\left(C\sin\sqrt{\frac{k}{m}}\,t\right) \qquad \text{(Equation B)}$$

$$-C\frac{k}{m}\exp\left(\sqrt{\frac{k}{m}}\,it\right) = -\frac{k}{m}\left[C\exp\left(\sqrt{\frac{k}{m}}\,it\right)\right] \qquad \text{(Equation C)}$$

Thus we see that all three proposed equations for x satisfy the differential equation. All that remains is to find a value for the constant C.

From Figure 15.17, we see that at $t = 0$, $x = A$. This is known as the *initial condition.* Our task is to find a value for C that satisfies this initial condition.

$$A = C\cos\sqrt{\frac{k}{m}}0 = C(1) \Rightarrow C = A \qquad \text{(Equation A)}$$

$$A = C\sin\sqrt{\frac{k}{m}}0 = C(0) \Rightarrow C = \frac{A}{0} = \infty \qquad \text{(Equation B)}$$

$$A = C\exp\left(\sqrt{\frac{k}{m}}i0\right) = C(1) \Rightarrow C = A \qquad \text{(Equation C)}$$

For Equation B, it was impossible to find a value for C that satisfies the initial conditions. Therefore, we have two equations that satisfy (or "solve") the differential equation given by Equation 15-19:

$$x = A\cos\sqrt{\frac{k}{m}}t \qquad \text{(Equation A)} \tag{15-20}$$

$$x = A\exp\sqrt{\frac{k}{m}}it \qquad \text{(Equation C)}$$

We will use Equation A for the solution, because it is easily calculated. Notice that in Figure 15.17, the position of the mass traces a cosine function with a maximum *amplitude* of A and a minimum amplitude of $-A$ as given by Equation 15-20.

Equation 15-20 shows that the mass will oscillate in a periodic manner. The *period* P is the time it takes to complete one cycle. One complete cycle around a circle is 2π radians; therefore,

$$\sqrt{\frac{k}{m}}P = 2\pi$$

$$P = 2\pi\sqrt{\frac{m}{k}} \tag{15-21}$$

The *frequency* f is the number of cycles that occur in a given length of time. Frequency is the inverse of the period; therefore,

$$f = \frac{1}{P} = \frac{1}{2\pi}\sqrt{\frac{k}{m}} \tag{15-22}$$

The spring-mass system shown in Figure 15.17 is relatively simple and hence is amenable to analytical solutions such as Equation 15-20. Many systems of engineering interest are not so simple and do not have analytical solutions. Therefore, a computer is required to calculate the position of the mass as a function of time.

To illustrate how a computer may be used to calculate the position of the mass as a function of time, we will use the spring-mass system shown in Figure 15.17. Because we know the analytical solution, we can compare the computer solution to the known analytical solution.

At any given time, we know the force F acting on the mass is

$$F = -kx \tag{15-23}$$

and from Newton's second law we know that

$$F = ma \tag{15-24}$$

We can solve Equation 15-24 explicitly for the acceleration a and substitute Equation 15-23 for F:

$$a = \frac{F}{m} = -\frac{k}{m}x \tag{15-25}$$

Acceleration is the time derivative of velocity, and velocity is the time derivative of position:

$$\frac{dv}{dt} = a \tag{15-26}$$

$$\frac{dx}{dt} = v \tag{15-27}$$

We can approximate each of these differential equations with difference equations:

$$\frac{dv}{dt} \approx \frac{\Delta v}{\Delta t} = \frac{v_{\text{new}} - v_{\text{old}}}{\Delta t} = a \tag{15-28}$$

$$\frac{dx}{dt} \approx \frac{\Delta x}{\Delta t} = \frac{x_{\text{new}} - x_{\text{old}}}{\Delta t} = v \tag{15-29}$$

From Equation 15-28, we can calculate the new velocity v_{new} from the old velocity v_{old} and from Equation 15-29, we can calculate the new position x_{new} from the old position x_{old},

$$v_{\text{new}} = v_{\text{old}} + a\Delta t \tag{15-30}$$

$$x_{\text{new}} = x_{\text{old}} + v\Delta t \tag{15-31}$$

where acceleration is given by Equation 15-25. Program 15.1 numerically calculates the acceleration, velocity, and position of the oscillating mass as a function of time. In addition, the program calculates the average error between the numerically calculated position and the analytical position (Equation 15-20). The program allows the user to specify Δt. It terminates when half a cycle has been completed, which occurs when the velocity changes from negative to positive. Table 15.2 shows a typical output. Figure 15.18 shows the average error decreases as Δt gets smaller. This occurs because the difference equations used by the computer better approximate the true derivative when the time step is small. However, when the time step is too small, so many computations are required that computer round-off errors become significant and the average error actually increases.

PROGRAM 15.1
Spring-mass program.

```
      program spring_mass
c
c     This program numerically calculates the position, velocity, and
c     acceleration of a spring/mass system given the starting amplitude
c     and k/m. It also compares the numerically computed values with
c     the known analytical solution.
c
      implicit none
      real amp, x_new, x_old, v_new, v_old, k_over_m, a, t, delta_t,
     & sum, error, ave_error
      integer count, skip
c
c     The output will be stored in a file.
c
      open (unit=7, file='spring.out', status='unknown')
c
c     Initialize variables.
c
      amp=10.0
      x_new=amp
      x_old=amp
      v_new=0.0
      v_old=0.0
      k_over_m=1.0
      a=-k_over_m*x_new
      t=0.0
      error=abs(x_new - amp*cos(sqrt(k_over_m)*t))
      count=0
      sum=0.0
      ave_error=0.0
c
c     From the screen, enter the delta t and number of lines to be
c     skipped between prints to the file.
c
      write (*,*) 'enter delta t '
      read (*,*) delta_t
      write (*,*) 'how many lines of output do you wish to skip between
     & line prints?'
      read (*,*) skip
c
c     Write the heading for the table.
c
```

```
      write (7,10) amp, k_over_m, delta_t, skip
   10 format (1x, 'amplitude = ', f5.2, 2x, 'k/m = ', f5.2, 2x, 'delta t = ',
     & e9.2, 2x, 'lines skipped = ',I5)
      write (7,20)
   20 format (1x, 'time',t15, 'acceleration',t30, 'velocity',t45,
     & 'position',t60, 'position error')
      write (7,30) t, a, v_new, x_new, error
   30 format (1x,f12.8,t15,f12.8,t30,f12.8,t45,f12.8,t60,e12.5)
c     Perform calculations as long as the velocity is negative. This
c     condition causes the program to calculate the acceleration,
c     velocity, and position for one half cycle.
c
      do while (v_new.le.0.0)
         a=-k_over_m*x_new
c
c        The first time through, calculate the new velocity based upon
c        half the delta t. All the other times, use the entire delta t.
c
         if (count.eq.0) then
           v_new=v_old + a*0.5*delta_t
         else
           v_new=v_old + a*delta_t
         endif
         x_new=x_old + v_new*delta_t
         t=t + delta_t
         error=abs(x_new - amp*cos(sqrt(k_over_m)*t))
         sum=sum + error
         count=count + 1
c
c        Write to the file after the specified number of lines are skipped.
c
         if (mod(count,skip + 1).eq.0) then
            write (7,30) t, a, v_new, x_new, error
         endif
         x_old=x_new
         v_old=v_new
      enddo
c
c     Calculate the average error.
c
      ave_error=sum/real(count)
      write (7,40) ave_error
   40 format (1x,t45 'average error = ',t60,e12.5)
      end
```

TABLE 15.2
Typical output from spring-mass program

amplitude = 10.00 k/m = 1.00 delta t = 0.10E-02 lines skipped = 99

Time	Acceleration	Velocity	Position	Position Error
0.00000000	−10.00000000	0.00000000	10.00000000	0.00000E+00
0.10000002	−9.95103455	−0.99335903	9.95004082	0.81172E−06
0.20000023	−9.80264759	−1.98179281	9.80066586	0.52698E−06
0.29999968	−9.55631733	−2.95042562	9.55336666	0.83625E−06
0.39999840	−9.21449852	−3.88957810	9.21060848	0.77015E−05
0.49999711	−8.78061485	−4.78986645	8.77582455	0.14931E−04
0.59999585	−8.25899410	−5.64229774	8.25335217	0.27408E−04
0.69999456	−7.65485811	−6.43835211	7.64841986	0.37035E−04
0.79999328	−6.97423553	−7.17007351	6.96706533	0.49990E−04
0.89999199	−6.22392845	−7.83015776	6.21609831	0.64125E−04
0.99999070	−5.41143370	−8.41200447	5.40302181	0.79489E−04
1.09999537	−4.54487181	−8.90979862	4.53596210	0.40331E−04
1.20000005	−3.63289952	−9.31857109	3.62358093	0.38323E−05
1.30000472	−2.68462706	−9.63424015	2.67499280	0.50000E−04
1.40000939	−1.70952928	−9.85364342	1.69967568	0.96820E−04
1.50001407	−0.71735072	−9.97459316	0.70737612	0.14442E−03
1.60001874	0.28199506	−9.99588013	−0.29199094	0.19160E−03
1.70002341	1.27852345	−9.91729069	−1.28844070	0.23641E−03
1.80002809	2.26227665	−9.73961258	−2.27201629	0.27817E−03
1.90003276	3.22342706	−9.46461487	−3.23289156	0.31410E−03
2.00003743	4.15236712	−9.09505081	−4.16146231	0.34642E−03
2.10003018	5.03981972	−8.63461494	−5.04845428	0.26731E−03
2.20002294	5.87691784	−8.08790398	−5.88500595	0.19066E−03
2.30001569	6.65529394	−7.46038294	−6.66275454	0.12266E−03
2.40000844	7.36717319	−6.75832081	−7.37393141	0.62756E−04
2.50000199	8.00544453	−5.98872995	−8.01143360	0.96884E−05
2.59999394	8.56372547	−5.15930128	−8.56888485	0.28534E−04
2.69998670	9.03644180	−4.27832413	−9.04071999	0.55424E−04
2.79997945	9.41886711	−3.35459757	−9.42222214	0.67578E−04
2.89997220	9.70718670	−2.39735389	−9.70958424	0.69099E−04
2.99996495	9.89851189	−1.41615522	−9.89992809	0.52592E−04
3.09995770	9.99093437	−0.42080745	−9.99135494	0.21035E−04
			Average error =	0.92951E−04

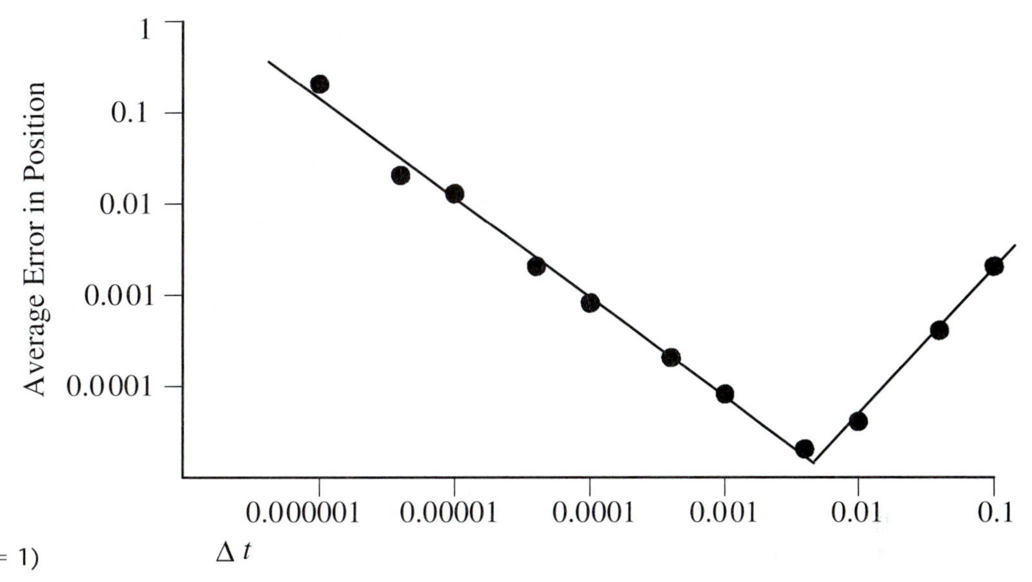

FIGURE 15.18
Effect of the time step on the average error in position. (A = 10, k/m = 1)

15.5 SUMMARY

Statics is the study of forces and moments acting on stationary objects. In order for the object to be stationary, Newton's first law requires the sum of all forces and moments to be zero. A moment is a twisting force about a reference point. The magnitude of the moment depends upon the reference point selected. A couple is a twisting force resulting when two equal-magnitude, parallel forces act in opposite directions. Unlike a moment, the magnitude of a couple does not depend upon the reference point.

Dynamics is the study of forces and/or moments acting on objects that cause them to accelerate. In order for an object to accelerate, Newton's second law requires the sum of all forces and/or moments to be nonzero.

Often, the motivation for making statics and dynamics calculations is to determine if the materials in an object will fail. Materials can fail in tension, compression, or shear.

Nomenclature

A area (m^2) or amplitude (m)
a multiplier (N/m)
b multiplier (N/m)
$\mathbf{C}$ position vector (m)
C arbitrary constant (m)
D diameter (m)
E modulus of elasticity (N/m^2)
$\mathbf{F}$ force vector (N)
F force scalar (N)
f frequency (s^{-1})

g acceleration due to gravity (m/s^2)
i unit vector in x-direction (dimensionless)
j unit vector in y-direction (dimensionless)
k unit vector in z-direction (dimensionless)
k spring constant (N/m)
L length (m)
M moment vector (N·m)
M moment scalar (N·m)
m mass (kg)
P period (s)
R position vector (m)
r position scalar (m)
t time (s)
v velocity (m/s)
x position vector (m)
x position scalar (m)
δ displacement (m)
ε strain (dimensionless)
σ stress (N/m^2)
τ stress (N/m^2)

Further Readings

Beer, F. P., and E. R. Johnston, *Mechanics for Engineers: Statics and Dynamics.* 4th ed. New York: McGraw-Hill, 1987.

PROBLEMS

15.1 Rank the following materials according to their strength-to-weight ratio (divide the ultimate strength by the density and rank them from highest to lowest): structural steel (A-7), stainless steel (18-8), cast iron (pearlitic malleable), aluminum alloy 7075-T6, titanium alloy (6A1-4V), brass, Douglas fir, glass (fused silica), concrete (high strength), polyethylene, E-glass, carbon fiber (high strength), Kevlar.

15.2 Four tugboats are having a contest to see which is strongest. Each tugboat attaches a rope to a very strong metal ring. Tugboat *A* heads due north and pulls with a force of 300,000 lb_f, Tugboat *B* heads due south and pulls with a force of 250,000 lb_f, Tugboat *C* heads due east and pulls with a force of 350,000 lb_f, and Tugboat *D* heads due west and pulls with a force of 275,000 lb_f. What is the magnitude and direction of the resultant force?

15.3 A buoy is attached to the ocean floor with a rope. The ocean currents exert a horizontal 1000-N force on the buoy. The rope angle is 75° with respect to the flat ocean floor. What is the tension in the rope?

15.4 A radio antenna is stabilized with three guide wires. All the guide wires are attached to the antenna at a height 100 ft from the ground and are fastened to the ground at a point 50 ft from the base of the antenna. Looking at the antenna from above, Wire *A* is pointed due north, Wire *B* is 120° clockwise from Wire *A*, and Wire *C* is 120° counterclockwise from Wire *A*. Calculate the tension force in each wire under the following wind conditions:

(a) A wind from the north that exerts a 50,000 ft·lb_f moment on the antenna.
(b) A wind from the south that exerts a 50,000 ft·lb_f moment on the antenna.
(c) A wind from the east that exerts a 50,000 ft·lb_f moment on the antenna.

15.5 A seesaw is 20 ft long with the fulcrum at the center. Three children want to get on at the same time. The masses of each child are: 50, 75, and 65 lb_m. The two heaviest children sit at opposite ends of the seesaw. Where should the lightest child sit to balance it?

15.6 A vertical, solid, round rod will support a 5,000-lb_m load in tension. What is the required rod diameter if it is made from structural steel? What is the required diameter if it is made from Kevlar? (*Note*: For safety, the maximum stress in the rod cannot exceed one-fourth of the ultimate strength of the material.)

15.7 Calculate the compression force in the arm of the crane shown in Figure 15.19. Also, calculate the tension in the support cable. If the support cable is made from steel with a yield strength of 35,000 psi, and a safety factor of 3 is specified, what is the minimum diameter of the cable?

15.8 Calculate the tension in the cable holding the coal car shown in Figure 15.20. If the cable is made from steel with a yield strength of 35,000 psi, and a safety factor of 3 is specified, what is the minimum diameter of the cable?

15.9 Draw a free body diagram numerically indicating the forces and moments acting on the 6.35-ft cantilever beam shown in Figure 15.21. (*Note:* Neglect the mass of the beam, vessel, and chain.)

15.10 Calculate the tension in the cable shown in Figure 15.22. If 3 is the safety factor, what is the minimum diameter of the steel cable if the yield strength is 35,000 psi?

15.11 Calculate the tension in the support cable shown in Figure 15.23 if the person is holding a 75-lb_m load. For a safety factor of 3, what is the minimum required cable diameter if the yield strength is 35,000 psi?

15.12 The person lifting the 75-lb_m load shown in Figure 15.23 jerks the cord, causing the load to accelerate upward at 53.2 ft/s^2. Calculate the maximum tension in the support cable. For a safety

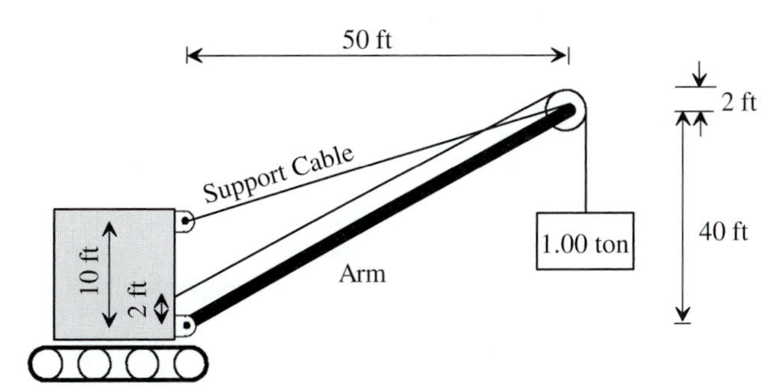

FIGURE 15.19
Crane holding a load.

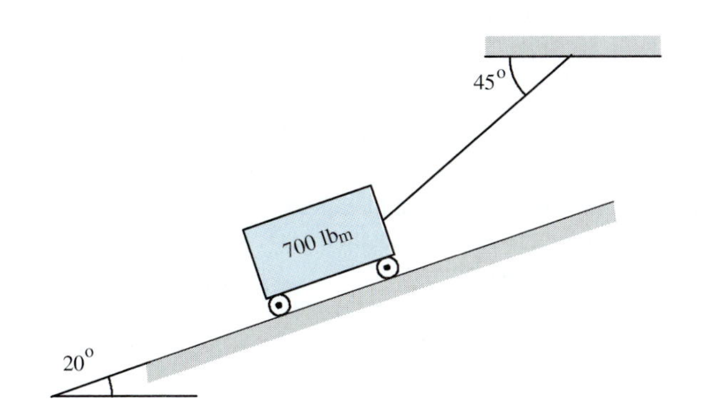

FIGURE 15.20
Cable supporting a coal car.

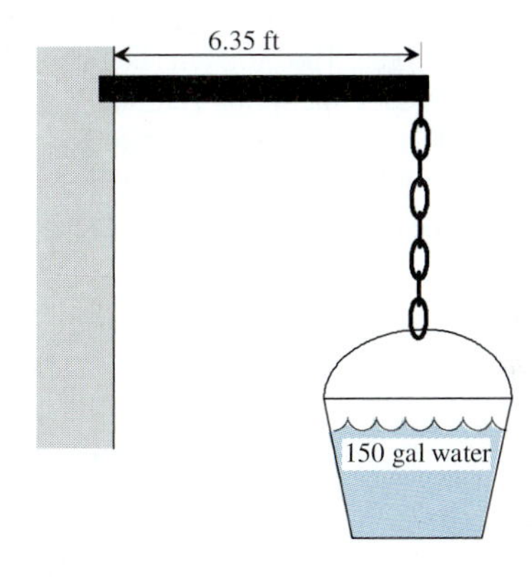

FIGURE 15.21
Cantilevered beam holding a large vessel of water.

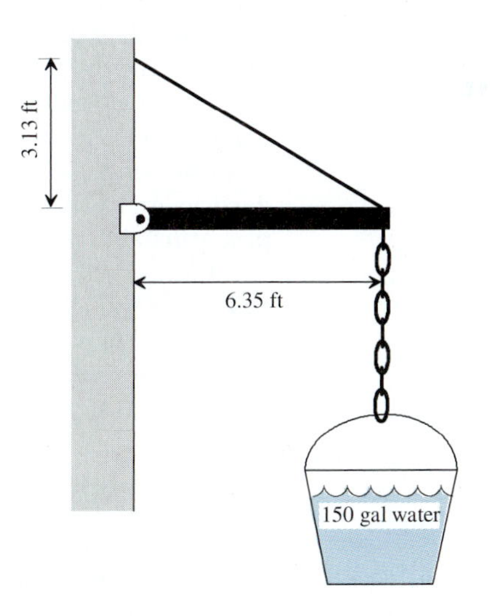

FIGURE 15.22
Support system for a large vessel of water.

factor of 3, what is the minimum required cable diameter if the yield strength is 35,000 psi?

15.13 The ideal spring-mass system described in the chapter will oscillate forever once set in motion. A real spring-mass system eventually stops oscillating due to drag forces. The drag force acting on a moving body is proportional to the square of the velocity and always acts in a direction opposite to the motion.

$$F_{drag} = k_{drag}v^2$$

Write a computer program that describes the acceleration, velocity, and position of a spring-mass system that has a drag force acting on it in addition to the spring force. The program should allow the user to specify the initial amplitude *A*, the spring constant *k*, the mass *m*, and the drag constant k_{drag}. The program should calculate the acceleration, velocity, and position for at least 10 cycles. Plot the position as a function of time to observe the decreasing amplitude due to the drag force.

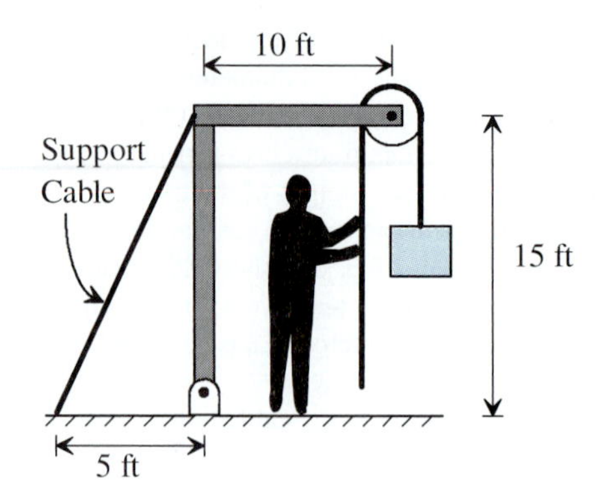

FIGURE 15.23
Person using a crane to hold a 75-lb_m load

Glossary

allowable stress The maximum force per unit area that may be safely applied.
compression force The force that causes the atoms of a structure to compact.
couple Two forces have the same magnitude and act parallel to each other, but in opposite directions
differential equation An equation that contains one or more derivatives.
dynamics The study of forces and moments on moving objects.
elastic limit The stress that initiates a permanent structural rearrangement.
free body diagram A drawing of a system with forces acting on it.
fulcrum The point or support on which a lever pivots.
mechanical advantage The ratio of a machine's output force to the applied input force.
modulus of elasticity The ratio of stress to strain.
moment A right-angle force exerted at a distance from a central pivot point.
particle An idealized object with mass, but no dimensions.
proportional limit The greatest stress that a material will sustain while maintaining a linear relationship between stress and strain.
rigid body An idealized object with both mass and dimensions, with the restriction that the components of the object do not move with respect to each other.
safety factor The ratio of the minimum load that would cause failure divided by the actual peak load.
shear force The force that tends to cause atoms of a structure to slide past each other.
statics The study of forces and moments on stationary objects.
strain The change in length per unit of original length caused by the application of a force.
tension force The force that tends to cause atoms of the structure to separate.
torque A twisting force resulting from a couple.
ultimate strength The highest stress obtained prior to failure.
yield strength The intersection of the stress-strain curve and a line that passes through 0.002 strain and is parallel to the linear portion of the stress-strain curve.

Mathematical Prerequisite

- Algebra (Appendix E, Mathematics supplement))
- Calculus (Appendix L, Mathematics supplement)

CHAPTER 16

Introduction to Electricity

Electricity is extremely important in our modern world. It is essential to communications (radio, television, Internet), information processing (computers), heavy industry (motors, welding, equipment control), homes (lights, appliances, air conditioning), and transportation (electric trains). Electricity is so versatile because it is a clean form of energy that can be exquisitely controlled. The fact that electrical and electronic engineers are the dominant engineering discipline—constituting more than a quarter of all engineers—attests to the importance of electricity. Because of its importance, all engineers benefit from understanding electricity.

16.1 FUNDAMENTALS OF ELECTRICITY

Figure 16.1 shows two spheres with electric charge. In Case A, the two spheres have like charges (two negatives), and repel each other. Mechanical forces (indicated by the hands) are needed to maintain the distance r between the spheres. In Case B, the two spheres have opposite charge, and attract each other. Again, mechanical forces are needed to maintain the distance r between the spheres.

The magnitude of the force of repulsion (or attraction) between the two spheres is

$$F = \left(\frac{1}{4\pi\varepsilon_0}\right)\frac{q_1 q_2}{r^2} \tag{16-1}$$

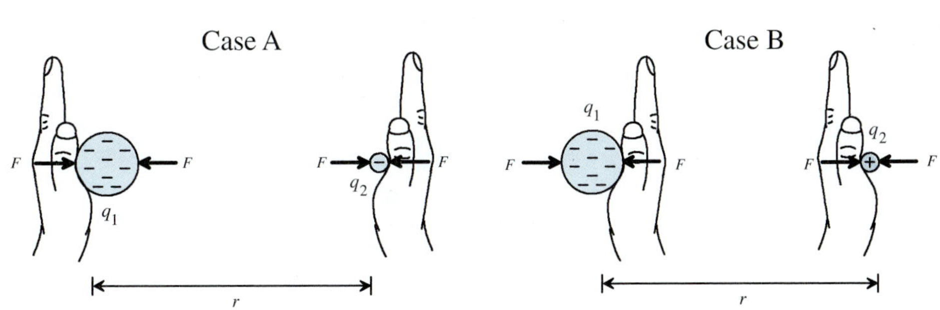

FIGURE 16.1 Charged spheres repel (Case A) and attract (Case B).

where r is the distance between the sphere centers (m), ε_0 is the permittivity of vacuum [$8.854187817 \times 10^{-12}$ $C^2/(N{\cdot}m^2)$], and q_1 and q_2 are the charges on each sphere (in units of *coulombs,* abbreviated C). A **coulomb** is a quantity of charge, analogous to the mole used to quantify atoms and molecules in chemistry. A coulomb of negative charge is equal to the charge of $6.24150636 \times 10^{18}$ electrons and a coulomb of positive charge is equal to the charge of $6.24150636 \times 10^{18}$ protons. (*Note:* Equation 16-1 is strictly valid only if the spheres are in a vacuum and the distance separating them is large relative to the sphere diameters.)

Figure 16.2(a) shows the force vectors **F** that result from placing Sphere 2 with positive charge q_2 at various radii r_i from central Sphere 1 with positive charge q_1. Sphere 2 may be viewed as a test sphere used to probe the amount of positive charge on Sphere 1. Unfortunately, the force on the test sphere depends upon its own charge, which may be arbitrarily selected. Simply reporting the force at various distances r_i from Sphere 1 does not give *unique* information because of the arbitrary amount of charge on the **test sphere.** To overcome this uniqueness problem, we invoke a concept called an **electric field,** the magnitude of which is defined as follows.

$$E \equiv \frac{F}{q_2} = \frac{\left(\frac{1}{4\pi\varepsilon_0}\right)\frac{q_1 q_2}{r^2}}{q_2} = \left(\frac{1}{4\pi\varepsilon_0}\right)\frac{q_1}{r^2} \tag{16-2}$$

(*Note:* The electric field is a vector that has the same direction as the force vector.) The electric field is independent of test charge q_2, which overcomes the uniqueness problem described above.

Figure 16.2(b) shows the **lines of force,** which are determined simply by connecting the force vectors shown in Figure 16.2(a). Notice that the lines of force are more dense in the regions where the force vectors are stronger. At any given point (for example, Point A), the vector that is tangent to the line of force gives the direction of the electric field **E.** Figure 16.2(c) shows the lines of force around a pair of oppositely charged spheres. The vector that is tangent to Point B gives the direction of the electric field **E** at that point.

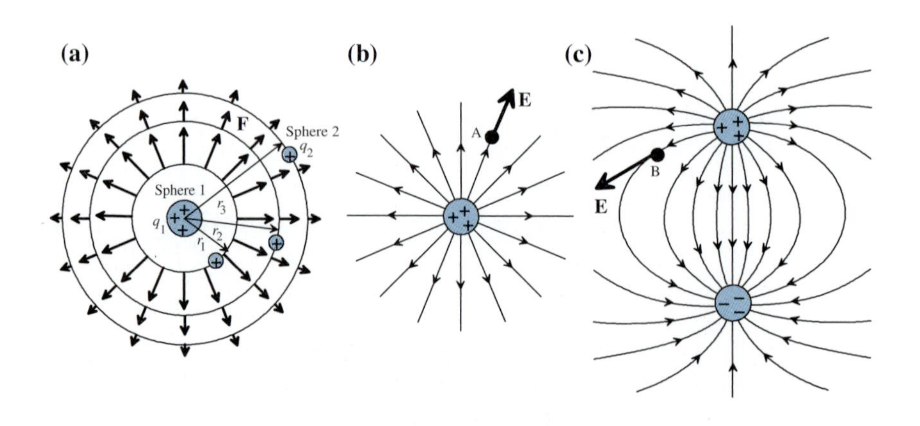

FIGURE 16.2
(a) Force vectors on the test sphere at various radii from the central sphere. (b) Lines of force around isolated positive charge. (c) Lines of force around a negative and a positive sphere.

EXAMPLE 16.1

Problem Statement: A test sphere with 0.500 C positive charge is located 3000 m from a central sphere with 10.0 C positive charge. (1) In a vacuum, what is the magnitude of the force of repulsion? (2) What is the magnitude of the electric field expressed in N/C? (3) What is the magnitude of the electric field expressed in V/m?

Solution:

1. Equation 16-1 gives the force of repulsion.

$$F = \left(\frac{1}{4\pi\varepsilon_0}\right)\frac{q_1 q_2}{r^2} = \left[\frac{1}{4\pi\left(8.85 \times 10^{-12}\frac{\text{C}^2}{\text{N}\cdot\text{m}^2}\right)}\right]\frac{(10.0\text{ C})(0.5\text{ C})}{(3000\text{ m})^2} = 5000\text{ N}$$

2. Equation 16-2 gives the electric field.

$$E = \left(\frac{1}{4\pi\varepsilon_0}\right)\frac{q_1}{r^2} = \left[\frac{1}{4\pi\left(8.85 \times 10^{-12}\frac{\text{C}^2}{\text{N}\cdot\text{m}^2}\right)}\right]\frac{(10.0\text{ C})}{(3000\text{ m})^2} = 9990\frac{\text{N}}{\text{C}}$$

3. Table 13.4 provides the necessary conversion factors.

$$E = 9990\frac{\text{N}}{\text{C}} \times \frac{\text{m}}{\text{m}} \times \frac{\text{J}}{\text{N}\cdot\text{m}} \times \frac{\text{V}}{\text{W/A}} \times \frac{\text{W}}{\text{J/s}} \times \frac{\text{C/s}}{\text{A}} = 9990\frac{\text{V}}{\text{m}}$$

Figure 16.3 shows the test sphere with charge q_2 initially an infinite distance away from the central sphere. According to Equation 16-1, at an infinite distance, the force on the test sphere is zero. As the test sphere is pushed toward the central sphere, the repulsive force increases, requiring that work be done to overcome this force. The **voltage V** at radius r is defined as the total amount of work input required to move the test sphere from an infinite distance to radius r per charge q_2

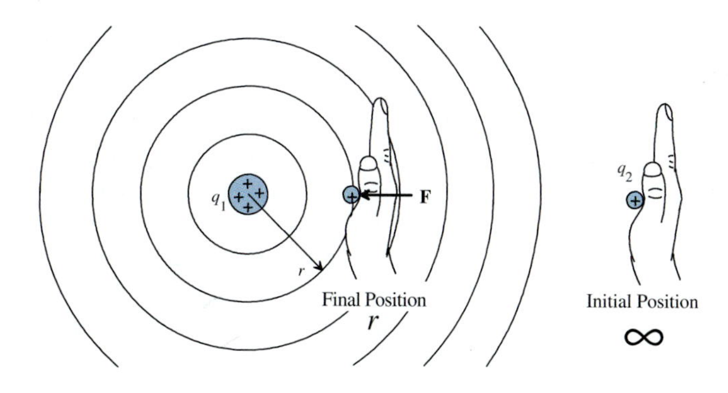

FIGURE 16.3
Illustration showing work required to move a test particle from an infinite distance to a position r from the central sphere. The lines show regions of constant voltage.

$$V \equiv \frac{W}{q_2} \quad (16\text{-}3)$$

From this equation, it is clear that a volt is a joule per coulomb.

EXAMPLE 16.2

Problem Statement: The central sphere has a negative charge of 10.0 C and the test sphere has a negative charge of 0.5 C. In a vacuum, what is the voltage at a distance r_1 of 3000 m?

Solution: The voltage is given by Equation 16-3:

$$V \equiv \frac{1}{q_2}W$$

The work input is calculated as the force exerted over a distance

$$W = \int_{\infty}^{r_1} F\,dr$$

$$V = \frac{1}{q_2}\int_{\infty}^{r_1} F\,dr = \frac{1}{q_2}\int_{\infty}^{r_1}\left[\left(\frac{1}{4\pi\varepsilon_0}\right)\frac{q_1 q_2}{r^2}\right]dr = \frac{q_1 q_2}{q_2}\left(\frac{1}{4\pi\varepsilon_0}\right)\int_{\infty}^{r_1}\left(\frac{1}{r^2}\right)dr$$

$$= q_1\left(\frac{1}{4\pi\varepsilon_0}\right)\int_{\infty}^{r_1}\left(\frac{1}{r^2}\right)dr = q_1\left(\frac{1}{4\pi\varepsilon_0}\right)\left[\frac{-1}{r}\right]_{\infty}^{r_1} = q_1\left(\frac{1}{4\pi\varepsilon_0}\right)\left[\frac{-1}{r_1} - \frac{-1}{\infty}\right]$$

$$= q_1\left(\frac{1}{4\pi\varepsilon_0}\right)\left[\frac{-1}{r_1}\right] = \frac{-q_1}{4\pi\varepsilon_0 r_1} = \frac{-(10.0\ \text{C})}{4\pi\left(8.85 \times 10^{-12}\frac{\text{C}^2}{\text{N}\cdot\text{m}^2}\right)(3000\ \text{m})}$$

$$= -3.00 \times 10^7\frac{\text{N}\cdot\text{m}}{\text{C}} \times \frac{\text{J}}{\text{N}\cdot\text{m}} \times \frac{\text{V}}{\text{J/C}} = -3.00 \times 10^7\ \text{V}$$

Note that the sign on the charge was ignored because we are interested in the magnitude of the voltage. By convention, the voltage near a negative charge is designated negative. Similarly, the voltage near a positive charge is designated positive.

In Example 16.2, we moved negatively charged objects through a vacuum. An example of such an object is a single electron. Although electrons do move through a vacuum (e.g., the cathode-ray tube used in some television and computer screens), it is more common for the electrons to move through a wire. Figure 16.4 shows an electric circuit in which a battery energizes electrons, causing them to flow clockwise through the circuit. If 1 coulomb of electrons is energized with 1 joule, we say these electrons are available with 1 volt of potential energy. Common batteries energize each coulomb of electrons with about 1.5 joules, so these electrons are available with 1.5 volts of potential energy.

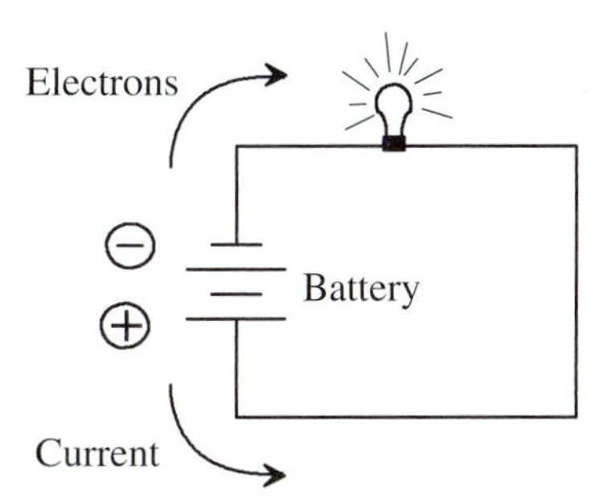

FIGURE 16.4
Electrons flow in a direction opposite to current.

Electrical phenomena were discovered thousands of years ago. To explain these phenomena, Benjamin Franklin (1706–1790) hypothesized that a mysterious "fluid" was responsible. Further, he hypothesized that this fluid flowed from one place to another. Logically, he designated the source of this fluid with a "+" sign, because it had an excess of the fluid. The sink for this fluid was designated with a "−" sign, because it had a deficiency of the fluid. The flow of this fluid was termed **current,** which flows counterclockwise through the circuit in Figure 16.4. It was not until 1897 that Sir Joseph Thomson (1856–1940) discovered that this fluid was actually a flow of electrons. Although Franklin had a 50 percent chance of correctly guessing the source of the fluid, he got it exactly backwards. By the time we learned of his mistake, it was too late—the convention of current flowing from "+" to "−" had already become well established.

16.2 FUNDAMENTALS OF MAGNETISM

Figure 16.5 shows two bar magnets. In Case A, the two magnets have like poles (two souths) facing each other. Because like poles repel each other, mechanical forces (indicated by the hands) are needed to maintain the distance r between the magnets. In Case B, the two magnets have opposite poles (one south and one north) facing each other. Because they attract each other, mechanical forces are needed to maintain the distance r between the magnets.

FIGURE 16.5
Magnets that repel (Case A) and attract (Case B).

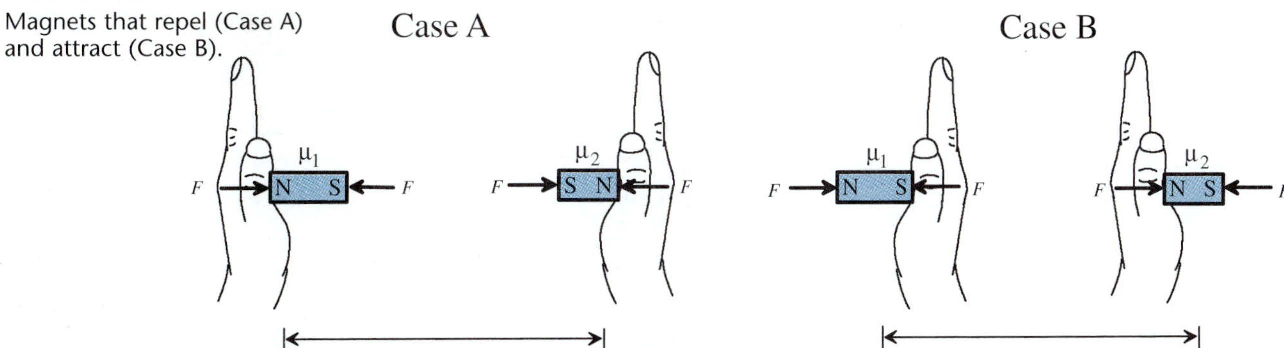

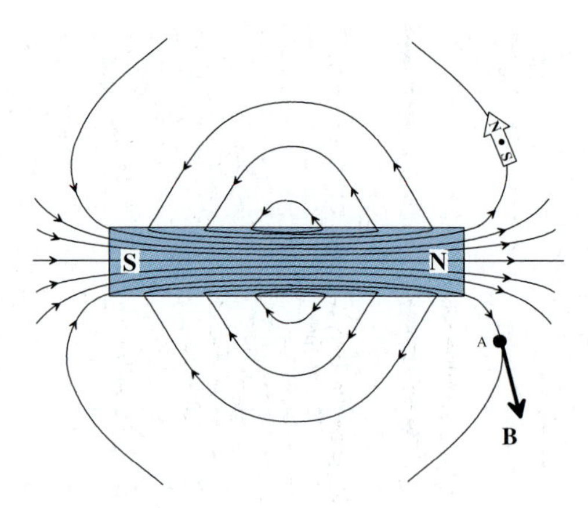

FIGURE 16.6
Lines of force surrounding a bar magnet.

The magnitude of the force of repulsion (or attraction) between the two magnets is

$$F = \left(\frac{3\mu_0}{2\pi}\right)\frac{\mu_1\mu_2}{r^4} \tag{16-4}$$

where μ_1 and μ_2 are the "magnetic dipole moments" of each magnet (a measure of magnetic strength with units A·m^2), r is the distance between the magnet centers (m), and μ_o is the permeability of vacuum ($4\pi \times 10^{-7}$ N/A^2). (*Note:* This equation is strictly valid only if the magnets are in a vacuum and the distance separating them is large relative to the magnet length.)

Figure 16.6 shows the lines of force that surround a bar magnet. By convention, the arrows on the lines of force are oriented away from the north pole and toward the south pole. If a compass is placed near the bar magnet, it orients itself along the lines of force, as shown. At Point A, the tangent to the line of force is the **magnetic field.** The magnetic field **B** is analogous to the electric field **E**.

Not only does a magnetic field **B** surround bar magnets, but it also surrounds a wire in which electric current flows (Figure 16.7). When a wire is wound into a coil, a **solenoid,** the magnetic fields from the current in each loop reinforce each other. Figures 16.8 and 16.9 show the lines of force surrounding a loosely wound and tightly wound solenoid, respectively.

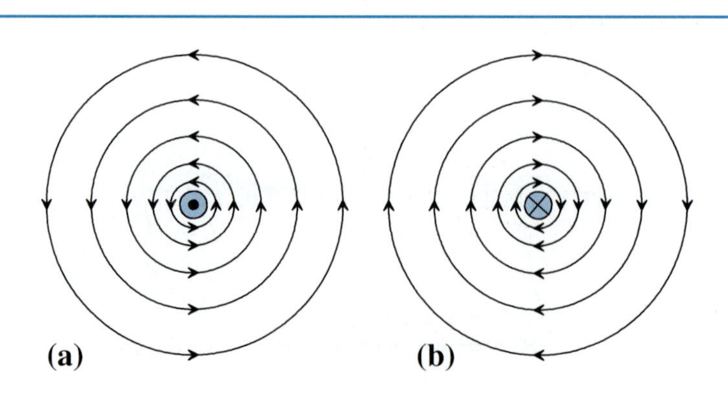

FIGURE 16.7
Lines of force surrounding wire with electric current emerging from the page (a) or going into the page (b).

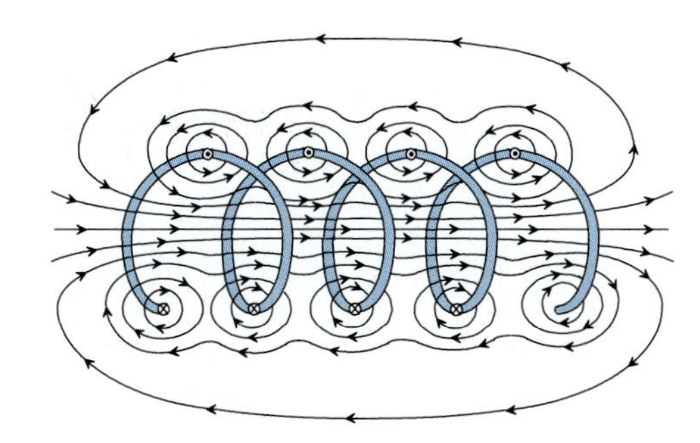

FIGURE 16.8
Lines of force surrounding loosely wound solenoid.

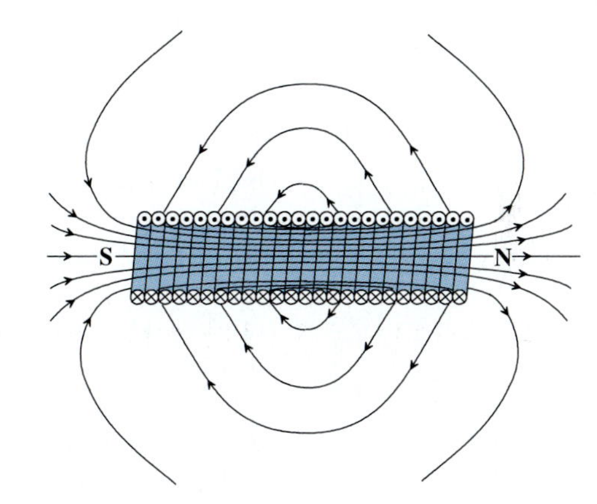

FIGURE 16.9
Lines of force surrounding tightly wound solenoid.

16.3 CONDUCTORS, SEMICONDUCTORS, AND INSULATORS

Figure 16.10 shows a schematic of four beryllium atoms aligned in a crystal. The nucleus of each atom contains four protons (+1 charge) and five neutrons (0 charge). Surrounding the nucleus of each atom are four electrons (−1 charge), two in the inner shell and two in the outermost (or *valence*) shell. Because opposite charges attract, the electrons are attracted to the nucleus.

If an electron is added to Atom 1, it displaces an electron from the valence shell, which jumps to Atom 2. This in turn causes an electron to jump to Atom 3, and so on. In

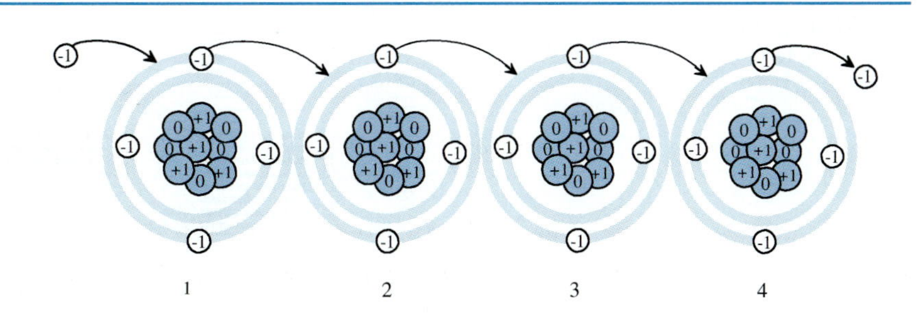

FIGURE 16.10
Schematic representation of four beryllium atoms illustrating the flow of electrons, which is electricity.

conductors, it is easy for electrons to jump to adjacent atoms. In contrast, in **insulators,** the electrons do not easily jump to adjacent atoms.

The electrons in the inner shells are closest to the nucleus. They are strongly attracted to the positively charged protons, and hence have a low energy level. In contrast, the electrons in the valence shell are shielded from the positively charged nucleus by the inner electron shells. These valence electrons primarily "see" other electrons, which have a negative charge. Because they are repelled by these other electrons, they have a high energy level. Interestingly, the energy levels of the electrons are not continuous, but occur in discrete increments or **quanta.** To use a concrete example, this means a given electron may have an energy level of exactly 4 or exactly 5, but none of the energy levels in between.

When energy—in the form of electric fields, heat, light, or particle impact—is added to an electron in a lower shell, it moves to a higher shell. Conversely, when an electron in a higher shell moves to a lower shell, it gives off energy in the form of heat or light.

Figure 16.11 shows a schematic representation of the energy levels in various atoms. The electrons in the lowest energy levels (nonconducting bands) are tightly bound to the nucleus and cannot jump to adjacent atoms. Electrons in the highest energy levels (conducting bands) are energetic enough to jump to adjacent atoms. In a *conductor,* the conducting and nonconducting bands are adjacent to each other without a **gap.** Adding just a little energy is sufficient to move electrons into the conducting band. In a conductor, the flow of electricity reduces as temperature increases because random atomic motion increases, making it difficult for electrons to flow to adjacent atoms.

In contrast, in an *insulator,* there is a large gap between the conducting and nonconducting bands. Because electrons are forbidden from occupying this gap, it is sometimes called the **forbidden zone.** A very large amount of energy is needed to move electrons from the nonconducting bands to the conducting bands. Under normal circumstances, not enough energy is available, so the electrons are confined to the nonconducting bands and are never able to jump to adjacent atoms.

A **semiconductor** has a small gap between the conducting and nonconducting bands. At low temperatures, very few of the electrons are energetic enough to occupy the conducting bands, hence the semiconductor conducts electricity almost as poorly as an insulator. At high temperatures, many electrons are energetic enough to occupy the conducting bands, hence the semiconductor conducts electricity almost as well as a conductor.

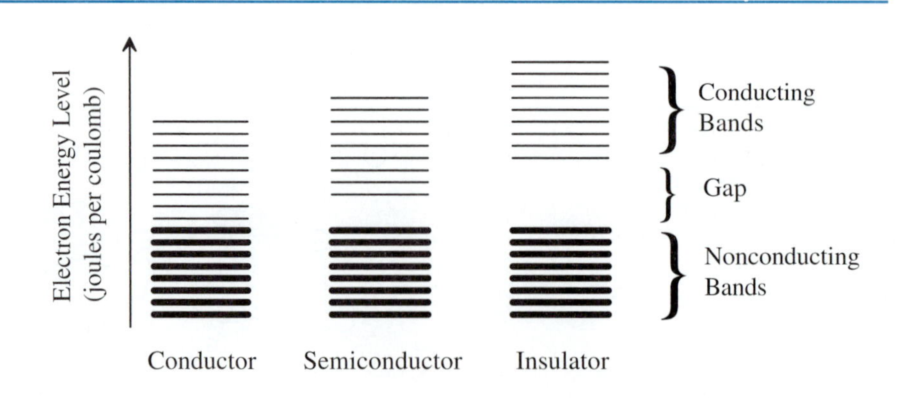

FIGURE 16.11
Schematic representation of electron energy levels in conductors, semiconductors, and insulators.

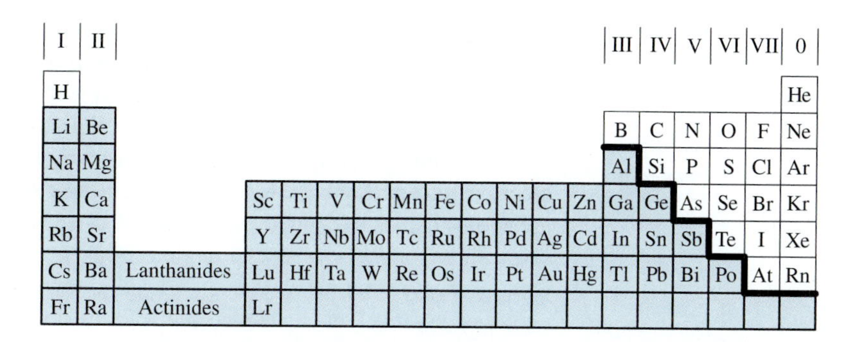

FIGURE 16.12
Periodic table showing metallic character of the elements. (*Note:* Symbols are defined in Table 11.2)

Figure 16.12 shows the periodic table of elements. The elements in the darkened boxes are *conductors* and the elements in light boxes are *insulators.* The heavy stair-step line separates the conductors from the insulators. The placement of this line is somewhat arbitrary as the elements near the line—such as silicon (Si) and germanium (Ge)—are semiconductors. At low temperature, they behave as insulators; and at high temperatures, they behave as conductors.

Silicon is a Group IV element, meaning each atom has four **valence electrons.** Figure 16.13(a) shows a silicon crystal. Each atom shares valence electrons with its neighbors, so that each atom effectively has eight valence electrons, making it stable. Figure 16.13(b) shows a silicon crystal **"doped"** with a small amount of phosphorus (P, a Group V element). There are nine valence electrons surrounding the phosphorus atom—one more than the stable eight. This extra electron is free to roam. Semiconductors with extra roaming electrons are called **n type**. Figure 16.13(c) shows a silicon crystal doped with a small amount of aluminum (Al, a Group III element). There are seven valence electrons surrounding the aluminum atom—one less than the stable eight. It needs an extra electron to make it stable. Semiconductors that are deficient in stabilizing electrons are called **p type.** The figure uses an open circle—a so-called **hole**—to represent the missing electron.

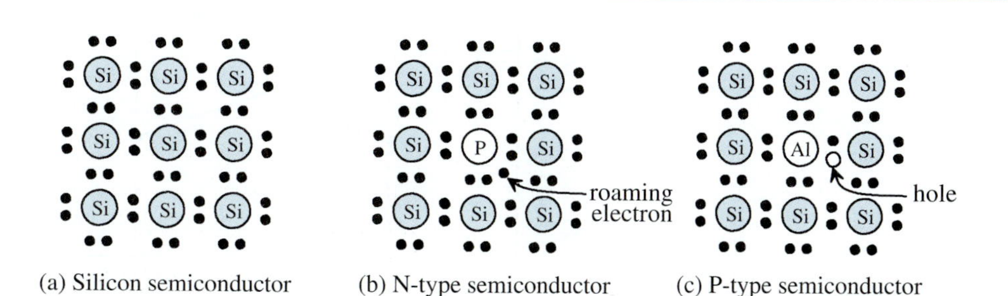

FIGURE 16.13
Semiconductor crystals.

Figure 16.14 shows a silicon crystal with an n-type region adjacent to a p-type region; the interface is called a **pn junction.** To simplify the diagram, only the roaming electrons and holes are shown. In the **depletion layer,** the extra roaming electrons from the phosphorus atom have moved rightward to fill the deficient aluminum atoms.

Deep into the n-type semiconductor away from the depletion layer, the numbers of electrons and protons are equal; however, the electron energy level is high because there are more than the required eight needed to fill each valence shell. Deep into the p-type semiconductor away from the depletion layer, the numbers of electrons and protons are equal; however, the electron energy level is low because there are fewer than the required eight needed to fill each valence shell. In the n side of the depletion layer, the phosphorus atoms have shed their extra electron, so there are more protons than electrons. In the p side of the depletion layer, the aluminum atoms have gained the electrons they need to make a stable valence shell, so there are more electrons than protons. In the depletion layer, the electron energy level varies with distance from the junction. In principle, the electron energy level throughout the semiconductor could be measured with a *voltmeter,* a device that reports the electron energy level in units of *volts* (joules per coulomb).

Figure 16.15 provides a side view of a schematic representation of a semiconductor in which a negative voltage is imposed on the right and a positive voltage is imposed on the left. Electrons in the conducting band move from right to left in response to the voltage

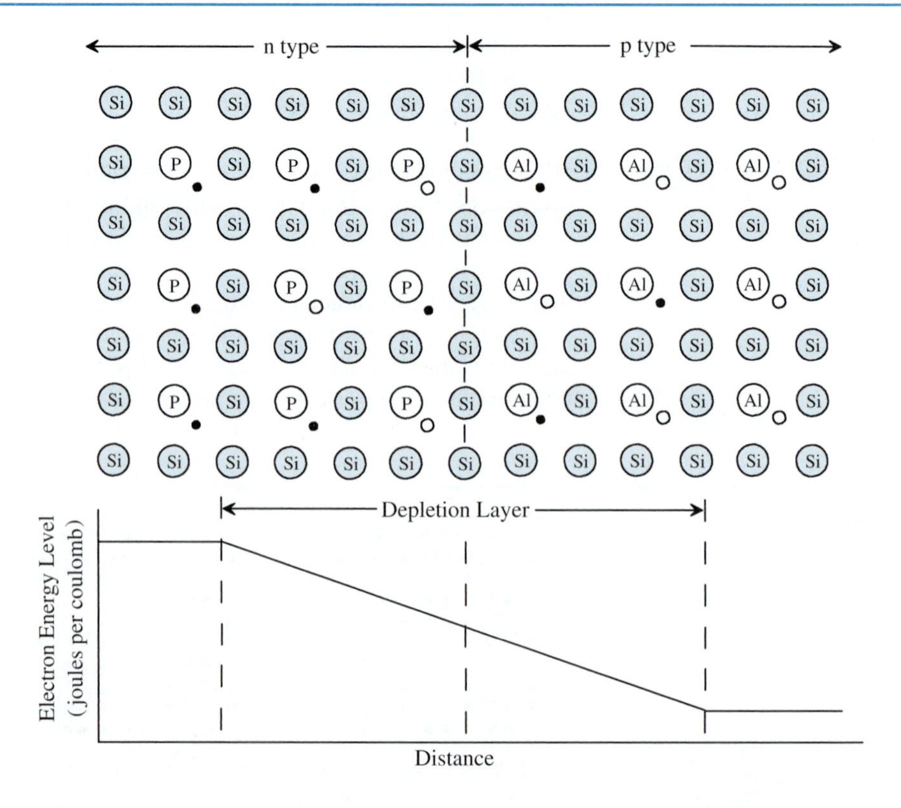

FIGURE 16.14
A pn junction.

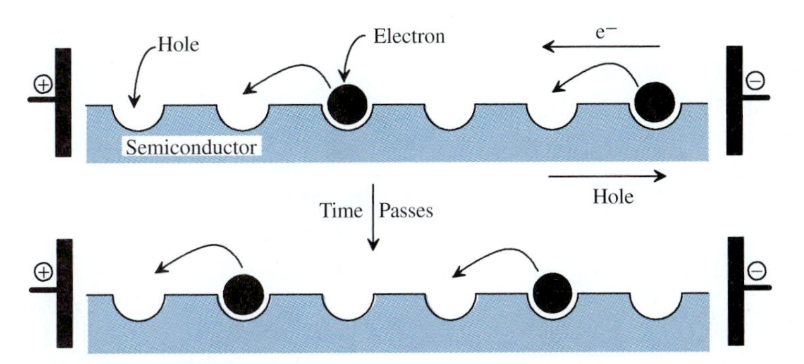

FIGURE 16.15
Schematic representation of electron travel through a semiconductor.

difference. Although holes do not physically move, it appears that they move in a direction opposite to the electron flow. Holes move in the same direction as current, so current may be visualized as a flow of holes.

16.4 ELECTROMAGNETIC RADIATION

Whenever a charged object accelerates, it emits **electromagnetic radiation.** A common example is electrons accelerating (oscillating) within a radio broadcast antenna. The electromagnetic radiation is emitted in the form of radio waves, which we use for communication.

Figure 16.16 shows an atom in which the outer electron is initially in a high-energy band. When the electron collapses to a low-energy band, it must accelerate, which causes it to emit electromagnetic radiation. The difference in energy between the low- and high-energy bands is discrete or *quantized*; therefore, the amount of energy in the electromagnetic radiation is also quantized.

As shown in Figure 16.16, the electromagnetic radiation leaves the atom as a "packet" of oscillating electric and magnetic fields called a **photon**. The energy of a single photon E (J) is

$$E = h\nu \tag{16-5}$$

where h is Planck's constant ($6.6260755 \times 10^{-34}$ J·s) and ν is the frequency that the electric and magnetic fields oscillate (s^{-1}). Equation 16-5 indicates that a high-energy photon has a rapid oscillation.

In a given medium, the speed of photon propagation u (m/s) is the wavelength λ (m) times the oscillating frequency of the electric (or magnetic) field ν (s^{-1}).

$$u = \lambda\nu \tag{16-6}$$

In a vacuum, the speed of a photon is designated c and has the value 2.99792458×10^{8} m/s. Substituting Equation 16-6 into 16-5 gives an alternate expression for the energy of a single photon in a vacuum

$$E = h\frac{c}{\lambda} \tag{16-7}$$

Equation 16-7 indicates that a high-energy photon has a short wavelength.

Figure 16.17 shows the wavelengths of various kinds of electromagnetic radiation.

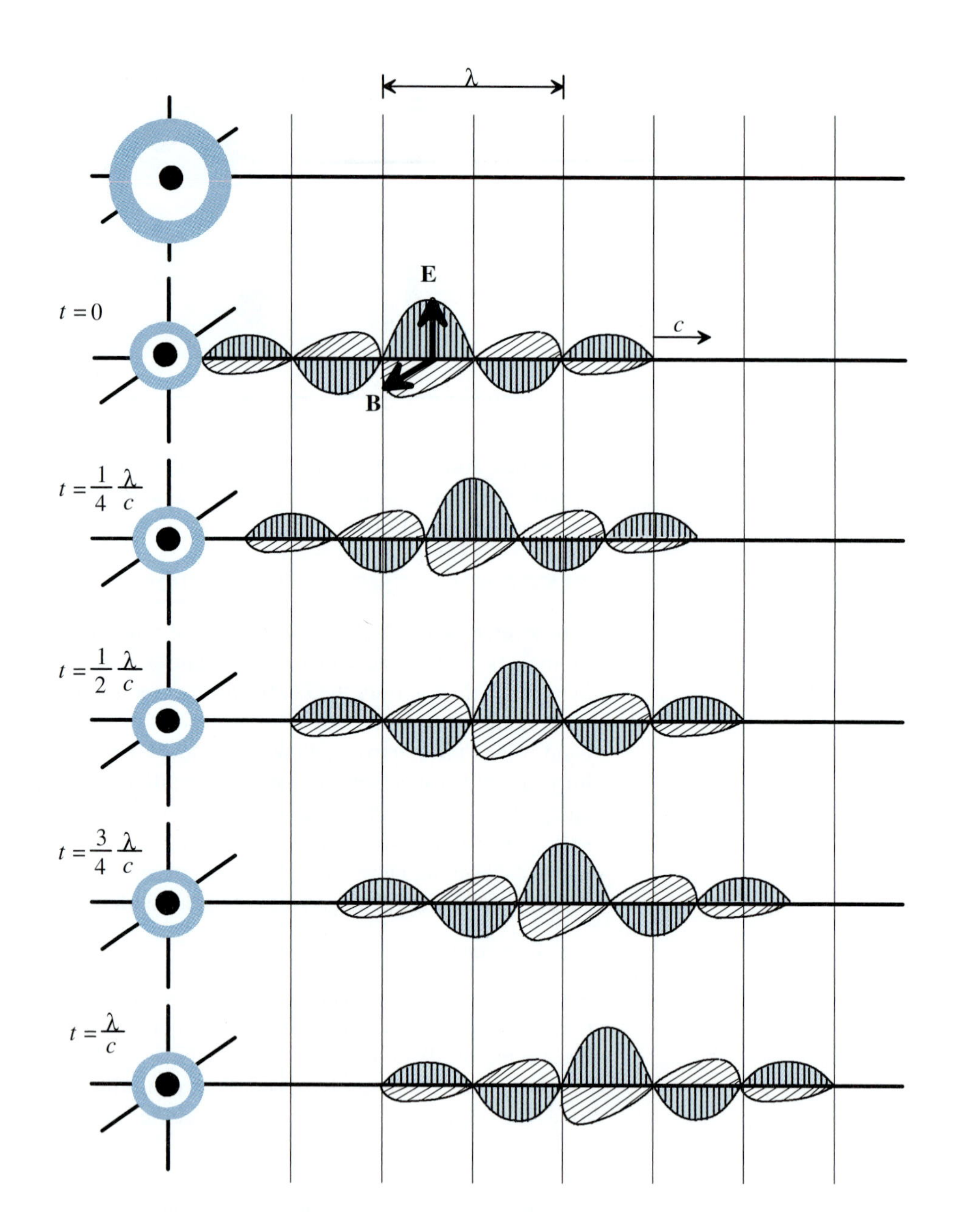

FIGURE 16.16
A photon emitted from an electron that transitions from a high-energy to a low-energy level. The electric field strength is shown in the vertical dimension and the magnetic field strength is shown in the horizontal dimension. The photon is traveling in a vacuum.

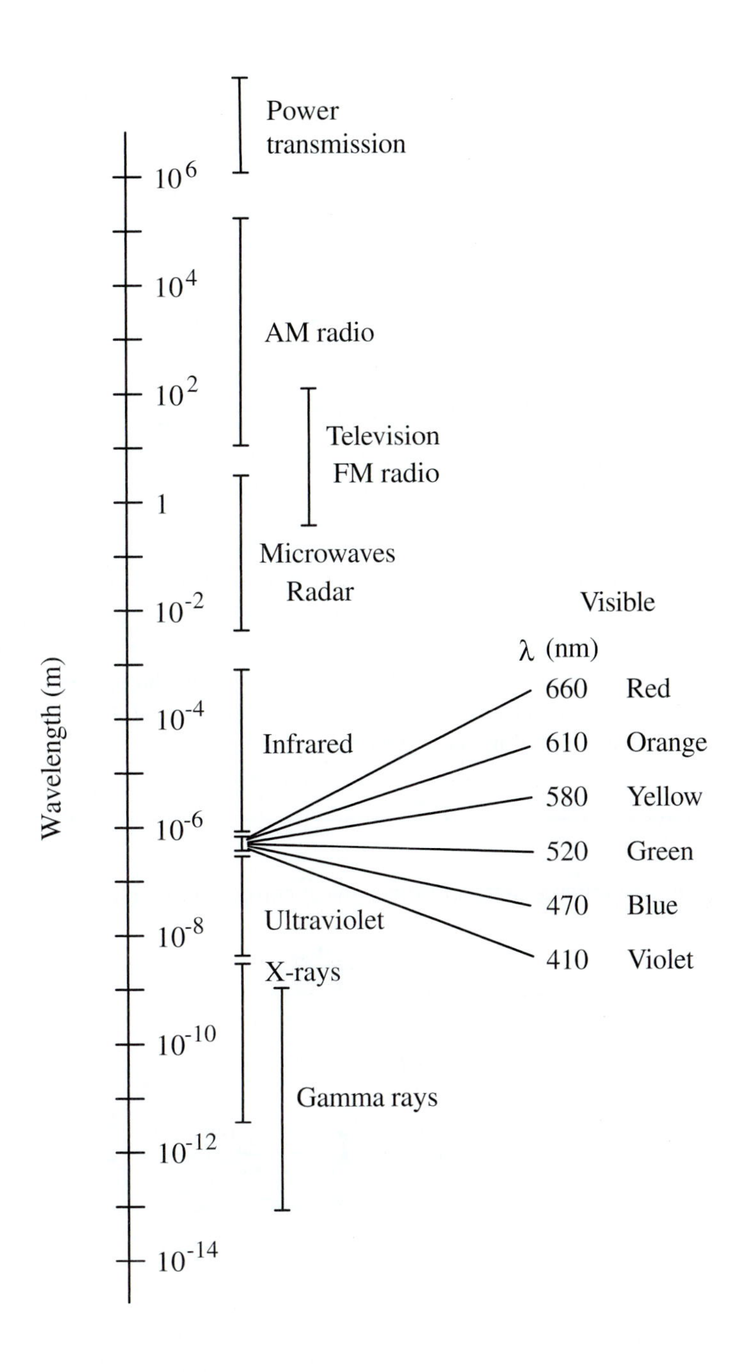

FIGURE 16.17
Wavelengths of electromagnetic radiation.

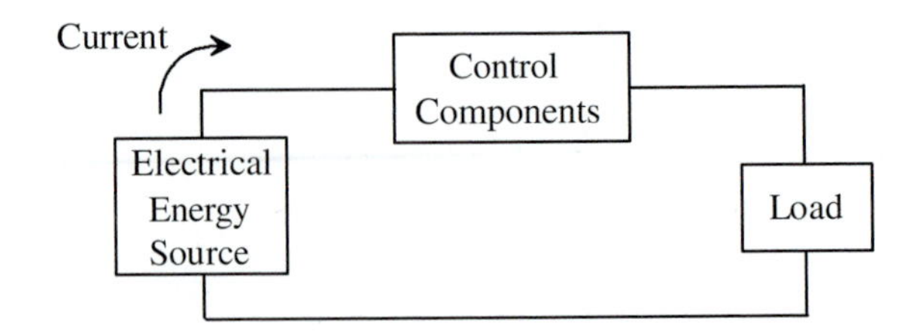

FIGURE 16.18
Generic electric circuit.

16.5 CIRCUIT COMPONENTS

Figure 16.18 shows a generic electric circuit. It is composed of three elements: an electrical energy source, a load, and control components. The **electrical energy source** provides the voltage required to push electrons through the circuit. The **load** transforms the electrical energy into other forms of energy (e.g., light, heat, motion). The **control components** modify or regulate the current coming from the energy source.

Because it is not possible to see electrons as they flow through electric circuits, electricity can be somewhat abstract. To help you visualize how circuit components work, occasionally we will make analogies to mechanical systems. For example, Figure 16.19(a) shows that electric current is analogous to fluid flow, and Figure 16.19(b) shows that voltage is analogous to pressure.

FIGURE 16.19
Analogies between electrical components (left) and mechanical components (right).

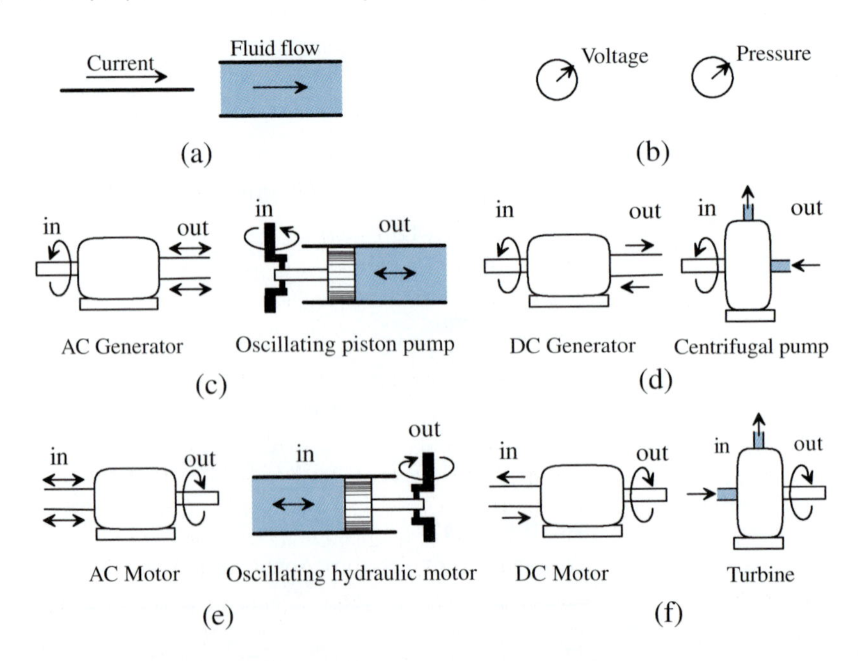

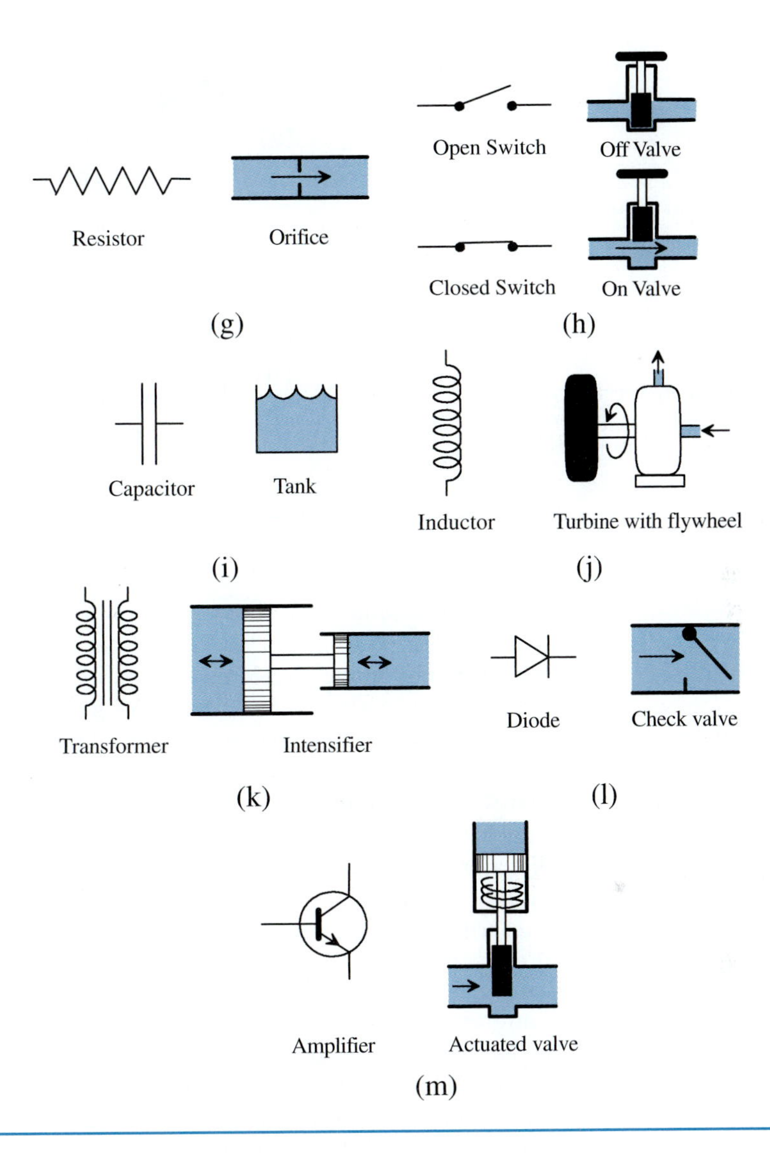

FIGURE 16.19
Continued.

16.5.1 Electrical Energy Sources

Methods for producing electrical energy from other energy sources are described below.

Generators.

Figure 16.20 illustrates two kinds of electric **generators,** devices that transform mechanical energy into electrical energy. A coil of wire rotates between the north and south poles of a magnet. A changing magnetic field induces current to flow in the wire, the reverse of the phenomenon illustrated in Figure 16.9. In Figure 16.20(a), the current flows through a **slip ring,** which consists of a spring-loaded stationary **brush**—typically

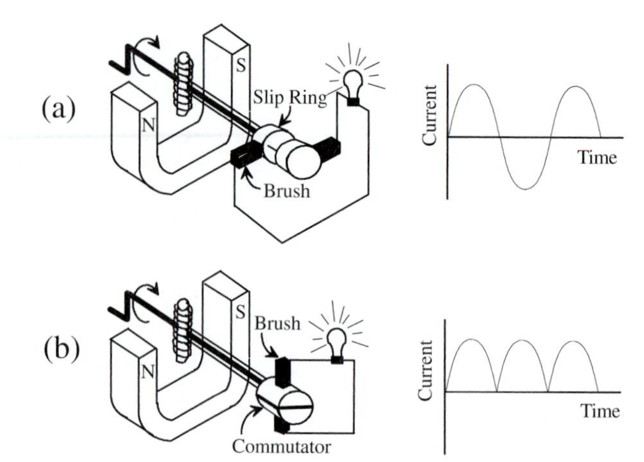

FIGURE 16.20
Electric generators convert mechanical energy into (a) alternating current electrical energy and (b) direct current electrical energy.

made from carbon—that slides against a rotating ring. Because the coil reverses direction with respect to the magnetic field, the direction of the current in the coil reverses. The electricity in the circuit also reverses direction, and is termed **alternating current** (AC). Figure 16.19(c) shows that an AC generator is analogous to an oscillating piston pump, a device with a rotating crank that drives a piston causing fluid to oscillate within a pipe. In Figure 16.20(b), the generator produces **direct current** (DC) in which the current flows in one direction only. This is accomplished by using a **commutator,** a slip ring that is split into two halves by a nonconductor. Just as the current is about to switch direction, each brush contacts the other portion of the commutator, allowing the current to flow in a single direction. Figure 16.19(d) shows that a DC generator is analogous to a centrifugal pump, a device in which fluid enters the center of a spinning disk and is flung to the periphery, causing the fluid to flow continuously.

Batteries.

Figure 16.21 shows a **battery,** a device that converts chemical energy into electrical energy. In this example, the battery has two chambers, one with an electrode (**anode**) of lithium (Li) metal immersed in water and the other with a wire electrode (**cathode**) immersed in an aqueous solution of silver chloride ions (Ag^+ and Cl^-). The silver is a positive ion, which attracts valence electrons from the lithium. The valence electrons from the lithium flow through the wire and join with the silver ions, causing them to plate onto the wire as neutral metal (Ag). The lithium becomes a positive ion that dissolves in the water. To neutralize the positive charge of the lithium ions, chloride ions transfer across the barrier forming a solution of lithium chloride ions (Li^+ and Cl^-). As the electrons flow from the lithium to the silver, they can power a load—in this case, a light bulb. The flow of electrons will continue until either the lithium electrode completely dissolves, or the dissolved silver ions completely plate onto the wire. When either of these occur, the battery is "dead." The chemistry shown in Figure 16.21 is but one example—there are many other possibilities. Some batteries must be disposed of once they are dead (*primary batteries*), whereas other batteries can be recharged using an electrical energy input (*secondary batteries*).

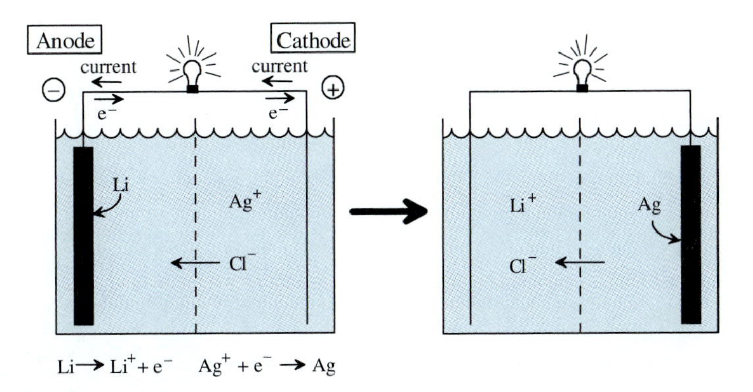

FIGURE 16.21
Battery.

Fuel cell.

Figure 16.22 shows a **fuel cell** which, like a battery, converts chemical energy into electrical energy. The major difference is that a battery is completely self-contained; no chemicals must be added while it is operating. In contrast, a fuel cell requires a continuous input of chemicals while it is operating. With few exceptions, fuel cells operate with an input of hydrogen and oxygen. At the anode, electrons are removed from the hydrogen to form protons (H^+) which diffuse through the porous membrane. The electrons flow through the wire to the cathode where they combine with protons and oxygen to form water. The reactions are

$$2\,H_2 \rightarrow 4\,H^+ + 4\,e^- \qquad \text{Anode}$$

$$4\,e^- + 4\,H^+ + O_2 \rightarrow 2\,H_2O \qquad \text{Cathode}$$

$$2\,H_2 + O_2 \rightarrow 2\,H_2O \qquad \text{Sum}$$

The summation reaction is simply the reaction of hydrogen and oxygen to form water. If this reaction were to occur in a combustor, the temperature would rise dramatically as the energy is released in the form of heat. When this reaction occurs in a fuel cell, the temperature is

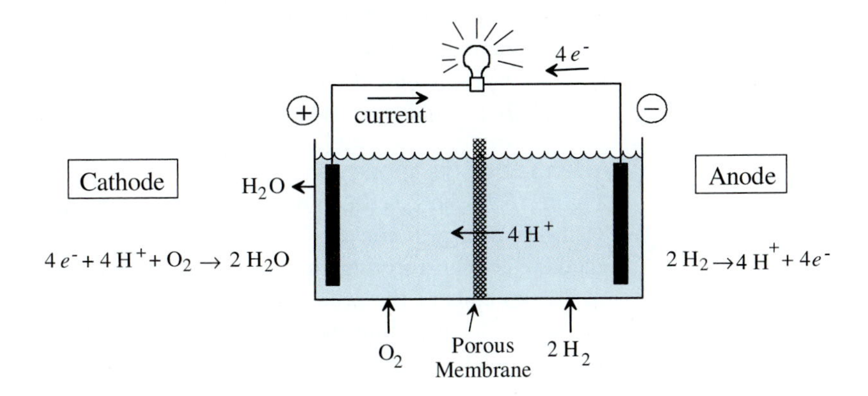

FIGURE 16.22
Fuel cell.

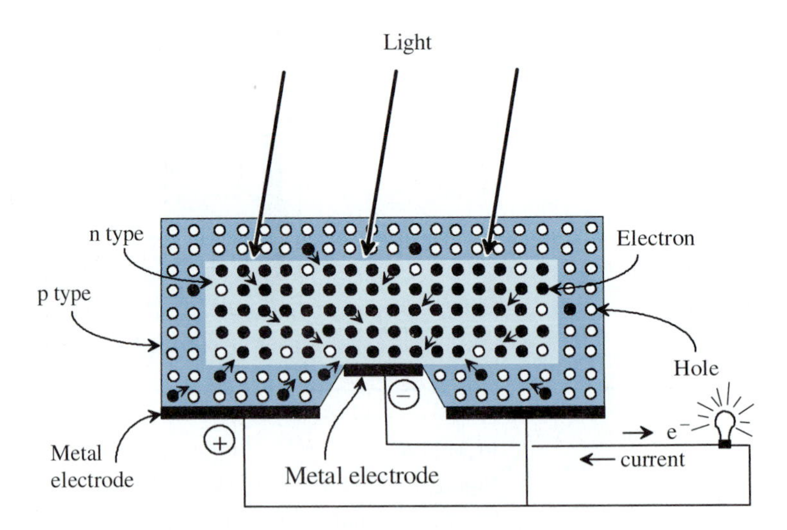

FIGURE 16.23
Photovoltaic cell.

close to room temperature, and the energy is released in the form of electricity. In theory, the **efficiency** of a fuel cell is 100 percent, meaning 100 percent of the input (chemical energy) can be converted to output (electrical energy). In practice, the efficiency is 40 to 80 percent, meaning 40 to 80 percent of the input chemical energy is converted to electrical energy output; the remaining input energy is lost as heat.

Photovoltaic cell.

Figure 16.23 shows a schematic representation of a **photovoltaic cell.** It consists of a pn junction with metal electrodes on the lower surface that connect to the p-type and n-type semiconductors. The pn junction produces a voltage difference between the electrodes, as illustrated in Figure 16.14. When light strikes the semiconductors, electrons jump into the conducting bands, are collected by the negative electrode, flow through the load, and return via the positive electrode. Less expensive photovoltaic cells use amorphous (i.e., noncrystalline) silicon in the pn junction; research cells convert up to about 13 percent of solar energy into electricity. More expensive photovoltaic cells use crystalline silicon in the pn junction; research cells convert up to about 30 percent of solar energy into electricity. In general, commercially available amorphous and crystalline photovoltaic cells are about half as efficient as research cells.

Thermoelectric generator.

Figure 16.24 shows a **thermoelectric generator,** a device that converts thermal energy into electrical energy. It consists of p-type and n-type semiconductors joined at the bottom by a metal plate, thus forming a pn junction. The bottom metal plate is heated and metal plates at the top ends of the semiconductors are cooled. The thermal energy at the pn junction energizes electrons into the conducting band, thus generating an electric current that can power a load. Thermoelectric generators are typically constructed of lead telluride with added impurities that make it a p-type or n-type semiconductor. Because it can make elec-

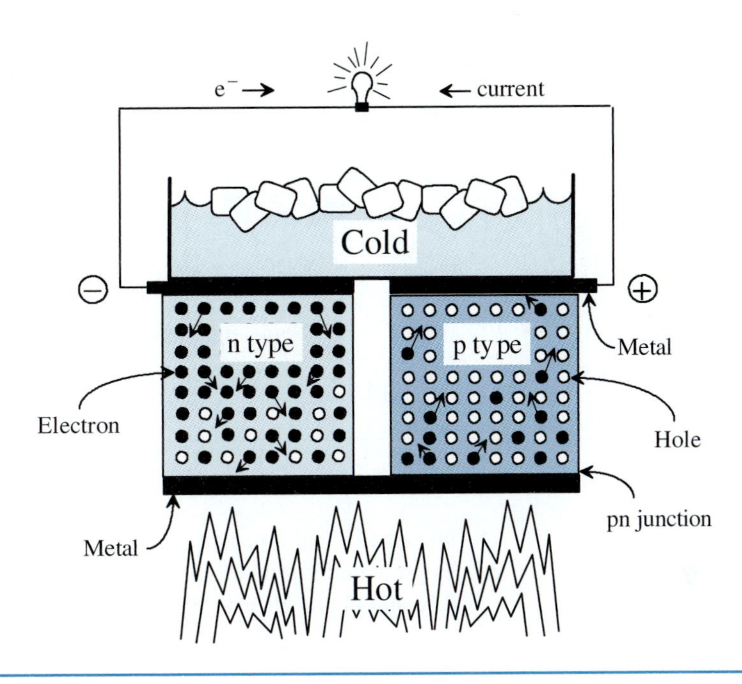

FIGURE 16.24
Thermoelectric generator.

tricity from thermal energy without moving parts, it is attractive for some specialized applications. For example, deep-space satellites can operate for decades using a radioisotope to provide the thermal energy. Unfortunately, the efficiency is very low (5 to 1 percent).

16.5.2 Loads

Methods for converting electrical energy into other types of energy are described below.

Motors.

Motors are devices that convert electrical energy into mechanical energy. Figures 16.25(a) and 16.25(b) show motors that produce rotary motion from alternating current (AC) and direct current (DC), respectively. A comparison to Figures 16.20(a) and 16.20(b) makes it clear that a motor is a generator operating in reverse. Figure 16.19(e) shows that an AC motor is analogous to an oscillating hydraulic motor, a device in which oscillating fluid pushes a piston that drives a crank shaft. Figure 16.19(f) shows that a DC motor is analogous to a turbine, a device in which continuously flowing fluid pushes against vanes causing a shaft to rotate.

Resistors.

Resistors are devices that convert electrical energy into thermal energy. Figure 16.26(a) shows that when an electric field from one of the previously described energy sources is imposed upon a conductor, electrons jump from atom to atom. Rather than traveling straight, their path is highly tortuous, which causes frictional heating. The heat can be used for a variety of practical purposes, such as cooking food or warming homes. The symbols for resistors [Figure 16.26(b)] emphasize the tortuous path of the electrons. Figure

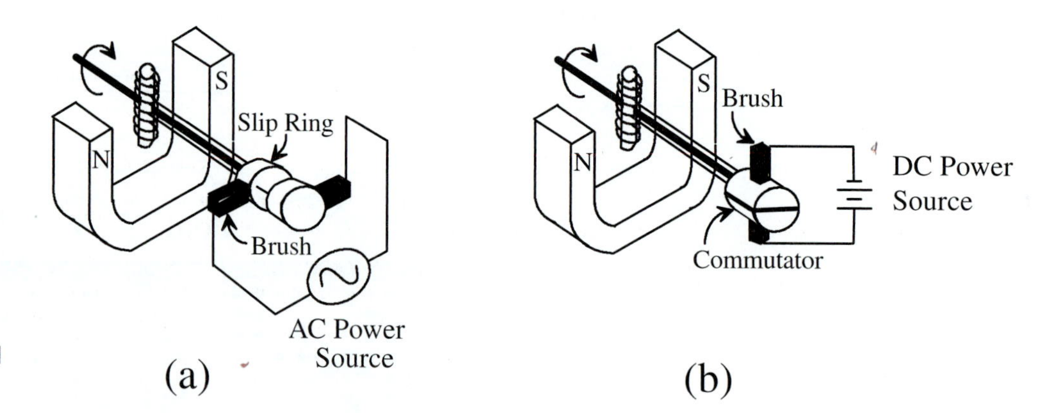

FIGURE 16.25
Electric motors that use (a) alternating current and (b) direct current.

16.19(g) shows that a resistor is analogous to an orifice, a device in which a constriction causes pressure drop when fluid flows through it.

The amount of electric power dissipated by the resistor is

$$P = iV \tag{16-8}$$

where P is the electric power dissipated (W or J/s), i is the current (A or C/s), and V is the voltage drop across the resistor (V or J/C). The voltage drop is proportional to the current

$$V = Ri \tag{16-9}$$

where the resistance R (Ω or V/A or J·s/C^2) is the proportionality constant. (*Note:* Equation 16-9 is called *Ohm's law* and is traditionally written $V = i\,R$.) Substituting Equation 16-9 into Equation 16-8 shows how resistance affects the electric power dissipation:

FIGURE 16.26
(a) Electron flow through a resistor. (b) Symbols for fixed and variable resistors.

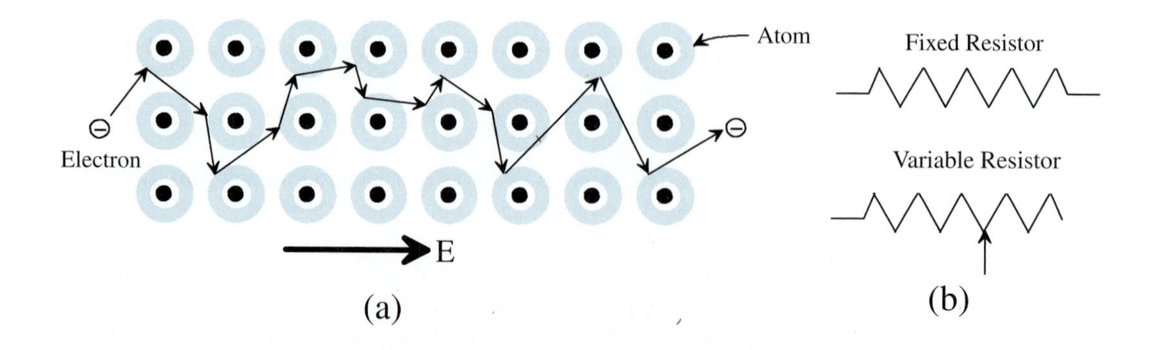

$$P = Ri^2 = \frac{V^2}{R} \quad (16\text{-}10)$$

Incandescent Light.

Figure 16.27(a) shows a schematic of an **incandescent light,** a device that converts electrical energy into light. As a result of an applied voltage, current flows through the thin tungsten wire in the bulb. Because of the wire resistance, the electrical energy is converted to thermal energy, which increases the wire temperature to the point that it radiates visible light. To prevent the tungsten wire from oxidizing and burning out, the wire must not contact oxygen in the air. Modern bulbs are filled with an inert gas (e.g., argon, nitrogen), whereas early bulbs used a vacuum. Incandescent lights are very inefficient; only 1.1 to 3.2 percent of the energy input is converted to visible light and the remainder is waste heat.

Fluorescent Light.

Figure 16.27(b) shows a schematic of a **fluorescent light.** It consists of a glass tube filled with low-pressure mercury (Hg) vapor plus an inert gas (e.g., argon). At opposite ends of the tubes are electrodes that apply a voltage. The voltage is sufficiently high to strip electrons off of the mercury to form a plasma, which can conduct electricity. The voltage provides energy to activate electrons into a higher energy level. When the electrons return to a lower energy level, they emit photons of ultraviolet light, which are not visible to humans. The ultraviolet light is absorbed by a fluorescent coating on the interior of the glass tube, which re-emits longer-wavelength visible light. External circuits (**starters**) are needed to preheat the lamp so the mercury vaporizes. Also, additional external circuits (**ballasts**) are needed to prevent the plasma current from getting too large. Because of their complexity, fluorescent lights are more expensive than incandescent lights, but their higher energy efficiency (4.4 to 12.2 percent) makes them cost-effective.

Neon Light.

A neon light is similar to a fluorescent light, except it is filled with neon gas rather than a mixture of mercury and argon. When the electrons in neon are excited to a higher energy level and return to a lower energy level, they emit visible light directly, so a fluorescent

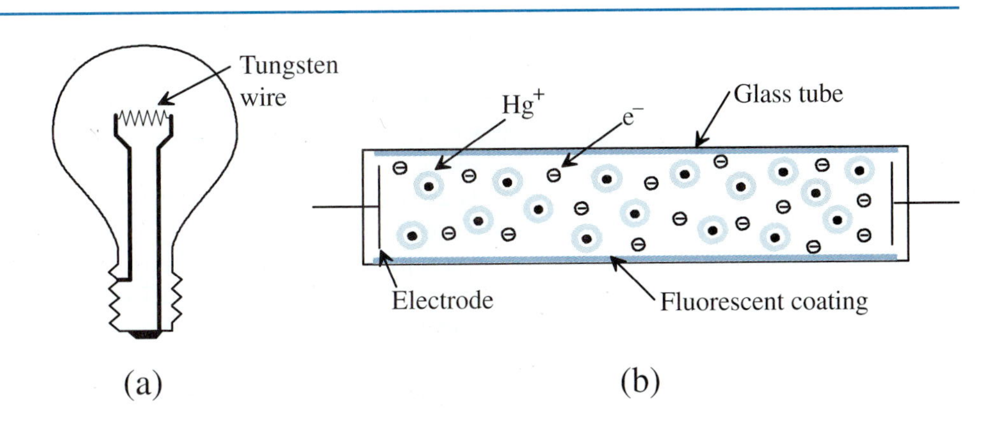

FIGURE 16.27
Schematic of (a) incandescent light bulb and (b) fluorescent bulb.

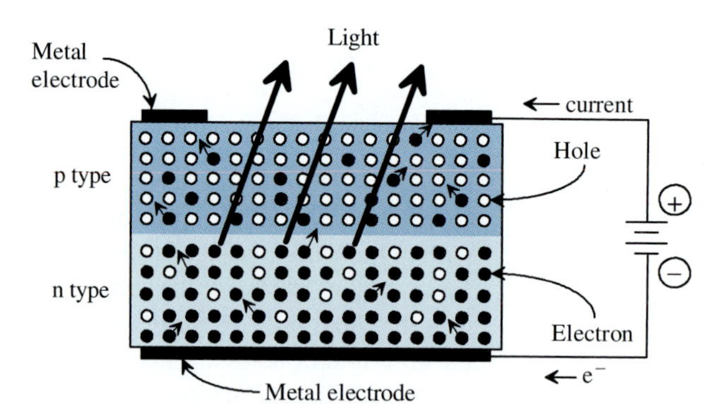

FIGURE 16.28
Light-emitting diode.

coating is not needed. The glass tube is often shaped into letters allowing neon lamps to be used for advertising signs. Instead of neon, other gases may be used allowing a variety of colors to be produced.

Light-Emitting Diode.

Figure 16.28 shows **a light-emitting diode** (LED). It consists of a pn junction in which negative voltage is applied to the n-type semiconductor and positive voltage is applied to the p-type semiconductor. As a result of the applied voltage, the electrons in the semiconductor are energized and emit light. (*Note:* The LED operates like a photovoltaic device in reverse.) Depending on the semiconductor chemistry, each **primary color** (red, blue, green) can be produced. Any color, or white light, can be formed by mixing LEDs. The efficiency depends on the color and is similar to, or slightly higher than, that of incandescent lights. The service life of LEDs is 5 to 10 years of continuous use. They are commonly used for indicator lights on electronic equipment. Also, they are being considered for illumination and flat-panel display screens.

Laser.

"Laser" is an acronym meaning "light amplification by stimulated emission of radiation." A laser can be constructed from many materials, including gases, liquids, and solids. Figure 16.29(a) shows that electrons in the laser material are energized to a high energy level by electrical, light, chemical, or nuclear energy input. Then, a passing photon stimulates the energized electrons to emit another photon, causing the electrons to return to a low energy level. Figure 16.29(b) shows that the laser material is located between two mirrors, one fully reflective and the other partially reflective. Photons bounce between the two mirrors and stimulate light emissions from the energized electrons in the laser material. Some of the bouncing photons exit from the partially reflective mirror as a parallel beam of **coherent** (in-phase), **monochromatic** (single-wavelength) light. A semiconductor laser can be constructed by placing a partially reflective mirror on the upper surface and a fully reflective mirror on the lower surface of the LED shown in Figure 16.28. The efficiency of some lasers is about 30%. High-power lasers are used for welding or nuclear fusion experiments. Low-power lasers are used for communicating through fiber optic cables and reading information encoded on compact discs.

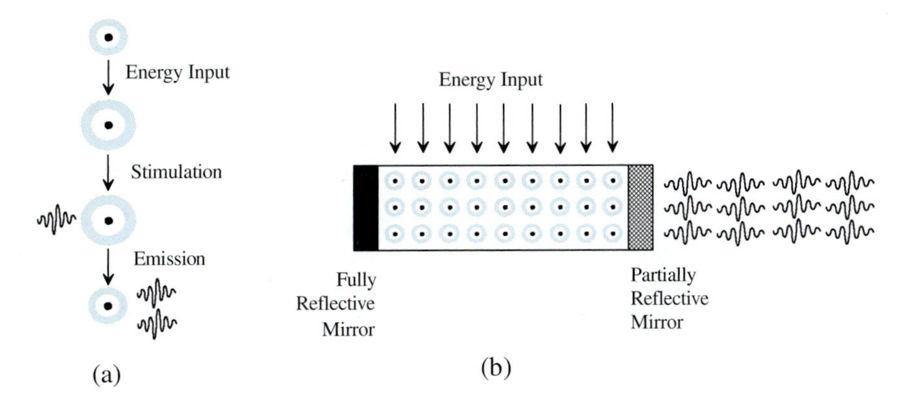

FIGURE 16.29
Laser. (a) Sequence leading to light emission.
(b) Generic laser device.

Cathode-Ray Tube.

Figure 16.30 shows a simplified schematic of a **cathode-ray tube,** which is used in television screens and computer monitors. A heater increases the temperature of the emitter to about 1050 K, causing electrons to "boil off" because of the **thermionic effect** (also called the *Edison effect*). The emitter has a negative voltage and the accelerator has a positive voltage, which accelerates the emitted electrons to a high velocity, forming an electron beam. Deflector electrodes alter the path of the beam by attracting it to the positive electrode and repelling it from the negative electrode. One pair of electrodes controls the east/west direction and the other controls the north/south direction. By carefully controlling the deflection electrodes, the electron beam is made to scan the entire screen. (*Note:* Some cathode-ray tubes use magnets rather than electrodes to control the electron beam.) The interior of the glass screen is coated with phosphors that emit light when struck by the electron beam; the phosphor chemistry determines the color that is emitted. In a black-and-white screen, the entire surface is coated with white-emitting phosphor. The intensity of the white light is controlled by the intensity of the electron beam. In a color screen, the screen is coated with tiny phosphor dots; each emits only one of the primary colors (red, green, blue). There are three electron beams, each directed to a single phosphor

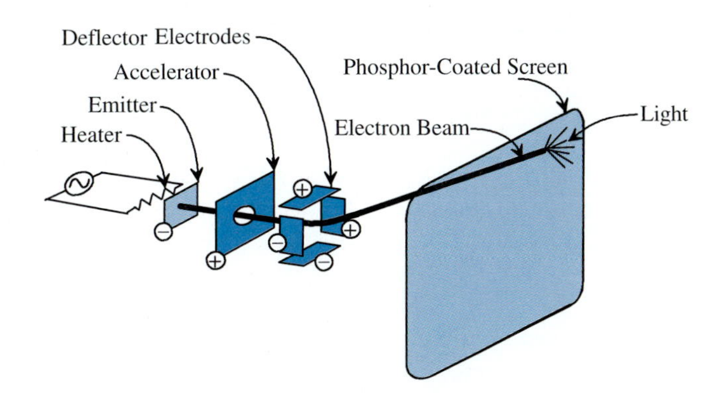

FIGURE 16.30
Simplified diagram of a cathode-ray tube.

Shuji Nakamura: Inventor of the Blue and Green LED

The potential impact of blue and green LEDs is remarkable. Compared to conventional LED lasers, blue LED lasers increase the amount of information stored on compact discs by about four times, and they increase the resolution of laser printers. Blue and green LEDs complete the primary colors (red, green, blue), thus allowing any color (including white) to be produced by LEDs. This opens the possibility of producing LED room lights that are more efficient and long-lived than incandescent lights. Also, LEDs can now be used in flat-panel color displays, traffic lights, and automotive taillights and turn signals.

Shuji Nakamura was born, raised, and educated on an isolated Japanese island. In 1979, he earned a master's degree and started working for Nichia, a small electronics company located on the island. Nichia made conventional red and yellowish green LEDs, but earned only a tiny share of the market dominated by giants such as Sanyo, Sharp, and Toshiba. Oddly, Nichia's sales department blamed Nakamura for their tiny market share, which made him angry. In 1988, he resolved to find the "holy grail," a blue LED. His direct boss refused to support his quest, so he bypassed him and spoke directly to the company CEO. He demanded a $3.3 million research budget and a year off to study a new semiconductor fabrication technique at the University of Florida. Amazingly, even though Nakamura had low seniority, the CEO granted his demands.

Lacking a PhD, at the University of Florida, Nakamura was treated as a "hands-on" engineer rather than a researcher. Working 7 days a week, 16 hours per day for 10 months, he constructed the semiconductor fabricator.

In 1989, when he returned to Japan to work on his blue LED, he had to select between two semiconductor types: gallium nitride and zinc selenide. The giants of industry and academia were focusing on zinc selenide. Unwilling to compete with them, Nakamura decided to study gallium nitride.

Nakamura's hands-on experience building semiconductor fabricators proved to be invaluable. He had the skills to customize his machine to grow gallium nitride crystals that were increasingly brighter and long lived. He documented his research achievements in 146 technical papers, six books, and 10 book chapters, all published without the knowledge of the CEO who feared that publishing might give competitors an advantage. In 1994, on the basis of his prodigious output, his alma matter awarded him a doctorate degree.

Although Nakamura's inventions allowed Nichia to increase its annual sales from $100 million to $400 million, his annual salary was only $100,000. Eventually, he sought opportunities outside of Nichia. Among his 17 job offers were some very lucrative industrial positions, but he decided to become a professor at the University of California in Santa Barbara, where he continues his inventive works.

Adapted from: Glenn Zorpette, *Scientific American,* pp. 30–31, August 2000, and M. G. Craford, N. Holonyak, Jr., and F. A. Kish, Jr., *Scientific American,* pp. 63–67, February 2001.

color. Adjusting the beam intensity regulates the color intensity. By mixing different intensities of each primary color, any other color can be produced. The entire apparatus is placed in a glass enclosure, which is evacuated so gases do not interfere with the electron beam.

Thermoelectric Cooler.

Figure 16.31 shows a schematic of a **thermoelectric cooler,** a thermoelectric generator operated in reverse. When DC power is connected to the p-type and n-type semiconduc-

FIGURE 16.31
Thermoelectric cooler.

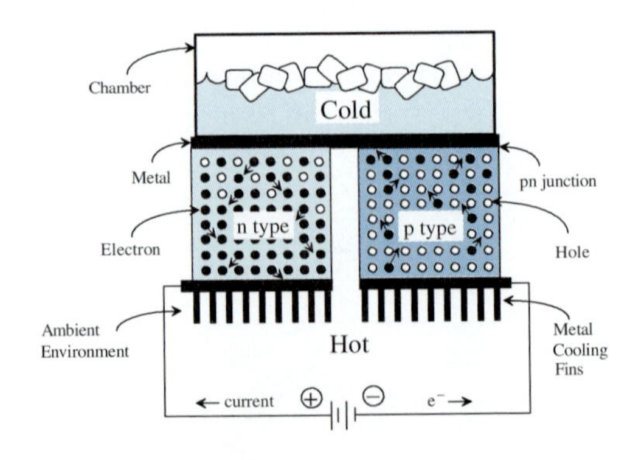

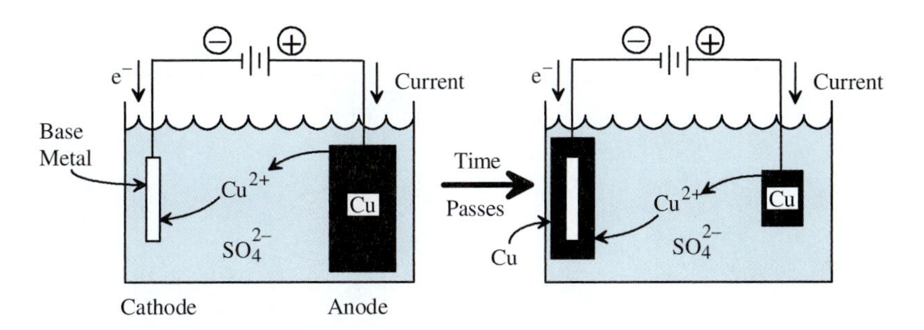

FIGURE 16.32
Example of electroplating.

tors as shown in Figure 16.31, thermal energy is "pumped" out of the cold chamber and rejected to the hot ambient environment through the cooling fins. Thermoelectric coolers have the advantage of no moving parts, but the efficiency is low compared to other cooler options. They are used in specialty applications, such as transporting medicine or picnic lunches.

Electroplating.

Figure 16.32 shows copper being electroplated onto a base metal. At the anode, electrons are removed from copper metal (Cu), allowing copper ions (Cu^{2+}) to dissolve into the aqueous solution. At the cathode, the electrons recombine with the copper ions to form copper metal, which plates onto the base metal. Common examples of **electroplating** include gold plating onto jewelry and chromium plating onto automobile components.

Electrochemistry.

Many chemical reactions that do not occur spontaneously can be performed by using electricity. For example, hydrogen and oxygen will combine spontaneously to form water, as shown in the fuel cell illustrated in Figure 16.22. This reaction can be reversed using an input of electrical energy (Figure 16.33). Many important industrial reactions are performed electrochemically, such as aluminum production from bauxite ore and chlorine production from salt.

Electroseparations.

Electricity can be used to separate chemical species. For example, Figure 16.34 shows **electrodialysis,** a process that separates salts from aqueous solutions. A DC energy source supplies the cathode with a negative voltage and the anode with a positive voltage. Positively charged ions (**cations**)—such as sodium (Na^+)—are attracted to the cathode,

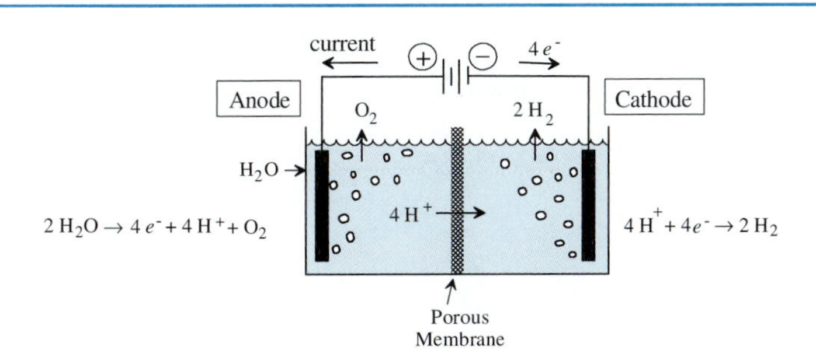

FIGURE 16.33
Electrolysis of water to form hydrogen and oxygen.

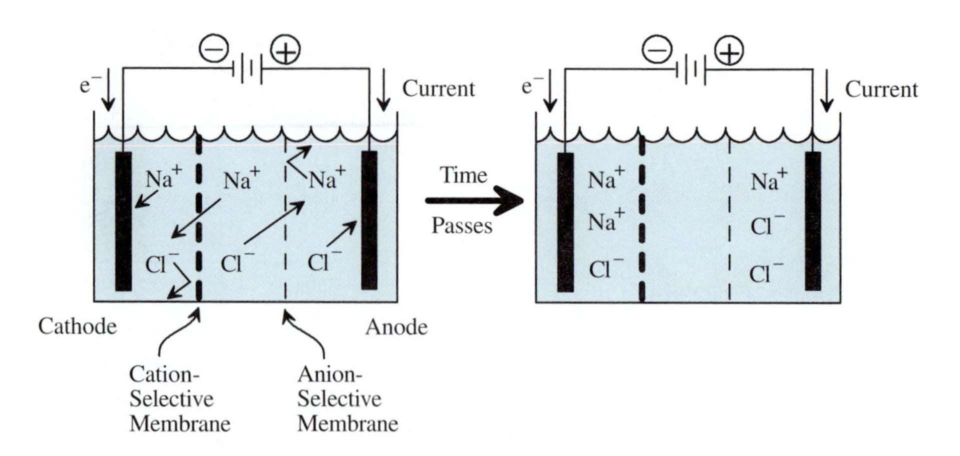

FIGURE 16.34
Electrodialysis.

and freely pass through the cation-selective membrane. Negatively charged ions (**anions**)—such as chloride (Cl^-)—are attracted to the anode, and freely pass through the anion-selective membrane. The net effect is that the salt concentration in the central chamber decreases, and the salt concentration in the cathode and anode chambers increases. Notice that the cathode chamber accumulates cations and that the anode chamber accumulates anions, which requires a great deal of energy. Mixing the contents of the cathode and anode chambers during electrodialysis reduces the energy requirements.

16.5.3 Control Components

Methods for controlling electrical energy are described below.

Switch.

Figure 16.35(a) shows a **switch,** a device that can disconnect a wire and stop current. A switch is a common device used to turn lights and equipment on or off. Figure 16.19(h) shows that a switch is analogous to an on/off valve, a device that can completely block a pipe and stop flowing fluid.

Relay.

Figure 16.35(b) shows a *normally open relay.* A spring (not shown) keeps the switch normally open, but when current flows through the solenoid, the magnetic field forces the switch to close. Figure 16.35(c) shows a *normally closed relay.* A spring (not

FIGURE 16.35
Switches and relays.

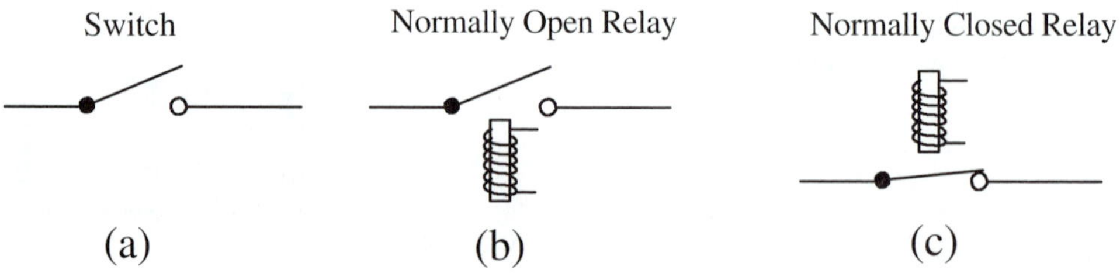

Switch (a)   Normally Open Relay (b)   Normally Closed Relay (c)

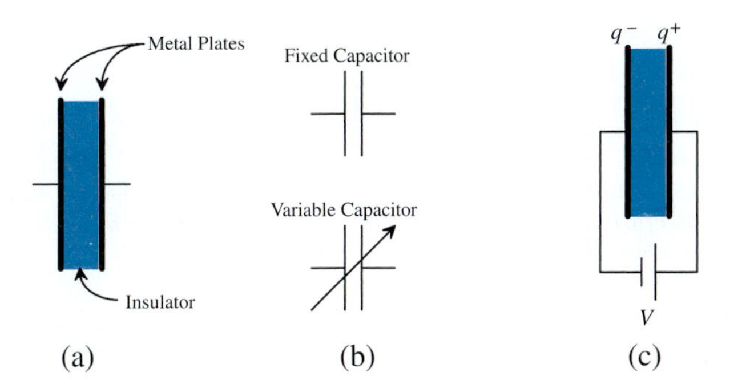

FIGURE 16.36
Capacitor. (a) Components. (b) Symbols. (c) DC voltage source coupled to a capacitor.

shown) keeps the switch normally closed, but when current flows through the solenoid, the magnetic field forces the switch to open. The purpose of a relay is to control the large amount of energy that can flow through the switch with the small amount of energy that flows through the solenoid.

Capacitor.

Figure 16.36(a) shows a **capacitor,** a device that stores electric charge. It consists of two metal plates separated by an electrical insulator. Figure 16.36(b) shows symbols for a capacitor, one fixed and one variable. Figure 16.36(c) shows a DC voltage source connected to a capacitor. The **capacitance** C (C/V or C^2/J or farad, F) is defined as follows:

$$C \equiv \frac{q}{V} \qquad (16\text{-}11)$$

where q is the charge (C) on one plate (either positive or negative) and V is the applied voltage (V). The capacitance is determined by the area and separation of the metal plates, and the electrical properties of the insulator. Figure 16.19(i) shows that a capacitor is analogous to a tank that contains fluid.

Equation 16-11 may be rearranged as follows:

$$q = CV \qquad (16\text{-}12)$$

Taking the derivative of each side with respect to time gives:

$$\frac{dq}{dt} = C\frac{dV}{dt} \qquad (16\text{-}13)$$

The charge changes as a result of a current flowing into, or out of, the capacitor; therefore,

$$i = \frac{dq}{dt} = C\frac{dV}{dt} \qquad (16\text{-}14)$$

Inductor.

Figure 16.37(a) shows the symbol for an **inductor,** a wire coil that produces a magnetic field when current flows through it. (This phenomenon was described earlier in Figures 16.8 and 16.9.) To understand how an inductor works, imagine we assemble the circuit

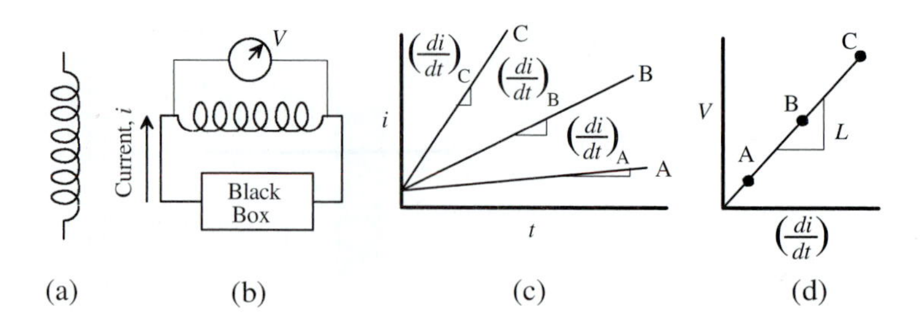

FIGURE 16.37 Inductor. (a) Symbol. (b) Circuit to test inductor. (c) Specified current from the black box. (d) Measured voltage across the inductor.

shown in Figure 16.37(b). It contains a voltmeter and a sophisticated "black box" that supplies any current that we specify. When we conduct our experiment, we specify that the current increase linearly [Figure 16.37(c)] with the following slopes:

$$\left(\frac{di}{dt}\right)_A \quad \left(\frac{di}{dt}\right)_B \quad \left(\frac{di}{dt}\right)_C$$

Figure 16.37(d) shows that when we plot our results, the measured voltage is linear and passes through the origin. These data satisfy the following equation:

$$V = L\frac{di}{dt} \tag{16-15}$$

where the slope L is defined as the ***inductance*** (V·s/A or J/A^2 or henry, H). Note this equation is similar to the one for a capacitor (Equation 16-14), except that voltage and current are interchanged.

Equation 16-15 is also similar to Newton's second law, which may be written as follows:

$$F = m\frac{dv}{dt} \tag{16-16}$$

Comparing these two equations reveals that inductance plays the same role as mass m and gives inertia to an electric circuit. This electrical inertia results because increasing the current increases the magnetic field in the inductor. To add energy to the magnetic field, the current must be more energetic; hence, the voltage must increase. Figure 16.19(j) shows that an inductor is analogous to a turbine with a flywheel attached to the rotating shaft. The flywheel gives the turbine inertia, and therefore the input fluid must have a higher pressure to increase the rotational speed.

Transformer.

Figure 16.38(a) shows a **transformer,** a device that transforms input voltage V_1 into output voltage V_2. It consists of two wire coils that loop around a core of iron, or other ferromagnetic material. An AC energy source is coupled to the primary coil and the load is coupled to the secondary coil. The alternating current in the primary coil produces an oscillating magnetic field in the iron core, which induces an alternating current in the secondary coil. In an ideal transformer with no energy loss, the voltages and currents are related as follows:

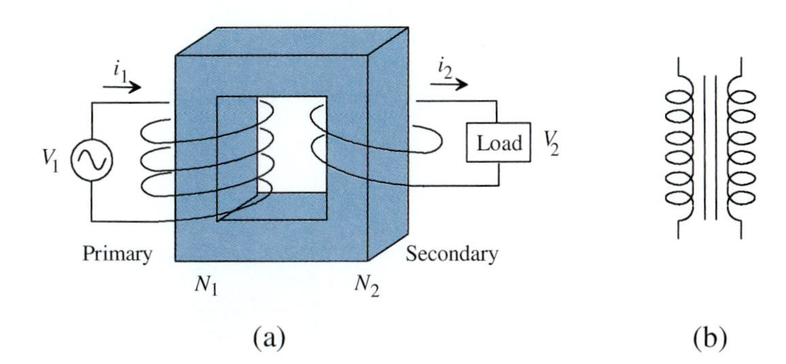

FIGURE 16.38
Transformer. (a) The primary and secondary coils loop around an iron core. (b) Transformer symbol.

$$\frac{V_2}{V_1} = \frac{N_2}{N_1} \tag{16-17}$$

$$\frac{i_2}{i_1} = \frac{N_1}{N_2} \tag{16-18}$$

where N is the number of turns in each coil. Figure 16.38(b) shows the symbol for a transformer. A common application of a transformer is to reduce the high voltage in the electricity distribution lines down to the low voltage used in a home. Figure 16.19(k) shows that a transformer is analogous to an **intensifier,** a device with two pistons that convert high-volume, low-pressure fluid into low-volume, high-pressure fluid.

Vacuum Tube Diode.

A **diode** is a device that allows current to flow in one direction only. Figure 16.39(a) shows a schematic of a vacuum tube diode. It has a heater that increases the temperature of the cathode, which "boils off" electrons because of the thermionic effect. The cathode has a negative voltage and the anode has a positive voltage, so the electrons emitted from the cathode are collected by the anode. The electrons can flow in one direction only: from the cathode to the anode [Figure 16.39(b)]. Although heavily used in the past, today vacuum tube diodes are used only in niche markets (e.g., high-power radio transmitters) because of the availability of solid-state diodes.

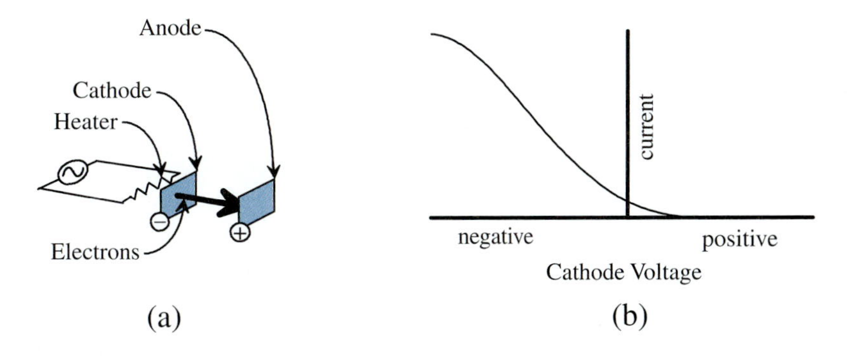

FIGURE 16.39
Vacuum tube diode.

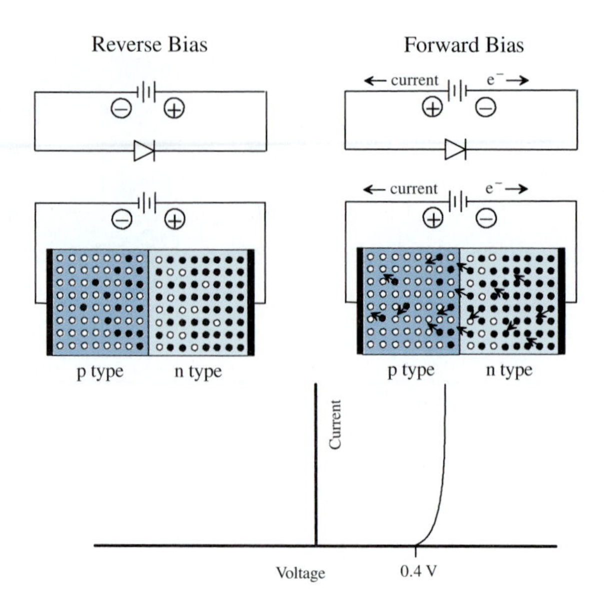

FIGURE 16.40
Solid-state diode.

Solid-State Diode.

Figure 16.40 shows a solid-state diode, which consists of a pn junction. When reverse-bias voltage is applied, no current flows; however, the depletion zone widens. When forward-bias voltage is applied, the depletion zone narrows and current flows, provided the applied voltage is above about 0.4 V. The symbol for the diode resembles an arrow showing the direction of current flow. Figure 16.19(l) shows that a diode is analogous to a **check valve,** a device with a flap that allows fluid to flow in one direction only.

Vacuum Tube Amplifier.

In an **amplifier**, a low-energy input controls a high-energy output, and thus "amplifies" the input. Figure 16.41 shows a schematic of a vacuum tube *amplifier.* It is similar to a vacuum tube diode except for the **grid**—located between the cathode and anode—which

FIGURE 16.41
Vacuum tube amplifier.

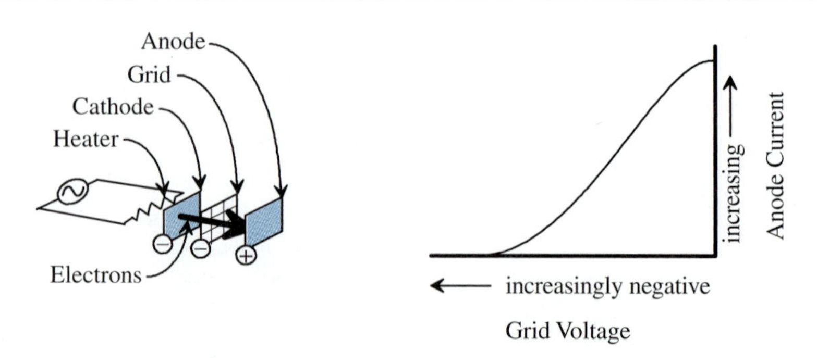

controls the flow of electrons from the cathode to the anode. When the grid voltage is strongly negative, it repels the emitted electrons and prevents them from reaching the anode. When the grid voltage is zero, electrons reach the anode unimpeded. A small amount of grid energy regulates a much larger amount of energy flowing from the cathode to the anode, and thus amplifies the input signal to the grid. Vacuum tubes are used in some niche markets, such as electric guitar amplifiers, high-end audio amplifiers, and high-power radio amplifiers.

Bipolar Transistor.

Figure 16.42(a) shows a **bipolar transistor,** an amplifier that regulates output current using an applied input voltage. Figure 16.42(b) shows its symbol. The bipolar transistor has a p-type semiconductor sandwiched between two n-type semiconductors. (This npn type is more common than the pnp type, which has an n-type semiconductor sandwiched between two p-type semiconductors.) The p-type semiconductor is very thin and "lightly doped," meaning it has a small concentration of Group III elements. In contrast, the n-type semiconductor is thick and "heavily doped," meaning it has a large concentration of Group V elements. Figure 16.42(c) shows that a positive voltage is applied to the **collector** and **base,** and a negative voltage is applied to the **emitter.** One pn junction has a reverse bias and the other has a forward bias. Electrons easily pass through the forward-bias pn junction into the p-type semiconductor. These "injected" electrons cannot easily combine with holes, because the p-type semiconductor is lightly doped. They are free to roam and can travel either to the base or collector. Because the p-type semiconductor is very thin compared to the collector, it is easier for the injected electrons to flow to the collector than the base. (For every electron that flows to the base, about 100 electrons flow to the collector, i.e., $i_c \approx 100\, i_b$.) Figure 16.42(d) shows that the collector current i_c varies approximately linearly with base voltage V_b but is not strongly affected by the collector voltage V_c. Figure 16.19(m) shows that an amplifier is analogous to an *actuated valve,* a device with a spring-loaded piston that opens and closes the valve. A small amount of fluid flowing into the piston regulates the much larger volume of fluid flowing through the valve.

Field-Effect Transistor.

Figure 16.43(a) shows a **field-effect transistor,** another amplifier that regulates output current by using an applied input voltage. Figure 16.43(b) shows its symbol. The field-effect transistor consists of an n-type semiconductor with a p-type semiconductor insert. The insert narrows a portion of the n-type semiconductor, making a channel. Figure 16.43(c) shows that a positive voltage is appled to the **drain,** and a negative voltage is applied to the **gate.** The pn junction has a reverse bias, so negligible current flows through the gate ($i_g \approx 0$). As increasingly negative voltage is applied to the gate, the depletion zone widens, which reduces the area for electrons to flow through the channel. When the gate voltage reaches V_{pinch}, the depletion zone is so wide that it blocks all electron flow. Figure 16.43(d) shows how the drain current i_d varies with gate voltage V_g; the drain current is not substantially affected by the drain voltage V_d.

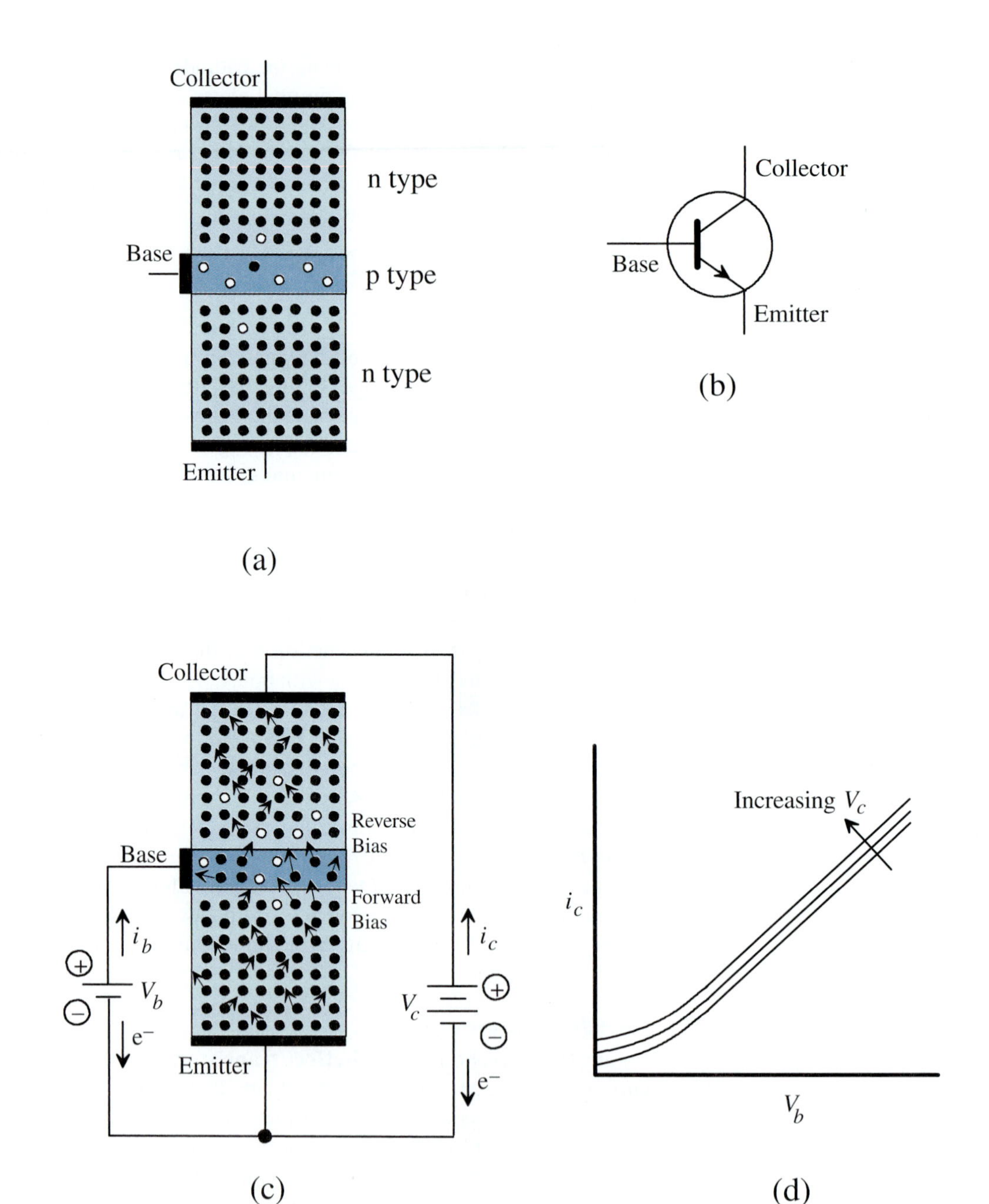

FIGURE 16.42

Bipolar transistor. (a) Schematic representation of npn bipolar transistor. (b) Symbol. (c) Voltage applied to npn bipolar transistor. (d) Current output resulting from applied voltages.

16.6 CIRCUIT ANALYSIS

A **circuit** is a series of electrical components connected by wires to accomplish a purpose. Figure 16.44 illustrates a generic circuit, which has the following features:

- A **node** is a point connecting two or more wires, shown as a dot in Figure 16.44.
- A **loop** is a closed path in a circuit.

When analyzing circuits, Kirchhoff's laws are enormously useful.

Kirchhoff's current law (*KCL*) may be stated as follows:

> *At a node, the sum of input currents equals the sum of output currents.*

The KCL may be understood by making an analogy to a pipe. Figure 16.45 shows a branched pipe. The inlet has 5 gal/min flowing into it. Clearly, because a pipe cannot store fluid, 5 gal/min must flow out. In this case, 2 gal/min flows out of one pipe and 3 gal/min flows out the other.

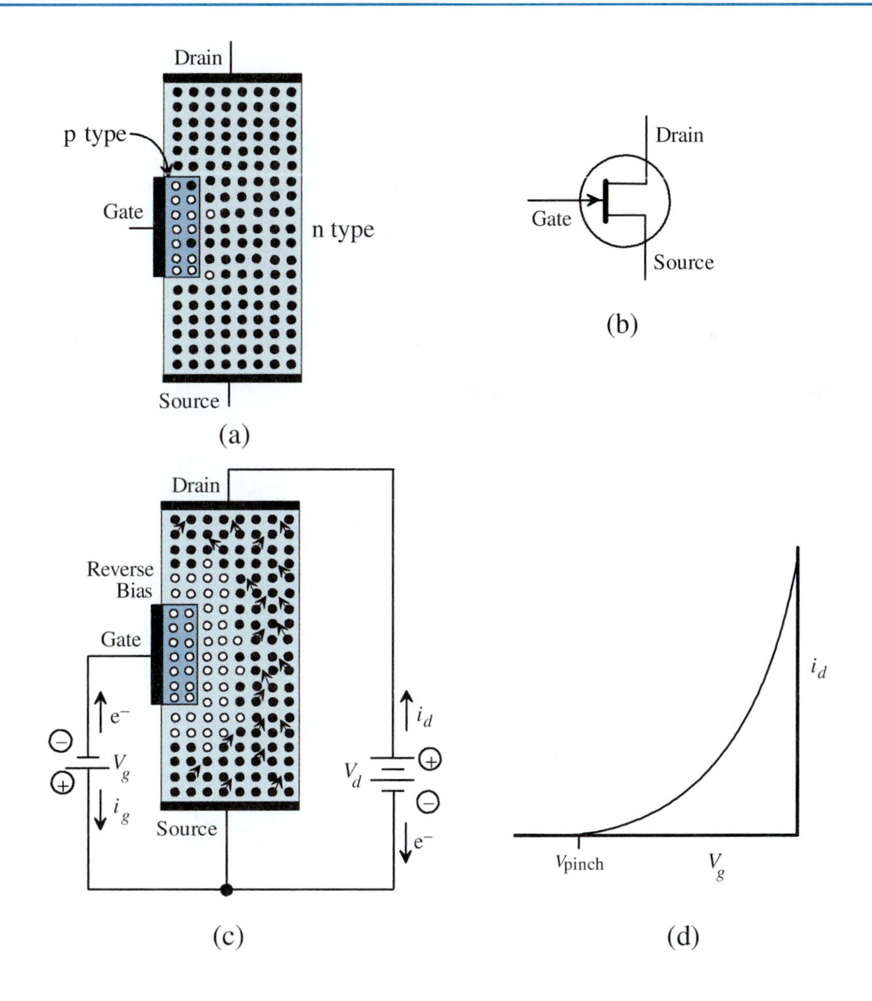

FIGURE 16.43
Field-effect transistor. (a) Schematic representation of field-effect transistor. (b) Symbol. (c) Voltage applied to field-effect transistor. (d) Current output resulting from applied voltage.

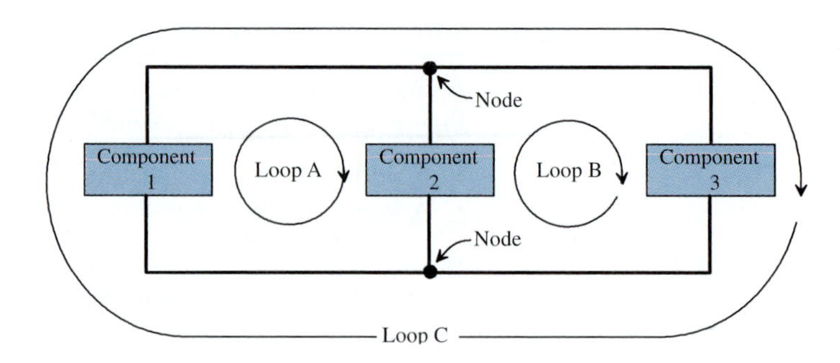

FIGURE 16.44
Illustration of a generic circuit.

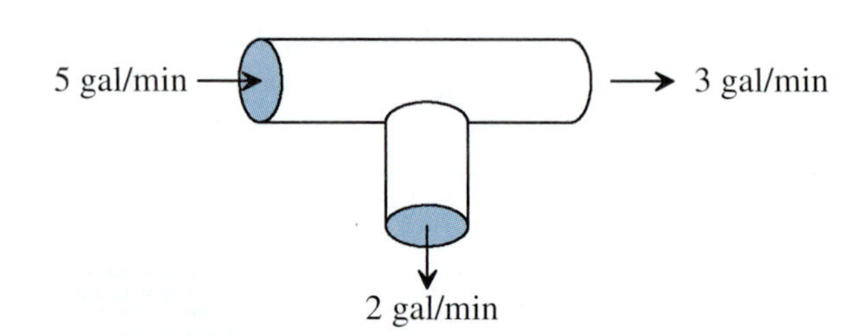

FIGURE 16.45
Pipe analogy to illustrate Kirchhoff's current law.

Kirchhoff's voltage law (*KVL*) may be stated as follows:

The sum of all voltage drops in a loop is zero.

The KVL may be understood by making an analogy to traveling in a loop on a mountain (Figure 16.46) in which height is analogous to voltage. The sum of all height changes while traveling in a loop is zero.

$$\Delta H_{\text{loop}} = \Delta H_{BA} + \Delta H_{CB} + \Delta H_{DC} + \Delta H_{AD}$$

$$= (H_B - H_A) + (H_C - H_B) + (H_D - H_C) + (H_A - H_D)$$

$$= (3 - 1) + (6 - 3) + (4 - 6) + (1 - 4)$$

$$= 0$$

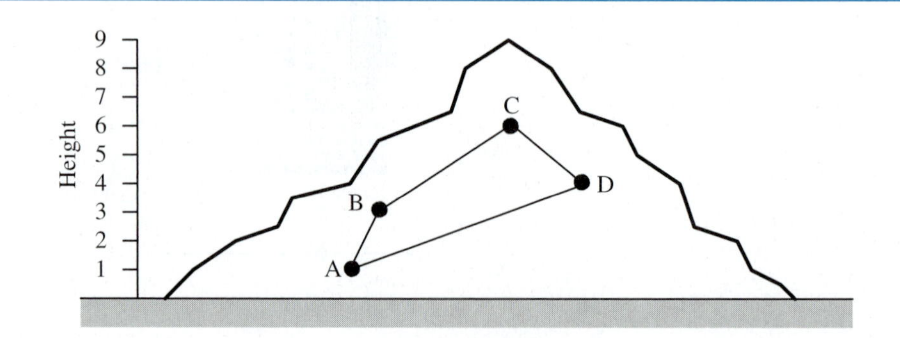

FIGURE 16.46
Mountain analogy to illustrate Kirchhoff's voltage law.

The Transistor—Invention of the Century

The **transistor** plays a central role in almost all electronic products. Some technologists consider it to be the major invention of the 20th century. Although many people laid the foundation for the transistor, its invention is credited to three individuals: William Shockley (the visionary), John Bardeen (the thinker), and Walter Brattain (the tinkerer).

William Shockley grew up mainly in Palo Alto, California. As a child, he was ill-tempered, spoiled, and uncontrollable—but he was also brilliant. In 1928, he became a physics major at the California Institute of Technology and later earned a PhD at the Massachusetts Institute of Technology. His first job was with Bell Labs, the research arm of the American Telephone and Telegraph Company (AT&T). He quickly developed a reputation as an ingenious problem solver. During World War II, he worked on a number of important problems, such as designing one of the first nuclear reactors, improving the destruction rate of enemy submarines by using depth charges, developing tactics for convoys to avoid enemy bombers, and training bomber crews.

John Bardeen was so bright, his parents moved him directly from third grade into junior high. In 1923, at age 15, he began his engineering studies at the University of Wisconsin. Later, he earned a PhD in mathematics from Princeton. During World War II, he developed techniques to protect U.S. ships from magnetic mines and torpedoes.

Walter Brattain grew up on a cattle ranch in Washington state. In 1920, he studied mathematics and physics at Whitman and later earned a PhD from the University of Minnesota. Growing up as a "cowboy," Brattain learned how to use his hands. This served him well; as a physicist, he had the reputation of being able to build anything.

The needs of AT&T spurred the invention of the transistor. They prospered by providing long-distance telephone service. Because the distances were long, the signal strength would decay as it traveled through the wires. Periodically, the signal was amplified by vacuum tubes; however, they were unreliable and wasted energy. To overcome this problem, Bell Labs was tasked with finding an improved amplifier. Young William Shockley was selected as the team leader.

In 1945, Shockley had devised a "field effect" semiconductor amplifier, but it failed to work as planned. Shockley challenged Brattain and Bardeen to determine why. They worked in the laboratory, while Shockley—something of a loner—preferred to work at home. Brattain had modest success when he dunked Shockley's field effect amplifier in a tub of water.

Taking a different approach, in December 1947 Brattain and Bardeen built a "point-contact" semiconductor amplifier. It was fashioned from a germanium crystal with fine gold contacts. Bardeen determined that the wires must be extremely tiny, but none were available in the required size. Ever resourceful, by hand, Brattain cut some gold foil with a razor blade and got the required dimensions. The gold contacts were pressed lightly against the germanium surface by using a spring reputedly fashioned from a paper clip. Amazingly, this crude device worked. When Brattain and Bardeen called Shockley to tell of their success, Shockley was pleased but also furious that he was not part of it.

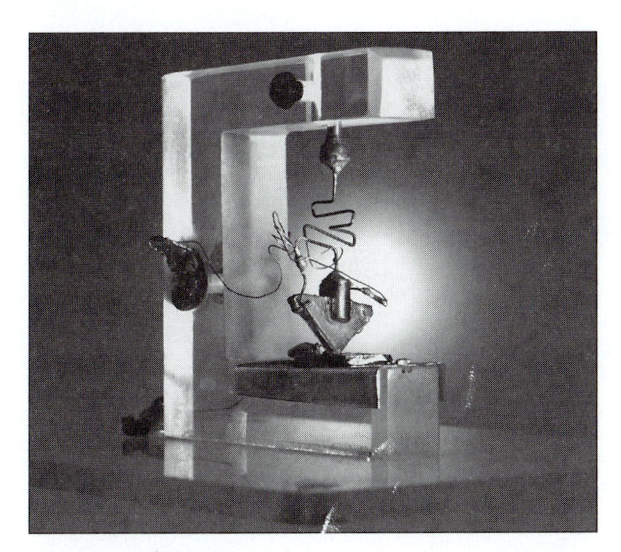

The first transistor.

Inventors of the transistor.

A few months later, when the press release was made, Bell Labs had to find a name. "Point-contact solid-state amplifier" was not very appealing. John Pierce—a Bell Labs employee and accomplished science fiction writer—noted that the new amplifier used *trans*resistance, and that a new electronic product called a therm*istor* had a catchy name. He coined the word "transistor," which stuck.

Brattain and Bardeen were recognized as the inventors of the point-contact transistor; in fact, their names and not Shockley's appear on the patent. The point-contact transistor was not robust, so Shockley set out to do them one better. During a 1-month period—in a flurry of creativity and anger, mostly in a Chicago hotel room—he invented the "sandwich" transistor, now called a bipolar transistor. Two years later, the device was actually built. Although modern transistors are not built quite the same as Shockley's original design, bipolar transistors are still used in some specialty applications.

Initially, the importance of the transistor was not recognized, but by 1956 its impact was apparent, so Shockley, Bardeen, and Brattain received the Nobel Prize that year.

Although Shockley's original field-effect transistor did not work, by 1960, John Atalla of Bell Labs built a functioning device. This is now the dominant type of transistor used today.

Eventually, Shockley left Bell Labs to found Shockley Semiconductor in the Santa Clara Valley, close to his childhood home. He assembled some top-notch talent, but his leadership style was so abrasive that key members of his company—the so-called "traitorous eight"—left to form Fairchild Semiconductor. Shockley never made any money from his invention, but he helped found "Silicon Valley," which allowed many others to become extremely wealthy.

Adapted from: The American Institute of Physics, www.pbs.org/transistor.

EXAMPLE 16.3

Problem Statement: In the circuit shown in Figure 16.47, determine the current through each resistor and determine the voltage at each node.

Solution: Apply the KVL to each loop.

Loop A:

$$\Delta V_{21} + \Delta V_{32} + \Delta V_{13} = 0$$

$$(V_2 - V_1) + (V_3 - V_2) + (V_1 - V_3) = 0$$

$$V_A - i_1 R_1 - i_3 R_3 = 0 \quad (16\text{-}19)$$

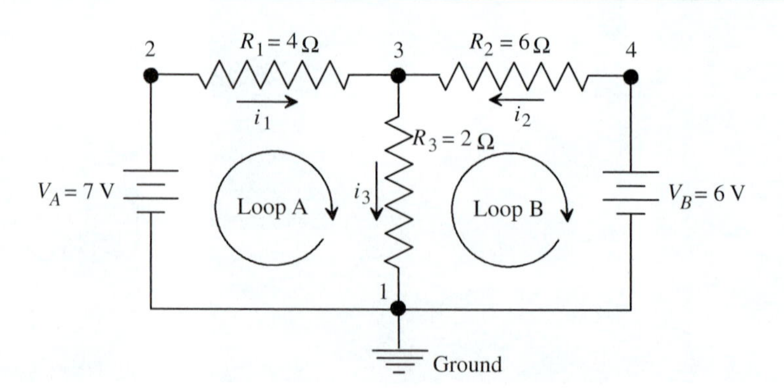

FIGURE 16.47
Circuit for Problem 16.3.

Loop B:

$$\Delta V_{31} + \Delta V_{43} + \Delta V_{14} = 0$$

$$(V_3 - V_1) + (V_4 - V_3) + (V_1 - V_4) = 0$$

$$i_3 R_3 + i_2 R_2 - V_B = 0 \tag{16-20}$$

where Ohm's law (Equation 16-9) is used to calculate the voltage differences. (*Note:* In Equations 16-19 and 16-20, a positive sign is used when the voltage becomes more positive and a negative sign is used when the voltage becomes more negative.)

Apply the KCL to Node 3

$$i_1 + i_2 = i_3 \tag{16-21}$$

Equations 16-19 to 16-21 can be written in the following standard form:

$$-R_1 i_1 + 0\ i_2 - R_3 i_3 = -V_A$$

$$0 i_1 + R_2 i_2 + R_3 i_3 = V_B$$

$$i_1 + i_2 - i_3 = 0$$

When this is solved for the voltages and resistances shown in Figure 16.47, the following currents result:

$$i_1 = 1.0 \text{ A}$$

$$i_2 = 0.5 \text{ A}$$

$$i_3 = 1.5 \text{ A}$$

Each node has the following voltages:

$V_1 = 0$ V (This is at ground voltage, which is defined as zero.)

$$V_2 = V_1 + V_A = 0\ V + 7\ V = 7\ V$$

$$V_3 = V_2 - i_1 R_1 = 7\ V - (1.0\text{ A})(4\ \Omega) \times \frac{1\text{ V}}{\text{A}\cdot\Omega} = 3\text{ V}$$

$$V_4 = V_1 + V_B = 0\ V + 6\ V = 6\ V$$

16.7 CIRCUITS

In this section, we describe a few circuits to show the effects of combining some of the circuit components described previously.

16.7.1 Resistors in Series

Figure 16.48(a) shows three resistors in series. Using the KVL,

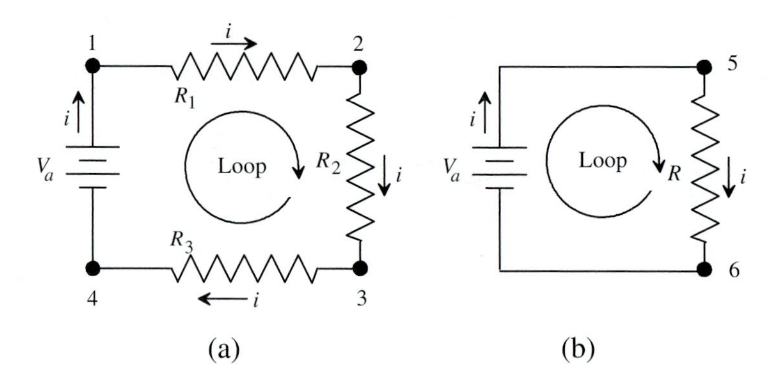

FIGURE 16.48
(a) Multiple resistors in series.
(b) An equivalent circuit with one resistor.

$$\Delta V_{14} + \Delta V_{21} + \Delta V_{32} + \Delta V_{43} = 0$$

$$V_a - R_1 i - R_2 i - R_3 i = 0$$

$$V_a = R_1 i + R_2 i + R_3 i = i(R_1 + R_2 + R_3) \tag{16-22}$$

Figure 16.48(b) shows a circuit with a single resistor. Using the KVL,

$$\Delta V_{56} + \Delta V_{65} = 0$$

$$V_a - iR = 0$$

$$V_a = iR \tag{16-23}$$

Comparing Equations 16-22 and 16-23,

$$R = R_1 + R_2 + R_3 \tag{16-24}$$

For n resistors in series, the equivalent resistance is determined as follows:

$$R = \sum_{i=1}^{n} R_i \tag{16-25}$$

16.7.2 Resistors in Parallel

Figure 16.49(a) shows three resistors in parallel. Using the KCL,

$$i = i_1 + i_a \qquad \text{Node 1} \tag{16-26}$$

$$i_a = i_2 + i_3 \qquad \text{Node 2} \tag{16-27}$$

Substituting Equation 16-27 into Equation 16-26 gives

$$i = i_1 + i_2 + i_3 \tag{16-28}$$

From Ohm's law, the current through each resistor relates to the voltage as follows:

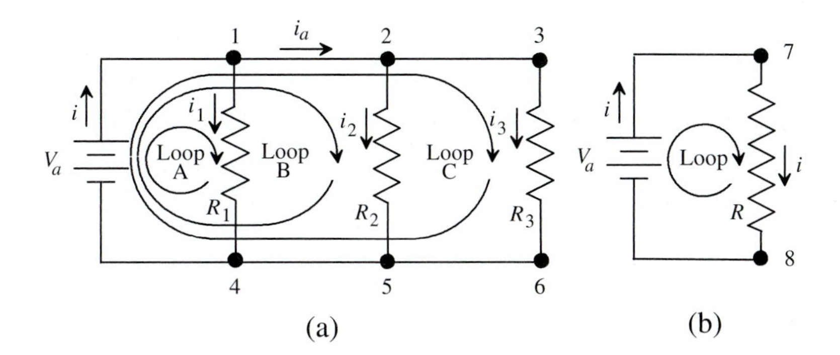

FIGURE 16.49
(a) Multiple resistors in parallel. (b) An equivalent circuit with one resistor.

$$i = \frac{V_{41}}{R_1} + \frac{V_{52}}{R_2} + \frac{V_{63}}{R_3} = \frac{V_a}{R_1} + \frac{V_a}{R_2} + \frac{V_a}{R_3} = \left(\frac{1}{R_1} + \frac{1}{R_2} + \frac{1}{R_3}\right)V_a \tag{16-29}$$

where the KVL applied to each loop reveals that the voltage drop across each resistor is equivalent to the voltage output of the battery.

The current for a circuit with a single resistor [Figure 16.49(b)] is

$$i = \frac{V_{87}}{R} = \frac{V_a}{R} = \left(\frac{1}{R}\right)V_a \tag{16-30}$$

where the KVL applied to the loop reveals that the voltage drop across the resistor is equivalent to the voltage output of the battery.

Comparing Equations 16-29 and 16-30 shows that the resistance of the individual resistors relates to a single equivalent resistor as follows:

$$\frac{1}{R} = \frac{1}{R_1} + \frac{1}{R_2} + \frac{1}{R_3} \tag{16-31}$$

For n parallel resistors, this can be generalized as follows:

$$\frac{1}{R} = \sum_{i=1}^{n} \frac{1}{R_i} \tag{16-32}$$

16.7.3 Resistor-Capacitor Circuits (Advanced Topic)

Figure 16.50 shows a circuit with a resistor and capacitor. A two-position switch charges the capacitor when in Position A and discharges the capacitor when in Position B.

Charging. The KVL is used to find the voltage drops across each node

$$\Delta V_{21} + \Delta V_{32} + \Delta V_{13} = 0 \tag{16-33}$$

The first voltage drop is from Ohm's law (Equation 16-9), the second is from the definition of capacitance (Equation 16-11), and the third is from the DC voltage source

$$-iR - \frac{q}{C} + V_a = 0 \tag{16-34}$$

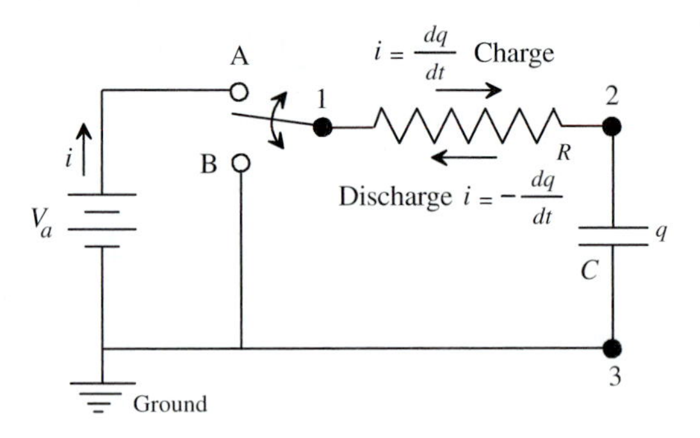

FIGURE 16.50
Resistor-capacitor circuit.

The current i is the rate that charge flows to the capacitor, $\frac{dq}{dt}$:

$$-\frac{dq}{dt}R - \frac{q}{C} + V_a = 0$$

$$\frac{dq}{dt} + \frac{1}{RC}q = \frac{V_a}{R} \tag{16-35}$$

This is a differential equation. To put this equation into words, we are seeking a mathematical function for q such that the function plus its derivative equals a constant V_a/R. (There is a minor issue associated with the RC term, but this is easily handled.) If we find such a function, we have "solved" the differential equation. In our search for this function for q, we propose the following:

$$q = k_1 - k_2 e^{-k_3 t} \tag{16-36}$$

The derivative of this function is

$$\frac{dq}{dt} = k_2 k_3 e^{-k_3 t} \tag{16-37}$$

There is a good chance this function for q will satisfy the differential equation because the function and its derivative have a similar form—we should be able to add them and be left with a constant. Now, our goal is to find the appropriate values of the constants k_1, k_2, and k_3. For this, we must identify some conditions that give additional information.

Condition 1: After an infinite time, the capacitor is fully charged ($q = CV_a$ at $t = \infty$).

$$CV_a = k_1 - k_2 e^{-k_3 \infty} = k_1 - k_2(0)$$

$$k_1 = CV_a \tag{16-38}$$

Condition 2: At the beginning, the capacitor has no charge ($q = 0$ at $t = 0$).

$$0 = k_1 - k_2 e^{-k_3 0} = k_1 - k_2(1)$$

$$k_2 = k_1 = CV_a \tag{16-39}$$

Condition 3: At the beginning, $V_2 = 0$, so the capacitor has no effect on the resistor ($i = V_a/R$ at $t = 0$).

$$i = \frac{dq}{dt} = \frac{V_a}{R} = k_2 k_3 e^{-k_3 0} = k_2 k_3 (1)$$

$$k_3 = \frac{V_a}{R}\frac{1}{k_2} = \frac{V_a}{R}\frac{1}{CV_a} = \frac{1}{RC} \tag{16-40}$$

(*Note: RC* has dimensions of time, and is known as the *capacitive time constant* τ_C.)

Now that we know the values of the three constants, we can write the following equations for q and i

$$q = k_1 - k_2 e^{-k_3 t} = CV_a - CV_a e^{-(1/RC)t} = CV_a(1 - e^{-(1/RC)t}) \tag{16-41}$$

$$i = \frac{dq}{dt} = k_2 k_3 e^{-k_3 t} = (CV_a)\left(\frac{1}{RC}\right)e^{-(1/RC)t} = \frac{V_a}{R}e^{-(1/RC)t} \tag{16-42}$$

We can verify that these are correct solutions to the differential equation by substituting Equations 16-41 and 16-42 into Equation 16-35

$$\frac{dq}{dt} + \frac{1}{RC}q = \frac{V_a}{R}$$

$$\frac{V_a}{R}e^{-(1/RC)t} + \frac{1}{RC}CV_a(1 - e^{-(1/RC)t}) = \frac{V_a}{R}e^{-(1/RC)t} + \frac{V_a}{R}(1 - e^{-(1/RC)t}) = \frac{V_a}{R}e^{-(1/RC)t} + \frac{V_a}{R} - \frac{V_a}{R}e^{-(1/RC)t} = \frac{V_a}{R}$$

The differential equation is satisfied, so Equation 16-41 is a proper solution.

Because we know the charge on the capacitor, we can determine the voltage at Node 2

$$V_2 - V_3 = \frac{q}{C}$$

The voltage at Node 3 is defined to be zero because it is grounded, so

$$V_2 = \frac{q}{C} = \frac{1}{C}q = \frac{1}{C}[CV_a(1 - e^{-(1/RC)t})] = V_a(1 - e^{-(1/RC)t}) \tag{16-43}$$

The capacitor charge q, current i, and Node 2 voltage V_2 are plotted in Figure 16.51.

Discharging. To discharge the capacitor, the switch connects to Position B. The KVL is used to find the voltage drops across each node

$$\Delta V_{21} + \Delta V_{32} + \Delta V_{13} = 0 \tag{16-44}$$

The first voltage drop is from Ohm's law (Equation 16-9), the second is from the definition of capacitance (Equation 16-11), and the third is zero because Nodes 1 and 3 have the same voltage

$$iR - \frac{q}{C} + 0 = 0 \tag{16-45}$$

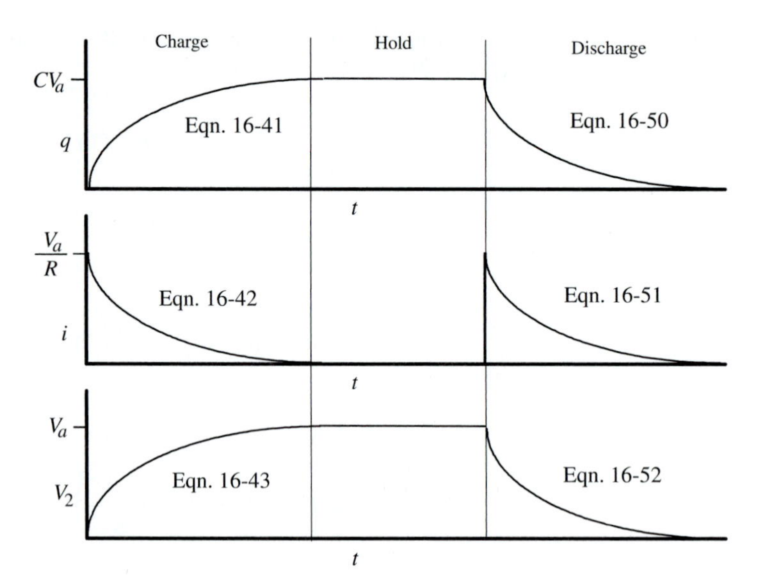

FIGURE 16.51
Charge, current, and voltage in a resistor-capacitor circuit.

The current i is the rate that charge flows from the capacitor, $-\frac{dq}{dt}$. (*Note:* The negative sign is needed because the capacitor charge is decreasing, but the current is positive.)

$$-\frac{dq}{dt}R - \frac{q}{C} + 0 = 0$$

$$\frac{dq}{dt} + \frac{1}{RC}q = 0 \qquad (16\text{-}46)$$

We will propose the same function for q as a solution:

$$q = k_1 - k_2 e^{-k_3 t}$$

$$\frac{dq}{dt} = k_2 k_3 e^{-k_3 t}$$

subject to the following conditions:

Condition 1: After an infinite time, the capacitor is completely discharged ($q = 0$ at $t = \infty$).

$$0 = k_1 - k_2 e^{-k_3 \infty} = k_1 - k_2(0) = k_1$$

$$k_1 = 0 \qquad (16\text{-}47)$$

Condition 2: At the beginning, the capacitor is fully charged ($q = CV_a$ at $t = 0$).

$$CV_a = k_1 - k_2 e^{-k_3 0} = 0 - k_2(1)$$

$$k_2 = -CV_a \qquad (16\text{-}48)$$

Condition 3: At the beginning, the capacitor voltage is the same as the DC voltage source ($V_2 = V_a$ at $t = 0$).

$$\frac{dq}{dt} = -i = -\frac{V_a}{R} = k_2 k_3 e^{-k_3 0} = k_2 k_3(1) = (-CV_a)k_3(1)$$

$$k_3 = -\frac{V_a}{R}\frac{1}{(-CV_a)} = \frac{1}{RC} \tag{16-49}$$

Now that we know the constants, we have the function for q:

$$q = k_1 - k_2 e^{-k_3 t} = 0 - (-CV_a)e^{-(1/RC)t} = CV_a e^{-(1/RC)t} \tag{16-50}$$

$$\frac{dq}{dt} = -i = k_2 k_3 e^{-k_3 t} = (-CV_a)\frac{1}{RC}e^{-(1/RC)t} = -\frac{V_a}{R}e^{-(1/RC)t} \tag{16-51}$$

By substituting Equations 16-50 and 16-51, we can verify that they solve the differential equation (Equation 16-46):

$$\frac{dq}{dt} + \frac{1}{RC}q = \frac{V_a}{R}e^{-(1/RC)t} + \frac{1}{RC}CV_a e^{-(1/RC)t} = -\frac{V_a}{R}e^{-(1/RC)t} + \frac{V_a}{R}e^{-(1/RC)t} = 0$$

Because we know the charge on the capacitor, we can determine the voltage at Node 2 (assuming $V_3 = 0$):

$$V_2 = \frac{q}{C} = \frac{1}{C}q = \frac{1}{C}(CV_a e^{-(1/RC)t}) = V_a e^{-(1/RC)t} \tag{16-52}$$

The capacitor charge q, current i, and Node 2 voltage V_2 are plotted in Figure 16.51.

16.7.4 Resistor-Inductor Circuits (Advanced Topic)

Figure 16.52 shows a circuit with a resistor and inductor. A two-position switch energizes the inductor when in Position A and de-energizes the inductor when in Position B.

Energizing

The KVL is used to find the voltage drops across each node.

$$\Delta V_{21} + \Delta V_{32} + \Delta V_{13} = 0 \tag{16-53}$$

The first voltage drop is from Ohm's law (Equation 16-9), the second is from the definition of inductance (Equation 16-15), and the third is from the DC voltage source.

$$-iR - L\frac{di}{dt} + V_a = 0$$

$$\frac{di}{dt} + \frac{R}{L}i = \frac{V_a}{L} \tag{16-54}$$

Again, we propose the following function to solve this equation, except this time we are finding a function for i rather than q:

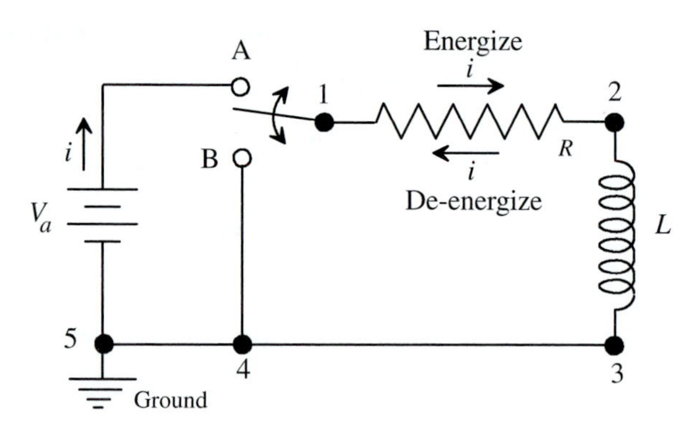

FIGURE 16.52
Resistor-inductor circuit.

$$i = k_1 - k_2 e^{-k_3 t} \tag{16-55}$$

$$\frac{di}{dt} = k_2 k_3 e^{-k_3 t} \tag{16-56}$$

To find the constants, we identify the following conditions:

Condition 1: After an infinite time, the inductor no longer resists the flow of current; only the resistor does ($i = V_a/R$ at $t = \infty$).

$$i = \frac{V_a}{R} = k_1 - k_2 e^{-k_3 \infty} = k_1 - k_2(0) = k_1$$

$$k_1 = \frac{V_a}{R} \tag{16-57}$$

Condition 2: Initially, the current is zero ($i = 0$ at $t = 0$).

$$i = 0 = k_1 - k_2 e^{-k_3 0} = k_1 - k_2(1) = k_1 - k_2$$

$$k_2 = k_1 = \frac{V_a}{R} \tag{16-58}$$

Condition 3: Initially, the voltage at Node 2 is V_a ($V_2 = V_a$ at $t = 0$).

$$\frac{di}{dt} = \frac{V_a}{L} = k_2 k_3 e^{-k_3 0} = k_2 k_3 (1)$$

$$k_3 = \frac{V_a}{L}\frac{1}{k_2} = \frac{V_a}{L}\frac{R}{V_a} = \frac{R}{L} \tag{16-59}$$

(*Note:* L/R has dimensions of time, and is known as the *inductive time constant* τ_L.)

Now that we know the values of the three constants, we can write the following equation for i.

$$i = k_1 - k_2 e^{-k_3 t} = \frac{V_a}{R} - \frac{V_a}{R} e^{-(R/L)t} = \frac{V_a}{R}(1 - e^{-(R/L)t}) \tag{16-60}$$

$$\frac{di}{dt} = k_2 k_3 e^{-k_3 t} = \frac{V_a}{R}\frac{R}{L}e^{-(R/L)t} = \frac{V_a}{L}e^{-(R/L)t} \tag{16-61}$$

Substituting Equations 16-60 and 16-61 into the differential equation (Equation 16-54) verifies that we have found the solution:

$$\frac{di}{dt} + \frac{R}{L}i = \frac{V_a}{L}$$

$$\frac{V_a}{L}e^{-(R/L)t} + \frac{R}{L}\frac{V_a}{R}\left(1 - e^{-(R/L)t}\right) = \frac{V_a}{L}e^{-(R/L)t} + \frac{V_a}{L}\left(1 - e^{-(R/L)t}\right) = \frac{V_a}{L}e^{-(R/L)t} + \frac{V_a}{L} - \frac{V_a}{L}e^{-(R/L)t} = \frac{V_a}{L}$$

Because we know the change in current through the inductor, we can determine the voltage at Node 2.

$$V_2 - V_3 = L\frac{di}{dt} \tag{16-62}$$

The voltage at Node 3 is defined to be zero because it is grounded, so

$$V_2 = L\frac{di}{dt} = L\frac{V_a}{L}e^{-(R/L)t} = V_a e^{-(R/L)t} \tag{16-63}$$

The current i and Node 2 voltage V_2 are plotted in Figure 16.53.

De-energizing.

To de-energize the inductor, the switch connects to Position B. The KVL is used to find the voltage drops across each node

$$\Delta V_{21} + \Delta V_{32} + \Delta V_{13} = 0 \tag{16-64}$$

The first voltage drop is from Ohm's law (Equation 16-9), the second is from the definition of inductance (Equation 16-15), and the third is zero because Nodes 1 and 3 have the same voltage.

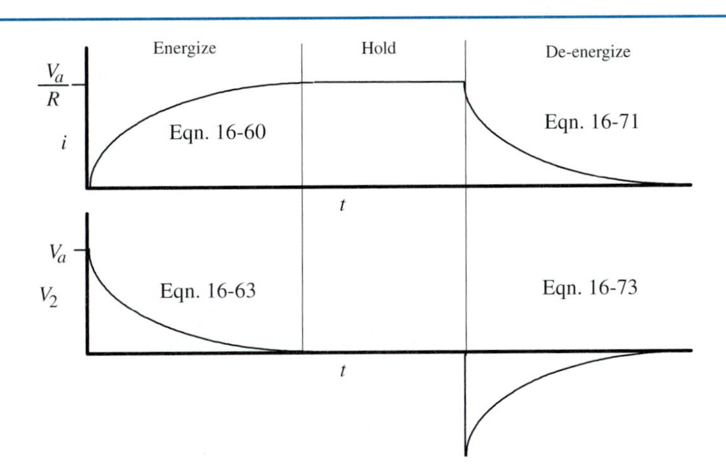

FIGURE 16.53
Current and voltage in a resistor-inductor circuit.

$$iR + L\frac{di}{dt} + 0 = 0$$

$$\frac{di}{dt} + \frac{R}{L}i = 0 \tag{16-65}$$

Again, we propose the following function to solve this equation

$$i = k_1 - k_2 e^{-k_3 t} \tag{16-66}$$

$$\frac{di}{dt} = k_2 k_3 e^{-k_3 t} \tag{16-67}$$

To find the constants, we identify the following conditions:

Condition 1: After an infinite time, the current is zero ($i = 0$ at $t = \infty$).

$$0 = k_1 - k_2 e^{-k_3 \infty} = k_1 - k_2(0)$$

$$k_1 = 0 \tag{16-68}$$

Condition 2: Initially, the current is limited only by the resistor ($i = V_a/R$ at $t = 0$).

$$\frac{V_a}{R} = k_1 - k_2 e^{-k_3 0} = 0 - k_2(1) = -k_2$$

$$k_2 = -\frac{V_a}{R} \tag{16-69}$$

Condition 3: At the beginning, the inductor voltage is the same as the DC voltage source ($V_2 = V_a$ at $t = 0$).

$$\frac{di}{dt} = -\frac{V_a}{L} = k_2 k_3 e^{-k_3 0} = -\frac{V_a}{R} k_3 \ (1)$$

$$k_3 = -\frac{V_a}{L}\left(-\frac{R}{V_a}\right) = \frac{R}{L} \tag{16-70}$$

Now that we know the values of the three constants, we can write the following equation for i:

$$i = k_1 - k_2 e^{-k_3 t} = 0 - \left(-\frac{V_a}{R}\right)e^{-(R/L)t} = \frac{V_a}{R}e^{-(R/L)t} \tag{16-71}$$

$$\frac{di}{dt} = k_2 k_3 e^{-k_3 t} = -\frac{V_a}{R}\frac{R}{L}e^{-(R/L)t} = -\frac{V_a}{L}e^{-(R/L)t} \tag{16-72}$$

Substituting Equations 16-71 and 16-72 into the differential equation (Equation 16-65) verifies that we have found the solution:

$$\frac{di}{dt} + \frac{R}{L}i = 0$$

$$-\frac{V_a}{L}e^{-(R/L)t} + \frac{R}{L}\left(\frac{V_a}{R}e^{-(R/L)t}\right) = -\frac{V_a}{L}e^{-(R/L)t} + \frac{V_a}{L}e^{-(R/L)t} = 0$$

Because we know the change in current through the inductor, we can determine the voltage at Node 2 (assuming $V_3 = 0$).

$$V_2 = L\frac{di}{dt} = L\left(-\frac{V_a}{L}e^{-(R/L)t}\right) = -V_a e^{-(R/L)t} \quad (16\text{-}73)$$

The current i and Node 2 voltage V_2 are plotted in Figure 16.53.

16.7.5 Inductor-Capacitor Circuits (Advanced Topic)

Figure 16.54 shows an inductor-capacitor circuit. The switch is briefly closed to provide energy to the circuit. After the switch is reopened, we apply the KVL giving

$$\Delta V_{12} + \Delta V_{21} = 0 \quad (16\text{-}74)$$

$$L\frac{di}{dt} + \frac{q}{C} = 0 \quad (16\text{-}75)$$

The current flowing to the capacitor changes the charge on the capacitor; therefore,

$$i = \frac{dq}{dt} \quad (16\text{-}76)$$

and

$$\frac{di}{dt} = \frac{d^2q}{dt^2} \quad (16\text{-}77)$$

Substituting Equation 16-77 into Equation 16-75 gives

$$L\frac{d^2q}{dt^2} + \frac{q}{C} = 0$$

$$\frac{d^2q}{dt^2} + \frac{1}{LC}q = 0 \quad (16\text{-}78)$$

FIGURE 16.54
Inductor-capacitor circuit.

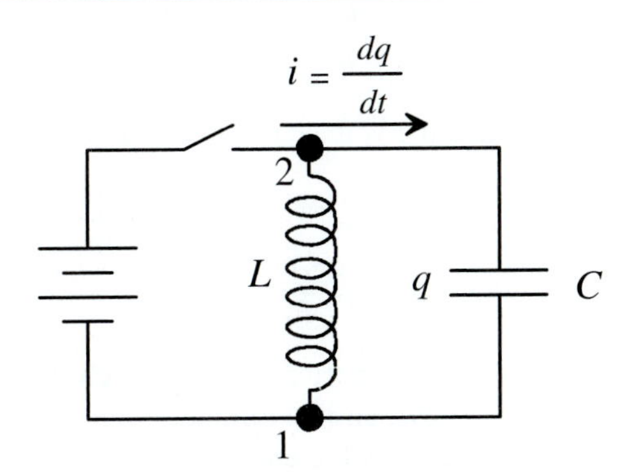

To solve this differential equation, we must find a function for q which, when differentiated twice and added to itself, gives zero. (There is a minor issue associated with the $1/(LC)$ constant, which is easily addressed.) A good candidate is the cosine function

$$q = q_m \cos \omega t \tag{16-79}$$

$$\frac{dq}{dt} = -q_m \omega \sin \omega t \tag{16-80}$$

$$\frac{d^2q}{dt^2} = -q_m\omega^2 \cos \omega t \tag{16-81}$$

where q_m is the maximum charge on the capacitor.

We can verify that Equation 16-79 is a solution by substituting Equations 16-79 and 16-81 into Equation 16-78:

$$-q_m\omega^2 \cos \omega t + \frac{1}{LC} q_m \cos \omega t = 0 \tag{16-82}$$

The following algebraic manipulations allow us to find the value for ω that satisfies this equation:

$$q_m\omega^2 \cos \omega t = \frac{1}{LC} q_m \cos \omega t$$

$$\omega^2 = \frac{1}{LC}$$

$$\omega = \sqrt{\frac{1}{LC}} \tag{16-83}$$

So, now that we know the value of ω, we can determine the charge

$$q = q_m \cos\left(\sqrt{\frac{1}{LC}}\, t\right) \tag{16-84}$$

the current

$$i = \frac{dq}{dt} = -q_m\sqrt{\frac{1}{LC}} \sin\left(\sqrt{\frac{1}{LC}}\, t\right) \tag{16-85}$$

and the voltage difference across the capacitor

$$\Delta V_{21} = \frac{q}{C} = \frac{q_m \cos\left(\sqrt{\frac{1}{LC}}\, t\right)}{C} = \frac{q_m}{C} \cos\left(\sqrt{\frac{1}{LC}}\, t\right) \tag{16-86}$$

Figure 16.55 shows the charge q, the current i, and the voltage difference ΔV_{21}.

For an ideal inductor-capacitor with no resistance, the oscillations would continue forever as the energy oscillates from the capacitor (electric) to the inductor (magnetic).

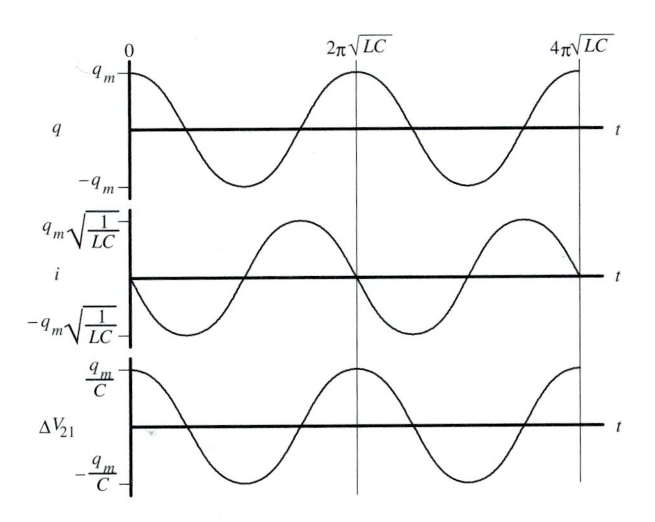

FIGURE 16.55
Charge, current, and voltage for an ideal inductor-capacitor circuit.

This is analogous to the ideal spring/mass system discussed in Chapter 15 in which the energy oscillates between potential and kinetic energy. In reality, each system will eventually stop oscillating because of resistance in the inductor-capacitor circuit or friction in the spring-mass system.

16.7.6 Low-Pass Filter

A *filter* is a device that separates desired frequencies from undesired frequencies. They have many applications. For example, when you "tune" in a radio station, you are separating the desired frequency (the station you want to listen to) from the undesired frequencies (all the other stations). Filters can also remove noise from signals.

Figure 16.56(a) shows a **"low-pass filter,"** which contains a resistor and capacitor. Voltmeters measure the input and output voltages, which are reported in Figure 16.56(b). The circuit is energized by a DC voltage source and a switch. If the switch energizes and

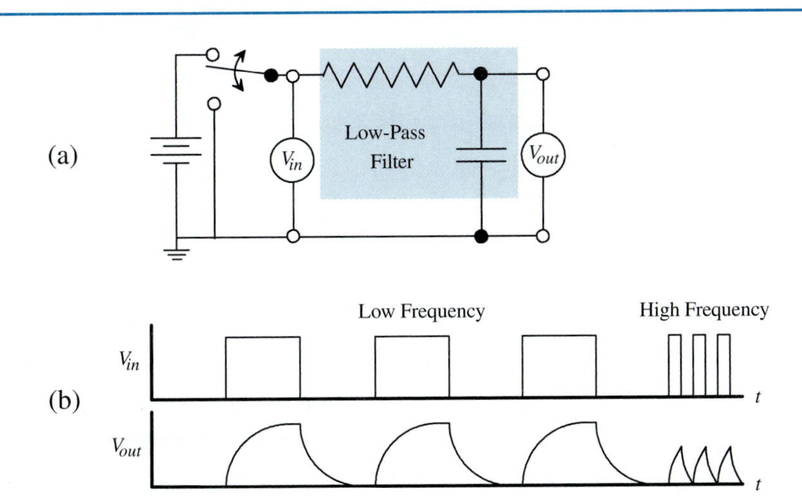

FIGURE 16.56
Low-pass filter.

de-energizes the circuit at a low frequency, there is time for the capacitor to charge and pass the input voltage to the output. However, if the switch energizes and de-energizes the circuit at a high frequency, there is not enough time for the capacitor to charge so little of the input voltage passes to the output.

16.7.7 High-Pass Filter

Figure 16.57(a) shows a **"high-pass filter,"** which contains a resistor and capacitor. Volt meters measure the input and output voltages, which are reported in Figure 16.57(b). The circuit is energized by a DC voltage source and a switch. When the switch energizes the circuit, the left and right plates of the capacitor instantly acquire a positive voltage. Over time, electrons flow through the resistor and reduce the positive charge on the right plate. If the switch energizes and de-energizes the circuit at a low frequency, the capacitor behaves as though it were an open switch so it is difficult for the input voltage to pass to the output. However, if the switch energizes and de-energizes the circuit at a high frequency, there is not enough time for many electrons to flow through the resistor and reduce the voltage on the right plate; therefore, the output voltage closely tracks the input voltage.

16.7.8 Rectifier

A **rectifier** converts AC electricity to DC electricity. Commonly, they are used to convert the AC electricity available at a wall socket to the DC electricity needed to operate electronic equipment, such as computers or battery chargers.

Figure 16.58(a) shows a "half-wave rectifier." In this case, a sinusoidal voltage source is connected to a single diode. Current flows through the load (a resistor, for example) only when the voltage source is positive. As shown in Figure 16.58(b), only the positive portion of the input voltage appears on the output.

Figure 16.58(c) shows a "full-wave rectifier." In this case, a sinusoidal voltage source is connected to four diodes. During the positive portion of the input voltage, current flows

FIGURE 16.57
High-pass filter.

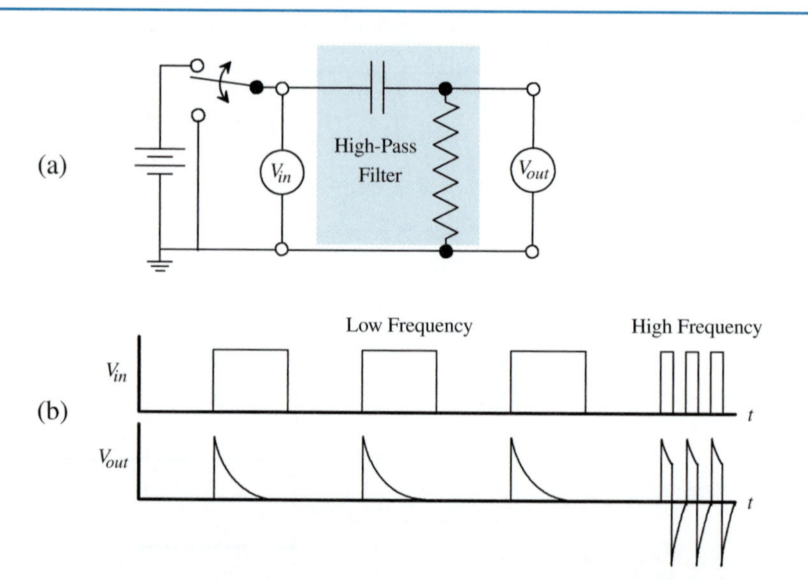

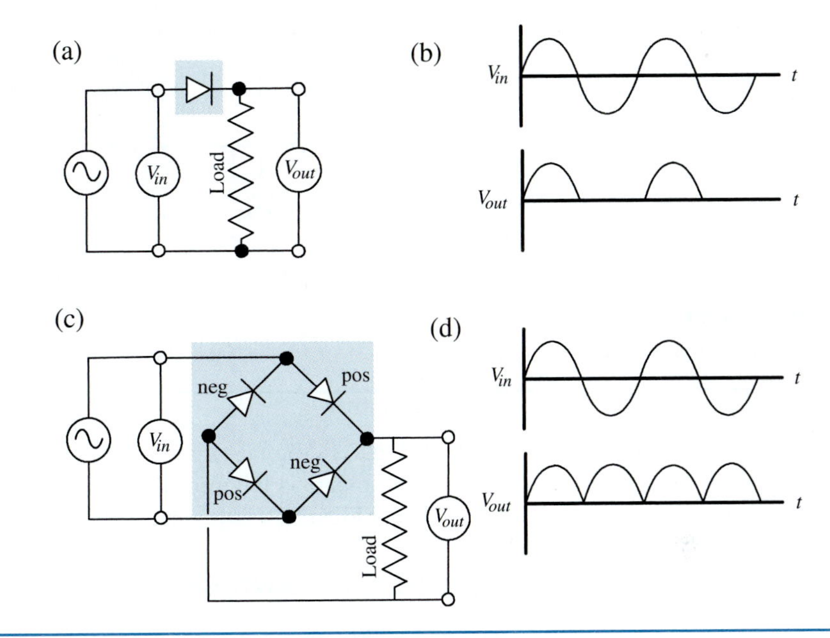

FIGURE 16.58
Rectifiers.

through the "neg" diodes. As shown in Figure 16.58(d), the entire portion of the input appears on the output; the negative portions of the input signal have been converted to positive voltage.

16.7.9 Logic OR Gate

Logic gates are used in digital circuits, such as those in computers (see Section 4.2.3). Figure 16.59(a) shows a logic OR gate, which consists of two diodes and a resistor. Switches A and B, which serve the logic gate, can be independently opened or closed. When a switch is closed, voltage from the DC power source causes current to flow through its corresponding diode. If current flows through either or both diodes, then voltage appears at Terminal C. Figure 16.59(b) shows the four voltage combinations that are possible with a

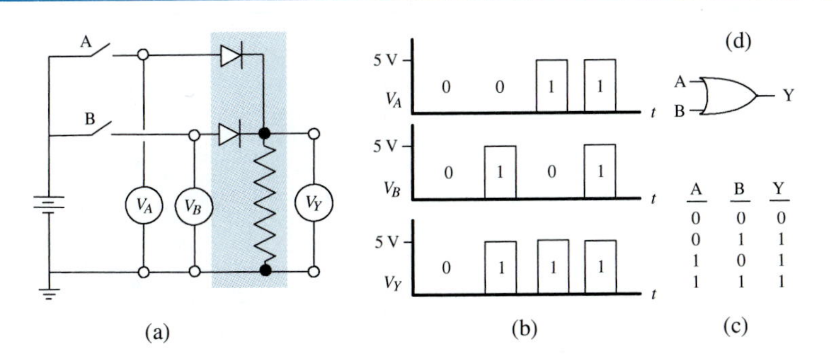

A	B	Y
0	0	0
0	1	1
1	0	1
1	1	1

FIGURE 16.59
Logic OR gate.

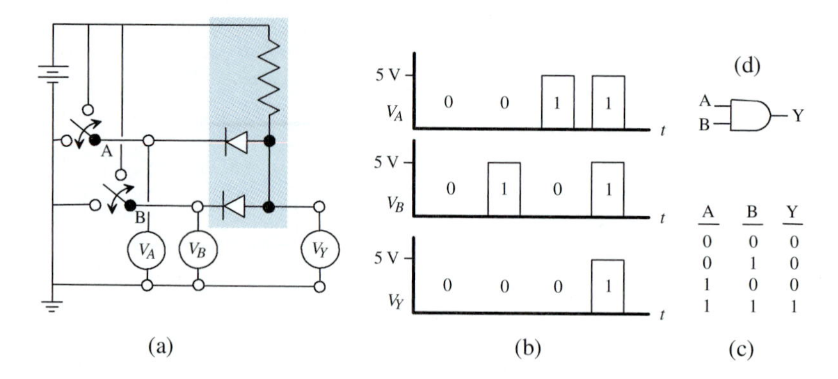

FIGURE 16.60
Logic AND gate.

logic OR gate. In digital circuits, the presence of a voltage is interpreted as a 1, and the absence of voltage is interpreted as a 0. Figure 16.59(c) shows the truth table for a logic OR gate and Figure 16.59(d) shows its symbol.

16.7.10 Logic AND Gate

Figure 16.60(a) shows a logic AND gate, which consists of two diodes and a resistor. Switches A and B, which serve the logic gate, can be independently connected to ground or the positive terminal of the DC power source. When one or both switches is connected to ground, current flows through the resistor and the output voltage V_Y drops to zero. Only when both switches are connected to the positive terminal does current stop flowing through the resistor and the output voltage V_Y becomes positive. Figure 16.60(b) shows the four voltage combinations that are possible with a logic AND gate. Figure 16.60(c) shows the truth table for a logic AND gate and Figure 16.60(d) shows its symbol.

16.7.11 Logic NOT Gate

Figure 16.61(a) shows a logic NOT gate, which consists of two resistors and a transistor. Switch A, which serves the logic gate, can be connected to ground or the positive terminal of the DC power source. When the switch is connected to ground, no voltage is applied to the base, the transistor is off, no current flows through Resistor 2, and the output voltage V_Y becomes positive. When the switch is connected to the positive terminal of the DC

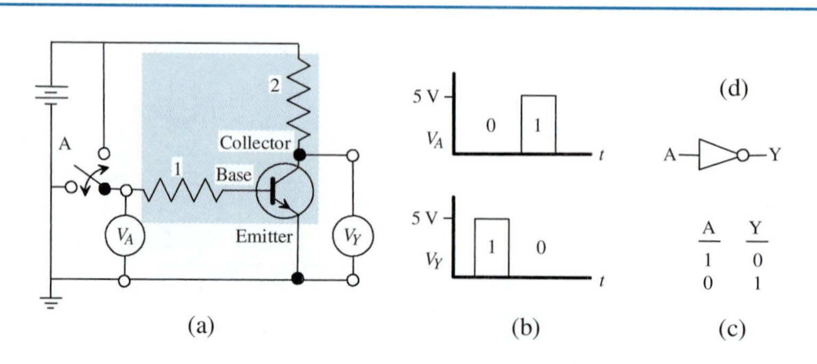

FIGURE 16.61
Logic NOT gate.

power source, voltage is applied to the base, the transistor is on, current flows through Resistor 2, and the output voltage V_Y becomes zero. Figure 16.61(b) shows the two voltage combinations that are possible with a logic NOT gate. Figure 16.61(c) shows the truth table for a logic NOT gate and Figure 16.61(d) shows its symbol.

16.8 ELECTRONIC CIRCUIT CONSTRUCTION

Electronic circuits can be constructed from **discrete components** such as resistors, capacitors, inductors, transistors, and diodes. In this case, each circuit element is packaged independently with wire leads that allow the component to become part of the circuit. For complex circuits, the discrete components are wired together by using a **printed circuit board** (Figure 16.62). The wiring pattern is a thin metal layer on the underside of a fiberglass laminate insulator board. Each discrete component is placed on the top of the printed circuit board. Holes in the insulator board allow the wire leads to be soldered to the metal pattern.

For extremely complex circuits, discrete components are too bulky, costly, and unreliable. This problem is overcome with **integrated circuits,** which are produced directly on a semiconductor crystal, typically silicon. Figure 16.63 shows the sequence of steps to make an integrated circuit, which are described below.

(a) A single crystal of silicon is doped to be a p-type semiconductor.
(b) A very thin layer (about 0.001 in thick) of n-type semiconductor is deposited on the surface using a gas-phase reaction.
(c) The surface of the semiconductor is oxidized to form an insulating layer of silicon oxide by placing it in an oven with high-temperature oxygen.
(d) Openings in the oxide layer are formed by a *photoengraving* technique: The oxide layer is coated with a light-sensitive polymer. With light, a pattern is imprinted onto the polymer. Via chemical baths, some portions of the polymer are dissolved away and other portions remain to protect the oxide layer, thus leaving the imprinted polymer pattern on the surface. With corrosive chemicals, the unprotected portions of the oxide layer are removed. Finally, the polymer pattern is removed as well.
(e) Gas-phase p-type chemicals are diffused into the openings in the oxide layer, thus doping regions of the silicon.
(f) The surface is oxidized again.
(g) Openings are again introduced into the oxide layer by photoengraving.

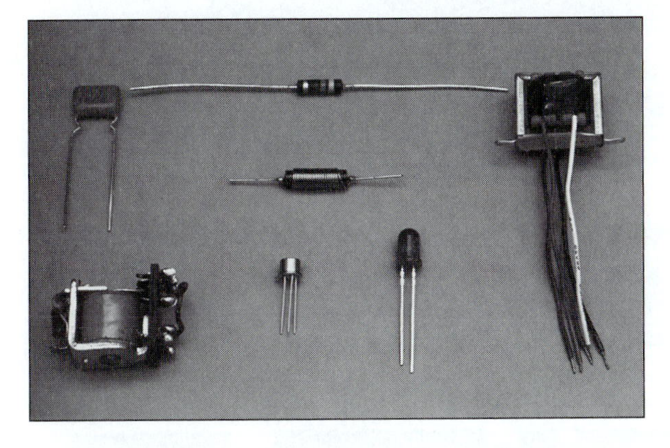

Discete electronic components.

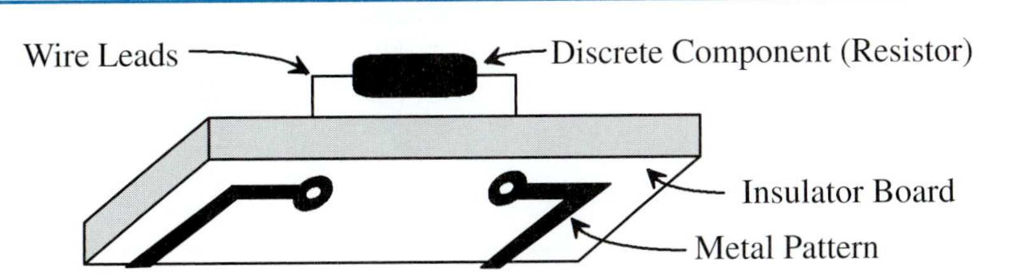

FIGURE 16.62
Portion of a printed circuit board.

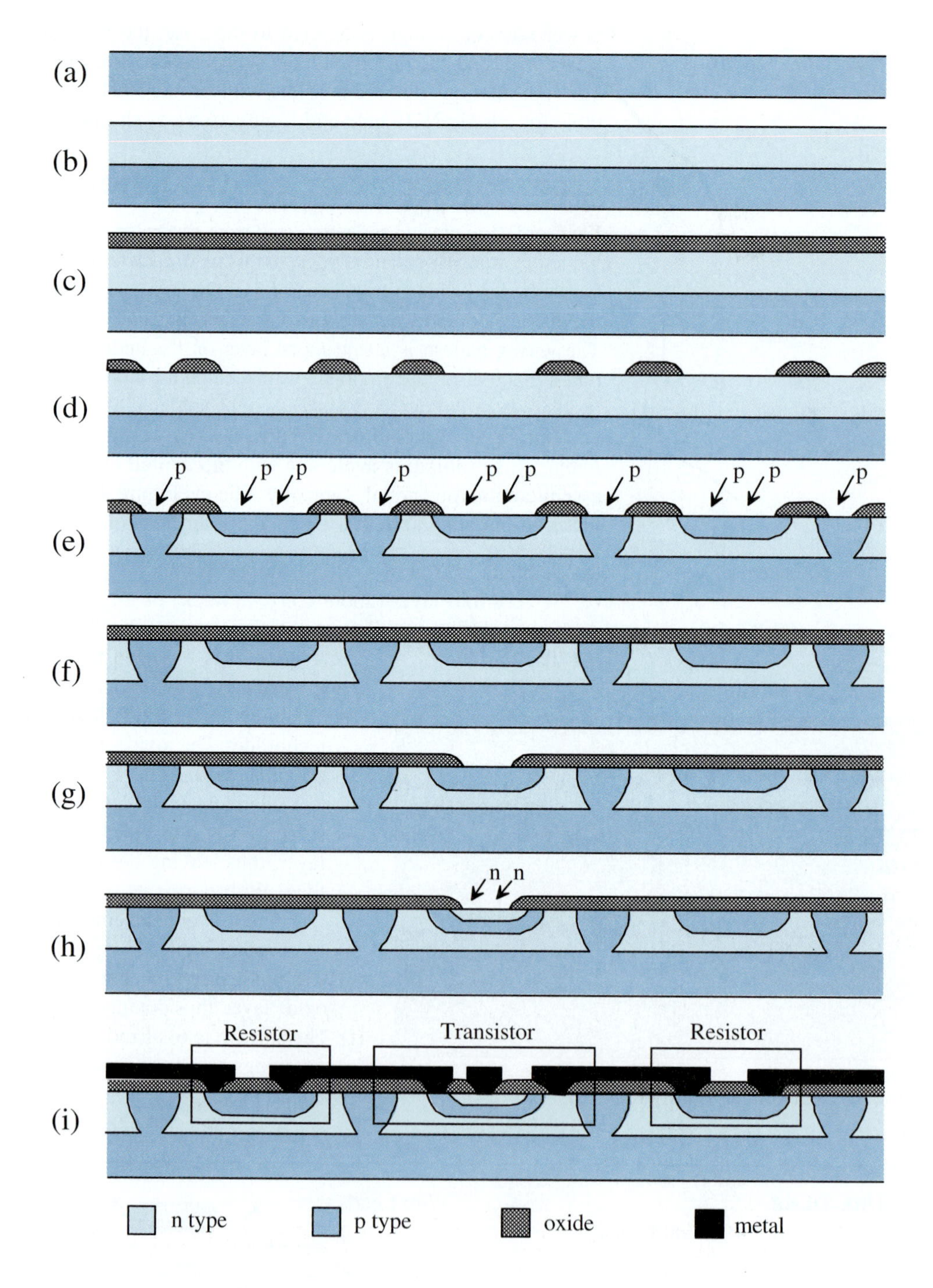

FIGURE 16.63
Fabrication of an integrated circuit.

(h) Gas-phase n-type chemicals are diffused into the openings in the oxide layer, thus doping regions of the silicon.
(i) More openings are introduced into the oxide layer. The surface is coated with a metal layer. Photoengraving produces a pattern of metal wires that connect the circuit elements together.

Integrated circuit.

With this technique, thousands of circuits can be placed on the silicon surface. Figure 16.63 shows that in this example, two resistors and a transistor were fabricated. Resistors, capacitors, transistors, and diodes are readily fabricated on integrated circuits, but inductors are difficult to fabricate.

Figure 16.64 shows how our integrated circuit functions as a logic NOT gate. Figure 16.64(a) shows the circuit in symbol notation. Figure 16.64(b) shows that when 0 V is applied to Point A, then no current flows through the circuit and Point C has 5 V. Figure 16.64(c) shows that when 5 V is applied to Point A, then current flows through the circuit and Point C has almost 0 V. This circuit performs the NOT logic described in Figure 16.61.

FIGURE 16.64
Logic NOT gate implemented on an integrated circuit.

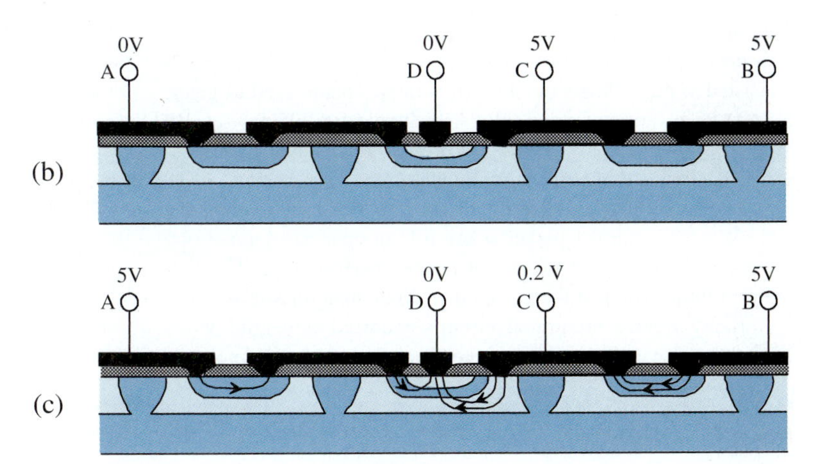

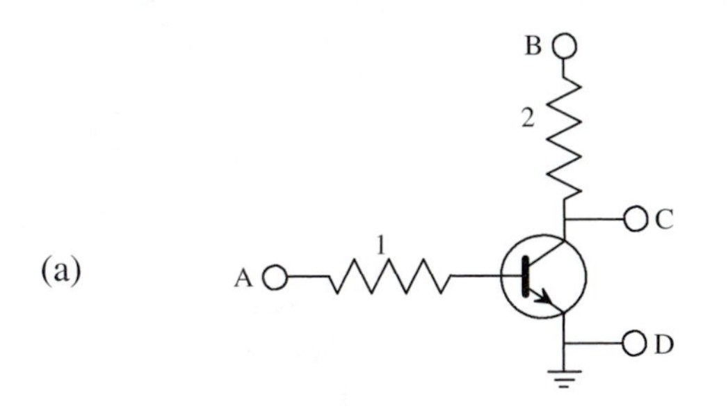

Pioneers of the Integrated Circuit

The integrated circuit is an essential component of modern products (e.g., computers, televisions, cellular telephones, and automobiles) and is used primarily to process information. Although many people played major roles in developing the integrated circuit, Jack Kilby is widely recognized to be the key inventor.

Kilby's interest in electronics began in high school. His father ran a small electric power company that served a wide region of Kansas. After a major ice storm knocked out telephone service, his father relied on a network of amateur radio operators to communicate with his workers. Kilby was fascinated and decided to study electrical engineering. His studies began just 4 months before the attack on Pearl Harbor; he interrupted them to serve in the military as a private.

After the war, in 1947, Kilby earned his BS degree in electrical engineering from the University of Illinois. His first job was with CentraLab, a pioneering company that purchased a license to the transistor developed by Bell Laboratories. Working during the day and attending graduate school in the evening, in 1950 he earned a master's degree in electrical engineering from the University of Wisconsin.

In May 1958, Kilby joined Texas Instruments (TI) in Dallas. Shortly after his arrival, many TI employees left for vacation. This gave him the freedom to play and address the "tyranny-of-numbers" problem. If complex circuits were constructed from discrete components, they would be too costly, heavy, and failure-prone. This could be avoided by constructing circuits containing resistors, capacitors, and transistors directly on a semiconductor surface. By September of 1958, he proved his concept by building a miniature circuit on the surface of germanium, and thus was born the integrated circuit. In March 1959, TI made a press announcement about its "solid circuit," a name that did not stick.

In October 1962, 4 years after the initial discovery, TI announced its first integrated circuit product line, the Series 51. It consisted of digital logic modules (flip-flops, counters, NOR gates, NAND gates, and exclusive ORs) which sold for $95 per set. By 1964, they were already selling Series 54, which contained transistor logic circuits that made a big impact in the market.

Integrated circuits were important weapons used to fight the Cold War. The military played a key role in their development by paying premium prices to get the needed performance. The first major military application was on the Minuteman missile system.

Kilby's early integrated circuits required tiny gold wires that rose above the plane and connected various components together, which was expensive and failure prone. Bob Noyce, a founder of Fairchild Semiconductor, improved the integrated circuit by constructing semiconductors in a single plane. Noyce's key innovation was to use oxide layers as insulators and metal coatings as wires, a development that was anticipated in Kilby's patent.

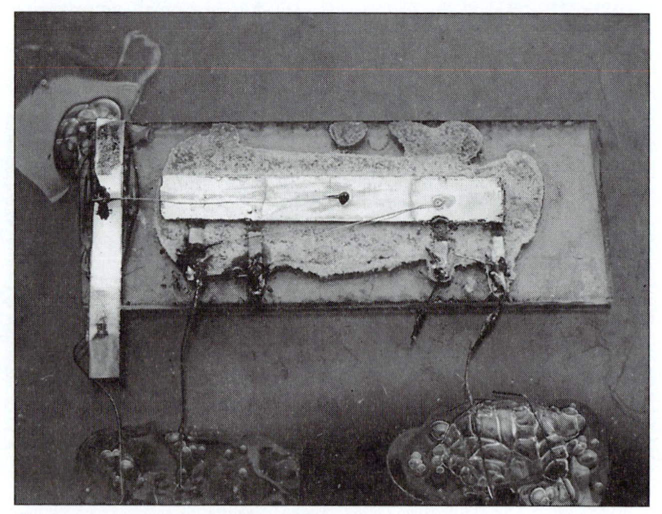

The first integrated circuit.

Legal battles between TI and Fairchild over integrated circuits wound up in the U.S. Supreme Court. The disputing companies agreed that both Kilby and Noyce should be given credit as coinventors of the integrated circuit.

Noyce was one of eight founders of Fairchild; Gordon Moore was another. They also helped found Intel, a giant manufacturer of integrated circuits. In 1965, Moore observed that as the industry progressed, the number of transistors on an integrated circuit doubled every 18 months, a trend that still continues. "Moore's law" is responsible for the ever-increasing performance of integrated circuits.

As time progressed, the role of integrated circuits became more prominent, and Jack Kilby received numerous accolades from society. He was appointed a distinguished professor of electrical engineering at Texas A&M University. His awards include five honorary doctorate degrees, the National Medal of Science, induction into the Inventors Hall of Fame, and the 2000 Nobel Prize in Physics.

Adapted from: Martha S. Polston, *The Bent of Tau Beta Pi*, pp. 13–15, Summer 2001; George Rostky, *Mechanical Engineering*, pp. 68–73, June 2000.

16.9 SUMMARY

Electricity is critical to the functioning of our technological society. Electricity is a convenient way to deliver energy through the flow of charge. There are two kinds of charge: positive and negative. Like charges repel, and unlike charges attract. The strength of charged Particle A can be quantified by the concept of an *electric field,* which is the force acting on test Particle B divided by the charge of the test particle. If the two charges are identical, it requires energy to push test Particle B toward charged Particle A; *voltage* is the push energy divided by the charge on test Particle B. The flow of positive charge is termed *current.* Current readily flows through a conductor, does not flow through an insulator, and can flow through a semiconductor if the temperature is high enough.

Magnetism is closely related to electricity. Magnets always have two poles: south and north. Like poles repel, and unlike poles attract. The strength of a magnet is quantified by the concept of a *magnetic field,* which is analogous to the electric field. Magnetism is produced by electrical current. A changing magnetic field induces electrical current in a wire.

Electromagnetic radiation is produced whenever a charged particle accelerates. The radiation is emitted in quantized "packets" (*photons*) of oscillating electric and magnetic fields. High-energy photons have rapidly oscillating fields and low-energy photons have slowly oscillating fields.

Electrical circuits have three components: electrical energy source, load, and control components. Electrical energy sources convert other forms of energy into electricity, and include generators, batteries, fuel cells, photovoltaic cells, and thermoelectric generators. Loads convert electrical energy into other forms of energy, and include motors, resistors, lights (incandescent, fluorescent, neon, light-emitting diode, laser), cathode-ray tubes, thermoelectric coolers, and processes that involve electroplating, electrochemistry, and electroseparations. Control components regulate electricity, and include switches, relays, capacitors, inductors, transformers, diodes, and amplifiers.

Kirchoff's current law (KCL) and Kirchoff's voltage law (KVL) are used to analyze electrical circuits. These laws can be used to analyze resistors in series, resistors in parallel, resistor-capacitor circuits, resistor-inductor circuits, and capacitor-inductor circuits. Some circuits are *filters,* which allow desired frequencies to be separated from undesired frequencies. Other circuits can perform logic functions, such as OR gates, AND gates, and NOT gates.

Electronic circuits can be constructed from discrete components, or fashioned in miniature integrated circuits.

Nomenclature

C capacitance (F or C/V or C^2/J)
c speed of light (m/s)
E electric field (N/C or V/m) or energy (J)
F force (N)
h Planck's constant (J·s)
i current (A)
k constant

L inductance (H or V·s/A or J/A^2)
m mass (kg)
N number of coils (dimensionless)
P power (W)
q charge (C)
R resistance (V/A or J·s/C^2 or Ω)
r distance (m)
t time (s)
u velocity (m/s)
V voltage (J/C or V)
v velocity (m/s)
W work (J)
ε_0 permittivity of vacuum [$C^2/(N \cdot m^2)$ or F/m]
λ wavelength (m)
μ magnetic dipole moment (A·m^2)
μ_0 permeability of vacuum (N/A^2 or H/m)
ν frequency (s^{-1})
τ time constant (s)
ω angular frequency (s^{-1})

Further Readings

Roadstrum, W. H., and D. H. Wolaver. *Electrical Engineering for All Engineers.* New York: Harper & Row, 1987.

Jones, M. H. *A practical introduction to electronic circuits.* Cambridge, England: Cambridge University Press, 1977.

Halliday, D., and R. Resnick. *Fundamentals of Physics.* New York: Wiley, 1974.

Culp, A. W. *Principles of Energy Conversion.* New York: McGraw-Hill, 1979.

PROBLEMS

16.1 A test sphere has a charge of +0.005 C and a central sphere has a charge of −0.1 C. The two spheres are separated by 100 m in a vacuum.
(a) What is the magnitude of the force acting on each sphere?
(b) Is the force attractive or repulsive?
(c) What is the magnitude of the electric field expressed in V/m?
(d) What is the voltage of the test sphere, assuming the voltage is zero an infinite distance away?

16.2 Two magnets each have a magnetic dipole moment of 5000 A·m^2. They are separated by a distance of 10 m in a vacuum. They are oriented so that the north poles are facing each other.
(a) What is the magnitude of the force acting on each magnet?
(b) Is the force attractive or repulsive?

16.3 Commonly, transistors are made from silicon doped with aluminum and phosphorus. Suggest some other elements that could be used to make transistors.

16.4 Calculate the energy content in a single photon of
(a) AM radio wave (1-km wavelength)
(b) FM radio wave (1-m wavelength)
(c) Red light
(d) Violet light
(e) X-ray (1-nm wavelength)
(f) Gamma ray (1-pm wavelength)

16.5 Using mechanical components, draw analogous circuits for the following electrical circuits:
(a) Three resistors in series
(b) Three resistors in parallel
(c) Resistor-capacitor
(d) Resistor-inductor
(e) Inductor-capacitor
(f) Low-pass filter
(g) High-pass filter

(h) Half-wave rectifier
(i) Full-wave rectifier
(j) Logic OR gate
(k) Logic AND gate
(l) Logic NOT gate

16.6 Using electrical components, design a device that provides cooling from a flame.

16.7 Design a circuit that converts 120-V AC house current to 12-V DC current. The DC current can have ripples.

16.8 Design a circuit that converts 120-V AC house current to 12-V DC current. The DC current cannot have significant ripples.

16.9 A solar cell is 20 percent efficient. Blue light is shining on it at a rate of 1.0×10^{23} photons per second.
(a) How much electric power is produced by the solar cell?
(b) If the voltage is 1.5 V, what is the current produced by the solar cell?

16.10 A blue LED is 5 percent efficient. It is provided with 80 mA of current at 5 V. At what rate (photons per second) are photons emitted?

16.11 A 100-W incandescent light is 2 percent efficient at producing visible light. At what rate (photons per second) are photons emitted, assuming that yellow is a typical wavelength for visible light?

16.12 A 25-W fluorescent light is 10 percent efficient at producing visible light. At what rate (photons per second) are photons emitted, assuming that yellow is a typical wavelength for visible light?

16.13 Calculate the required current for the following devices operating on 120-V household current:
(a) Hair dryer that produces 1 kW of heat
(b) 100-W light bulb
(c) Motor that produces 1 hp of shaft work (assume the motor is 70 percent efficient)

16.14 A cathode-ray tube is operated with the emitter at -1000 V and the accelerator at $+2000$ V.
(a) What is the voltage difference between the emitter and accelerator?
(b) At the emitter, what is the potential energy of a single electron?
(c) At the accelerator, what is the kinetic energy of a single electron?
(d) At the accelerator, what is the speed of a single electron?

16.15 Copper is electroplated at a rate of 5 kg/h. What is the required current?

16.16 A fuel cell is fed hydrogen at a rate of 10 kg/h.
(a) At what rate (kg/h) must oxygen be supplied?
(b) What current is produced?

16.17 An electrolyzer is fed water at a rate of 10 kg/h.
(a) At what rate (kg/h) is hydrogen produced?
(b) At what rate (kg/h) is oxygen produced?
(c) How much current is required?

16.18 A 100-pF capacitor is charged with 5 V.
(a) How much charge is stored?
(b) How many electrons are stored on the negative plate?

16.19 On occasion, the electricity grid loses electric power during lightning storms. When the grid power is re-established, there can be damaging voltage spikes. To avoid damage to sensitive equipment, a protective circuit can be constructed from a relay and a switch that closes only while the button is actively pushed. When the grid is operating properly, the switch of the protective circuit is closed temporarily, which energizes the relay and supplies electric power to the sensitive equipment. However, when the grid goes down, the relay is de-energized. Once the grid is repaired and comes back on line, the relay stays de-energized, preventing voltage spikes from damaging the sensitive equipment. The switch is pushed only when the grid voltage is returned to the normal range.
(a) What type of relay should be used, normally open or normally closed?
(b) Design the protective circuit that accomplishes the task described above.

16.20 At time zero, a 5-H inductor has 10 A of current flowing through it. The current increases linearly so that 2 s later, the current is 20 A. What is the required voltage difference across the inductor?

16.21 A 50,000-V high-voltage line supplies current to an industrial park. The industrial park requires 1000 A at 440 V. A transformer will be used to reduce the voltage. (Assume that the transformer is "ideal," meaning it does not degrade any of the electrical energy into heat.)
(a) How much power is used by the industrial park?
(b) To supply the industrial park, how much current must be drawn from the high-voltage line?
(c) The secondary coil has 10 turns. How many turns are needed in the primary coil?

16.22 Using an inductor and resistor, design the following circuits:
(a) High-pass filter.
(b) Low-pass filter.

16.23 For the circuit shown in Figure 16.65, calculate the following:
(a) Equivalent resistance of the five resistors
(b) Voltages at nodes A and B
(c) Current through each of the five resistors

16.24 Show that the following quantities have dimensions of time:
(a) RC
(b) L/R
(c) $\sqrt{LC}$

16.25 The low-pass filter circuit shown in Figure 16.56 has a 30-Ω resistor, a 5-mF capacitor, and a 9-V battery. The switch energizes and de-energizes the circuit periodically. Using a spreadsheet, plot the input and output voltages for the following cases:
(a) $t_{energize} = 150\ \mu s$, $t_{de\text{-}energize} = 150\ \mu s$
(b) $t_{energize} = 150$ ms, $t_{de\text{-}energize} = 150$ ms
(c) $t_{energize} = 150$ s, $t_{de\text{-}energize} = 150$ s

16.26 The high-pass filter circuit shown in Figure 16.57 has a 30-Ω resistor, a 5-mF capacitor, and a 9-V battery. The switch energizes

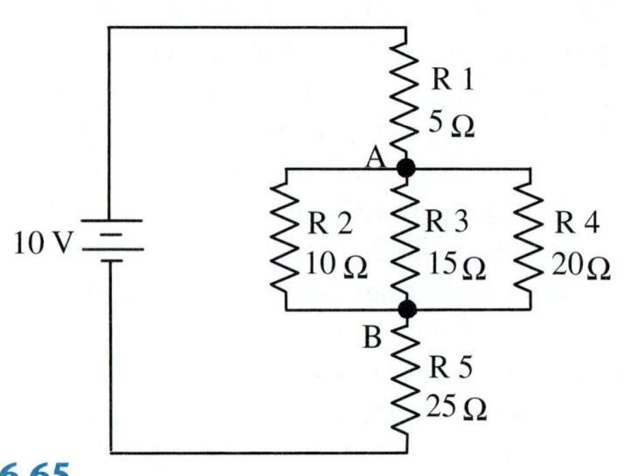

FIGURE 16.65
Circuit for Problem 16.23.

and de-energizes the circuit periodically. Using a spreadsheet, plot the input and output voltages for the following cases:

(a) $t_{\text{energize}} = 150\ \mu\text{s}$, $t_{\text{de-energize}} = 150\ \mu\text{s}$
(b) $t_{\text{energize}} = 150\ \text{ms}$, $t_{\text{de-energize}} = 150\ \text{ms}$
(c) $t_{\text{energize}} = 150\ \text{s}$, $t_{\text{de-energize}} = 150\ \text{s}$

16.27 The low-pass filter circuit shown in Figure 16.56 is modified by replacing the capacitor with an inductor. It now has a 30-Ω resistor, a 4.5-H inductor, and a 9-V battery. The switch energizes and de-energizes the circuit periodically. Using a spreadsheet, plot the input and output voltages for the following cases:

(a) $t_{\text{energize}} = 150\ \mu\text{s}$, $t_{\text{de-energize}} = 150\ \mu\text{s}$
(b) $t_{\text{energize}} = 150\ \text{ms}$, $t_{\text{de-energize}} = 150\ \text{ms}$
(c) $t_{\text{energize}} = 150\ \text{s}$, $t_{\text{de-energize}} = 150\ \text{s}$

16.28 The high-pass filter circuit shown in Figure 16.57 is modified by replacing the capacitor with an inductor. It now has a 30-Ω resistor, a 4.5-H inductor, and a 9-V battery. The switch opens and closes periodically. Using a spreadsheet, plot the input and output voltages for the following cases:

(a) $t_{\text{energize}} = 150\ \mu\text{s}$, $t_{\text{de-energize}} = 150\ \mu\text{s}$
(b) $t_{\text{energize}} = 150\ \text{ms}$, $t_{\text{de-energize}} = 150\ \text{ms}$
(c) $t_{\text{energize}} = 150\ \text{s}$, $t_{\text{de-energize}} = 150\ \text{s}$

16.29 The inductor-capacitor circuit is analogous to the spring-mass system described in Section 15.4. Create a table showing how the inductor-capacitor variables (q_m, L, C, V) relate to the spring-mass variables (k, A, x, m).

16.30 Draw electrical circuits that accomplish the following logical functions:

(a) NOR (Figure 4.4)
(b) NAND (Figure 4.4)
(c) adder (Figure 4.5)

Glossary

alternating current Electric current that reverses direction periodically.
amplifier A device that allows a small amount of electrical energy to regulate a large amount of electrical energy.
amporphous Noncrystalline.
anion Negatively charged ion.
anode Positively charged electrode.
ballast A device that regulates current in a fluorescent light.
base Electrode that regulates current flow through a bipolar transistor.
battery Device that converts chemical energy into electrical energy by reacting electrodes themselves.
bipolar transistor A transistor that contains two pn junctions.
brush Electric contact that slides against a rotating component.
capacitance A measure of the amount of charge stored in a capacitor for a given voltage.
capacitor A device that stores electric charge.
cathode Negatively charged electrode.
cathode-ray tube A device that produces visual images by bombarding a phosphor screen with high-velocity electrons.
cation Positively charged ions.
check valve A device that allows fluid to flow in one direction only.
circuit Collection of electrical components wired together.
coherent Photons that are in phase.

collector Electrode that receives electrons in a bipolar transistor.
commutator Split slip ring that converts alternating current to direct current.
conductor Substance through which current readily flows.
control components Devices that regulate current flow in an electrical circuit.
coulomb Quantity of charge.
crystalline Molecules or atoms ordered into a repeating structure.
current Flow of charge opposite to electron flow.
depletion layer Zone of a pn junction where electrons from the n-type semiconductor occupy holes in the p-type semiconductor.
diode A device that allows current to flow in one direction only.
direct current Electric current that flows in a single direction.
discrete components Individual circuit elements.
dope Introduction of substances into semiconductor crystal, typically by diffusion.
drain Electrode that receives electrons in a field-effect transistor.
efficiency Ratio of output to input.
electric field Region surrounding an electric charge that describes the force on a test sphere.
electricity Study of charge and related phenomena.
electrochemistry Chemical reactions driven by electrical energy.
electrodialysis A process for separating ions by electrical energy.
electromagnetic radiation Electric/magnetic energy emitted as photons.
electroplating A process that coats a surface with metal by using electrical energy.
electroseparations Separation process driven directly by electrical energy.
emitter Source of electrons in a bipolar transistor.
energy efficiency Ratio of energy output to energy input.
energy source A device produces electrical energy from other forms of energy.
field-effect transistor A solid-state amplifier that regulates current flow by controlling the size of the depletion layer.
fluorescent light A device that converts electrical energy to light using fluorescence.
forbidden zone Energy level that is forbidden to electrons.
fuel cell A device that converts the chemical energy of added reactants into electrical energy.
gap Energy level that is forbidden to electrons.
gate Electrode that regulates the size of the depletion zone in a field-effect transistor.
generator A device that produces electrical energy from a mechanical energy input.
grid A device that controls the flow of electrons in a tube amplifier.
high-pass filter A circuit that passes high-frequency signals, but not low-frequency signals.
hole Absence of an electron.
incandescent light A device that converts electrical energy to light energy by heating a metal to very high temperatures.
inductance A measure of the tendency for an inductor to resist increases in current.
inductor A coil that resists current increases.
insulator Substance that does not readily conduct electricity.
integrated circuit A device in which circuit components are mounted directly onto a semiconductor crystal.
intensifier A device that increases or decreases the pressure of a pulsating fluid.
laser A device that emits monochromatic, coherent light.
light-emitting diode A pn junction that emits light when current flows through it.
lines of force Imaginary lines surrounding an electric charge that are parallel to the force on a test sphere.
load Component of an electric circuit that converts electrical energy into another form of energy.
loop Closed path in a circuit.

low-pass filter A circuit that passes low-frequency signals, but not high-frequency signals.
magnetic field Region surrounding a magnet that describes the force on a test magnet.
monochromatic Single color.
motor A device that converts electrical energy to mechanical energy.
node Point connecting two or more wires in a circuit.
n-type semiconductor Semiconductor doped with Group V elements.
photons Packets of electromagnetic radiation.
photovoltaic cell A device that directly converts light energy into electrical energy.
pn junction Interface of p-type and n-type semiconductors.
primary colors Red, blue, and green colors from which all other colors are produced.
printed circuit board An electrically insulating board onto which electrical components are assembled and wired together.
p-type semiconductor Semiconductor doped with Group III elements.
quanta Discrete increment.
rectifier A circuit that converts alternating current to direct current.
relay A device in which electrical energy opens or closes a switch.
resistor A device that converts electrical energy directly to heat.
semiconductor Substance through which current flows moderately well.
slip ring Rotating component through which electrical contact is made.
solenoid Wire coil.
starter A circuit that vaporizes mercury in a fluorescent light.
switch A device that mechanically opens or closes a circuit.
test sphere Charged sphere used to test electric fields.
thermionic effect The phenomenon that allows electrons to be emitted from a high-temperature surface.
thermoelectric cooler A heat pump that is powered by electrical energy, without moving parts.
thermoelectric generator A device that converts thermal energy directly to electrical energy.
transformer A device that increases or decreases the voltage of alternating current.
transistor A solid-state amplifier.
valence electrons Electrons in the outermost shell of an atom.
voltage Work required to move a test sphere toward a charge, expressed per unit of charge on the test sphere.

SECTION FOUR

ENGINEERING ACCOUNTING

Compared with other engineering texts, this section is unique and arguably the most important. The concept of "accounting"—the idea that things can be counted—underlies all the conservation equations used in science and engineering. This concept is lost on most students until they are juniors or seniors; only after looking at many trees does the brightest student discern a pattern in the forest. Our goal in this section is to lay a framework within which almost all engineering problems can be analyzed. We discuss accounting "rules," how to establish a system, how to track changes that occur at the system boundary, and how to treat the objects that are counted. Careful study and understanding of this section will make your future engineering coursework much easier.

The list of things that can be counted is large, but the rules are consistent, whether tracking mass, money, or entropy. The scope of this section is enormous, because these are the overarching principles of engineering. The ride is fun. Enjoy!

Mathematical Prerequisite

- Algebra (Appendix E, Mathematics Supplement)

CHAPTER 17

Accounting

You may be surprised to learn that accounting is a major component of engineering. Normally, we think that accounting is the realm of business majors, who count money; however, because engineers manage big-budget projects, they also must count money. In addition, engineers must count mass, charge, linear momentum, angular momentum, energy, and **entropy.** (Later, we discuss these topics in more detail.) As shown in Table 17.1, each engineering discipline has particular accounting specialties. Because modern engineering is becoming increasingly interdisciplinary, it should be emphasized that every engineer must be familiar with all accounting topics.

Proper accounting is a prerequisite to organized, rational thought. As such, we must be very careful in our approach. We start our accounting process by first defining the **system,** meaning, the subset of the universe we wish to study.

TABLE 17.1
Accounting specialties for engineering disciplines

Engineering Discipline	Mass	Charge	Linear Momentum	Angular Momentum	Energy	Entropy	Money
Electrical		X			X		X
Civil	X		X	X			X
Mechanical	X	X	X	X	X	X	X
Chemical	X	X	X	X	X	X	X
Industrial	X	X	X	X	X		X
Aerospace	X	X	X	X	X	X	X
Materials	X	X	X	X	X	X	X
Agricultural	X	X	X	X	X	X	X
Nuclear	X	X	X	X	X	X	X
Architectural	X		X	X			X
Biomedical	X	X	X	X	X	X	X
Computer		X			X		X

17.1 SYSTEMS: SUBSETS OF THE UNIVERSE

The *universe* is defined as everything that exists, including the planets, the stars, and the space in between. Engineers rarely study something as large as the universe. Instead, we study a subset of the universe called the *system* (Figure 17.1). The **surroundings** are everything in the universe **except** the system. A system can be anything of interest, such as an individual solar system, a planet, a rocket ship, a car, an engine, or an individual atom. The specification of the system is completely arbitrary and is at the discretion of the engineer; however, there are rules the engineer must follow:

Rule 1: *Once a system is specified, it cannot be changed midway through a calculation.*
Rule 2: *The system boundary can be any shape, but it must be a closed surface.*
Rule 3: *The system boundary can be rigid, thus defining a volume of space; or it can be flexible, thus defining an object.*

Beyond these three rules, the engineer is free to define the system and its boundaries as she wishes. Often, a difficult problem can be made easier by altering the definition of the system. The selection of a system and its boundary is part of the "art of engineering."

When defining a system, it is necessary to use **space-time coordinates** (Figure 17.2). The spatial coordinates describe where the system exists and the time coordinates describe when the system exists. Generally, engineers study systems that change with respect to time; therefore, the contents of the system would be described at time t_1 and later at time t_2.

Figure 17.3 shows a **hydraulic lift,** a device used to lift automobiles at service stations. Hydraulic fluid (oil) is pumped into a cylinder that lifts the piston and the automobile. An engineer studying this device would have a number of questions such as: (1) What pressure is required to lift the automobile? (2) How much work does the pump do when

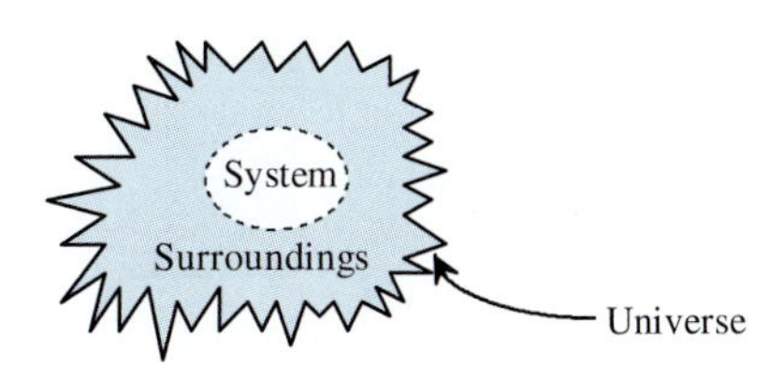

FIGURE 17.1
A system is a subset of the universe.

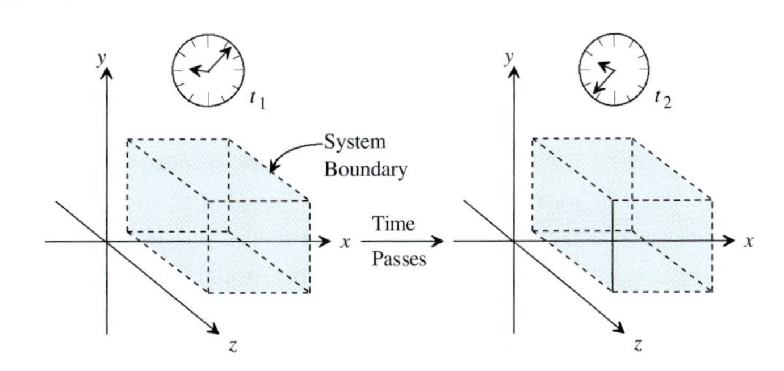

FIGURE 17.2
Space-time coordinates required to define a system.

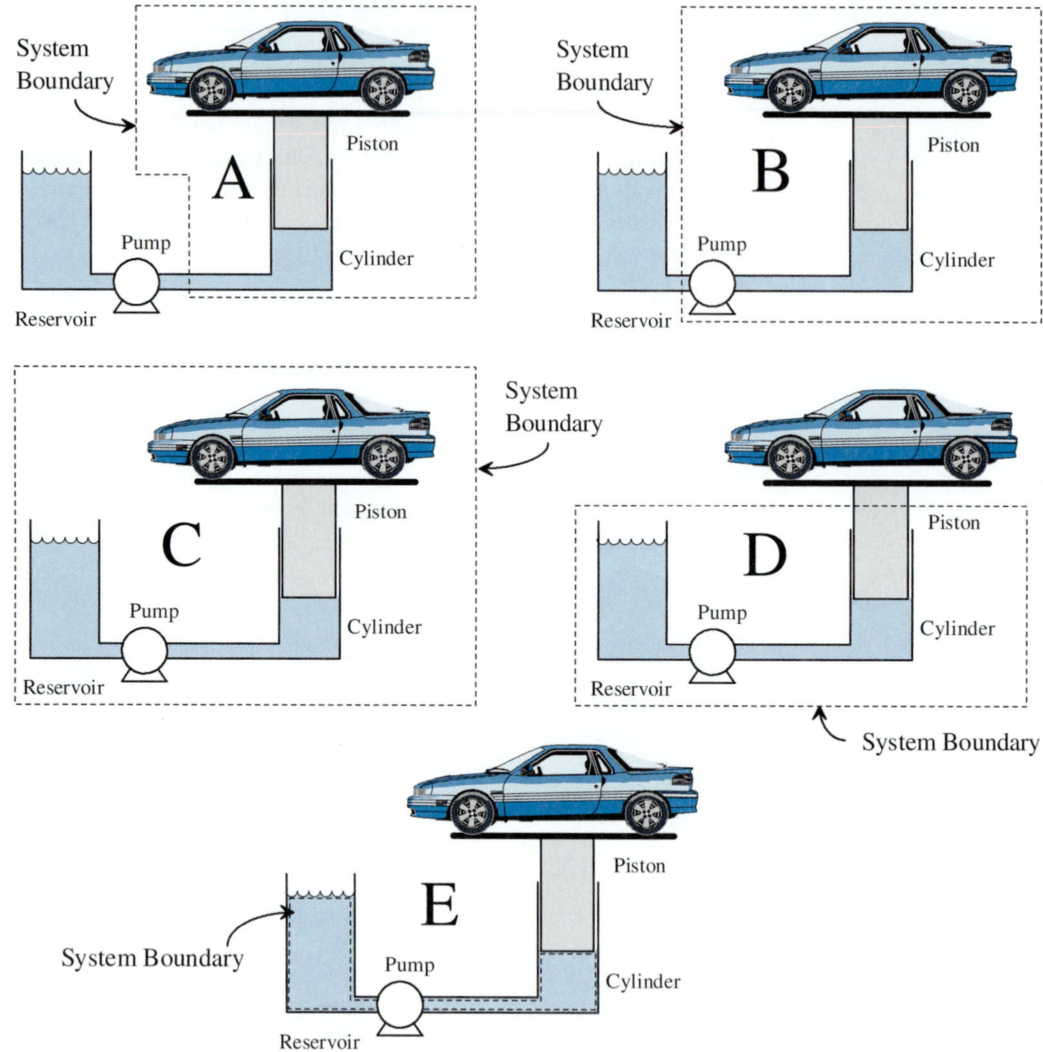

FIGURE 17.3
Five possible systems for analyzing a hydraulic lift.

lifting the automobile? (3) What volume of hydraulic fluid must be pumped to lift the automobile a given height? To address these practical questions, the engineer must first specify a system, the subset of the universe she wishes to study. Figure 17.3 shows five possible systems (out of many more) for analyzing this device. Systems A through D use rigid system boundaries encompassing various components of the device, whereas System E uses a "rubber boundary" that encompasses just the hydraulic fluid as it flows from the reservoir to the cylinder. Each of these systems will yield the same answer to the practical questions listed above; however, some systems will be easier to analyze.

17.2 INTENSIVE AND EXTENSIVE QUANTITIES

As the system increases in size, **extensive quantities** increase, but **intensive quantities** remain the same. As an illustration of the difference between intensive and extensive quantities, imagine that you are an engineer working for a chemical company. Your assignment is to develop a new process to make a high-strength polymer. You go into the laboratory and, using laboratory beakers, you determine the temperature, pressure, and reactant concentration that work best. You present your results to management and they decide to scale up the process by 100,000 times to serve an industrial market. Clearly, you would not increase the temperature, pressure, and reactant concentration by 100,000 times! However, the reactor volume, reactant mass, and product mass would all scale up by 100,000 times. Those quantities that were independent of scale (temperature, pressure, concentration) were all intensive, whereas those quantities that did depend on scale (volume, reactant mass, product mass) are extensive.

Extensive quantities (e.g., mass, moles, volume, and money) can be counted, but intensive quantities (e.g., color, temperature, pressure) cannot. The following rules apply to intensive and extensive quantities.

Rule 4: *The ratio of two extensive quantities is intensive.* For example, mass m is extensive and volume V is extensive, but density ρ is intensive.

$$\text{Intensive} = \rho = \frac{m}{V} = \frac{\text{extensive}}{\text{extensive}} \tag{17-1}$$

Rule 5: *An equation with an extensive quantity on the left-hand side must also have an extensive quantity on the right-hand side.* As an illustration, consider the perfect gas equation:

$$PV = nRT \quad \text{(Extensive form)} \tag{17-2}$$

The left-hand side contains the extensive quantity volume V and the right-hand side contains the extensive quantity moles n.

Rule 6: *The extensive form of an equation can be transformed into an intensive form by dividing both sides by an extensive variable.* For example, the extensive form of the perfect gas equation (Equation 17-2) can be transformed into an intensive form by dividing both sides by moles n,

$$P\frac{V}{n} = RT$$

$$P\hat{V} = RT \quad \text{(Intensive form)} \tag{17-3}$$

where $\hat{V} \equiv V/n$ is the *specific volume.*

TABLE 17.2
Examples of specific quantities

Quantity	Extensive Quantity	Specific Quantity (Intensive)
Volume	V (m^3)	$\hat{V}$ ($\frac{m^3}{kg}$ or $\frac{m^3}{mol}$)
Internal energy	U (J)	$\hat{U}$($\frac{J}{kg}$ or $\frac{J}{mol}$)
Enthalpy	H (J)	$\hat{H}$ ($\frac{J}{kg}$ or $\frac{J}{mol}$)

Specific quantities are formed by dividing an extensive quantity by mass or moles (Table 17.2). An extensive quantity results when the specific quantity is multiplied by mass (or moles).

EXAMPLE 17.1

Problem Statement: At 244.2°C and 36 × 10^5 Pa, steam has the following specific quantities: $\hat{V}$ = 0.0554 m^3/kg, $\hat{U}$ = 2602.2 kJ/kg, and $\hat{H}$ = 2801.7 kJ/kg. What is the total volume *V*, internal energy *U*, and **enthalpy** *H* of 2 kg steam at this temperature and pressure?

Solution:

$$V = m\hat{V} = 2\text{ kg}\left(0.0554\,\frac{\text{m}^3}{\text{kg}}\right) = 0.111\text{ m}^3$$

$$U = m\hat{U} = 2\text{ kg}\left(2602.2\,\frac{\text{J}}{\text{kg}}\right) = 5204.4\text{ J}$$

$$H = m\hat{H} = 2\text{ kg}\left(2801.7\,\frac{\text{J}}{\text{kg}}\right) = 5603.4\text{ J}$$

17.3 STATE AND PATH QUANTITIES

State quantities are independent of path and **path** quantities are dependent on path. A simple analogy is shown in Figure 17.4. Suppose we wished to travel from the state of California to the state of New York. We could take a very direct path (Path A), a southern path (Path B), or a northern path (Path C). Clearly, the distance traveled depends on whether Path A, B, or C is taken, so distance traveled is a path quantity. On the other hand, the state we occupy (California or New York) does not depend on the path we took, so it is a state quantity (pardon the pun).

As an example of state and path quantities, consider Figure 17.5. The bucket is defined as the system. Initially, the bucket contains 4 L of water and finally, it contains 2 L of water. Three different paths for changing the state of the system are presented: Path A drains 1 L and then drains another 1 L, Path B drains 3 L and adds 1 L, and Path C drains

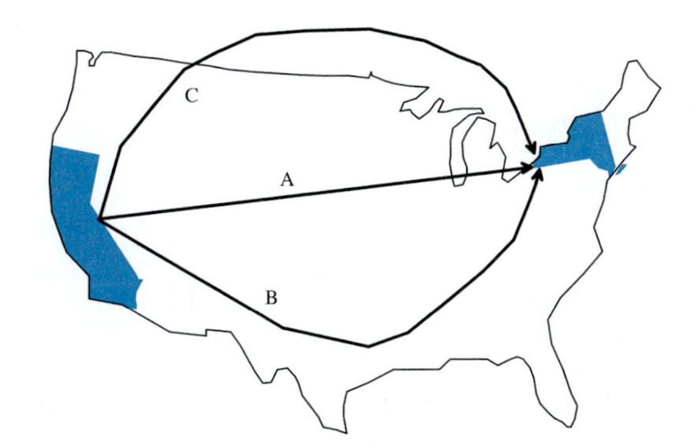

FIGURE 17.4
Analogy showing state and path quantities using a map.

4 L and adds 2 L. From this simple example, we can conclude that the water in the bucket is a state quantity (independent of path) and the water added or drained is a path quantity (dependent on path).

The following rules pertain to state and path quantities:

Rule 7: *Extensive state quantities are contained within the system boundary, whereas extensive path quantities pass through the system boundary and have the ability to change the state of the system.* For example, Figure 17.5 shows that the water volume in the system (extensive state quantity) is changed by adding or draining water volume (extensive path quantities).

Rule 8: *Algebraic combinations of state quantities are also state quantities.* A good example of this rule is specific enthalpy $\hat{H}$, which is defined as the sum of specific internal energy $\hat{U}$ and $P\hat{V}$.

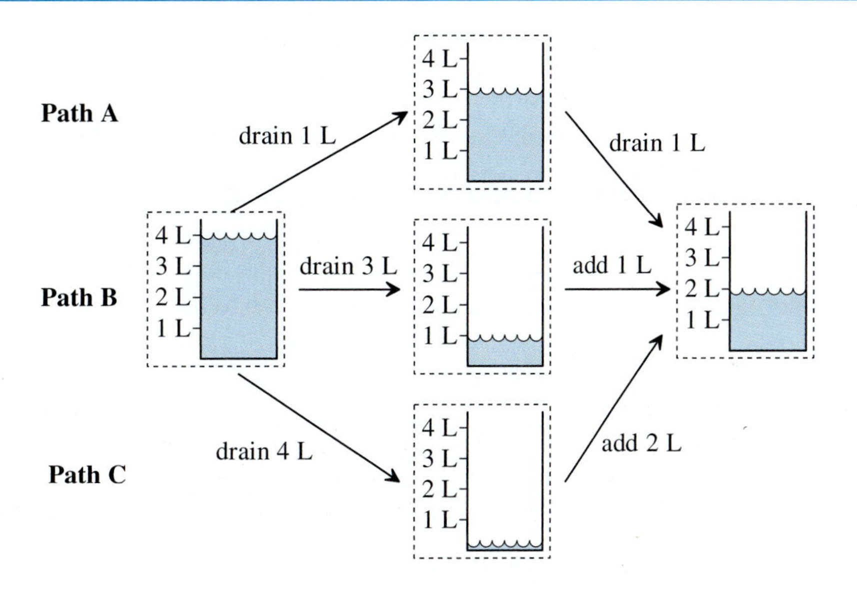

FIGURE 17.5
Illustration of state and path quantities using water buckets. (*Note:* The density of the water is assumed to be constant.)

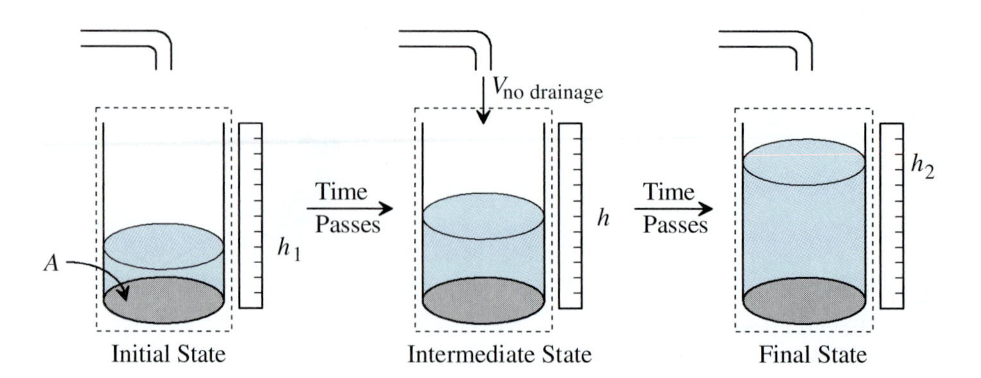

FIGURE 17.6
Illustration showing that an algebraic combination of state quantities and well-defined path quantities is a state quantity.

$$\hat{H} \equiv \hat{U} + P\hat{V} \tag{17-4}$$

Because $\hat{U}$, P, and $\hat{V}$ are all state quantities, then $\hat{H}$ must also be a state quantity.

Rule 9: *Algebraic combinations of state quantities and well-defined path quantities are state quantities.* To illustrate this rule, consider a cylindrical tank with cross-sectional area A being filled with water (Figure 17.6). Initially, the tank is filled to a height h_1. When a volume of water $V_{\text{no drainage}}$ is added to the tank in the absence of drainage, the new height h_2 is

$$h_2 = h_1 + \frac{V_{\text{no drainage}}}{A} \tag{17-5}$$

The initial height h_1 and cross-sectional area A are both state quantities. They are algebraically combined with the path quantity $V_{\text{no drainage}}$ to give the state quantity h_2. Notice that the path quantity $V_{\text{no drainage}}$ must be well defined by specifying that there is no drainage. Without this careful description, Equation 17-5 would be invalid in general.

17.4 CLASSIFICATION OF QUANTITIES

In our discussion so far, we described a system that must first be defined in terms of space-time coordinates. Then, we enumerated rules that relate to intensive, extensive, state, and path quantities. When using a quantity, it is necessary to properly classify it so we know which rules apply. To help classify quantities, we can think of their relationship to the system:

1. *Space-time coordinates* can be used to define system boundaries.
2. *State quantities* define the system contents.
3. *Path quantities* define things that affect the system by passing through the system boundary.

Table 17.3 summarizes the classification scheme and gives some examples of each type of quantity.

TABLE 17.3
Classification scheme for quantities

Quantity	Space-Time Coordinate	Intensive	Extensive	State	Path
Time	X			X	
Position	X			X	
Distance traveled	X				X
Volume of the system itself	X			X	
Volume of liquid in the system			X	X	
Volume of liquid added to the system			X		X
Temperature		X		X	
Pressure		X		X	
Heat			X		X
Specific heat[1]		X			X
Work			X		X
Specific work[2]		X			X

[1] For example, heat needed per pound of water to raise its temperature by 1 Fahrenheit degree.

[2] For example, work needed per pound of water to lift it 5 feet.

17.5 STATE PROPERTIES OF MATTER

Up to this point in our discussion, we have referred to state quantities and have not been very specific about whether these "quantities" were intensive or extensive; in fact, we have allowed state quantities to be either intensive or extensive. If we were to restrict our discussion to intensive state quantities, we would be referring to state **properties of matter,** such as temperature, pressure, specific volume, specific internal energy, and specific enthalpy.

To illustrate the use of state properties, study Figure 17.7, which shows a piston/cylinder filled with gas. The accompanying graph indicates the temperature and pressure. Path A consists of two steps: (1) work compresses the gas (at constant temperature), and (2) heat (at constant pressure) raises its temperature. Path B also consists of two steps: (1) heat warms the gas (at constant pressure), and (2) work compresses it (at constant temperature). Regardless of the path, the gas started at State 1 and ended at State 2; thus, the state properties at State 1 and State 2 do **not** depend upon the path taken. As specified by Rule 7, path quantities (i.e., heat and work) were able to effect a change in state properties.

17.6 SPECIAL TYPES OF SYSTEMS

When analyzing problems, engineers may find it necessary to classify the system they are studying. The following classification schemes are often used: (1) closed and open, (2) isolated and nonisolated, (3) homogeneous and heterogeneous, and (4) steady state and unsteady state. Let us discuss each of these classifications further.

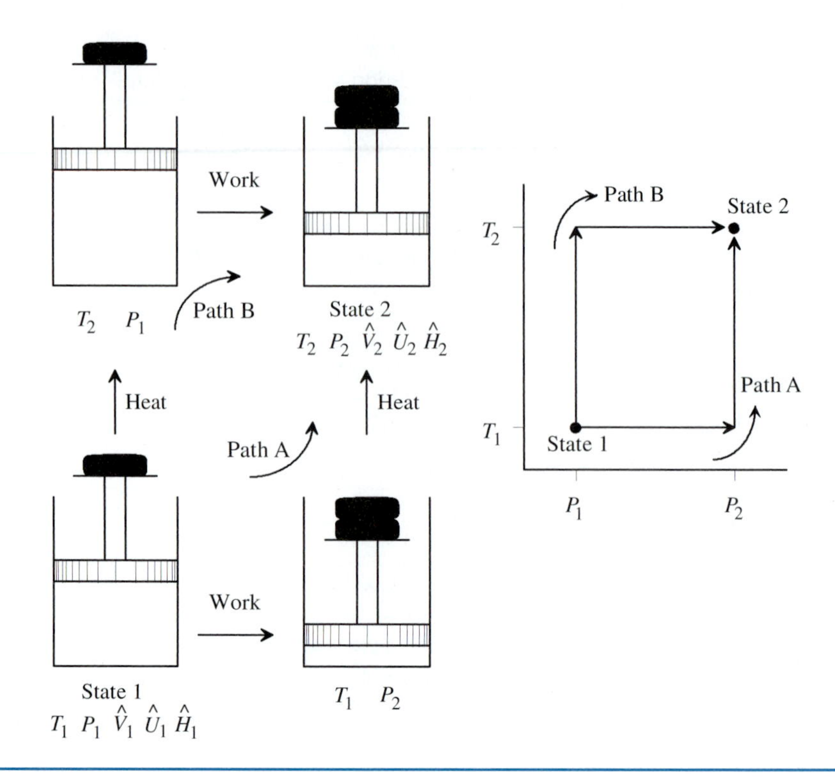

FIGURE 17.7
Illustration of state properties.

17.6.1 Open and Closed Systems

An **open system** is one in which mass **does** cross the boundary and a **closed system** is one in which mass **does not** cross the boundary. An automobile engine is an example of an open system: air and fuel enter the engine and exhaust leaves. A lightbulb is an example of a closed system; it is filled with an inert gas that ideally does not leak out of the bulb; therefore, no mass crosses the boundary.

Figure 17.8 shows a system that contains mass m_{sys}. Many properties of matter (e.g., specific volume, specific enthalpy, and specific internal energy) accompany mass. Therefore, once the mass in the system is known, we know other quantities as well. For example, knowing m_{sys} (kg) and the specific volume $\hat{V}_{sys}$ (m^3/kg), we can calculate the volume in the system V_{sys} (m^3).

Figure 17.8 shows that when mass flows into a system, properties (e.g., specific volume, specific enthalpy, and specific internal energy) enter the system with the mass. If mass flows into the system at rate $\dot{m}_{in}$ (kg/s) and the mass has specific volume $\hat{V}_{in}$ (m^3/kg), then volume enters the system at rate $\dot{V}_{in}$ (m^3/s). Similarly, if mass flows out of the system at rate $\dot{m}_{out}$ (kg/s) and the mass has specific volume $\hat{V}_{out}$ (m^3/kg), then volume exits the system at rate $\dot{V}_{out}$ (m^3/s).

Our earlier discussion emphasized the importance of being able to classify a given quantity as state or path so we could apply the proper accounting rules. Table 17.4 classifies the quantities shown in Figure 17.8. Notice that in all cases, intensive quantities are strictly state *or* path quantities. However, in the case of open systems, there is pos-

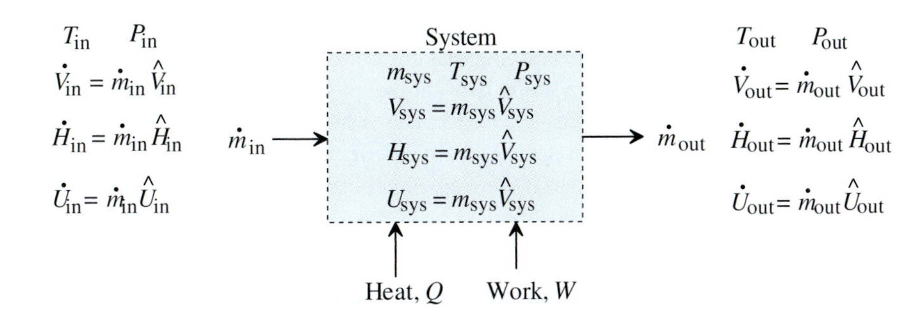

FIGURE 17.8
An open system has mass enter and exit the system. Notice that properties accompany the mass in the system, mass entering the system, and mass exiting the system.

sible confusion regarding whether a given extensive quantity is a state or path quantity. In open systems, extensive quantities have a dual nature and can be classified as state *and* path quantities. Take volume as an example. The volume **in** the system is a state quantity, whereas the volume **entering** or **leaving** the system is a path quantity. Similar arguments can be made for enthalpy and internal energy.

Although open systems have potential for confusion because of the dual nature of extensive quantities, no such confusion exists with closed systems. In closed systems, all quantities are strictly classified as either state *or* path quantities. For this reason, and others, thermodynamicists often prefer to study closed systems.

17.6.2 Isolated and Nonisolated Systems

An **isolated system** has nothing crossing the boundary: no mass, no charge, no linear momentum, no angular momentum, and no energy. In contrast, a *nonisolated system* has at least one of the above-mentioned quantities crossing the boundary.

TABLE 17.4
Analysis of the quantities shown in Figure 17.8

Quantity			Open System		Closed System	
	Intensive	Extensive	Path Quantity	State Quantity	Path Quantity	State Quantity
Heat		X	X		X	
Specific heat (Q/m_{sys})	X		X		X	
Work		X	X		X	
Specific work (W/m_{sys})	X		X		X	
Temperature	X			X		X
Pressure	X			X		X
Specific volume	X			X		X
Volume		X	X	X		X
Specific enthalpy	X			X		X
Enthalpy		X	X	X		X
Specific internal energy	X			X		X
Internal energy		X	X	X		X

The only truly isolated system is the universe itself. By definition, the universe contains everything that exists, so nothing can cross the universe system boundary. In practice, many systems that engineers study closely approximate isolated systems. For example, a well-insulated container of high-temperature, high-pressure gas approximates an isolated system. It deviates from a perfectly isolated system if there is some heat transfer across the insulation and if there are small leaks that allow the gas to slowly escape.

17.6.3 Homogeneous and Heterogeneous Systems

A **homogeneous system** has the same properties everywhere within the system, whereas a **heterogeneous system** has different properties depending upon what portion of the system is examined. Figure 17.9 shows an example of two such systems. In the left figure, salt has been added to water without mixing. As a result, the undissolved salt stays on the bottom. The composition of this system depends on where the sample is taken; a sample taken from the top will be rich in water and a sample taken from the bottom will be rich in salt. When the contents are stirred, the salt dissolves. In this case, it becomes a homogeneous system. Regardless of where the sample is taken, the composition is everywhere the same.

The determination of whether a system is homogeneous or heterogeneous depends on the scale at which it is examined. For example, if the system contains a tank of water, a pile of stone, a pile of sand, and a pile of cement, we would clearly understand this to be a heterogeneous system. Once these components are mixed together, they form concrete. If we examined a cubic meter of concrete, we would find that each cubic meter is pretty much the same as any other, so on this scale, we would say our system is homogeneous. If, however, we examined each cubic millimeter of concrete, the system would be heterogeneous; in some locations we would examine a grain of sand; in other locations, we would examine a stone, and so forth. A decision to treat the concrete as homogeneous or heterogeneous depends upon the question we are trying to answer. If we want to know how thick to pour the concrete on a road that carries heavy trucks, we would consider it to be a homogeneous material because the road will be very thick relative to the size of an individual stone. However, if we want to know how the strength of a thin layer of concrete depends upon the size of the stones, we may want to treat the system as heterogeneous.

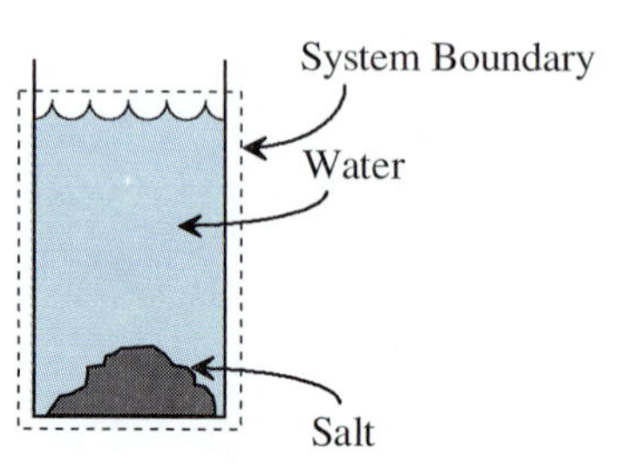

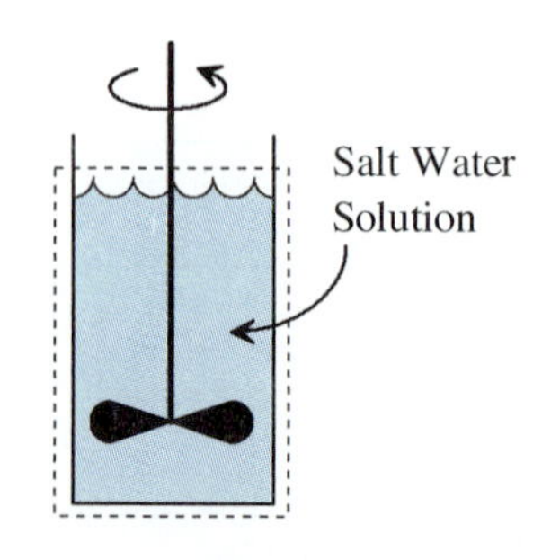

FIGURE 17.9
Example of a heterogeneous and homogeneous system.

17.6.4 Steady-State and Unsteady-State Systems

A **steady-state system** does not change with respect to time, whereas an **unsteady-state system** does change. To understand this concept, imagine that you are an industrial engineer studying a tire warehouse that services a tire factory. Once the tire is made by the factory, it is placed in the warehouse until it is sold. If tires are made at a rate of 1000/h and they are shipped to customers at a rate of 1000/h, the number of tires stored in the warehouse (our system) will never change. This warehouse is said to be at "steady state." If the factory installs a new machine that doubles production to 2000/h, but customer demand stays constant at 1000/h, then tires will accumulate in the warehouse. Because the number of tires stored in the warehouse will change with respect to time, then, in this case, the warehouse is an unsteady-state system.

17.7 UNIVERSAL ACCOUNTING EQUATION

Now that we have described fundamental concepts such as systems, intensive/extensive quantities, and state/path quantities, you have the background to understand one of the most powerful weapons in the engineering arsenal, the **universal accounting equation.** This equation is an **axiom,** meaning it cannot be derived from a more fundamental equation. The reputed source of this equation is shown in Figure 17.10. It is so broad and general, many nonengineering disciplines use it also, because it forms the foundation of rational thought. Although it is extremely important, it is not universally understood. For example, politicians often promise the electorate something for nothing. If the electorate understood this equation, they would be able to see through the politician's nonsense.

Final amount – initial amount = input – output + generation – consumption

FIGURE 17.10
Reputed source of the universal accounting equation. (Drawing kindly provided by A. Robert Holtzapple.)

The *universal accounting equation* requires that the engineer do the following:

1. Establish a system, i.e., a subset of the universe must be identified.
2. Determine what will be counted.
3. Define a time interval during which the counting will be performed.

During this time interval, the definition of the system cannot change.

Suppose an industrial engineer is studying the inventory of a warehouse that stores electrical appliances. First, he would define the warehouse as the system. Second, he would decide if he is going to count refrigerators, ovens, stoves, or dishwashers. Third, he would determine the time period of interest.

The engineer specifies the initial state of the system using extensive state quantities (e.g., the number of refrigerators, ovens, stoves, and dishwashers in the warehouse). As time passes, extensive path quantities (e.g., appliances added or removed) act on the system, causing it to arrive at a final state (Figure 17.11). Mathematically, this is stated as

$$\underbrace{\text{Final amount} - \text{initial amount}}_{\text{state quantities}} = \underbrace{\text{input} - \text{output} + \text{generation} - \text{consumption}}_{\text{path quantities}} \tag{17-6}$$

where the left side accounts for all state quantities and the right side accounts for all path quantities.

It is very important to be careful about the exact definition of each term in the universal accounting equation:

Final amount	Specifies the amount of the counted quantity in the system *at the end* of the time period.
Initial amount	Specifies the amount of the counted quantity in the system *at the beginning* of the time period.
Input	Specifies the amount of the counted quantity passing through the system boundary *into* the system during the time interval.
Output	Specifies the amount of the counted quantity passing through the system boundary *out of* the system during the time interval.
Generation	Specifies the amount of the counted quantity *produced* during the time interval within the system boundary.
Consumption	Specifies the amount of the counted quantity *destroyed* during the time interval within the system boundary.

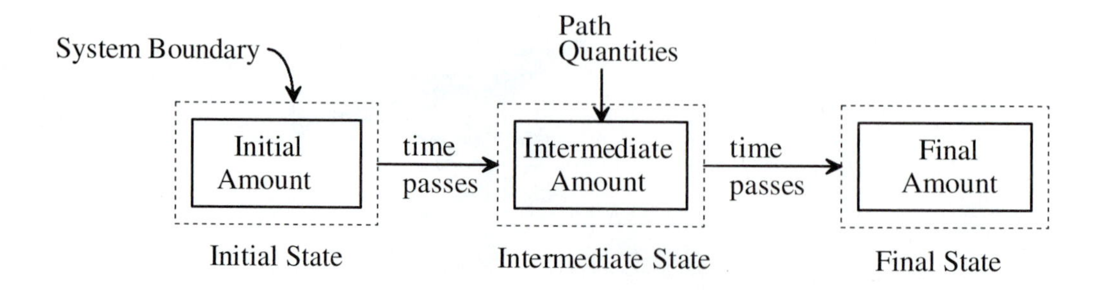

FIGURE 17.11 General approach of the universal accounting equation.

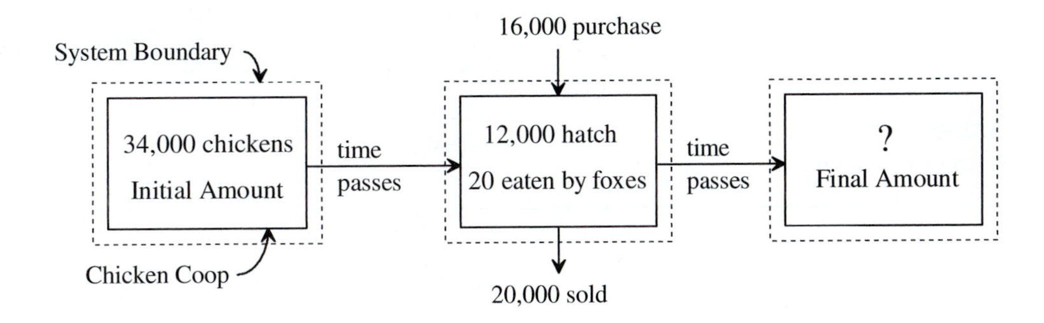

FIGURE 17.12
Accounting for chickens in a chicken coop.

You should note that any input, output, generation, or consumption occurring before or after the time period is completely ignored.

Using the following definitions,

$$\text{Accumulation} \equiv \text{final amount} - \text{initial amount} \tag{17-7}$$

$$\text{Net input} \equiv \text{input} - \text{output} \tag{17-8}$$

$$\text{Net generation} \equiv \text{generation} - \text{consumption} \tag{17-9}$$

the *universal accounting equation* may be more compactly written as

$$\text{Accumulation} = \text{net input} + \text{net generation} \tag{17-10}$$

The left side of the equation describes the change in state caused by the path quantities on the right side of the equation.

EXAMPLE 17.2

Problem Statement: A chicken coop starts the year with 34,000 chickens. During the year, 16,000 chicks are purchased, 20,000 chickens are sold, 12,000 eggshatch, and 20 chickens are eaten by foxes (Figure 17.12). How many chickens are present at the end of the year?

Solution: A chicken is an extensive quantity, so the universal accounting equation is applicable. The system is the chicken coop and the time interval is one year. Equation 17-6 may be solved explicitly for the final state (i.e., the final number of chickens).

$$\begin{aligned}\text{Final amount} &= \text{initial amount} + \text{input} - \text{output} + \text{generation} - \text{consumption}\\ &= 34{,}000 + 16{,}000 - 20{,}000 + 12{,}000 - 20\\ &= 41{,}980 \text{ chickens}\end{aligned}$$

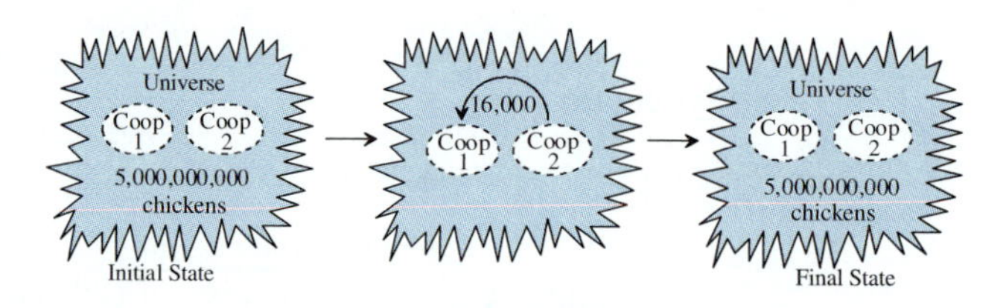

FIGURE 17.13
Illustration of input.

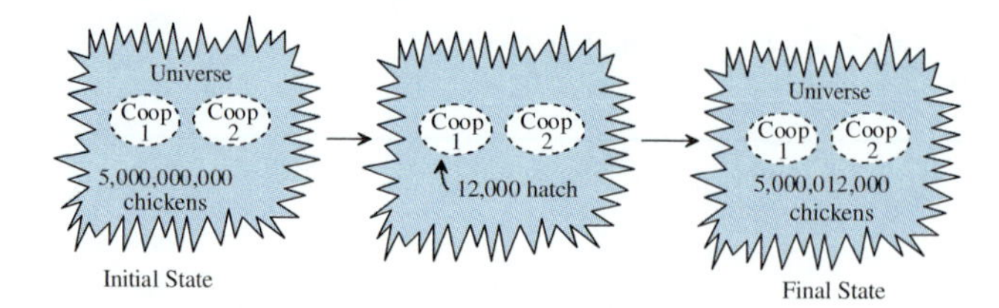

FIGURE 17.14
Illustration of generation.

17.7.1 Input or Generation? Output or Consumption?

In the chicken coop example, we had to decide if the 16,000 purchased chicks are an input or a generation. We intuitively feel they should be counted as input, but how do we know this is proper? The answer comes by analyzing this transaction from the perspective of the entire universe. If the total number of chickens in the universe increases as a result of the transaction, it would be counted as generation. If the total number of chickens in the universe stays the same as a result of the transaction, it would be counted as input. Figure 17.13 shows that the purchased chickens were simply transferred from another chicken coop, so the total number of chickens in the universe stays the same. Therefore, the purchased chickens were properly counted as input.

In the case of the 12,000 hatchlings, the total number of chickens in the universe increased (Figure 17.14). Therefore, the hatchlings are properly counted as generation.

Similar arguments can be made for the sale of 20,000 chickens. They were transferred from one coop to another, so the total number of chickens in the universe did not change. Therefore, they are properly counted as output. In the case of the 20 chickens eaten by foxes, the universe lost 20 chickens, so they were properly counted as consumption.

17.7.2 Steady State

A system is at *steady state* when there is no accumulation. In this case, the universal accounting equation becomes

$$0 = (\text{input} - \text{output}) + (\text{generation} - \text{consumption}) \quad (17\text{-}11)$$

$$0 = \text{net input} + \text{net generation} \quad (17\text{-}12)$$

EXAMPLE 17.3

Problem Statement: A farmer wishes to operate his chicken coop at steady state. During the year, 16,000 chicks are purchased, 12,000 eggs hatch, and 20 chickens are eaten by foxes. How many chickens should be sold that year?

Solution: As previously stated, chickens are extensive quantities, so the universal accounting equation is applicable. The system is the chicken coop and the time interval is one year. Equation 17-11 may be solved explicitly for the number of chickens that should be sold.

$$\text{Output} = \text{input} + \text{generation} - \text{consumption}$$

$$= 16{,}000 + 12{,}000 - 20$$

$$= 27{,}980$$

17.7.3 Conserved Quantities

Many engineering quantities (e.g., mass, charge, momentum, energy) are **conserved,** meaning they are neither generated nor consumed. For conserved quantities, the universal accounting equation reduces to

$$\text{Final amount} - \text{initial amount} = (\text{input} - \text{output}) \qquad (17\text{-}13)$$

$$\text{Accumulation} = \text{net input} \qquad (17\text{-}14)$$

For a system that is at steady state and for a quantity that is conserved, the universal accounting equation reduces to

$$0 = \text{input} - \text{output} \qquad (17\text{-}15)$$

$$0 = \text{net input} \qquad (17\text{-}16)$$

EXAMPLE 17.4

Problem Statement: During the month of September, Carol (an engineering student) makes a \$600 deposit to her checking account and writes a \$300 check to her sorority for the month's rent and food. The bank charges her a \$1 service charge for writing the check and she earns \$3 interest. At the beginning of the month, her checking account balance was \$2000. What is her balance at the end of the month?

Solution: Dollars are extensive quantities, so the universal accounting equation applies. The system is Carol's checking account and the time interval is one month. For ordinary people, dollars are a conserved quantity, so Equation 17-13 may be used.

$$\text{Final amount} = \text{initial amount} + \text{input} - \text{output}$$

$$= \$2000 + (\$600 + \$3) - (\$300 + \$1)$$

$$= \$2302$$

Clearly, the \$600 deposit and \$300 check are inputs and outputs, respectively. Notice that the \$3 interest is counted as input, not generation, and the \$1 service charge is counted as output, not consumption. This can be better understood by considering Figure 17.15. The interest and service charge are simply transferred between two accounts. The total number of dollars in the universe is unaffected by this transaction; therefore, it would be improper to count them as generation or consumption.

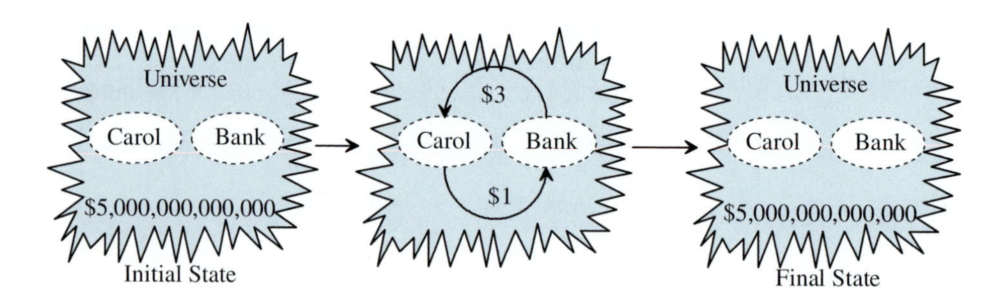

FIGURE 17.15
Service charges and interest are transferred between two accounts.

As stated previously, for most of us, dollars are a conserved quantity. When we make a purchase, our account decreases and the store's account increases by the same amount; hence, the total number of dollars in the universe is unaffected. There are some minor exceptions, however. If we were to have a fire in our home that burned cash, the number of dollars in the universe would diminish. Therefore, this loss would be counted as consumption. Counterfeiters who are able to successfully produce realistic dollars are able to generate dollars; that is, the total number of dollars in the universe has increased because of their illegal activity. The only legal method for increasing the number of dollars in the universe is through the U.S. Federal Reserve Bank. As U.S. economic activity increases due to larger population and greater standard of living, the Federal Reserve Bank generates dollars. If they generate too many dollars, we experience inflation and if they generate too few dollars, we experience deflation.

17.8 SUMMARY

Accounting is an important engineering activity; in fact, it forms the basis of rational thought. Engineers typically count money, mass, charge, linear momentum, angular momentum, energy, and entropy. First, engineers specify the system, then they determine what will be counted, and finally, they specify the time period during which the accounting will be performed.

Only extensive quantities can be counted. Extensive quantities are affected by the system size, whereas intensive quantities are not. Those quantities that characterize the state of the system at any given time are termed state quantities. Path quantities can act on the system to change its state.

Nomenclature

A area (m^2)
H enthalpy (J)
$\hat{H}$ specific enthalpy (J/kg)
$\dot{H}$ enthalpy rate (J/s)
h height (m)
m mass (kg)
$\dot{m}$ mass rate (kg/s)
n moles (mol)

P pressure (Pa)
Q heat rate (J/s)
R universal gas constant [Pa·m^3/(mol·K)]
T temperature (K)
U internal energy (J)
$\hat{U}$ specific internal energy (J/kg)
$\dot{U}$ internal energy rate (J/s)
V volume (m^3)
$\hat{V}$ specific volume (m^3/kg)
$\dot{V}$ volume rate (m^3/s)
W work rate (J/s)
ρ density (kg/m^3)

Further Readings

Felder, R. M., and R. W. Rousseau. *Elementary Principles of Chemical Processes.* New York: Wiley, 1986.

Glover, C. J., and H. Jones. *Conservation Principles of Continuous Media.* New York: McGraw-Hill College Custom Series, 1994.

Glover, C. J.; K. M. Lunsford; and J. A. Fleming. *Conservation Principles and the Structure of Engineering.* New York: McGraw-Hill College Custom Series, 1992.

PROBLEMS

17.1 In Table 17.5, classify the indicated quantities.

17.2 In the 1980 census, Booneville had 3428 residents, and in the 1990 census, it had 2783. There were 178 births and 87 deaths during the decade between 1980 and 1990. What was the net input of people moving into Booneville? Realtors report that 187 people moved into Booneville between 1980 and 1990. How many people left?

17.3 Into a piston/cylinder is placed 3.00 kg of liquid water at 25°C and 1.013×10^5 Pa. At these initial conditions, the specific volume is 0.00100 m^3/kg. Heat is added, converting the liquid into steam at 100°C and 1.013×10^5 Pa. At these final conditions, the specific volume is 1.673 m^3/kg. How much volume accumulates (i.e., what is the change in volume) from this process? Did this volume change result from input, output, generation, or consumption?

17.4 Into a steam boiler is placed 3.00 kg/s of liquid water at 25°C and 1.013×10^5 Pa. At these inlet conditions, the specific volume is 0.00100 m^3/kg. Steam emerges at 100°C and 1.013×10^5 Pa. At these outlet conditions, the specific volume is 1.673 m^3/kg. The boiler is operated at steady state so that no mass accumulates in the boiler tubes. At what rate (m^3/s) is water volume generated by the boiler?

17.5 The initial liquid volume in a tank is 5 m^3. During the day, 10 m^3 is added and 3 m^3 is withdrawn. What is the final volume? (*Note:* The liquid temperature is constant throughout the day.)

17.6 Liquid flows into a tank at 10 m^3/s and flows out of the tank at 3 m^3/s. At what rate (m^3/s) does volume accumulate in the tank? Initially, the tank has 5 m^3. After 10 s, what is the volume in the tank?

17.7 Liquid water has its maximum density at 3.98°C (1.00000 g/cm^3). At 100°C, its density is 0.95838 g/cm^3. The system contains 100.00000 g of water initially at 3.98°C. Then the temperature is raised to 100°C. How much liquid volume is generated in the system?

17.8 When water freezes, it expands. The density of liquid water at 0°C is 0.99987 g/cm^3 and the density of ice is 0.917 g/cm^3. When liquid water gets into the crack of a road and then freezes, it damages the road due to the expansion. A crack in the road (our system) initially contains 23.689 g of liquid water at 0°C. Later, the water freezes. How much volume was generated?

17.9 The following substances have the indicated densities at 20°C: water = 0.9982 g/cm^3, ethanol = 0.7893 g/cm^3, ethanol/ water (50% by mass) = 0.9139 g/cm^3. Initially, 100.0000 g of 20°C water is added to the system. Then, 100.0000 g of 20°C ethanol is gently added so there is a layer of water at the bottom and a layer of ethanol on top. Finally, the ethanol and water are mixed together and brought to 20°C. How much liquid volume was generated (consumed) by this mixing process?

17.10 The following substances have the indicated densities at 20°C: water = 0.9982 g/cm^3, acetic acid = 1.0477 g/cm^3, acetic

TABLE 17.5
Classification of quantities

Quantity	Space-Time Coordinate	Extensive	Intensive	State	Path
Altitude					
Longitude					
Latitude					
Displacement					
Distance traveled					
Heat					
Work					
Temperature					
Pressure					
Volume of the system					
Liquid volume within system					
Liquid volume added to system					
Liquid volume removed from system					
Mass in the system					
Net mass added to the system					
Specific volume					
Specific enthalpy					
Enthalpy within system					
Enthalpy added to system					
Enthalpy removed from system					
Specific internal energy					
Internal energy within system					
Internal energy added to system					
Internal energy removed from system					
Density					
Concentration					
Color					
Bank account balance					
Bank deposits					
Bank withdrawals					
Bank service charge					
Bank interest					
Momentum					
Velocity (momentum per unit mass)					
Force (momentum addition to system)					

acid/water (50% by mass) = 1.0562 g/cm^3. Initially, 100.0000 g of 20°C acetic acid is added to the system. Then, 100.0000 g of 20°C water is gently added so there is a layer of acetic acid at the bottom and a layer of water on top. Finally, the acetic acid and water are mixed together and brought to 20°C. How much liquid volume was generated (consumed) by this mixing process?

17.11 A candle factory makes Santa Claus candles during the year and stores them in a warehouse. In September, the factory sells them to retail stores so they can be on the shelves by Christmas. The candles are produced every day at a rate of 450 per day. On the morning of January 1, there are 5493 candles in the warehouse. On the evening of February 16, there is a fire in the warehouse that destroys some candles. Once the fire is extinguished, the factory does an inventory and finds that there are 17,895 unburned candles. How many candles were consumed by the fire?

17.12 The "optimist version" of the universal accounting equation (Equations 17-6 and 17-10) uses accumulation to account for state quantities. Reformulate it in the "pessimist version" using depletion to account for state quantities.

17.13 An isolated 1000-acre nature preserve supports a population of rabbits and foxes. The land supports 10 breeding pairs of rabbits per acre. Each year, a breeding pair produces 100 rabbits. Of these rabbits, 40% are eaten by foxes; the remaining rabbits die from disease, are eaten by owls, or live to breed. In order to survive, a fox must eat one rabbit every 5 days. At steady state, how many foxes live on the nature preserve? (Assume the land surrounding the nature preserve has no rabbits or foxes.)

17.14 An extinct volcano has a crater that is approximately the shape of an inverted right circular cone (Figure 17.16). The base of the cone is 0.500 mile wide and the tip makes an angle of 45°. Each year, it rains 5 inches. Each day, the evaporation rate is 500 gallons per acre of surface. At steady state, how deep (in feet) is the lake?

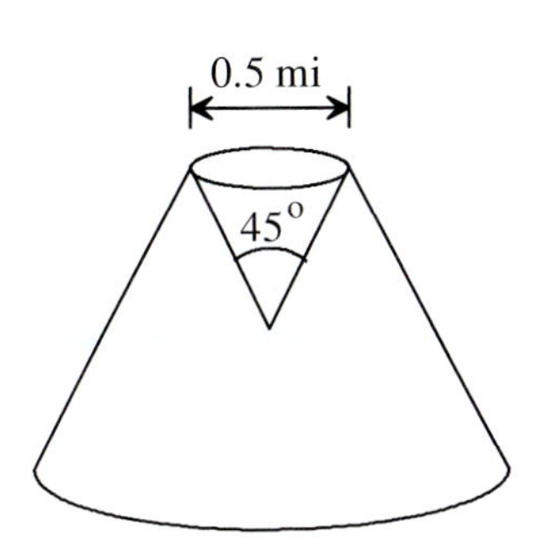

FIGURE 17.16
Volcano lake.

Glossary

axiom An assumption that cannot be rigorously proved to be true, but seems to be true from experience or observation.

closed system A system in which mass does not cross the boundary.

conserved A quantity that is neither generated or consumed.

enthalpy $H = U + PV$.

entropy A measure of disorder.

extensive quantities Quantities that depend on scale.

heterogeneous system A system that has different properties, depending on what portion of the system is examined.

homogeneous system A system that has the same properties everywhere within the system.

hydraulic lift A device used to lift automobiles at service stations.

intensive quantities Quantities that are independent of scale.

isolated system A system in which nothing crosses the boundary.

open system A system in which mass crosses the boundary.

path quantities Quantities that depend on path.

properties of matter Descriptions of materials, such as temperature, pressure, specific volume, specific internal energy, and specific enthalpy.

space-time coordinates A coordinate system that describes where and when.

specific quantities Quantities formed by dividing an extensive quantity by mass or moles.

state quantities Quantities that are independent of path.

steady-state system A system that does not change with respect to time.

surroundings Everything in the universe, except the system.

system The subset of the universe we wish to study.

universal accounting equation Accumulation = net input + net generation.

unsteady-state system A system that changes with respect to time.

Fire Stuff

One of the most powerful conservation laws is the conversation of mass. Simply, this law states that mass can neither be created nor destroyed. If this seems perfectly obvious to you, congratulations. You are far ahead of many great scientists from the past. As an example, consider burning wood in your fireplace. From breaking your back lugging in the wood, you know that a great deal of mass went into the fireplace, but afterwards the ashes are light enough to be scattered by the wind. Mass is conserved?

Around 1650, on the basis of observations like the fireplace example, a German scientist named Johann Joachim Becher proposed that burnable substances contained *phlogiston,* or "fire stuff." He proposed that when something burns, the phlogiston leaves in the form of a flame. The ash left after burning is the non-phlogiston part of a particular substance. Readily burnable things obviously contained much phlogiston, and hard-to-burn substances contained little. It took some real brainwork to realize that during combustion, most of the mass went out the chimney.

Near the time of the American Revolution, the French scientist Antoine Laurent Lavoisier (1743–1794) conducted a series of experiments to study combustion. In one experiment (shown below), he demonstrated that the combustion gases (carbon dioxide and water) have more mass than the original candle. As the candle burned, the balance pan slowly moved downward because the carbon dioxide was captured by the potassium hydroxide according to the reaction

$$KOH + CO_2 \rightarrow KHCO_2$$

Within a few years, Lavoisier demonstrated that combustion was the chemical combination of burnable substance with oxygen from the air.

EXAMPLE 18.3

Problem Statement: Into a heated kettle is placed 100 kg of fresh maple tree sap that contains 5% sugar and 95% water. Water is boiled away to produce maple syrup containing 80% sugar and 20% water. What is the final mass T of maple syrup? How much water S evaporated? How much water W is in the maple syrup?

Solution: The sugar is not volatile, so its mass does not change. Therefore, Figure 18.3 shows that the final amount of sugar is the same as the initial amount.

$$\text{Sugar balance: Final amount} - \text{initial amount} = \text{input} - \text{output}$$

$$\text{Final amount} - \text{initial amount} = 0 - 0 = 0$$

$$\text{Final amount} = \text{initial amount} = 5 \text{ kg}$$

$$\text{Water balance: Final amount} - \text{initial amount} = \text{input} - \text{output}$$

$$W - 95 \text{ kg} = 0 - S$$

$$W + S = 95 \text{ kg}$$

$$\text{Total balance: Final amount} - \text{initial amount} = \text{input} - \text{output}$$

$$T - 100 \text{ kg} = 0 - S$$

$$T + S = 100 \text{ kg}$$

$$0.8 = \frac{5 \text{ kg}}{T} \text{ (because final solution is 80\% sugar)}$$

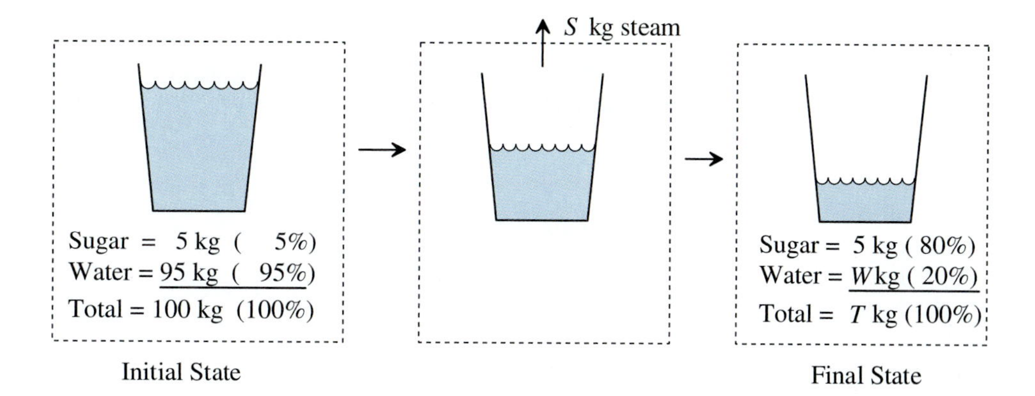

FIGURE 18.3
Water is evaporated from maple sap to make maple syrup.

$$T = \frac{5 \text{ kg}}{0.8} = 6.25 \text{ kg}$$

$$S = 100 \text{ kg} - T = 100 \text{ kg} - 6.25 \text{ kg} = 93.75 \text{ kg}$$

$$W = 95 \text{ kg} - S = 95 \text{ kg} - 93.75 \text{ kg} = 1.25 \text{ kg}$$

18.2 GENERATION AND CONSUMPTION OF MASS (ADVANCED TOPIC)

In the special cases of nuclear reactions or particle accelerators, conservation of mass is violated. In these situations, the entire universal accounting equation is required to count mass:

$$\text{Final amount} - \text{initial amount} = \text{input} - \text{output} + \text{generation} - \text{consumption} \quad (18\text{-}6)$$

In nuclear reactors, mass is literally consumed and converted to energy.[1] In particle accelerators, mass is literally generated from energy.

Einstein developed the famous equation that relates energy to mass:

$$E = mc^2 \quad (18\text{-}7)$$

where E is the amount of energy created (destroyed), m is the amount of mass destroyed (created), and c is the speed of light in a vacuum (2.99792458×10^8 m/s).

Fission is the process whereby a heavy element (e.g., uranium) is broken into smaller fragments. The mass of the fragments is actually less than the original heavy element.

$$^{235}\text{U} + \text{n} \rightarrow \text{fission products} + 2.43\ \text{n} \qquad E = 3.2 \times 10^{-11}\ \text{J/atom} \quad (18\text{-}8)$$

This fission reaction is initiated when uranium-235 is impacted by a slow neutron (indicated with the symbol "n"). In a nuclear reactor (e.g., nuclear power plant, nuclear submarine), fission is done slowly in a controlled manner. In an atomic bomb, fission occurs very rapidly.

1. Mass and energy are related according to Einstein's $E = mc^2$. Some people prefer to say "mass-energy is conserved; therefore, the total amount of mass-energy in the universe does not change with time." Although we agree with this statement, we prefer to account for mass and energy separately because their properties are so different.

EXAMPLE 18.4

Problem Statement: If 1.00000 kg of uranium-235 is fissioned, how much energy is generated and how much mass is consumed? What is the final mass of the products?

Solution:

Energy generated

$$= 1.00000 \text{ kg} \times \frac{1000 \text{ g}}{1 \text{ kg}} \times \frac{\text{gmol}}{235 \text{ g}} \times \frac{6.022 \times 10^{23} \text{ atoms}}{\text{gmol}} \times \frac{3.2 \times 10^{-11} \text{ J}}{\text{atom}}$$

$$= 8.2 \times 10^{13} \text{ J}$$

$$\text{Mass consumed} = m = \frac{E}{c^2} = \frac{8.2 \times 10^{13} \text{ J}}{(3.0 \times 10^8 \text{ m/s})^2} \times \frac{\text{kg·m}^2/\text{s}^2}{J} = 0.00091 \text{ kg}$$

$$\begin{aligned}\text{Final mass} &= \text{final amount} \\ &= \text{initial amount} + \text{input} - \text{output} + \text{generation} - \text{consumption} \\ &= 1.00000 \text{ kg} + 0 - 0 + 0 - 0.00091 \text{ kg} \\ &= 0.99909 \text{ kg}\end{aligned}$$

The amount of energy released by fissioning 1 kg of uranium-235 is roughly the same as is released by combusting 2,000,000 kg of coal! Notice that less than a gram of mass (the approximate mass of a paper clip) was consumed to create this enormous amount of energy.

Fusion is a process whereby lighter elements are combined to form heavier elements. The mass of the products is less than the mass of the reactants. For example, if two deuteriums (hydrogen-2) are fused together to make helium-4,

$$\begin{array}{ccccc} ^{2}\text{H} & + & ^{2}\text{H} & \rightarrow & ^{4}\text{He} \\ 2.014 \text{ kg} & & 2.014 \text{ kg} & & 4.00260 \text{ kg} \end{array} \tag{18-9}$$

the helium has less mass than the two hydrogens. The lost mass is converted to energy. Fusion is employed in hydrogen bombs and is the process that powers the sun. The sun loses 4.5 million tons each second through fusion. Physicists and engineers are attempting to build fusion reactors to make electricity, but have been unable to generate more power than is consumed by the reactor.

EXAMPLE 18.5

Problem Statement: If 1.00000 kg of hydrogen-2 is fused to make helium-4, what is the final mass of the helium-4? How much mass is consumed and how much energy is generated?

Solution: The final mass of helium-4 is determined from Reaction 18-9:

$$\text{Mass } ^4\text{He} = 1.00000 \text{ kg } ^2\text{H} \times \frac{4.00260 \text{ kg } ^4\text{He}}{(2.014 + 2.014) \text{ kg } ^2\text{H}} = 0.9937 \text{ kg } ^4\text{He}$$

$$\begin{aligned}\text{Consumption} &= \text{initial amount} - \text{final amount} + \text{input} - \text{output} + \text{generation} \\ &= 1.00000 \text{ kg} - 0.9937 \text{ kg} + 0 - 0 + 0 \\ &= 0.0063 \text{ kg}\end{aligned}$$

$$E = mc^2 = 0.0063 \text{ kg } (3.00 \times 10^8 \text{ m/s})^2 \times \frac{\text{J}}{\text{kg·m}^2/\text{s}^2} = 5.7 \times 10^{14} \text{ J}$$

Notice that the energy released from the fusion reaction was about 7 times greater than the energy released from the fission reaction. This is why hydrogen (fusion) bombs are much more powerful than atomic (fission) bombs.

Particle accelerators accelerate protons (or other particles) to extremely high velocities that approach the speed of light. (It is impossible to exceed the speed of light.) Enormous amounts of energy are required to accelerate particles to these velocities. When these particles impact stationary objects, or when two particles (each moving in opposite directions) collide, the energy of these particles is converted into mass. New particles [with strange names such as positrons (positive electrons), antiprotons, bosons, fermions, leptons, hadrons, classons, neutrinos, and muons] are generated. Physicists use these particle accelerators to probe the structure of matter. One physicist likened the process to studying the workings of a watch by smashing it with a hammer and observing the pieces fly out.

18.3 CHEMICAL REACTORS (ADVANCED TOPIC)

Unlike nuclear reactors, chemical reactors are not able to destroy mass (at least, we have been unable to measure losses of mass in chemical reactors). However, chemical reactors **are** able to convert chemical species (**reactants**) into other chemical species (products). Therefore, when we account for chemical species, we use the entire universal accounting equation

$$\text{Final amount} - \text{initial amount} = \text{input} - \text{output} + \text{generation} - \text{consumption} \quad (18\text{-}10)$$

For example, a natural gas furnace (Figure 18.4) is a chemical reactor in which methane and oxygen reactants are converted to carbon dioxide and water products,

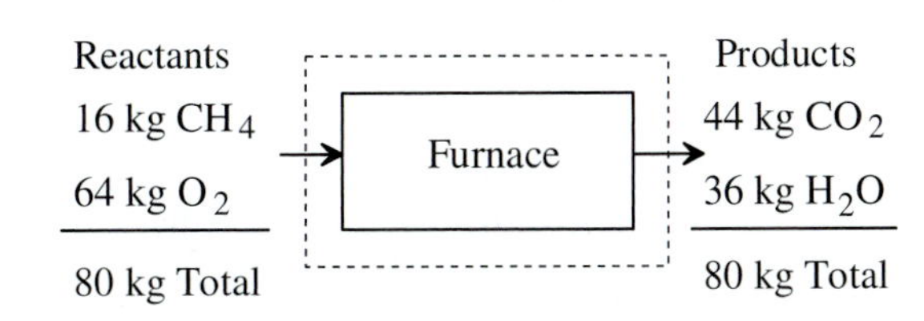

FIGURE 18.4
A natural gas furnace is a chemical reactor in which methane and oxygen are converted to carbon dioxide and water. Notice that the mass of reactants is identical to the mass of products.

$$\begin{aligned} CH_4 + 2\,O_2 &\rightarrow CO_2 + 2\,H_2O \\ 16\text{ kg} + 64\text{ kg} &= 44\text{ kg} + 36\text{ kg} \\ 80\text{ kg} &= 80\text{ kg} \end{aligned} \quad (18\text{-}11)$$

The chemical species methane and oxygen are consumed and the chemical species carbon dioxide and water are generated. However, the gas furnace is unable to destroy mass. If 80 kg of methane and oxygen are fed to the furnace, 80 kg of carbon dioxide and water will be produced.

In Equation 18-11, we had to balance it so that the mass of the reactants equals the mass of the products. Balancing a chemical reaction is certainly a problem you have encountered in your chemistry class. Most students use the "by-guess-and-by-golly" method, where they look at the reactants and products, fiddle with the numbers, and hopefully emerge with a balanced equation. This approach is not recommended because it is time consuming and prone to error. A more systematic approach is desired.

Unlike nuclear reactors in which elements are consumed and generated, chemical reactors merely rearrange elements; none are generated or consumed. Therefore, in chemical reactors, we can say that elements are conserved and the universal accounting equation becomes

$$\text{Final amount} - \text{initial amount} = \text{input} - \text{output} \quad (18\text{-}12)$$

If the reactor is run at steady state (i.e., there is no accumulation, so the final amount and initial amount are identical), then the universal accounting equation for elements in a chemical reactor becomes

$$\begin{aligned} 0 &= \text{input} - \text{output} \\ \text{input} &= \text{output} \end{aligned} \quad (18\text{-}13)$$

To balance Equation 18-11, we specify that a single mole of methane reacts with a moles of oxygen and produces b moles of carbon dioxide and c moles of water. By applying Equation 18-13 to each element, we can solve for a, b, and c, thus balancing our equation. (*Note:* If you are uncomfortable with the concept of a "**mole,**" please review Chapter 11 on thermodynamics.)

$$\begin{array}{lcccccl} & \text{Input} & & \text{Output} & & & \\ & 1\,CH_4 & + a\,O_2 & \rightarrow b\,CO_2 & + c\,H_2O & & \\ \text{Carbon (C):} & 1(1) & + a(0) & = b(1) & + c(0) & \Rightarrow & b = 1 \\ \text{Hydrogen (H):} & 1(4) & + a(0) & = b(0) & + c(2) & \Rightarrow & c = 2 \\ \text{Oxygen (O):} & 1(0) & + a(2) & = b(2) & + c(1) & \Rightarrow & a = 2 \end{array}$$

If the chemical reactor is operated so that exactly two moles of oxygen are added per mole of methane, we say that the feed is **stoichiometric.** If three moles of oxygen were added per mole of methane, we would say the oxygen was added in **excess** and the methane was **limiting.**

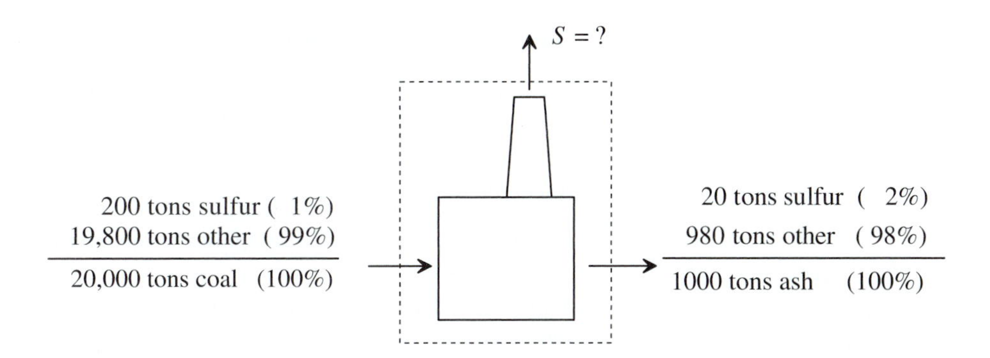

FIGURE 18.5
Electric power plant burning sulfur-laden coal.

EXAMPLE 18.6

Problem Statement: Each day, a power plant burns 20,000 tons of coal that contains 1% sulfur. The plant produces 1000 tons per day of ash containing 2% sulfur. The power plant is operated in a steady-state manner so that no ash accumulates in the furnace. How much sulfur is emitted up the stack?

Solution: Because sulfur is an element, it is conserved in this chemical reaction. Because the furnace (Figure 18.5) is operated in a steady-state manner, we use Equation 18-13:

$$\text{Input} = \text{output}$$

$$\begin{matrix}\text{Mass of sulfur} \\ \text{in coal feed}\end{matrix} = (\text{mass of sulfur in ash}) + (\text{mass of sulfur in stack gas})$$

$$(0.01)20{,}000 \text{ ton} = (0.02)1000 \text{ ton} + S$$

$$200 \text{ ton} = 20 \text{ ton} + S$$

$$S = 200 \text{ ton} - 20 \text{ ton} = 180 \text{ ton}$$

EXAMPLE 18.7

Problem Statement: Pure methane is fed to a furnace at a rate of 1.0 kg/s. A stoichiometric amount of air is fed to the furnace, which operates in a steady-state manner. The methane completely reacts, forming carbon dioxide and water. What is the required feed rate of oxygen and air? What is the flow rate of carbon dioxide, water, and nitrogen (and other inerts) exiting the furnace? What is the total mass of feed and the total mass of exit gases? [*Note:* Air is 21 mole % oxygen, 79 mole % nitrogen (and other inerts). It has an average molecular weight of 29.0 g/mol or 29.0 kg/kmol.]

Solution: Unlike the previous example in which chemical elements (which are conserved) were counted, here, we are counting chemical species—methane (CH_4), dioxygen (O_2), carbon dioxide (CO_2), and water (H_2O)—which are not conserved. Therefore, we need the full universal accounting equation,

$$\text{Final amount} - \text{initial amount}$$
$$= \text{Accumulation} = \text{input} - \text{output} + \text{generation} - \text{consumption}$$

Because the furnace is operated at steady state (Figure 18.6), there is no accumulation and this equation simplifies to

$$0 = \text{Input} - \text{output} + \text{generation} - \text{consumption}$$

This equation is applied to each chemical species. (Notice that the reaction stoichiometry and molecular weights are featured prominently in the calculations. Also, notice that "kmol" is used rather than "mol." A "kmol" is 1000 times larger than a "mol" and is the appropriate unit when measuring mass in kilograms rather than grams.)

CH_4: $0 = \text{input} - 0 + 0 - \text{consumption}$

$$\text{Consumption} = \text{input} = 1.0\,\frac{\text{kg CH}_4}{\text{s}}$$

O_2: $0 = \text{input} - 0 + 0 - \text{consumption}$

$$\text{Input} = \text{consumption}$$

$$= 1.0\,\frac{\text{kg CH}_4}{\text{s}} \times \frac{\text{kmol CH}_4}{16\text{ kg CH}_4} \times \frac{2\text{ kmol O}_2}{1\text{ kmol CH}_4} \times \frac{32\text{ kg O}_2}{\text{kmol O}_2} = 4.0\,\frac{\text{kg O}_2}{\text{s}}$$

Air: $\text{Input} = 4.0\,\frac{\text{kg O}_2}{\text{s}} \times \frac{\text{kmol O}_2}{32\text{ kg O}_2} \times \frac{1.00\text{ kmol air}}{0.21\text{ kmol O}_2} \times \frac{29\text{ kg air}}{\text{kmol air}} = 18.3\,\frac{\text{kg air}}{\text{s}}$

CO_2: $0 = 0 - \text{output} + \text{generation} - 0$

$$\text{Output} = \text{generation}$$

$$= 1.0\,\frac{\text{kg CH}_4}{\text{s}} \times \frac{\text{kmol CH}_4}{16\text{ kg CH}_4} \times \frac{1\text{ kmol CO}_2}{1\text{ kmol CH}_4} \times \frac{44\text{ kg CO}_2}{\text{kmol CO}_2} = 2.75\,\frac{\text{kg O}_2}{\text{s}}$$

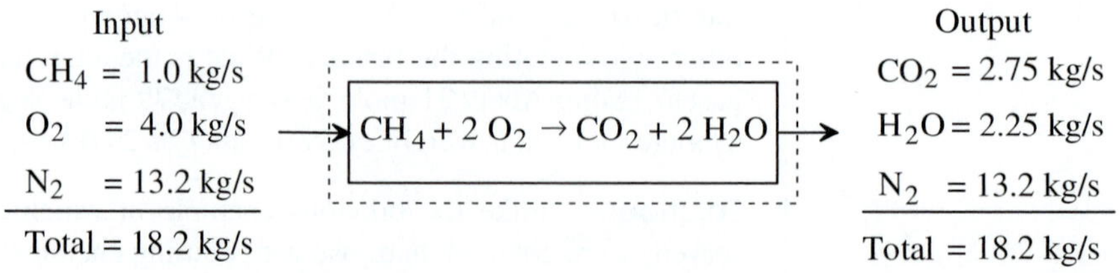

FIGURE 18.6
Combustion of methane in a furnace.

H_2O: $0 = 0 - \text{output} + \text{generation} - 0$

$$\text{Output} = \text{generation}$$

$$= 1.0\,\frac{\text{kg CH}_4}{\text{s}} \times \frac{\text{kmol CH}_4}{16\ \text{kg CH}_4} \times \frac{2\ \text{kmol H}_2\text{O}}{1\ \text{kmol CH}_4} \times \frac{18\ \text{kg H}_2\text{O}}{\text{kmol H}_2\text{O}} = 2.25\,\frac{\text{kg H}_2\text{O}}{\text{s}}$$

N_2: $0 = \text{input} - \text{output} + 0 - 0$

$$\text{Output} = \text{input}$$

$$= 17.3\,\frac{\text{kg air}}{\text{s}} \times \frac{\text{kmol air}}{29\ \text{kg air}} \times \frac{0.79\ \text{kmol N}_2}{\text{kmol air}} \times \frac{28\ \text{kg N}_2}{\text{kmol N}_2} = 13.2\,\frac{\text{kg N}_2}{\text{s}}$$

Total: $0 = \text{input} - \text{output} + 0 - 0$

$$\text{Input} = \text{output}$$

$$(\text{CH}_4) + (\text{O}_2) + (\text{N}_2) = (\text{CO}_2) + (\text{H}_2\text{O}) + (\text{N}_2)$$

$$1.0\ \text{kg/s} + 4.0\ \text{kg/s} + 13.2\ \text{kg/s} = 2.75\ \text{kg/s} + 2.25\ \text{kg/s} + 13.2\ \text{kg/s}$$

$$= 18.2\ \text{kg/s}$$

18.4 SUMMARY

In general, mass is a conserved quantity, meaning it can be neither created nor destroyed. An exception is nuclear reactions in which a measurable amount of matter is created or destroyed. Very few engineers work with nuclear reactions (except for nuclear engineers), so for most engineers, mass is a conserved quantity.

Although total mass is conserved, chemical species are not. In chemical reactors, chemical species may be either created or destroyed.

Nomenclature

c speed of light (m/s)
E energy (J)
m mass (kg)

Further Readings

Eide, A. R.; R. D. Jenison; L. H. Mashaw; and L. L. Northrup. *Engineering Fundamentals and Problem Solving.* 3rd ed. New York: McGraw-Hill, 1997.

Felder, R. M., and R. W. Rousseau. *Elementary Principles of Chemical Processes.* New York: Wiley, 1986.

PROBLEMS

18.1 Balance the following chemical reactions:

(a) $C_2H_5OH + O_2 \rightarrow CO_2 + H_2O$

(b) $CO + H_2 \rightarrow CH_4 + H_2O$

(c) $CO + H_2 \rightarrow H_3COH$

(d) $CO + H_2 \rightarrow C_2H_5OH + H_2O$

(e) $C_6H_{12}O_6 + O_2 \rightarrow CO_2 + H_2O$

(f) $CO + H_2 \rightarrow H_3CCH_2CH_2CH_2CH_3 + H_2O$

(g) $C_8H_{18} + O_2 \rightarrow CO_2 + H_2O$

(h) $C_6H_{12}O_6 \rightarrow C_2H_5OH + CO_2$

18.2 How much water must be added to 1.36 lb_m of salt to give an 8.3% salt solution?

18.3 What is the final salt concentration resulting when 1.03 kg of Solution A (2.3% salt) is mixed with 3.04 kg of Solution B (4.8% salt)?

18.4 A grain drier is fed 100 ton/d of wet corn (50% moisture), which dries it to 9% moisture. How much dried corn (containing 9% moisture) is produced each day? How much water is removed from the corn each day?

18.5 An alcohol solution contains 6% (by mass) ethanol. This solution is fed to a distillation column at a rate of 1000 kg/h. The top of the distillation column produces highly concentrated ethanol product (90% ethanol by mass), and the bottom of the distillation column is essentially ethanol free (0.005% ethanol by mass). At what rate is liquid produced from the top of the column? At what rate is liquid produced from the bottom of the column?

18.6 A 1.000-kg sugar solution initially contains 2.3% sugar with the remainder water. How much dry sugar must be added to obtain a solution that is 26.3% sugar?

18.7 A military vehicle burns gasoline at a rate of 2 gallons (13.2 lb_m) per hour. Although gasoline is a mixture of many chemicals, for simplicity assume it has the formula C_8H_{18}. What is the required air flow rate (lb_m/h) assuming it is added stoichiometrically? For desert operations, this military vehicle is fitted with a special device that condenses moisture out of the engine exhaust. What is the maximum amount of water (lb_m/h) that could be recovered from the engine exhaust?

18.8 In Brazil, many of the cars are designed to run on ethanol (C_2H_5OH). An ethanol-powered car is on a highway trip burning 8 liters (6.4 kg) of pure ethanol per hour. What is the required air flow rate (kg/h) assuming it is added stoichiometrically? At what rate (kg/h) are carbon dioxide and water produced by the car?

18.9 An evaporator has the capacity to remove 1000 kg/h of pure water from a brine solution. The brine solution is 10.0% salt and is fed to the evaporator at a rate of 5000 kg/h. What is the concentration of salt exiting the evaporator?

18.10 Air conditioners not only cool the air but remove moisture as well. Each hour, an air conditioner removes 10.0 gal liquid water from the air. The moisture content of the incoming air is 0.015 lb_m water/lb_m dry air and the moisture content of the outgoing air is 0.010 lb_m water/lb_m dry air. What is the air flow rate (in lb_m dry air/h) through the air conditioner?

18.11 A chemist has two ethanol/water stock solutions: one is 10.0% (by mass) ethanol (density = 0.9819 g/cm^3) and the other is 80.0% (by mass) ethanol (density = 0.8436 g/cm^3). What volume of each stock solution is required to make a 1.000 L solution of ethanol/water that is 50.0% (by mass) ethanol (density = 0.9139 g/cm^3)? Does the total volume of the two stock solutions equal the volume of the final solution?

18.12 Zeolite is used to remove moisture from methane. A vertical column is filled with 1000.0 kg of dry zeolite. The zeolite has the capacity to hold 0.100 kg water/kg dry zeolite. Once the zeolite becomes saturated with moisture, it must be regenerated by heating. The inlet moisture content of the methane is 7.00% (by mass) and the outlet moisture content is 0.05% (by mass). How much methane (kg) will be produced before the zeolite must be regenerated? (*Hint:* Calculate the ratio of the water to methane at the inlet and the outlet.)

18.13 An electric power plant is designed to burn coal that is 0.0300 lb_m sulfur/lb_m dry coal. An eastern coal is available that is 0.0525 lb_m sulfur/lb_m dry coal, and a western coal is available that is 0.0136 lb_m sulfur/lb_m dry coal. The eastern coal contains 0.12 lb_m water/lb_m dry coal, and the western coal contains 0.17 lb_m water/lb_m dry coal. In what ratio should the coal be fed to the power plant? (Express your answer as ton wet eastern coal/ton wet western coal.)

18.14 A city of 100,000 people decides to install a seawater desalination plant to supply their water. Each person requires 100 gallons of water per day; all of their water requirements will be supplied by the desalination plant. The seawater enters the desalination plant as a 3.5% salt solution. A concentrated brine (10.0% salt solution) is discharged to the ocean as waste. The treated water is essentially salt free. What is the required flow rate (gallons per day) of seawater to feed the desalination plant?

18.15 Fresh orange juice is about 12.0% solids (by mass). Because it is very expensive to ship water from Florida (where the juice is produced) to New York (where the juice is consumed), evaporators are used to remove water. The concentrated orange juice (42.0% solids by mass) is frozen and shipped north where it is reconstituted back to the original 12.0% solids. The city of New York has about 15 million people, each of whom drinks one 10-oz glass of orange juice each day, on average. To supply the New Yorkers with juice, how much water (in gallons/h) must be removed from the fresh orange juice by the evaporators?

18.16 Strawberries cost \$0.50/lb_m and are 85% water and 15% solids. Dry sugar costs \$0.10/lb_m. To make strawberry jam, crushed strawberries and sugar are added in a 45:55 mass ratio.

This mixture is heated to remove water until the final solids content is 66%. What is the cost of the raw materials used to make 1.00 lb_m of jam? Some consumers are concerned about their intake of sugar. If the jam is made from strawberries alone, without the addition of sugar, what is the cost of the raw materials used to make 1.00 lb_m of jam if the sugar-free jam is also 66% solids?

18.17 When one gmole of hydrogen is burned with oxygen, it releases 285.84 kJ of energy. A scientist proposes that there is a loss of mass associated with chemical reactions, just as there is with nuclear reactions. She attempts to measure this loss of mass by charging a chemical reactor with 1.000000 kg of room-temperature hydrogen and a stoichiometric amount of pure, room-temperature oxygen. She then initiates the reaction with a spark, cools the product to room temperature, and weighs the water product. Assuming the loss of mass can be calculated by Einstein's famous equation $E = mc^2$, how much mass is lost? Do you think the scientist can measure this lost mass?

18.18 Write a computer program that balances chemical reactions. The program asks the user how many reactants and how many products are involved in the reaction. It then asks the user about the elemental composition of each reactant and each product; that is, it asks how many moles of carbon, hydrogen, and oxygen are contained in each mole of reactant and product. It then computes the balanced equation and prints it out. To test your program, use the chemical reactions described in Problem 18.1.

Glossary

conservation of mass Mass can be neither created nor destroyed; also called conservation of matter.

excess reactant A reactant that remains after other reactants are completely converted to product.

fission The process whereby a heavy element is broken into lighter elements.

fusion The process whereby lighter elements are combined to form heavier elements.

limiting reactant A reactant that is completely consumed before other reactants are converted to product.

mole The base unit of the amount of substance; Avogadro's number.

particle accelerator A machine that accelerates particles to extremely high velocities that approach the speed of light.

reactants Chemical species that are converted into other chemical species.

stoichiometric Quantitative relationship that describes the amount of products resulting from the complete conversion of reactants.

universal accounting equation Final amount − initial amount = input − output + generation − consumption

Mathematical Prerequisite

- Algebra (Appendix E, Mathematics Supplement)

CHAPTER 19

Accounting for Charge

Charge is a property of matter, just as mass is a property of matter. As you have undoubtedly learned in your chemistry class, ordinary matter is composed of the three fundamental particles shown in the following table:

Fundamental Particle	Mass (g)	Charge
Electron	$9.1093897 \times 10^{-28}$	−1
Proton	$1.6726231 \times 10^{-24}$	+1
Neutron	$1.6749286 \times 10^{-24}$	0

Because charge is a property of matter, and matter is an extensive quantity, charge also must be an extensive quantity. Therefore, charge can be counted by using the universal accounting equation.

19.1 ACCOUNTING FOR CHARGE

We have two types of charge to count, one positive and the other negative. The symbol q^+ will be used to count positive charge and q^- will be used to count negative charge. The universal accounting equation for each type of charge is

$$q^+_{\text{final}} - q^+_{\text{initial}} = q^+_{\text{in}} - q^+_{\text{out}} + q^+_{\text{gen}} - q^+_{\text{cons}} \quad \text{(General)} \tag{19-1}$$

$$q^-_{\text{final}} - q^-_{\text{initial}} = q^-_{\text{in}} - q^-_{\text{out}} + q^-_{\text{gen}} - q^-_{\text{cons}} \quad \text{(General)} \tag{19-2}$$

Ordinarily, charge is neither generated nor consumed, so we can say that under usual conditions, charge is a **conserved quantity;** therefore, the accounting equation becomes

$$q^+_{\text{final}} - q^+_{\text{initial}} = q^+_{\text{in}} - q^+_{\text{out}} \quad \text{(Ordinarily)} \tag{19-3}$$

$$q^-_{\text{final}} - q^-_{\text{initial}} = q^-_{\text{in}} - q^-_{\text{out}} \quad \text{(Ordinarily)} \tag{19-4}$$

Equations 19-3 and 19-4 are not necessarily valid in nuclear reactions where mass is converted to energy, and vice versa. In this case, charge also can be created or consumed. For example, consider the reaction of an electron e^- with a positron e^+ (i.e., a particle that is identical to an electron, except it has positive charge), which forms pure energy with no mass left over:

$$e^- + e^+ \rightarrow \text{energy} \tag{19-5}$$

In this reaction, positive and negative charge is "consumed"; that is, the amount of positive and negative charge in the universe is now less.

The above reaction can be run in reverse, energy being converted to an electron and a positron:

$$\text{Energy} \rightarrow e^- + e^+ \tag{19-6}$$

In this reaction, positive and negative charge is "generated"; that is, the amount of positive and negative charge in the universe is now greater.

If we subtract Equation 19-2 from Equation 19-1, we obtain

$$(q^+ - q^-)_{\text{final}} - (q^+ - q^-)_{\text{initial}}$$
$$= (q^+ - q^-)_{\text{in}} - (q^+ - q^-)_{\text{out}} + (q^+ - q^-)_{\text{gen}} - (q^+ - q^-)_{\text{cons}} \tag{19-7}$$

In all the years physicists have studied matter, they have always observed that when a positive charge is generated, a negative charge is generated with it (Equation 19-6). Similarly, whenever a positive charge is consumed, a negative charge is consumed with it (Equation 19-5). Mathematically, we can state this observation as

$$q^+_{\text{gen}} = q^-_{\text{gen}} \tag{19-8}$$

$$q^+_{\text{cons}} = q^-_{\text{cons}} \tag{19-9}$$

When Equations 19-8 and 19-9 are substituted into Equation 19-7, we obtain

$$(q^+ - q^-)_{\text{final}} - (q^+ - q^-)_{\text{initial}} = (q^+ - q^-)_{\text{in}} - (q^+ - q^-)_{\text{out}} \tag{19-10}$$

If we define net positive charge $q^{\text{net}+}$ as

$$q^{\text{net}} \equiv q^+ - q^- \tag{19-11}$$

Equation 19-10 becomes

$$q^{\text{net}+}_{\text{final}} - q^{\text{net}+}_{\text{initial}} = q^{\text{net}+}_{\text{in}} - q^{\text{net}+}_{\text{out}} \qquad \text{(General)} \tag{19-12a}$$

As derived here, this equation counts net positive charge. It can also be used to count net negative charge,

$$q^{\text{net}-}_{\text{final}} - q^{\text{net}-}_{\text{initial}} = q^{\text{net}-}_{\text{in}} - q^{\text{net}-}_{\text{out}} \qquad \text{(General)} \tag{19-12b}$$

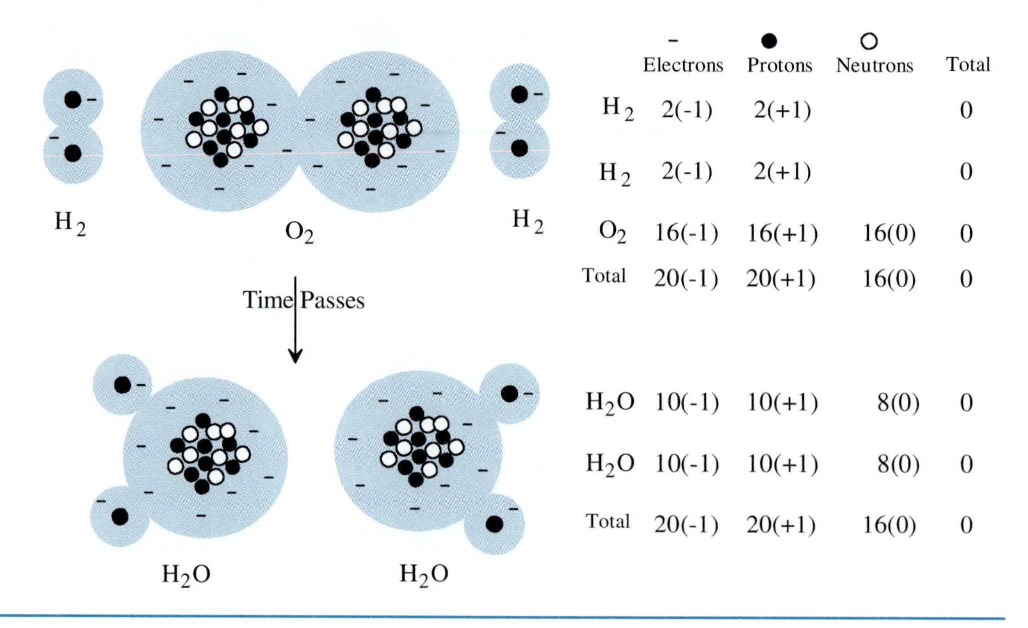

FIGURE 19.1
Chemical reaction in which oxygen and hydrogen form water.

19.2 ACCOUNTING FOR CHARGE IN CHEMICAL REACTIONS

Hydrogen can react with oxygen according to the following chemical reaction

$$2\ H_2 + O_2 \rightarrow 2\ H_2O$$

Figure 19.1 illustrates this reaction. If we count the positive and negative charge before and after the reaction, there is no change. Chemical reactions are "ordinary" in the sense that they do not generate or consume charge; therefore, chemical reactions obey accounting Equations 19-3 and 19-4.

EXAMPLE 19.1

Problem Statement: Into a reactor are placed 2.0 mol of hydrogen and 1.0 mol of oxygen. A reaction occurs in which all of the hydrogen and oxygen react to form water (Figure 19.2). (a) Initially, how many moles of positive, negative, and net positive charge are in the reactor? (b) At the end of the reaction, how many moles of positive, negative, and net positive charge are in the reactor?

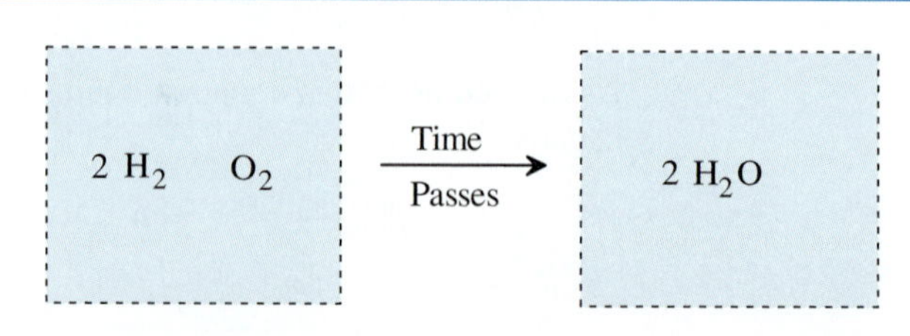

FIGURE 19.2
Reaction of hydrogen and oxygen to form water.

Solution: (a)

$$q^{+}_{\text{initial}} = 2 \text{ mol H}_2 \frac{2 \text{ mol } (+)}{\text{mol H}_2} + 1 \text{ mol O}_2 \frac{16 \text{ mol } (+)}{\text{mol O}_2} = 20 \text{ mol } (+)$$

$$q^{-}_{\text{initial}} = 2 \text{ mol H}_2 \frac{2 \text{ mol } (-)}{\text{mol H}_2} + 1 \text{ mol O}_2 \frac{16 \text{ mol } (-)}{\text{mol O}_2} = 20 \text{ mol } (-)$$

$$q^{\text{net}+}_{\text{initial}} = q^{+}_{\text{initial}} - q^{-}_{\text{initial}} = 20 \text{ mol } (+) - 20 \text{ mol } (-) = 0$$

(b) Equations 19-3, 19-4, and 19-12a are used to account for the charge. There is no input or output of charge during the time period.

$$q^{+}_{\text{final}} - q^{+}_{\text{initial}} = q^{+}_{\text{in}} - q^{+}_{\text{out}} = 0 - 0 \quad \Rightarrow \quad q^{+}_{\text{final}} = q^{+}_{\text{initial}} = 20 \text{ mol } (+)$$

$$q^{-}_{\text{final}} - q^{-}_{\text{initial}} = q^{-}_{\text{in}} - q^{-}_{\text{out}} = 0 - 0 \quad \Rightarrow \quad q^{+}_{\text{final}} = q^{-}_{\text{initial}} = 20 \text{ mol } (-)$$

$$q^{\text{net}+}_{\text{final}} - q^{\text{net}+}_{\text{initial}} = q^{\text{net}+}_{\text{in}} - q^{\text{net}+}_{\text{out}} = 0 - 0 \quad \Rightarrow \quad q^{\text{net}+}_{\text{final}} = q^{\text{net}+}_{\text{initial}} = 0 \text{ mol}$$

EXAMPLE 19.2

Problem Statement: **Fuel cells** are very efficient devices for producing electricity and have been used in the U.S. space program. A fuel cell (Figure 19.3) is a device in which hydrogen and oxygen react to form water in such a way that some of the electrons involved in the reaction flow through a wire as electricity. The two "half reactions" are

$$2 \text{ H}_2 \rightarrow 4 \text{ H}^+ + 4 e^- \qquad \text{Anode}$$

$$4 e^- + 4 \text{ H}^+ + \text{O}_2 \rightarrow 2 \text{ H}_2\text{O} \qquad \text{Cathode}$$

Into the fuel cell, 0.040 mol of hydrogen is fed. How much charge flows through the wire?

FIGURE 19.3
Fuel cell.

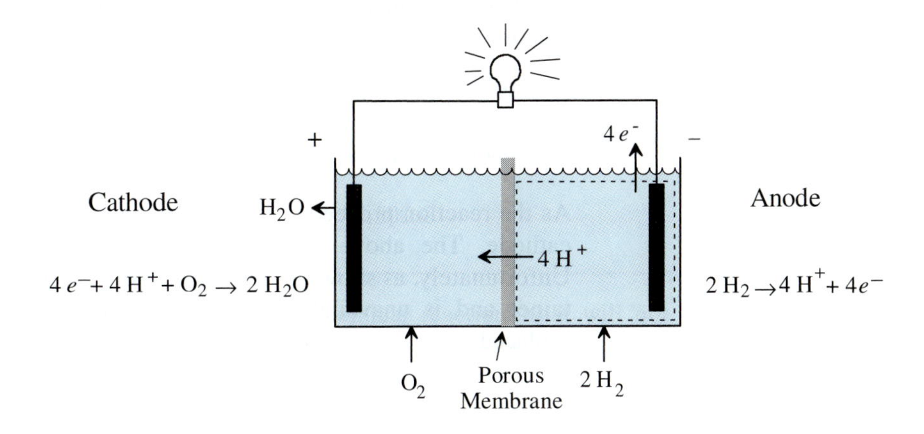

aluminum metal from bauxite (hydrated aluminum oxide). Another important electrolysis reaction is chlorine/sodium hydroxide (caustic soda) production from sodium chloride (salt). The two "half reactions" are

$$2\ (Na^+\ Cl^-) \rightarrow 2\ Na^+ + Cl_2 + 2\ e^- \qquad \text{Anode}$$

$$2\ H_2O + 2\ e^- \rightarrow 2\ OH^- + H_2 \qquad \text{Cathode}$$

$$2\ NaCl + 2\ H_2O \rightarrow 2\ NaOH + Cl_2 + H_2 \qquad \text{Summation reaction}$$

Figure 19.5 illustrates the chlorine/sodium hydroxide electrolysis reaction.

EXAMPLE 19.4

Problem Statement: A 1000-A electric current flows through the sodium hydroxide/chlorine electrolysis reactor illustrated in Figure 19.5. What is the chlorine production rate (g/s)?

Solution: The anode will be defined as the system. The system is assumed to operate at steady state; therefore, there is no accumulation of net negative charge. We write

$$q_{final}^{net-} - q_{initial}^{net-} = q_{in}^{net-} - q_{out}^{net-}$$

$$0 = q_{in}^{net-} - q_{out}^{net-}$$

$$q_{in}^{net-} = q_{out}^{net-}$$

The net output of negative charge flows with the electrons:

$$q_{out}^{net-} = q_{in}^{net-} = 1000\ \text{A} \times \frac{\text{C}}{\text{A·s}} \times \frac{\text{mol}\ (-)}{96{,}485\ \text{C}} = 0.010364\ \frac{\text{mol}\ (-)}{\text{s}}$$

Using reaction stoichiometry, the flow rate of chlorine can be determined from q_{in}^{net-} as

$$F_{Cl_2} = 0.010364\ \frac{\text{mol}\ (-)}{\text{s}} \times \frac{1\ \text{mol}\ Cl^-}{1\ \text{mol}\ (-)} \times \frac{\text{mol}\ Cl_2}{2\ \text{mol}\ Cl^-} \times \frac{70.906\ \text{g}\ Cl_2}{\text{mol}\ Cl_2}$$

$$= 0.367\ \frac{\text{g}\ Cl_2}{\text{s}}$$

Electroplating is a process in which metal ions react at a cathode to form elemental metals. Generally, the cathode is an inexpensive metal onto which a more precious metal is deposited. Examples of electroplating include gold plating onto jewelry, silver plating onto tableware, chromium plating onto automobile bumpers, and gold plating onto electronic terminals to prevent corrosion. Some common plating reactions are:

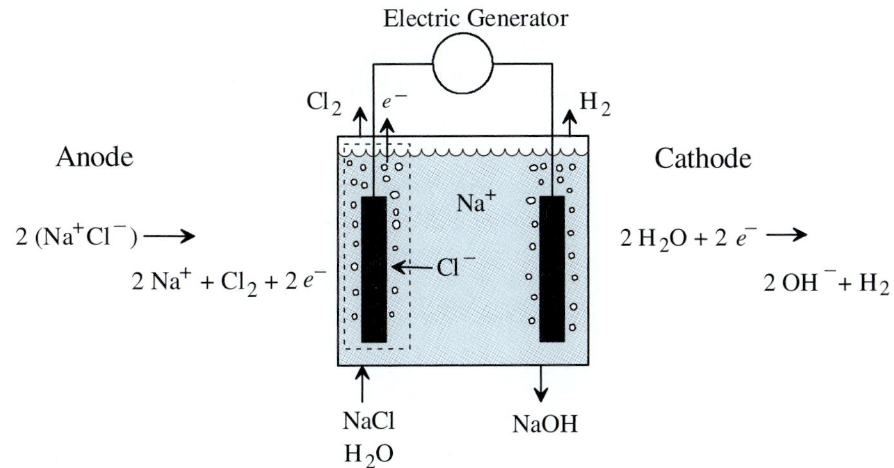

FIGURE 19.5
The electrolysis of salt and water to form sodium hydroxide, chlorine, and hydrogen.

$$\text{Silver:} \quad Ag^+ + e^- \rightarrow Ag$$

$$\text{Gold:} \quad Au^{3+} + 3\,e^- \rightarrow Au$$

$$\text{Copper:} \quad Cu^{2+} + 2\,e^- \rightarrow Cu$$

$$\text{Chromium:} \quad Cr^{3+} + 3\,e^- \rightarrow Cr$$

FIGURE 19.6
Electroplating of copper onto a base metal (e.g., steel).

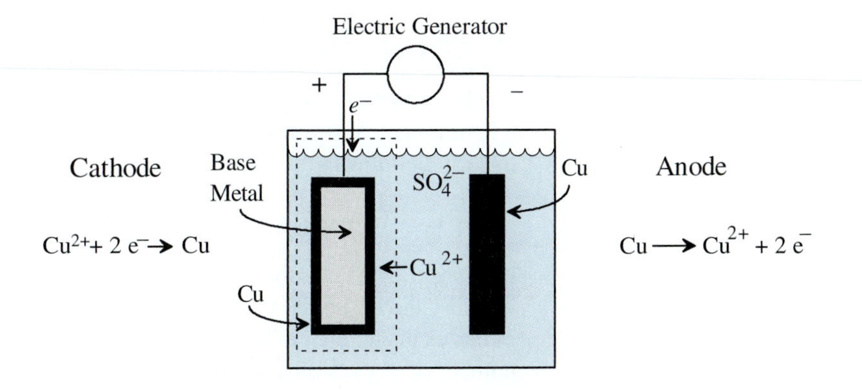

EXAMPLE 19.5

Problem Statement: A steel bar with a 500-cm^2 surface area is to be coated with 0.300 cm of copper. How much current (in coulombs) must flow?

Solution: We will choose the cathode as our system. Copper metal accumulates within the system, but copper is a neutral species with no net charge; therefore, there is no accumulation of net negative charge within the system:

$$q_{final}^{net-} - q_{initial}^{net-} = q_{in}^{net-} - q_{out}^{net-}$$

$$0 = q_{in}^{net-} - q_{out}^{net-}$$

$$q_{in}^{net-} = q_{out}^{net-}$$

Here, we are counting net negative charge. It relates to the flow of net positive charge as

$$q_{in}^{net-} = q_{out}^{net-} = -q_{out}^{net+} = -(-q_{in}^{net+}) = q_{in}^{net+}$$

By calculating the net flow of positive charge into the system, we can calculate the net negative charge into the system as

$$q_{in}^{net-} = q_{in}^{net+}$$

$$= (500\ \text{cm}^2 \times 0.300\ \text{cm}) \times \frac{8.9\ \text{g Cu}}{\text{cm}^3} \times \frac{\text{mol Cu}}{63.546\ \text{g Cu}} \times \frac{1\ \text{mol Cu}^{2+}}{\text{mol Cu}}$$

$$\times \frac{2\ \text{mol}\ (+)}{\text{mol Cu}^{2+}} \times \frac{96{,}485\ \text{C}}{\text{mol}\ (+)}$$

$$= 4.05 \times 10^6\ \text{C}$$

(*Note:* Both the density of copper and its atomic mass were obtained from Chapter 11 on thermodynamics.)

19.3 ACCOUNTING FOR CHARGE IN NUCLEAR REACTIONS

Nuclear reactions occur when the nuclei of atoms fuse or break apart. Figure 19.7 shows a fusion nuclear reaction in which deuterium (hydrogen-2) and tritium (hydrogen-3) react to form helium-4 plus a neutron; that is,

$${}^{2}_{1}\text{H} + {}^{3}_{1}\text{H} \rightarrow {}^{4}_{2}\text{He} + \text{n} + \text{Energy}$$

Because the mass of the helium and neutron is lower than the deuterium and tritium, mass was lost and converted to energy. (*Note:* The number in the lower left of the element symbol indicates the number of protons in the nucleus, and the number in the upper left indicates the number of protons plus neutrons in the nucleus.) As shown in the figure, in this nuclear reaction charge is conserved—the amount of charge before and after the reaction is the same.

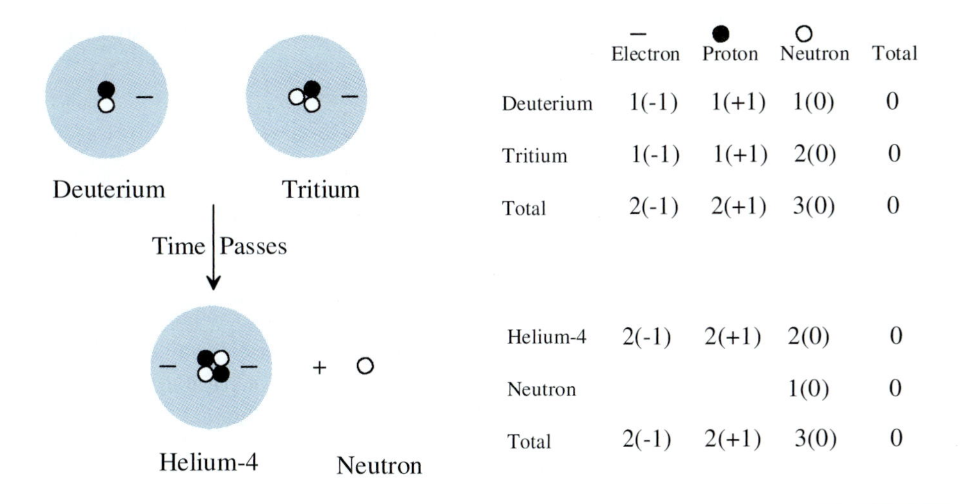

	Electron (–)	Proton (●)	Neutron (○)	Total
Deuterium	1(-1)	1(+1)	1(0)	0
Tritium	1(-1)	1(+1)	2(0)	0
Total	2(-1)	2(+1)	3(0)	0
Helium-4	2(-1)	2(+1)	2(0)	0
Neutron			1(0)	0
Total	2(-1)	2(+1)	3(0)	0

FIGURE 19.7
Nuclear reaction.

19.4 ELECTRIC CIRCUITS

A **capacitor** (Figure 19.8) is a device constructed from two conducting plates separated by an insulating material. Energy—supplied by a battery, fuel cell, or generator—transfers electrons from one plate to the other.

EXAMPLE 19.6

Problem Statement: A capacitor plate is made from 0.050 mol of copper (atomic number 29). A battery transfers 5000 C of electrons onto the plate. How much total negative charge is on the plate when the transfer is complete?

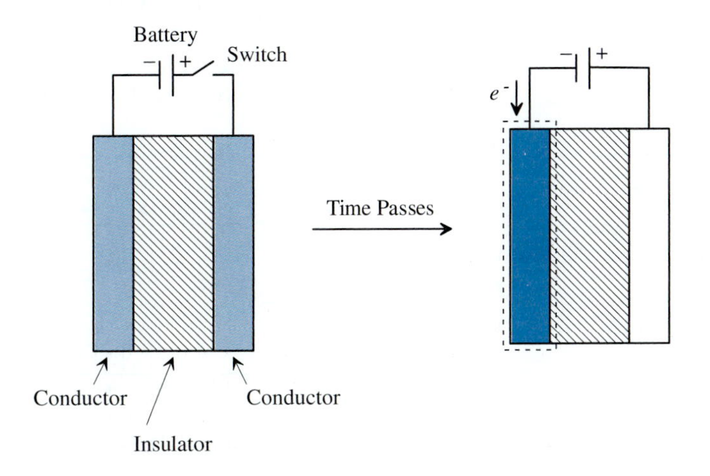

FIGURE 19.8
A capacitor is a device in which electrons are transferred from one plate to the other.

Solution: This is an "ordinary" process, so we can use accounting Equation 19-4:

$$q^-_{\text{final}} - q^-_{\text{initial}} = q^-_{\text{in}} - q^-_{\text{out}}$$

$$q^-_{\text{final}} = q^-_{\text{initial}} + q^-_{\text{in}} - q^-_{\text{out}}$$

$$= \left[0.050 \text{ mol Cu} \times \frac{29 \text{ mol } (-)}{\text{mol Cu}}\right] + \left[5000 \text{ C} \times \frac{\text{mol } (-)}{96{,}485 \text{ C}}\right] - 0$$

$$= 1.5 \text{ mol } (-)$$

Figure 19.9 shows an electric circuit in which a battery causes charge (i.e., electrons) to flow through two resistors in parallel. This is an "ordinary" process, so we can use Equation 19-4 to count charge, thus

$$q^-_{\text{final}} - q^-_{\text{initial}} = q^-_{\text{in}} - q^-_{\text{out}} \tag{19-4}$$

At each junction point, the wires have no capacity to store charge, so charge cannot accumulate. Therefore, the final and initial amount of charge in the system are identical,

$$q^-_{\text{final}} = q^-_{\text{initial}} \tag{19-13}$$

When we substitute Equation 19-13 into Equation 19-4, we obtain

$$q^-_{\text{final}} - q^-_{\text{final}} = 0 = q^-_{\text{in}} - q^-_{\text{out}}$$

$$q^-_{\text{in}} = q^-_{\text{out}} \tag{19-14}$$

Equation 19-14 indicates that the charge q^-_{in} (C) input to the junction point equals the charge q^-_{out} (C) output from the junction. We can modify Equation 19-14 to state that the **rate** that charge enters $\dot{q}^-_{\text{in}}$ (C/s) is equal to the **rate** that it exits $\dot{q}^-_{\text{out}}$ (C/s):

$$\dot{q}^-_{\text{in}} = \dot{q}^-_{\text{out}} \tag{19-15}$$

Electrical phenomena were mathematically described before atomic theory was developed. Benjamin Franklin (1706–1790), an early experimenter with electricity, guessed that electric current (i) flows from the positive terminal of a battery to the negative terminal. Although he had a 50 percent chance of being correct, he was wrong. We

FIGURE 19.9
Electron flow and current through two parallel resistors.

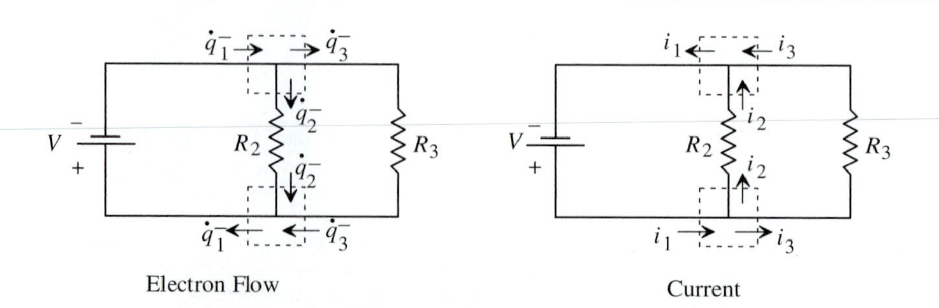

now know that electron flow $(\dot{q}^-)$ is from the negative terminal to the positive terminal. Unfortunately, the notation with current flowing from positive to negative is well entrenched, so we must adopt that notation as well.[1] Because current i flows in the opposite direction of electrons $\dot{q}^-$, we can think of current i as a flow of positive charge or "holes."

Because current i always flows exactly opposite to electron flow $\dot{q}^-$, Equation 19-15 can be adapted to account for currents at a junction:

$$i_{in} = i_{out} \tag{19-16}$$

This relationship is so useful, it is given the name **Kirchhoff's current law** (*KCL*).

We can use Kirchhoff's current law to analyze the circuit shown in Figure 19.9. At each junction, we know that

$$i_1 = i_2 + i_3 \tag{19-17}$$

From Chapter 12 on rate processes, we know that

$$i_2 = \frac{V}{R_2} \qquad i_3 = \frac{V}{R_3} \tag{19-18}$$

Therefore, knowing the voltage of the battery, we can calculate the total current flowing out of the battery by substituting Equations 19-18 in Equation 19-17:

$$i_1 = \frac{V}{R_2} + \frac{V}{R_3} = V\left(\frac{1}{R_2} + \frac{1}{R_3}\right) \tag{19-19}$$

If we were to replace resistors R_2 and R_3 with a single resistor R_1 that gives the same current flow i_1, this resistor R_1 would have to satisfy the following relationship

$$i_1 = \frac{V}{R_1} \tag{19-20}$$

Comparing Equations 19-19 and 19-20, we conclude that the equivalent resistor R_1 is related to the two parallel resistors as

$$\frac{1}{R_1} = \frac{1}{R_2} + \frac{1}{R_3} \tag{19-21}$$

Other useful relationships for resistors in series and parallel are described in Chapter 12.

EXAMPLE 19.7

Problem Statement: A 9-V battery is used in the circuit shown in Figure 19.9 with $R_2 = 20\ \Omega$ and $R_3 = 10\ \Omega$. How much current flows from the battery? What resistance would be required if a single resistor were to replace the two parallel resistors?

1. The U.S. Navy developed a number of manuals and books in which current flowed from negative to positive, but their notation has not been widely adopted.

Solution: From Equation 19-19, we can find the current

$$i_1 = V\left(\frac{1}{R_2} + \frac{1}{R_3}\right) = (9\text{ V})\left(\frac{1}{20\ \Omega} + \frac{1}{10\ \Omega}\right) \times \frac{1\text{ J/C}}{\text{V}} \times \frac{1\ \Omega}{\text{J·s/C}^2} \times \frac{1\text{ A}}{\text{C/s}}$$

$$= 1.35\text{ A}$$

(Notice the three conversion factors that convert the answer to the desired unit.) From Equation 19-21, we find the equivalent resistance.

$$\frac{1}{R_1} = \frac{1}{R_2} + \frac{1}{R_3} = \frac{1}{20\ \Omega} + \frac{1}{10\ \Omega} = 0.15\ \Omega^{-1} \Rightarrow R_1 = \frac{1}{0.15\ \Omega^{-1}} = 6.7\ \Omega$$

19.5 SUMMARY

Many important problems, particularly those involving electric circuits, require engineers to count charge. Charge can be negative or positive. In electric circuits, the flow of positive charge is termed current *i*.

Ordinarily, in the absence of nuclear reactions, charge is a conserved quantity, meaning it is neither created nor destroyed. In those situations where charge is created (destroyed), both positive and negative charge are created (destroyed) in equal amounts, thus the net charge of the universe is constant.

Facts about Fax

Sending graphical images over a long distance has been a human need for centuries. This need was first met in 1843 by Scottish clockmaker Alexander Bain. The image to be sent was prepared on a raised metal block (see schematic below). A pendulum stylus passed over the image, thus making and breaking electrical connections. The electrical pulses were sent over the wires to a receiving pendulum stylus that reproduced the image in another medium. To generate a proper image, it was very important that the sending stylus and receiving stylus move in synchrony. After a single stylus pass, the sending and receiving images were incrementally moved to "scan" the next line.

The first improvements were to replace the raised metal block with tin foil onto which the image was drawn in ink. The ink was not electrically conductive, so the interrupted electrical signal generated the image at the receiver. In 1865, Abbé Caselli installed the first commercial fax between Paris and Lyons. Even Thomas Edison got involved by improving the means for synchronizing the two pendulums.

In the late 1800s, a photoelectric cell was developed. In 1902, Germany's Arthur Korn incorporated the photocell into a

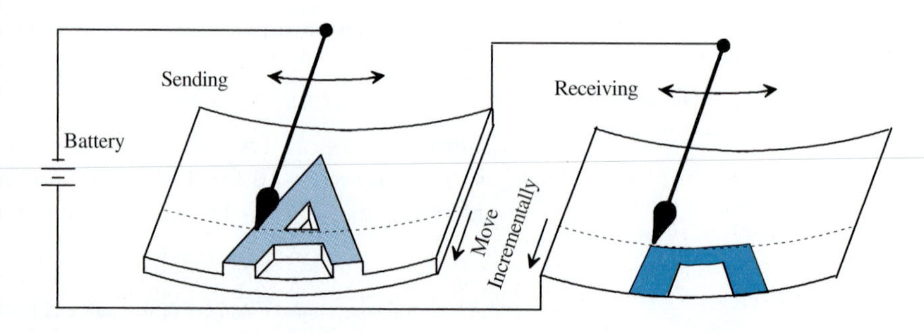

fax machine that allowed photographs to be sent electronically. In 1911, newspaper offices in Berlin, London, and Paris were connected with a photowire circuit.

Transmission of faxes over the telephone lines was hampered by AT&T, which decided not to pursue fax technology during the 1930s. Because they held a monopoly on phone lines, this stifled innovation. However, in the late 1960s, the Federal Communications Commission allowed non-AT&T equipment to be connected to the phone lines. This spread innovation, but there were no standards, so equipment from different manufacturers could not communicate. To overcome this problem, in 1968, the United Nations set the first international standards for fax communications.

Because of the difficulty of transmitting Japanese characters by telegraph, Japan was highly motivated to develop fax technology. In the 1970s, the largest fax research center in the world was established in Japan's phone company. The early fax machines printed on thermal paper, but later as copier technology became less expensive, plain-paper fax became common. From 1980 to 1992, the cost of fax machines fell by 30 times. The reduced cost allowed U.S. sales to increase from 0.5 million in 1985 to 6 million in 1991. Modern computers have fax cards that allow images to be transmitted without paper.

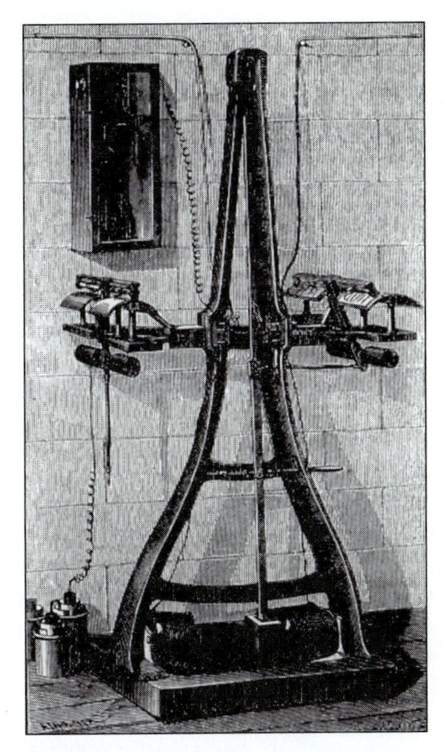

A 19th-century fax machine.

Adapted from: H. Petroski, "Fax and Content," *American Scientist* 84, No. 6 (November–December 1996), pp. 527–531.

Pioneers in Electricity

About 1672, Otto von Guericke (1602–1686) devised the first frictional electric machine. He placed a globe of sulfur on a shaft; one hand was placed on the globe and the other turned the crank, causing the globe to rotate. The friction allowed sizable electric sparks to be generated.

In 1745, while at the University of Leyden, Pieter van Musschenbroek (1692–1761) developed the *Leyden jar,* the first capacitor (also called a *condenser*). It consisted of a water-filled metal jar suspended with insulating silk cords. A brass wire was placed in the water and was held in place by a cork. He charged the Leyden jar with a frictional electric machine. Not knowing how much charge was stored, his laboratory assistant received an astonishing shock when he touched the metal jar and brass wire at the same time.

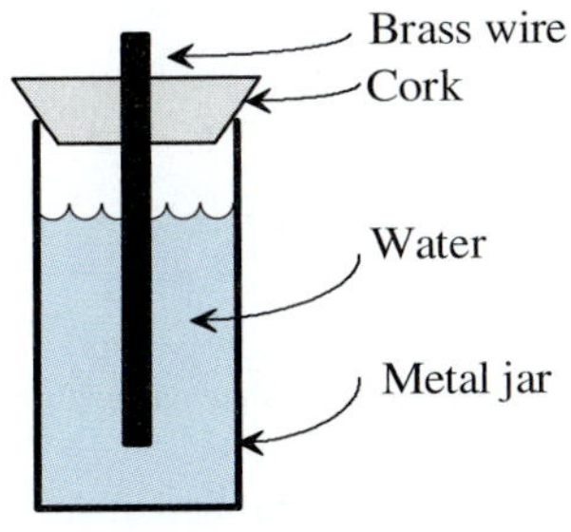

Schematic of a Leyden jar.

Benjamin Franklin (1706–1790), noting the sparks and crackles coming from Leyden jars, surmised that lightning might be the same phenomenon on a larger scale.

In 1752, he built a kite modified with a pointed wire and a silk thread attached to a key. During a thunderstorm, he put his hand to the key and obtained a spark. (He was lucky; two others who attempted this experiment were killed.) He also was able to charge a Leyden jar with the key, proving that lightning was the same as the electricity produced by the frictional electric machines.

Franklin knew there were two types of electric charge. When electrified, two amber rods repel each other. Also, two electrified glass rods repel each other. However, an electrified amber rod and an electrified glass rod attract. He hypothesized that a "fluid" was in excess in one rod and deficient in the other. If both rods had excess "fluid," they repelled. Similarly, if both rods were deficient in "fluid," they repelled. However, if one had excess and the other was deficient, they attracted. He further hypothesized the excess "fluid" would flow to fill the deficiency. Logically, he designated the object with excess fluid as "+" and the deficient one with "−." It was not until 1897 that Sir Joseph Thomson (1856–1940) showed that this "fluid" was electrons. Properly, electrons flow from negative to positive, but the convention has existed so long, we continue to say electric current flows from positive to negative.

Benjamin Franklin, with son William, proved that lightning and electricity are identical by flying a kite in a thunderstorm in June 1752.

Adapted from: Issaac Asimov, *Asimov's Biographical Encyclopedia of Science and Technology* (Garden City, NY: Doubleday, 1982).

Nomenclature

- F flow rate (kg/s or mol/s)
- i current (A)
- q charge (C)
- R resistance (V/A)
- V voltage (V)

Further Readings

Glover, C. J., and H. Jones. *Conservation Principles of Continuous Media.* New York: McGraw-Hill College Custom Series, 1994.

Glover, C. J., K. M. Lunsford, and J. A. Fleming. *Conservation Principles and the Structure of Engineering.* New York: McGraw-Hill College Custom Series, 1992.

Whitten, K. W., and K. D. Gailey. *General Chemistry.* 2nd ed. Philadelphia: Saunders, 1984.

PROBLEMS

19.1 For each circuit shown in Figure 19.10, find the current coming from the 1.5-V battery. Use the following resistances:
(a) $R_1 = 50\ \Omega, R_2 = 10\ \Omega, R_3 = 100\ \Omega$
(b) $R_1 = 30\ \Omega, R_2 = 20\ \Omega, R_3 = 10\ \Omega$
(c) $R_1 = 100\ \Omega, R_2 = 30\ \Omega, R_3 = 200\ \Omega$
(d) $R_1 = 5\ \Omega, R_2 = 1\ \Omega, R_3 = 10\ \Omega$
(e) $R_1 = 75\ \Omega, R_2 = 20\ \Omega, R_3 = 300\ \Omega$

19.2 For each circuit shown in Figure 19.11, find the current coming from the 1.5-V battery. Use the following resistances:
(a) $R_1 = 50\ \Omega, R_2 = 10\ \Omega, R_3 = 100\ \Omega, R_4 = 75\ \Omega$
(b) $R_1 = 30\ \Omega, R_2 = 20\ \Omega, R_3 = 10\ \Omega, R_4 = 5\ \Omega$
(c) $R_1 = 100\ \Omega, R_2 = 30\ \Omega, R_3 = 200\ \Omega, R_4 = 200\ \Omega$
(d) $R_1 = 5\ \Omega, R_2 = 1\ \Omega, R_3 = 10\ \Omega, R_4 = 3\ \Omega$
(e) $R_1 = 75\ \Omega, R_2 = 20\ \Omega, R_3 = 300\ \Omega, R_4 = 75\ \Omega$

19.3 A capacitor plate is constructed from a sheet of aluminum that measures 1.0 cm × 5.0 cm × 0.1 cm. Initially, the plate is neutral, meaning there are an equal number of electrons and protons. How many protons and electrons are there in the plate? A current of 2.0 A flows for 5.0 s. When the current stops flowing, how many electrons accumulated on the plate? When the current stops flowing, how many protons and electrons are there in the plate? (*Note:* Tables in Chapter 11 will be helpful to you.)

FIGURE 19.10
Three-resistor circuits.

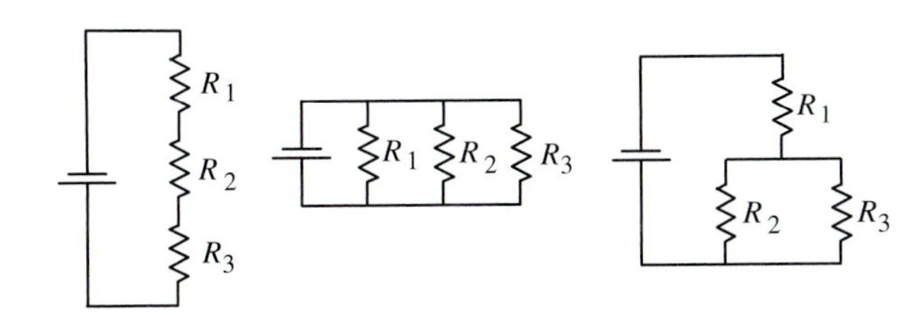

FIGURE 19.11
Four-resistor circuits.

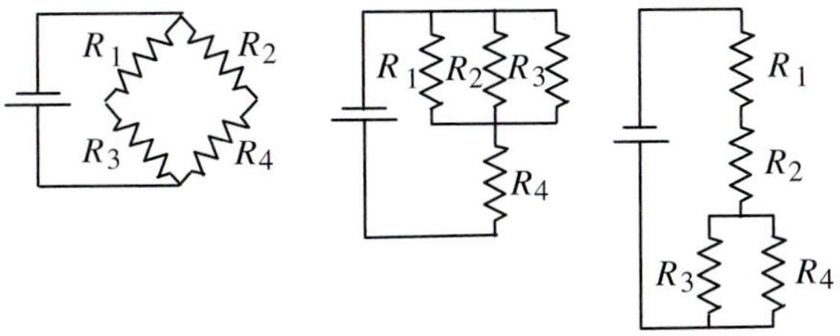

19.4 A hydrogen/oxygen fuel cell is fed 5.0 kg/min of hydrogen and a stoichiometric amount of oxygen (i.e., there is exactly enough oxygen to react with all the hydrogen). What is the electric current produced by the fuel cell?

19.5 The lead-acid battery is commonly used to store electrical energy in automobiles. At the anode, lead reacts with sulfate ions to release electrons and form lead sulfate, which precipitates on the anode. At the cathode, lead oxide reacts with sulfate ions, protons, and electrons to form water and lead sulfate, which precipitates on the cathode. How much charge is produced (in coulombs) when 4.56 g of lead sulfate precipitates in the battery?

19.6 The alkaline dry cell is a primary battery, meaning it cannot be recharged. It has the following reactions occurring during discharge:

$$Zn + 2\,OH^- \rightarrow Zn(OH)_2 + 2\,e^- \qquad \text{Anode}$$

$$2\,MnO_2 + 2\,H_2O + 2\,e^- \rightarrow 2\,MnO(OH) + 2\,OH^- \qquad \text{Cathode}$$

A dry cell has 5.13 g of zinc, which is the limiting reactant (i.e., all other reactants are in excess). How much charge (in coulombs) will be produced if the zinc completely reacts to zinc hydroxide?

19.7 The nickel-cadmium (nicad) battery is a secondary battery, meaning it can be recharged. It has the following reactions occurring during discharge:

$$Cd + 2\,OH^- \rightarrow Cd(OH)_2 + 2\,e^- \qquad \text{Anode}$$

$$NiO_2 + 2\,H_2O + 2\,e^- \rightarrow Ni(OH)_2 + 2\,OH^- \qquad \text{Cathode}$$

A nicad battery has 0.87 g of cadmium, which is the limiting reactant (all other reactants are in excess). How much charge (in coulombs) will be produced if the cadmium completely reacts to cadmium hydroxide?

19.8 An electronics manufacturer makes corrosion-resistant, gold-plated connectors. Each connector is cylindrical in shape with a diameter of 0.100 cm and a length of 1.00 cm. Each connector is to be coated with a 10-μm gold layer. The connectors enter the plating bath at a rate of 1000 connectors per hour. What current (in amperes) will be required for the plating operation?

19.9 An automobile company operates a process that puts a 15-μm chrome plating onto automobile bumpers. Each bumper has 2.79 ft^2 of plated surface area. The bumpers are plated at a rate of

100/h. What current (in amperes) will be required for the plating operation? (*Note:* The density of chromium is 7.2 g/cm^3.)

19.10 A chemical company decides to build a plant that produces 400,000 tons of sodium hydroxide (caustic soda) each year. The plant is designed to operate 8000 hours per year; the remaining hours are used for plant maintenance. How much chlorine and hydrogen (each in tons per year) will be produced as coproducts with the sodium hydroxide? What current (in amperes) will be needed to operate the electrolysis unit?

19.11 Aluminum is manufactured from bauxite ($Al_2O_3 \cdot xH_2O$) according to the following reaction:

$$4\ Al^{3+} + 12\ e^- \rightarrow 4\ Al \qquad \text{Cathode}$$

$$6\ O^{2-} \rightarrow 3\ O_2 + 12\ e^- \qquad \text{Anode}$$

$$2\ Al_2O_3 \rightarrow 4\ Al + 3\ O_2 \qquad \text{Net Reaction}$$

A company decides to produce 500,000 tons of aluminum each year. The plant operates 8000 hours each year, with the remaining time used for equipment maintenance. What current (in amperes) is required?

19.12 A family wishes to preserve their child's baby shoes by coating them with copper. First, the shoes are immersed in salt water to make them electrically conductive. Then they are placed into the electroplating bath and connected to an electric generator. The shoes have a total surface area of 210 cm^2. The copper coating is to be 0.15 cm thick. How much charge (in coulombs) is required?

19.13 A fusion reactor is constructed in which two deuteriums react to form tritium and hydrogen according to the following reaction:

$${}^2_1H + {}^2_1H \rightarrow {}^3_1H + {}^1_1H$$

The fusion reactor is fed 3.0 mol/s of deuterium, all of which reacts to form tritium and hydrogen. The fusion reactor contains plasma and is designed so that one electron is discharged from the reactor per tritium produced. At what rate (C/s) do electrons leave the reactor? At what rate (C/s) does positive charge accumulate in the reactor?

19.14 An earth-orbiting satellite has a particle beam accelerator used to shoot down enemy ballistic missiles. The accelerator strips the electrons off of hydrogen and accelerates the resulting protons toward earth to hit the missile. The proton beam flows at a rate of 3.00 kg/s. At what rate (C/s) is negative charge accumulating on the satellite?

Glossary

anode A positively charged electrode.

battery A device that produces electricity from the chemical energy in the anodes and cathodes.

capacitor A charge storage device that is constructed from two conducting plates separated by an insulating material.

cathode A negatively charged electrode.

conserved quantities Quantities that are neither generated nor consumed.

coulomb For a 1-ampere current, the number of electrons that flow past a point during 1 second.

electrolysis reactions Chemical reactions that require an input of electrical energy.

electroplating A process in which metal ions react at a cathode to form elemental metals.

Faraday's constant The ratio of Avogadro's number to the coulomb.

fuel cell A device that directly converts the chemical energy in added reactants to electrical energy.

Kirchhoff's current law The total current flowing into a junction is equal to the total current flowing out of a junction.

lead-acid battery A secondary battery in which the anode is constructed of lead and the cathode is constructed of lead oxide.

primary battery A battery that cannot be recharged.

secondary battery A battery that can be recharged.

Mathematical Prerequisites

- Algebra (Appendix E, Mathematics Supplement)
- Vectors (Appendix I, Mathematics Supplement)
- Calculus (Appendix L, Mathematics Supplement)

CHAPTER 20

Accounting for Linear Momentum

Previously, we described how to account for mass m. Mass is a **scalar** quantity, meaning it does not depend on direction. In a closed system, mass is a **state** quantity, meaning it is independent of path.

Velocity v is also a state quantity. However, unlike mass, velocity is a **vector** quantity, meaning it does depend on direction. The vector nature of velocity is indicated with bold type.

According to Rule 8 in Chapter 17 on accounting, when two state quantities are algebraically combined, a new state quantity is formed. Mass and velocity may be multiplied together to form a new state quantity[1] called **linear momentum p.**

$$\mathbf{p} = m\mathbf{v} \tag{20-1}$$

Linear momentum is also a vector quantity, as indicated by the bold type. Linear momentum was a concept developed by Newton.

20.1 REVIEW OF NEWTON'S LAWS

Newton's first law states that a body in its "natural state" (i.e., no forces are imposed upon it) has a constant linear momentum.

Newton's second law states that when a net force is imposed upon a body, its linear momentum changes.

Newton's third law states that forces *always* exist by the interaction of two (or more) bodies. The force on one body is equal and opposite to the force acting on the other body.

1. Momentum is only a state quantity in a closed system. In an open system, momentum is both a state and a path quantity.

20.2 CONSERVATION OF LINEAR MOMENTUM

Newton's three laws are consistent with linear momentum being a conserved quantity; that is, the total amount of linear momentum in the universe is constant and cannot be changed. When accounting for linear momentum, the generation and consumption terms are always zero, so the universal accounting equation reduces to

$$\text{Final amount} - \text{initial amount} = \text{input} - \text{output} \tag{20-2}$$

which can also be written as

$$\text{Accumulation} = \text{net input} \tag{20-3}$$

There are two ways that the linear momentum of a system can change: (1) change of mass and (2) unbalanced forces.[2]

20.2.1 Linear Momentum Change by Mass Transfer

Figure 20.1 shows a system with initial linear momentum $\mathbf{p}_i$. The system has only one body, so the initial linear momentum is

$$\mathbf{p}_i = m_1\mathbf{v}_1 \tag{20-4}$$

As time passes, mass enters the system. Each body is moving at a velocity, so each body has its own linear momentum. The final linear momentum $\mathbf{p}_f$ is found by summing the linear momentum of each body, or

$$\mathbf{p}_f = m_1\mathbf{v}_1 + m_2\mathbf{v}_2 + m_3\mathbf{v}_3 \tag{20-5}$$

Recall that linear momentum is a vector quantity, meaning it has both a magnitude and a direction. The initial linear momentum can be described using components,

$$p_i\mathbf{i} = \mathbf{p}_i = m_1\mathbf{v}_1 = m_1(v_1\mathbf{i}) = m_1v_1\mathbf{i} \tag{20-6}$$

where $\mathbf{i}$ is the unit vector. By comparing the components of the left and right side of Equation 20-6, we can write the following scalar equation:

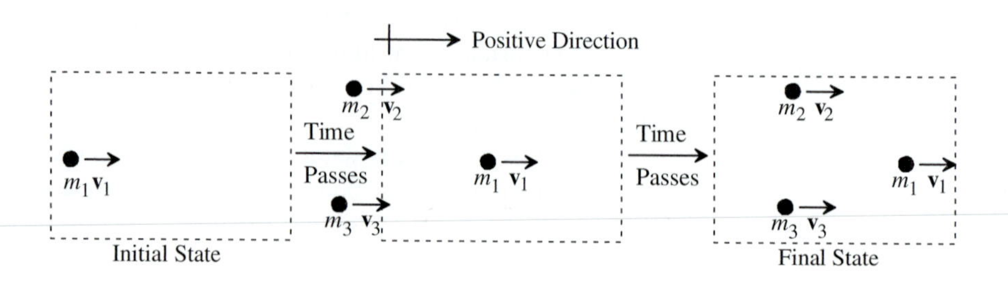

FIGURE 20.1 Linear momentum change by mass transfer.

2. Photons also carry momentum, but this is beyond the scope of this book.

$$p_i = m_1 v_1 \tag{20-7}$$

Through similar arguments, we can write the following scalar equation for the final momentum:

$$p_f = m_1 v_1 + m_2 v_2 + m_3 v_3 \tag{20-8}$$

EXAMPLE 20.1

Problem Statement: An inventor develops a "baseball-powered cart" (Figure 20.2). The 20-lb_m cart has a "sail" that catches baseballs thrown by a baseball gun similar to those used for batting practice. The ball drops into the cart after it is caught by the sail. Each baseball has a mass of 1.0 lb_m and travels at a velocity v_{ball} of 90 mph (132 ft/s). Initially, the cart is at rest. What is its velocity after 10 balls impact the cart?

Solution: Final amount = initial amount + input − output

$$(m_{cart} + 10 m_{ball}) v_f = m_{cart} v_i + 10(m_{ball} v_{ball}) - 0$$

$$v_f = \frac{m_{cart} v_i + 10(m_{ball} v_{ball})}{m_{cart} + 10 m_{ball}}$$

$$= \frac{(20\ lb_m)(0\ ft/s) + 10(1.0\ lb_m \cdot 132\ ft/s)}{20\ lb_m + 10(1.0\ lb_m)} = 44\ ft/s$$

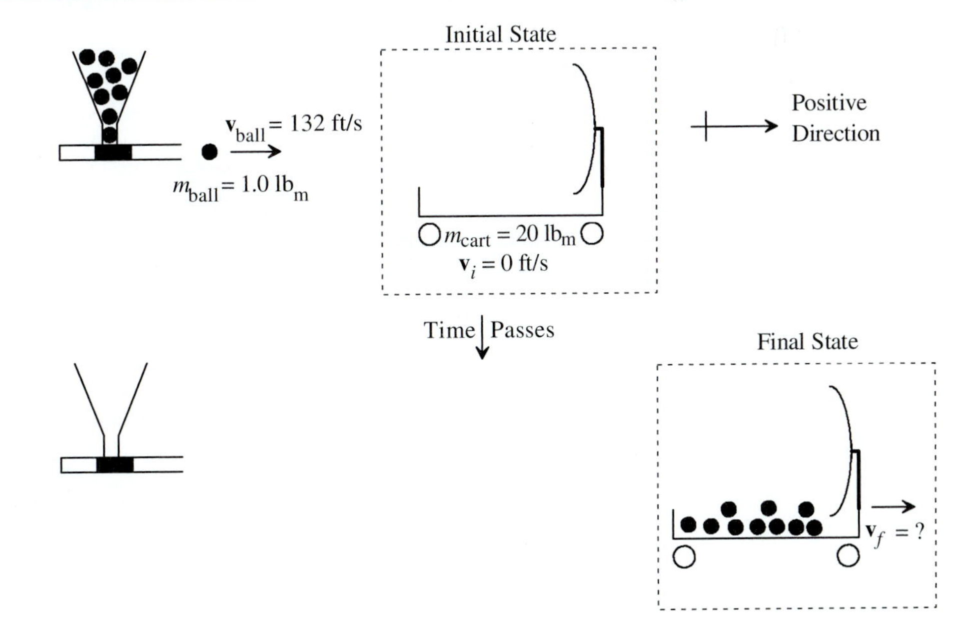

FIGURE 20.2
Baseball-powered cart.

EXAMPLE 20.2

Problem Statement: The old steam locomotives you have seen in movies (or perhaps ridden on) use open-cycle steam engines. The steam that pushes the drive piston is vented to the atmosphere after it has completed its work. For the engine to keep operating, water must be replaced. To avoid stopping, some trains were designed to scoop water from an open trough "on the fly."

A locomotive has a mass of 50.0 tons and travels 60.0 mph. It scoops up 2.00 tons of water, which causes it to slow down (Figure 20.3). What is its velocity immediately after scooping the water?

Solution: Final amount = initial amount + input − output

$$(m_{\text{loco}} + m_{\text{water}})v_f = m_{\text{loco}}v_{\text{loco}} + m_{\text{water}}v_{\text{water}} - 0$$

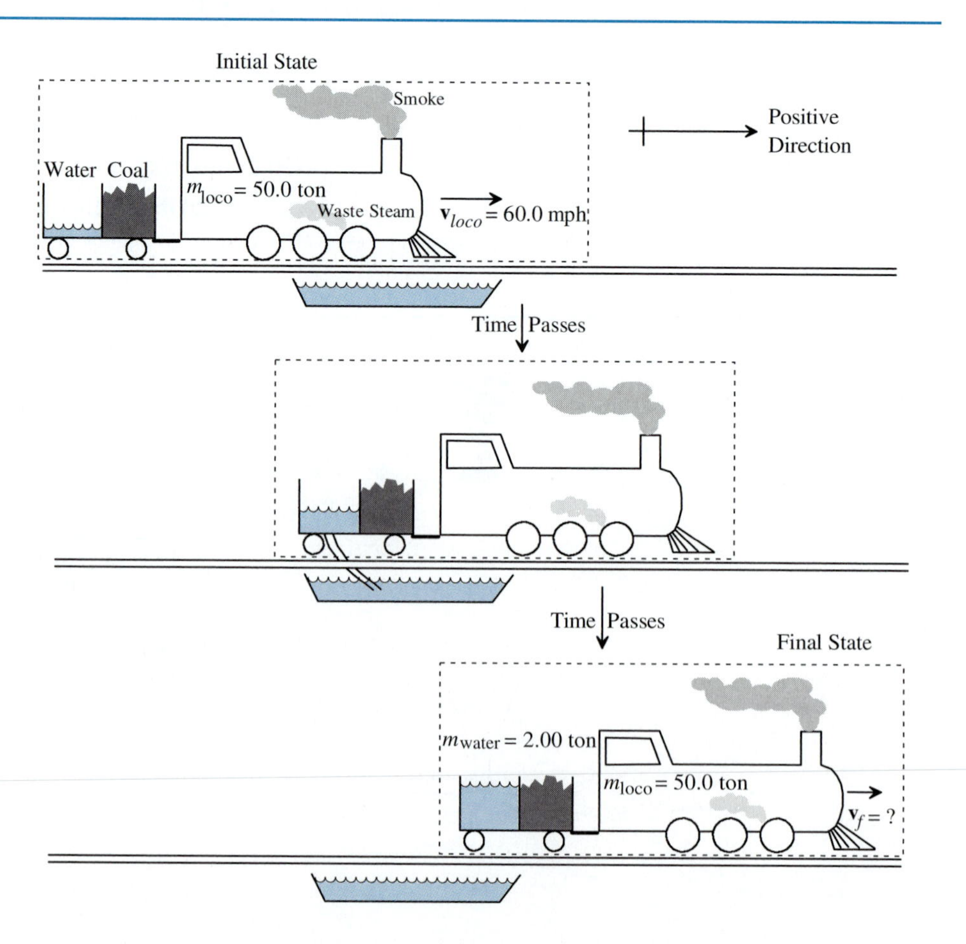

FIGURE 20.3
Steam locomotive scooping up water "on the fly."

$$v_f = \frac{m_{\text{loco}} v_{\text{loco}} + m_{\text{water}} v_{\text{water}}}{m_{\text{loco}} + m_{\text{water}}}$$

$$= \frac{(50.0 \text{ ton})(60.0 \text{ mph}) + (2.00 \text{ ton})(0 \text{ mph})}{50.0 \text{ ton} + 2.00 \text{ ton}} = 57.7 \text{ mph}$$

20.2.2 Linear Momentum Change by Unbalanced Forces

As previously discussed, the accounting equation for momentum is

$$\text{Final amount} - \text{initial amount} = \text{input} - \text{output} \tag{20-2}$$

Forces are a means by which momentum can be transported into, or out of, a system. If this seems a little unusual to you, carefully study the following discussion.

The SI units of force are

$$\text{Units of force} = \frac{\text{kg·m}}{\text{s}^2} = \frac{\frac{\text{kg·m}}{\text{s}}}{\text{s}} \tag{20-9}$$

and the SI units of momentum are

$$\text{Units of momentum} = \frac{\text{kg·m}}{\text{s}} \tag{20-10}$$

Comparing Equations 20-9 and 20-10, it is easy to see that the units of force correspond to "momentum per second"; that is, **force** is a rate of momentum flow:

$$\text{Units of force} = \frac{\text{units of momentum}}{\text{s}} = \text{units of momentum flow rate} \tag{20-11}$$

If we indicate the rate of momentum flow as $\dot{\mathbf{p}}$ [with units (kg·m/s)/s or kg·m/s^2], then we can say

$$\dot{\mathbf{p}} = \mathbf{F} \tag{20-12}$$

which is the mathematical way of saying that momentum flow across a system boundary can be caused by forces.

Because it may have been awhile since you studied rate processes in Chapter 12, it is worthwhile to review the idea of *rates* using a simple example. Figure 20.4 shows a water tank that sits on a scale. Two pipes are connected to the tank; one pipe delivers water at rate $\dot{m}_{\text{in}}$ (kg/s) and the other removes water at rate $\dot{m}_{\text{out}}$ (kg/s). In a given time interval Δt, the mass of water added is

$$m_{\text{in}} = \dot{m}_{\text{in}} \Delta t \tag{20-13}$$

Similarly, the mass of water out is

$$m_{\text{out}} = \dot{m}_{\text{out}} \Delta t \tag{20-14}$$

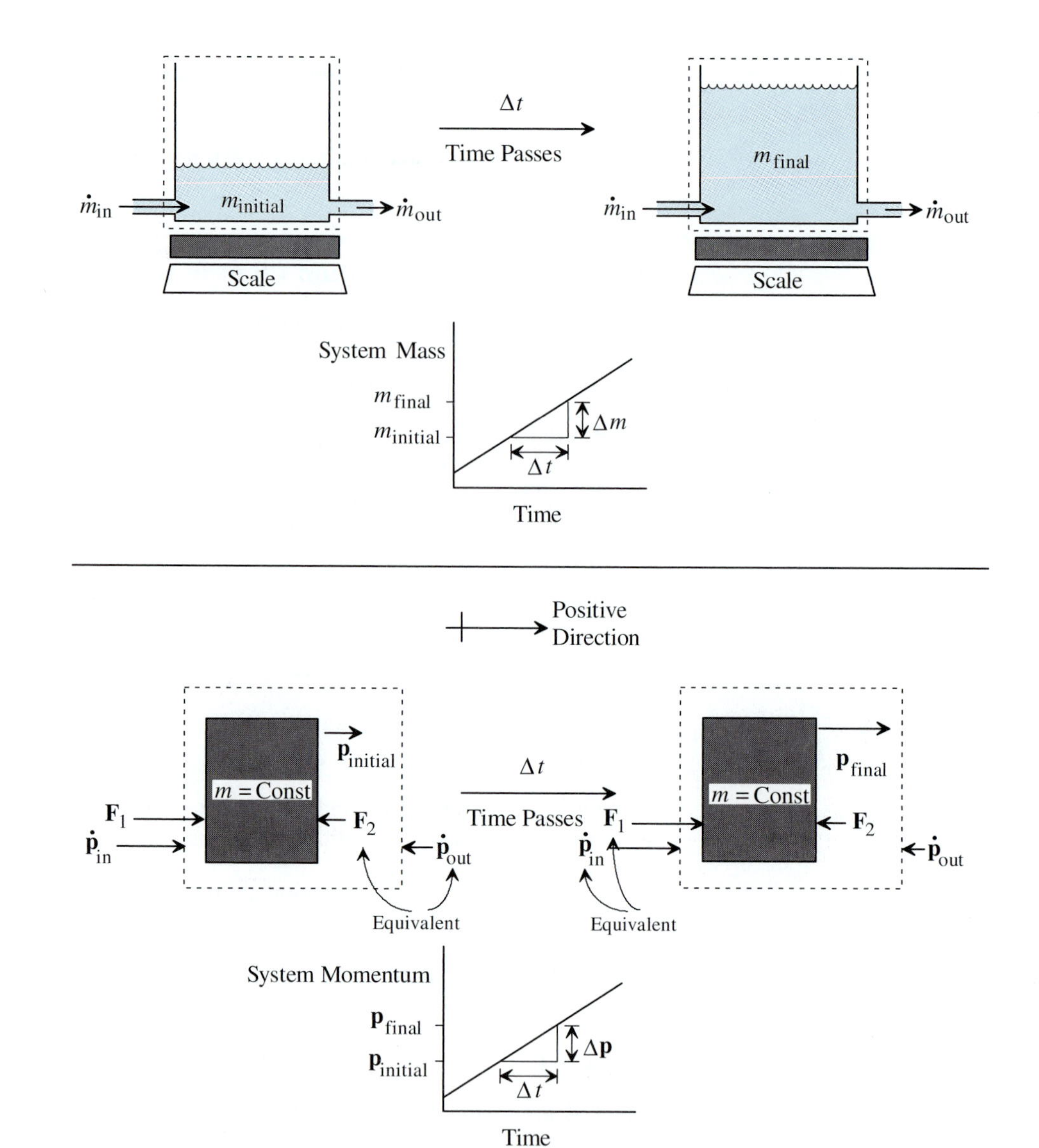

FIGURE 20.4
Analogy between mass accumulation in a tank and momentum accumulation.

Mass is a conserved quantity, so it also obeys Equation 20-2. Thus,

$$m_{\text{final}} - m_{\text{initial}} = m_{\text{in}} - m_{\text{out}} \tag{20-15}$$

Substituting Equations 20-13 and 20-14 gives

$$m_{\text{final}} - m_{\text{initial}} = \dot{m}_{\text{in}}\Delta t - \dot{m}_{\text{out}}\Delta t \tag{20-16}$$

Both sides may be divided by Δt, giving

$$\frac{m_{\text{final}} - m_{\text{initial}}}{\Delta t} = \frac{\Delta m}{\Delta t} = \dot{m}_{\text{in}} - \dot{m}_{\text{out}} \tag{20-17}$$

The above equation is valid only if the mass flow rates are constant (which they are in Figure 20.4). We can generalize this relationship by evaluating Equation 20-17 at an instant in time, that is, in the limit as Δt goes to 0,

$$\lim_{\Delta t \to 0} \frac{\Delta m}{\Delta t} = \frac{dm}{dt} = \dot{m}_{\text{in}} - \dot{m}_{\text{out}} \tag{20-18}$$

which is valid even if the mass flow rates vary with time.

By defining $\dot{m}_{\text{net}} = \dot{m}_{\text{in}} - \dot{m}_{\text{out}}$, we can write

$$\frac{dm}{dt} = \dot{m}_{\text{net}} \tag{20-19}$$

Figure 20.4 also shows a constant-mass system that has two forces acting on it, each in opposite directions. By analogy to Equation 20-19, the rate at which the momentum accumulates in the system is

$$\frac{d\mathbf{p}}{dt} = \dot{\mathbf{p}}_{\text{net}} \tag{20-20}$$

Based upon the relationship given in Equation 20-12, we can substitute forces for the rate of momentum flow

$$\frac{d\mathbf{p}}{dt} = \mathbf{F}_{\text{net}} \tag{20-21}$$

By substituting the definition of momentum, and assuming the mass of the system is constant, we obtain

$$\frac{d(m\mathbf{v})}{dt} = m\frac{d\mathbf{v}}{dt} = \mathbf{F}_{\text{net}} \tag{20-22}$$

By definition, acceleration is the rate of change for velocity:

$$m\mathbf{a} = \mathbf{F}_{\text{net}} \tag{20-23}$$

This equation should be very familiar to you; it is Newton's second law! Thus, Newton's second law is a momentum accounting statement for a closed system, that is, one in which there is no mass crossing the system boundary.

In summary, according to Newton's second law, a net force acting on a body changes the body's linear momentum. In addition, Newton's third law states that forces always occur between two (or more) bodies and that the forces act in opposite directions. Therefore, the linear momentum of one body changes the same as the other body, but in opposite directions, thus conserving the linear momentum of the universe. However, there is no requirement that the system contain both bodies. (Recall that the system designation is arbitrary and at the discretion of the engineer.) Therefore, it is possible that a given system could have unbalanced forces, thus changing the system linear momentum (Figure 20.5).

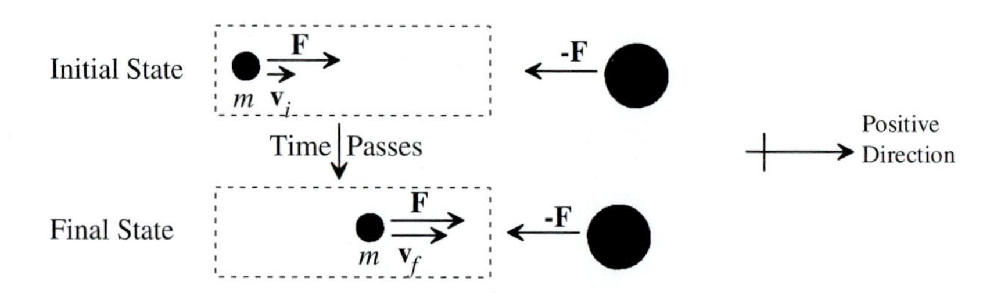

FIGURE 20.5
Change in linear momentum resulting from unbalanced forces.

EXAMPLE 20.3

Problem Statement: A 50-kg cannonball is dropping toward the earth with an initial velocity $v_{\text{ball},i}$ of 5.0 m/s. After 1.0 s, what is its velocity? Using the cannonball as the system (Figure 20.6), what is the net input of linear momentum to the system?

Solution: The magnitude of the gravitational attractive force between the cannonball and the earth is given by the gravitational force equation[3]

$$F = G\frac{m_{\text{ball}}m_{\text{earth}}}{(r_{\text{earth}})^2} = \left(6.67 \times 10^{-11}\,\frac{\text{N}\cdot\text{m}^2}{\text{kg}^2}\right)\frac{(50\text{ kg})(6.0 \times 10^{24}\text{ kg})}{(6.38 \times 10^6\text{m})^2} = 490\text{ N}$$

The **acceleration** of the cannonball toward the earth is found from Newton's second law, so

$$a_{\text{ball}} = \frac{F}{m_{\text{ball}}} = \frac{490\text{ N}}{50\text{ kg}} \times \frac{1\text{ kg}\cdot\text{m/s}^2}{\text{N}} = 9.8\text{ m/s}^2$$

The final velocity is found using an equation of motion

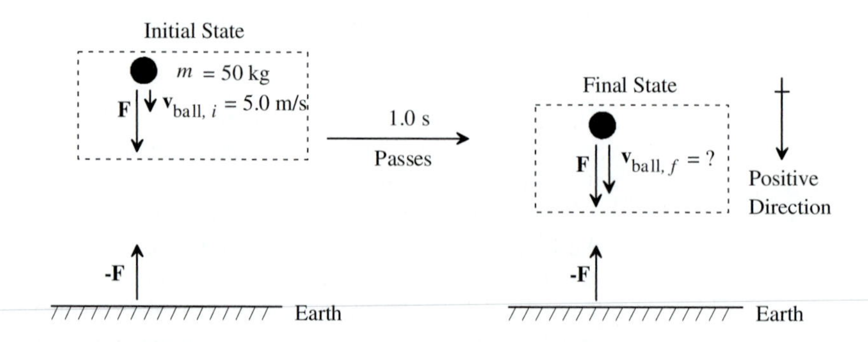

FIGURE 20.6
Cannonball falling to earth. The system contains only the cannonball.

3. This equation is valid for two infinitesimally small masses separated by distance r. It can be shown that the force of attraction on earth behaves as though (1) the entire mass of the earth were shrunken to a point located at the earth's center, and (2) the entire mass of the object were shrunken to a point located at the object's center.

$$v_{\text{ball},f} = v_{\text{ball},i} + a_{\text{ball}}t = 50 \text{ m/s} + (9.8 \text{ m/s}^2)(1.0 \text{ s}) = 14.8 \text{ m/s}$$

The net linear momentum input is

$$\begin{aligned}\text{Net input} &= \text{final amount} - \text{initial amount}\\ &= m_{\text{ball}}v_{\text{ball},f} - m_{\text{ball}}v_{\text{ball},i}\\ &= (50 \text{ kg})(14.8 \text{ m/s}) - (50 \text{ kg})(5.0 \text{ m/s}) = 490 \text{ kg·m/s}\end{aligned}$$

Note that the net linear momentum input during the 1.0-s time period is identical to the force acting on the system for 1.0 s. That is,

$$\text{Net input} = F\,\Delta t$$

$$490 \text{ kg·m/s} = \left(490 \text{ N} \times \frac{\text{kg·m/s}^2}{\text{N}}\right)(1.0 \text{ s})$$

This is consistent with the assertion that force may be viewed as a rate of momentum addition to (or removal from) the system.

EXAMPLE 20.4

Problem Statement: A 50-kg cannonball is dropping toward the earth with an initial velocity $v_{\text{ball},i}$ of 5.0 m/s. After 1.0 s, what is the velocity of the earth toward the cannonball? Using the cannonball and the earth as the system (Figure 20.7), what is the net input of linear momentum to the system?

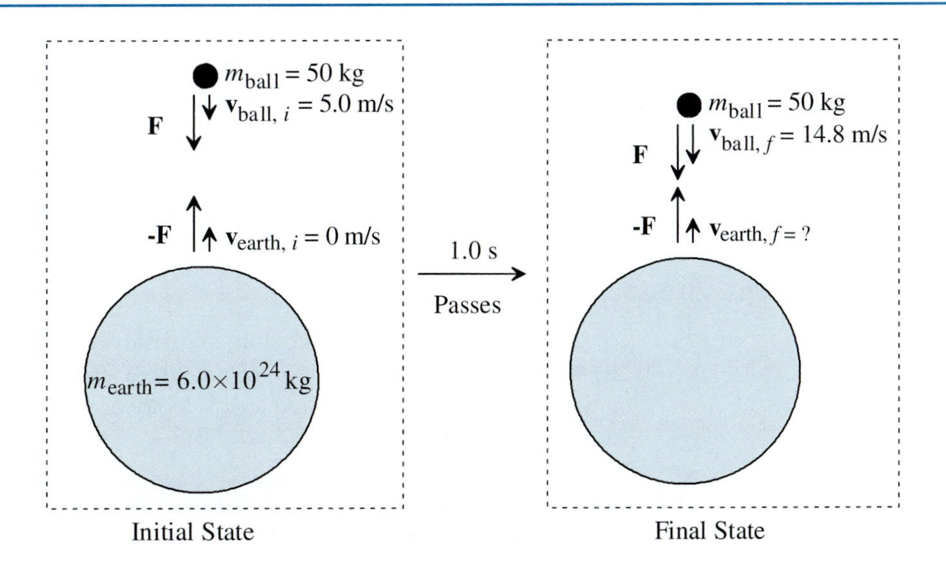

FIGURE 20.7
Cannonball falling to earth. The system contains the cannonball and the earth.

Solution: This problem is identical to the previous example, except we are changing our system definition to include both the cannonball and the earth. The acceleration of the earth toward the cannonball is found from Newton's second law by writing

$$a_{earth} = -\frac{F}{m_{earth}} = -\frac{490\ \text{N}}{6.0 \times 10^{24}\text{kg}} = -8.2 \times 10^{-23}\text{m/s}^2$$

The magnitude of the attractive force is identical to the previous example, except the direction is opposite.

The final velocity of the earth is found using an equation of motion.

$$v_{earth,f} = v_{earth,i} + a_{earth}t$$

$$= (0\ \text{m/s}) + (-8.2 \times 10^{-23}\ \text{m/s}^2)(1.0\ \text{s}) = -8.2 \times 10^{-23}\ \text{m/s}$$

The net linear momentum input is

$$\text{Net input} = \text{final amount} - \text{initial amount}$$

$$= (m_{ball}v_{ball,f} + m_{earth}v_{earth,f}) - (m_{ball}v_{ball,i} + m_{earth}v_{earth,i})$$

$$= m_{ball}(v_{ball,f} - v_{ball,i}) + m_{earth}(v_{earth,f} - v_{earth,i})$$

$$= 50\ \text{kg}(14.8\ \text{m/s} - 5.0\ \text{m/s}) + 6.0 \times 10^{24}\ \text{kg}\ (-8.2 \times 10^{-23}\ \text{m/s} - 0)$$

$$= 0$$

Because the cannonball/earth system had no change in mass and there were no *unbalanced* forces, there was no net input of linear momentum.

20.3 SYSTEMS WITHOUT NET LINEAR MOMENTUM INPUT

Many systems of interest have no unbalanced forces, nor do they transfer mass across the system boundary. In these cases, the accounting equation reduces to

$$\text{Final amount} - \text{initial amount} = 0$$

$$\text{Final amount} = \text{initial amount} \qquad (20\text{-}24)$$

EXAMPLE 20.5

Problem Statement: A white 2-lb_m billiard ball is traveling at 2 ft/s and hits a black 1-lb_m billiard ball that is stationary (Figure 20.8). The white billiard ball continues moving in the same direction, but at a reduced velocity of 1.5 ft/s. What is the final velocity of the black billiard ball?

Solution: There are no unbalanced forces nor is there mass transfer across the boundary, so Equation 20-24 may be used.

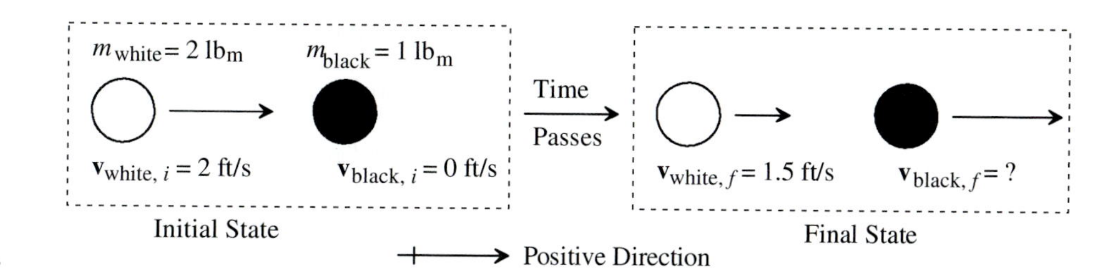

FIGURE 20.8
Collision of two billiard balls.

$$\text{Final amount} = \text{initial amount}$$

$$m_{black}v_{black,f} + m_{white}v_{white,f} = m_{black}v_{black,i} + m_{white}v_{white,i}$$

$$m_{black}v_{black,f} = m_{black}v_{black,i} + m_{white}v_{white,i} - m_{white}v_{white,f}$$

$$= m_{black}v_{black,i} + m_{white}(v_{white,i} - v_{white,f})$$

$$v_{black,f} = v_{black,i} + \frac{m_{white}}{m_{black}}(v_{white,i} - v_{white,f})$$

$$= 0 \text{ ft/s} + \frac{2 \text{ lb}_m}{1 \text{ lb}_m}(2 \text{ ft/s} - 1.5 \text{ ft/s}) = 1 \text{ ft/s}$$

EXAMPLE 20.6

Problem Statement: A 2500-lb_m automobile traveling at 60 mph hits a parked 3500-lb_m truck (Figure 20.9). Upon impact, the bumpers lock together. What is the velocity of the locked-together automobile and truck?

Solution: There are no unbalanced forces, nor does any mass transfer across the system boundary, so Equation 20-24 may be used.

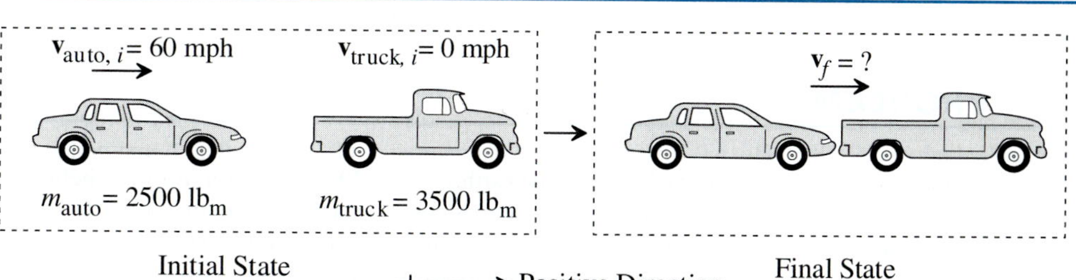

FIGURE 20.9
Impact of an automobile and truck.

1.0 s, how fast (ft/s) is the fireman traveling? (Neglect the friction of the firehose on the ground and the mass of the firehose.)

20.22 A 150-lb_m student buys a CO_2 fire extinguisher at the local hardware store. The container has a mass of 3.25 lb_m and it holds 5.00 lb_m of CO_2. The student puts on roller blades, straps the tank to his back, and points the nozzle toward the rear. The CO_2 exits at 20.0 ft/s. The student is initially at rest. When the tank is exhausted, how fast is the student traveling?

Glossary

acceleration The rate at which velocity changes.

force The rate that momentum enters or leaves the system; the influence on a body that causes it to accelerate in the absence of any other counteracting forces.

linear momentum A vector resulting from multiplying a mass by its velocity vector.

Newton's first law In the absence of a net force the momentum of an object does not change.

Newton's second law A net force changes the momentum of an object.

Newton's third law The force on one body is equal and opposite to the force on the second body; isolated forces cannot exist.

scalar A quantity that does not depend on direction.

state A quantity that is independent of path.

vector A quantity with a magnitude and a direction.

velocity A vector describing the time rate of change for the position of a body.

Mathematical Prerequisites

- Algebra (Appendix E, Methematics Supplement)
- Vectors (Appenndix I, Mathematics Supplement)
- Calculus (Appendix L, Mathematics Supplement)

CHAPTER 21

Accounting for Angular Momentum

In Chapter 20, we described the motion of objects that move in a straight line. Here, we describe the motion of objects that *rotate* (e.g., wheels, gyroscopes) and *orbit* (e.g., satellites). Whereas *linear momentum* is useful to calculate straight-line motion, **angular momentum** is useful to calculate rotary motion.

The equations that describe rotary motion are consistent with Newton's laws. However, Newton's laws have been reformulated to use convenient measures of rotary motion that are described as follows.

21.1 MEASURES OF ROTARY MOTION

Unlike linear motion, in which movement is easily described by linear position, angular motion is most easily described with angles.

21.1.1 Angular Measures of Rotary Motion

Figure 21.1 shows a rock tied to a string. The rock travels around a central point in a circular motion. Its position can be described by angle θ, which is the arc length s divided by the radius r,

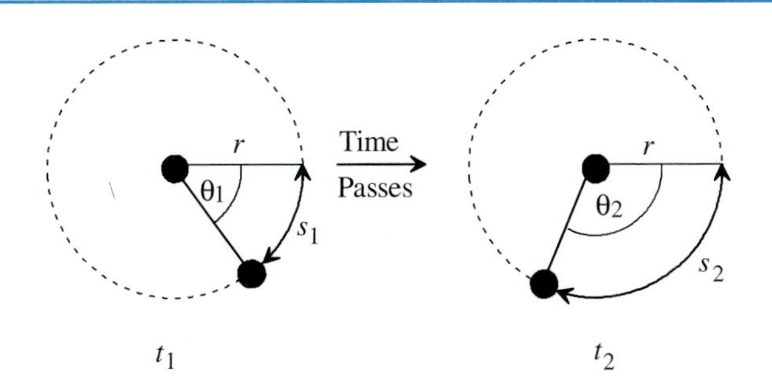

FIGURE 21.1
A rock tied to a string travels in a circular motion about a central point.

$$\theta = \frac{s}{r} \tag{21-1}$$

where θ is in radians. The **angular displacement** $\Delta\theta$ is

$$\Delta\theta = \theta_2 - \theta_1 \tag{21-2}$$

The **average angular speed** $\overline{\omega}$ is

$$\overline{\omega} = \frac{\theta_2 - \theta_1}{t_2 - t_1} = \frac{\Delta\theta}{\Delta t} \tag{21-3}$$

The **instantaneous angular speed** ω is

$$\omega = \lim_{\Delta t \to 0} \frac{\Delta\theta}{\Delta t} = \frac{d\theta}{dt} \tag{21-4}$$

where typical units are radians per second or radians per minute.

If the rotation rate changes, we can define the **average angular acceleration** $\overline{\alpha}$ as

$$\overline{\alpha} = \frac{\omega_2 - \omega_1}{t_2 - t_1} = \frac{\Delta\omega}{\Delta t} \tag{21-5}$$

and the **instantaneous angular acceleration** α is

$$\alpha = \lim_{\Delta t \to 0} \frac{\Delta\omega}{\Delta t} = \frac{d\omega}{dt} \tag{21-6}$$

Table 21.1 compares these angular measures of rotary motion to analogous measures of linear motion.

TABLE 21.1
Comparison of linear and rotary measures of motion

Linear Motion	Rotary Motion
$x = x_0 + v_0 t + \frac{1}{2} a_0 t^2$	$\theta = \theta_0 + \omega_0 t + \frac{1}{2}\alpha_0 t^2$
Integrate ↑↓ Differentiate	Integrate ↑↓ Differentiate
$\frac{dx}{dt} = v = v_0 + a_0 t$	$\frac{d\theta}{dt} = \omega = \omega_0 + \alpha_0 t$
Integrate ↑↓ Differentiate	Integrate ↑↓ Differentiate
$\frac{dv}{dt} = a = a_0$	$\frac{d\omega}{dt} = \alpha = \alpha_0$

21.1.2 Common Measures of Rotation

Ordinarily, people do not measure rotation rates as ω radians per second because a radian is hard to visualize. Instead, it is much more common to measure rotation rates as **frequency** in revolutions per second or revolutions per minute. Frequency f is defined as

$$f = \lim_{\Delta t \to 0} \frac{\Delta n}{\Delta t} = \frac{dn}{dt} \tag{21-7}$$

where n is the number of revolutions. One revolution corresponds to an angle of 2π radians; therefore, a proportion can be established to relate arbitrary n revolutions to arbitrary angle θ

$$\frac{1 \text{ revolution}}{2\pi \text{ radians}} = \frac{n}{\theta}$$

$$n = \frac{\theta}{2\pi} \tag{21-8}$$

By putting Equation 21-8 into Equation 21-7,

$$f = \frac{d}{dt}\left(\frac{\theta}{2\pi}\right) = \frac{1}{2\pi}\frac{d\theta}{dt} = \frac{1}{2\pi}\,\omega \tag{21-9}$$

we can relate frequency f to instantaneous angular speed ω.

Another way to describe rotation rate is by the **period,** that is, the time it takes to complete one revolution. The period P is the inverse of frequency:

$$P = \frac{1}{f} = \frac{2\pi}{\omega} \tag{21-10}$$

21.1.3 Relationships between Angular Momentum and Linear Motion

As previously stated, angular motion is entirely consistent with Newton's laws, which are formulated for linear motion. Therefore, it is possible to relate angular equations of motion to linear equations of motion.

Figure 21.2 shows the **velocity** and position **vectors** for the rock as it travels in a circle around a fixed point. The **velocity magnitude** v is the speed that the rock travels around the circle, and it is

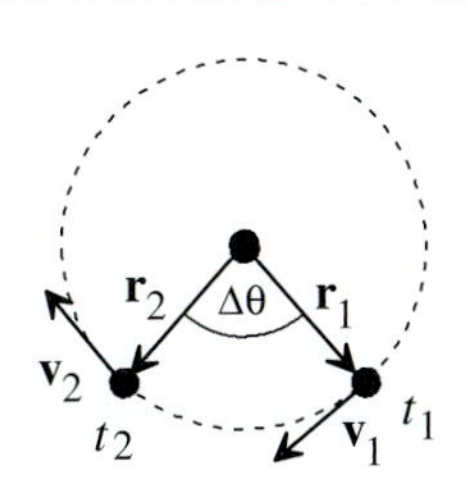

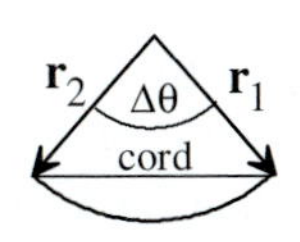

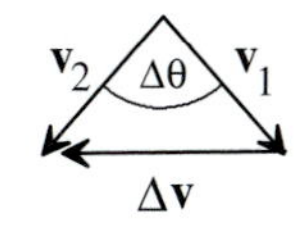

FIGURE 21.2
The velocity and position vectors of a rock tied to a string that travels in a circle around a fixed point.

$$\text{Speed} = v = |\mathbf{v}_1| = |\mathbf{v}_2| \tag{21-11}$$

During the time interval from t_1 to t_2, the position vectors $\mathbf{r}_1$ and $\mathbf{r}_2$ trace arc length Δs. The arc length is the speed v multiplied by the time interval

$$\Delta s = v(t_2 - t_1) = v\Delta t \tag{21-12}$$

which may be written in differential form as

$$ds = v\,dt$$

$$v = \frac{ds}{dt} \tag{21-13}$$

From Equation 21-1, we know that

$$s = r\theta \tag{21-14}$$

which can be differentiated with respect to time as

$$\frac{ds}{dt} = \frac{d}{dt}(r\theta) = r\frac{d\theta}{dt} = r\omega \tag{21-15}$$

(*Note:* r is constant.) By comparing Equation 21-13 to 21-15, we determine that

$$v = r\omega \tag{21-16}$$

which relates the speed v to the instantaneous angular speed ω.

Figure 21.2 shows that position vectors $\mathbf{r}_1$ and $\mathbf{r}_2$ trace arc length Δs during the time interval Δt. During a differential time interval dt, the arc length and cord length are identical:

$$\text{Cord length} = \text{arc length} = ds = v\,dt \tag{21-17}$$

Figure 21.2 shows that the velocity vector $\mathbf{v}_2$ is the sum of vectors $\mathbf{v}_1$ and $\Delta\mathbf{v}$. During a differential time interval, $\mathbf{v}_2$ is the sum of vectors $\mathbf{v}_1$ and $d\mathbf{v}$. The position vectors and velocity vectors form similar triangles; therefore

$$\frac{ds}{|\mathbf{r}_1|} = \frac{d|\mathbf{v}|}{|\mathbf{v}_1|}$$

$$\frac{v\,dt}{r} = \frac{dv}{v} \tag{21-18}$$

which can be rearranged as

$$a = \frac{dv}{dt} = \frac{v^2}{r} \tag{21-19}$$

The direction of this acceleration is toward the center pivot point. This acceleration is called the **centripetal acceleration,** meaning "seeking a center." It may seem odd that an object that is traveling at a constant speed v is accelerating; however, it is important to realize that velocity is a vector with both a magnitude **and** a direction. The inward acceleration is caused by changes in direction, **not** changes in magnitude.

By substituting Equation 21-16 for v, we can derive an alternate expression for the centripetal acceleration

$$a = \frac{v^2}{r} = \frac{(r\omega)^2}{r} = r\omega^2 \tag{21-20}$$

EXAMPLE 21.1

Problem Statement: A rock is tied to a 2.00-m string. A student swings the rock in a complete circle twice every second. What is the frequency, period, angular velocity, speed (i.e., velocity magnitude), and centripetal acceleration?

Solution:

$$\text{Frequency} = f = 2 \text{ rev/s}$$

$$\text{Period} = P = \frac{1}{f} = \frac{\text{s}}{2 \text{ rev}} = 0.5 \text{ s/rev}$$

$$\text{Angular velocity} = \omega = 2\pi f = \left(\frac{2\pi \text{ rad}}{\text{rev}}\right)\left(\frac{2 \text{ rev}}{\text{s}}\right) = 12.6 \text{ rad/s}$$

$$\text{Speed} = v = r\omega = (2.00 \text{ m})\left(12.6 \frac{\text{rad}}{\text{s}}\right) = 25.1 \text{ m/s}$$

$$\text{Centripetal acceleration} = a = \frac{v^2}{r} = \frac{(25.1 \text{ m/s})^2}{2.00 \text{ m}} = 316 \text{ m/s}^2$$

21.2 CENTRIPETAL AND CENTRIFUGAL FORCES

We have seen that a rock that rotates about a fixed central point has a centripetal acceleration toward the center (Figure 21.3). According to Newton's second law, there must be a corresponding **centripetal force** causing this inward acceleration,

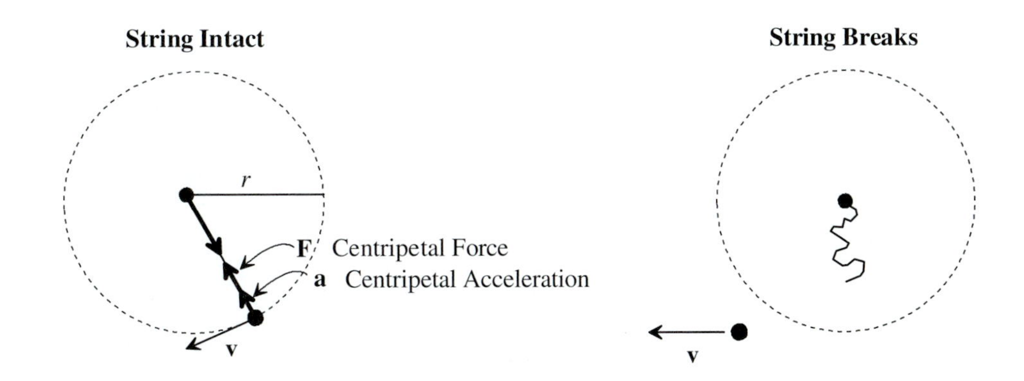

FIGURE 21.3
Centripetal force maintains the rock in a circular orbit.

$$F = ma = m\frac{v^2}{r} = mr\omega^2 \tag{21-21}$$

where m is the mass of the rock. The centripetal force is applied by the tension in the string attached to the rock. If the string were to break, the centripetal force could no longer be applied, so the rock would fly off in a direction tangential to the circle, as in the figure.

Figure 21.4 shows a box tethered to a cable rotating about a fixed center. In the box is an observer and a cannonball mounted on a spring. (*Note:* We assume the spring has negligible mass compared to the cannonball.) As the tethered box rotates at speed v, a centripetal force F_1 is required to maintain the box and its contents in a circular orbit,

$$F_1 = m_1\frac{v^2}{r}$$

where m_1 is the total mass of the box (including the cannonball, spring, and observer). The centripetal force required to maintain the cannonball in the circular orbit is

$$F_2 = m_2\frac{v^2}{r}$$

where m_2 is the cannonball mass.

Figure 21.4 shows an analysis of the forces acting on the cannonball and spring. The force $\mathbf{F}_2$ is required to keep the cannonball rotating in a circular direction. Because of Newton's third law, there must be an equal and opposite force $-\mathbf{F}_2$ acting on the left end of the spring. The box supplies a force $\mathbf{F}_2$ to the right end of the spring. (*Note:* If the spring had a significant mass, the force acting on the right of the spring must be slightly greater than the force acting on the left of the spring. The extra force from the right would be needed to maintain the spring in a circular orbit.) These two forces act to compress the spring. The observer inside the box may be unaware that he is in a rotating system (after all, the earth rotates and you are probably unaware that you are on a rotating system). If the box were to increase its rotational speed, the observer would notice that the spring compresses. He would have to conclude that a force $-\mathbf{F}_2$ was acting to compress the spring and push the cannonball toward the floor. In

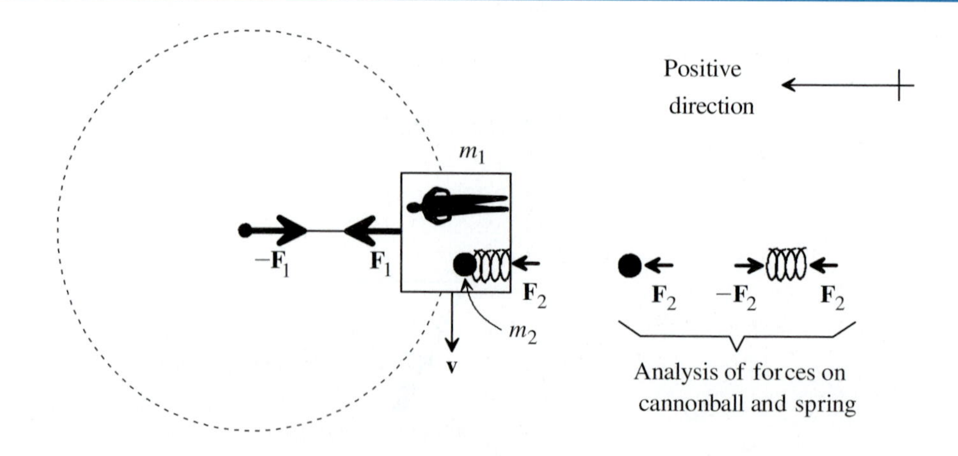

FIGURE 21.4
Illustration of centrifugal force.

actuality, there is no force pushing the cannonball to the floor; in fact, there is a force acting in the opposite direction that keeps the cannonball in a circular orbit. Thus, the observer would hypothesize a force that does not exist. This nonexistent force is called the **centrifugal force,** meaning "fleeing from the center." The observer's invalid conclusion about the existence of a centrifugal force results because the observer is in a noninertial frame of reference.

The tendency for heavy objects to fly outward in rotating systems is very useful to engineers. For example, if an engineer wishes to separate a slightly more dense solid from a slightly less dense liquid, she could employ a "centrifuge." The centrifuge spins the liquid/solid slurry so the generated forces separate the solid from the liquid much faster than could be achieved by gravity alone.

From the perspective of these riders, there is a centrifugal force pushing them outward. In actuality, there is a centripetal force acting inward to keep them in circular motion.

EXAMPLE 21.2

Problem Statement: Travel to Mars takes approximately 1 year. Because gravity is needed to maintain strong bones and muscles, there is concern that astronauts would weaken during their 1-year trip and be unable to function on Mars. To solve this problem, an engineer proposes that the space capsule be given "artificial gravity" by tethering two space capsules together with a 100-m cord (Figure 21.5). At what speed v must the capsules rotate to simulate 1 "g" of gravity (i.e., an acceleration of 9.8 m/s^2)? What are the required angular velocity, period, and frequency?

Solution:

$$\text{Acceleration} = g = \frac{v^2}{r} = 9.8 \text{ m/s}^2$$

$$\text{Speed} = v = \sqrt{gr} = \sqrt{(9.8 \text{ m/s}^2)(50 \text{ m})} = 22.1 \text{ m/s}$$

$$\text{Angular velocity} = \omega = \frac{v}{\text{r}} = \frac{22.1 \text{ m/s}}{50 \text{ m}} = 0.442 \text{ rad/s}$$

$$\text{Period} = P = \frac{2\pi}{\omega} = \frac{2\pi \text{ rad/rev}}{0.442 \text{ rad/s}} = 14.2 \text{ s/rev}$$

$$\text{Frequency} = f = \frac{1}{P} = \frac{\text{rev}}{14.2 \text{ s}} = 0.0703 \text{ rev/s}$$

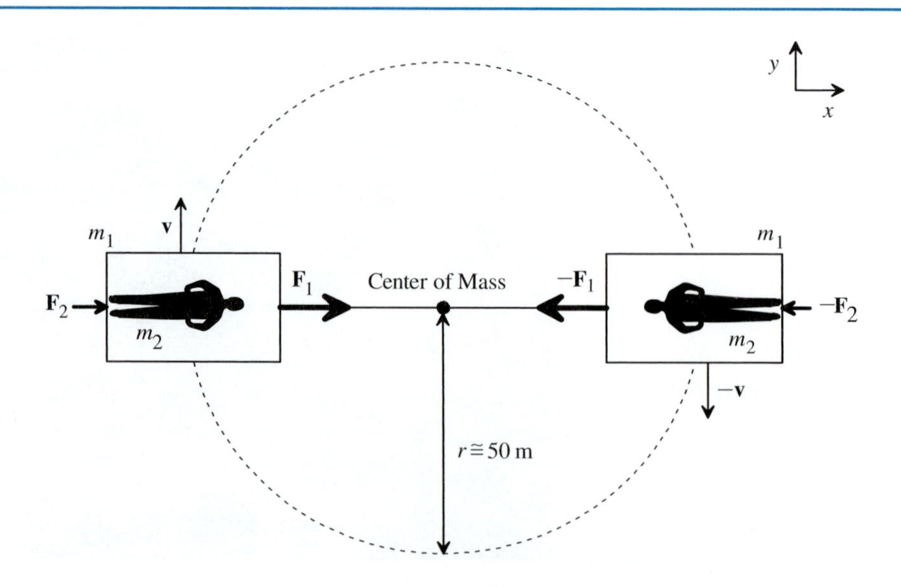

FIGURE 21.5
Two space capsules tethered together and rotating to simulate gravity.

EXAMPLE 21.3

Problem Statement: We know the earth takes 365.24 days to orbit the sun (i.e., its period is 365.24 days). On the basis of this knowledge, calculate the distance from the sun to the earth, assuming a circular orbit.

Solution: The centripetal force that maintains the earth in a circular orbit is supplied by gravity. Thus,

$$F = G\frac{m_{\text{sun}}m_{\text{earth}}}{r^2} = m_{\text{earth}}a_{\text{centripetal}} = m_{\text{earth}}\frac{v^2}{r}$$

$$Gm_{\text{sun}} = v^2\text{r}$$

$$v = \frac{\text{circumference}}{\text{period}} = \frac{2\pi r}{P}$$

$$Gm_{\text{sun}} = \left(\frac{2\pi r}{P}\right)^2 \text{r} = \left(\frac{2\pi}{\text{P}}\right)^2 \text{r}^3$$

$$r = \sqrt[3]{\left(\frac{P}{2\pi}\right)^2 Gm_{\text{sun}}}$$

$$r = \sqrt[3]{\left(\frac{(365.34\text{ d})\left(\frac{24\text{ h}}{\text{d}}\right)\left(\frac{3600\text{ s}}{\text{h}}\right)}{2\pi}\right)^2\left(6.673\times10^{-11}\frac{\text{N·m}^2}{\text{kg}^2}\right)(1.9891\times10^{30}\text{ kg})\left(\frac{\frac{\text{kg·m}}{\text{s}^2}}{\text{N}}\right)}$$

$$= 1.496\times10^{11}\text{ m}\times\frac{\text{km}}{1000\text{ m}} = 1.496\times10^{8}\text{ km}$$

21.3 ANGULAR MOMENTUM FOR PARTICLES

A **particle** is an object that has mass confined to a single point with no volume. The objects we have described so far (stone, earth, space capsules) can be approximated as particles because the dimensions of the object are small compared to the circle radius.

As described in the previous chapter, linear momentum **p** is a vector given by

$$\mathbf{p} = m\mathbf{v} \tag{21-22}$$

The magnitude of the momentum vector p is

$$p = |\mathbf{p}| = m|\mathbf{v}| = mv \tag{21-23}$$

The position of a particle can be described by the position vector **r** (Figure 21.6). The position vector magnitude r is

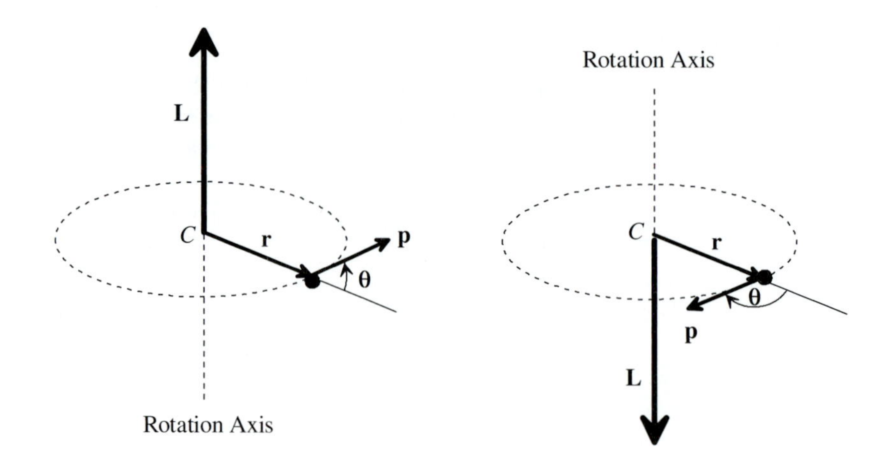

FIGURE 21.6
Angular momentum vectors **L** are determined by the position vectors **r** and momentum vectors **p**.

$$r = |\mathbf{r}| \tag{21-24}$$

which for a circular orbit is just the circle radius.

Angular momentum **L** is a vector determined by the position vector **r** and linear momentum vector **p.** The direction of the angular momentum vector **L** is determined by the conventions shown in Figure 21.6. Angular momentum magnitude is given by

$$L = rp \sin \theta = r(mv)\sin \theta \tag{21-25}$$

where θ is the angle shown in Figure 21.6. For a circular orbit, the angle is always 90°. Because $\sin 90° = 1$, Equation 21-25 reduces to

$$L = rp = rmv \qquad \text{(Circular orbit)} \tag{21-26}$$

We can develop an alternative expression by substituting Equation 21-16 for v:

$$L = rm(r\omega) = mr^2\omega \qquad \text{(Circular orbit)} \tag{21-27}$$

The **mass moment of inertia** I about the vertical axis through the center C is defined as

$$I \equiv mr^2 \qquad \text{(Orbiting particle)} \tag{21-28}$$

Using this definition, we may write Equation 21-27 as

$$L = I\omega \qquad \text{(Angular momentum)} \tag{21-29}$$

The corresponding equation for linear momentum is

$$p = mv \qquad \text{(Linear momentum)} \tag{21-30}$$

Term-by-term comparison shows that L corresponds with p, I corresponds with m, and ω corresponds with v.

EXAMPLE 21.4

Problem Statement: What is the angular momentum magnitude of the earth as it circles the sun? (Assume a circular orbit.)

Solution: $L = rmv$

$$v = \frac{\text{circumference}}{\text{period}} = \frac{2\pi r}{P} = \frac{2\pi(1.496 \times 10^{11}\ \text{m})}{(365.26\ \text{d})\left(\frac{24\ \text{h}}{\text{d}}\right)\left(\frac{3600\ \text{s}}{\text{h}}\right)} = 29{,}780\ \text{m/s}$$

$$L = (1.496 \times 10^{11}\ \text{m})(5.9785 \times 10^{24}\ \text{kg})(29{,}780\ \text{m/s}) = 2.664 \times 10^{40}\ \text{kg·m}^2/\text{s}$$

21.4 ANGULAR MOMENTUM FOR RIGID BODIES (ADVANCED TOPIC)

Unlike a particle that has mass but no dimensions (i.e., volume), a **rigid body** is an object that has both mass and dimensions (i.e., length, area, or volume). A rigid body moves as a whole; the components of the rigid body do not move with respect to each other.

Unlike a particle that has all the mass at a given radius, a rigid body has mass distributed across many radii. Consider, for example, the cylinder with density ρ shown in Figure 21.7. The mass of the differential volume element is

$$dm = \rho\, dV = \rho[(r\, d\theta)dr\, dh] \tag{21-31}$$

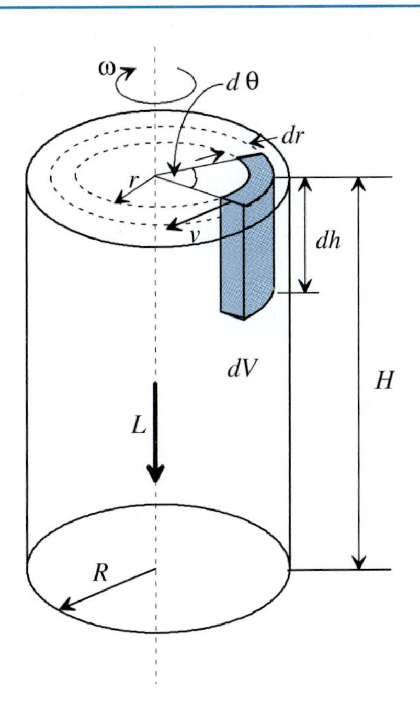

FIGURE 21.7
Solid cylinder rotating about its axis.

From Equation 21-26, a differential amount of angular momentum contributed by the differential volume element traveling in a circular orbit is

$$dL = rv\,dm \tag{21-32}$$

Substituting Equation 21-16 for v and Equation 21-31 for dm gives

$$dL = r(r\omega)(\rho r\,d\theta\,dr\,dh) = \omega\rho r^3\,dr\,d\theta\,dh \tag{21-33}$$

The total angular momentum can be determined by performing a triple integration:

$$\int_0^L dL = \omega\rho\int_0^H\int_0^{2\pi}\int_0^R r^3\,dr\,d\theta\,dh$$

$$[L]_0^L = \omega\rho\int_0^H\int_0^{2\pi}\left[\frac{1}{4}r^4\right]_0^R d\theta\,dh$$

$$[L - 0] = \omega\rho\int_0^H\int_0^{2\pi}\left[\left(\frac{1}{4}R^4\right) - \left(\frac{1}{4}\right)0^4\right]d\theta\,dh$$

$$L = \omega\rho\int_0^H\int_0^{2\pi}\frac{1}{4}R^4\,d\theta\,dh$$

$$= \omega\rho\frac{1}{4}R^4\int_0^H[\theta]_0^{2\pi}\,dh$$

$$= \omega\rho\frac{1}{4}R^4\int_0^H[2\pi - 0]\,dh$$

$$= \omega\rho\frac{1}{4}R^4\int_0^H 2\pi\,dh$$

$$= \omega\rho\frac{1}{4}R^4 2\pi\int_0^H dh$$

$$= \omega\rho\frac{1}{4}R^4 2\pi[h]_0^H$$

$$= \omega\rho\frac{1}{4}R^4 2\pi[H - 0]$$

$$= \omega\rho\frac{1}{4}R^4 2\pi H$$

$$= \omega\rho\frac{1}{2}R^4 \pi H$$

$$= \omega\frac{1}{2}R^2(\rho\pi R^2 H) \tag{21-34}$$

The total mass m of the cylinder is its density times its volume.

$$m = \rho V = \rho \pi R^2 H \tag{21-35}$$

Equation 21-34 then becomes

$$L = \omega \frac{1}{2} R^2(m) = \left(\frac{mR^2}{2}\right)\omega \tag{21-36}$$

It is customary to write the angular momentum as

$$L = I\omega \tag{21-37}$$

where I is the mass moment of inertia. By comparing Equations 21-36 and 21-37, we see that the mass moment of inertia for a solid cylinder rotating about its center axis is

$$I = \frac{mR^2}{2} \qquad \text{(Solid cylinder)} \tag{21-38}$$

Table 21.2 gives mass moments of inertia for other geometries.

EXAMPLE 21.5

Problem Statement: A solid-cylinder flywheel is constructed from steel. Its radius is 0.500 m and it is 0.100 m thick. It rotates at 5000 rpm. Calculate the angular momentum about its axis of rotation.

Solution: $L = \text{I}\omega$

$$I = m\tfrac{1}{2}R^2$$

$$m = \rho V = \rho \pi R^2 H = \left(7900\,\frac{\text{kg}}{\text{m}^3}\right)\pi(0.500\text{ m})^2(0.100\text{ m}) = 620\text{ kg}$$

$$I = (620\text{ kg})\,\tfrac{1}{2}\,(0.500\text{ m})^2 = 77.6\text{ kg·m}^2$$

$$\omega = 2\pi f = \left(\frac{2\pi\text{ rad}}{\text{rev}}\right)\left(\frac{5000\text{ rev}}{\text{min}}\right)\left(\frac{\text{min}}{60\text{ s}}\right) = 524\,\frac{\text{rad}}{\text{s}}$$

$$L = (77.6\text{ kg·m}^2)\left(524\,\frac{\text{rad}}{\text{s}}\right) = 40{,}600\text{ kg·m}^2/\text{s}$$

21.5 TORQUE

Torque is a twisting force. Like angular momentum, it is also a vector quantity. Torque **T** is determined by the position vector **r** and the force vector **F.** The direction of the torque vector **T** is given by the conventions shown in Figure 21.8. The magnitude of the torque is

$$T = rF\sin\theta \tag{21-39}$$

TABLE 21.2
Mass moments of inertia* (Reference)

Particle

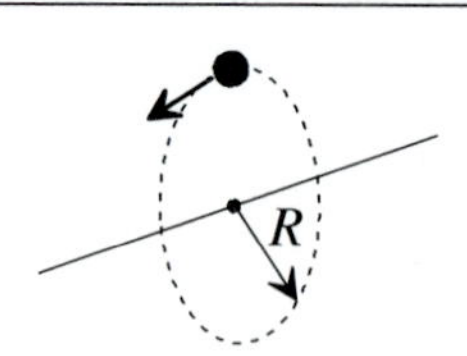

$I = mR^2$

Thin Hoop

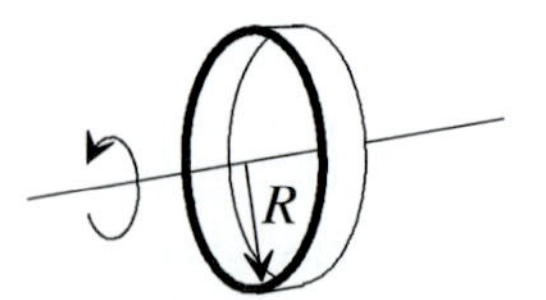

$I = mR^2$

Solid Cylinder

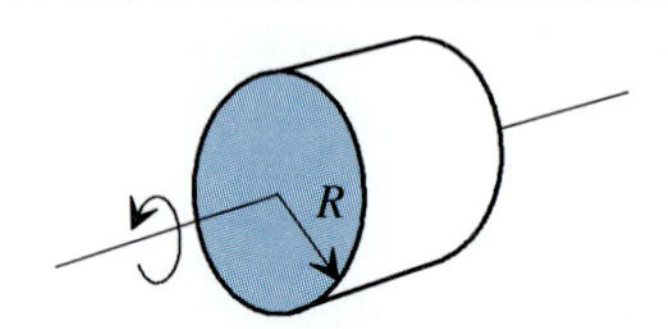

$I = m\frac{1}{2}R^2$

Annular Cylinder

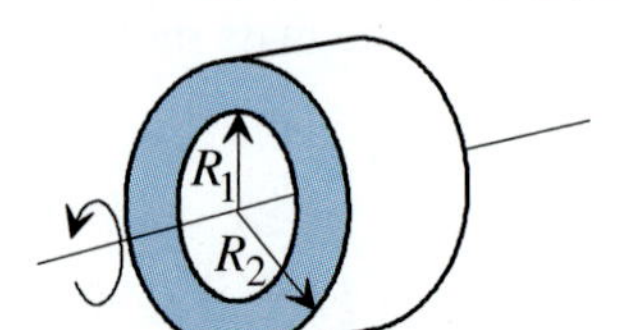

$I = m\frac{1}{2}(R_1^2 + R_2^2)$

Solid Sphere

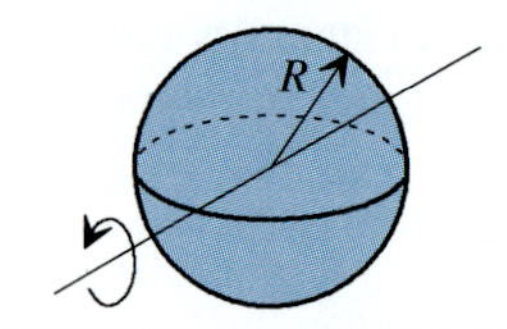

$I = m\frac{2}{5}R^2$

Thin Spherical Shell

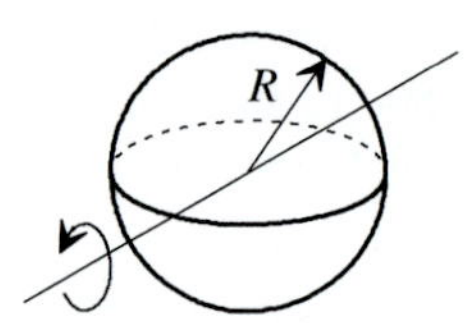

$I = m\frac{2}{3}R^2$

Solid Cylinder

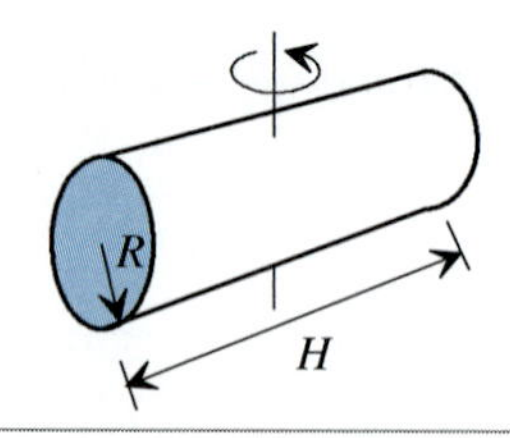

$I = m\left(\frac{1}{4}R^2 + \frac{1}{12}H^2\right)$

Solid Rectangular Parallelepiped

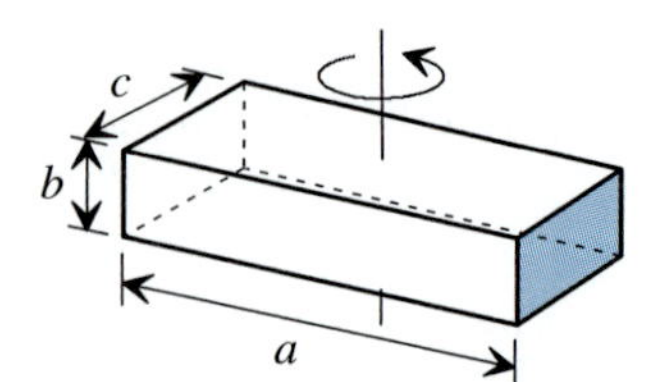

$I = m\frac{1}{12}(a^2 + c^2)$

Hoop

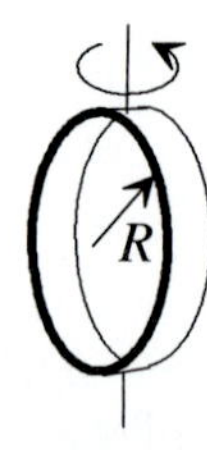

$I = m\frac{1}{2}R^2$

Cone

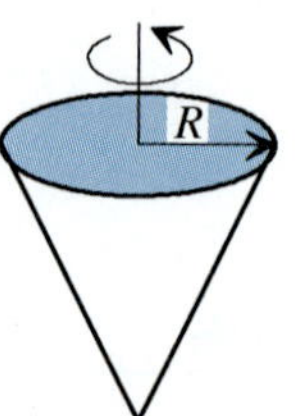

$I = m\frac{3}{10}R^2$

*$I = \int r^2\, dm$

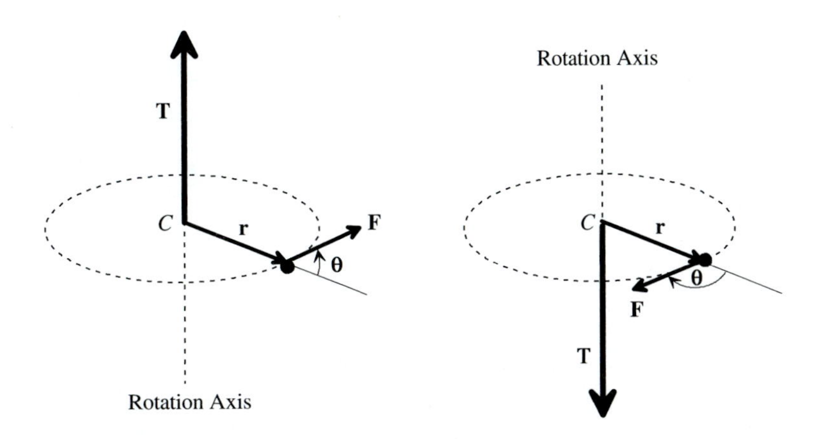

FIGURE 21.8
Torque vectors **T** are determined by the position vectors **r** and force vectors **F**.

where θ is the angle shown in Figure 21.8. For a force applied at a right angle to the radius, the torque becomes

$$T = rF \qquad \text{(Right-angle force)} \tag{21-40}$$

21.6 CONSERVATION OF ANGULAR MOMENTUM

Just as linear momentum is a conserved quantity, so is angular momentum. Because it is conserved, the total angular momentum in the universe is neither generated nor consumed, so the universal accounting equation reduces to

$$\text{Final amount} - \text{initial amount} = \text{input} - \text{output} \tag{21-41}$$

which can also be written as

$$\text{Accumulation} = \text{net input} \tag{21-42}$$

There are two ways that the angular momentum of a system can change: (1) change of mass and (2) unbalanced torque.

21.6.1 Angular Momentum Change by Mass Transfer

Figure 21.9 shows a system with initial angular momentum $\mathbf{L}_i$. The system has only one body, so the initial angular momentum is

$$\mathbf{L}_i = \mathbf{L}_1 \tag{21-43}$$

As time passes, mass enters the system. Each body has its own moment of inertia and angular velocity. The final angular momentum is found by summing the angular momentum of each body:

$$\mathbf{L}_f = \mathbf{L}_1 + \mathbf{L}_2 + \mathbf{L}_3 \tag{21-44}$$

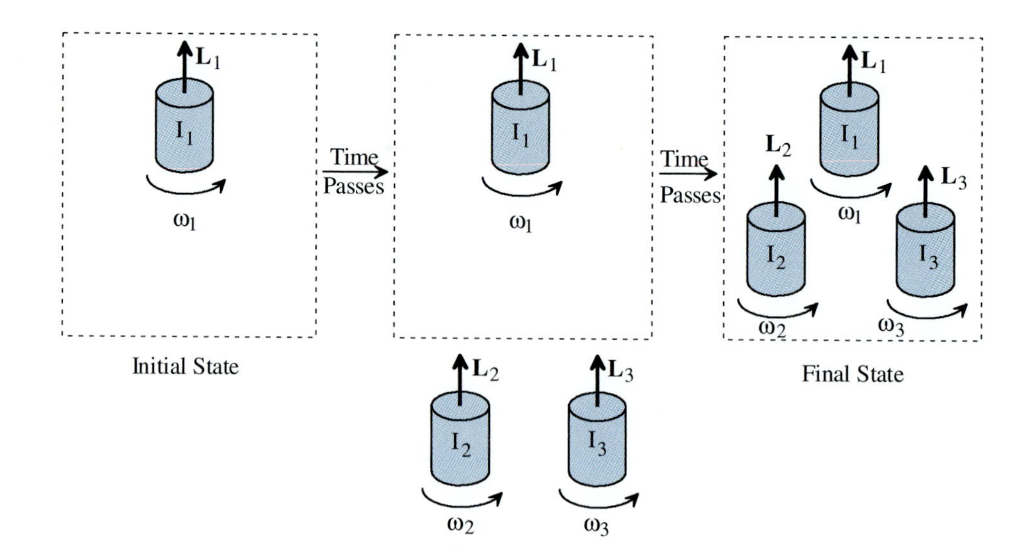

FIGURE 21.9 Angular momentum change by mass transfer.

EXAMPLE 21.6

Problem Statement: Mary is playing with spinning tops. She draws a circle on the sidewalk with chalk. Her game is to see how many tops she can have simultaneously spinning in the circle. Each top has a mass moment of inertia equal to 0.00025 kg·m^2 and spins at 2000 revolutions per minute (209 rad/s). Mary is left-handed, so all tops spin counterclockwise when viewed from above. At the beginning, she has no tops in the circle, but after 3 min, she has five tops spinning simultaneously within the circle. What is the final angular momentum of the system (i.e., the contents of the circle)?

Solution: Final amount = initial amount + input − output

$$= 0 + 5(I\omega) - 0$$

$$= 0 + 5(0.00025 \text{ kg·m}^2)(209 \text{ rad/s}) - 0$$

$$= 0.261 \text{ kg·m}^2/\text{s}$$

(*Note:* All tops are spinning in the same direction and are pointed vertically, so the vector nature of angular momentum does not affect the calculations.)

EXAMPLE 21.7

Problem Statement: John decides to play tops with Mary. John is right-handed, so his tops spin clockwise when viewed from above. All tops have a mass moment of inertia equal to 0.00025 kg·m^2 and spin at 2000 rpm (209 rad/s). At the beginning, there are no tops in the circle, but after 3 min, there are five spinning clockwise and five spinning counterclockwise. What is the final angular momentum of the system (i.e., the contents of the circle)?

Solution: Final amount = initial amount + input − output

$$= 0 + [5(I\omega_1) + 5\,(I\omega_2)] - 0$$

$$= 0 + [5(0.00025\ \text{kg·m}^2)(209\ \text{rad/s}) + 5(0.00025\ \text{kg·m}^2)(-209\ \text{rad/s})] - 0$$

$$= 0$$

[*Note:* All tops have identical angular momentum magnitude. However, because half spin clockwise (downward angular momentum vector) and half spin counterclockwise (upward angular momentum vector), the vectors cancel when added.]

21.6.2 Angular Momentum Change by Unbalanced Torque

When torque T is applied to a body, it changes the body's angular momentum L:

$$T = \frac{dL}{dt} = \frac{d(I\omega)}{dt} = I\frac{d\omega}{dt} = I\alpha \qquad \text{(Angular momentum)} \qquad (21\text{-}45)$$

This equation is entirely analogous to force F changing the linear momentum p of a body:

$$F = \frac{dp}{dt} = \frac{d(mv)}{dt} = m\frac{dv}{dt} = ma \qquad \text{(Linear momentum)} \qquad (21\text{-}46)$$

According to Newton's third law, a force or torque results from the interaction of two (or more) bodies. Further, the force (or torque) on one body is equal and opposite to the force (or torque) on the other body. By Newton's third law, the angular momentum of the *universe* is constant and cannot change. This means that if torque is applied to an object, somewhere in the universe there must be an equal and opposite torque.

However, Newton's third law does not guarantee that the angular momentum of a given *system* is constant. Recall that the system may be arbitrarily specified by the engineer. Therefore, there is no requirement that the system contain both the body and its pair.

EXAMPLE 21.8

Problem Statement: An electric motor is used to accelerate a solid-cylinder flywheel (Figure 21.10). The flywheel is constructed from steel and has a radius of 0.500 m and is 0.100 m thick. Initially, the flywheel is stationary. Then, the electric motor applies a torque of 500 N·m for 60 s. What is the final angular momentum? How fast does the flywheel rotate?

Solution: Angular momentum accumulation = net input

$$\frac{dL}{dt} = T$$

$$dL = T\,dt$$

$$\int_0^{L_f} dL = \int_0^{t_f} T\,dt = T\int_0^{t_f} dt$$

The angular momentum of this skater is approximately constant. When her arms and leg are extended, she has a greater momentum of inertia, so she spins more slowly. When her arms and leg are drawn in, she has less moment of inertia, so she spins more rapidly.

$$[L]_0^{L_f} = T[t]_0^{t_f}$$

$$[L_f - 0] = T[t_f - 0]$$

$$L_f = Tt_f = (500\ \text{N}\cdot\text{m})(60\ \text{s}) = 30{,}000\ \text{N}\cdot\text{m}\cdot\text{s} \times \frac{\text{kg}\cdot\text{m/s}^2}{1\ \text{N}} = 30{,}000\ \frac{\text{kg}\cdot\text{m}^2}{\text{s}}$$

$$\omega = \frac{L}{I} = \frac{30{,}000\ \frac{\text{kg}\cdot\text{m}^2}{\text{s}}}{77.6\ \text{kg}\cdot\text{m}^2} = 387\ \frac{\text{rad}}{\text{s}}$$

$$f = \frac{1}{2\pi}\omega = \frac{1\ \text{rev}}{2\pi\ \text{rad}}\left(\frac{387\ \text{rad}}{\text{s}}\right) = 61.5\ \frac{\text{rev}}{\text{s}} = 61.5\ \text{Hz}$$

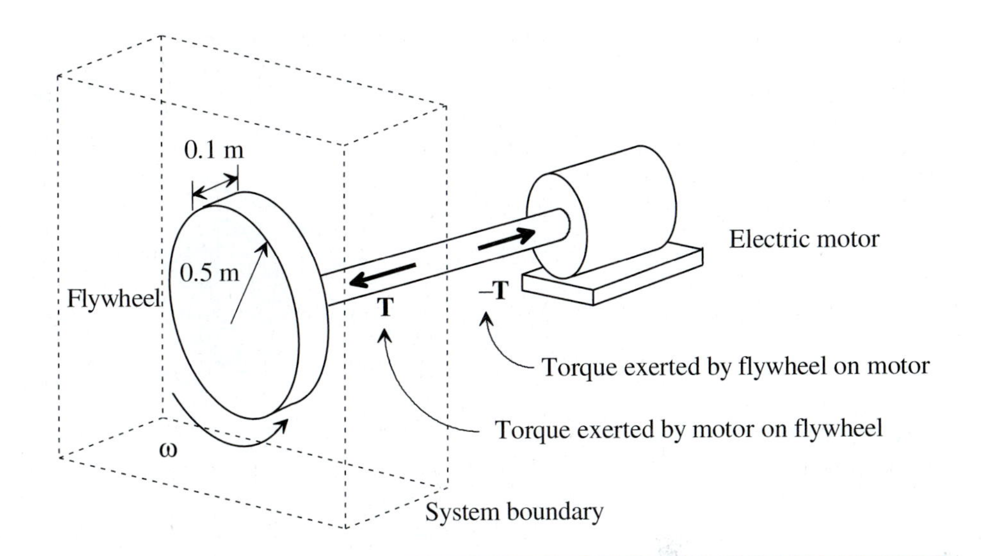

FIGURE 21.10
The torque produced by the electric motor causes the flywheel angular momentum to increase. Note that according to Newton's third law, an equal and opposite torque is exerted by the flywheel on the electric motor. The motor is attached to the earth; therefore, this torque changes the angular momentum of the earth. However, because the earth mass is so large, the change in the earth rotation is imperceptibly small.

(*Note:* We calculated the flywheel mass moment of inertia in Example 21.5.)

21.7 SYSTEMS WITHOUT NET MOMENTUM INPUT

Many systems of interest have no unbalanced torque, nor do they transfer mass across the system boundary. In these cases, the universal accounting equation reduces to

$$\text{Final amount} - \text{initial amount} = 0$$

$$\text{Final amount} = \text{initial amount}$$

EXAMPLE 21.9

Problem Statement: An engineer is designing a space telescope. The telescope must be able to point at any desired star, so a pointing mechanism is required. Initially, the engineer proposes to use small rockets to point the telescope. However, he soon realizes that the amount of propellant needed during the life of the telescope would be exceedingly large. Instead, he decides to mount a flywheel in the telescope (Figure 21.11). The electric motor that drives the flywheel is coupled to the space telescope. The electricity needed to drive the motor comes from solar panels mounted on the space telescope. According to Newton's third law, any torque applied to the flywheel is accompanied by an equal and opposite torque on the electric motor. Because the motor is coupled to the space telescope, this torque will cause an equal and opposite change in the angular momentum of the telescope.

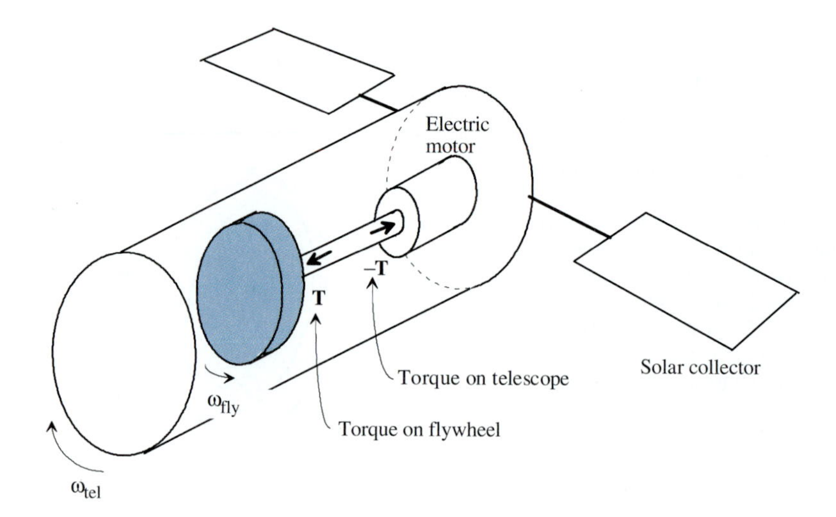

FIGURE 21.11
Space telescope with flywheel directional control.

The flywheel mass moment of inertia is 10.0 kg·m², and the space telescope mass moment of inertia is 10,000 kg·m² with the solar panels fully extended. Initially, the space telescope and flywheel are stationary. Then, the electric motor applies torque to the flywheel, causing it to rotate at 5.00 rpm. At what frequency is the space telescope rotating?

Solution: Final amount = initial amount

$$I_{tel}\omega_{tel,f} + I_{fly}\omega_{fly,f} = i_{tel}\omega_{tel,i} + I_{fly}\omega_{fly,i}$$

$$I_{tel}2\omega f_{tel,f} + I_{fly}2\pi f_{fly,f} = I_{tel}2\pi f_{tel,i} + I_{fly}2\pi f_{fly,i}$$

$$I_{tel}f_{tel,f} + I_{fly}f_{fly,f} = I_{tel}f_{tel,i} + I_{fly}f_{fly,i}$$

$$f_{tel,f} = \frac{I_{tel}f_{tel,i} + I_{fly}f_{fly,i} - I_{fly}f_{fly,f}}{I_{tel}}$$

$$= \frac{10{,}000 \text{ kg·m}^2 \text{ (0 rpm)} + 10 \text{ kg·m}^2 \text{ (0 rpm)} - 10 \text{ kg·m}^2 \text{ (5.00 rpm)}}{10{,}000 \text{ kg·m}^2}$$

$$= -0.00500 \text{ rpm}$$

(*Note:* The negative sign indicates that the telescope rotates in a direction opposite to the flywheel.)

EXAMPLE 21.10

Problem Statement: The space telescope described in the previous example has a mass moment of inertia equal to 10,000 kg·m² with the solar collectors fully extended. When the solar collectors are retracted against the sides of the telescope, the mass moment of inertia reduces to 8000 kg·m^2. Although the telescope mass is unchanged by this action, the mass is located nearer the rotation axis which reduces the mass moment of inertia. Initially, with the solar collectors fully extended, the telescope is rotating at −0.00500 rpm. What is its rotation rate when the solar collectors are retracted?

Solution: Final amount = initial amount

$$I_{tel,f}\omega_{tel,f} = I_{tel,i}\omega_{tel,i}$$

$$I_{tel,f}2\pi f_{tel,f} = I_{tel,i}\,2\pi f_{tel,i}$$

$$I_{tel,f}f_{tel,f} = I_{tel,i}f_{tel,i}$$

$$f_{tel,f} = \frac{I_{tel,i}}{I_{tel,f}} = f_{tel,i}$$

$$= \frac{10{,}000 \text{ kg·m}^2}{8000 \text{ kg·m}^2}(-0.00500 \text{ rpm}) = -0.00625 \text{ rpm}$$

21.8 SUMMARY

Angular momentum is completely analogous to linear momentum. Table 21.3 shows the corresponding equations for linear and angular motion.

Angular momentum is a conserved quantity, meaning it can be neither created nor destroyed. The amount of angular momentum in the system can be altered by mass crossing the boundary, or by unbalanced torques acting on the system. An unbalanced torque represents a net input of angular momentum into the system, causing an angular acceleration.

TABLE 21.3
Corresponding equations for linear and angular motion

	Linear Motion	Angular Motion
Displacement	$x = x_0 + v_0t + \frac{1}{2}a_0t^2$	$\theta = \theta_0 + \omega_0t + \frac{1}{2}\alpha_0t^2$
Velocity	$v = v_0 + a_0t$	$\omega = \omega_0 + \alpha_0t$
Acceleration	$a = a_0$	$\alpha = \alpha_0$
Momentum	$p = mv$	$L = I\omega$
Force	$F = \frac{dp}{dt} = \frac{d}{dt}(mv) = m\frac{dv}{dt} = ma$	$T = \frac{dL}{dt} = \frac{d}{dt}(I\omega) = I\frac{d\omega}{dt} = I\alpha$
Kinetic energy	$E = \frac{1}{2}mv^2$	$E = \frac{1}{2}I\omega^2$

Mathematical Prerequisites

- Algebra (Appendix E, Mathematics Supplement)
- Calculus (Appendix L, Mathematics Supplement)

CHAPTER 22

Accounting for Energy

Energy is often defined as "the capacity to do work." This is a rather circular definition, because work is a type of energy. Energy is an abstract concept best described as a unit of exchange (much like money) used by the scientific and engineering community to equate various physical phenomena (see Chapter 11 on thermodynamics). Energy is a **scalar** quantity.

22.1 ENERGY ACCOUNTING

The universal accounting equation may be applied to a system to count energy.

$$\text{Final amount} - \text{initial amount} = \text{input} - \text{output} + \text{generation} - \text{consumption} \tag{22-1}$$

As described in Chapter 18 on accounting for mass, energy can be generated in fusion and fission reactions and consumed in particle accelerators.[1] Because very few engineers work with these systems, we can generally say that energy is conserved, meaning it is neither generated nor consumed. Therefore, the universal accounting equation for energy reduces to

$$\text{Final amount} - \text{initial amount} = \text{input} - \text{output} \tag{22-2}$$

which can also be written as

$$\text{Accumulation} = \text{net input} \tag{22-3}$$

Forms of energy that can accumulate are state quantities, and forms of energy that are net input are path quantities. Table 22.1 categorizes the types of energy frequently encountered in engineering. All of these are discussed in more detail in subsequent sections.

1. Mass and energy are related according to Einstein's $E = mc^2$. Some people prefer to say, "mass-energy is conserved; therefore, the total amount of mass-energy in the universe does not change with time." Although we agree with this statement, we prefer to count mass and energy separately, because their properties are so different.

TABLE 22.1
Classification of energies commonly encountered in engineering

Path		State		
Work	**Heat**	**Kinetic**	**Potential**	**Internal**
Mechanical Shaft Hydraulic Electrical Chemical Monochromatic radiation (e.g., laser light)	Conduction Blackbody radiation	Kinetic	Gravity Spring Electrical Magnetic	Sensible Phase change Chemical reaction

22.2 PATH ENERGIES

Path quantities depend on the path taken. Path energies may be further classified as heat or work.

22.2.1 Work

Work is energy flow across a system boundary resulting from any driving force (e.g., pressure, torque, voltage, concentration) other than temperature.

Mechanical Work.

Mechanical work results when a force F is exerted over a linear distance x (Figure 22.1).

$$\text{Mechanical work} = \int F\,dx \tag{22-4}$$

If the force is constant as a function of distance, then it may be brought out of the integral

$$\text{Mechanical work} = F\int dx = F\Delta x \tag{22-5}$$

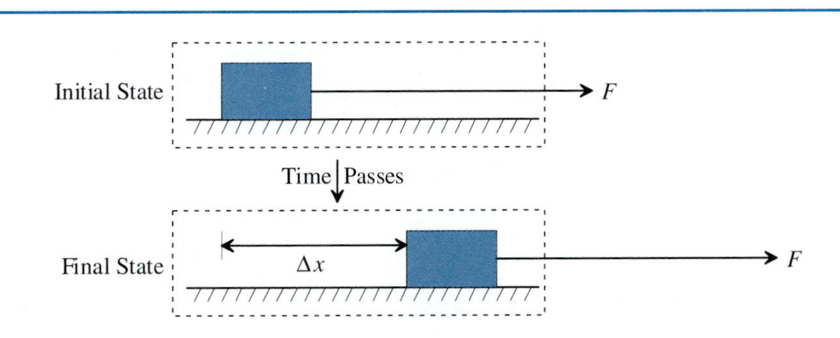

FIGURE 22.1
Illustration of mechanical work.

EXAMPLE 22.1

Problem Statement: An airplane jet engine has a thrust of 2000 lb_f. The jet flies horizontally for 300 miles. How much mechanical work was performed?

Solution:

$$\text{Mechanical work} = F\,\Delta x = (2000\ \text{lb}_f)(300\ \text{mi})\left(\frac{5280\ \text{ft}}{\text{mi}}\right) = 3.2 \times 10^9\ \text{ft}\cdot\text{lb}_f$$

Shaft Work.

Whereas mechanical work referred to force exerted over a linear distance, **shaft work** refers to torque T (twisting force) exerted over a circular distance (Figure 22.2). The definition of shaft work is the same as that of mechanical work:

$$\text{Shaft work} = \int F\,dx \tag{22-6}$$

The distance traveled x by the periphery of a shaft of radius r is

$$x = r\theta \tag{22-7}$$

where θ is the twist angle measured in radians. The derivative of Equation 22-7 is

$$dx = r\,d\theta \tag{22-8}$$

because r is a constant. This may be substituted into Equation 22-6 to give

$$\text{Shaft work} = \int Fr\,d\theta \tag{22-9}$$

Assuming the force is constant, it and the radius r may be removed from the integral to give

$$\text{Shaft work} = Fr \int d\theta = Fr\theta = T\theta \tag{22-10}$$

where $T = Fr$.

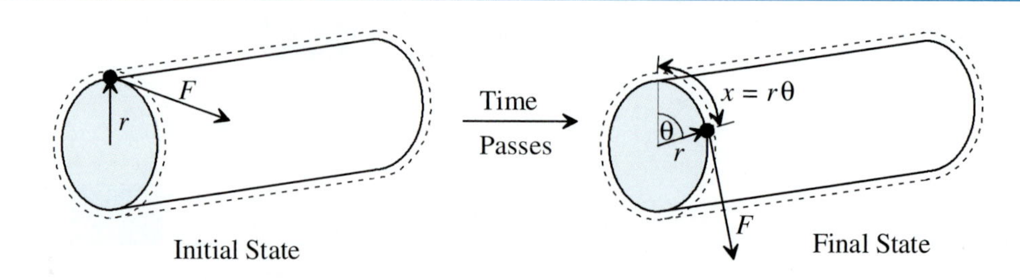

FIGURE 22.2
Shaft work.

EXAMPLE 22.2

Problem Statement: A vehicle engine produces 80 $ft \cdot lb_f$ of torque while traveling a distance of 1.0 mile. During the trip, the drive shaft rotated 2000 times. What was the shaft work produced by the engine?

Solution:

$$\text{Shaft work} = T\theta = (80 \text{ ft} \cdot \text{lb}_f)(2000 \text{ rev})\left(\frac{2\pi \text{rad}}{\text{rev}}\right) = 1.0 \times 10^6 \text{ ft} \cdot \text{lb}_f$$

Normally, rotating shafts are used to power devices such as automobiles, pumps, and fans. The **power** (i.e., energy per unit time) transmitted by the shaft is

$$\text{Shaft power} = \frac{\text{shaft work}}{t} = \frac{T\theta}{t} = T\left(\frac{\theta}{t}\right) = T\omega \tag{22-11}$$

where t is time, ω is rotational speed (in radians per unit time), and T is torque.

EXAMPLE 22.3

Problem Statement: An automobile engine shaft is rotating at 3000 revolutions per minute (rpm) and has 80 $ft \cdot lb_f$ of torque. How much shaft power (hp) is produced by the engine?

Solution:

$$\text{Shaft power} = T\omega$$

$$= (80 \text{ ft} \cdot \text{lb}_f)\left(3000 \frac{\text{rev}}{\text{min}}\right)\left(\frac{2\pi \text{ rad}}{\text{rev}}\right)\left(\frac{60 \text{ min}}{\text{h}}\right)\left(\frac{\text{hp} \cdot \text{h}}{1.98 \times 10^6 \text{ ft} \cdot \text{lb}_f}\right)$$

$$= 45.7 \text{ hp}$$

Hydraulic Work.

Figure 22.3 shows a piston with cross-sectional area A pressurizing incompressible hydraulic fluid to pressure P_2. Pressure P_1 exists on the other side of the piston. A very thin shaft with negligible cross-sectional area applies a force F to the piston. As the piston moves a distance Δx, fluid volume V (m^3) flows through the valve at a volumetric flow rate Q (m^3/s).

The **hydraulic work** is

$$\text{Hydraulic work} = \int F\, dx \tag{22-12}$$

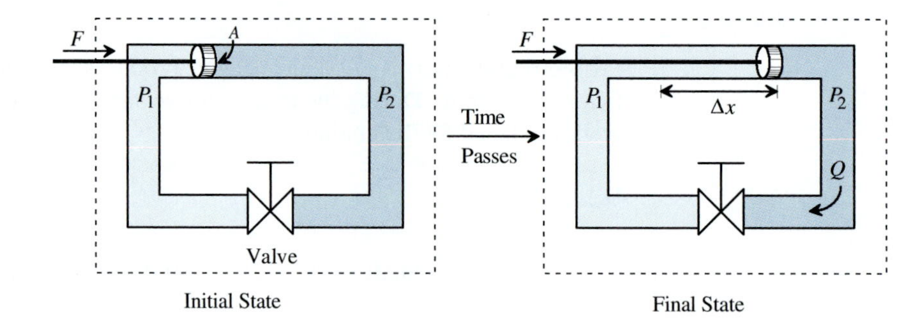

FIGURE 22.3
Hydraulic work.

Assuming the force is constant, the force can be removed from the integral, thus

$$\text{Hydraulic work} = F\int dx = F\Delta x \tag{22-13}$$

The force required to move the piston is

$$F = A(P_2 - P_1) \tag{22-14}$$

We may substitute Equation 22-14 into Equation 22-13, giving

$$\text{Hydraulic work} = A(P_2 - P_1)\Delta x \tag{22-15}$$

The volume V of hydraulic fluid that flows is

$$V = A\Delta x \tag{22-16}$$

which we may substitute into Equation 22-15, giving

$$\text{Hydraulic work} = V(P_2 - P_1) \tag{22-17}$$

Recall from our study of thermodynamics in Chapter 11 that pressure has dimensions of energy per unit volume. With this interpretation of pressure, Equation 22-17 is very easy to remember.

Commonly, hydraulic fluid is reported as a flow rate Q (m^3/s) rather than as a volume V (m^3). The volume that flows can be easily calculated by multiplying the flow rate by time

$$V = Qt \tag{22-18}$$

We may substitute this expression into Equation 22-17, giving

$$\text{Hydraulic work} = Qt\,(P_2 - P_1) \tag{22-19}$$

EXAMPLE 22.4

Problem Statement: Water is pumped at a rate of 20 ft^3/min from a pressure of 15 psia to 100 psia (*Note:* See Section 14.6 for a description of "psia."). After 10 min, how much hydraulic work was performed?

Solution:

$$\text{Hydraulic work} = Qt(P_2 - P_1)$$

$$= \left(\frac{20\ \text{ft}^3}{\text{min}}\right)(10\ \text{min})\left(\frac{100\ \text{lb}_f}{\text{in}^2} - \frac{15\ \text{lb}_f}{\text{in}^2}\right)\left(\frac{144\ \text{in}^2}{\text{ft}^2}\right)$$

$$= 2.4 \times 10^6\ \text{ft·lb}_f$$

Electrical Work.

Figure 22.4 shows an electrical circuit in which an electric generator applies a force F that causes a charge q to change its voltage from V_1 to V_2. As you can imagine, it is very hard to grab onto an electron and push it. Fortunately, it can be done with magnetism. Provided the magnetic strength is constantly increasing, it can push electrons through a wire. Conceptually, the increasing magnetic strength could be generated by flowing an increasingly large current through a coil that surrounds the wire. (In real generators, the varying magnetic strength is produced mechanically by rotating armatures.)

The **electrical work** is the force exerted over the distance.

$$\text{Electrical work} = \int F\,dx \tag{22-20}$$

Assuming the force is constant, it may be removed from the integral to yield

$$\text{Electrical work} = F\int dx = F\Delta x \tag{22-21}$$

From electrical theory, the force required to move the charge q is

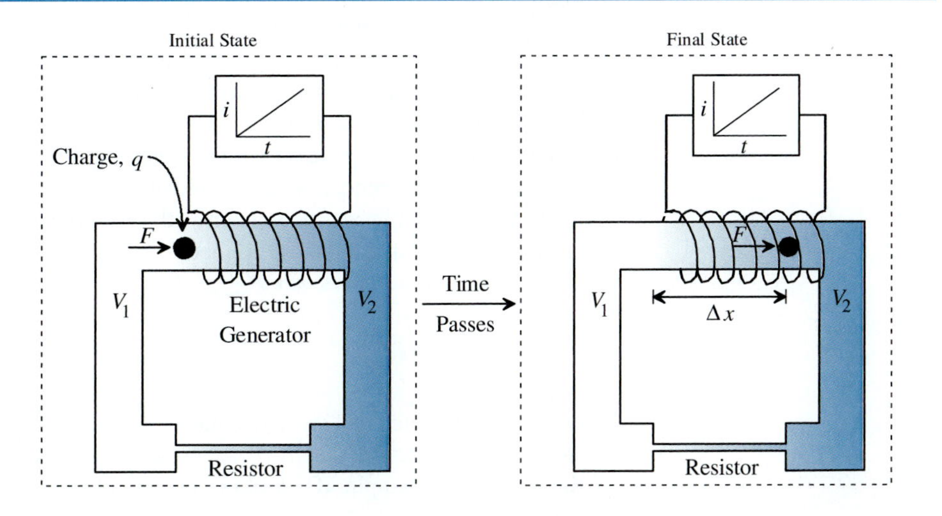

FIGURE 22.4 Electrical work. (*Note:* The coil is conceptual.)

$$F = q\frac{V_2 - V_1}{\Delta x} \tag{22-22}$$

This expression may be substituted into Equation 22-21 to give the electrical work as

$$\text{Electrical work} = \left(q\frac{V_2 - V_1}{\Delta x}\right)\Delta x = q(V_2 - V_1) \tag{22-23}$$

Recall from Chapter 12, on rate processes, that voltage has dimensions of energy per coulomb; thus, Equation 22-23 is very easy to remember.

The charge is current times time,

$$q = it \tag{22-24}$$

which may be substituted into Equation 22-23, giving

$$\text{Electrical work} = it(V_2 - V_1) \tag{22-25}$$

Notice the similarity between the final equations for electrical work and hydraulic work.

EXAMPLE 22.5

Problem Statement: An electric generator produces 10 A of current against a voltage difference of 110 V. The generator runs for 2 h. How much electrical work was performed?

Solution:

$$\text{Electrical work} = it(V_2 - V_1) = (10\text{ A})(2\text{ h})(110\text{ V})\left(\frac{1\text{ W}}{\text{A}\cdot\text{V}}\right) = 2200\text{ Wh}$$

Chemical Work (Advanced Topic).

Figure 22.5 shows a cylinder filled with brine (salt water) and pure water. A semipermeable membrane separates the pure water and brine. The semipermeable membrane is able to pass water, but not salt. A piston pushes on the brine and another piston retains the pure water. A vacuum is on the dry side of each piston.

The salt in the brine displaces water; therefore, the water concentration (with units mol water/m^3) in the brine is less than the water concentration in the pure water. Because of the concentration difference, water should diffuse across the semipermeable membrane and dilute the brine. However, the piston with cross-sectional area A has an applied force F that resists the inflow of water. The applied force causes the brine to be pressurized; this pressure is called the **osmotic pressure** P_{osmotic}. For dilute salt concentrations, the osmotic pressure is

$$P_{\text{osmotic}} = RT\frac{s}{V} \tag{22-26}$$

where R is the universal gas constant [$R = 8.314$ J/(mol·K)], T is the absolute temperature (K), V is the water volume (m^3) added to the salt, and s is the moles of salt ions added to the water. For example, if 1 mol of NaCl is added to the water, it dissociates into 1 mol of

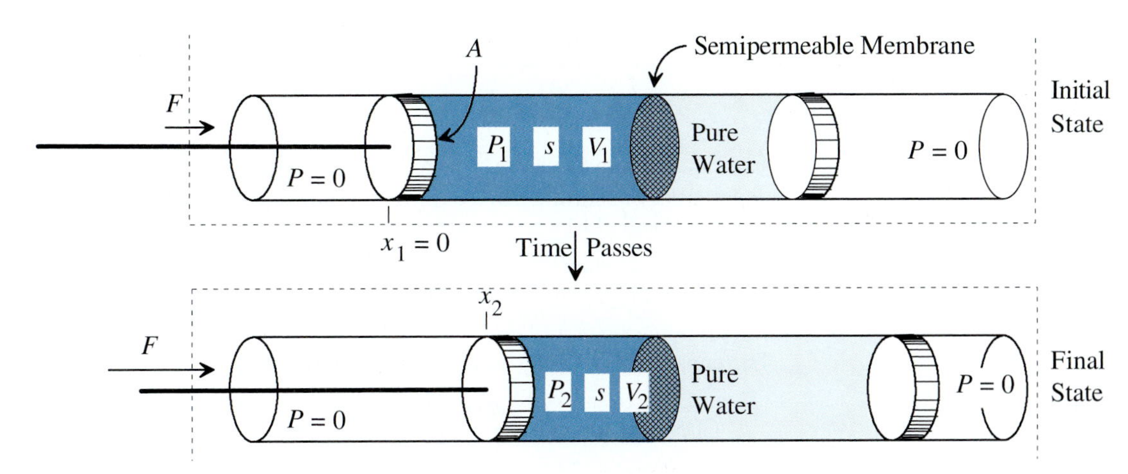

FIGURE 22.5
Chemical work. (*Note: V* is the volume of water in the brine, not the total brine volume.)

Na^+ and 1 mol of Cl^-, so $s = 2$ mol of ions. (It should be noted that this equation is general and applies to other species besides salt. For example, it also applies to sugar solutions where s is the moles of sugar added to the water. However, sugar does not dissociate into ions; therefore, if 1 mol of sugar is added to the water, $s = 1$ mol.)

The **chemical work** required to push water through the membrane is

$$\text{Chemical work} = \int_{x_1}^{x_2} F\,dx \tag{22-27}$$

In this case, the force is not constant because the osmotic pressure increases as the salt concentration increases. Therefore, the force cannot be removed from the integral. We must find an expression for the applied force in terms of position x so we can perform the integration.

The applied force needed to push against the osmotic pressure is

$$F = AP_{\text{osmotic}} \tag{22-28}$$

because there is a vacuum on the other face of the piston. Equation 22-26 may be substituted for the osmotic pressure

$$F = ART\frac{s}{V} \tag{22-29}$$

As the piston moves rightward, water is squeezed out of the brine, but salt remains. The volume of water in the brine is related to the piston position as

$$V = V_1 - Ax = V_1\left(1 - \frac{A}{V_1}x\right) \tag{22-30}$$

where V_1 is the initial water volume in the brine and Ax is the volume swept by the piston. Substituting Equation 22-30 into Equation 22-29 gives

$$F = \frac{ARTs}{V_1\left(1 - \frac{A}{V_1}x\right)} \tag{22-31}$$

This equation gives us the required force as a function of position so we may perform the integration. Substituting this expression into Equation 22-27 gives

$$\text{Chemical work} = \int_{x_1=0}^{x_2} \frac{ARTs}{V_1\left(1 - \frac{A}{V_1}x\right)}\,dx \tag{22-32}$$

The constant terms may be removed from the integral:

$$\text{Chemical work} = \frac{ARTs}{V_1}\int_{x_1=0}^{x_2} \frac{dx}{1 - \frac{A}{V_1}x} \tag{22-33}$$

The solution to this integral is

$$\begin{aligned}
\text{Chemical work} &= \frac{ARTs}{V_1}\left[-\frac{V_1}{A}\ln\left(1 - \frac{A}{V_1}x\right)\right]_{x_1=0}^{x_2} \\
&= -RTs\left[\ln\left(1 - \frac{A}{V_1}x\right)\right]_{x_1=0}^{x_2} \\
&= -RTs\left[\ln\left(1 - \frac{A}{V_1}x_2\right) - \ln\left(1 - \frac{A}{V_1}0\right)\right] \\
&= -RTs\left[\ln\left(1 - \frac{A}{V_1}x_2\right) - 0\right] \\
&= -RTs\ln\left(1 - \frac{Ax_2}{V_1}\right)
\end{aligned} \tag{22-34}$$

The volume of pure water that flows through the semipermeable membrane ΔV is

$$\Delta V = Ax_2 \tag{22-35}$$

This may be substituted into Equation 22-34, giving

$$\begin{aligned}
\text{Chemical work} &= -RTs\ln\left(1 - \frac{\Delta V}{V_1}\right) \\
&= -RTs\ln\left(\frac{V_1}{V_1} - \frac{\Delta V}{V_1}\right) \\
&= -RTs\ln\left(\frac{V_1 - \Delta V}{V_1}\right) \\
&= -RTs\ln\left(\frac{V_2}{V_1}\right)
\end{aligned} \tag{22-36}$$

$$\text{Chemical work} = RTs \ln\left(\frac{V_1}{V_2}\right) \tag{22-37}$$

where V_2 is the final water volume in the brine.

EXAMPLE 22.6

Problem Statement: Salt ions (1.00 mol total) are dissolved in 2.00 m³ of water. A pump pressurizes this mixture and pumps it through a semipermeable membrane that allows 0.50 m³ of pure water to pass, but no salt passes. The water temperature is 300 K. How much chemical work is required?

Solution:

$$\text{Chemical work} = -RTs \ln\left(1 - \frac{\Delta V}{V_1}\right)$$

$$= -\left(8.314 \frac{\text{J}}{\text{mol}\cdot\text{K}}\right)(300\text{ K})(1.00\text{ mol}) \ln\left(1 - \frac{0.50\text{ m}^3}{2.00\text{ m}^3}\right) = 718\text{ J}$$

Laser Light Work.

Light travels in little "packets" called **photons** (Figure 22.6). Photon energy is both electrical and magnetic. The energy oscillates as the photon travels at the speed of light c (2.99792458×10^8 m/s). (Figure 22.6 shows just the electrical portion of the energy.) The distance between the electrical energy peaks is called the wavelength λ. The energy E in a single photon is

$$E = h\frac{c}{\lambda} \tag{22-38}$$

where h is **Planck's constant** ($6.6260755 \times 10^{-34}$ J·s).

Laser light is emitted at a single wavelength; therefore, laser light is a form of monochromatic radiation. Further, the electrical energy peaks align, so the light is **coherent.** Lasers are commonly powered by electricity, although some are chemically powered. In each case, the driving force that creates the laser light is not temperature, so laser light is classified as work, not heat.

FIGURE 22.6
Laser light.

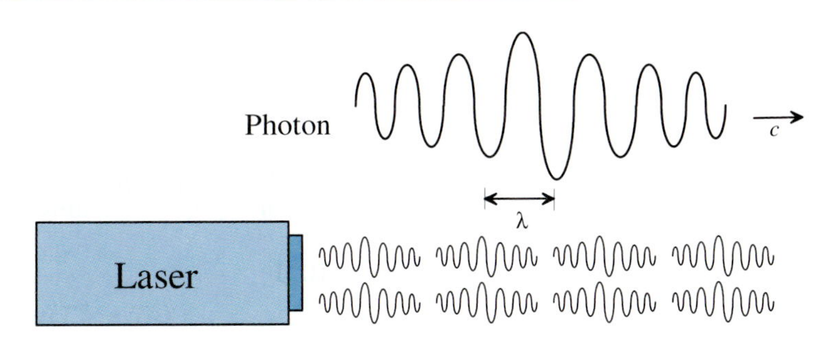

$$= (2.0\text{ h})\left(\frac{3600\text{ s}}{\text{h}}\right)\left[\frac{\pi}{4}(0.0010\text{ m})^2\right]\left(5.67 \times 10^{-8}\frac{\text{J}}{\text{s}\cdot\text{m}^2\cdot\text{K}^4}\right)\cdot$$

$$[(600\text{ K})^4 - (500\text{ K})^4] = 22\text{ J}$$

22.3 STATE ENERGIES

State energies do not depend on path. State energy is often specified by other state quantities such as temperature, pressure, position, and velocity.

22.3.1 Kinetic Energy

Kinetic energy is energy associated with motion. Figure 22.9 shows a rigid object of mass m being accelerated from an initial state v_1 to a final state v_2 resulting from an applied force F. An input of mechanical work causes the object to change the state of its kinetic energy E_k according to Equation 22-2.

$$E_{k,\text{final}} - E_{k,\text{initial}} = \Delta E_k = \text{mechanical work input} - 0 = \int F\,dx \quad (22\text{-}41)$$

Because the applied force is constant, this becomes

$$\Delta E_k = F\Delta x \quad (22\text{-}42)$$

According to Newton's second law for a constant force applied to a rigid body,

$$F = ma = m\left(\frac{v_2 - v_1}{t}\right) \quad (22\text{-}43)$$

The position of an object at any time t starting at position x_1 with velocity v_1 is

$$x_2 = x_1 + v_1 t + \tfrac{1}{2}at^2 \quad (22\text{-}44)$$

$$x_2 - x_1 = v_1 t + \frac{1}{2}\left(\frac{v_2 - v_1}{t}\right)t^2$$

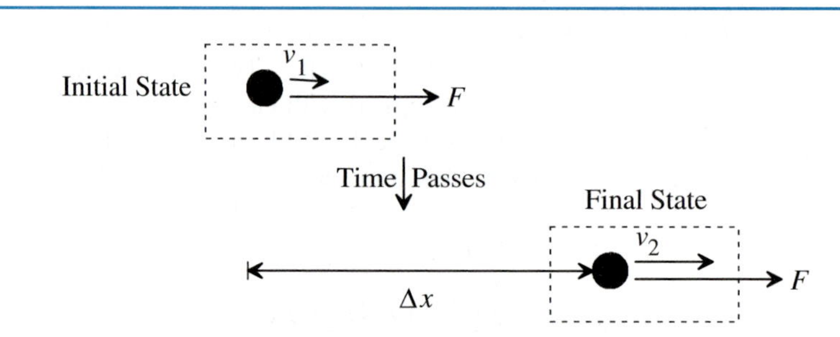

FIGURE 22.9
Kinetic energy.

$$\Delta x = v_1 t + \tfrac{1}{2}(v_2 - v_1)t$$

$$\Delta x = (v_1 + \tfrac{1}{2}v_2 - \tfrac{1}{2}v_1)t$$

$$\Delta x = \tfrac{1}{2}(v_2 + v_1)t \tag{22-45}$$

Substituting Equations 22-43 and 22-45 into Equation 22-42 gives

$$\Delta E_k = m\left(\frac{v_2 - v_1}{t}\right)\left[\frac{1}{2}(v_2 + v_1)t\right] \tag{22-46}$$

which algebraically simplifies to

$$\Delta E_k = \tfrac{1}{2}mv_2^2 - \tfrac{1}{2}mv_1^2 \tag{22-47}$$

For an object that starts from rest ($v_1 = 0$), this simplifies to

$$\Delta E_k = \tfrac{1}{2}mv_2^2 \tag{22-48}$$

EXAMPLE 22.10

Problem Statement: A 2000-lb_m automobile is stopped at a red light. When the light changes to green, the auto accelerates to 55 mph. What is the change in kinetic energy?

Solution:

$$\Delta E_k = \frac{1}{2}mv^2 = \frac{1}{2}(2000\ \text{lb}_\text{m})\left(55\frac{\text{mi}}{\text{h}} \times \frac{\text{h}}{3600\ \text{s}} \times \frac{5280\ \text{ft}}{\text{mi}}\right)^2 \frac{\text{lb}_\text{f}\text{·s}^2}{32.174\ \text{lb}_\text{m}\text{·ft}}$$

$$= 2.0 \times 10^5\ \text{ft·lb}_\text{f}$$

(Note the role of the conversion factor g_c in this calculation.)

22.3.2 Potential Energy

Potential energy is energy associated with position.

Gravitational Potential Energy.

Figure 22.10 shows a mass being lifted against the force of gravity. The input of mechanical work causes the mass to change its potential energy E_p according to Equation 22-2.

$$E_{p,\text{final}} - E_{p,\text{initial}} = \Delta E_p = \text{mechanical work input} - 0 = \int F\,dx \tag{22-49}$$

The applied force is constant, so this becomes

$$\Delta E_p = F\Delta x \tag{22-50}$$

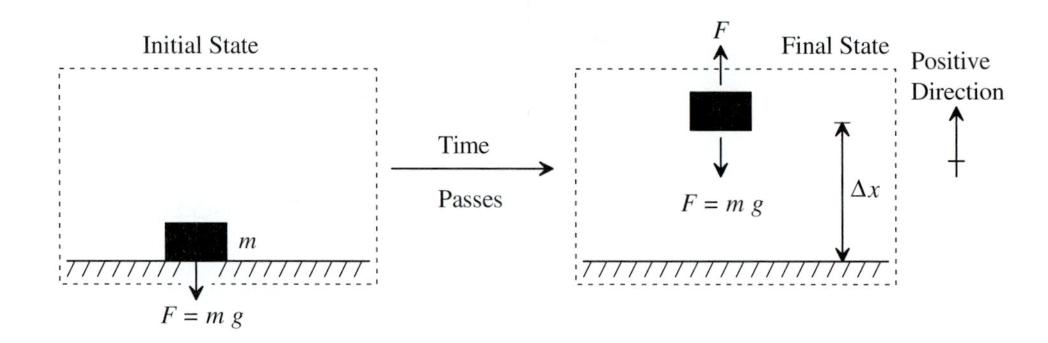

FIGURE 22.10
Potential energy due to gravity.

The force of gravity is mg, where g is the acceleration of an object in earth's gravity. (We know that mg must be the force due to gravity, because when dropped, an object will accelerate at g, assuming there are no other external forces.) This may be substituted into Equation 22-50, giving

$$\Delta E_p = mg\Delta x \tag{22-51}$$

EXAMPLE 22.11

Problem Statement: A grandfather clock is driven by a 3.00-kg mass that is raised 1.00 m. What is the potential energy of the raised mass? The mass is raised once each week. At what rate is the potential energy decreasing as the mass falls during the week?

Solution:

$$\Delta E_p = mg\Delta x = (3.00\text{ kg})\left(9.8\,\frac{\text{m}}{\text{s}^2}\right)(1.00\text{ m})\frac{1\text{ J}}{\text{kg}\cdot\text{m/s}^2} = 29.4\text{ J}$$

$$\text{Rate} = \frac{29.4\text{ J}}{(1\text{ wk})\left(\dfrac{7\text{ d}}{\text{wk}}\right)\left(\dfrac{24\text{ h}}{\text{d}}\right)\left(\dfrac{3600\text{ s}}{\text{h}}\right)}$$

$$= 4.86 \times 10^{-5}\,\frac{\text{J}}{\text{s}} \times \frac{1\text{ W}}{\text{J/s}} = 4.86 \times 10^{-5}\text{ W}$$

Spring Potential Energy.

Figure 22.11 shows a force compressing a spring to change its position from x_1 to x_2, where x_1 is the uncompressed state. The input of mechanical work changes the potential energy E_p according to Equation 22-2.

$$E_{p,\text{final}} - E_{p,\text{initial}} = \Delta E_p = \text{mechanical work input} - 0 = \int F\,dx \tag{22-52}$$

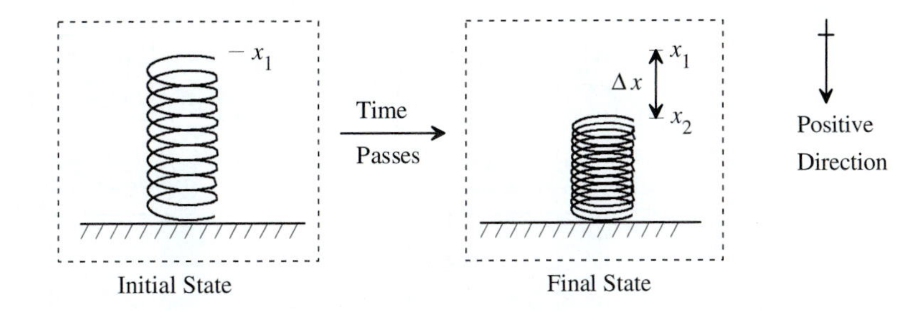

FIGURE 22.11
Spring potential energy.

In this case, the applied force is not constant. Hook's law provides the relationship between force and position,

$$F = kx \tag{22-53}$$

where k is the spring constant and $x = 0$ corresponds to the uncompressed state. Substituting Hook's law into Equation 22-52 gives

$$\begin{aligned} \Delta E_p &= \int_{x_1}^{x_2} kx\,dx \\ &= k\left[\tfrac{1}{2}x^2\right]_{x_1}^{x_2} \\ &= \tfrac{1}{2}k\left[x_2^2 - x_1^2\right] \end{aligned} \tag{22-54}$$

If the initial uncompressed position x_1 is zero, this equation becomes

$$\Delta E_p = \tfrac{1}{2}kx_2^2 \tag{22-55}$$

EXAMPLE 22.12

Problem Statement: A spring with a spring constant of 80.0 lb_f/in is compressed 4.00 in from the uncompressed position. What is the potential energy stored in the spring?

Solution:

$$\Delta E_p = \frac{1}{2}kx_2^2 = \frac{1}{2}\left(80.0\,\frac{\text{lb}_\text{f}}{\text{in}}\right)(4.00\text{ in})^2 \times \frac{\text{ft}}{12\text{ in}} = 53.3\text{ ft}\cdot\text{lb}_\text{f}$$

Electrical Potential Energy (Advanced Topic).

Figure 22.12 shows a **capacitor**—two parallel metal plates sandwiching an electrical insulator. A capacitor is a device for storing electrical charge. As electrons are transferred from one plate to the other, a voltage difference develops between the plates. The size of a capacitor is indicated by the *capacitance C*, defined as

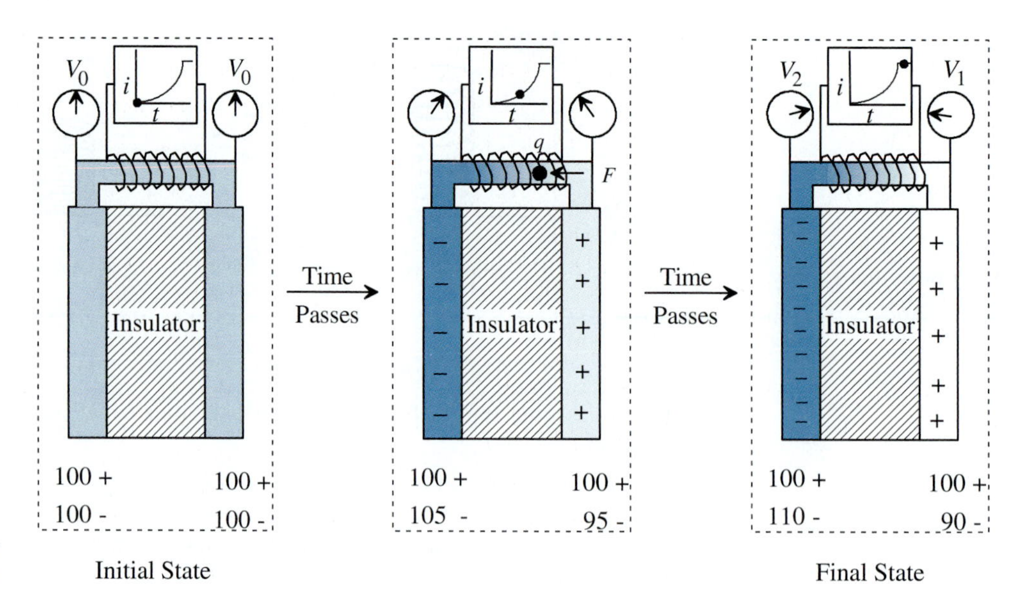

FIGURE 22.12
Electrical potential energy in a capacitor. (*Note:* This coil is conceptual.)

$$C \equiv \frac{q}{V_2 - V_1} \tag{22-56}$$

where q is the charge transferred from one plate to the other, and $V_2 - V_1$ is the voltage difference between the two plates. The capacitance is determined by the area of the plates, the spacing between the plates, and the properties of the electrical insulator. The unit of capacitance is the farad (F), that is, C/V or C^2/J.

Equation 22-56 can be easily understood by making an analogy to a water tank (Figure 22.13). The tank cross-sectional area A is analogous to capacitance C, the change in liquid volume ΔV is analogous to the charge transferred q, and the liquid height x is analogous to the voltage V; thus, we write

$$A = \frac{\Delta V}{x_2 - x_1} \tag{22-57}$$

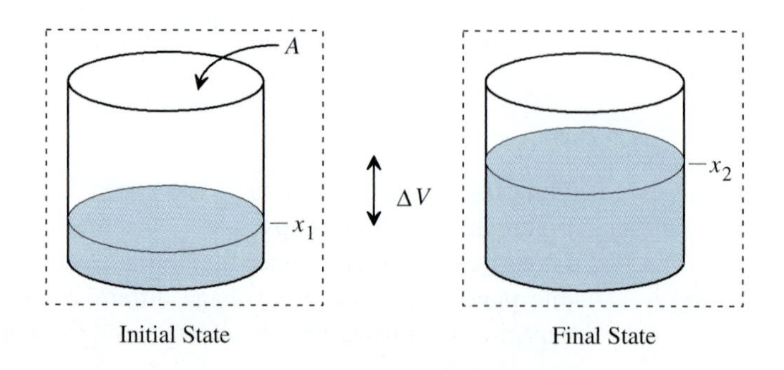

FIGURE 22.13
Analogy between a capacitor and a water tank.

In Figure 22.12, a wire connects the two plates. A generator (conceptually illustrated as a coil surrounding the wire) is able to push electrons from a low-voltage region to a high-voltage region using magnetism produced by current flowing through the coil. As the current increases, the magnetic strength also increases. Provided the magnetic strength is increasing, it will impose a force on the charge, causing it to flow from a region of low voltage to high voltage.

Initially, in each plate, the ratio of electrons to protons is identical. For example, using simple numbers, we could say there are 100 electrons and 100 protons in each plate. Then, a force is imposed upon the charge in the wire when current flows through the coil. A charge q (say 10 electrons) is transferred from the positive plate to the negative plate. The positive plate has more protons (100) than electrons (90), and the negative plate has more electrons (110) than protons (100). The electrons are more tightly packed on the negative plate. Because electrons repel each other, it takes energy provided by the generator to push them together. The voltage change indicates how many joules of energy were required to transfer a coulomb of electrons from the positive plate to the negative plate.

As previously derived (Equation 22-23), the amount of electrical work required to transfer electrons from a region of low voltage to a region of high voltage is

$$\text{Electrical work} = W_{\text{elec}} = q(V_2 - V_1) \tag{22-58}$$

where q is the charge transferred (coulombs) and V is voltage (joules per coulomb). Initially, there is no voltage difference between the two plates, so very little electrical work will be required to move a charge q. However, as charge builds up on the capacitor, the voltage difference increases. Therefore, the later phases of the charge transfer will require more work. This situation is best handled using calculus.

At any instant of time, a differential amount of charge dq is transferred from one plate to the other, and a differential amount of work dW_{elec} is needed to push the charge against a voltage difference $(V_2 - V_1)$.

$$dW_{\text{elec}} = (V_2 - V_1)\, dq \tag{22-59}$$

The voltage difference changes as charge is transferred; the capacitance given in Equation 22-56 tells us how. It may be substituted into Equation 22-59, yielding

$$dW_{\text{elec}} = \left(\frac{q}{C}\right) dq \tag{22-60}$$

This may be integrated, giving

$$\int_0^{W_{\text{elec}}} dW_{\text{elec}} = \frac{1}{C}\int_0^q q\, dq$$

$$[W_{\text{elec}}]_0^{W_{\text{elec}}} = \frac{1}{C}\left[\frac{1}{2}q^2\right]_0^q$$

$$[W_{\text{elec}} - 0] = \frac{1}{2C}[q^2 - 0]$$

$$W_{\text{elec}} = \frac{q^2}{2C} \tag{22-61}$$

Again substituting Equation 22-56,

$$W_{\text{elec}} = \tfrac{1}{2}\,q(V_2 - V_1) \tag{22-62}$$

By accounting for the energy flow into the capacitor using Equation 22-2, we can calculate its change in potential energy:

$$E_{p,\text{final}} - E_{p,\text{initial}} = \Delta E_p = \text{electrical work input} - 0 = \tfrac{1}{2}\,q(V_2 - V_1) \tag{22-63}$$

EXAMPLE 22.13

Problem Statement: How much energy is stored in a 10.0-μF capacitor that has a 5.00-V difference across the plates?

Solution:

$$\Delta E_p = \tfrac{1}{2}\,q(V_2 - V_1) = \tfrac{1}{2}\,[C(V_2 - V_1)](V_2 - V_1) = \tfrac{1}{2}\,C(V_2 - V_1)^2$$

$$= \tfrac{1}{2}(10.0 \times 10^{-6}\,\text{F})\left(\frac{\text{C}^2/\text{J}}{\text{F}}\right)(5.00\,\text{V})^2\left(\frac{\text{J/C}}{\text{V}}\right)^2 = 1.25 \times 10^{-4}\,\text{J}$$

Magnetic Potential Energy (Advanced Topic).

Figure 22.14 shows an electrical circuit containing an **inductor,** that is, a coil of wire. When electric current flows through a coil, a magnet is generated. (You can see this for yourself by making a coil from ordinary wire and connecting each end to the terminals of

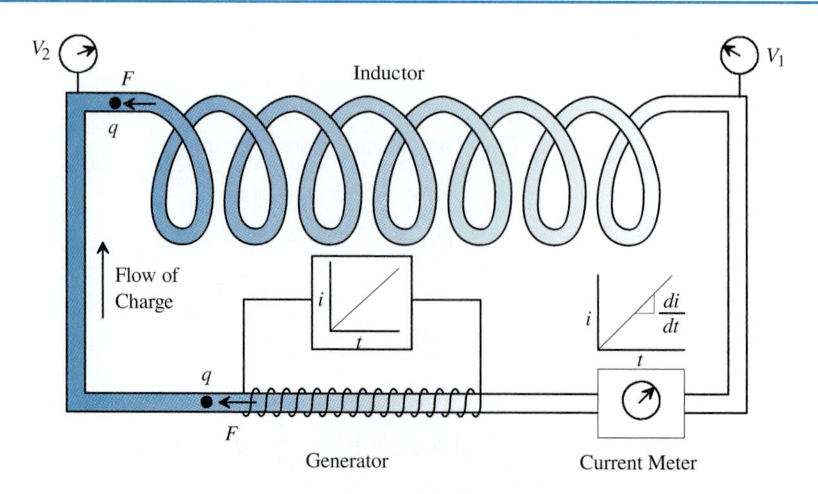

FIGURE 22.14
Magnetic potential energy stored in an inductor. (*Note:* The generator coil is conceptual.)

a battery.) The circuit in Figure 22.14 also has a current meter that indicates the charge flow rate in the wire, and an electric generator, conceptually indicated by the coil encircling the wire. (Recall that a generator is a device that uses changing magnetic strengths to push charge through a wire. Here, the changing magnetic strength is produced by flowing an increasingly large current through a coil. In most generators, the changing magnetic strength is produced by mechanically rotating the armature near magnets.)

As current flows through the generator, the magnetic strength increases, which allows it to push a charge q from a low-voltage region to a high-voltage region. As charge flows through the inductor, another magnet is formed that pushes the charge in the opposite direction. Provided the current in the wire is continuously increasing (as indicated by the current meter), a voltage difference will be produced in this circuit. The voltage difference in the circuit ($V_2 - V_1$) is proportional to the increasing current (di/dt), or

$$V_2 - V_1 = L\frac{di}{dt} \tag{22-64}$$

where the proportionality constant L is called the **inductance.** The inductance is determined by geometric factors, such as the volume of the coil and its number of turns. The unit of inductance is the henry (H), defined as V·s/A or $J \cdot s^2/C^2$.

This inductor electrical circuit resists change. Figure 22.15 shows an analogous hydraulic circuit that resists change. The pump is analogous to the generator and the inductor is analogous to a turbine coupled to a flywheel. Provided the pump and turbine are spinning at the same rate, the two pressures P_1 and P_2 will be identical. However, if the pump speeds up, causing the fluid flow rate to increase (analogous to di/dt), its discharge pressure must increase to make the turbine spin faster. The greater the mass of the flywheel, the more the turbine resists the change, and hence, the pressure difference must be greater to get the turbine speed to match the pump speed.

As previously derived (Equation 22-23), the amount of electrical work required to push charge q against a voltage difference is

$$\text{Electrical work} = W_{\text{elec}} = q(V_2 - V_1) \tag{22-65}$$

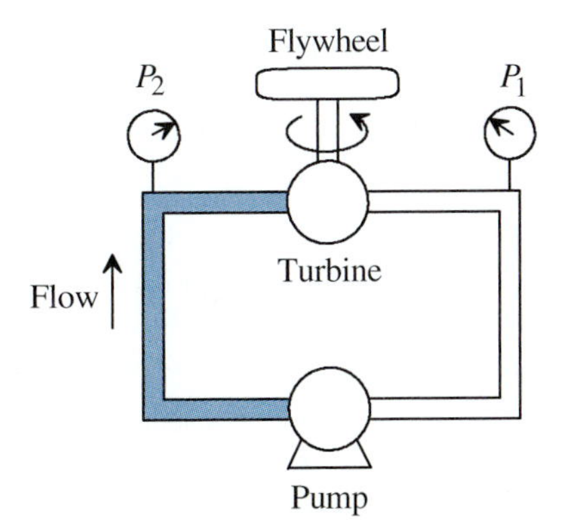

FIGURE 22.15
Analogous hydraulic circuit to the inductor circuit shown in Figure 22.14.

Equation 22-64 may be substituted for the voltage difference, giving

$$W_{\text{elec}} = qL\frac{di}{dt} \tag{22-66}$$

If a differential amount of charge dq flows against the voltage difference, a differential amount of electrical work dW_{elec} is required to push it:

$$dW_{\text{elec}} = dq\, L\frac{di}{dt} \tag{22-67}$$

Current i is defined as the charge flow rate:

$$i \equiv \frac{dq}{dt} \tag{22-68}$$

By solving this equation for dq, it can be substituted into Equation 22-67:

$$dW_{\text{elec}} = (i\, dt)L\frac{di}{dt} \tag{22-69}$$

The dt's cancel, giving

$$dW_{\text{elec}} = iL\, di \tag{22-70}$$

This equation may be integrated to show how much work is required to build up the inductor magnet assuming zero initial current.

$$\int_0^{W_{\text{elec}}} dW_{\text{elec}} = L\int_0^i i\, di$$

$$[W_{\text{elec}}]_0^{W_{\text{elec}}} = L[\tfrac{1}{2}i^2]_0^i$$

$$[W_{\text{elec}} - 0] = \tfrac{1}{2}L[i^2 - 0]$$

$$W_{\text{elec}} = \tfrac{1}{2}Li^2 \tag{22-71}$$

By accounting for the energy flow into the inductor using Equation 22-2, we can calculate its change in potential energy as

$$E_{p,\text{final}} - E_{p,\text{initial}} = \Delta E_p = \text{electrical work input} - 0 = \tfrac{1}{2}Li^2 \tag{22-72}$$

EXAMPLE 22.14

Problem Statement: How much energy is stored in a 5.00-H inductor with 10.0 A of current flowing through it?

Solution:

$$\Delta E_p = \frac{1}{2}Li^2 = \frac{1}{2}(5.00\text{ H})\left(\frac{\text{J}\cdot\text{s}^2/\text{C}^2}{\text{H}}\right)(10.0\text{ A})\left(\frac{\text{C/s}}{\text{A}}\right) = 250\text{ J}$$

22.3.3 Internal Energy/Enthalpy

Internal energy results from translational, rotational, and vibrational kinetic energies and the electronic potential energy of atoms or molecules (Figure 22.16). Although we could conceivably classify these energies as the previously discussed kinetic and potential energies, it is not practical (or even possible) to separate them because it is difficult to measure individual atoms or molecules. Therefore, we lump these energies together and call it *internal energy.*

Sensible Energy.

If path energy (heat or work) is added to a collection of atoms or molecules, the temperature increases (provided there is no phase change). If the atoms (molecules) are confined to a constant-volume container (Figure 22.17), the added energy changes the internal energy U. The input of path energy changes the internal energy according to Equation 22-2:

$$\begin{aligned} U_{\text{final}} - U_{\text{initial}} &= \Delta U \\ &= \text{path energy input} - \text{path energy output} \\ &= \text{net path energy input} \end{aligned} \tag{22-73}$$

The internal energy change ΔU can be determined from the temperature change,

$$\Delta U = mC_v(T_2 - T_1) = \text{net path energy input} \tag{22-74}$$

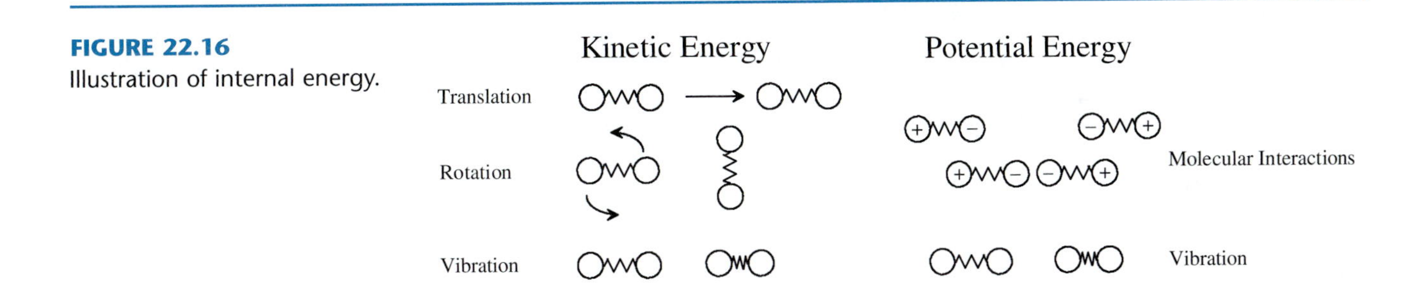

FIGURE 22.16
Illustration of internal energy.

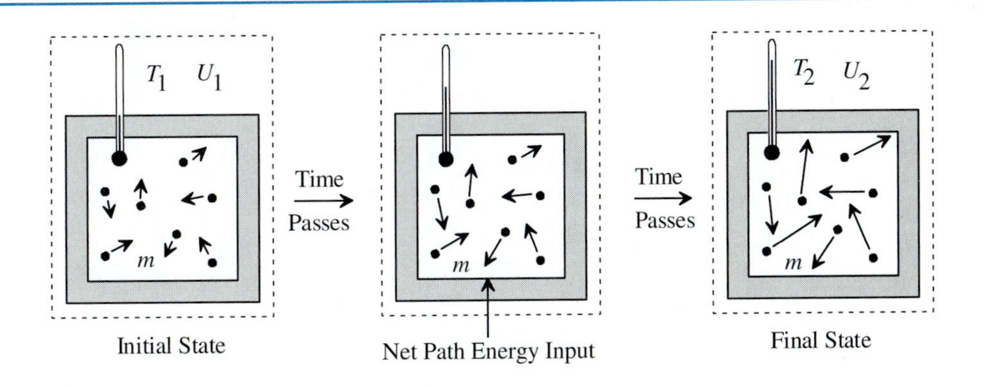

FIGURE 22.17
Constant-volume sensible energy change. (*Note:* Only translational energy is shown.)

where m is the mass of atoms or molecules and C_v is termed the **constant-volume heat capacity.** (*Note*: Although this is the standard term for C_v, it is very misleading. It incorrectly implies that heat is the only path energy that can change internal energy. C_v would be better termed *internal energy capacity,* but this is not standardly used.) Equation 22-74 is valid provided C_v is constant over the temperature range from T_1 to T_2.

When path energy—heat or work—is added to a collection of atoms or molecules in a constant-pressure container (Figure 22.18), the temperature increases (assuming no phase change). The path energy input changes the internal energy U according to Equation 22-2:

$$U_{\text{final}} - U_{\text{initial}} = \Delta U = \text{path energy input} - \text{path energy output} \tag{22-75}$$

In this case, there is an output of path energy in the form of work, because the piston pushes with a force F to move the piston mass from position x_1 to x_2:

$$\Delta U = \text{path energy input} - \int F\,dx \tag{22-76}$$

Because the force is constant, we can remove F from the integral, giving

$$\begin{aligned} \Delta U &= \text{path energy input} - F \int dx \\ &= \text{path energy input} - F\Delta x \end{aligned} \tag{22-77}$$

The upward force is the pressure times the piston frontal area:

$$F = PA \tag{22-78}$$

As the piston of cross-sectional area A moves a distance Δx, it sweeps a volume ΔV:

$$\Delta V = A\Delta x$$

$$\Delta x = \frac{\Delta V}{A} \tag{22-79}$$

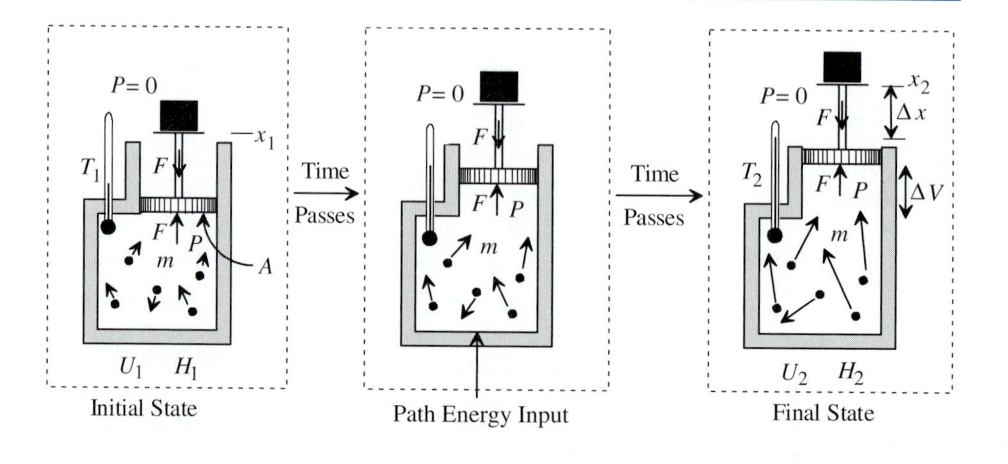

FIGURE 22.18
Constant-pressure sensible energy change. (*Note:* Only translational internal energy is shown.)

Substituting Equations 22-78 and 22-79 into Equation 22-77 gives

$$\Delta U = \text{path energy input} - (PA)\left(\frac{\Delta V}{A}\right)$$

$$\Delta U = \text{path energy input} - P\Delta V \quad (22\text{-}80)$$

Because pressure P and volume V are state properties of the system, we may move them to the left side of the equation:

$$\Delta U + P\Delta V = \text{path energy input} \quad (22\text{-}81)$$

The left side appears frequently in engineering problems. Because it is an algebraic combination of state quantities (see Rule 8 in Chapter 17, "Accounting"), it may be defined as a new state quantity called **enthalpy** H:

$$\Delta H = \Delta U + P\Delta V = \text{path energy input} \quad (22\text{-}82)$$

(Equation 22-82 is a valid relationship for ΔH provided pressure is constant, which is true in this case. For more explanation, see the footnote on page 319.)

The enthalpy change ΔH can be determined from the temperature change,

$$\Delta H = mC_p(T_2 - T_1) = \text{path energy input} \quad (22\text{-}83)$$

where m is the mass of atoms or molecules and C_p is termed the **constant-pressure heat capacity**. (*Note:* For previously discussed reasons, it should be called the *enthalpy capacity,* but this is not standardly used.) Equation 22-83 is valid provided C_p is constant over the temperature range from T_1 to T_2.

Because it is very difficult to measure heat capacities in a constant-volume apparatus, generally constant-pressure heat capacities are reported (Table 22.2).

EXAMPLE 22.15

Problem Statement: Liquid water (2.00 lb_m) is heated from 40°F to 50°F at constant pressure. How much heat is required?

Solution:

$$\text{Heat input} = \Delta H = mC_p(T_2 - T_1)$$

$$= (2.00\ lb_m)\left(1.00\ \frac{Btu}{lb_m \cdot °F}\right)(50°F - 40°F) = 20\ Btu$$

Phase Change.

When path energy is added to a material, it may change phase. Figure 22.19 shows liquid changing to vapor (**vaporization**) at constant pressure. As path energy is input to the liquid, the atoms (molecules) separate and move more rapidly, thus increasing the internal energy. In addition, the volume changes by ΔV. Therefore, in accounting for the energy in this constant-pressure process, enthalpy is the appropriate state quantity, and

TABLE 22.2
Constant-pressure heat capacities (Reference)

Solids	$C_p\left(\frac{\text{Btu}}{\text{lb}_m\cdot°\text{F}}\right)$ or $\left(\frac{\text{cal}}{\text{g}\cdot°\text{C}}\right)$	Liquids	$C_p\left(\frac{\text{Btu}}{\text{lb}_m\cdot°\text{F}}\right)$ or $\left(\frac{\text{cal}}{\text{g}\cdot°\text{C}}\right)$
Asbestos	0.20	Acetic acid	0.51
Brick	0.22	Acetone	0.51
Chalk	0.215	Ethanol	0.58
Coal	0.30	Gasoline	0.50
Concrete	0.156	Glycerol	0.58
Cork	0.485	Kerosene	0.50
Glass	0.199	Naphthalene	0.31
Granite	0.195	Machine oil	0.40
Soil	0.44	Mercury	0.03
Ice (32°F)	0.487	Sulfuric acid	0.33
Sand	0.195	Sea water	0.94
Wood (oak)	0.57	Water	1.00
Gases (1 atm)	$C_p\left(\frac{\text{Btu}}{\text{lb}_m\cdot°\text{F}}\right)$ or $\left(\frac{\text{cal}}{\text{g}\cdot°\text{C}}\right)$	**Gases (1 atm)**	$C_p\left(\frac{\text{Btu}}{\text{lb}_m\cdot°\text{F}}\right)$ or $\left(\frac{\text{cal}}{\text{g}\cdot°\text{C}}\right)$
Air (80°F)	0.240	He (80°F)	1.24
Cl_2 (80°F)	0.115	N_2 (80°F)	0.249
CO_2 (80°F)	0.204	O_2 (80°F)	0.220
CO (80°F)	0.249	SO_2 (100°F)	0.149
H_2 (80°F)	3.42	Steam (212°F)	0.493

Adapted from: J. R. Welty, C. E. Wicks, and R. E. Wilson, *Fundamentals of Momentum, Heat, and Mass Transfer,* 2nd ed. (New York: Wiley, 1976); T. Baumeister, E. A. Avallone, and T. Baumeister III, *Marks' Standard Handbook for Mechanical Engineers*, 8th ed. (New York: McGraw-Hill, 1978).

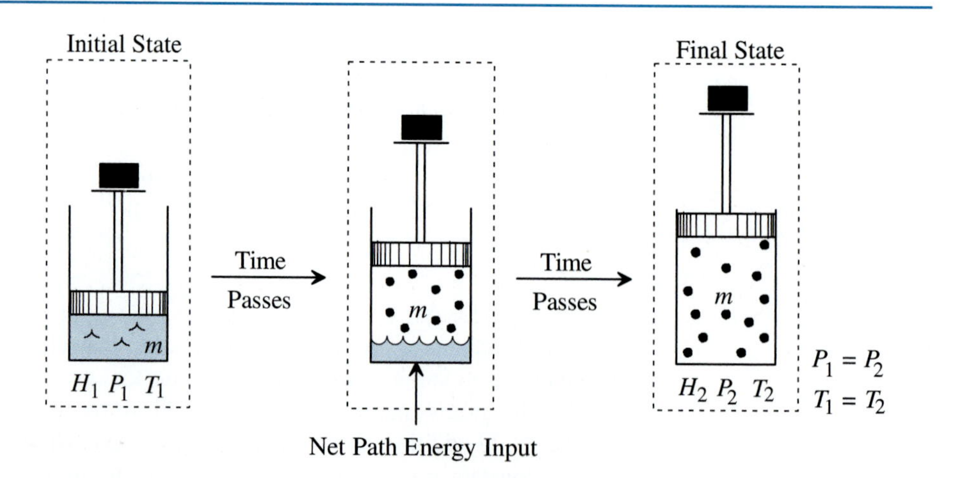

FIGURE 22.19
Vaporization of a liquid.

$$H_2 - H_1 = \Delta H_{vap} = \text{net path energy input} \tag{22-84}$$

where ΔH_{vap} is often called the **heat of vaporization** (although it should properly be called the *enthalpy of vaporization*).

If the pressure is constant during the phase change, the temperature does not change even though path energy (heat or work) is added. Because the effect of the added heat is hidden (latent), the enthalpy change is sometimes called the *latent heat of vaporization.* In contrast, no-phase-change systems increase temperature when path energy (heat or work) is added, which is "sensible."

The *specific enthalpy of vaporization* is defined as

$$\Delta \hat{H}_{vap} = \frac{\Delta H_{vap}}{m} \tag{22-85}$$

A similar definition is for the *specific enthalpy of fusion:*

$$\Delta \hat{H}_{fus} = \frac{\Delta H_{fus}}{m} \tag{22-86}$$

where the phase change is from a solid to a liquid. The *specific enthalpy of sublimation* is

$$\Delta \hat{H}_{sub} = \frac{\Delta H_{sub}}{m} \tag{22-87}$$

where the phase change is from a solid to a vapor. Table 22.3 lists some phase-change enthalpies.

EXAMPLE 22.16

Problem Statement: How much heat is required to vaporize 2.0000 kg of liquid water?

Solution:

$$\text{Heat input} = \Delta H_{vap} = m\Delta \hat{H}_{vap} = (2.0000 \text{ kg})\left(2256.7 \frac{\text{kJ}}{\text{kg}}\right) = 4513.4 \text{ kJ}$$

Chemical Reaction.

A chemical reaction rearranges the atoms to form new molecules. For example, the reaction

$$CH_4 + 2\,O_2 \rightarrow CO_2 + 2\,H_2O \tag{22-88}$$

involves carbon, hydrogen, and oxygen atoms. The hydrogen is joined to the carbon originally, but after the reaction, the hydrogen is joined to oxygen.

Figure 22.20 shows a chemical reaction in which methane and oxygen reactants are placed in a constant-pressure reactor. A spark initiates the reaction to form CO_2 and H_2O. This is an exothermic reaction, so the products are very hot. Because no energy has been withdrawn from the system, the internal energy of the products is the same as the original reactants. The original atomic configuration had a high potential (electronic)

TABLE 22.3
Phase-change enthalpies (Reference)

Compound	Melting Temp (°C)†	Latent Heat of Fusion (kJ/kg)	Boiling Temp (°C)†	Latent Heat of Vaporization (kJ/kg)
Acetic acid	16.6	201.3	118.2	406.2
Acetone	−95.0	98.0	56.0	520
Ammonia	−77.8	331.9	−33.43	1371
Benzene	5.53	125.9	80.10	393.9
Bromine	−7.4	67.6	58.6	194
n-Butane	−138.3	80.20	−0.6	383.77
Carbon disulfide	−112.1	57.7	46.25	352
Carbon monoxide	−205.1	29.9	−191.5	215.7
Carbon tetrachloride	−22.9	16.3	76.7	195
Chlorine	−101.00	90.34	−34.06	288
Copper	1083	204.8	2595	4793
Diethyl ether	−116.3	98.5	34.6	351.5
Ethane	−183.3	95.08	−88.6	489.5
Ethanol	−114.6	109.0	78.5	837.4
Ethylene	−169.2	119.4	−103.7	482.7
Ethylene glycol	−13	180.9	197.2	917
Helium	−269.7	5	−268.9	21
Hydrogen	−259.19	59	−252.76	448
Iron	1535	270	2800	6338
Lead	327.4	24.6	1750	868.2
Magnesium	650	378	1120	5419
Methane	−182.5	59	−161.5	509.9
Methanol	−97.9	98.85	64.7	1100
Nitrogen	−210.0	25.7	−195.8	199.0
Oxygen	−218.75	13.9	−182.97	213
Propane	−187.69	79.8	−42.07	425.7
Sulfur (rhombic)	113	39.14	444.6	326
Water	0.00	333.56	100.00	2256.7

† $P = 1$ atm

Adapted from: R. M. Felder and R. W. Rousseau, *Elementary Principles of Chemical Processes,* 2nd ed. (New York: Wiley, 1986).

energy, which was converted into kinetic (translational, rotational, and vibrational) energy of the products. Path energy is then withdrawn from the hot products to cool them to the initial temperature T_1.

Because the volume changes, the appropriate state quantity in the energy accounting is enthalpy:

$$H_2 - H_1 = \Delta H_{\text{reaction}} = \text{path energy input} - \text{path energy output} \tag{22-89}$$

Because there is no path energy input, this equation simplifies to

$$\Delta H_{\text{reaction}} = -\text{path energy output} \tag{22-90}$$

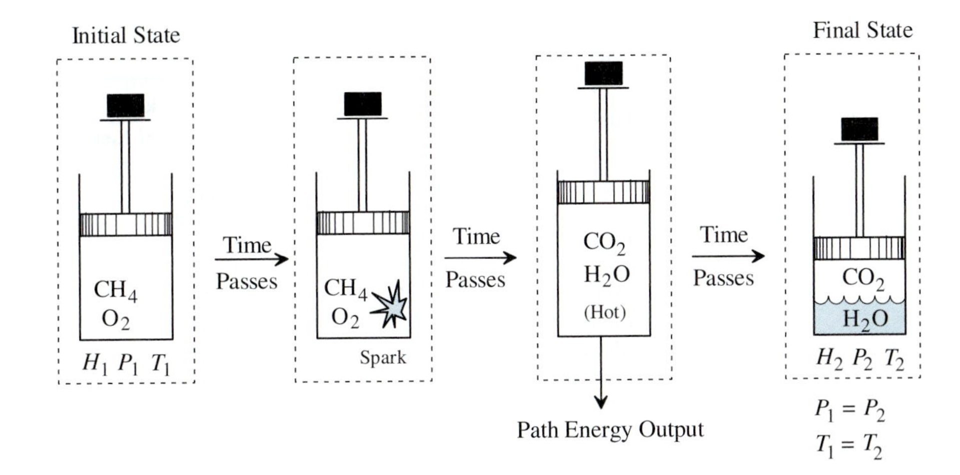

FIGURE 22.20
Chemical reaction enthalpy change.

In engineering, combustion reactions (i.e., reactions between fuel and oxygen) are common. The **specific heat of combustion** (more properly called *specific enthalpy of combustion*) is

$$\Delta\hat{H}_{\text{combustion}} = \frac{\Delta H_{\text{reaction}}}{n_{\text{fuel}}} \tag{22-91}$$

where n_{fuel} is the moles or mass of fuel that actually react. Table 22.4 shows some specific heats of combustion for some common fuels.

EXAMPLE 22.17

Problem Statement: How much heat can be removed from a furnace by completely combusting 2.00 mol of methane with a stoichiometric amount of oxygen? (*Note*: The products are returned to the original temperature of the reactants and water product is present as a liquid.)

Solution:

$$\text{Path heat output} = -n\,\Delta\hat{H}_{\text{combustion}} = -(2.00\text{ mol})\left(-890.36\,\frac{\text{kJ}}{\text{mol}}\right) = 1780\text{ kJ}$$

22.4 TOTAL ENERGY CONSERVATION

So far in our discussion of energy, we have focused on many specific types of energy (mechanical work, heat, kinetic energy, potential energy, internal energy, etc.). As we focused upon each of these energies, we used the universal accounting equation

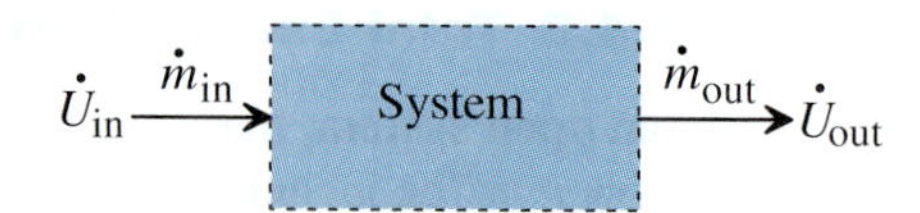

FIGURE 22.23
The flow of internal energy across the system boundary.

where $\dot{m}$ is the mass flow rate (kg/s) and v is the velocity (m/s).

Internal Energy Flow.

Figure 22.23 shows internal energy flowing across the system boundary. The internal energy flow rate $\dot{U}$ (J/s) is

$$\dot{U}_{in} = \dot{m}_{in}\hat{U}_{in} \qquad \dot{U}_{out} = \dot{m}_{out}\hat{U}_{out} \tag{22-96}$$

where $\dot{m}$ is the mass flow rate (kg/s) and $\hat{U}$ is the specific internal energy (J/kg).

Fluid PV Flow Work.

Fluids include liquids, gases, vapors, and supercritical fluids. The flow of fluids is extremely important in engineering. For example, it is involved with liquid flow through pipelines, airflow over automobiles and airplanes, oil flow out of a reservoir, and water flow past a boat hull.

When fluid flows into (or out of) a system, the fluid pressure and fluid volume can change. This allows for **PV flow work** to occur (Figure 22.24). The *PV* flow work input is

$$W_{flow,in} = F_{in}\Delta x_{in} = (P_{in}A_{in})\Delta x_{in} = P_{in}(A_{in}\Delta x_{in}) = P_{in}V_{in} \tag{22-97a}$$

and the *PV* flow work output is

$$W_{flow,out} = F_{out}\Delta x_{out} = (P_{out}A_{out})\Delta x_{out} = P_{out}(A_{out}\Delta x_{out}) = P_{out}V_{out} \tag{22-97b}$$

When the mass flow rate is known, the *PV* flow work $\dot{W}_{flow}$ (J/s) can be calculated

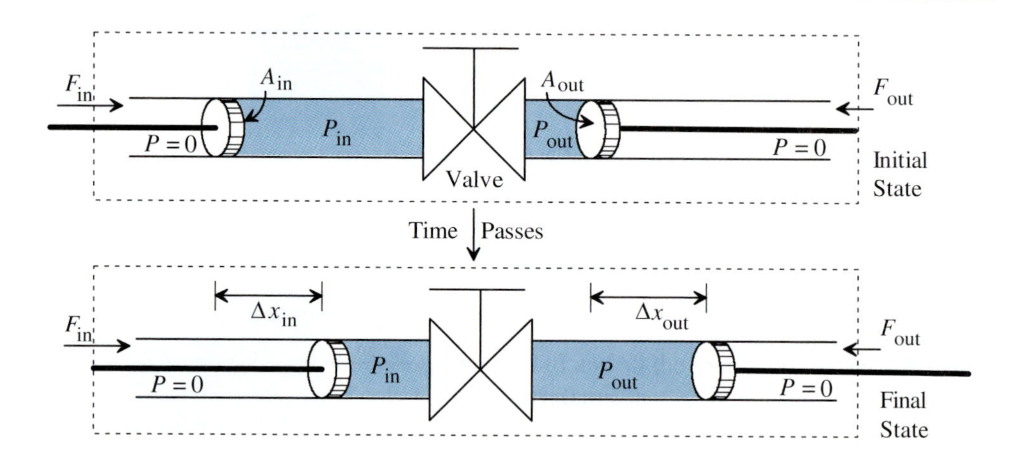

FIGURE 22.24
Flow work.

$$\dot{W}_{\text{flow,in}} = \dot{m}_{\text{in}} P_{\text{in}} \hat{V}_{\text{in}} \qquad \dot{W}_{\text{flow,out}} = \dot{m}_{\text{out}} P_{\text{out}} \hat{V}_{\text{out}} \tag{22-98}$$

where $\dot{m}$ is the mass flow rate (kg/s), P is the pressure (N/m^2), and $\hat{V}$ is the specific volume (m^3/kg).

Enthalpy Flow.

Enthalpy H is defined as

$$H \equiv U + PV \tag{22-99}$$

and specific enthalpy is defined as

$$\hat{H} \equiv \hat{U} + P\hat{V} \tag{22-100}$$

Figure 22.25 shows enthalpy flow across the system boundary. The enthalpy flow rate $\dot{H}$ (J/s) is

$$\dot{H}_{\text{in}} = \dot{m}_{\text{in}} \hat{H}_{\text{in}} = \dot{m}_{\text{in}} (\hat{U}_{\text{in}} + P_{\text{in}} \hat{V}_{\text{in}}) \tag{22-101a}$$

$$\dot{H}_{\text{out}} = \dot{m}_{\text{out}} \hat{H}_{\text{out}} = m_{\text{out}} (\hat{U}_{\text{out}} + P_{\text{out}} \hat{V}_{\text{out}}) \tag{22-101b}$$

where $\dot{m}$ is the mass flow rate (kg/s) and $\hat{H}$ is the specific enthalpy (J/kg). Enthalpy flow accounts for both internal energy and PV flow work crossing the system boundary. Although a separate accounting could be made for internal energy and PV flow work, these two appear together so often in engineering calculations that many handbooks report enthalpy in preference to internal energy.

22.5 FLUID FLOW (ADVANCED TOPIC)

Figure 22.26 shows a system through which fluid flows. The fluid enters with mass flow rate $\dot{m}_{\text{in}}$ and with the state quantities velocity v_{in}, height z_{in}, density ρ_{in}, pressure P_{in}, and specific internal energy $\hat{U}_{\text{in}}$. The fluid exits with mass flow rate $\dot{m}_{\text{out}}$ and with the state quantities v_{out}, z_{out}, ρ_{out}, P_{out}, and $\hat{U}_{\text{out}}$. For many systems of practical interest, mass does not accumulate in the system ($\dot{m}_{\text{in}} = \dot{m}_{\text{out}} = \dot{m}$). Also, the inlet and outlet densities are often the same ($\rho_{\text{in}} = \rho_{\text{out}} = \rho$), which is true for all liquids and also for gases, vapors, and supercritical fluids, provided $P_{\text{in}} \cong P_{\text{out}}$.

As previously described, the total energy accounting equation for an open system is

$$\Delta E_k + \Delta E_p + \Delta U = W_{\text{in}} - W_{\text{out}} + Q_{\text{in}} - Q_{\text{out}} + M_{\text{in}} - M_{\text{out}} \tag{22-93}$$

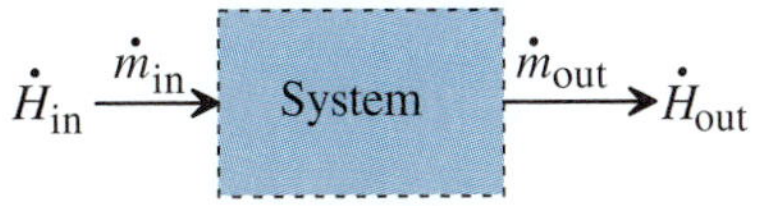

FIGURE 22.25
Enthalpy flow across the system boundary.

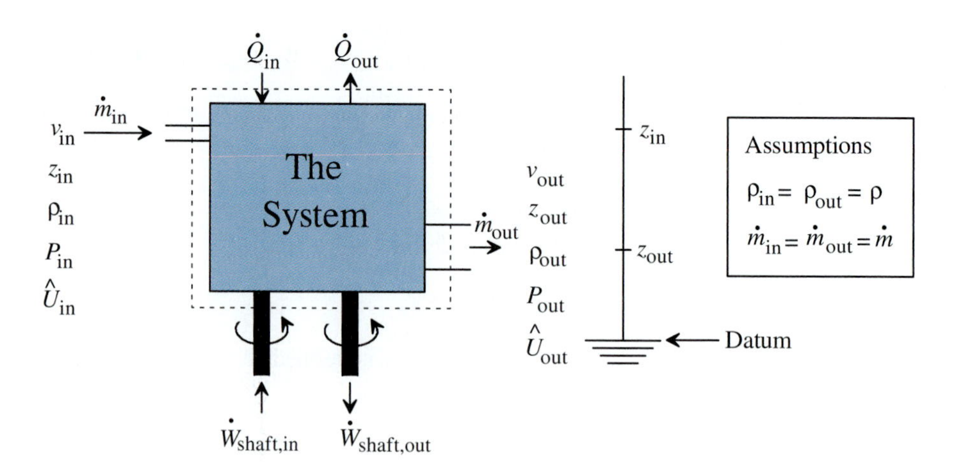

FIGURE 22.26
Fluid flow through a system.

If we assume the system is at steady state ($\Delta E_k = \Delta E_p = \Delta U = 0$) and consider only shaft work, this equation becomes

$$0 = W_{shaft,in} - W_{shaft,out} + Q_{in} - Q_{out} + M_{in} - M_{out} \tag{22-102}$$

Gravitational potential energy, kinetic energy, internal energy, and PV flow work cross the system boundary with the flowing mass. Thus,

$$0 = \dot{W}_{shaft,in} - \dot{W}_{shaft,out} + \dot{Q}_{in} - \dot{Q}_{out} + \dot{E}_{p,in} - \dot{E}_{p,out} + \dot{E}_{k,in} - \dot{E}_{k,out}$$
$$+ \dot{U}_{in} - \dot{U}_{out} + \dot{W}_{flow,in} - \dot{W}_{flow,out} \tag{22-103}$$

where the raised dots indicate a flow of energy.

Substituting Equations 22-94 to 22-96, and 22-98 gives

$$0 = \dot{W}_{shaft,in} - \dot{W}_{shaft,out} + \dot{Q}_{in} - \dot{Q}_{out}$$
$$+ \dot{m}_{in} g z_{in} - \dot{m}_{out} g z_{out} + \tfrac{1}{2}\dot{m}_{in} v_{in}^2 - \tfrac{1}{2}\dot{m}_{out} v_{out}^2$$
$$+ \dot{m}_{in}\hat{U}_{in} - \dot{m}_{out}\hat{U}_{out} + \dot{m}_{in} P_{in}\hat{V}_{in} - \dot{m}_{out} P_{out}\hat{V}_{out} \tag{22-104}$$

Because we are assuming that no mass accumulates in the system ($\dot{m}_{in} = \dot{m}_{out} = \dot{m}$), this simplifies to

$$0 = \dot{W}_{shaft,in} - \dot{W}_{shaft,out} + \dot{Q}_{in} - \dot{Q}_{out} + \dot{m}[g(z_{in} - z_{out})$$
$$+ \tfrac{1}{2}(v_{in}^2 - v_{out}^2) + (\hat{U}_{in} - \hat{U}_{out}) + (P_{in}\hat{V}_{in} - P_{out}\hat{V}_{out})] \tag{22-105}$$

The specific volume $\hat{V}$ is the inverse of density ρ; therefore,

$$0 = \dot{W}_{\text{shaft,in}} - \dot{W}_{\text{shaft,out}} + \dot{Q}_{\text{in}} - \dot{Q}_{\text{out}} + \dot{m}\left[g(z_{\text{in}} - z_{\text{out}}) + \frac{1}{2}(v_{\text{in}}^2 - v_{\text{out}}^2) + (\hat{U}_{\text{in}} - \hat{U}_{\text{out}}) + \left(\frac{P_{\text{in}}}{\rho_{\text{in}}} - \frac{P_{\text{out}}}{\rho_{\text{out}}}\right)\right] \quad (22\text{-}106)$$

Because we are assuming that the density does not change ($\rho_{\text{in}} = \rho_{\text{out}} = \rho$), this becomes

$$0 = \dot{W}_{\text{shaft,in}} - \dot{W}_{\text{shaft,out}} + \dot{Q}_{\text{in}} - \dot{Q}_{\text{out}} + \dot{m}\left[g(z_{\text{in}} - z_{\text{out}}) + \tfrac{1}{2}(v_{\text{in}}^2 - v_{\text{out}}^2) + (\hat{U}_{\text{in}} - \hat{U}_{\text{out}}) + \left(\frac{P_{\text{in}} - P_{\text{out}}}{\rho}\right)\right] \quad (22\text{-}107)$$

Algebraic rearrangement gives

$$\frac{P_{\text{out}} - P_{\text{in}}}{\rho} + \frac{1}{2}(v_{\text{out}}^2 - v_{\text{in}}^2) + g(z_{\text{out}} - z_{\text{in}}) + \left[(\hat{U}_{\text{out}} - \hat{U}_{\text{in}}) + \frac{\dot{Q}_{\text{out}} - \dot{Q}_{\text{in}}}{\dot{m}}\right] = \frac{\dot{W}_{\text{shaft,in}} - \dot{W}_{\text{shaft,out}}}{\dot{m}} \quad (22\text{-}108)$$

The terms in the square brackets are zero if the fluid has negligible viscosity (i.e., no viscous heating) and if the system is well insulated. Using these assumptions gives

$$\frac{P_{\text{out}} - P_{\text{in}}}{\rho} + \frac{1}{2}(v_{\text{out}}^2 - v_{\text{in}}^2) + g(z_{\text{out}} - z_{\text{in}}) = \frac{\dot{W}_{\text{shaft,in}} - \dot{W}_{\text{shaft,out}}}{\dot{m}} \quad (22\text{-}109)$$

If there is no shaft work being done on the system (e.g., pump) or by the system (e.g., turbine), then the right term is zero and we have the **Bernoulli equation**

$$\frac{P_{\text{out}} - P_{\text{in}}}{\rho} + \frac{1}{2}(v_{\text{out}}^2 - v_{\text{in}}^2) + g(z_{\text{out}} - z_{\text{in}}) = 0 \quad (22\text{-}110a)$$

$$\frac{P_{\text{out}} - P_{\text{in}}}{\rho g} + \frac{1}{2g}(v_{\text{out}}^2 - v_{\text{in}}^2) + (z_{\text{out}} - z_{\text{in}}) = 0 \quad (22\text{-}110b)$$

In the second version of the Bernoulli equation, the dimension of each term is length. Therefore, pressure differences, velocities, and heights can all be expressed as length or **head.** The Bernoulli equation is one of the most famous in engineering. It has many uses, such as describing the lift of an airplane wing.

EXAMPLE 22.18

Problem Statement: Assuming the pump is 100% efficient and there is no fluid friction, how much shaft work is required to pump 100 lb_m/s of water from an open reservoir in a valley to an open reservoir at the top of a 3000-ft mountain? The pipe diameter is constant. The pump is capable of generating a 3000-ft head.

Solution: Because the reservoirs are open to atmospheric pressure, $P_{\text{in}} = P_{\text{out}}$. Because the pipe diameter is constant, $v_{\text{in}} = v_{\text{out}}$. Equation 22-109 simplifies to

$$0 + 0 + g(z_{\text{out}} - z_{\text{in}}) = \frac{\dot{W}_{\text{shaft,in}} - 0}{\dot{m}}$$

$$\dot{W}_{\text{shaft,in}} = \dot{m}g(z_{\text{out}} - z_{\text{in}})$$

$$= \left(100\,\frac{\text{lb}_\text{m}}{\text{s}}\right)\left(32.174\,\frac{\text{ft}}{\text{s}^2}\right)(3000\text{ ft} - 0\text{ ft}) \times \frac{\text{lb}_\text{f}\cdot\text{s}^2}{32.174\text{ lb}_\text{m}\cdot\text{ft}}$$

$$= 300{,}000\,\frac{\text{ft}\cdot\text{lb}_\text{f}}{\text{s}} \times \frac{3600\text{ s}}{\text{h}} \times \frac{5.05 \times 10^{-7}\text{ hp}\cdot\text{h}}{\text{ft}\cdot\text{lb}_\text{f}} = 545\text{ hp}$$

EXAMPLE 22.19

Problem Statement: A horizontal 6.00-in pipe has a venturi with a 1.00-in throat (Figure 22.27). The upstream pressure is 18 psia with a water flow rate of 100 gal/min. What is the pressure at the throat, assuming there is negligible friction loss (i.e., no viscous heating losses)?

Solution: There is no shaft work ($W_{\text{shaft}} = 0$); therefore, we may use the Bernoulli equation. Because the pipe is horizontal, $z_{\text{in}} = z_{\text{out}}$.

$$\frac{P_{\text{out}} - P_{\text{in}}}{\rho} + \frac{1}{2}(v_{\text{out}}^2 - v_{\text{in}}^2) + 0 = 0$$

$$P_{\text{out}} = P_{\text{in}} + \frac{\rho}{2}(v_{\text{in}}^2 - v_{\text{out}}^2)$$

$$v_{\text{in}} = \frac{Q}{A_{\text{in}}} = \frac{Q}{\frac{\pi}{4}D_{\text{in}}^2} = \frac{100\text{ gal/min}}{\frac{\pi}{4}\left(6.00\text{ in} \times \frac{\text{ft}}{12\text{ in}}\right)^2} \times \frac{0.1337\text{ ft}^3}{\text{gal}} \times \frac{\text{min}}{60\text{ s}} = 1.14\,\frac{\text{ft}}{\text{s}}$$

$$v_{\text{out}} = \frac{Q}{A_{\text{out}}} = \frac{Q}{\frac{\pi}{4}D_{\text{out}}^2} = \frac{100\text{ gal/min}}{\frac{\pi}{4}\left(1.00\text{ in} \times \frac{\text{ft}}{12\text{ in}}\right)^2} \times \frac{0.1337\text{ ft}^3}{\text{gal}} \times \frac{\text{min}}{60\text{ s}} = 40.9\,\frac{\text{ft}}{\text{s}}$$

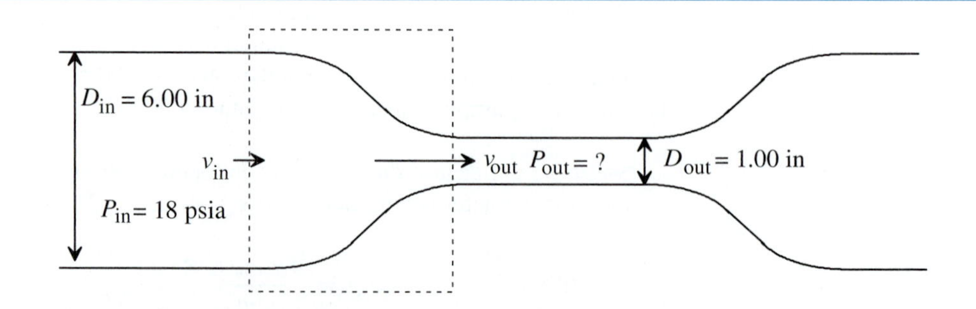

FIGURE 22.27
A pipe with a venturi.

$$P_{out} = 18\,\frac{\text{lb}_f}{\text{in}^2} + \frac{62.4\,\dfrac{\text{lb}_m}{\text{ft}^3}}{2}\left[\left(1.14\,\frac{\text{ft}}{\text{s}}\right)^2 - \left(40.9\,\frac{\text{ft}}{\text{s}}\right)^2\right]$$

$$\times \frac{\text{lb}_f\cdot\text{s}^2}{32.174\ \text{lb}_m\cdot\text{ft}} \times \frac{\text{ft}^2}{144\ \text{in}^2}$$

$$= 6.77\,\frac{\text{lb}_f}{\text{in}^2} = 6.77\ \text{psia}$$

Note: A vacuum is created at the throat. This principle is used in laboratory aspirators to create a vacuum and in automobile carburetors to suck fuel into the incoming air.

22.6 ENERGY CONVERSION

Energy may be converted from one form to another. In fact, designing, constructing, and maintaining energy conversion devices is one of the most important functions of engineers. Electric power plants, automobiles, planes, chemical plants, motors, lightbulbs, televisions, and pumps are all examples of energy conversion devices. We will illustrate some energy conversion processes with examples.

EXAMPLE 22.20

Problem Statement: A man lifts a 50.0-lb_m load 100 ft using the pulley system shown in Figure 22.28. How much work is performed?

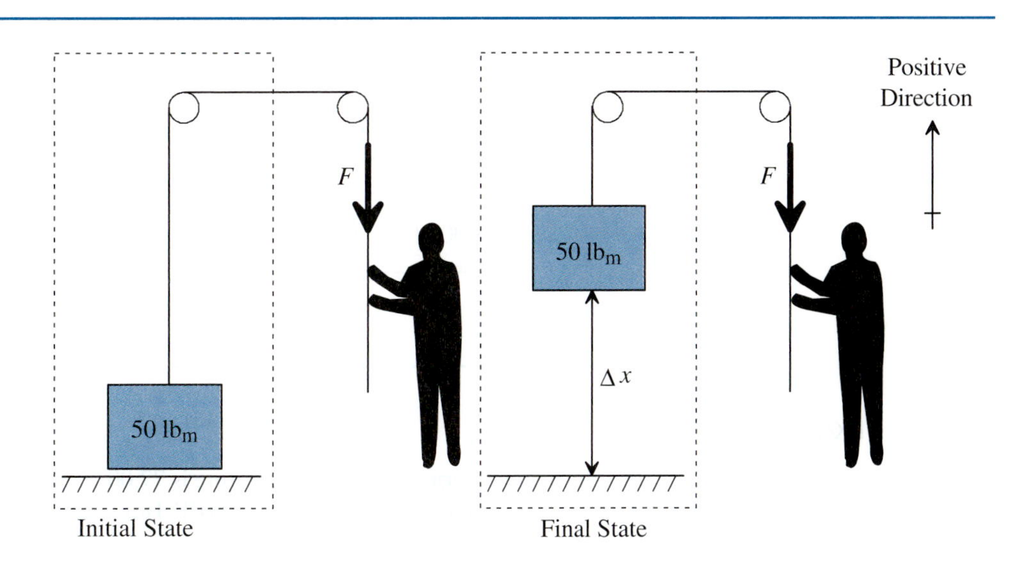

FIGURE 22.28
Man lifting a load with a pulley.

$$\dot{m} = -\frac{30.0 \text{ hp} + 77.1 \text{ hp}}{\left(-45.7\,\frac{\text{MJ}}{\text{kg}}\right)\left(\frac{\text{kg}}{2.2 \text{ lb}_\text{m}}\right)\left(\frac{\text{Btu}}{1054 \text{ J}}\right)\left(\frac{10^6 \text{ J}}{\text{MJ}}\right)} \times \frac{2542 \text{ Btu}}{\text{hp}\cdot\text{h}} = 13.8\,\frac{\text{lb}_\text{m}}{\text{h}}$$

$$\text{Fuel volume} = \frac{\dot{m}}{\rho_\text{fuel}} = \frac{13.8 \text{ lb}_\text{m}/\text{h}}{6.6 \text{ lb}_\text{m}/\text{gal}} = 2.0\,\frac{\text{gal}}{\text{h}}$$

22.7 SEQUENTIAL ENERGY CONVERSION

Many useful energy conversions involve sequences of steps. For example, a power plant burns fuel that generates steam that turns a turbine shaft that drives a generator that makes electricity. Each step has an efficiency η defined as

$$\eta = \frac{\text{desired energy}}{\text{energy input}} \tag{22-111}$$

Figure 22.29 shows sequential energy conversion. The efficiency for the entire process is the efficiency of each step multiplied together:

$$\eta_\text{overall} = \frac{E_4}{E_1} = \left(\frac{E_2}{E_1}\right)\left(\frac{E_3}{E_2}\right)\left(\frac{E_4}{E_3}\right) = \eta_1\eta_2\eta_3 = \prod_{i=1}^{n} \eta_i \tag{22-112}$$

Of course, no energy is consumed in the sequential process because energy is conserved. However, only a fraction of the energy input is converted to the desired energy; the rest is converted to waste energy (usually heat). Figure 22.30 shows some typical energy conversion efficiencies.

EXAMPLE 22.24

Problem Statement: Calculate the overall efficiency of an electric power plant that has the following steps: (1) coal-fired boiler (chemical reaction → sensible/phase change), (2) steam turbine (sensible/phase change → shaft work), and (3) large electric generator (shaft work → electricity).

Solution: The efficiencies are found in Figure 22.30, and

$$\eta_\text{overall} = \eta_1\eta_2\eta_3 = (0.82)(0.43)(0.99) = 0.35$$

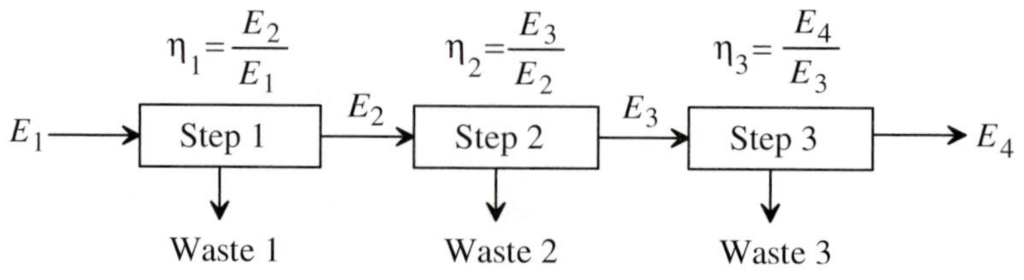

FIGURE 22.29
Sequential energy conversion.

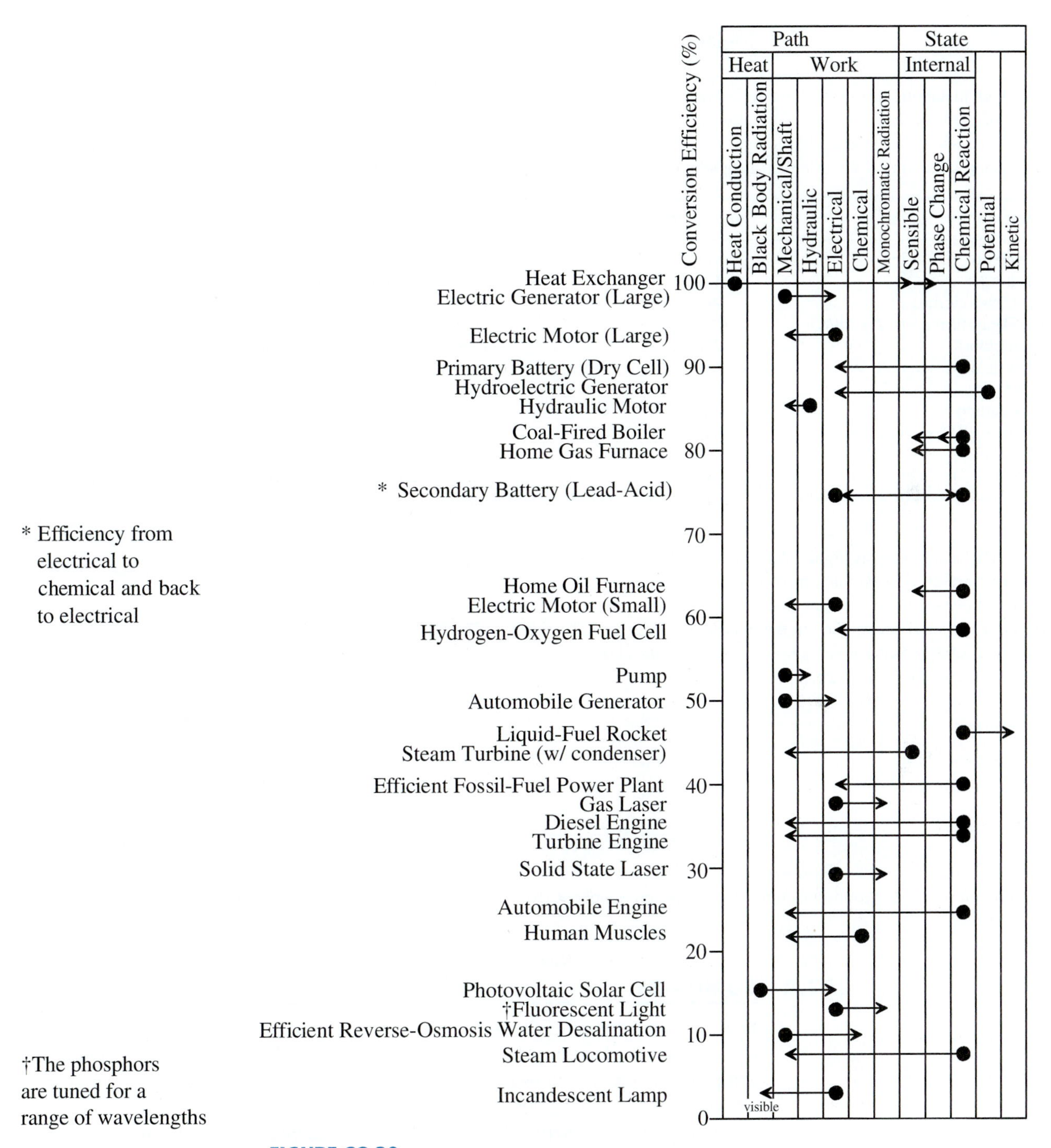

FIGURE 22.30

Typical conversion efficiencies. (*Note:* Energy classification based upon closed system.)

Metabolic Energy

Through metabolic processes, humans convert the chemical energy of food into mechanical work. The efficiency of this conversion process is about 22 percent with a range from 10 to 30 percent. The following table shows the rate that metabolic energy is expended for various activities; that is, for each activity, it shows the rate that food is processed by the cells in the body. The portion of the metabolic energy not converted into mechanical work exits the body as heat.

Activity	Power (W)	Activity	Power (W)
Sleeping	84	Shoveling sand, moderate rate	470
Writing	125	Swimming, breaststroke, 1 mi/h	490
Washing clothes	260	Chopping wood	520
Walking, firm road, 2.5 mi/h	270	Digging	620
Scrubbing	330	Climbing stairs, 116 steps/min	680
Bicycle ergometer, 70 W	360	Cycling, race, 23 mi/h	680
Golf	380	Basketball	800
Gardening	400	Wrestling	910
Walking, 45° slope,1.5 mi/h	410	Endurance march	1030
Crawling	430	Swimming, breaststroke, 2 mi/h	2020
Tennis	440	Swimming, breaststroke, 3 mi/h	6770

The accompanying figure shows that, in the short term, humans are able to produce mechanical power in excess of 1 horsepower. This rate is not sustainable, as waste products (e.g., lactic acid) accumulate in the muscle. For long periods of time, some humans are able to sustain a mechanical power output of about 0.2 to 0.3 hp.

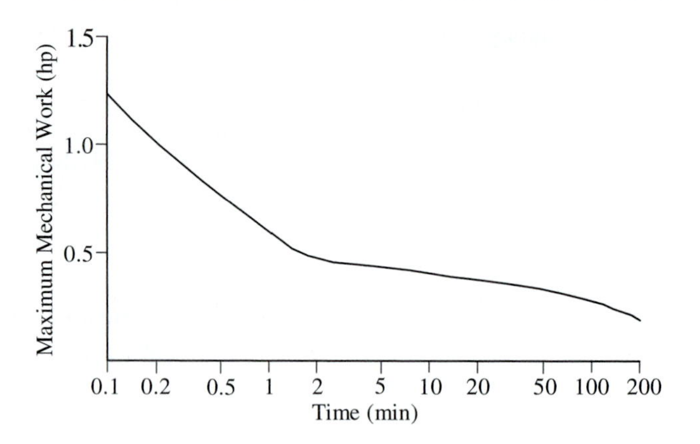

Adapted from: W. E. Woodson, *Human Factors Design Handbook* (New York: McGraw-Hill, 1981).

22.8 SUMMARY

Energy is the "unit of exchange" used by scientists and engineers to quantify many phenomena such as electricity, light, fluid flow, motion, fuel combustion, temperature change, and phase change. As an engineer, you will undoubtedly have to analyze energy transformation processes during your career.

Accounting for energy is complex because it comes in many forms. The state of the system is described by its potential, kinetic, and internal (sensible, phase change, chemical reaction) energies. The energy state of the system can be altered by path energies, that is, heat, work, and energy that enter the system via mass transfer. Heat is energy flow across the system boundary resulting from a temperature driving force. Work is energy flow across the system boundary resulting from any other driving force (force, torque, voltage, pressure, concentration, electromagnetic fields). Energy that enters the system via mass transfer includes potential, kinetic, internal, and *PV* flow work.

When energy conversion occurs sequentially, the efficiency of the entire process is found by multiplying the efficiency of each individual step.

Nomenclature

A area (m^2)
a acceleration (m/s^2)
C capacitance (C/V)
C_p constant-pressure heat capacity [J/(kg·K)]
C_v constant-volume heat capacity [J/(kg·K)]
c speed of light (m/s)
D diameter (m)
E energy (J)
$\dot{E}$ energy rate (J/s)
E_k kinetic energy (J)
E_p potential energy (J)
F force (N)
g acceleration due to gravity (m/s^2)
H enthalpy (J)
$\hat{H}$ specific enthalpy (J/kg)
$\dot{H}$ enthalpy flow (J/s)
h Planck's constant (J·s)
i current (A)
k thermal conductivity [J/(s·m·K)] or spring constant (N/m)
L inductance (V·s/A)
M energy associated with mass flow (J)
m mass (kg)
$\dot{m}$ mass flow (kg/s)
n mass (kg) or moles (mol)
P pressure (N/m^2 or Pa)
Q volumetric flow rate (m^3/s) or heat (J)
$\dot{Q}$ heat flow (J/s)
q charge (C)
R universal gas constant [J/(mol·K) or Pa·m^3/(mol·K)]
r radius (m)
s moles of salt ions (mol)
T torque (N·m) or temperature (K)
t time (s)
U internal energy (J)
$\hat{U}$ specific internal energy (J/kg)
$\dot{U}$ internal energy flow (J/s)
V volume (m^3) or voltage (V)

$\hat{V}$ specific volume (m^3/kg)
v velocity (m/s)
W work (J)
$\dot{W}$ work flow (J/s)
x distance (m)
z height (m)
η efficiency (dimensionless)
θ angle (rad)
λ wavelength (m)
ρ density (kg/m^3)
σ Stefan-Boltzmann constant [J/(s·m^2·K^4)]
ω rotation rate (rad/s)

Further Readings

Felder, R. M., and R. W. Rousseau. *Elementary Principles of Chemical Processes.* 2nd ed. New York: Wiley, 1986.

Fenn, J. B. *Engines, Energy, and Entropy: A Thermodynamics Primer.* San Francisco: W. H. Freeman, 1982.

Glover, C. J.; and H. Jones. *Conservation Principles of Continuous Media.* New York: McGraw-Hill College Custom Series, 1994.

Glover, C. J., K. M. Lunsford, and J. A. Fleming. *Conservation Principles and the Structure of Engineering.* New York: McGraw-Hill College Custom Series, 1992.

Serway, R. A. *Principles of Physics.* Fort Worth: Saunders, 1994.

Smith, J. M., and H. C. Van Ness. *Introduction to Chemical Engineering Thermodynamics.* 3rd ed. New York: McGraw-Hill, 1975.

PROBLEMS

22.1 A 2.00-ton automobile is traveling at 55 mph. What is its kinetic energy expressed in ft·lb_f, Btu, and J?

22.2 A 2.00-ton automobile is traveling at 55 mph and brakes to a complete stop to park. It does not move until the next day. How much heat (Btu) was released as the brakes cooled?

22.3 A 2.00-ton automobile is parked at the top of a hill that is 100 ft high and inclined 25° with respect to the horizon. The parking brake fails and the automobile rolls down the hill. Neglecting air resistance and rolling resistance, what is the automobile velocity (ft/s) at the bottom of the hill? Solve this problem using the following methods: (a) find the component of earth's gravity force acting to accelerate the automobile, and (b) convert potential energy to kinetic energy.

22.4 A spherical 3-mi diameter comet impacts the earth with a velocity of 20 mi/s. The average density of the comet is 7 g/cm^3. Calculate the amount of heat released from the impact site. Express the answer as megatons of TNT explosive equivalent.

22.5 An electric water heater can heat 2.00 gal/min from 60.0°F to 120°F. How much electrical power (kW) is required? (*Note:* Electricity can be converted to heat with 100% efficiency. Because of insulation losses, the unit is 98% efficient.)

22.6 A natural-gas water heater can heat 2.00 gal/min from 60.0°F to 120°F. The water heater is 70% efficient. At what rate (lb_m/h) must the natural gas be burned? (*Note:* Natural gas is primarily methane.)

22.7 A portable kerosene heater can supply 30,000 Btu/h to a room. At what rate (lb_m/h) must the kerosene be combusted? (*Note:* The fuel is burned with 99.0% efficiency. All exhaust gases are vented directly into the room, so the heat transfer efficiency approaches 100%.)

22.8 A fuel-oil-fired steam generator produces 100 kg/min of steam. Liquid water enters at 100°C and exits as 100°C steam. The steam generator is 70% efficient. At what rate (kg/min) must the fuel oil be combusted?

22.9 An automobile has a fuel economy of 25 mi/gal when combusting gasoline. If the engine is modified to burn pure ethanol, but the engine efficiency stays the same, what is the expected fuel economy? (*Note:* Densities are found in Chapter 11.) Ethanol has a high octane rating and can be burned in high-compression, high-efficiency engines. If the ethanol-powered engine is 10 percent more efficient than the gasoline engine, what is the expected fuel economy?

22.10 A computer uses 100 W of electricity. It is cooled by a "muffin fan" that brings in room-temperature air at 25°C and discharges it at 35°C. What is the required flow rate in ft^3/min?

22.11 A laboratory aspirator has an inlet diameter of 0.500 in and a throat diameter of 0.100 in. The inlet water pressure is 70 psig. What water flow rate (in gal/min) is needed to form a 3.0-psia vacuum?

22.12 A pump flows 100 gal/min of water. The inlet pressure is 0 psig and the outlet pressure is 20 psig. What is the required shaft power (hp) of the motor, assuming the pump is 60% efficient? The motor is 85% efficient? How much electricity (kW) is required to power the motor?

22.13 An electric motor drives a pulley that lifts a 1000-lb_m load 20 ft into the air. How much work (ft·lb_f) is done lifting the load? If the motor is 60% efficient and the pulley is 95% efficient, how much electrical energy (J) is required? If the mass is lifted in 1 min, how much electrical power (W) is required?

22.14 An electric car can be powered by a battery; however, it takes a long time to charge a battery. A clever engineer realizes that capacitors can be charged very quickly and suggests that a capacitor be used to store energy instead of a battery. Most automobiles require about 15 hp of shaft power to travel at highway speeds. Assuming the combined efficiency of the capacitor and electric motor is 85%, what is the required capacitance (F) to operate the automobile for 2 h on a highway, assuming it is initially charged with 110 V?

22.15 An electric car can be powered by a battery; however, it takes a long time to charge a battery. A clever engineer realizes that superconducting inductors can be charged very quickly and suggests that an inductor be used to store energy instead of a battery. Most automobiles require about 15 hp of shaft power to travel at highway speeds. Assuming the combined efficiency of the inductor and electric motor is 85%, what is the required inductance (H) to operate the automobile for 2 h on a highway, assuming it is initially charged with 100 A of electric current?

22.16 An electric car can be powered by a battery; however, it takes a long time to charge a battery. A clever engineer realizes that a flywheel coupled to an electric generator can be charged very quickly and suggests that a flywheel be used to store energy instead of a battery. When fully charged, the flywheel will spin at 100,000 rpm. The flywheel is a thin hoop with a 2.00-ft diameter. Most automobiles require about 15 hp of shaft power to travel at highway speeds. Assuming the combined efficiency of the flywheel and electric motor is 85%, what is the required flywheel mass (kg) to operate the automobile for 2 h on a highway? (*Note:* The kinetic energy of a flywheel is given in Table 21.3.)

22.17 The water level behind a dam is 150 ft. A turbine is located at the bottom of the dam through which the water flows at a rate of 10,000 gal/min. The water is discharged through a 2.00-ft-diameter pipe into the stream below the dam. Assuming the turbine is 100 percent efficient, what is the shaft power (hp) output of the turbine? The turbine is coupled to an electric generator. Assuming the generator is 100 percent efficient, how much electricity (W) will be produced? In actuality, the turbine and generator are not 100 percent efficient. Using the typical efficiency for a hydroelectric generator reported in Figure 22.30, estimate the actual electric power production.

22.18 Seawater is about 3.5% salt by weight. Although there are many minerals dissolved in it, assume that sodium chloride is the dominant salt.

Reverse osmosis is a process that uses semipermeable membranes that pass water, but not salt. Estimate the theoretical energy (J) required to remove 10 L of pure water from 100 L of seawater at 25°C. (*Note:* Sodium chloride dissociates into sodium ions and chloride ions when dissolved in water. One mole of sodium chloride produces 2 moles of ions when dissolved.)

As a result of various inefficiencies, reverse osmosis requires much more energy than the theoretical energy requirement (Figure 22.30). How much actual energy (J) is required to produce 10 L of pure water from 100 L of seawater?

A city decides to produce drinking water from seawater because surface water is not available. They decide to produce 100 m^3/s of drinking water from 1000 m^3/s of seawater. What is the required electrical power (kW)? (*Note:* Take into account the inefficiencies determined above.)

Electricity costs the city $0.07/kWh. What is the cost to produce 1.00 m^3 of drinking water? How much is the city going to spend for electricity each month for drinking water?

22.19 An efficient power plant burns coal to make steam that rotates a turbine that powers an electric generator that makes electricity. What is a typical overall efficiency for such a plant? (See Figure 22.30.) At what rate must lignite coal be burned (ton/h) to make 1000 MW of electricity?

22.20 A photovoltaic solar cell will be used to charge a lead-acid battery that runs the television, lights, etc. of a Charleston, South Carolina, home. What is the overall conversion efficiency from sunlight to delivered electricity? (See Figure 22.30 to account for the efficiencies of the solar cell and battery.)

On an average 24-h day, the home uses electricity at a rate of 1 kW. In one day, how much energy (J) is used by the home?

During a clear December day, 9.03 MJ/m^2 strikes Charleston, South Carolina. Typically, the light intensity is only 62% of this because of cloud cover. How much solar cell surface area is required to keep the batteries charged during a typical December?

22.21 A spring with a spring constant of 50 lb_f/in is compressed 3.00 in. Onto it is placed a 1.00-lb_m baseball. When the spring is released, it thrusts the baseball vertically into the air. What is the highest height obtained by the baseball? What is its speed just before it impacts the ground?

22.22 On a flat highway, a 1500-lb_m automobile traveling at 55 mi/h requires 15 hp of shaft power to overcome air resistance and rolling resistance. For high-speed highways, the maximum allowable *grade* (i.e., rise over run) is 5 percent. How much shaft power is required for the automobile to maintain the 55 mi/h speed while climbing a 5 percent grade?

22.23 Because it is expensive to make and store electricity, an inventor decides to use a windmill to make hot water rather than electricity. The device is very simple; it consists of fan blades exposed to the wind and paddles that churn water in a tank. The tank is insulated to store the hot water until it is needed. The fan blade is spinning at 300 revolutions per minute (rpm). The torque in the shaft is 100 ft·lb_f. What is the shaft power (hp)?

Water is fed at 60°F and withdrawn at 180°F. At what rate (lb_m/h) is hot water produced by this invention?

22.24 The ultimate source of energy for hydroelectric power is the sun. Sunlight vaporizes water from a body of water that travels into the upper cold atmosphere where it condenses to form rain. Assume the rain is captured behind a 300-ft dam with a turbine at the bottom that is coupled to an electric generator. What is a typical conversion efficiency for converting the potential energy of the water to electricity? (See Figure 22.30.) What is the conversion efficiency for the entire process, that is, what percentage of the solar energy that evaporated the water is ultimately converted into electricity?

22.25 An engineer wishes to heat a chemical reactor with 100,000 J of energy. She is trying to decide whether to heat it electrically or with steam.

The engineer has steam available at 100°C and 1 atm pressure with specific enthalpy 48.168 kJ/mol. When the steam gives up its energy to the reactor, it condenses. The condensate temperature is 100°C, and its pressure is 1 atm with a specific enthalpy of 7.544 kJ/mol. How many gmoles of steam must be used to heat the reactor?

The engineer has electric power available at 220 V (i.e., 220 joules per coulomb). After the electric current flows through a resistor to heat the reactor, it is discharged at "ground potential" (i.e., 0 V). How many coulombs are required to heat the reactor?

22.26 A 60-kg person climbs stairs at a rate of 116 steps/min. Each step is 15 cm high. While climbing, work is required to increase the potential energy of the person. What is the efficiency with which a person performs this task; that is, how much work is performed per unit of metabolic energy expended?

22.27 An engineer connects an electric generator to a stationary bicycle. To lose weight, he rides the bicycle and uses the electrical output to power a 100-W television. The bicycle efficiency is 95 percent and the electric generator efficiency is 50 percent . Every day, he rides the bicycle for 20 minutes. While on this exercise program, he does not increase his daily food intake. In one year, approximately how many kilograms of body fat would be burned off? (*Note:* Fat has a metabolic energy content of 9 kcal/g, whereas carbohydrates and protein have a metabolic energy content of 4 kcal/g.)

22.28 An inventor develops a 10-lb_m wagon with a sail located on the front. As an object is thrown at the sail, it drops into the wagon. As a result of conservation of momentum, the wagon is propelled by the impact of the object. A 1.00-lb_m baseball is traveling at 20 ft/s and hits the stationary wagon. Using conservation of momentum, calculate the final velocity of the wagon with its baseball payload. Calculate the kinetic energy of the system (i.e., baseball + wagon) before the impact and after the impact. Where did the loss of kinetic energy go? What is the efficiency of this process; that is, what percentage of the original kinetic energy is retained by the system?

The baseball is removed from the wagon and the process is repeated with a 5-lb_m rock traveling at the same velocity (20 ft/s). The rock is removed from the wagon and the process is repeated with a 15-lb_m bowling ball traveling at the same velocity (20 ft/s). Calculate the efficiency of the process using the 5-lb_m rock and 15-lb_m bowling ball. Discuss the implications of your calculations in terms of *reversibility,* a topic described in Chapter 11.

22.29 Write a computer program that calculates the efficiency of the process described in Problem 22.28. Prepare a table that lists the efficiencies for a given combination of mass and velocity. Vary the mass from 5 to 50 lb_m in 5-lb_m increments and vary the velocity from 20 to 100 ft/s in 20-ft/s increments. Review your table and report your conclusions.

Glossary

Bernoulli equation The equation that describes the interconversion of pressure energy, gravitational potential energy, and kinetic energy for an incompressible fluid flowing at steady state.

blackbody An object that completely absorbs light.

capacitor A device that stores electrical charge.

chemical work Work required to change chemistry or concentration.

closed system A system in which mass does not cross the boundary.

coherent light Light in which the electrical energy peaks of the photons align.
constant-pressure heat capacity The amount of heat that causes a 1° temperature change in a material maintained at constant pressure.
constant-volume heat capacity The amount of heat that causes a 1° temperature change in a material maintained at constant volume.
electrical work Energy required to move a charge across a voltage difference.
energy A unit of exchange used by the scientific and engineering community to equate various phenomena.
enthalpy $H = U + PV$
fluids Materials that flow when pressure is applied, including liquids, gases, vapors, and supercritical fluids.
head Pressure differences, velocities, and heights expressed as length.
heat Energy flow resulting from a temperature driving force.
heat of vaporization The heat required to convert a liquid to a vapor.
hydraulic work Energy required to flow a fluid across a pressure difference.
incoherent light Light in which the electrical energy peaks of the photons do not align.
inductance The quantitative factor that describes the "inertia" of an electrical circuit to resist changes in current.
inductor A coil of wire.
internal energy The energy contained within matter resulting from translational, rotational, and vibrational kinetic energies and the electronic potential energy of atoms and molecules.
kinetic energy The energy associated with a moving mass.
laser light work The energy required to produce laser light.
mechanical work The energy associated with applying a force over a distance.
open system A system in which mass crosses the boundary.
osmotic pressure The pressure required to force the solvent through a semipermeable membrane separating two solutions with different concentrations of solute.
Planck's constant $6.6260755 \times 10^{-34}$ J·s
photon A "packet" of light.
potential energy Energy associated with position.
power Energy per unit time.
***PV* flow work** The energy that enters or leaves a system resulting from the volume and pressure of a fluid.
scalar A quantity completely described by its magnitude; has no direction.
shaft work Energy from a rotating shaft.
specific heat of combustion Energy released from combusting a unit quantity of fuel.
state energy Energy that does not depend on path.
Stefan-Boltzmann constant 5.67051×10^{-8} J/(s·m^2·K^4)
vaporization The change of liquid to vapor.
work The energy flow across a system boundary resulting from any driving force other than temperature.

- A process is *reversible* if, when the system is returned to its initial state, there *are no* changes in the surroundings.
- A process is *irreversible* if, when the system is returned to its initial state, there *are* changes in the surroundings.

Consider the cannon example. In principle, the cannon (our first system) could be returned to its initial state. A person could find the cannonball at its final destination and return it to the cannon. A large vacuum cleaner could be operated near the cannon to gather up the smoke and, through various chemical processes, convert it back into gunpowder. Although this could be done, it would cause *many* changes in the surroundings, therefore firing a cannon is a highly irreversible process.

In contrast, the hypothetical frictionless pulley (our second system) is a reversible process because once it is returned to its initial state, there are no detectable changes in the surroundings. A real pulley, of course, has some friction. Because of the friction, there would be some slight changes in the surroundings. As the pulley got warm and rejected heat to the surroundings, the temperature of the surroundings would rise as well. With good bearings and lubrication, the amount of heat generated by the pulley is very small; therefore, a real pulley (with friction) is only a slightly irreversible process.

To become more comfortable with our new quantifiable definitions of reversibility and irreversibility, consider the system shown in Figure 23.2. It consists of two frictionless pulleys mounted on each end of a table. One end of each rope is attached to a wooden block that slides on the table surface. Pans are attached to the other end of each rope that extend over the side of the table. Weights can be placed on the pans as well as on the wooden block in order to introduce various amounts of friction into our system. To keep the numbers simple, imagine that the coefficient of friction between the wooden block and the table is 1.0; that is, if a 1-lb_f weight is placed on top of the wooden block, it will cause a frictional force of 1.0 lb_f that resists the direction of motion.

Initially, the system (the components within the dotted line) has the right pan in the up position and the left pan in the down position. Each pan starts with a 1-lb_f block on it. Also, there is a 1-lb_f block sitting on the wooden block. Because of friction, neither pan can move. Then, an additional 1-lb_f weight sitting on a shelf is placed on the right pan. This extra weight is able to overcome friction, allowing the right pan to move downward to the floor and raising the left pan upward. The extra weight on the right is slid off onto the floor and an extra weight sitting on a shelf is added to the left pan. There is now sufficient weight in the left pan to drop it to the floor, which raises the right pan. The extra weight on the left is slid off onto the floor, thus returning the system to its original state. Although the system has returned to its original state, the surroundings have not. Two weights that were previously on shelves are now on the floor. Further, the temperature of the wooden block increased due to friction. Later, as the block cools, heat is rejected into the surroundings, causing a change in the surroundings. If there were no friction in the system, it would be reversible and behave the same as the pulley shown in Figure 23.1(b); therefore, we can say that friction causes irreversibility.

Although the system shown in Figure 23.2 is irreversible, it still conserved energy. The potential energy of the weights sitting on the shelves was converted into internal energy of the blocks and later into heat. Here, we see a system in which potential energy was degenerated into internal energy and then into heat, summarized as

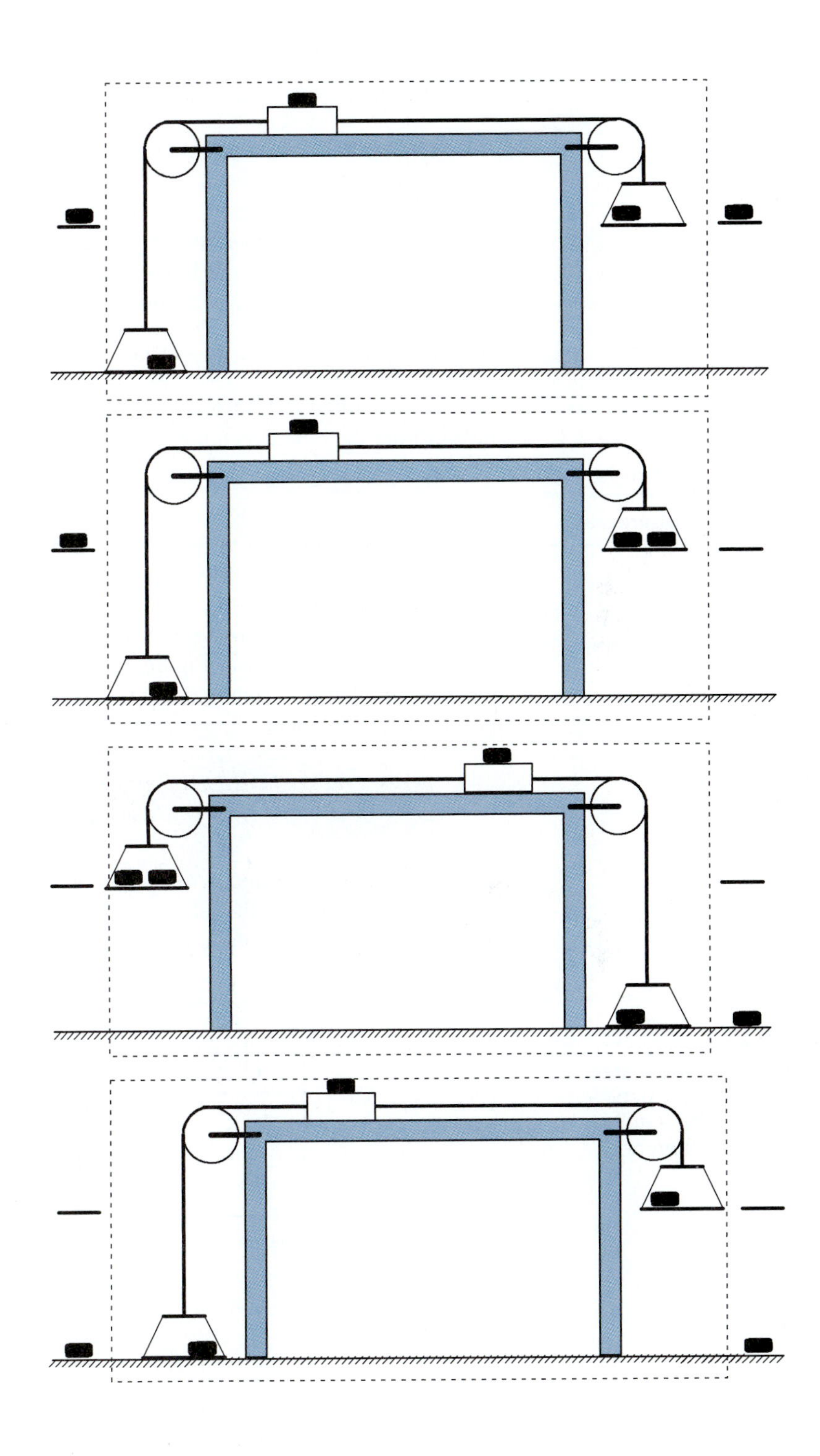

FIGURE 23.2
A system in which friction causes irreversibility.

Potential energy $\rightarrow$ internal energy $\rightarrow$ heat

Recall that the designation of the system boundary is completely arbitrary and up to the discretion of the engineer. If we had changed the designation of our system to include only the wooden block, then we would have seen work crossing the system boundary as the rope pulled the block across the table. However, there would have been no change in potential energy because the block did not change its height. Thus, we can describe the energy changes in this newly designated system as

Work $\rightarrow$ internal energy $\rightarrow$ heat

As we learned in Chapter 22, on accounting for energy, for a closed system (i.e., no mass crosses the boundary), the energy accounting equation is

$$\Delta E_k + \Delta E_p + \Delta U = W_{\text{in}} - W_{\text{out}} + Q_{\text{in}} - Q_{\text{out}} \quad (23\text{-}1)$$

where ΔE_k is the change in kinetic energy of the system, ΔE_p is the change in potential energy of the system, ΔU is the change in internal energy of the system, W is the work that crosses the system boundary, and Q is the heat that crosses the system boundary. Table 23.1 shows these various forms of energy classified as *ordered* energy and *disordered* energy. Thus, the system in Figure 23.2 transformed ordered energy (potential or work) into disordered energy (internal and heat). We can generalize these observations and make the following statement:

TABLE 23.1
Classification of energies

Ordered	Disordered
Kinetic	Internal
Potential	Heat
Work	

Observation 1: Irreversibilities occur when *ordered* energy is converted to *disordered* energy.

Figure 23.3 shows another example of an irreversible system. A piston/cylinder filled with steam is located at the left of a copper rod and an ice bath is located at the right. As heat is transferred down the copper rod, the steam condenses to liquid water, and the ice

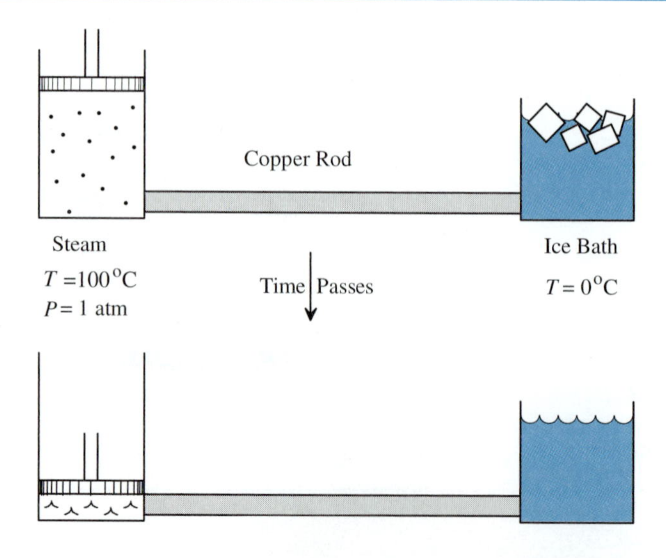

FIGURE 23.3
Irreversibilities associated with heat transfer between two bodies with a finite temperature difference.

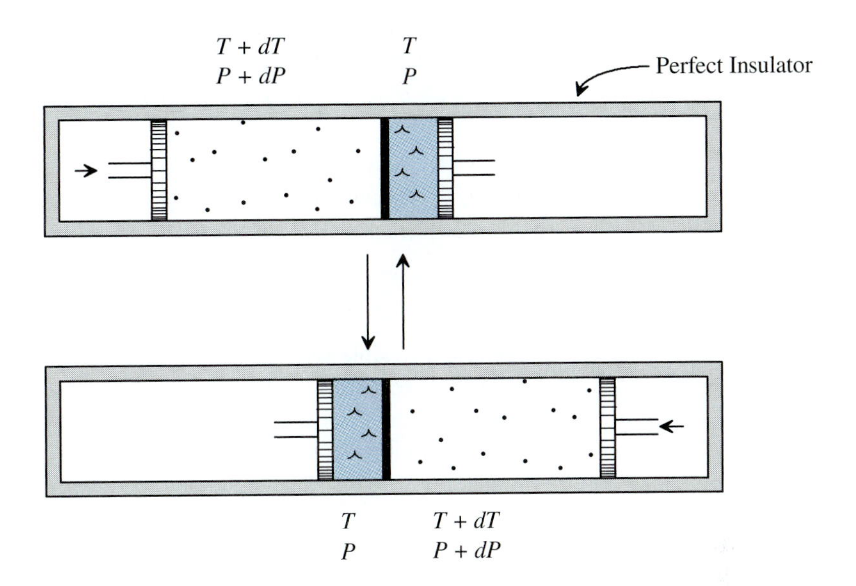

FIGURE 23.4
Reversible heat transfer using a differential temperature difference dT.

melts. This is clearly an irreversible process. Heat will not spontaneously flow from the ice bath to regenerate the steam. Although we could regenerate the steam using a flame and refreeze the ice using a freezer, these extra steps will change the surroundings. This leads to a second observation about irreversibility.

> **Observation 2:** Heat transfer from a high-temperature body to a low-temperature body is an irreversible process.

Observation 2 indicates that heat transfer over a finite ΔT is irreversible. However, heat transfer across a differential temperature difference (dT) is reversible. To illustrate this point, consider Figure 23.4 showing two piston/cylinders, the left one filled with steam and the right one filled with liquid water. The temperatures and pressures in each piston/cylinder are approximately the same. The two piston/cylinders are assumed to be in perfect thermal contact and the system is assumed to be perfectly insulated (i.e., *adiabatic*) so there are no heat losses to the surroundings. Initially, a slight extra pressure (dP) is applied to the cylinder on the left. According to the phase diagram for water, its temperature will be slightly higher (dT) than the piston/cylinder on the right. This slightly higher temperature allows heat transfer, so the steam on the left will condense, causing the liquid water on the right to vaporize. Later, this process could be reversed by applying a slightly higher pressure (dP) to the right piston/cylinder, causing its temperature to increase slightly (dT). Thus, the system can return to its original state—which means that heat transfer caused by a differential temperature difference (dT) is reversible. Recall from the chapters on thermodynamics and rate processes that a temperature difference is an example of a driving force. We can generalize our observation about differential temperature differences being reversible by stating that *any* differential driving force (e.g., temperature, pressure, voltage, concentration) allows the system to be reversible.

Observation 3: Systems that have only differential driving forces are reversible.

Of course, systems that have differential driving forces move infinitely slowly because the driving forces are so small.

Observation 4: Reversible systems are infinitely slow.

Because reversible systems are infinitely slow, they are idealizations that do not exist in reality. A few exceptions to Observations 3 and 4 are electrical superconductors and superfluids, which have no electrical resistance or fluid flow resistance, respectively. Superconductors and superfluids are systems that have finite, measurable rates but are entirely reversible because they do not degenerate ordered energy into disordered energy (see Observation 1).

Figure 23.5 shows gas inside a piston/cylinder undergoing expansion and then compression. The system is defined to be the gas only. The gas is maintained at constant temperature by immersing the piston/cylinder in a constant-temperature water bath. The upper portion of the figure describes an irreversible expansion/compression and the lower portion

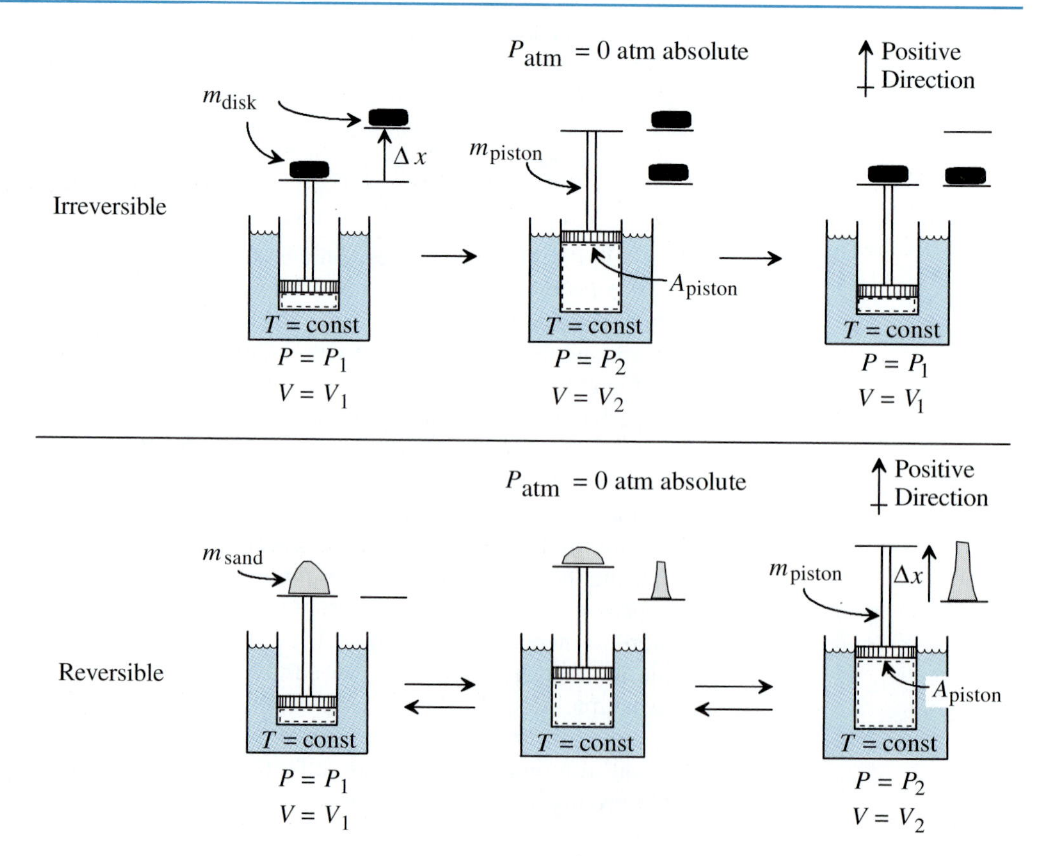

FIGURE 23.5
An irreversible and reversible expansion/compression of a gas.

of the figure describes a reversible expansion/compression. In both cases, the piston is idealized to have no friction and no leakage, but it does have mass m_{piston}. The atmospheric pressure of the surroundings is assumed to be a perfect vacuum (P_{atm} = 0 atm absolute).

In the irreversible process shown, the gas starts at an initial volume V_1.[1] The initial pressure P_1 is determined by the force of the piston and heavy disk sitting on the piston,

$$P_1 = \frac{F_{piston} + F_{disk}}{A_{piston}} = \frac{(m_{piston} + m_{disk})g}{A_{piston}} \tag{23-2}$$

where g is the acceleration due to gravity (m/s^2). (*Note:* g may also be viewed as the specific force due to gravity (N/kg).) The heavy disk is removed from the piston by sliding it onto an adjacent shelf. The piston immediately shoots up until the pressure P_2 in the cylinder is balanced solely by the force exerted by the piston weight:

$$P_2 = \frac{F_{piston}}{A_{piston}} = \frac{m_{piston}\, g}{A_{piston}} \tag{23-3}$$

This process irreversibly expands the gas. To complete the cycle, a heavy disk from a higher shelf is slid onto the piston, causing it to rapidly move downward and irreversibly compress the gas back to P_1. Note that although the gas was returned to its initial state, the surroundings were changed: a heavy disk was transferred from the higher shelf to the lower shelf. In addition, because this process occurred so rapidly, we knew it would be irreversible based upon Observation 4.

In the reversible process, the piston has a pile of sand on it. The gas starts at the same initial state as the irreversible case, that is, V_1 and P_1. The piston mass is identical to the piston mass in the irreversible case and the sand mass is identical to the heavy disk. One by one, grains of sand are moved laterally from the piston to the adjacent shelf, slowly allowing the piston to rise. When all the grains of sand are removed, the pressure in the cylinder will equal P_2 given by Equation 23-3. Then, the grains of sand are slowly added back to the piston until the gas is compressed to its original pressure P_1. The system has returned to its original state, and so have the surroundings; therefore, this is a reversible process. (*Note:* In actuality, this only approximates a reversible process. A truly reversible process would need an infinite number of sand grains that are infinitely light, but for our purposes, a grain of sand is effectively "infinitely light.")

During the irreversible expansion of the gas, the piston with mass m_{piston} moves a distance Δx. The gas does work on the piston by increasing its potential energy; therefore

$$W_{out,irrev} = m_{piston}\, g \Delta x \tag{23-4}$$

Equation 23-3 allows the following substitution,

$$W_{out,irrev} = P_2 A_{piston} \Delta x = P_2 \Delta V = P_2(V_2 - V_1) \tag{23-5}$$

where the volume change is equal to the piston cross-sectional area A_{piston} times the linear displacement. This work is shown as the shaded area in the graph at the upper left corner of Figure 23.6. As a reference line, this graph also shows the **isotherm** that relates P and V at constant temperature. (*Note:* You can calculate this isotherm from the perfect gas equation.)

1. *Note:* There are many possible irreversible processes; this is but one example.

For the reversible expansion, the work output can be determined by integrating the force exerted over a distance:

$$W_{\text{out,rev}} = \int F\,dx \tag{23-6}$$

The force constantly changes; at the beginning with most of the sand on the piston, more force is required. Later, when most of the sand is gone, less force is required. Because this process is occurring very slowly, the downward force of the piston/sand is exactly balanced by the upward force of the gas; therefore,

$$W_{\text{out,rev}} = \int PA_{\text{piston}}\,dx \tag{23-7}$$

The term $A_{\text{piston}}\,dx$ can be replaced by dV (if this does not seem plausible to you, just look at the units, area × length = volume); therefore,

$$W_{\text{out,rev}} = \int P\,dV \tag{23-8}$$

The pressure term cannot be removed from the integral because pressure depends on volume. We need to find a relationship between pressure and volume. Assuming the gas is perfect (sometimes called "ideal"), we can express the pressure in terms of volume:

$$PV = nRT \tag{23-9}$$

$$P = \frac{nRT}{V} \tag{23-10}$$

By substituting Equation 23-10 into Equation 23-8, we get

$$W_{\text{out,rev}} = \int \frac{nRT}{V}\,\text{dV} \tag{23-11}$$

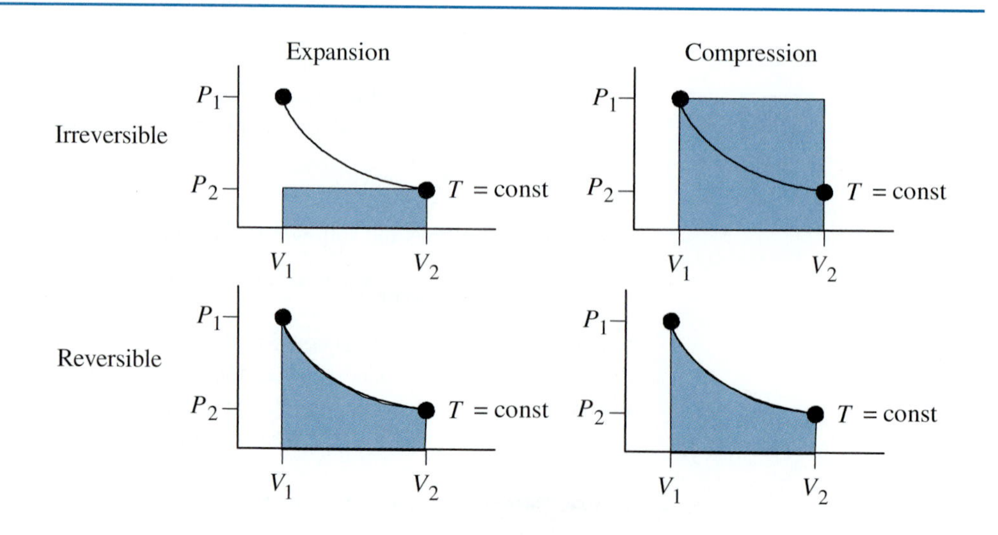

FIGURE 23.6
Work output during expansion and work input during compression.

The terms in the numerator are all constant, so they can be removed from the integral:

$$W_{\text{out,rev}} = nRT \int \frac{dV}{V} \tag{23-12}$$

This integral may be performed from the initial volume to the final volume:

$$W_{\text{out,rev}} = nRT \int_{V_1}^{V_2} \frac{dV}{V} = nRT[\ln V]_{V_1}^{V_2} = nRT[\ln V_2 - \ln V_1] = nRT\frac{V_2}{V_1} \tag{23-13}$$

This result is shown as the area under the isotherm in the graph at the lower left corner of Figure 23.6. It is obvious by comparing the two graphs for the reversible and irreversible expansions that the reversible expansion produces more work than the irreversible expansion. Although it was shown here for a specific example, this result is true in general. This leads us to our next observation.

Observation 5: A reversible process has more work output than an irreversible process.

We can analyze the work required to compress the gas using a very similar analysis. For the irreversible compression, the piston and disk do work on the gas when they change their potential energy. Therefore, the work input to irreversibly compress the gas is

$$W_{\text{in,irrev}} = -(m_{\text{piston}} + m_{\text{disk}})g\Delta x = -P_1 A_{\text{piston}}\Delta x$$

$$= -P_1\Delta V = -P_1(V_1 - V_2) = P_1(V_2 - V_1) \tag{23-14}$$

(*Note:* The negative sign is needed for the potential energy term, because Δx is negative according to the sign convention in Figure 23.5, but clearly, $W_{\text{in,irrev}}$ is positive.) This result is shown as the shaded area on the upper right graph in Figure 23.6.

For the reversible compression, we integrate the force exerted over the distance just as before:

$$W_{\text{in,rev}} = -\int F\,dx \tag{23-15}$$

(*Note:* The negative sign is needed because dx is negative according to the sign convention in Figure 23.5, but $W_{\text{in,rev}}$ is clearly a positive quantity.) Using the same substitutions and derivations as for the reversible expansion, we obtain

$$W_{\text{in,rev}} = -nRT \int_{V_2}^{V_1} \frac{dV}{V} = -nRT[\ln V]_{V_2}^{V_1} = -nRT[\ln V_1 - \ln V_2]$$

$$= nRT[\ln V_2 - \ln V_1] = nRT \ln \frac{V_2}{V_1} \tag{23-16}$$

By comparing the reversible and irreversible work inputs needed to compress the gas, it is clear that the irreversible process requires more work, thus leading us to another general observation.

Observation 6: A reversible process requires less work input than an irreversible process.

By now, you can see that reversible processes have many advantages over irreversible processes. Because work (e.g., electricity) is very precious, engineers should constantly strive to remove irreversibilities from processes. This can be accomplished by eliminating friction and specifying equipment that requires small driving forces (e.g., temperature, pressure, voltage, concentration). Although clever designs at times are able to minimize driving force requirements, sometimes the only way to eliminate large driving forces is to specify large pieces of equipment (e.g., extra large heat exchangers). Thus, engineers are often presented with interesting trade-offs between capital equipment and energy costs. For example, doubling a heat exchanger will reduce irreversibilities and save energy, but at some point, the further energy savings cannot justify yet another doubling of the heat exchanger. The economic optimum depends on the cost of energy and the cost of the equipment.

By comparing the work output from the reversible expansion to the work input required for the reversible compression (Equations 23-13 and 23-16), we see that they are identical. This leads us to yet another observation.

Observation 7: A reversible process that has a given work *output* when run in the forward direction requires the same work *input* when run in the reverse direction.

Observation 7 is very useful to engineers because the same equations developed by analyzing the forward direction can be used in the reverse direction.

In Figure 23.5, the gas was defined as the "system." Because no mass crosses the boundary, it is a **closed system.** Therefore, the following equation can be used to perform an energy accounting

$$\Delta E_k + \Delta E_p + \Delta U = W_{\text{in}} - W_{\text{out}} + Q_{\text{in}} - Q_{\text{out}} \tag{23-1}$$

(If this equation is unfamiliar to you, please review Chapter 22 on accounting for energy.) There are no changes in kinetic or potential energy of the system (i.e., the gas) nor is there a change in the internal energy because the temperature is constant. (Recall, from Chapter 11 on thermodynamics, that $\Delta U = mC_v\Delta T$. Because the temperature is constant, $\Delta T = 0$ and thus $\Delta U = 0$.)

For the expansions, there is no work input or heat output; therefore Equation 23-1 becomes

$$0 + 0 + 0 = 0 - W_{\text{out}} + Q_{\text{in}} - 0 \tag{23-17}$$

which reduces to

$$Q_{\text{in}} = W_{\text{out}} \tag{23-18}$$

This equation says that the work output by the gas that raised the piston was supplied as heat from the water bath. For the irreversible expansion, substituting Equation 23-5 for the work output, we obtain

$$Q_{\text{in,irrev}} = P_2(V_2 - V_1) \qquad \text{(Irreversible expansion)} \tag{23-19}$$

and for the reversible expansion, substituting Equation 23-13 for the work output, we obtain

$$Q_{\text{in,rev}} = nRT \ln \frac{V_2}{V_1} \qquad \text{(Reversible expansion)} \tag{23-20}$$

For the compression, there is no change in the state of the system for the reasons previously described, but there is work input (to compress the gas) and heat output. Thus, the energy accounting equation is

$$0 + 0 + 0 = W_{\text{in}} - 0 + 0 - Q_{\text{out}} \tag{23-21}$$

which reduces to

$$Q_{\text{out}} = W_{\text{in}} \tag{23-22}$$

For the irreversible compression, we may substitute Equation 23-14 to get

$$Q_{\text{out}} = P_1(V_2 - V_1) \qquad \text{(Irreversible compression)} \tag{23-23}$$

and for the reversible compression, we may substitute Equation 23-16 to get

$$Q_{\text{out}} = nRT \ln \frac{V_2}{V_1} \qquad \text{(Reversible compression)} \tag{23-24}$$

23.2 ENTROPY

Thermodynamics was a science developed primarily in the 1800s as people tried to explain the workings of steam engines. It quickly became apparent that heat available at high temperatures is much more valuable than heat available at low temperatures. For example, 1000 Btu of heat taken from a large water reservoir at 400°F can be used to make about 1 lb_m of 250-psia steam, which may be fed to a steam engine that produces useful shaft work to run trains or factories. In contrast, 1000 Btu taken from a large water reservoir at 100°F may be used to heat a home, and that's about it. Heat available at low temperatures just is not worth much.

In thermodynamics, we wish to quantify this notion that high-temperature heat is valuable, but low-temperature heat is not. One question we must deal with is, What heat are you talking about? As shown by Equations 23-19 and 23-20, which describe the heat required for the irreversible and reversible expansions, there are a number of heat inputs that allow the gas to go from volume V_1 to volume V_2. In fact, there are an infinite number of possible irreversible heat inputs; the irreversible path shown in Figure 23.5 is only one of an infinite number of possible irreversible expansions. However, there is **only one** reversible path from V_1 to V_2 at constant temperature. Thus, when quantifying this notion that high-temperature heat is more valuable than low-temperature heat, it is logical to specify "reversible heat," that is, the heat that passes the system boundary for a reversible process. We can propose a number of algebraic combinations of reversible heat Q_{rev} and temperature T to help us quantify our notion:

$$\cdots \frac{Q_{rev}}{T^3} \quad \frac{Q_{rev}}{T^2} \quad \frac{Q_{rev}}{T} \quad Q_{rev}T \quad Q_{rev}T^2 \quad Q_{rev}T^3 \cdots$$

When heat is delivered to a reversible process at higher temperatures, the first three proposed quantities get smaller whereas the second three get larger. Of these proposed quantities, the one that has proved most useful is Q_{rev}/T. In fact, the quantity is so useful, it is given a special name, *entropy S:*

$$\Delta S \equiv \frac{Q_{rev}}{T} \tag{23-25}$$

where T must be the absolute temperature. Q_{rev} is defined to mean the *net input* of reversible heat into a system. Entropy is an interesting quantity because it combines a path quantity (heat) with a state quantity (temperature). This raises the natural question, Is entropy a state quantity or a path quantity? Although it is true that heat is a path quantity, we have been careful to define a very specific path—the reversible path. According to Rule 9 in Chapter 17, on accounting, algebraic combinations of state quantities and *well-defined* path quantities are state quantities. Thus, we conclude that entropy is a state quantity.

Observation 8: Entropy is a state quantity.

The (Δ) in Equation 23-25 emphasizes that net reversible heat input to a system causes a "change" in the state of its entropy.

To become more comfortable with this newly defined quantity, entropy, let's calculate the entropy change associated with the reversible expansion of gas from V_1 to V_2. Equation 23-20 provides an expression for Q_{rev} so that

$$\Delta S = \frac{Q_{rev}}{T} = \frac{nRT \ln \frac{V_2}{V_1}}{T} = nR \ln \frac{V_2}{V_1} \tag{23-26}$$

This equation can be manipulated by dividing both sides by the number of moles n. Similarly, the numerator and denominator of the logarithm may be divided by n:

$$\frac{\Delta S}{n} = \Delta \hat{S} = R \ln \frac{\frac{V_2}{n}}{\frac{V_1}{n}} = R \ln \frac{\hat{V}_2}{\hat{V}_1} \tag{23-27}$$

Here, we have shown that not only is entropy a state quantity, it also can be defined as a *specific* state quantity much like specific volume, specific internal energy, and specific enthalpy. Thus, entropy can be a property of matter.

Observation 9: Entropy can be a property of matter.

By inspecting Equation 23-26, we see that as the volume increases, so does the entropy. When the volume increases, the gas molecules have more possible places to go, so we can interpret the gas in the large-volume system to be more disordered. This leads us to a famous observation about entropy.

Observation 10: Entropy is a measure of the disorder of a system.

EXAMPLE 23.1

Problem Statement: Calculate the entropy change of the gas when one mole of perfect gas *reversibly* expands from 1 m^3 to 2 m^3.

Solution:

$$\Delta S_{exp} = S_2 - S_1 \equiv \frac{Q_{rev}}{T} = \frac{nRT \ln \frac{V_2}{V_1}}{T}$$

$$= nR \ln \frac{V_2}{V_1} = (1 \text{ mol})\left(8.314 \frac{\text{J}}{\text{mol·K}}\right) \ln \frac{2 \text{ m}^3}{1 \text{ m}^3} = 5.76 \frac{\text{J}}{\text{mol}}$$

EXAMPLE 23.2

Problem Statement: Calculate the entropy change of the gas when one mole of perfect gas *irreversibly* expands from 1 m^3 to 2 m^3 using the path described in Figure 23.5.

Solution:

$$\Delta S_{exp} = S_2 - S_1 \equiv \frac{Q_{rev}}{T} = \frac{nRT \ln \frac{V_2}{V_1}}{T}$$

$$= nR \ln \frac{V_2}{V_1} = (1 \text{ mol})\left(8.314 \frac{\text{J}}{\text{mol·K}}\right) \ln \frac{2 \text{ m}^3}{1 \text{ m}^3} = 5.76 \frac{\text{J}}{\text{mol}}$$

Note that the entropy change of the gas was the same for both the reversible and irreversible expansions! This unexpected result occurs because entropy is a state quantity and does not depend on the specific path taken. Although the actual heat transfer during the irreversible expansion is given by Equation 23-19, the definition of entropy does not ask for the actual heat transfer; it asks for the heat transfer that *would* have occurred *if* the process were reversible.

Observation 11: Entropy changes for a real (i.e., irreversible) process that goes from State 1 to State 2 are calculated by finding a reversible path that goes from State 1 to State 2 and determining the heat transfer across the system boundary during this hypothetical reversible path.

Entropy and Disorder

Entropy is a measure of the disorder of a system. To help you visualize this, imagine we have a perfect (or ideal) gas in which the volume of a single molecule is V_m. It is placed in a container with initial volume V_1. In the following figure, the volume V_1 is only 4 times larger than V_m. There are 4 possible arrangements if there is one molecule placed in the container, 12 possible arrangements with two molecules, and 24 possible arrangements with three molecules.

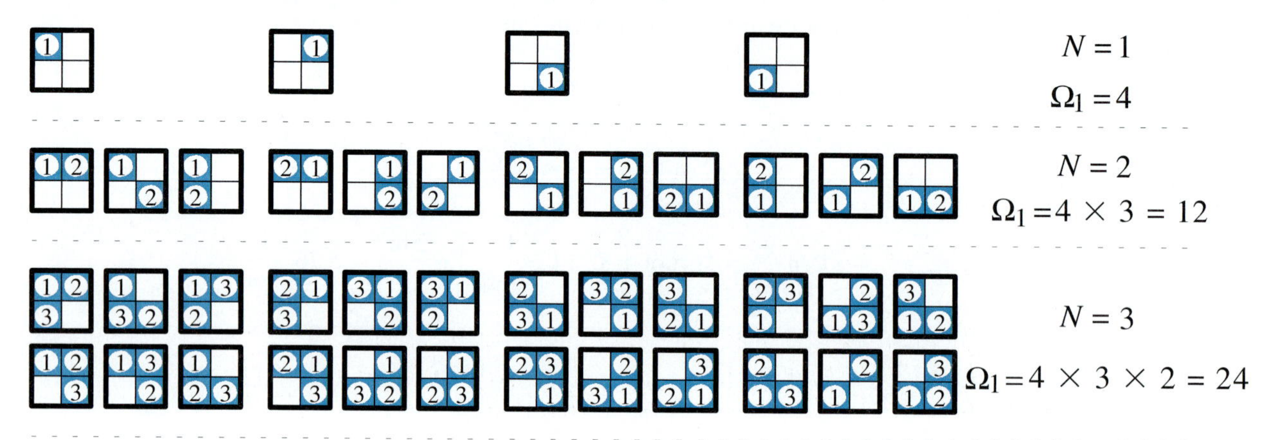

Appendix G (Mathematics Supplement) shows that the number of possible arrangements can be calculated according to the following formula:

$$\Omega_1 = \frac{\left(\dfrac{V_1}{V_m}\right)!}{\left(\dfrac{V_1}{V_m} - N\right)!}$$

where Ω_1 is the number of possible arrangements and N is the number of molecules. If the container volume is **much** larger than the total volume of the molecules (which is true for a perfect gas), then the number of possible arrangements is given by

$$\Omega_1 \cong \left(\frac{V_1}{V_m}\right)^N \quad \text{where} \quad \frac{V_1}{V_m} >> N$$

This is illustrated in the following figure:

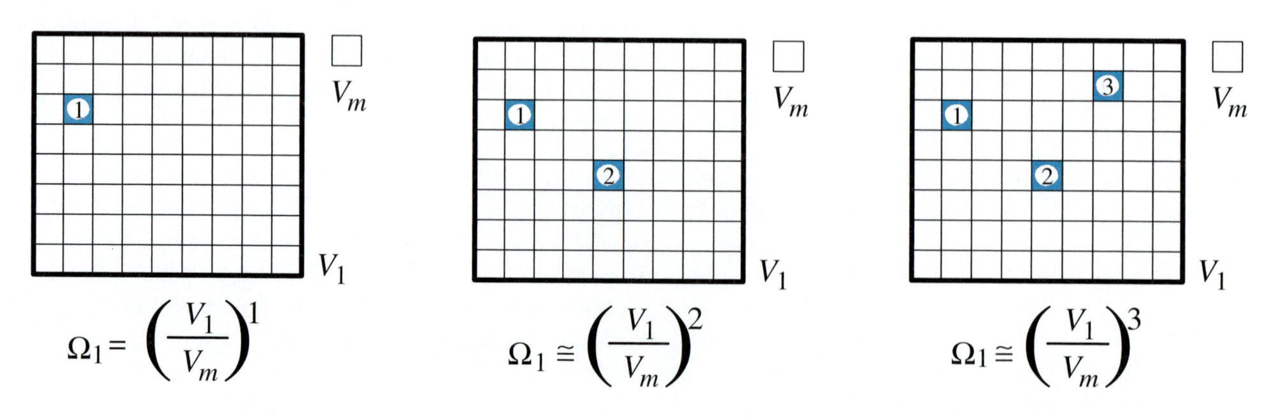

If the container volume is increased to V_2, then the number of possible arrangements is

$$\Omega_2 \cong \left(\frac{V_2}{V_m}\right)^N \quad \text{where} \quad \frac{V_2}{V_m} >> N$$

The ratio of the number of possible arrangements is

$$\frac{\Omega_2}{\Omega_1} = \frac{\left(\frac{V_2}{V_m}\right)^N}{\left(\frac{V_1}{V_m}\right)^N} = \left(\frac{V_2}{V_1}\right)^N$$

If the natural logarithm is taken of both sides, we have

$$\ln \frac{\Omega_2}{\Omega_1} = \ln \left(\frac{V_2}{V_1}\right)^N = N \ln \left(\frac{V_2}{V_1}\right)$$

We can express the total number of molecules as the number of moles n times Avogadro's number N_A:

$$\ln \frac{\Omega_2}{\Omega_1} = nN_A \ln \left(\frac{V_2}{V_1}\right)$$

Both sides of the equation may be multiplied by a constant, in this case R/N_A where R is the universal gas constant:

$$\frac{R}{N_A} \ln \frac{\Omega_2}{\Omega_1} = \frac{R}{N_A} nN_A \ln \left(\frac{V_2}{V_1}\right)$$

$$\frac{R}{N_A} \ln \frac{\Omega_2}{\Omega_1} = nR \ln \left(\frac{V_2}{V_1}\right)$$

This equation should look familiar to you; the right side is identical to Equation 23-26.

$$\frac{R}{N_A} \ln \frac{\Omega_2}{\Omega_1} = \Delta S = S_2 - S_1$$

The constant R/N_A is called Boltzmann's constant, k_B.

$$k_B \ln \frac{\Omega_2}{\Omega_1} = \mathrm{S}_2 - \mathrm{S}_1$$

Using a property of logarithms, the left side can be expressed as

$$k_B \ln \Omega_2 - k_B \ln \Omega_1 = S_2 - S_1$$

Term-by-term comparison shows that

$$S = k_B \ln \Omega$$

This equation says that entropy S is proportional to the number of possible arrangements Ω. As the number of possible arrangements increases, so does the "disorder" of the system; therefore, entropy is a measure of the system's disorder. This equation is so important that Boltzmann had it engraved on his tombstone.

EXAMPLE 23.3

Problem Statement: Calculate the entropy change of the gas when one mole of ideal gas is *reversibly* compressed from 2 m^3 to 1 m^3.

Solution:

$$\Delta S_{\text{comp}} = S_1 - S_2 \equiv \frac{Q_{\text{rev}}}{T} = \frac{nRT \ln \frac{V_1}{V_2}}{T}$$

$$= nR \ln \frac{V_1}{V_2} = (1 \text{ mol})\left(8.314 \frac{\text{J}}{\text{mol·K}}\right) \ln \frac{1 \text{ m}^3}{2 \text{ m}^3} = -5.76 \frac{\text{J}}{\text{K}}$$

In the above examples, the gas was expanded from V_1 to V_2 and then compressed back to V_1. It was a **single-cycle process,** meaning that the system was returned to its original state. The total entropy change for this single-cycle process is

$$\Delta S_{\text{cycle}} = \Delta S_{\text{exp}} + \Delta S_{\text{comp}} = \left(5.76 \frac{\text{J}}{\text{K}}\right) + \left(-5.76 \frac{\text{J}}{\text{K}}\right) = 0$$

Thus, for a single-cycle process, the entropy *of the system* does not change. We would reach this conclusion regardless of whether we were studying a reversible or irreversible system. This is an expected result because entropy is a state quantity; if the system returns to its original state, then the entropy (a state quantity) of the system does not change.

Observation 12: In all single-cycle processes, the entropy *of the system* does not change.

23.2.1 Some More Entropy Changes at Constant Temperature

So far in our discussion of entropy, we have looked only at processes that occur at constant temperature. In this section, we will add a few more constant-temperature entropy changes to our repertoire.

Melting.

Figure 23.7 shows a solid in which heat is applied, causing it to melt. Because the volume changes slightly upon melting, this process is performed in a piston/cylinder. A mass is placed upon the piston to maintain a constant pressure in the system. The system is defined as the contents within the piston/cylinder. The piston/cylinder and its solid contents are placed in a water bath that has a temperature differentially higher than the temperature of the piston/cylinder contents.

An energy accounting of the contents shows that there is no change in the kinetic or potential energy, but there certainly is a change in the internal energy as the relatively motionless solid atoms (molecules) become highly mobile liquid atoms (molecules). A heat input across the system boundary was necessary to cause the melting. Because the mass was raised (in this case where the solid expands upon melting), the contents of the piston/cylinder performed work on the surroundings. Thus, the energy accounting of the contents gives us

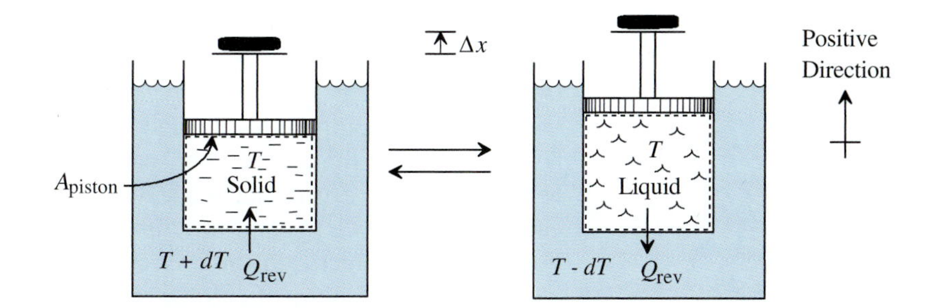

FIGURE 23.7
Melting and freezing.

$$\Delta E_k + \Delta E_p + \Delta U = W_{in} - W_{out} + Q_{in} - Q_{out} \tag{23-1}$$

$$0 + 0 + \Delta U = 0 - W_{out} + Q_{in} - 0 \tag{23-28}$$

The work output by the contents is

$$W_{out} = \int F\,dx = F \int dx = F\Delta x = (PA_{piston})\Delta x = P\Delta V \tag{23-29}$$

Substituting Equation 23-29 into Equation 23-28 and rearranging gives

$$\Delta U + P\Delta V = Q_{in} \tag{23-30}$$

Entropy and Information

Imagine you are an engineer designing a laser printer. The image on the paper is constructed from *pixels,* tiny dots that are either black or white. The pixels are controlled by sending a string of ones and zeros from the computer to the printer. For simplicity, imagine that the printer has only four pixels. We could generate the following images by sending the following numbers to the printer:

The number of possible messages, M, can be calculated by the following formula:

$$M = s^n$$

where s is the number of "letters" in our alphabet (two: a one or a zero) and n is the length of our message (four). By taking the base-2 logarithm of both sides, we obtain

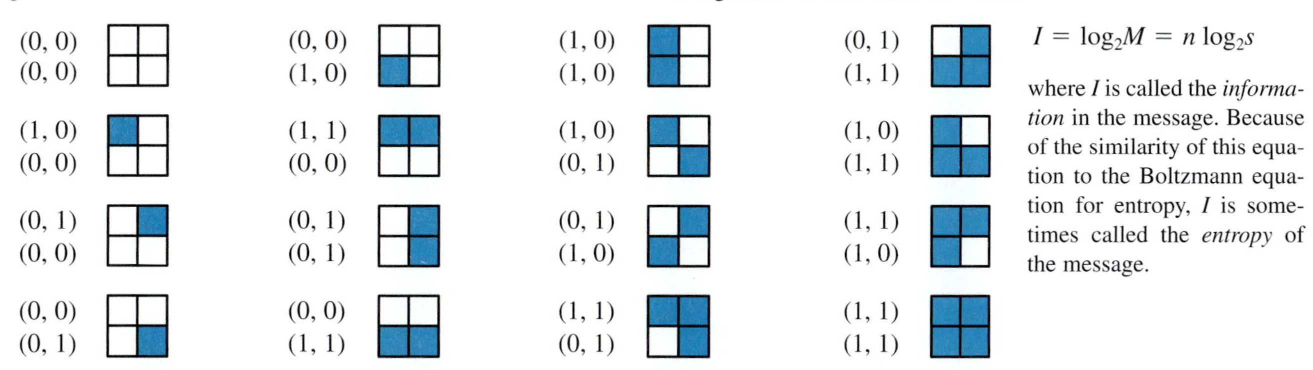

$$I = \log_2 M = n \log_2 s$$

where I is called the *information* in the message. Because of the similarity of this equation to the Boltzmann equation for entropy, I is sometimes called the *entropy* of the message.

Adapted from: P. P. Panter, *Modulation, Noise, and Spectral Analysis* (New York: McGraw-Hill, 1965).

The terms on the left are the change in enthalpy (at constant pressure),

$$\Delta H_{fus} = Q_{in} \tag{23-31}$$

where the subscript indicates **"fusion,"** that is, melting. This equation is valid whether the heat is supplied from a reversible heat source (as in Figure 23.7) or from an irreversible heat source (e.g., a very hot flame). Recall from Observation 11 that when calculating entropy changes, we hypothesize the existence of a reversible path and base our calculations on this hypothetical path, even if our actual path is not reversible.

Using the definition of entropy and substituting Equation 23-31, we obtain

$$\Delta S_{fus} \equiv \frac{Q_{rev}}{T} = \frac{Q_{in}}{T} = \frac{\Delta H_{fus}}{T} \qquad \text{(Constant pressure)} \tag{23-32}$$

EXAMPLE 23.4

Problem Statement: Calculate the entropy change when 2.00 kg of ice are melted at 0°C. The latent heat of fusion for water (taken from the table in Chapter 22) is 333.56 kJ/kg.

Solution:

$$\Delta S_{fus} = \frac{\Delta H_{fus}}{T} = \frac{m\Delta\hat{H}}{T} = \frac{(2.00 \text{ kg})(333.56 \text{ kJ/kg})}{(273.15 + 0) \text{ K}} = 2.44 \text{ kJ/K}$$

Freezing.

Freezing occurs when a liquid is transformed into a solid and is the reverse process shown in Figure 23.7. The energy accounting for freezing is nearly identical to that for melting, except that heat must be removed and work is done on the contents (in this case where the liquid contracts upon freezing). Thus, our energy accounting becomes

$$\Delta E_k + \Delta E_p + \Delta U = W_{in} - W_{out} + Q_{in} - Q_{out} \tag{23-1}$$

$$0 + 0 + \Delta U = W_{in} - 0 + 0 - Q_{out} \tag{23-33}$$

W_{in} is given by Equation 23-29, except that a negative sign is needed because now the piston is moving in the negative direction:

$$W_{in} = -P\Delta V \tag{23-34}$$

Substituting Equation 23-34 into 23-33 and rearranging gives

$$\Delta U + P\Delta V = \Delta H_{freeze} = -\Delta H_{fus} = -Q_{out} \tag{23-35}$$

Note that the enthalpy change of freezing is exactly opposite the enthalpy change of fusion (i.e., melting). This equation is valid whether the heat is removed from the system and reversibly rejected to a heat sink that is differentially cooler than the piston/cylinder contents (as in Figure 23.7) or whether it is irreversibly rejected to a very cold heat sink.

Recall from Observation 11 that when calculating entropy changes, we hypothesize the existence of a reversible path, and base our calculations on this hypothetical path, even if our actual path is not reversible.

We can calculate the entropy change resulting from freezing by starting with the definition of entropy

$$\Delta S_{\text{freeze}} \equiv \frac{Q_{\text{rev}}}{T} \tag{23-36}$$

where Q_{rev} is the *net input* of reversible heat. Equation 23-35 gives us the *output* of heat; therefore, we must introduce a negative sign when we substitute it into Equation 23-36, which becomes

$$\Delta S_{\text{freeze}} = \frac{-Q_{\text{out}}}{T} = \frac{-\Delta H_{\text{fus}}}{T} \qquad \text{(Constant pressure)} \tag{23-37}$$

EXAMPLE 23.5

Problem Statement: Calculate the entropy change when 2.00 kg of liquid water are frozen at 0°C.

Solution:

$$\Delta S_{\text{freeze}} = \frac{-\Delta H_{\text{fus}}}{T} = \frac{-m\Delta \hat{H}}{T} = \frac{-(2.00\text{ kg})(333.56\text{ kJ/kg})}{(273.15 + 0)\text{ K}} = -2.44\text{ kJ/K}$$

Thus, the entropy change of freezing is identical to the entropy change of melting, except for a sign change. This indicates that in a single-cycle process where a solid substance is melted and then is returned to its original solid state, there is no change in entropy. This is consistent with Observations 8 and 12.

Vaporization and Condensation.

The same arguments used to derive the entropy changes from melting and freezing may be used to calculate the entropy changes from vaporization and condensation:

$$\Delta S_{\text{vap}} = \frac{\Delta H_{\text{vap}}}{T} \qquad \text{(Constant pressure)} \tag{23-38}$$

$$\Delta S_{\text{cond}} = \frac{-\Delta H_{\text{vap}}}{T} \qquad \text{(Constant pressure)} \tag{23-39}$$

EXAMPLE 23.6

Problem Statement: Calculate the entropy change when 2.00 kg of water are vaporized at 100°C. The latent heat of vaporization for water (taken from the table in Chapter 22) is 2256.7 kJ/kg.

Solution:

$$\Delta S_{vap} = \frac{\Delta H_{vap}}{T} = \frac{m\Delta \hat{H}_{vap}}{T} = \frac{(2.00 \text{ kg})(2256.7 \text{ kJ/kg})}{(273.15 + 100) \text{ K}} = 12.1 \text{ kJ/K}$$

Chemical Reaction.

Chemical reactions also cause a change in enthalpy, which allows the entropy change to be calculated:

$$\Delta S_{react,forward} = \frac{\Delta H_{react,forward}}{T} \quad \text{(Constant pressure)} \quad (23\text{-}40)$$

$$\Delta S_{react,reverse} = \frac{-\Delta H_{react,forward}}{T} \quad \text{(Constant pressure)} \quad (23\text{-}41)$$

The subscript "forward" indicates the reaction goes as written, and "reverse" indicates the reaction goes opposite as written.

EXAMPLE 23.7

Problem Statement: Hydrogen reacts with oxygen according to the reaction

$$H_2 + \tfrac{1}{2} O_2 \rightarrow H_2O$$

The reaction occurs on a catalyst at room temperature (25°C). What is the change in entropy when 2.00 mol of hydrogen react with oxygen to form water? The enthalpy change for this reaction (as taken from the heats of combustion listed in Chapter 22) is −285.84 kJ/mol hydrogen.

Solution:

$$\Delta S_{react,forward} = \frac{\Delta H_{react,forward}}{T}$$

$$= \frac{n\Delta \hat{H}}{T} = \frac{(2.00 \text{ mol})(-285.84 \text{ kJ/mol})}{(273.15 + 25) \text{ K}} = -1.92 \text{ kJ/K}$$

If this reaction were performed in reverse (water was split into hydrogen and oxygen by electrolysis, for example), then the entropy change would be +1.92 kJ/K.

Gas Mixing.

Figure 23.8 shows two cylinders immersed in a water bath that holds the temperature constant. One cylinder holds Species A (white spheres) and the other cylinder holds Species B (black spheres). The valve connecting each cylinder is opened, allowing the contents of each cylinder to mix. Species A is undergoing an isothermal expansion from V_1 to $(V_1 + V_2)$. Equation 23-26 allows us to calculate the change in entropy for Species A:

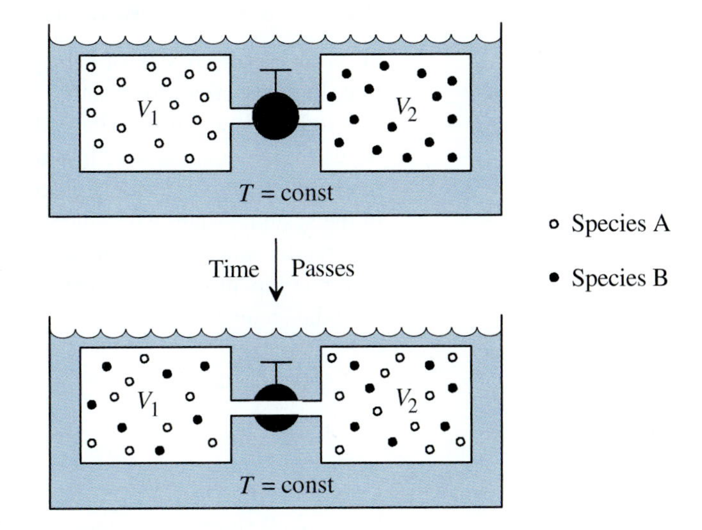

FIGURE 23.8
Mixing of gases.

$$\Delta S_A = n_A R \ln \frac{V_1 + V_2}{V_1}$$

Species B undergoes an expansion from V_2 to $(V_1 + V_2)$. Equation 23-26 allows us to calculate the change in entropy for Species B:

$$\Delta S_B = n_B R \ln \frac{V_1 + V_2}{V_2}$$

The total entropy change for this mixing process is

$$\Delta S_{tot} = \Delta S_A + \Delta S_B = n_A R \ln \frac{V_1 + V_2}{V_1} + n_B R \ln \frac{V_1 + V_2}{V_2} \tag{23-42}$$

EXAMPLE 23.8

Problem Statement: The gases within two tanks are allowed to mix. One tank contains 3.0 mol of methane within a 100-L volume, and the other tank contains 5.0 mol of ethane within a 50-L volume. What is the entropy change of the two gases when they mix?

Solution:

$$\Delta S_{tot} = (3.0 \text{ mol})\left(8.314 \frac{\text{J}}{\text{mol·K}}\right) \ln \frac{100 \text{ L} + 50 \text{ L}}{100 \text{ L}}$$

$$+ (5.0 \text{ mol})\left(8.314 \frac{\text{J}}{\text{mol·K}}\right) \ln \frac{100 \text{ L} + 50 \text{ L}}{50 \text{ L}} = 56 \frac{\text{J}}{\text{K}}$$

23.2.2 Entropy Changes at Nonconstant Temperature

So far in our discussion of entropy, we have performed all processes at constant temperature. In this section, we discuss some processes that occur at nonconstant temperature.

If the temperature of the system changes, we cannot use Equation 23-25. What would we use as *the* temperature? Instead, we imagine a reversible process in which we add a differential amount of heat at a given temperature, which causes a differential entropy change in the system. Then, when the system comes to a new temperature, we add an additional differential amount of heat, which causes an additional differential entropy change. We continue this process until the new temperature is reached. In this case, we define a differential change in entropy as

$$dS \equiv \frac{dQ_{\text{rev}}}{T} \tag{23-43}$$

Constant-Volume Temperature Change.

Figure 23.9 shows a substance within a constant-volume container that is immersed in a temperature bath. In this case, the bath is **not** at constant temperature; rather, it has a differentially higher temperature than the contents of the container, which allows a differential amount of reversible heat to be added.

There is no change in kinetic or potential energy in the system, which is defined to be the contents of the container. There is no heat output, nor is there work input or output. Therefore, the energy accounting gives

$$\Delta E_k + \Delta E_p + \Delta U = W_{\text{in}} - W_{\text{out}} + Q_{\text{in}} - Q_{\text{out}} \tag{23-1}$$

$$0 + 0 + \Delta U = 0 - 0 + Q_{\text{in}} - 0$$

$$\Delta U = Q_{\text{in}} \tag{23-44}$$

As discussed in Chapter 11 on thermodynamics, the constant-volume heat capacity of the substance in the container allows us to calculate the change in internal energy as

$$\Delta U = nC_v \Delta T \tag{23-45}$$

where n is the moles of the substance in the container.

At an instant in time, the change in internal energy is

FIGURE 23.9
Constant-volume temperature change.

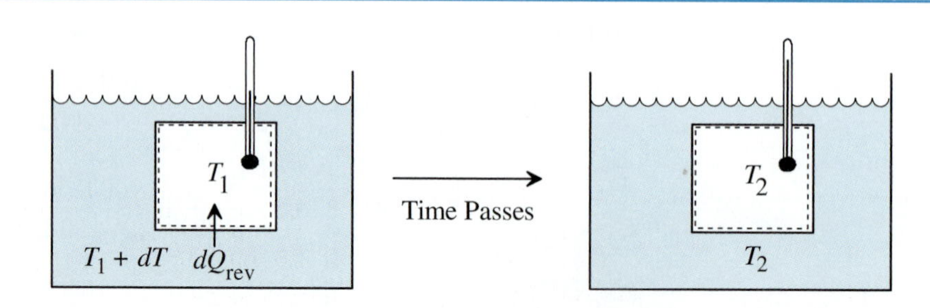

$$dU = nC_v\,dT \tag{23-46}$$

The differential change in internal energy is caused by a differential amount of heat addition, which in this case is reversibly added:

$$dU = dQ_{\text{rev}} = nC_v\,dT \tag{23-47}$$

From the definition of entropy given in Equation 23-43, we can calculate the differential entropy change of the substance in the constant-volume container as

$$dS \equiv \frac{dQ_{\text{rev}}}{T} = \frac{nC_v\,dT}{T} \tag{23-48}$$

To find the total entropy change of the system as it is heated from T_1 to T_2, we integrate the above expression:

$$\int_{S_1}^{S_2} dS = [S]_{S_1}^{S_2} = S_2 - S_1 = \Delta S = \int_{T_1}^{T_2} \frac{nC_v\,dT}{T}$$

Assuming C_v does not change with temperature, then

$$\Delta S = nC_v \int_{T_1}^{T_2} \frac{dT}{T} = nC_v[\ln T]_{T_1}^{T_2} = nC_v[\ln T_2 - \ln T_1] = nC_v \ln \frac{T_2}{T_1}$$

Thus, we have derived a simple relationship for the total entropy change of a substance when its temperature changes while maintained at a constant volume:

$$\Delta S = nC_v \ln \frac{T_2}{T_1} \tag{23-49}$$

Although this equation was derived by assuming the heat was added reversibly, it applies even for irreversible heat addition (recall Observations 8 and 11) at constant volume.

EXAMPLE 23.9

Problem Statement: While maintaining a constant volume, 1.0 mol of helium is heated from 20°C to 30°C. What is its change in entropy? [*Note:* The constant-volume heat capacity for helium is 12.5 J/(mol·K).]

Solution:

$$\Delta S = nC_v \ln \frac{T_2}{T_1} = (1.0\text{ mol})\left(12.5\,\frac{\text{J}}{\text{mol·K}}\right)\ln \frac{(30 + 273.15)\text{ K}}{(20 + 273.15)\text{ K}} = 0.42\text{ J/K}$$

Constant-Pressure Temperature Change.

Figure 23.10 shows a substance within a constant-pressure container that is immersed in a temperature bath. In this case, the bath is **not** at constant temperature; rather, it has a differentially higher temperature than the contents of the container, which allows a differential amount of reversible heat to be added.

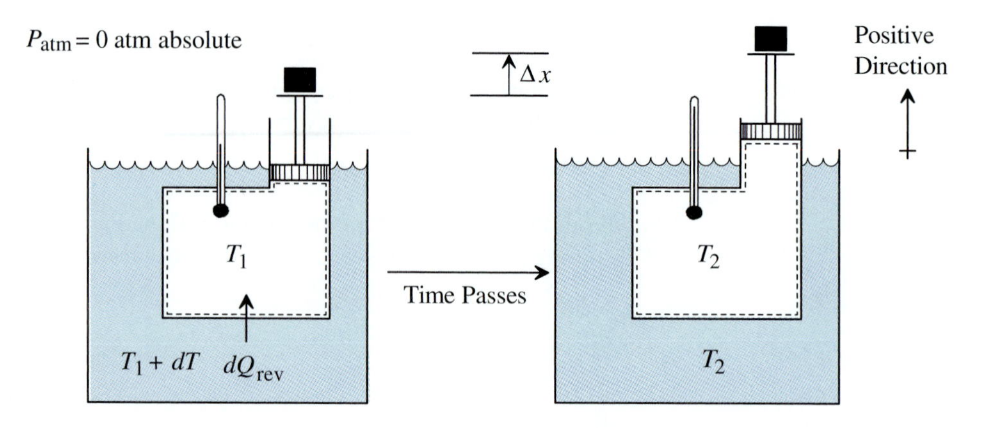

FIGURE 23.10
Temperature change at constant pressure.

There is no change in kinetic or potential energy in the system, which is defined to be the contents of the container. There is no heat output, nor is there work input, but there is work output because a weight is lifted as the substance expands. Therefore, the energy accounting gives

$$\Delta E_k + \Delta E_p + \Delta U = W_{\text{in}} - W_{\text{out}} + Q_{\text{in}} - Q_{\text{out}} \tag{23-1}$$

$$0 + 0 + \Delta U = 0 - W_{\text{out}} + Q_{\text{in}} - 0$$

$$\Delta U = -W_{\text{out}} + Q_{\text{in}} \tag{23-50}$$

As we previously derived (Equation 23-29), the work output of the gas that lifts the mass is

$$W_{\text{out}} = P\Delta V$$

which may be substituted into Equation 23-50, giving

$$\Delta U = -P\Delta V + Q_{\text{in}} \tag{23-51}$$

The work term may be brought over to the left, giving

$$\Delta U + P\Delta V = Q_{\text{in}} \tag{23-52}$$

The left side of the equation is defined as change in enthalpy, provided the pressure is constant; thus,

$$\Delta H = Q_{\text{in}} \tag{23-53}$$

As discussed in Chapter 11 on thermodynamics, the constant-pressure heat capacity of the substance in the container allows us to calculate the change in enthalpy as

$$\Delta H = nC_p\Delta T \tag{23-54}$$

At an instant in time, a differential change in temperature causes a differential change in enthalpy

$$dH = nC_p\, dT \tag{23-55}$$

The differential change in enthalpy is caused by a differential amount of heat addition, which in this case is reversibly added:

$$dH = dQ_{\text{rev}} = nC_p\,dT \tag{23-56}$$

Based upon the definition of entropy given in Equation 23-43, we can calculate the differential entropy change of the substance in the constant-pressure container to be

$$dS = \frac{dQ_{\text{rev}}}{T} = \frac{nC_p\,dT}{T} \tag{23-57}$$

To find the total entropy change of the system as it is heated from T_1 to T_2, we integrate the above expression, writing

$$\int_{S_1}^{S_2} dS = [S]_{S_1}^{S_2} = S_2 - S_1 = \Delta S = \int_{T_1}^{T_2} \frac{nC_p\,dT}{T}$$

Assuming C_p does not change with temperature, then

$$\Delta S = nC_p \int_{T_1}^{T_2} \frac{dT}{T} = nC_p[\ln T]_{T_1}^{T_2} = nC_p[\ln T_2 - \ln T_1] = nC_p \ln \frac{T_2}{T_1}$$

Thus, we have derived a simple expression for the entropy change of a substance that changes temperature from T_1 to T_2 while the pressure is maintained constant:

$$\Delta S = nC_p \ln \frac{T_2}{T_1} \tag{23-58}$$

Although this equation was derived by assuming the heat was added reversibly, it applies even for irreversible heat addition (recall Observations 8 and 11).

EXAMPLE 23.10

Problem Statement: While maintaining a constant pressure, 1.0 mol of helium is heated from 20°C to 30°C. What is its change in entropy? [*Note:* The constant-pressure heat capacity for helium obtained from Table 22.2 is 20.8 J/(mol·K).]

Solution:

$$\Delta S = nC_p \ln \frac{T_2}{T_1} = (1.0 \text{ mol})\left(20.8 \frac{\text{J}}{\text{mol·K}}\right) \ln \frac{(30 + 273.15)\text{ K}}{(20 + 273.15)\text{ K}} = 0.70 \text{ J/K}$$

23.3 ACCOUNTING FOR ENTROPY

Heat is an extensive quantity; therefore, based upon Rule 5 in Chapter 17 on accounting, so is entropy. Because entropy is an extensive quantity, it may be counted using the universal accounting equation. As with energy, we must develop separate accounting equations for closed and open systems.

23.3.1 Accounting for Entropy in Closed Systems

So far, all the systems we have studied were *closed,* meaning no mass crosses the system boundary. The proposed accounting statement for the entropy change of a closed system may be written as

$$\Delta S = S_{\text{in}} - S_{\text{out}} + S_{\text{gen}} - S_{\text{cons}} \qquad \text{(Proposed)} \tag{23-59}$$

Here, we are being very general and allowing for the possibility of generating and consuming entropy, because we do not assert that entropy is a conserved quantity.

To test this proposed accounting equation, let's use it to evaluate the entropy changes in a rod that transfers heat from a constant-temperature water bath at $(T_1 + dT)$ to a constant-temperature water bath at $(T_2 + dT)$ (Figure 23.11). The rod is insulated, except where it touches the water baths, so that heat is conducted from the hot bath to the cold bath only, and not out the sides. The temperature of the left end of the rod is T_1 and the temperature of the right end is T_2. Because the temperature of the hot bath is differentially higher than the left end of the rod, reversible heat Q_{rev} enters the left end of the rod. Because the temperature of the cold water bath is differentially colder than the right end of the rod, reversible heat Q_{rev} flows out the right end of the rod.

We define our system to be the rod itself. Although the rod temperature does vary spatially (one end is hot and the other is cold), it does not vary its temperature with time (the hot end stays a constant temperature and the cold end stays a constant temperature, regardless of how much time has elapsed). Also, the rod does not change its volume or pressure. Because none of these other state properties have changed, there will be no change in the state of the rod's entropy. Therefore,

$$\Delta S = 0 \tag{23-60}$$

The entropy input can be calculated from the definition of entropy

$$S_{\text{in}} \equiv \frac{Q_{\text{rev,in}}}{T} \tag{23-61}$$

and the flow of entropy output can be determined from the definition of entropy

$$S_{\text{out}} \equiv \frac{Q_{\text{rev,out}}}{T} \tag{23-62}$$

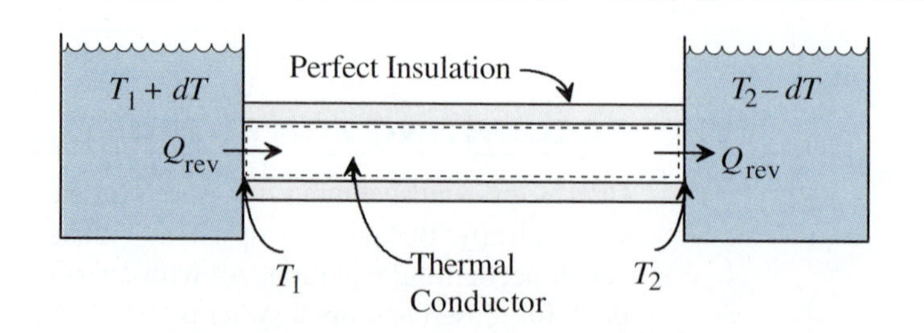

FIGURE 23.11
Heat transfer from a high-temperature bath to a low-temperature bath.

(*Note:* Previously, we have emphasized that entropy is a state quantity, which it is. Here, we are showing that entropy can flow across the system boundary, so in this case, it is also a path quantity. This issue of a quantity being both state and path was discussed in Chapter 17 on accounting. If this confuses you, please refer back to that chapter.)

Equations 23-60, 23-61, and 23-62 may be substituted into our proposed accounting equation, yielding

$$0 = \frac{Q_{\text{rev,in}}}{T_1} - \frac{Q_{\text{rev,out}}}{T_2} + S_{\text{gen}} - S_{\text{cons}} \tag{23-63}$$

Because the rod is perfectly insulated and the rod temperature does not change with time, then $Q_{\text{rev,in}} = Q_{\text{rev,out}}$, and

$$0 = \frac{Q_{\text{rev}}}{T_1} - \frac{Q_{\text{rev}}}{T_2} + S_{\text{gen}} - S_{\text{cons}} = Q_{\text{rev}}\left(\frac{1}{T_1} - \frac{1}{T_2}\right) + S_{\text{gen}} - S_{\text{cons}} \tag{23-64}$$

Rearranging this equation slightly gives

$$S_{\text{cons}} - S_{\text{gen}} = Q_{\text{rev}}\left(\frac{1}{T_1} - \frac{1}{T_2}\right) \tag{23-65}$$

Because heat only flows from a higher temperature to a lower temperature, then $T_1 > T_2$. Therefore, $1/T_1 < 1/T_2$, which means the right side of Equation 23-65 is negative. This is consistent with entropy being generated; the right side of the equation would have to be positive for entropy to be consumed. Therefore,

$$S_{\text{gen}} = Q_{\text{rev}}\left(\frac{1}{T_2} - \frac{1}{T_1}\right) \tag{23-66}$$

Of course, from Observation 11, we can generalize this equation to include any heat flow, even if it were not reversibly added or removed from the rod.

$$S_{\text{gen}} = Q\left(\frac{1}{T_2} - \frac{1}{T_1}\right) \tag{23-67}$$

The flow of heat from a high-temperature body to a low-temperature body is a classic irreversible process. We can generalize our observation that entropy is generated in this heat transfer process as follows:

> **Observation 13:** Irreversible processes generate entropy.

It is important that you fully comprehend Observation 13. In Chapter 17 on accounting, we were careful to distinguish between the word **input** and the word *generate.* When a quantity is generated within a system, this means that there is more of it in the universe. In contrast, when a quantity is input to a system, it must be taken from the surroundings, so there is no net change of the quantity in the universe. (Review Chapter 17 on accounting if the distinction between *input* and **generation** is unfamiliar to you.) Thus, based upon Observation 13, we understand that entropy is **not** a conserved quantity, and the amount of it in the universe increases as irreversible processes occur.

As another example of an irreversible process, let's return to the irreversible expansion and compression depicted in Figure 23.5. To simplify the situation, let's imagine that the universe contains only the water bath (the surroundings) and the piston/cylinder (the system). In addition, we will include the weight and the shelves in our definition of the system. (*Note:* These objects were not included in our original definition of the system when we studied this process earlier.) Because the water bath is assumed to occupy most of the universe, it obviously is very large. Therefore, if heat flows into, or out of, the water bath, its temperature T will not change by a measurable amount.

In Figure 23.5, when the gas irreversibly expanded, heat was drawn out of the water bath (i.e., surroundings) into the system as given by Equation 23-19. Thus,

$$Q_{\text{out,surr}} = P_2(V_2 - V_1) \tag{23-68}$$

When the gas was irreversibly compressed, heat was added to the water bath (surroundings) as given by Equation 23-23; thus,

$$Q_{\text{in,surr}} = P_1(V_2 - V_1) \tag{23-69}$$

If we perform an entropy accounting on the water bath (surroundings),

$$\Delta S = S_{\text{in}} - S_{\text{out}} = \frac{Q_{\text{in,surr}}}{T} - \frac{Q_{\text{out,surr}}}{T}$$

$$= \frac{P_1(V_2 - V_1)}{T} - \frac{P_2(V_2 - V_1)}{T} = \frac{V_2 - V_1}{T}(P_1 - P_2) \tag{23-70}$$

Because $V_2 > V_1$ and $P_1 > P_2$, then ΔS must be positive, meaning the entropy in the surroundings increased. The expansion/compression is a single-cycle process, meaning the system returned to its original state. Thus, we can make the following observation:

> **Observation 14:** Single-cycle irreversible processes increase the entropy of the surroundings.

In a single-cycle process, the entropy of the system does not change. Because the single-cycle irreversible process increased the entropy of the surroundings, we can conclude that the entropy of the universe must have increased, which confirms Observation 13.

For the reversible expansion/compression depicted in Figure 23.5, the heat removed from the water bath (surroundings) during the expansion is given by Equation 23-20, which we can write as

$$Q_{\text{out,surr}} = nRT \ln \frac{V_2}{V_1} \tag{23-71}$$

and the heat added to the water bath (surroundings) during the compression is given by Equation 23-24,

$$Q_{\text{in,surr}} = nRT \ln \frac{V_2}{V_1} \tag{23-72}$$

By performing an entropy accounting of the water bath (surroundings) for this single-cycle process, we obtain

$$\Delta S = S_{in} - S_{out} = \frac{Q_{in,surr}}{T} - \frac{Q_{out,surr}}{T} = \frac{nRT \ln \frac{V_2}{V_1}}{T} - \frac{nRT \ln \frac{V_2}{V_1}}{T} = 0 \qquad (23\text{-}73)$$

This is a very interesting result. It says that a reversible single-cycle process does **not** change the entropy of the surroundings.

Observation 15: Single-cycle reversible processes do not increase the entropy of the surroundings.

Of course, the entropy of a single-cycle process does not change (Observation 12) and we just observed that the entropy of the surroundings does not change when a reversible single-cycle process is performed; therefore, we can conclude that the entropy of the universe does not change when reversible processes occur.

Observation 16: Reversible processes do not increase the entropy of the universe.

Observations 15 and 16 are two of the most important observations in all of thermodynamics. We can summarize them as

$$\Delta S_{univ.} > 0 \qquad \text{(Irreversible process)} \qquad (23\text{-}74)$$

$$\Delta S_{univ} = 0 \qquad \text{(Reversible processes)} \qquad (23\text{-}75)$$

These observations have been immortalized as the second law of thermodynamics, which can be stated in words as: The entropy of the universe can stay the same, or it can increase, but it can never decrease. This is equivalent to saying that entropy cannot be consumed, so we must revise our entropy accounting equation for a closed system to

$$\Delta S = S_{in} - S_{out} + S_{gen} \qquad \text{(Actual)} \qquad (23\text{-}76)$$

The universe is an example of an isolated system—a system in which *nothing* crosses the boundary (see Chapter 17 on accounting for a discussion of isolated systems). Equations 23-74 and 23-75 may be generalized for any isolated system:

$$\Delta S_{isolated\ system.} > 0 \qquad \text{(Irreversible process)} \qquad (23\text{-}77)$$

$$\Delta S_{isolated\ system} = \qquad \text{(Reversible processes)} \qquad (23\text{-}78)$$

Observation 17: In an isolated system, irreversible processes increase the system entropy, whereas reversible processes do not change the system entropy.

In Scenario 1, the entropy (disorder) of the isolated system did not change; this is allowed by nature. In Scenario 2, the entropy (disorder) of the isolated system increased; this is allowed by nature. In Scenario 3, the entropy (disorder) of the isolated system decreased; this is **not** allowed by nature.

Accounting for entropy is complicated by the fact that it is **not** conserved; entropy can be generated, but not consumed. The increased entropy in Scenario 2 resulted not because entropy had been added to the system (remember, it is isolated, so *nothing* is added). Instead, entropy was generated within the system.

From Chapter 22, we know that energy is both a state and a path quantity. The state forms of energy include kinetic, potential, and internal energies and the path forms include heat, work, and energy that enters with mass transferred across the system boundary. Similarly, entropy is both a state and path quantity. When we describe the entropy within our system, we are using the state form of entropy. The path forms of entropy include entropy generated within the system and entropy that crosses the system boundary. Entropy can cross the system boundary in two ways: (1) heat transfer and (2) mass transfer. Entropy is not added to the system when work crosses the system boundary.

The concept of entropy is closely tied to concepts of reversibility and irreversibility. Irreversible processes are those that occur rapidly due to large driving forces (e.g., force, torque, voltage, pressure, concentration, electromagnetic fields). Except for a few cases (e.g., superconductors and superfluids), most real processes are irreversible. Reversible processes are an idealization in which the process is envisioned to occur infinitely slowly because the driving forces are infinitely small. Irreversible processes generate entropy, whereas reversible processes do not.

Entropy is among the most abstract quantities used by engineers. It is unlikely that you can fully grasp entropy from a single reading of this chapter. We recommend that you read it again to capitalize on the perspective gained from the first reading.

Nomenclature

A area (m^2)
C_p constant-pressure heat capacity [J/(mol·K)]
C_v constant-volume heat capacity [J/(mol·K)]
E_k kinetic energy (J)
E_p potential energy (J)
F force (N)
g acceleration due to gravity (m/s^2)
k_B Boltzmann's constant [J/(molecule·K)]
H enthalpy (J)
$\hat{H}$ specific enthalpy (J/mol or J/kg)
M energy associated with mass flow (J) or number of possible messages (dimensionless)
m mass (kg)
$\dot{m}$ mass flow (kg/s)
N number of molecules (dimensionless)
n moles (mol) or length of message (dimensionless)
P pressure (Pa)
Q heat (J)
$\dot{Q}$ heat rate (J/s)

R universal gas constant [Pa·m^3/(mol·K) or J/(mol·K)]
S entropy (J/K)
$\hat{S}$ specific entropy [J/(mol·K) or J/(kg·K)]
$\dot{S}$ entropy rate *(W/K)*
s number of letters in the alphabet (dimensionless)
T temperature (K)
U internal energy (J)
$\hat{U}$ specific internal energy (J/mol or J/kg)
V volume (m^3)
$\hat{V}$ specific volume (m^3/mol or m^3/kg)
W work (J)
x distance (m)
Ω number of possible molecular arrangements (dimensionless)

Further Readings

Fenn, J. B. *Engines, Energy, and Entropy: A Thermodynamics Primer.* San Francisco: W.H. Freeman, 1982.

Glover, C. J., and H. Jones. *Conservation Principles of Continuous Media.* New York: McGraw-Hill College Custom Series, 1994.

Glover, C. J.; K. M. Lunsford; and J. A. Fleming. *Conservation Principles and the Structure of Engineering.* New York: McGraw-Hill College Custom Series, 1992.

Serway, R. A. *Principles of Physics.* Fort Worth: Saunders, 1994.

Smith, J. M., and H. C. Van Ness. *Introduction to Chemical Engineering Thermodynamics.* 3rd ed. New York: McGraw-Hill, 1975.

PROBLEMS

23.1 Indicate whether the following statements are true or false:

(a) If a movie shows an irreversible process, it is impossible to determine if the movie is being run in the forward or reverse direction.

(b) When a system is returned to its original state, irreversible processes cause a change in the surroundings.

(c) A system that has friction is reversible.

(d) Molasses quickly flowing through a small-diameter pipe is an example of an irreversible process.

(e) Electricity flowing through a resistor is an example of a reversible process.

(f) A spinning satellite is sent into deep space in which there is a perfect vacuum. Because this is a reversible process, the satellite will never stop spinning.

(g) Using a gas furnace and an air-conditioning system, the temperature of a storage room is maintained at exactly 20°C. A container of gas is stored in the room at exactly 20°C. This temperature is about 300 K greater than absolute zero, so the gas molecules are in rapid motion. Because of frictional irreversibilities, the gas molecules slow down; therefore, the container of gas must be wrapped with heating tape to constantly supply the energy needed to maintain the gas temperature at 20°C.

(h) An ice cube melts while sitting on the kitchen counter. This is an example of an irreversible process because heat is transferred from a higher temperature (the room air) to a lower temperature (the ice).

(i) The entropy of the universe increases when an ice cube melts while sitting on the kitchen counter.

(j) An ice tray filled with liquid water is placed in the freezer. A few hours later, the liquid water freezes and becomes ice. An increase in entropy is used to determine time's arrow. Because this process occurred as time progressed, obviously, the entropy of the water increased due to the freezing process.

(k) When a refrigerator/freezer is operated, the entropy of the universe increases.

(l) A single cycle of a cyclic reversible operation is completed. The entropy of the system did not change.

(m) A single cycle of a cyclic irreversible operation is completed. The entropy of the system did not change.

(n) A single cycle of a cyclic reversible operation is completed. The entropy of the surroundings did not change.
(o) A single cycle of a cyclic irreversible operation is completed. The entropy of the surroundings did not change.
(p) A single cycle of a cyclic reversible operation is completed. The entropy of the universe did not change.
(q) A single cycle of a cyclic irreversible operation is completed. The entropy of the universe did not change.
(r) A complex process is analyzed by an engineer. He models the system as being reversible. The real process produces more work than the model predicts.
(s) A complex process is analyzed by an engineer. He models the system as being reversible. The real process requires more work than the model predicts.
(t) The system contains a mixture of 1-atm gases that are separated into their individual species at 1 atm. The entropy of the system increases.
(u) The system contains a mixture of 1-atm gases that are separated into their individual species at 1 atm. The entropy of the surroundings increases.
(v) An isolated system has two 1.00-kg objects. Initially, each object has an entropy of 1000 kJ/K. Within the system, a process occurs that causes one object to have 500 kJ/K and the other to have 700 kJ/K. This process is possible according to the second law of thermodynamics.
(w) An isolated system has two 1.00-kg objects. Initially, each object has an entropy of 1000 kJ/K. Within the system, a process occurs that causes one object to have 500 kJ/K and the other to have 1500 kJ/K. This process is possible according to the second law of thermodynamics.
(x) An isolated system has two 1.00-kg objects. Initially, each object has an entropy of 1000 kJ/K. Within the system, a process occurs that causes one object to have 500 kJ/K and the other to have 2000 kJ/K. This process is possible according to the second law of thermodynamics.
(y) Two engineering students are having a philosophical discussion about the existence of God. One student argues that life would be impossible without Divine intervention. According to her argument, life is highly ordered; such order does not spontaneously arise out of disorder. Life violates the second law of thermodynamics.

23.2 Perform the following calculations:
(a) Calculate the change in entropy of 10.0 mol of nitrogen gas that reversibly expands from 2.0 m^3 to 5.0 m^3 at 300 K.
(b) Calculate the change in entropy of 10.0 mol of nitrogen gas that irreversibly expands from 2.0 m^3 to 5.0 m^3 at 300 K. What do you conclude?
(c) Calculate the change in entropy of 10.0 mol of nitrogen gas that reversibly expands from 2.0 m^3 to 5.0 m^3 at 500 K. What do you conclude?
(d) Calculate the change in entropy of 10.0 mol of oxygen gas that reversibly expands from 2.0 m^3 to 5.0 m^3 at 300 K. What do you conclude?
(e) Calculate the change in entropy of 10.0 mol of nitrogen gas that reversibly compresses from 5.0 m^3 to 2.0 m^3 at 300 K. The nitrogen has returned to the original state described in Part (a). What do you conclude?

23.3 Calculate the entropy change of the two gases when 5.0 mol of nitrogen in a 3.0 m^3 tank are mixed with 3.0 mol of oxygen in a 2.0 m^3 tank. The final mixing volume is 5.0 m^3. The temperature is maintained at a constant 300 K.

23.4 Air consists of 21% oxygen by volume and about 79% nitrogen by volume. What is the entropy change when 1.0 mol of air at 1.0 atm of pressure is separated into two tanks, one with pure oxygen and the other with pure nitrogen? The pure gases are also at 1.0 atm of pressure. The temperature is maintained at a constant 300 K.

23.5 One pound mass of the following substances is heated from 70°F to 90°F at a constant 1.0-atm pressure. What is the entropy change of each substance? (*Note:* Constant-pressure heat capacities are available in Chapter 22 on energy accounting.)
(a) Air
(b) Nitrogen
(c) Oxygen
(d) Hydrogen
(e) Carbon dioxide

23.6 One kilogram of the following substances is melted at 1.0-atm pressure. What is the entropy change of each substance? (*Note:* Latent heats of fusion are available in Table 22.3.)
(a) Ammonia
(b) Benzene
(c) Copper
(d) Helium
(e) Hydrogen
(f) Propane
(g) Water

23.7 One kilogram of the following substances is frozen at 1.0-atm pressure. What is the entropy change of each substance? (*Note:* Latent heats of fusion are available in Table 22.3.)
(a) Ammonia
(b) Benzene
(c) Copper
(d) Helium
(e) Hydrogen
(f) Propane
(g) Water

23.8 One kilogram of the following substances is vaporized at 1.0-atm pressure. What is the entropy change of each substance? (*Note:* Latent heats of vaporization are available in Table 22.3.)
(a) Ammonia
(b) Benzene

(c) Copper
(d) Helium
(e) Hydrogen
(f) Propane
(g) Water

23.9 One kilogram of the following substances is condensed at 1.0-atm pressure. What is the entropy change of each substance? (*Note:* Latent heats of vaporization are available in Table 22.3.)
(a) Ammonia
(b) Benzene
(c) Copper
(d) Helium
(e) Hydrogen
(f) Propane
(g) Water

23.10 An *automobile differential* is a device that connects to the drive shaft and distributes motive power to the rear wheels. When making a turn, the differential allows the wheel at the outer edge of the turn to spin faster than the wheel at the inner edge. A particular differential has an efficiency of 97%, meaning 97% of the drive-shaft power is delivered to the wheels. The remaining power (3%) is dissipated as heat. The drive shaft power input to the differential is 15 hp. What is the power output (hp) to the wheels? How much heat (Btu/h) is produced in the differential? The differential temperature is 150°F; at what rate does it generate entropy [Btu/(h·°R)]?

23.11 An electrical resistor has a voltage drop of 5.00 V when 0.300 A of current flows. The resistor is at a steady-state temperature of 55°C. At what rate (W) is heat dissipated? At what rate is entropy generated (W/K)?

23.12 Fluid is flowing through a pipe at a rate of 100 gal/min. The inlet pressure is 20 psig and the outlet pressure is 0 psig. The pipe is submerged in a lake that maintains the fluid at a constant temperature of 50°F. At what rate (Btu/h) is heat dissipated from the pipe? At what rate [Btu/(h·°R)] is entropy generated?

23.13 Water flows over a 100-ft waterfall at a rate of 50,000 gal/min. In this process, the potential energy of the water is converted to thermal energy at the bottom of the waterfall. Normally, the water temperature would increase, but some water evaporates and maintains the water temperature at a constant 60°F. At what rate (lb_m/min) does the water evaporate? At what rate is entropy generated [Btu/(min·°R)]?

23.14 The surface of the sun is about 5780 K and the surface of the earth is about 300 K. When 1.0 J of heat is radiated from the sun to the earth, how much entropy is generated? The temperature of deep space is 2.73 K. If 1.0 J of heat is radiated from the earth to a comet that is at the temperature of deep space, how much entropy is generated?

23.15 Because of the friction of tides, the earth rotates more slowly as time progresses. Each year, the time it takes for the earth to complete one rotation (the length of a day) increases by 20 μs. The mass of the earth is 5.9785×10^{24} kg and the radius is 6.37×10^6 m. Assuming it has a uniform density, what is the rotational kinetic energy (J) of the earth? (See Table 21.3 on accounting for angular momentum.) At what rate (J/yr) is kinetic energy being lost due to friction? Assuming the earth temperature is 15°C, at what rate [J/(yr·K)] is entropy being generated due to tidal friction?

Glossary

closed system A system in which mass does not cross the boundary.

energy A unit of exchange used by the scientific and engineering community to equate various phenomena.

entropy A measure of the disorder of a system.

fusion The act of melting a solid.

generation The amount of the counted quantity produced during the time interval within the system boundary.

input The amount of the counted quantity passing through the system boundary into the system during the time interval.

irreversible A system which, when returned to its original state, causes a change in the surroundings.

isotherm Constant temperature.

reversible A system which, when returned to its original state, does not cause a change in the surroundings.

single-cycle process A process that causes a system to return to its original state.

surroundings Everything in the universe except the system.

system A subset of the universe we wish to study.

Mathematical Prerequisite

- Algebra (Appendix E, Mathematics Supplement)
- Calculus (Appendix L, Mathematics Supplement)

CHAPTER 24

Accounting for Money

Money! It's a hit. . . .

—*Pink Floyd, "Dark Side of the Moon"*

Once you've had your creative spark, designed your new idea, satisfied all of the laws of physics and those of man, only one thing remains—you have to create it cheaply enough for someone to buy and less expensively than the next guy. Money is, of course, the driving force behind nearly all engineering. The money aspect of engineering can be graceful, allowing an easy choice between two nearly equal engineering schemes, or brutal, denying you any way to make a salable product. So many people have been concerned with money (bankers, merchants, accountants, businesspeople, economists, etc.) for so long (since swinging from the trees) that whole vocabularies and a dozen mind-sets exist when it comes to money. We can do little but provide a glimpse of some ways to look at money.

The term describing the mind-set used by engineers is **engineering economics,** the discipline that translates engineering technology into a form that permits evaluation by businesses or investors. Engineering economics provides the means to answer such questions as:

- How much capital will be needed to build a new plant?
- Which, among the various process alternatives, generates the most profit?
- Should a larger heat exchanger be installed to save energy?
- How much should be spent to avoid equipment failures?

The ability of the engineer to provide meaningful answers to these questions is one of the principal (all puns intended) ways we differ from scientists.

24.1 MICRO- AND MACROECONOMICS

Since giving up the barter system, societies have used tokens called *money* to ease the flow of goods and services. The value of money is controlled both by market forces and by government institutions.

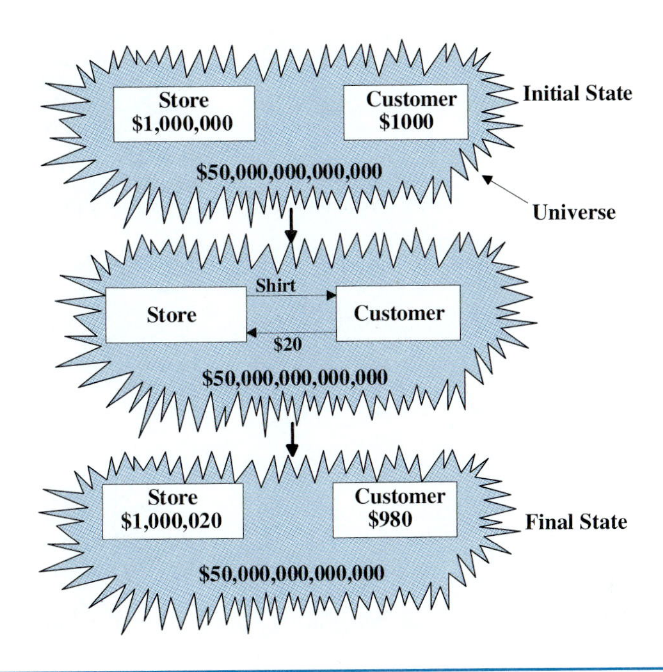

FIGURE 24.1
Money conservation for ordinary citizens.

24.1.1 Microeconomics

The universal accounting equation can be applied to money because it is an extensive quantity:

$$\underbrace{\text{Final amount} - \text{initial amount}}_{\text{Accumulation}} = \text{input} - \text{output} + \text{generation} - \text{consumption} \tag{24-1}$$

To the ordinary citizen, money is a conserved quantity; that is, it is neither created nor destroyed during use. Because generation and consumption are usually zero (unless engaged in counterfeiting or you forgot where you buried your money), accumulation of money is the difference between input and output. The schematic in Figure 24.1 shows that money is swapped for goods and services, but the amount of money in the universe does not change because of these transactions.

24.1.2 Macroeconomics

If wealth were truly a conserved quantity, there would be a finite wealth in the universe and nothing could be done to change this. This would also mean that everyone plays a "zero sum" game in which the gain by one person must always result in a loss by others. Fortunately, wealth can be "created," and engineers are integrally involved in this process. Consider a pool of crude oil beneath the surface of the earth. Through the efforts of petroleum engineers, chemical engineers, industrial engineers, and so forth, this pool of oil is tapped and refined into gasoline, lubricants, asphalt, petrochemicals, pharmaceuticals,

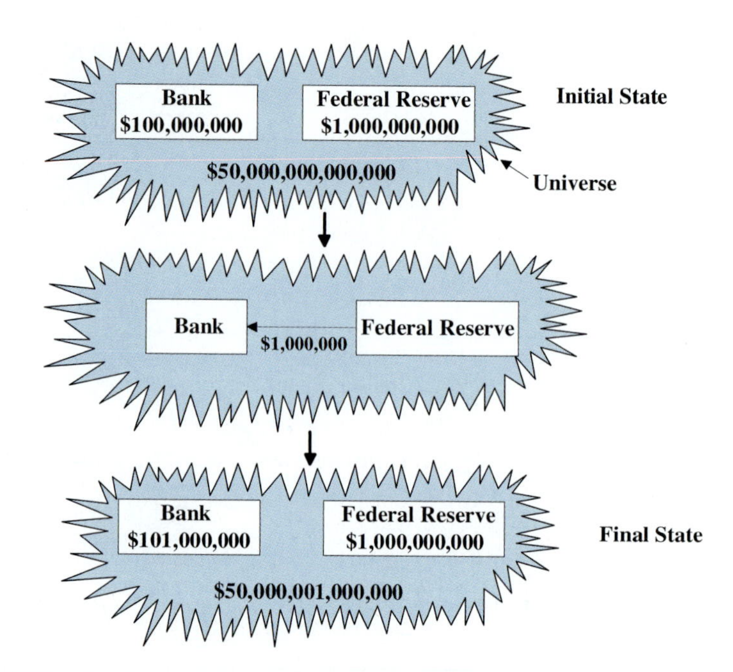

FIGURE 24.2
Creation of money. Although the actual process is more complex, we can visualize it as follows: The bank borrows money from the Federal Reserve. The bank's account increases but the Federal Reserve's account stays the same.

dyes, and so on, the worth of which is more than the cost to find the oil and convert it into these products. So now there is more wealth in the universe than before; wealth can be generated. Now what?

If money were conserved while wealth is generated, there would be more wealth than dollars, and dollars would become more valuable. If this were done in a disastrous fashion, a recession or depression of the economy would occur. To prevent this, the Federal Reserve creates money to match the increase in wealth (Figure 24.2). If the government "manufactures" more money than there is wealth, then the value of the currency falls. This is called **inflation.** In pathological cases, runaway inflation is followed by depression of the economy. The trick is for governments to keep the amount of currency at least roughly equal to the amount of wealth in the country.

24.2 INTEREST AND INVESTMENT COSTS

"What a capital idea!" said Tom with interest.

Just as banks and credit card companies charge you interest on the money you use, so do businesses pay for the use of borrowed money. The amount of money borrowed is termed the **principal P**, and the **interest I** is the "rent" paid for use of the principal.

Generally, interest payments are based upon a legal contract between the lender and the borrower. As in any legal contract, it is necessary to define terms.

An **interest period** t_I is the length of time after which interest is due. For example, a contract may be written as follows: "On January 1, $1000 in principal will be transferred from the lender to the borrower. Thereafter, each December 31, the borrower agrees to pay interest of $100. At any time, the contract may be terminated when the borrower returns the $1000 principal to the lender." In this contract, the interest period was 1 year.

The *number of interest periods n* is defined as

$$n = \frac{t}{t_I} \tag{24-2}$$

where t is the elapsed time. Both t and t_I have dimensions of time, so n is dimensionless. In the contract described above, if the borrower returns the principal after five years, then five interest periods would have elapsed.

The **interest rate** i is defined as the amount of interest paid in a single interest period I_p divided by the principal:

$$i = \frac{I_p}{P} \tag{24-3}$$

Both I_p and P are expressed in dollars, so i is dimensionless. In the above contract, the interest rate is 0.1 or 10%. An alternate way to express the interest rate is per unit of time

$$\hat{i} = \frac{I_p}{Pt_I} \tag{24-4}$$

In this definition of the interest rate, the dimension of $\hat{i}$ is inverse time. In the above contract, the interest rate is 0.1 per year, or 10% per year. Alternatively, it could be said that the annual interest rate is 0.1 or 10%.

Interest may be calculated in many ways, as we discuss in the following sections.

24.2.1 Simple Interest

The easiest form of interest is termed **simple interest;** it specifies an interest payment based only on the original principal. For example, suppose you borrow \$1000 for three years. The contract defines the interest period as one year and states that you must pay simple interest at a rate of 12%. After three years, three interest periods have elapsed. You would owe \$360 in interest plus the original \$1000 of principal for a sum of \$1360. Figure 24.3 illustrates how simple interest is determined.

We can generalize simple interest calculations as follows. From Equation 24-3, the interest I_p due after each interest period is

$$I_p = Pi \tag{24-5}$$

If the money is borrowed for n periods, the total interest I is

FIGURE 24.3
Illustration of simple interest.

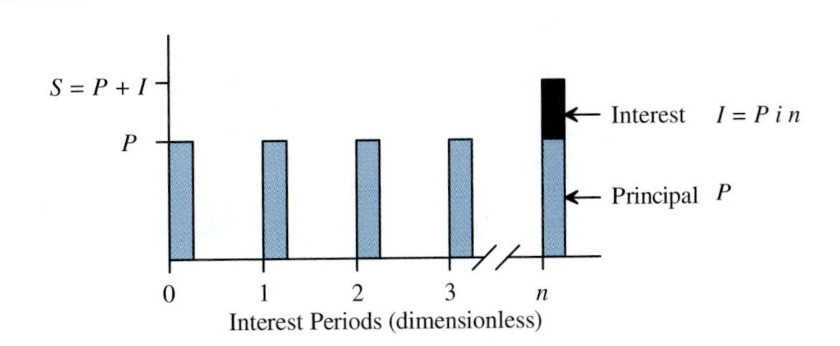

$$I = I_p n = Pin \tag{24-6}$$

Of course, the lender also wants his principal returned at the end of the loan period, so the sum S to be repaid is

$$S = P + I = P + Pin = P(1 + in) \tag{24-7}$$

Using our example, we can calculate the sum to be repaid as

$$S = \$1000[1 + (0.12)(3)] = \$1360$$

In this contract, the *interest period* was 1 year so three interest periods elapsed. Although most contracts are written with interest periods of 1 year, there is nothing sacred about this. Just as easily, we could have chosen another interest period. For example, the contract could have defined the interest period to be 1 month. Further, the contract could specify an interest rate of 1% with the loan to be repaid after 36 months. We can still use Equation 24-7 to calculate the sum S to be repaid

$$S = \$1000[1 + (0.01)(36)] = \$1360$$

24.2.2 Compound Interest

With simple interest, the interest is not paid until the loan is terminated. This leaves the lender deprived of the interest money that he could have loaned out in the interim. There are two ways to overcome this: (1) specify that the interest payments are to be made at the end of each interest period, or (2) the unpaid interest is treated as unpaid principal and collects interest at the same rate as the principal. Figure 24.4 illustrates this second option.

Imagine you borrow $1000 from a credit card company. They define the interest period to be 1 month and they charge an interest rate of 1%. Further, they **compound** the interest monthly, meaning if each month's interest is unpaid, it is added to the principal. If no payments were made for 1 year, Table 24.1 shows what you would owe. Notice that when the interest is compounded, the sum to be repaid is higher.

When compounding, the formula for calculating the sum to be repaid is straightforward to derive. For an initial principal P_0, after the first interest period, the sum S to be repaid is the same as for simple interest after a single interest period:

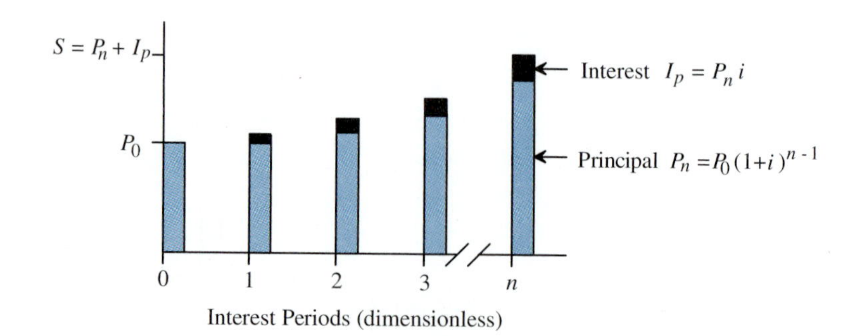

FIGURE 24.4
Illustration of compound interest.

TABLE 24.1
Principal and interest on $1000 borrowed at 1% monthly interest

	Simple Interest		Monthly Compounding	
Month	Unpaid Principal	Interest	Unpaid Principal	Interest
1	$1000.00	$10.00	$1000.00	$10.00
2	$1000.00	$10.00	$1010.00	$10.10
3	$1000.00	$10.00	$1020.10	$10.20
4	$1000.00	$10.00	$1030.30	$10.30
5	$1000.00	$10.00	$1040.60	$10.41
6	$1000.00	$10.00	$1051.01	$10.51
7	$1000.00	$10.00	$1061.52	$10.62
8	$1000.00	$10.00	$1072.14	$10.72
9	$1000.00	$10.00	$1082.86	$10.83
10	$1000.00	$10.00	$1093.69	$10.94
11	$1000.00	$10.00	$1104.62	$11.05
12	$1000.00	$10.00	$1115.67	$11.16
Sum S to be repaid	$1120.00		$1126.83	

$$S = P_0 + I = P_0 + P_0 i = P_0(1 + i) \tag{24-8}$$

After the second interest period, the sum to be repaid is

$$S = P_0 + I = P_0(1 + i) + P_0(1 + i)i = P_0(1 + i)(1 + i) = P_0(1 + i)^2 \tag{24-9}$$

In general, for n interest periods, the sum to be repaid is

$$S = P_0 + I = P_0(1 + i)^n \tag{24-10}$$

For the credit card example shown in Table 24.1, the sum to be repaid is

$$S = P_0 + I = P_0(1 + 0.01)^{12} = 1000(1.01)^{12} = \$1126.83$$

24.2.3 Continuous Compound Interest

An alternative formulation of Equation 24-10 is

$$S = P_0\left(1 + \frac{\hat{i}}{m}\right)^{mt} \tag{24-11}$$

where $\hat{i}$ is the *annual* interest rate, m is the number of interest periods *per year,* and t is the number of *years*. For the credit card example,

$$S = \$1000\left(1 + \frac{0.12\ \text{yr}^{-1}}{12\ \text{yr}^{-1}}\right)^{(12\ \text{yr}^{-1})(1\ \text{yr})} = \$1126.83$$

we get the same answer.

Rather than compound once a month, once a week, or daily, we could envision a process in which the compounding would occur continuously. In this case, there would be an infinite number of compoundings m per year

$$S = \lim_{m \to \infty} P_0 \left(1 + \frac{\hat{i}}{m}\right)^{mt} \tag{24-12}$$

By defining

$$\frac{1}{k} \equiv \frac{\hat{i}}{m} \tag{24-13}$$

and substituting into Equation 24-12, we obtain

$$S = \lim_{k \to \infty} P_0 \left(1 + \frac{1}{k}\right)^{k\hat{i}t} = P_0 \left[\lim_{k \to \infty} \left(1 + \frac{1}{k}\right)^{k}\right]^{\hat{i}t} \tag{24-14}$$

The term in the square brackets is the Euler number $e = 2.7182818284 \ldots$, the base of the natural logarithms. (This is a topic of calculus. If you are unfamiliar with it, test it with your calculator. Choose $k = 20{,}000$ and see how close to e the answer is.)

$$S = P_0 e^{\hat{i}t} \tag{24-15}$$

Figure 24.5 illustrates this relationship between S and time.

Although **continuous interest** rates are not widely used in business (probably because calculus is not widely appreciated among the general populus), they are routinely used in engineering economics calculations. This is so because the tools of calculus can be brought to bear on the problem and the answers derived using continuous interest rates are often close to those for discrete payments.

24.2.4 Summary of Interest Rates

For a given interest rate i, the sum S that must be repaid depends upon the type of compounding. When taking out a loan, be sure to read the fine print to determine the compounding method used by the bank.

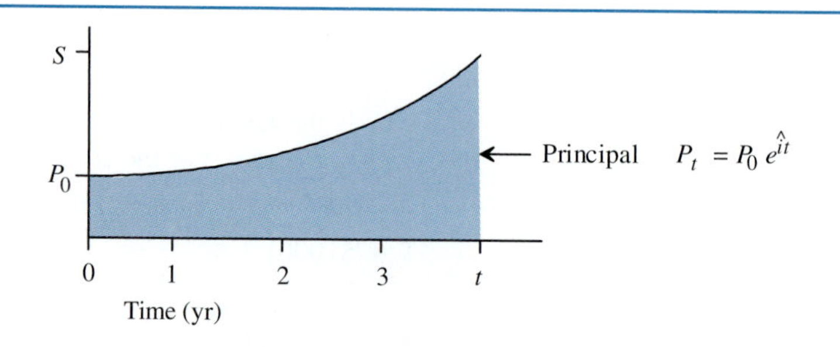

FIGURE 24.5
Illustration of continuous interest.

EXAMPLE 24.1

Problem Statement: You borrow \$1000 at an annual interest rate of 10% for two years. Determine the amount to be repaid after two years using (a) simple interest, (b) yearly compounding, (c) monthly compounding, (d) daily compounding, and (e) continuous compounding.

Solution:

(a) Simple interest:

$$S = P(1 + in) = \$1000[1 + (0.10)(2)] = \$1200.00$$

(b) Yearly compounding:

$$S = P_0\left(1 + \frac{\hat{i}}{m}\right)^{mt} = \$1000\left(1 + \frac{0.1\ \text{yr}^{-1}}{1\ \text{yr}^{-1}}\right)^{(1\ \text{yr}^{-1})(2\ \text{yr})} = \$1210.00$$

(c) Monthly compounding:

$$S = P_0\left(1 + \frac{\hat{i}}{m}\right)^{mt} = \$1000\left(1 + \frac{0.1\ \text{yr}^{-1}}{12\ \text{yr}^{-1}}\right)^{(12\ \text{yr}^{-1})(2\ \text{yr})} = \$1220.39$$

(d) Daily compounding:

$$S = P_0\left(1 + \frac{\hat{i}}{m}\right)^{mt} = \$1000\left(1 + \frac{0.1\ \text{yr}^{-1}}{365\ \text{yr}^{-1}}\right)^{(365\ \text{yr}^{-1})(2\ \text{yr})} = \$1221.37$$

(e) Continuous compounding:

$$S = P_0 e^{\hat{i}t} = \$1000 e^{(0.1\ \text{yr}^{-1})(2\ \text{yr})} = \$1221.40$$

Note that there is little difference between the value calculated by daily compounding and continuous compounding.

24.3 PRESENT WORTH AND DISCOUNT

The cornerstone of engineering economics is to use interest rates to relate present money to future or past money. Given an annual interest rate of 10% with annual compounding, Example 24.1 shows that \$1000 today is worth \$1210 two years from now. This could be restated as "\$1000 is the *present worth* of \$1210 two years from now." This ability to relate future and present moneys is a surprisingly powerful tool.

24.3.1 Present Worth

The present worth of a future dollar is the amount of principal that must be invested today at a given interest rate to yield 1 dollar at the given future date. For annually compounded interest, we can rearrange Equation 24-10 to give

$$\text{Present worth} = P_0 = S\frac{1}{(1+i)^n} \tag{24-16}$$

With m interest periods per year, we can rearrange Equation 24-11 to give

$$\text{Present worth} = P_0 = S\frac{1}{(1+\frac{\hat{i}}{m})^{mt}} \tag{24-17}$$

Continuous interest can also be used to calculate present worth. We can rearrange Equation 24-15 to give

$$\text{Present worth} = P_0 = S\frac{1}{e^{\hat{i}t}} \tag{24-18}$$

The difference between future worth and present worth is known as the *discount,* particularly for bonds.

$$\text{Discount} = S - P_0 \tag{24-19}$$

EXAMPLE 24.2

Problem Statement: A 10-year bond with \$1000 face value is sold with a \$400 discount, meaning it can be purchased today for \$600 and the borrower will pay \$1000 10 years from now. Determine (a) the annual interest rate with annual compounding and (b) the annual interest rate with continuous compounding.

Solution:

(a) Annual compounding:

$$P_0 = S\frac{1}{(1+i)^n} \qquad \text{(Equation 24-16)}$$

$$(1+i)^n = \frac{S}{P_0}$$

$$(1+i)^n = \left(\frac{S}{P_0}\right)^{1/n}$$

$$i = \left(\frac{S}{P_0}\right)^{1/n} - 1 = \left(\frac{\$1000}{\$600}\right)^{1/10} - 1 = 0.0524 = 5.24\%$$

(b) Continuous compounding:

$$P_0 = S\frac{1}{e^{\hat{i}t}} \qquad \text{(Equation 24-18)}$$

$$e^{\hat{i}t} = \frac{S}{P_0}$$

$$\ln(e^{\hat{i}t}) = \ln\left(\frac{S}{P_0}\right)$$

$$\hat{i}t = \ln\left(\frac{S}{P_0}\right)$$

$$\hat{i} = \frac{1}{t}\ln\left(\frac{S}{P_0}\right) = \frac{1}{10\text{ yr}}\ln\left(\frac{\$1000}{\$600}\right) = 0.0511\text{ yr}^{-1} = 5.11\%\text{ per year}$$

24.4 ANNUITIES

An **annuity** is a series of equal payments occurring at equal time periods that last for a finite length of time. Annuities are of three types:

- Sinking funds used to build capital (e.g., for planned equipment replacement, for planned maintenance).
- Installment loans (e.g., mortgage, car payment).
- Retirement plans (e.g., from a life insurance policy or retirement fund).

24.4.1 Sinking Fund Annuity

A sinking fund annuity is used to accumulate capital. Figure 24.6 illustrates a 4-year sinking fund annuity with annual compound interest and annual payment R. Table 24.2 shows the sum value S resulting from each annual payment to the 4-year sinking fund annuity.

In general, the sum value S for an n-period annuity is given by the expression

$$S = R(1+i)^{n-1} + R(1+i)^{n-2} + \cdots + R(1+i) + R \tag{24-20}$$

If we multiply both sides of Equation 24-20 by $(1 + i)$, we obtain

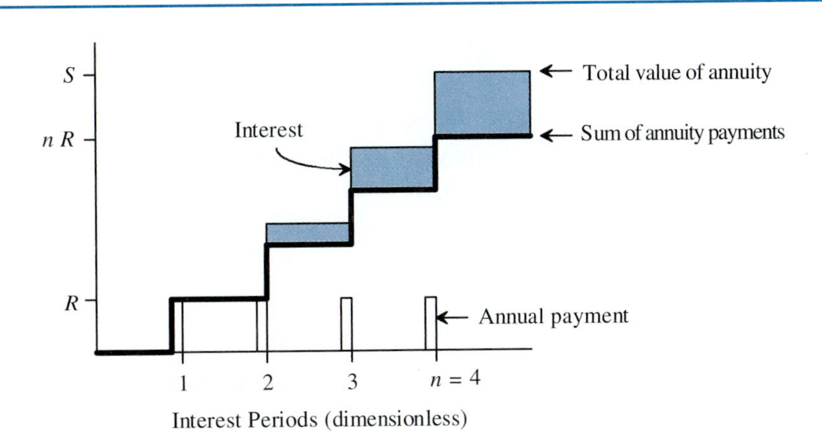

FIGURE 24.6 Illustration of a 4-year sinking fund annuity with a 1-year interest period and annual payments using annual compound interest.

TABLE 24.2
Value of a 4-period annuity resulting from each payment.

Period Deposited	Annuity Sum Value
1	$S_1 = R(1 + i)^3$
2	$S_2 = R(1 + i)^2$
3	$S_3 = R(1 + i)^1$
4	$S_4 = R(1 + i)^0$
Total	$S = R(1 + i)^3 + R(1 + i)^2 + R(1 + i)^1 + R(1 + i)^0$

$$S(1 + i) = R(1 + i)^n + R(1 + i)^{n-1} + \cdots + R(1 + i)^2 + R(1 + i) \tag{24-21}$$

Subtracting Equation 24-20 from 24-21, we get

$$S(1 + i) - S = R(1 + i)^n - R$$

$$S[(1 + i) - 1] = R[(1 + i)^n - 1]$$

$$Si = R[(1 + i)^n - 1]$$

$$S = R\frac{(1 + i)^n - 1}{i} \tag{24-22}$$

If payments are made more frequently than once a year, then

$$S = \hat{R}\frac{\left(1 + \frac{\hat{i}}{m}\right)^{mt} - 1}{\hat{i}} \tag{24-23}$$

where m is the number of payments made per year, $\hat{i}$ is the annual interest rate, and t is the time during which the capital is accumulated. The annual payment $\hat{R}$ is calculated as

$$\hat{R} = mr \tag{24-24}$$

where r is the individual payment.

If payments to the annuity are made continuously, then there are an infinite number of annual payments m:

$$S = \lim_{m \to \infty} \hat{R}\frac{\left(1 + \frac{\hat{i}}{m}\right)^{mt} - 1}{\hat{i}} \tag{24-25}$$

By substituting k as defined by Equation 24-13, we obtain

$$S = \lim_{k \to \infty} \hat{R} \frac{\left(1 + \frac{1}{k}\right)^{k\hat{i}t} - 1}{\hat{i}} = \hat{R} \frac{\left[\lim_{k \to \infty} \left(1 + \frac{1}{k}\right)\right]^{k\hat{i}t} - 1}{\hat{i}} = \hat{R} \frac{e^{\hat{i}t} - 1}{\hat{i}} \tag{24-26}$$

EXAMPLE 24.3

Problem Statement: An investor puts \$1000 per year into an annuity with 10% annual interest. After 20 years, what is the annuity worth if the payment is made (a) annually, (b) monthly, and (c) continuously?

Solution:

(a) Annually:

$$S = R \frac{(1 + i)^n - 1}{i} = \$1000 \frac{(1 + 0.1)^{20} - 1}{0.1} = 57{,}275.00$$

(b) Monthly:

$$S = \hat{R} \frac{(1 + \frac{\hat{i}}{m})^{mt} - 1}{\hat{i}} = (\$1000 \text{ yr}^{-1}) \frac{\left(1 + \frac{0.1 \text{ yr}^{-1}}{12 \text{ yr}^{-1}}\right)^{(12 \text{ yr}^{-1})(20 \text{ yr})} - 1}{0.1 \text{ yr}^{-1}} = \$63{,}280.74$$

(*Note:* The monthly payment is \$1000/12 or \$83.33.)

(c) Continuously:

$$S = \hat{R} \frac{e^{\hat{i}t} - 1}{\hat{i}} - (\$1000 \text{ yr}^{-1}) \frac{e^{(0.1 \text{ yr}^{-1})(20 \text{ yr})} - 1}{0.1 \text{ yr}^{-1}} = \$63{,}890.56$$

24.4.2 Installment Loans

If you borrow a lump sum of money and agree to make equal periodic payments until the loan is paid off, this is an **installment loan.** Common examples of installment loans are mortgages and car loans. Figure 24.7 illustrates a 4-year installment loan with annual payments R. Initially, most of the payment goes to pay interest and a relatively small portion goes to reduce the principal. When the last payment is made, the principal is zero and the loan is complete.

Table 24.3 shows how the initial principal P_0 is calculated for a 4-year installment loan. In general, P_0 is given by

$$P_0 = R \frac{1}{(1 + i)} + R \frac{1}{(1 + i)^2} + \cdots + R \frac{1}{(1 + i)^{n-1}} + R \frac{1}{(1 + i)^n} \tag{24-27}$$

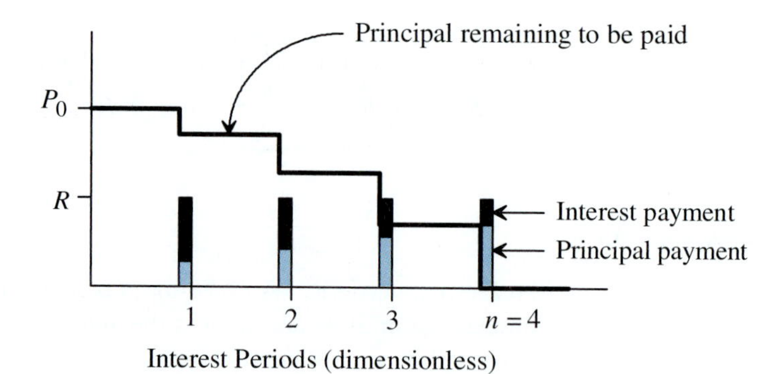

FIGURE 24.7
Illustration of a 4-year installment loan with a 1-year interest period and annual payments using annual compound interest.

If we multiply both sides of Equation 24-27 by $(1 + i)$, we obtain

$$P_0(1 + i) = R + R\frac{1}{(1 + i)} + \cdots + R\frac{1}{(1 + i)^{n-2}} + R\frac{1}{(1 + i)^{n-1}} \tag{24-28}$$

Subtracting Equation 24-27 from 24-28, we get

$$P_0(1 + i) - P_0 = R - R\frac{1}{(1 + i)^n}$$

$$P_0[(1 + i) - 1] = R\left[1 - \frac{1}{(1 + i)^n}\right]$$

$$P_0 i = R[1 - (1 + i)^{-n}]$$

$$P_0 = R\frac{1 - (1 + i)^{-n}}{i} \tag{24-29}$$

If payments are made more frequently than once a year, then

$$P_0 = \hat{R}\frac{1 - \left(1 + \frac{\hat{i}}{m}\right)^{-mt}}{\hat{i}} \tag{24-30}$$

where m is the number of payments per year, $\hat{i}$ is the annual interest rate, t is the number of years it takes to pay off the loan, and $\hat{R}$ is the annual loan payment given by Equation 24-24.

If loan payments are made continuously, then there are an infinite number of annual payments m:

$$P_0 = \lim_{m \to \infty} \hat{R}\frac{1 - \left(1 + \frac{\hat{i}}{m}\right)^{-mt}}{\hat{i}} \tag{24-31}$$

TABLE 24.3
Principal of a 4-year installment loan resulting from each payment

Payment Year	Principal
1	$P_1 = R\frac{1}{(1+i)}$
2	$P_2 = R\frac{1}{(1+i)^2}$
3	$P_3 = R\frac{1}{(1+i)^3}$
4	$P_4 = R\frac{1}{(1+i)^4}$
Total	$P_0 = R\frac{1}{(1+i)} + R\frac{1}{(1+i)^2} + R\frac{1}{(1+i)^3} + R\frac{1}{(1+i)^4}$

By substituting k as defined by Equation 24-13, we obtain

$$P_0 = \lim_{k\to\infty} \hat{R}\frac{1-\left(1+\frac{1}{k}\right)^{-k\hat{i}t}}{\hat{i}} = \hat{R}\frac{1-\left[\lim_{k\to\infty}\left(1+\frac{1}{k}\right)\right]^{-k\hat{i}t}}{\hat{i}} = \hat{R}\frac{1-e^{-\hat{i}t}}{\hat{i}} \qquad (24\text{-}32)$$

EXAMPLE 24.4

Problem Statement: What is the annual payment for a \$100,000 mortgage with a 20-yr term at 10% compounded annual interest if the payments are made (a) annually, (b) monthly, and (c) continuously?

Solution:

(a) Annually:

$$P_0 = R\frac{1-(1+i)^{-n}}{i} \qquad \text{(Equation 24-29)}$$

$$R = P_0\frac{i}{1-(1+i)^{-n}} = \$100{,}000\frac{0.1}{1-(1+0.1)^{-20}} = \$11{,}745.96$$

(b) Monthly:

$$P_0 = \hat{R}\frac{1-\left(1+\frac{\hat{i}}{m}\right)^{-mt}}{\hat{i}} \qquad \text{(Equation 24-30)}$$

EXAMPLE 24.5

Problem Statement: An initial principal of $100,000 is placed in a perpetuity at 10% annual interest. A periodic payment R is received once every 3 years. What is the periodic payment if the perpetuity is compounded (a) annually, (b) monthly, (c) continuously?

Solution:

(a) Annually:

$$P_0 = R\frac{1}{(1+i)^n - 1} \qquad \text{(Equation 24-34)}$$

$$R = P_0[(1+i)^n - 1] = \$100{,}000[(1+0.1)^3 - 1] = \$33{,}100.00$$

(b) Monthly:

$$P_0 = R\frac{1}{(1+\frac{\hat{i}}{m})^{mt} - 1} \qquad \text{(Equation 24-35)}$$

$$R = P_0\left[\left(1+\frac{\hat{i}}{m}\right)^{mt} - 1\right]$$

$$= \$100{,}000\left[\left(1+\frac{0.1\ \text{yr}^{-1}}{12\ \text{yr}^{-1}}\right)^{(12\ \text{yr}^{-1})(3\ \text{yr})} - 1\right] = \$34{,}818.18$$

(c) Continuously:

$$P_0 = R\frac{1}{e^{\hat{i}t}} - 1 \qquad \text{(Equation 24-37)}$$

$$R = P_0(e^{\hat{i}t} - 1) = \$100{,}000(e^{(0.1\ \text{yr}^{-1})(3\ \text{yr})} - 1) = \$34{,}985.88$$

24.5.2 Capitalized Costs

Capitalized costs are used to choose between competing investments for equipment that must be periodically replaced after it becomes worn out. The periodic payment R needed to replace old equipment is equal to its initial purchase price C_i minus the scrap value C_s of the old equipment,

$$R = C_i - C_s \qquad (24\text{-}38)$$

Capitalized cost K is the initial cost of the equipment C_i plus the initial principal P_0 of the perpetuity established to replace the equipment,

$$K = C_i + P_0 \qquad (24\text{-}39)$$

If the interest is compounded annually, we can substitute Equation 24-34 for P_0:

$$K = C_i + R\frac{1}{(1+i)^n - 1} = C_i + (C_i - C_s)\frac{1}{(1+i)^n - 1} \tag{24-40}$$

If the interest is compounded m times per year, then we substitute Equation 24-35 for P_0:

$$K = C_i + R\frac{1}{\left(1+\frac{\hat{i}}{m}\right)^{mt} - 1} = C_i + (C_i - C_s)\frac{1}{1+\left(\frac{\hat{i}}{m}\right)^{mt} - 1} \tag{24-41}$$

If the interest is compounded continuously, then we substitute Equation 24-37 for P_0:

$$K = C_i + R\frac{1}{e^{\hat{i}t} - 1} = C_1 + (C_i - C_s)\frac{1}{e^{\hat{i}t} - 1} \tag{24-42}$$

EXAMPLE 24.6

Problem Statement: A radio station has two choices: buy a standard power supply for \$50,000 with a useful life of 10 years and a scrap value of \$2000, or buy a "gold-plated" model for \$80,000 with a useful life of 20 years and a scrap value of \$5000. Which should the station buy, assuming 8% annual interest with annual compounding?

Solution:

(a) Standard model:

$$K = C_i + (C_i - C_s)\frac{1}{(1+i)^n - 1}$$

$$= \$50{,}000 + (\$50{,}000 - \$2000)\frac{1}{(1+0.08)^{10} - 1} = \$91{,}418$$

(b) "Gold-plated" model:

$$K = C_i + (C_i - C_s)\frac{1}{(1+i)^n - 1}$$

$$= \$80{,}000 + (\$80{,}000 - \$5000)\frac{1}{(1+0.08)^{20} - 1} = \$100{,}486$$

The standard model has a lower capitalized cost, so it is the more economical choice.

24.6 DISCOUNT FACTORS AND COMPOUNDING FACTORS

If someone offered you \$1000 today or \$1000 one year from now, which would you choose? Assuming there is no deflation in the economy, any rational person would choose the \$1000 today. This money could be invested at 10% interest, so a year from now it would be worth \$1100. Money now is worth more than money in the future. This simple concept is often described as "the time value of money."

To relate future money S to present money P we use discount factor F_d:

$$P = F_d S \qquad (F_d < 1) \tag{24-43}$$

To relate present money P to future money S we use compounding factor F_c:

$$S = F_c P \qquad (F_c > 1) \tag{24-44}$$

Equation 24-44 also relates past money P to current money S. By comparing Equations 24-43 and 24-44, we see that the compounding factor is simply the inverse of the discount factor.

The discount factor and compounding factor depend upon:

- Interest rates.
- Time.
- Payment type (lump sum or continuous).
- Type of compounding (simple interest, annual, monthly, daily, continuous).

For simplicity in our discussion, we will use continuous compounding only.

24.6.1 Lump-Sum Payments

The simplest discount or compounding factor is for single lump-sum payments in the future or in the past (Figure 24.9). Equation 24-18 allows us to relate lump-sum payment S made t years in the future with an annual interest rate $\hat{\imath}$, to the present worth P,

$$P = (e^{\hat{i}t})S = F_d S \tag{24-45}$$

where F_d is the discount factor for this scenario. Likewise, Equation 24-18 relates present lump-sum payment P to future worth S, t years in the future with annual interest rate $\hat{\imath}$,

$$S = (e^{\hat{i}t})P = F_c P \tag{24-46}$$

where F_c is the compounding factor for this scenario. Note that the compounding factor is simply the inverse of the discount factor.

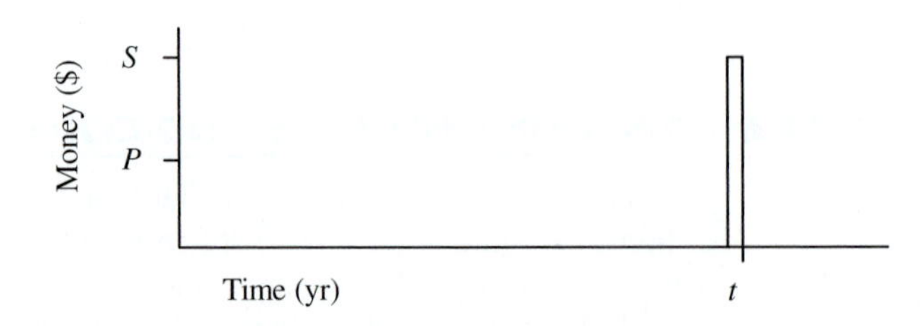

FIGURE 24.9
Lump-sum payment made t years in the future.

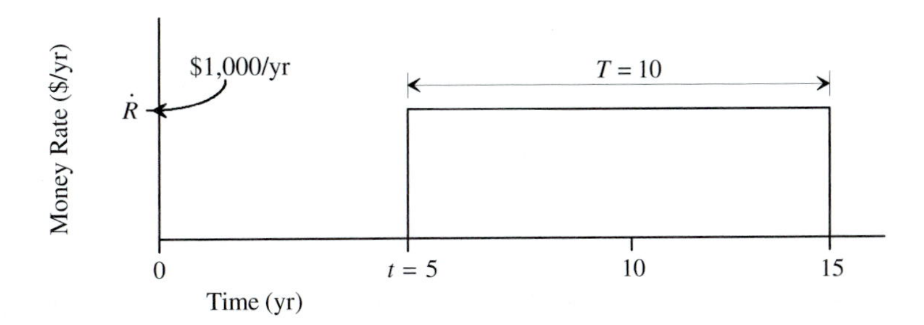

FIGURE 24.10
Illustration of continuous payment (for text example).

24.6.2 Constant Continuous Payment

Figure 24.10 illustrates another payment scenario in which continuous payments are made at a constant rate $\dot{R}$ ($/yr) for T years, starting t years from now. This scenario allows us to answer the question, "What is the present worth of an annuity having 10% annual interest compounded continuously, having annual payments of $1000, but the payments start 5 years from now and continue for 10 years?"

If we look t years into the future, Equation 24-32 tells us that the present worth of the annuity *at that time* would be

$$P_t = \dot{R}\,\frac{1 - e^{-\hat{i}T}}{\hat{i}} \tag{24-47}$$

Using Equation 24-45, we can discount P_t to its present value P_0:

$$P_0 = e^{-\hat{i}t}P_t = e^{-\hat{i}t}\dot{R}\,\frac{1 - e^{-\hat{i}T}}{\hat{i}} \tag{24-48}$$

From this expression, we can now determine the present worth of the annuity illustrated in Figure 24.10.

$$P_0 = e^{-(0.1\ \mathrm{yr}^{-1})(5\ \mathrm{yr})}(\$1000\ \mathrm{yr}^{-1})\,\frac{1 - e^{-(0.1\ \mathrm{yr}^{-1})(10\ \mathrm{yr})}}{0.1\ \mathrm{yr}^{-1}} = \$3834$$

24.6.3 Linearly Declining Continuous Payment

Figure 24.11 illustrates a scenario in which payments start at an initial rate of $\dot{R}_0$ ($/yr) but linearly decline to zero after T years. This scenario allows us to answer the following question: "What is the present value of an investment that initially pays at a rate of $10,000 per year and decreases linearly to zero after five years assuming 10% interest and continuous compounding?"

The instantaneous payment rate $\dot{R}$ at any time t is

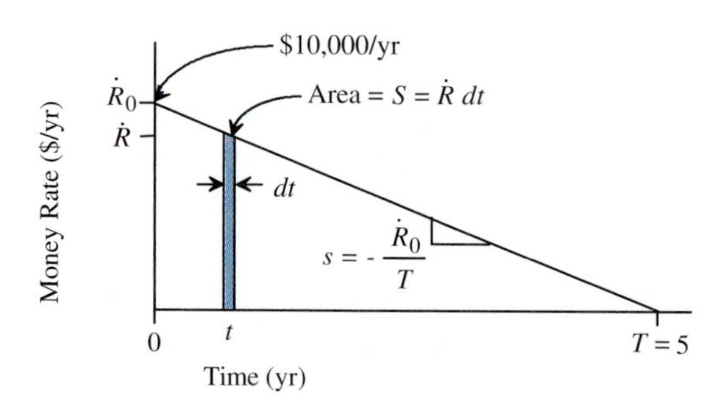

FIGURE 24.11
Illustration of linearly declining continuous payment (for text example)

$$\dot{R} = \dot{R}_0 - st = \dot{R}_0 - \frac{\dot{R}_0}{T}t = \dot{R}_0\left(1 - \frac{t}{T}\right) \tag{24-49}$$

where s is the slope. At time t, during time increment dt, the future sum of money S will be $\dot{R}\ dt$. We can discount this "lump sum" of money to present worth using Equation 24-45,

$$P_0 = e^{-\hat{i}t}S = e^{\hat{i}t}\dot{R}\,dt \tag{24-50}$$

where $\hat{i}$ is the annual interest rate. We can replace $\dot{R}$ with Equation 24-49, giving

$$P_0 = e^{-\hat{i}t}\dot{R}_0\left(1 - \frac{t}{T}\right)dt = \dot{R}_0\left(1 - \frac{t}{T}\right)e^{-\hat{i}t}\,dt \tag{24-51}$$

This expression can be integrated from 0 to T, giving

$$P_0 = \int_0^T \dot{R}_0\left(1 - \frac{t}{T}\right)e^{-\hat{i}t}\,dt$$

$$= \dot{R}_0\int_0^T\left(1 - \frac{t}{T}\right)e^{-\hat{i}t}\,dt = \dot{R}\left\{\int_0^T e^{-\hat{i}t}\,dt + \int_0^T -\frac{t}{T}e^{-\hat{i}t}\,dt\right\}$$

$$= \dot{R}_0\left\{\int_0^T e^{-\hat{i}t}\,dt - \frac{1}{T}\int_0^T te^{-\hat{i}t}\,dt\right\} = \dot{R}_0\left\{\left(-\frac{1}{\hat{i}}e^{-\hat{i}t}\right)_0^T - \frac{1}{T}\left[\frac{e^{-\hat{i}t}}{(-\hat{i})^2}(-\hat{i}t - 1)\right]_0^T\right\}$$

$$= \dot{R}_0\left\{\left[\left(-\frac{1}{\hat{i}}e^{-\hat{i}T}\right) - \left(-\frac{1}{\hat{i}}e^{-\hat{i}0}\right)\right]\right.$$

$$\left. - \frac{1}{T}\left[\left(\frac{e^{\hat{i}T}}{\hat{i}^2}(-\hat{i}T - 1)\right) - \left(\frac{e^{-\hat{i}0}}{\hat{i}^2}(-\hat{i}0 - 1)\right)\right]\right\}$$

$$= \dot{R}_0\left\{\left[\left(-\frac{1}{\hat{i}}e^{-\hat{i}T}\right) - \left(-\frac{1}{\hat{i}}1\right)\right] - \frac{1}{T}\left[\left(\frac{e^{-\hat{i}T}}{\hat{i}^2}(-\hat{i}T - 1)\right) - \left(\frac{1}{\hat{i}^2}(0 - 1)\right)\right]\right\}$$

$$= \dot{R}_0 \left\{ \left[\left(-\frac{1}{\hat{i}} e^{\hat{i}T} \right) + \frac{1}{\hat{i}} \right] - \frac{1}{T} \left[\left(\frac{e^{-\hat{i}T}}{\hat{i}^2} (-\hat{i}T - 1) \right) + \frac{1}{\hat{i}^2} \right] \right\}$$

$$= \dot{R}_0 \left\{ \frac{1}{\hat{i}} [1 - e^{-\hat{i}T}] - \frac{1}{T\hat{i}^2} [(e^{-\hat{i}T}(-\hat{i}T - 1)) + 1] \right\}$$

$$= \frac{\dot{R}_0}{\hat{i}} \left\{ [1 - e^{\hat{i}T}] - \frac{1}{T\hat{i}} [(e^{\hat{i}T}(-\hat{i}T - 1)) + 1] \right\}$$

$$= \frac{\dot{R}_0}{\hat{i}} \left\{ [1 - e^{-\hat{i}T}] - \frac{1}{T\hat{i}} [-\hat{i}Te^{-\hat{i}T} - e^{-\hat{i}T} + 1] \right\}$$

$$= \frac{\dot{R}_0}{\hat{i}} \left\{ [1 - e^{-\hat{i}T}] - \frac{1}{T\hat{i}} [-\hat{i}Te^{-\hat{i}T} + (1 - e^{-\hat{i}T})] \right\}$$

$$= \frac{\dot{R}_0}{\hat{i}} \left\{ [1 - e^{-\hat{i}T}] - \left[\frac{-\hat{i}Te^{-\hat{i}T}}{T\hat{i}} + \frac{(1 - e^{-\hat{i}T})}{T\hat{i}} \right] \right\}$$

$$= \frac{\dot{R}_0}{\hat{i}} \left\{ [1 - e^{-\hat{i}T}] - \left[-e^{-\hat{i}T} + \frac{(1 - e^{-\hat{i}T})}{T\hat{i}} \right] \right\}$$

$$= \frac{\dot{R}_0}{\hat{i}} \left\{ 1 - e^{-\hat{i}T} + e^{-\hat{i}T} - \frac{(1 - e^{-\hat{i}T})}{T\hat{i}} \right\}$$

$$= \frac{\dot{R}_0}{\hat{i}} \left\{ 1 - \frac{(1 - e^{-\hat{i}T})}{T\hat{i}} \right\} \tag{24-52}$$

From this expression, we can now determine the present value of the investment illustrated in Figure 24.11:

$$P_0 = \frac{\$10{,}000 \text{ yr}^{-1}}{0.1 \text{ yr}^{-1}} \left\{ 1 - \frac{[1 - e^{-(0.1 \text{ yr}^{-1})(5 \text{ yr})}]}{(5 \text{ yr})(0.1 \text{ yr}^{-1})} \right\} = \$21{,}306$$

24.7 INFLATION

Previously, we used the annual interest rate $\hat{i}$ that might be offered by a bank, annuity company, or other financial institution. In order for invested money to make an actual return, the interest rate must be higher than the rate of inflation $\hat{i}_{\text{inflation}}$. The real rate of return $\hat{i}_{\text{real}}$ is calculated as

$$\hat{i}_{\text{real}} = \frac{1 + \hat{i}}{1 + \hat{i}_{\text{inflation}}} - 1 \quad (24\text{-}53)$$

For example, if the bank pays 8% interest, and the inflation rate is 4%, then the real interest rate is only 3.85%. (*Note:* As an approximation, $\hat{i}_{\text{real}} \approx \hat{i} - \hat{i}_{\text{inflation}}$.) When performing economic evaluations of situations that occur over many years (e.g., capitalized cost analysis of long-lived equipment), it is best to use i_{real} rather than i. When you use the real interest rate in your analysis, the relationships you develop between future money and present money will be expressed in terms of current, noninflated money.

24.8 A QUICK LOOK AT OTHER TOPICS

There are many interesting topics in engineering economics, but the details go beyond the scope of this text. Because economics is critical to almost all engineering, further study in this area is advised.

24.8.1 Depreciation

When buying equipment for an engineering process, you might think that because money has been spent, this represents a loss to the company. This is not true. In principle, the new equipment could be returned for the same amount that was invested, so no loss has occurred. The loss to the company happens over the operating lifetime of the equipment. Just as when someone buys a new car, its value depreciates over the years, so equipment declines in value until it is worth nothing; or at most, its value is the scrap metal or salvageable parts. As equipment depreciates, a company can use this loss to help offset taxes on profits. There are many ways to compute **depreciation.** Although we could spend a lot of time on this aspect of economics alone, we will just note that the form of allowed depreciation affects the overall cost of a project.

24.8.2 Taxes

The tax laws influence the choices within a company and therefore often impact engineering decisions. The depreciation of equipment is usually an allowable tax write-off against company profits, and, indeed, prescribed depreciation schedules are part of business tax laws. Taxes influence businesses and engineering in both negative and positive ways. Cities, communities, or states that wish to increase engineering operations will often grant tax incentives to lure businesses to their location. On the other hand, particular types of businesses can be discouraged by levying increased taxes on their operations.

24.8.3 Cash Flow Management; Inventory Control

A challenge for many engineers is to manage an engineering project or a particular task within a project. This requires managing the people, cash, and materials to bring a product to completion within a specific time. There are many tools for managing engineering projects, including critical path analysis, time and material charts, and task and subtask milestone management. There are many excellent texts on project management and, if working for an engineering company, you will be taught whatever system the company uses.

For those projects that assemble or use parts made elsewhere, often batches of parts will be bought and stored until used. This *inventory* of parts represents a cash outlay drawing no interest. Tax laws usually count idle inventories as assets to the company and tax them. A current trend in engineering project management is the use of *just-in-time* part delivery. Obviously, the overall project must be controlled closely, but the savings from a reduced inventory can be appreciable. A careful analysis of the savings versus the risk of delaying the project must be considered.

24.9 SUMMARY

In ordinary business transactions, money is a conserved quantity. It is neither generated nor consumed and is merely transferred from one party to another. In rare instances, money may be consumed by a fire, for example. Ordinary citizens do not have the legal authority to generate money; those who do are counterfeiters and are subject to imprisonment. Only the federal government has the legal right to generate money through the Federal Reserve. If it generates too much money relative to the growth in the economy, then we have inflation. Conversely, if it generates too little money, then we have deflation.

The key idea in this chapter is "the time value of money." Stated another way, \$100 today has more value than \$100 next year, even if there were no inflation. Because money can be invested and put to work, its value grows as time progresses.

When a lender loans money, the money transferred to the borrower is called the *principal*. The principal belongs to the lender and is only temporarily in the possession of the borrower. As a fee for the use of this loaned money, the lender charges *interest*. When specifying the method for calculating interest, the lender and borrower can establish any contractual relationship they deem to be mutually fair.

The easiest method to calculate interest is called *simple interest* in which the same interest is paid each year regardless of the length of the loan. At the end of the loan, the borrower returns the principal (after all, it belongs to the lender) plus all the interest payments.

A slightly more complex method of calculating the fee for using the principal is called *compounding*. Compounding is a process in which the lender periodically offers the borrower the opportunity to pay the interest due. If the borrower does not pay the interest at that point, then the unpaid interest is added to the principal. Compounding can occur as frequently as the contract between lender and borrower specifies. Here, we used annual, monthly, daily, and continuous compounding.

An *annuity* is a constant payment occurring at equal time intervals during a finite time period. Annuities are used to accumulate capital (i.e., sinking fund), to pay installment loans, or as retirement plans. The terms of the annuity contract can specify compounding at any mutually agreeable interval.

A *perpetuity* is a constant payment occurring at equal time intervals for an infinite length of time. Perpetuities have a number of purposes. For example, an endowment is a perpetuity in which only the interest from the investment is used; the principal stays untouched, allowing the interest to be produced forever. An endowment might be used to pay student scholarships, fund research, support a symphony orchestra, or maintain a building.

Capitalized costs are used by engineers to assess different options when making purchasing decisions. Capitalized cost is the amount of money needed to purchase the equipment plus the principal needed to establish a perpetuity to replace the equipment when it wears out.

Discount factors relate future money to present money and *compounding factors* relate present money to future money. The compounding factor is the inverse of the discount factor. The discount factor (and compounding factor) depends upon the interest rate, the compounding method, and the rate at which payments are made.

Nomenclature

C_i initial price ($)
C_s scrap value ($)
F_c compound factor (dimensionless)
F_d discount factor (dimensionless)
I total interest owed ($)
I_p interest owed after an interest period ($)
i interest rate (dimensionless)
$\hat{i}$ interest rate (yr^{-1})
K capitalized cost ($)
k constant (dimensionless)
m number of interest periods per year (yr^{-1})
n number of interest periods (dimensionless)
P principal ($)
R payment per period ($)
$\dot{R}$ payment per year ($/yr)
r individual payment ($)
S sum owed ($)
T time (yr)
t time (yr)
t_I interest period (yr)

Further Readings

Bullinger, C. E. *Engineering Economy.* New York: McGraw-Hill, 1958.

Eide, A. R.; R. D. Jenison; L. H. Mashaw; and L. L. Northup. *Engineering Fundamentals and Problem Solving.* 3rd ed. New York: McGraw-Hill, 1997.

Grant, E. L.; W. G. Ireson; and R. S. Leavenworth. *Principles of Engineering Economy.* 6th ed. New York: Ronald, 1976.

Peters, M. S., and K. D. Timmerhaus. *Plant Design and Economics for Chemical Engineers.* 4th ed. New York: McGraw-Hill, 1991.

Smith, G. W. *Engineering Economy: Analysis of Capital Expenditures.* Ames, IA: Iowa State University Press, 1973.

Thuesen, H. G.; W. J. Fabrycky; and G. L. Thuesen. *Engineering Economy.* Englewood Cliffs, NJ: Prentice Hall, 1977.

PROBLEMS

24.1 You borrow $10,000 at an annual interest rate of 8%. You repay the loan after 4 years. How much must you repay using (a) simple interest, (b) yearly compounding, (c) monthly compounding, (d) daily compounding, and (e) continuous compounding?

24.2 If your fairy godmother had invested $1000 for you when you were born, and the investment had averaged 15% annual interest, how much money would you have at age 18 if the investment were compounded (a) annually, (b) monthly, (c) daily, and (d) continuously?

24.3 When you were born, your fairy godfather invested $1000 for you at 15% annual interest with continuous compounding. Rather than allow you to redeem this investment in a lump sum, he specifies that when you reach age 18, you must establish a perpetuity with his investment. Assuming 15% annual interest with continuous compounding, what is your annual check from this perpetuity?

24.4 You have a $15,000, 5-yr installment loan for your car. The bank charges you 8% annual interest with monthly compounding. What is your monthly payment?

24.5 An automobile company is offering a special program. To new graduates who have a job offer, they will sell a car with no money down; the entire purchase price of the automobile can be financed through the automobile company with a 5-yr loan at 3% annual interest and monthly compounding. Alternatively, if you pay cash, they will knock $750 off the selling price. As a new graduate who meets the company's criteria, you are eyeing a $15,000 beauty and are trying to decide how to finance it because you have no money. You could finance the car through the automobile company, or you could take a 5-yr $14,250 loan from a bank at 8% annual interest with monthly compounding and pay "cash" for the car. For each scenario, what is your monthly payment? What is the most economical option?

24.6 You are going to take a $100,000 mortgage on a house at 8% annual interest compounded monthly. What are your monthly payments if you repay the loan in 15 years? During the life of the loan, what was the sum total of all your payments? How much interest did you pay? Repeat these calculations using a 30-yr loan.

24.7 A 25-year-old engineer is trying to establish fiscally responsible habits. He determines that if he eats at a restaurant each working day, he will spend $5/day for lunch; but if he brings lunch from home (i.e., he brown-bags it), it costs only $2/day. During his 40-year career, if he were to invest his daily savings in an annuity that pays 10% interest, what would be the cash value of the annuity when he retires at age 65? (*Note:* Assume payments are made monthly to the annuity.)

24.8 A 65-year-old retiree has saved $500,000 during her life and wishes to provide monthly income for her "golden years." She is in good health and is trying to assess her options. She consults with an annuity company that offers 10% annual interest compounded monthly. What will her monthly check be if she establishes a 20-yr annuity, 30-yr annuity, or a perpetuity?

24.9 A 25-year-old engineer is planning for his retirement at age 65. During his working life, he decides to save $500 each month and invest it in an annuity that pays 10% annual interest compounded monthly. At retirement, what is the annuity worth? When he retires, he decides to take his principal and start a new annuity that pays him a monthly check for 20 years. Assuming this new annuity pays 10% annual interest compounded monthly, what is his monthly check during retirement?

24.10 In 5 years, a bond can be cashed in for $5000. It earns an interest rate of 10% per year compounded continuously. What should you pay for the bond? What is the discount?

24.11 A 25-year-old smoker decides to quit smoking and take the $5/day savings and put it into an annuity that pays 10% annual interest compounded monthly. When she retires at 65, what is the cash value of the annuity? (*Note:* Assume payments are made monthly to the annuity.)

24.12 A piece of equipment will be purchased for $50,000. It has a life of 15 years. Its estimated scrap value is $5000. What is the capitalized cost of this equipment assuming 10% annual interest rate with annual compounding?

24.13 A carbon-steel reactor costs $10,000 and has an estimated life of 3 years in a corrosive environment. A stainless-steel reactor costs $30,000 and has an estimated life of 10 years in the same corrosive environment. The scrap value of the carbon-steel reactor is $2000 and the scrap value of the stainless-steel reactor is $6000. Using capitalized costs with 10% annual interest and annual compounding, which reactor should be purchased?

24.14 As a well-known capitalist, an oil company offers you an investment opportunity. Their surveys indicate that if they drill in a particular formation, the well is guaranteed to produce profit at an initial rate of $50,000 per year. In their experience with this formation, the profit is expected to linearly decline to zero after 10 years. You expect a 15% annual interest rate on your investments with continuous compounding. What is the most you should pay for this well?

Glossary

annuity A series of equal payments occurring at equal time periods that last a finite length of time.

capitalized costs The amount of money needed to purchase equipment plus the principal needed to establish a perpetuity to replace the equipment when it wears out.

compound interest The interest accumulated on unpaid interest as well as the original principal.

continuous compound interest A method in which the interest on unpaid interest and principal is assessed continuously rather than after finite time periods, such as monthly or annually.

depreciation A decrease in value because of age, wear, or market conditions.

engineering economics The discipline that translates engineering technology into a form that permits evaluation by businesses or investors.

inflation An increase in the money supply relative to goods and services that results in a continuing rise in prices.

installment loan The payment of equal periodic amounts to repay a lump sum of money.

interest The "rent" paid for borrowed money.

interest period The length of time after which interest is due.

interest rate The "rent" for borrowed principal expressed as a percentage of the principal per unit time.

perpetuity A constant payment occurring at equal time intervals for an infinite length of time.

principal The amount of money borrowed.

simple interest The interest payment is based only on the original principal and not unpaid interest.

APPENDIXES

APPENDIX A

UNIT CONVERSION FACTORS

A.1 PLANE ANGLE

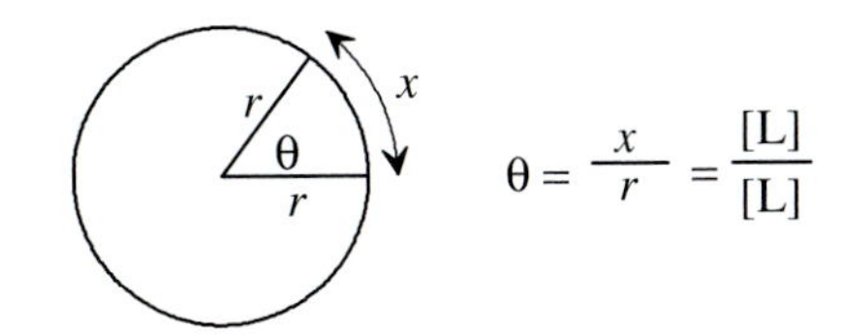

TABLE A.1
Plane angle conversion factors (Reference)

	°	′	″	rad	rev
1 degree =	1	60	3600	$\pi/180$	1/360
1 minute =	1/60	1	60	$\pi/10{,}800$	1/21,600
1 second =	1/3600	1/60	1	$\pi/648{,}000$	1/1,296,000
1 radian =	$180/\pi$	$10{,}800/\pi$	$648{,}000/\pi$	1	$1/(2\pi)$
1 revolution =	360	21,600	1,296,000	2π	1

90° = 100 grade [*a*] = 100^g = 100 gon   90° = 1000 angular mil [*b*]

[*a*] All grade subdivisions are indicated with decimals, so there are no equivalent units of minutes or seconds. This system is not widely used except in France.

[*b*] During World War II, the U.S. artillery divided a right angle into 1000 parts called *angular mil.*

An angle θ is defined by

$$\theta \equiv \frac{x}{r} \tag{A-1}$$

where the angle is measured in radians. Because the perimeter around a circle is $2\pi r$, one complete revolution is

$$\theta = \frac{2\pi r}{r} = 2\pi\left(\frac{r}{r}\right) = 2\pi \text{ rad} \tag{A-2}$$

The perimeter may also be divided into 360 equally spaced divisions called *degrees*. Therefore,

$$2\pi \text{ rad} = 360° \tag{A-3}$$

The degree may be further subdivided into 60 divisions called *minutes,* and the minutes may be subdivided into 60 divisions called *seconds*. This is a fractional system of measuring angles that dates back to the Babylonians.

A.2 SOLID ANGLE

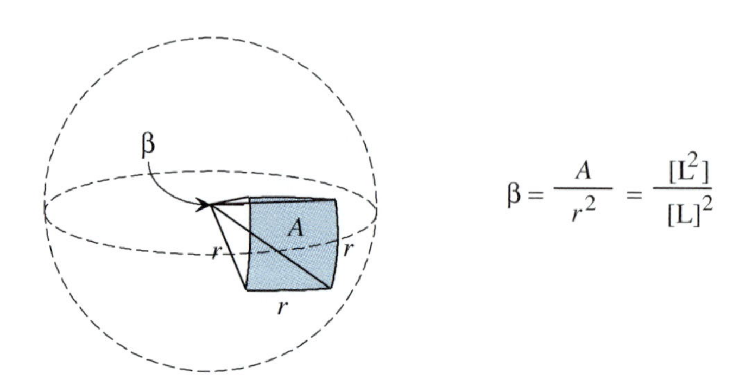

TABLE A.2
Solid angle conversion factors (Reference)

	Square Degree	Square Minute	Square Second	Steradian	Sphere
1 square degree =	1	$(60)^2$	$(3600)^2$	$(\pi/180)^2$	$(\pi/4)(180)^{-2}$
1 square minute =	$(1/60)^2$	1	$(60)^2$	$(\pi/10{,}800)^2$	$(\pi/4)(10{,}800)^{-2}$
1 square second =	$(1/3600)^2$	$(1/60)^2$	1	$(\pi/648{,}000)^2$	$(\pi/4)(648{,}000)^{-2}$
1 steradian =	$(180/\pi)^2$	$(10{,}800/\pi)^2$	$(648{,}000/\pi)^2$	1	$(4\pi)^{-1}$
1 sphere =	$(4/\pi)(180)^2$	$(4/\pi)(10{,}800)^2$	$(4/\pi)(648{,}000)^2$	4π	1

1 sphere = 2 hemisphere   1 sphere = 8 spherical right angles

A solid angle β is defined as the surface area on the sphere A divided by the radius r squared:

$$\beta \equiv \frac{A}{r^2} \tag{A-4}$$

The surface can be defined by projecting four radii from the center of a sphere and connecting the ends of adjacent radii with circumference segments. If the angle between adjacent radii is one radian, then a square is defined on the sphere surface that has circumference segments of length r. This solid angle is a *steradian,* given by the formula

$$\beta = \frac{A}{r^2} = \frac{r^2}{r^2} = 1 \text{ steradian} \tag{A-5}$$

If the angle between adjacent radii is 1 degree, then the solid angle is a *square degree*; if the angle between adjacent radii is 1 minute, then the solid angle is a *square minute*; and if the angle between adjacent radii is 1 second, then the solid angle is a *square second.* Table A.2 shows the relationship between these various solid angle measurements.

If a sphere is divided into two parts, then the solid angle is a *hemisphere.* If a hemisphere is divided into four equal parts, the solid angle formed is a *spherical right angle.*

A.3 LENGTH

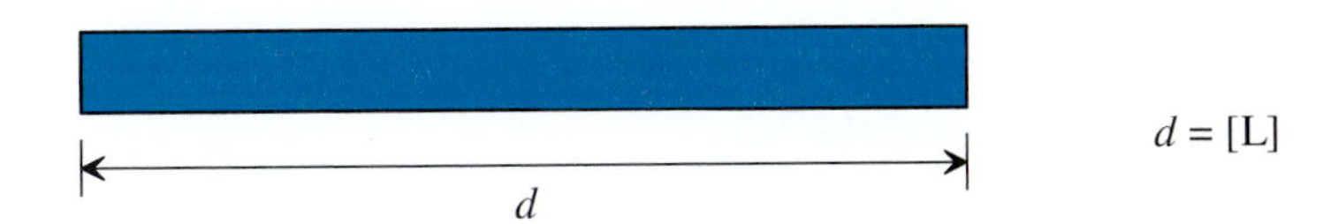

TABLE A.3
Length conversion factors (Reference)

	cm	m	km	in	ft	mi [e]
1 centimeter =	1	0.01	1.0000 E–05	0.3937	0.03281	6.214 E–06
1 meter =	100	1	0.001	39.37	3.281	6.214 E–04
1 kilometer =	1.00 E+05	1000	1	3.937 E+04	3281	0.6214
1 inch [b] =	2.54000	0.02540	2.540 E–05	1	0.08333	1.578 E–05
1 foot [a] =	30.48000	0.304800	3.048 E–04	12	1	1.894 E–04
1 U.S. statute mile =	1.609 E+05	1609	1.609	6.336 E+04	5280	1

1 nautical mile (n. mile) [f]=1852 m=1.151 mi=6076 ft	1 rod (rd) = 1 pole = 1 perch = 16.5 ft	1 fermi (fm) [j] = 1.00 E–15 m
1 ångström (Å) [k] = 1.00 E–10 m	1 yard (yd) = 3 ft	1 micron (μ) [l] = 1.00 E–06 m
1 light-year (ly) [g] = 9.4606 E+12 km	1 bolt of cloth = 120 ft	1 printer's pica = 0.16604 in
1 parsec (pc) [h] = 3.086 E+13 km	1 mil [d] = 1 thou = 0.001 in	1 printer's pica = 12 points
1 astronomical unit (i) = 1.496 E+08 km	1 pace = 30 in	1 fathom (fath) [c] = 6 ft
1 statute league = 2640 fathoms	1 cable [m] = 120 fathoms	1 cubit = 18 in
1 chain (ch) = 66 ft = 100 Gunter's links (li)	1 palm = 3 in	1 span = 9 in
1 furlong (fur) = 660 ft = 1/8 mi	1 hand = 4 in	1 skein = 360 ft

[a] The *foot* has been used in England for over 1000 years and is approximately equal to the length of a man's foot.
[b] The *inch* is derived from "ynce," the Anglo-Saxon word for twelfth part.
[c] A *fathom* is used to describe the depth of the sea. It is approximately the distance between the hands when the arms are outstretched; its name is derived from the Anglo-Saxon word for "embrace."
[d] The *mil* is equal to one thousandth of an inch and is not to be confused with the millimeter. It is commonly used in metal machining.
[e] The *mile* traces to the Romans and is about equal to 1000 double paces (about 5 ft).
[f] The *nautical mile* is the average meridian length of 1 minute of latitude, a definition that makes navigation easier.
[g] The *light-year* is the distance light travels in 1 year.
[h] The *parsec* is the height of an isosceles triangle of which the base is equal to the diameter of the earth's orbit around the sun, and the angle opposite that base is 1″.
[i] An *astronomical unit* is approximately equal to the mean distance from the earth to the sun.
[j] The *fermi* is used to measure nuclear distances.
[k] The *ångström* is used to measure atomic distances (a hydrogen atom is approximately 1 Å).
[l] The *micron* is slang for "micrometer" and is not SI.
[m] The *cable* is used to measure lengths at sea and dates back to the middle of the 16th century.

A.4 AREA

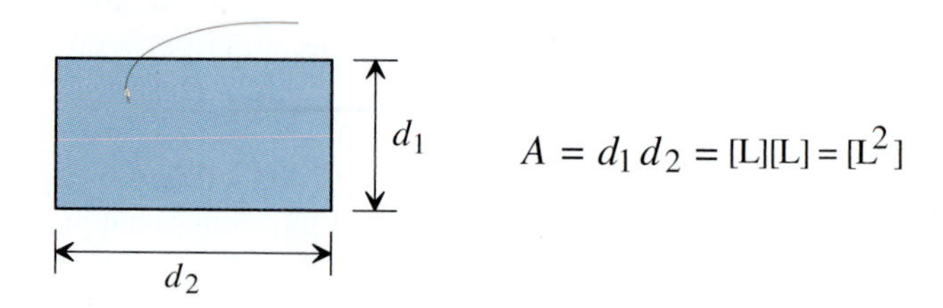

TABLE A.4
Area conversion factors (Reference)

	m^2	cm^2	ft^2	in^2
1 square meter =	1	1.000 E+04	10.76	1550
1 square centimeter =	1.000 E–04	1	0.001076	0.1550
1 square foot =	0.09290	929.0	1	144
1 square inch =	6.452 E–04	6.452	0.006944	1

1 square mile = 2.788 E+07 ft^2 = 640 acres
1 yd^2 = 9 ft^2
1 square rod = 30.25 yd^2 = 272.25 ft^2
1 rood = 40 square rod
1 acre [*b*] = 4 roods = 160 square rods = 43,560 ft^2

1 are (a) [*a*] = 100 m^2
1 hectare (ha) [*a*] = 100 are = 1000 m^2 = 2.471 acres
1 barn (b) [*d*] = 1.0000 E–28 m^2
1 circular mil (cir mils) [*c*] = $(0.001\ \text{in})^2\pi/4$ = 7.854 E–07 in^2
1 U.S. township = 36 mi^2 = 36 sections

[*a*] An area 10 m on a side is an *are* and an area 100 m on a side is a *hectare* (i.e., 100 are). Both the are and hectare are used in international agriculture to measure land area.
[*b*] The *acre,* which has been in existence since about 1300, is the approximate area that a yoke of oxen could plow in a day.
[*c*] A *circular mil* is the cross-sectional area of a circle that is 1 mil (0.001 in) in diameter. It was first used to measure the cross-sectional area of wire.
[*d*] The *barn* is used to measure the effective target area of atomic nuclei when bombarded with particles. The unit was invented in 1942 as Manhattan Project code; it probably derives from the expression "I bet you couldn't hit the broadside of a barn."

A.5 VOLUME

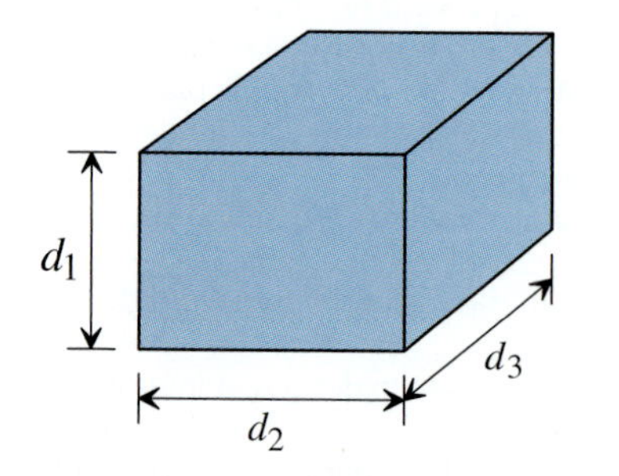

$$V = d_1 d_2 d_3 = [L][L][L] = [L^3]$$

TABLE A.5
Volume conversion factors (Reference)

	m^3	cm^3	L	ft^3	in^3
1 cubic meter =	1	1.000 E+06	1000	35.31	6.102 E+04
1 cubic centimeter =	1.000 E–06	1	0.001	3.531 E–05	0.06102
1 liter =	0.001000	1000	1	0.03531	61.02
1 cubic foot =	0.02832	2.832 E+04	28.32	1	1728
1 cubic inch =	1.639 E–05	16.39	1.639 E–02	5.787 E–04	1

1 acre-foot [*d*] = 43,560 ft^3
1 board-foot (fbm or bd-ft) [*e*] = 144 in^3
1 barrel (bbl) [*h*] = 42 gal
1 U.K. gallon = 1 Imperial gallon = 1.2009 U.S gallon [*c*]

1 stere (st) [*a*] = 1 m^3
1 yd^3 = 27 ft^3
1 masonry perch = 24.75 ft^3

1 λ = 1 μL [*b*]
1 cord (wood) [*f*] = 128 ft^3
1 cord-foot [*g*] = 16 ft^3
1 ft^3 = 7.4805195 U.S. gallon (liq)

[*a*] The *stere* is no longer recommended.
[*b*] 1 μL is sometimes called 1 λ, but the use of this unit is not recommended.
[*c*] The *gallon* was first mentioned in 1342 and was given legal status in 1602. The U.S. gallon originated with the old English wine gallon in Colonial times. The *Imperial gallon* (which is about 20% larger than the U.S. gallon) is defined by a 1963 British law as the volume occupied by 10 lb_m of distilled water provided the water has a density of 0.998859 g/mL weighed in air with a density of 0.001217 g/mL against weights with a density of 8.136 g/mL.
[*d*] An *acre-foot* is the volume when one acre is covered by water with 1 ft depth. This unit is commonly used in agricultural irrigation.
[*e*] A *board-foot* corresponds to the volume occupied by a board that is 1 ft × 1 ft × 1 in.
[*f*] The *cord* describes the volume of a wood stack that measures 4 ft × 4 ft × 8 ft.
[*g*] A *cord-foot* describes the volume of a wood stack that measures 4 ft × 4 ft × 1 ft.
[*h*] U.S. petroleum barrel.

TABLE A.6
Customary units of volume (Reference)

United Kingdom (Liquids and Solids)

20 minims (min)	= 1 scruple	= 1.1838 E–06 m^3
3 scruples	= 1 fluid drachm	= 3.5515 E–06 m^3
8 fluid drachms	= 1 fluid ounce (fl oz)	= 2.8413 E–05 m^3
5 fluid ounces	= 1 gill or noggin	= 1.4207 E–04 m^3
4 gills	= 1 pint (pt)	= 5.6825 E–04 m^3
2 pints	= 1 quart (qt)	= 1.1365 E–03 m^3
2 quarts	= 1 pottle or quartern (dry)	= 2.2730 E–03 m^3
2 quarterns (dry)	= 1 gallon (gal)	= 4.5461 E–03 m^3
2 gallons	= 1 peck (pk)	= 9.0919 E–03 m^3
4 pecks	= 1 bushel (bu)	= 3.6368 E–02 m^3
9 gallons	= 1 firkin	= 4.0914 E–02 m^3
9 pecks	= 1 kilderkins	= 8.1830 E–02 m^3
3 bushels	= 1 sack or bag	= 1.0910 E–01 m^3
36 gallons	= 1 barrel (bbl)	= 1.6365 E–01 m^3
8 bushels	= 1 quarter or seam	= 2.9094 E–01 m^3
640 gallons	= 1 lasts	= 2.9094 m^3

United States (Liquid)

60 minims (min)	= 1 fluid dram (fl dr)	= 3.6967 E–06 m^3
3 teaspoons (t or tsp)	= 1 tablespoon (T or Tbsp)	= 1.4787 E–05 m^3
2 tablespoons	= 1 fluid ounce (fl oz)	= 2.9574 E–05 m^3
8 fluid drams	= 1 fluid ounce (fl oz)	= 2.9574 E–05 m^3
4 fluid ounces	= 1 gill	= 1.1829 E–04 m^3
2 gills	= 1 cup	= 2.3659 E–04 m^3
2 cups	= 1 liquid pint (pt)	= 4.7318 E–04 m^3
2 liquid pints	= 1 liquid quart (qt)	= 9.4635 E–04 m^3
4 liquid quarts	= 1 gallon (gal)	= 3.7854 E–03 m^3
9 gallons	= 1 firkin	= 3.4068 E–02 m^3
31.5 gallons	= 1 barrel (bbl)*	= 1.1924 E–01 m^3
63 gallons	= 1 hogshead (hhd)	= 2.3847 E–01 m^3
84 gallons	= 1 puncheon	= 3.1797 E–01 m^3
126 gallons	= 1 U.K. butt	= 4.7696 E–01 m^3
252 gallons	= 1 tun	= 9.5392 E–01 m^3

United States (Dry)

2 dry pints	= 1 dry quart (qt)	= 1.1012 E–03 m^3
4 dry quarts	= 1 dry gallon (gal)	= 4.4049 E–03 m^3
2 dry gallons	= 1 peck (pk)	= 8.8098 E–03 m^3
4 pecks	= 1 bushel (bu)	= 3.5239 E–02 m^3
105 dry quarts	= 1 dry barrel (bbl)*	= 1.1563 E–01 m^3

*Not to be confused with a U.S. petroleum barrel.

A.6 MASS

TABLE A.7
Mass unit conversions (Reference)

	g	kg	lb_m	slug
1 gram-mass =	1	0.001	0.002205	6.852 E–05
1 kilogram-mass =	1000	1	2.205	0.06852
1 pound-mass [c] =	453.6	0.4536	1	0.03108
1 slug =	1.4594 E+04	14.594	32.174	1

1 grain [*b*] = 6.479891 E–05 kg
1 short hundred weight = 100 lb_m
1 short ton = 2000 lb_m
1 tonne (t) = 1 metric ton = 1000 kg
1 metric carat [*d*] = 2.000 E–04 kg
1 point = 0.01 metric carat
1 γ [*a*] = 1 μg = 1.000 E–09 kg

1 glug = 980.665 g = 0.980665 kg
1 mug = 1 metric slug = 1 par = 1 TME = 9.80665 kg
1 unified atomic mass unit (u) [*e*] = 1 dalton = 1.6605402 E–27 kg
1 atomic mass unit, chem. (amu) [*e*] = 1.66024 E–27 kg
1 atomic mass unit, phys. (amu) [*e*] = 1.65979 E–27 kg
1 eV of equivalent mass [*f*] = 1.7827 E–36 kg

[*a*] The symbol "γ" is used to represent 1 μg, but its use is discouraged.
[*b*] The *grain* dates back to the 16th century and is thought to be equal to the weight of a wheat grain.
[*c*] The *pound* originated with the Roman Libra (327 g). The Imperial Standard Pound was defined in 1855 as the mass of platinum with given dimensions. In 1963, the pound was defined as 0.45359237 kg exactly, a number chosen because it is evenly divided by seven to ease the conversion from grains to grams.
[*d*] Precious stones are measured in *metric carats,* which correspond to 200 mg.
[*e*] The *atomic mass unit* was originally intended to be the mass of a single hydrogen atom, the lightest element. In 1885, it was suggested that more elements would have integer numbers for their atomic weights if the atomic mass unit were defined using 1/16 the mass of oxygen. Chemists used oxygen in its natural abundance (2480:5:1 ^{16}O:^{18}O:^{17}O) whereas physicists used isotopically pure ^{16}O for their standard. Thus, there was a slight discrepancy between the scales used by chemists and physicists (272 parts per million). It was later found that expressing the atomic mass unit as 1/12th the mass of a single carbon-12 atom allowed even more elements to have masses that were integer numbers. Thus, the *unified atomic mass unit* was established, which had the added benefit of eliminating the discrepancy between the chemist and physicist scales.
[*f*] The famous Einstein relationship $E=mc^2$ showed that when mass is destroyed, energy is produced (and vice versa). The amount of energy E is found by multiplying the destroyed mass m by the speed of light c squared. Thus, physicists and nuclear engineers sometimes express mass in energy units, such as electron volts (eV).

TABLE A.8
Customary units of mass (Reference)

Avoirdupois Weights		Apothecaries Weights [b]		Troy Weights [c]	
1 pound (lb avdp) [a] = 7000 grains [d]		1 pound (lb ap) = 5760 grains [d]		1 pound (lb t) = 5760 grains [d]	
16 drams (dr avdp)	= 1 ounce (oz)	20 grains	= 1 scruple (s ap)	24 grains	= penny weight (dwt)
16 ounces	= 1 pound (lb avdp)	3 scruples	= 1 U.K. drachm (dr ap)	20 penny weights	= 1 ounce (oz t)
14 pounds	= 1 stone	3 scruples	= 1 U.S. dram (dr ap)	12 ounce (oz t)	= 1 pound (lb t)
28 pounds	= 1 quarter	8 drachm or dram	= 1 ounce (oz ap)		
112 pounds	= 1 long hundred weight (cwt)	12 ounce (oz ap)	= 1 pound (lb ap)		
252 pounds	= 1 wey				
2240 pounds	= 1 long ton				

[a] The common pound with which we are familiar (and the pound indicated by the symbol "lb_m") is the avoirdupois pound.
[b] The apothecary scale is not used anymore.
[c] The troy scale is used in the United States for weighing precious metals.
[d] The grain is the same in all systems.

A.7 DENSITY

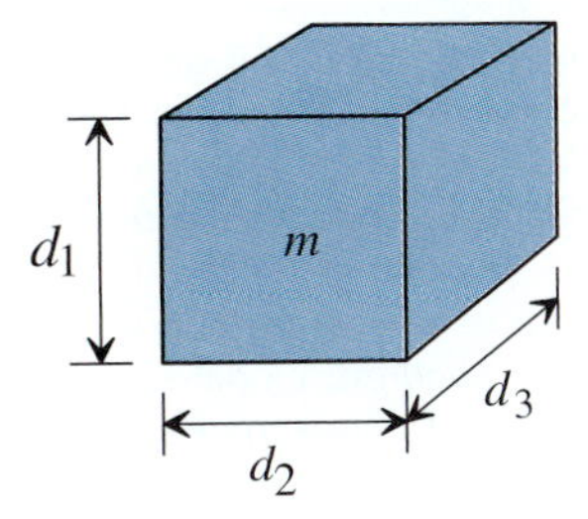

$$\rho = \frac{m}{d_1\, d_2\, d_3} = \frac{[M]}{[L][L][L]} = \frac{[M]}{[L^3]}$$

TABLE A.9
Density conversion factors (Reference)

	g/cm³	kg/m³	lb_m/ft^3	lb_m/in^3	slug/ft³
1 gram per cubic centimeter =	1	1000	62.43	0.03613	1.940
1 kilogram per cubic meter =	0.001	1	0.06243	3.613 E–05	0.001940
1 pound-mass per cubic foot =	0.01602	16.02	1	5.787 E–04	0.03108
1 pound-mass per cubic inch =	27.68	2.768 E+04	1728	1	53.71
1 slug per cubic foot =	0.5154	515.4	32.174	0.01862	1

Density can also be expressed by *specific gravity* SG, a dimensionless number formed by dividing the density of substance A ρ_A by the density of a reference substance ρ_R:

$$SG = \frac{\rho_A}{\rho_R} \tag{A-6}$$

Although any reference may be used, the most common reference substance is water at its maximum density (4°C, 1.000 g/cm³).

A.8 TIME

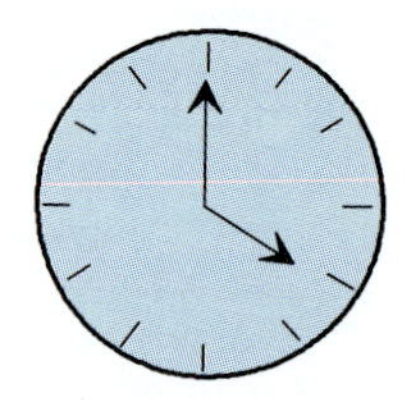

TABLE A.10
Time conversion factors (Reference)

	yr	d	h	min	s
1 year [*a*] =	1	365.24	8.766 E+03	5.259 E+05	3.1557 E+07
1 day [*c*] =	2.738 E–03	1	24	1440	8.640 E+04
1 hour [*d*] =	1.141 E–04	4.167 E–02	1	60	3600
1 minute [*e*] =	1.901 E–06	6.944 E–04	1.667 E–02	1	60
1 second [*e*] =	3.169 E–08	1.157 E–05	2.778 E–04	1.667 E–02	1

1 year = 365.24 solar days [*c*]
1 mean solar day [*c*] = 86,400 s

1 year = 366.24 sidereal days [*b*]
1 sidereal day [*b*] = 86,164 s

1 week = 7 days
1 fortnight = 2 weeks

[*a*] A *year* is the time required for the earth to return to a given position as it orbits the sun. Our calendar is adjusted to the *tropical year*, the time it takes for the earth to orbit the sun between successive vernal equinoxes (March 21, the spring date in which light and dark are equal).

[*b*] A *sidereal day* is the mean time taken for the earth to complete one revolution as determined by comparing the earth's position to distant stars.

[*c*] A *solar day* is the mean time required for the sun to return to a fixed position (e.g., overhead) in the sky. The solar day and sidereal day differ. The solar day is slightly longer because the sun is viewed from a different position as the earth orbits the sun. In common parlance, we refer to a solar day, not a sidereal day. It has been known since the Egyptians and Babylonians that there are $365\frac{1}{4}$ solar days per year.

[*d*] In ancient times, the day was divided into 24 time fractions which we call *hours*. Light and darkness were each divided into 12 equal time fractions regardless of the time of year. According to the season, the length of the dark-hour differed from the light-hour. When mechanical clocks were invented, the length of the hour was standardized. In England, each community kept its own local time; each community was completely independent of the others. In 1880, Greenwich mean time was established as the official time throughout England. Today, most of the world has agreed to standardize on Greenwich mean time.

[*e*] The *minute* and *second* of time trace to the Babylonians, who used units of 60. Efforts to decimalize time have proved unsuccessful.

A.9 SPEED/VELOCITY

$$\text{Speed, Velocity} = \frac{d}{t_2 - t_1} = \frac{[L]}{[T]} = [L/T]$$

TABLE A.11
Speed/velocity conversion factors (Reference)

	m/s	cm/s	ft/s	km/h	mi/h (mph)	knot
1 meter per second =	1	100	3.281	3.6	2.237	1.944
1 centimeter per second =	0.01	1	0.03281	0.036	0.02237	0.01944
1 foot per second =	0.3048	30.48	1	1.097	0.6818	0.5925
1 kilometer per hour =	0.2778	27.78	0.9113	1	0.6214	0.5400
1 mile per hour =	0.4470	44.70	1.467	1.609	1	0.8690
1 nautical mile per hour =	0.5144	51.44	1.688	1.852	1.151	1

1 knot = 1 nautical mile per hour   1 mi/min = 88.00 ft/s = 60.00 mi/h

A.10 FORCE

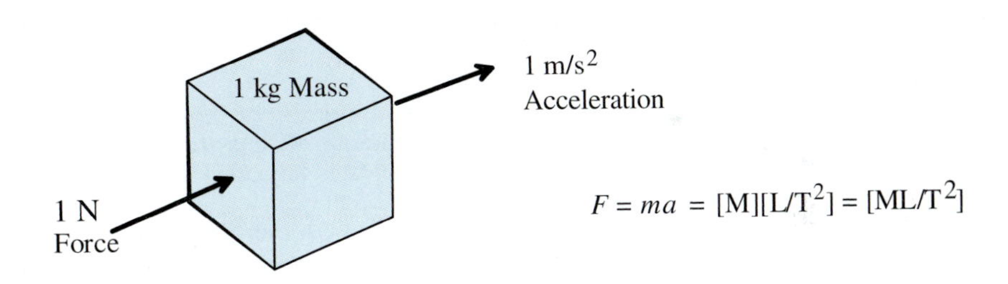

TABLE A.12

	N	dyne	pdl	kg_f	g_f	lb_f
1 newton =	1	1.00 E+05	7.233	0.1020	102.0	0.2248
1 dyne =	1.00 E-05	1	7.233 E-05	1.020 E-06	0.001020	2.248 E-06
1 poundal =	0.1383	1.383 E+04	1	0.01410	14.10	0.03108
1 kilogram-force =	9.807	9.807 E+05	70.93	1	1000	2.205
1 gram-force =	0.009807	980.7	0.07093	0.001	1	0.002205
1 pound-force =	4.448	4.448 E+05	32.174	0.4536	453.6	1

1 pound-force = 16 ounce-force
1 ton-force = 2000 lb_f
1 kilopond [*b*] = 1 kg_f
1 fors [*c*] = 1 g_f
1 kip [*a*] = 1000 lb_f

[*a*] The *kip* (for *K*ilo *I*mperial *P*ound) is sometimes used to describe the load on a structure.
[*b*] The *kilopond* is used in Germany for "kilogram-force."
[*c*] The *fors* (Latin for "force") was proposed in 1956 as an alternate name for "gram-force."

A.11 PRESSURE

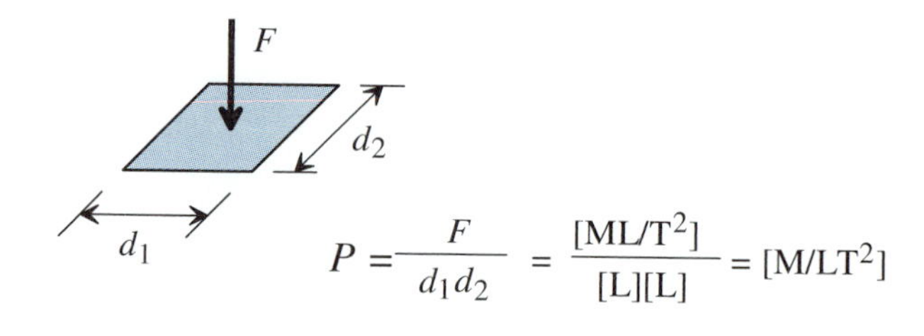

$$P = \frac{F}{d_1 d_2} = \frac{[\mathrm{ML/T^2}]}{[\mathrm{L}][\mathrm{L}]} = [\mathrm{M/LT^2}]$$

TABLE A.13
Pressure conversion factors (Reference)

	Pa	dyne/cm^2	lb$_f$/ft^2	lb$_f$/in^2 (psi)	atm	cm-Hg	in-H_2O
1 newton per square meter =	1	10	0.02089	1.450 E–04	9.869 E–06	7.501 E–04	0.004015
1 dyne per square centimeter =	0.1	1	0.002089	1.450 E–05	9.869 E–07	7.501 E–05	4.015 E–04
1 pound-force per square foot =	47.88	478.8	1	0.006944	4.725 E–04	0.03591	0.1922
1 pound-force per square inch =	6895	6.895 E+04	144	1	0.06805	5.171	27.68
1 standard atmosphere [*c*] =	1.013 E+05	1.013 E+06	2116	14.696	1	76	406.8
1 centimeter [*d*] of mercury at 0°C =	1333	1.333 E+04	27.84	0.1934	0.01316	1	5.353
1 inch [*d*] of water at 4°C =	249.1	2491	5.202	0.03613	0.002458	0.1868	1

1 kg$_f$/m^2 = 9.806650 Pa
1 atm = 2.493 ft-Hg = 33.90 ft-H_2O = 27,714 ft-air (1 atm, 60°F) [*d*]
1 bar [*a*] = 1 barye = 1.00 E+06 dyne/cm^2 = 0.1 MPa = 100 kPa ≈ 1 atm
1 millibar (mb) = 1.00 E+03 dyne/cm^2 = 1000 microbar (μb) [*b*]
1 torr [*e*] = (101325/760) Pa ≈ 1 μm-Hg = 0.1 cm-Hg

1 kip/in^2 (ksi) = 1000 lb$_f$/in^2
1 technical atmosphere [*c*] = 1 kg$_f$/cm^2
1 micron pressure = 1 μm-Hg [*d*]
1 g$_f$/cm^2 = 980.665 dyne/cm^2

[*a*] The bar is most commonly employed in meteorology because it is approximately equal to the atmospheric pressure on earth. Although the bar is not properly SI, its use is temporarily tolerated because it is so widespread. The *barye* was the original name given to this unit of pressure in 1900, but it has been shortened to "bar."
[*b*] Although there is no proper abbreviation for the bar, the *millibar* (mb) and *microbar* (μb) abbreviations are sometimes used.
[*c*] Because the atmospheric pressure changes (in fact, meteorologists measure it to predict weather changes), a *standard atmosphere* P° has been defined as 101,325.0 Pa. The *technical atmosphere* is defined as 1 kg$_f$/cm^2. Unless otherwise specified, an "atmosphere" is generally the "standard atmosphere." The use of *atmosphere* for pressure measurements is discouraged by SI, but its use will probably continue because it is easily visualized.
[*d*] The simplest way to measure pressure is with a *manometer*, a U-shaped tube filled with liquid. Differences in pressure acting on each liquid column change the liquid levels, which are then easily read using a meterstick. For accurate work, the conversion factors in Table A.13 may be used only if the temperature of the liquid is controlled [4°C for water, 0°C for mercury (Hg)]. (Alternatively, tables listing the liquid density as a function of temperature may be used to correct the reading, provided the manometer temperature is known.) Also, the local acceleration due to gravity (g) affects the reading. The values given in Table A.13 use the standard acceleration due to gravity (g°).
[*e*] The *torr* differs from a mm-Hg by less than 1 part in 7 million. The use of *torr* is discouraged by SI.

A.12 ENERGY

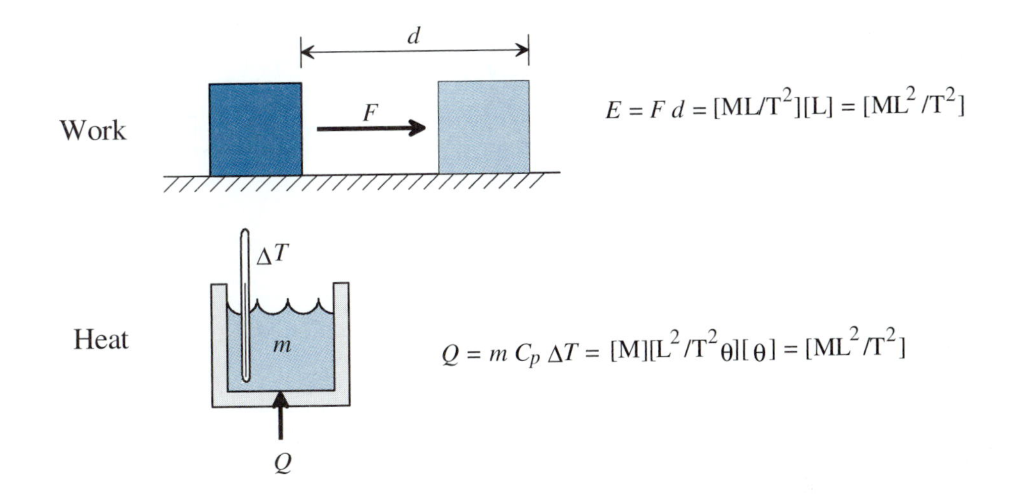

TABLE A.14
Energy conversion factors (Reference)

	J	erg	ft·lb_f	cal	Btu	kW·h	hp·h
1 joule =	1	1.000 E+07	0.7376	0.2390	9.485 E–04	2.778 E–07	3.725 E–07
1 erg =	1.000 E–07	1	7.376 E–08	2.390 E–08	9.485 E–11	2.778 E–14	3.725 E–14
1 foot $pound_f$ =	1.356	1.356 E+07	1	0.3240	0.001286	3.766 E–07	5.051 E–07
1 calorie [b] =	4.184	4.184 E+07	3.086	1	0.003968	1.162 E–06	1.559 E–06
1 Brit. thermal unit [c] =	1054	1.054 E+10	777.6	252.0	1	2.929 E–04	3.928 E–04
1 kilowatt hour [d] =	3.600 E+06	3.600 E+13	2.655 E+06	8.606 E+05	3414	1	1.341
1 horsepower hour [d] =	2.685 E+06	2.685 E+13	1.980 E+06	6.414 E+05	2545	0.7457	1

1 electron volt (eV) [f] = 1.60217733 E–19 J
1 kg_f·m = 9.806650 J
1 V·C = 1 J
1 (dyne/cm^2)·cm^3 = 1 erg [e]
1 atm·ft^3 = 2116 ft·lb_f [e]
1 ton (nuclear equivalent TNT) = 4.184 E+09 J

1 kcal [a] = 1 calorie (kg)
1 g_f·cm = 980.6650 erg
1 V·A·s = 1 J
1 atm·L = 101.3 J [e]
1 psia·ft^3 = 144 ft·lb_f [e]

1 W·h = 3600 J [d]
1 W·s = 1 J [d]
1 Pa·m^3 = 1 J [e]
1 atm·cm^3 = 0.1013 J [e]
1 bar·cm^3 = 0.1 J [e]

[a] *Kilocalorie* is the heat required to raise 1 kg water by 1 K. Because the heat capacity of water is not constant, a variety of kilocalories are defined. This is the *thermochemical* kilocalorie, the most commonly used.
[b] *Calorie* is the heat required to raise 1 g water by 1 K. In diet books, the energy content in food is usually expressed in *calories*, but actually *kilocalories* are meant. Sometimes dietitians use *Calorie* to mean *kilocalorie*. Because the heat capacity of water is not constant, a variety of calories are defined. This is the *thermochemical* calorie, the most commonly used.
[c] *British thermal unit* is the heat required to raise 1 lb_m water by 1 F°. Because the heat capacity of water is not constant, a variety of Btus are defined. This is the *thermochemical* Btu, the most commonly used.
[d] Energy = power × time. These units can be visualized as answering the question "how much energy is expended if a 1-kW (1-hp) motor operates for 1 hour?"
[e] Energy = pressure × volume
[f] The energy required to move a single electron through a vacuum with 1 volt of potential is an *electron volt.*

A.13 POWER

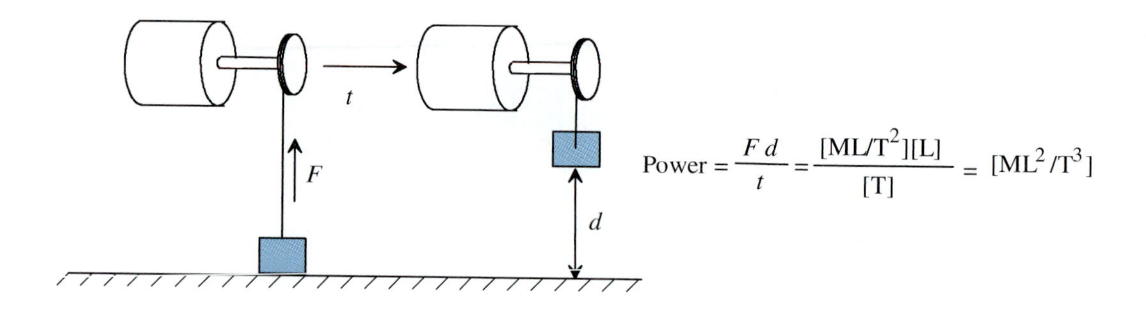

TABLE A.15
Power conversion factors (Reference)

	W	kW	ft·lb$_f$/s	hp	cal/s	Btu/h
1 watt [a] =	1	0.001	0.7376	0.001341	0.2390	3.414
1 kilowatt =	1000	1	737.6	1.341	239.0	3414
1 foot-pound$_f$ per second =	1.356	0.001356	1	0.001818	0.3240	4.629
1 horsepower [b] =	745.7	0.7457	550	1	178.2	2546
1 calorie per second =	4.184	0.004184	3.086	0.005611	1	14.29
1 British thermal unit per hour =	0.2929	2.929 E–04	0.2160	3.928 E–04	0.07000	1

1 W = 1.00 E+07 erg/s 1 ft·lb$_f$/s = 60 ft·lb$_f$/min = 3600 ft·lb$_f$/h 1 hp = 33,000 ft·lb$_f$/min = 550 ft·lb$_f$/s
1 hp (electric) ≡ 746 W 1 ton of refrigeration [d] = 12,000 Btu/h
1 hp = 0.0760181 hp (boiler) = 0.999598 hp (electric) = 1.01387 hp (metric) = 0.999540 hp (water) [c]

[a] A *watt* is a J/s.
[b] In 1782, James Watt (1736–1819) devised the *horsepower* to help him sell steam engines. He assumed a horse could pull with a force of 180 lb$_f$ and, when harnessed to a capstan, would walk a 24-ft diameter circle $2\frac{1}{2}$ times each minute. This was a work expenditure of 32,400 ft·lb$_f$/min, which he rounded to 33,000 ft·lb$_f$/min (550 ft·lb$_f$/s). Engines are often rated in *brake horsepower* (bhp) or *shaft horsepower,* which is the power available at the turning drive shaft.
[c] A *boiler horsepower* (bhp) is the amount of heat needed to evaporate 34.5 lb$_m$/h of water at 212°F. A *metric horsepower* is the power required to raise 75 kg 1 meter per second.
[d] A *ton of refrigeration* is a U.S. term that describes the amount of refrigeration required to freeze 1 ton (2000 lb$_m$) per day of water at 32°F.

A.14 AMOUNT OF SUBSTANCE

The *mole* is the number of atoms in 0.012 kg (12 g) of carbon-12. This number is given special recognition as *Avogadro's constant* N_A, which is

$$N_A = 6.0221367 \times 10^{23}\ \text{atoms/mol} \tag{A-7}$$

The *coulomb* C is the number of electrons that flow in a 1-ampere current in 1 second. The number of electrons in a coulomb N_C is

$$N_C = 6.24150636 \times 10^{18}\ \text{electrons/C} \tag{A-8}$$

The ratio of Avogadro's number to the coulomb is called the *Faraday constant F*

$$F = \frac{N_A}{N_C} = 96{,}485.309 \text{ C/mol} \tag{A-9}$$

The *mol* is sometimes called the *gram-mole.* The *kilogram-mole* (kmol) is the number of atoms in 12 kg of carbon-12, the *pound-mole* (lbmol) is the number of atoms in 12 lb_m of carbon-12, and the *ton-mole* is the number of atoms in 12 tons of carbon-12. The number of atoms in each of these units is calculated as:

$$\frac{6.022 \times 10^{23} \text{ atoms}}{\text{mol}} \times \frac{1000 \text{ mol}}{\text{kmol}} = 6.022 \times 10^{26} \frac{\text{atoms}}{\text{kmol}}$$

$$\frac{6.022 \times 10^{23} \text{ atoms}}{\text{mol}} \times \frac{453.6 \text{ mol}}{\text{lbmol}} = 2.732 \times 10^{26} \frac{\text{atoms}}{\text{lbmol}}$$

$$\frac{6.022 \times 10^{23} \text{ atoms}}{\text{mol}} \times \frac{453.6 \text{ mol}}{\text{lbmol}} \times \frac{2000 \text{ lbmol}}{\text{ton-mole}} = 5.463 \times 10^{29} \frac{\text{atoms}}{\text{ton-mol}}$$

Further Readings

Jerrard, H. G., and D. B. McNeill. *A Dictionary of Scientific Units.* Englewood, NJ: Franklin Publishing, Inc., 1964.

Klein, H. A. *The World of Measurements.* New York: Simon and Schuster, 1974.

National Institute of Standards and Technology. *The International System of Units (SI).* NIST Special Publication 330, U.S. Department of Commerce, 1991.

Weast, R. C. *CRC Handbook of Chemistry and Physics.* 58th ed. West Palm Beach: CRC Press, 1978.

decisions with respect to professional services solicited or provided by them or their organizations in private or public engineering practice.

e. Engineers shall not solicit or accept a professional contract from a government body on which a principal or officer of their organization serves as a member.

5. Engineers shall avoid deceptive acts in the solicitation of professional employment.

a. Engineers shall not falsify or permit misrepresentation of their, or their associates', academic or professional qualifications. They shall not misrepresent or exaggerate their degree of responsibility in or for the subject matter of prior assignments. Brochures or other presentations incident to the solicitation of employment shall not misrepresent pertinent facts concerning employers, employees, associates, joint ventures or past accomplishments with the intent and purpose of enhancing their qualifications and their work.

b. Engineers shall not offer, give, solicit or receive, either directly or indirectly, any political contribution in an amount intended to influence the award of a contract by public authority, or which may be reasonably construed by the public of having the effect or intent to influence the award of a contract. They shall not offer any gift, or other valuable consideration in order to secure work. They shall not pay a commission, percentage or brokerage fee in order to secure work except to a bona fide employee or bona fide established commercial or marketing agencies retained by them.

III. Professional Obligations

1. Engineers shall be guided in all their professional relations by the highest standards of integrity.

a. Engineers shall admit and accept their own errors when proven wrong and refrain from distorting or altering the facts in an attempt to justify their decisions.

b. Engineers shall advise their clients or employers when they believe a project will not be successful.

c. Engineers shall not accept outside employment to the detriment of their regular work or interest. Before accepting any outside employment they will notify their employers.

d. Engineers shall not attempt to attract an engineer from another employer by false or misleading pretenses.

e. Engineers shall not actively participate in strikes, picket lines, or other collective coercive action.

f. Engineers shall avoid any act tending to promote their own interest at the expense of the dignity and integrity of the profession.

2. Engineers shall at all times strive to serve the public interest.

a. Engineers shall seek opportunities to be of constructive service in civic affairs and work for the advancement of the safety, health and well-being of their community.

b. Engineers shall not complete, sign or seal plans and/or specifications that are not of a design safe to the public health and welfare and in conformity with accepted engineering standards. If the client or employer insists on such unprofessional conduct, they shall notify the proper authorities and withdraw from further service on the project.

c. Engineers shall endeavor to extend public knowledge and appreciation of engineering and its achievements and to protect the engineering profession from misrepresentation and misunderstanding.

3. Engineers shall avoid all conduct or practice which is likely to discredit the profession or deceive the public.
   a. Engineers shall avoid the use of statements containing a material misrepresentation of fact or omitting a material fact necessary to keep statements from being misleading or intended or likely to create an unjustified expectation, or statements containing prediction of future success.
   b. Consistent with the foregoing, Engineers may advertise for recruitment of personnel.
   c. Consistent with the foregoing, Engineers may prepare articles for the lay or technical press, but such articles shall not imply credit to the author for work performed by others.
4. Engineers shall not disclose confidential information concerning the business affairs or technical processes of any present of former client or employer without his consent.
   a. Engineers in the employ of others shall not without the consent of all interested parties enter promotional efforts or negotiations for work or make arrangements for other employment as a principal or to practice in connection with a specific project for which the Engineer has gained particular and specialized knowledge.
   b. Engineers shall not, without the consent of all interested parties, participate in or represent an adversary interest in connection with a specific project or proceeding in which the Engineer has gained particular specialized knowledge on behalf of a former client or employer.
5. Engineers shall not be influenced in their professional duties by conflicting interests.
   a. Engineers shall not accept financial or other considerations, including free engineering designs, from material or equipment suppliers for specifying their product.
   b. Engineers shall not accept commissions or allowances, directly or indirectly, from contractors or other parties dealing with clients or employers of the Engineer in connection with work for which the Engineer is responsible.
6. Engineers shall uphold the principle of appropriate and adequate compensation for those engaged in engineering work.
   a. Engineers shall not accept remuneration from either an employee or employment agency for giving employment.
   b. Engineers, when employing other engineers, shall offer a salary according to professional qualifications.
7. Engineers shall not attempt to obtain employment or advancement or professional engagements by untruthfully criticizing other engineers, or by other improper or questionable methods.

   a. Engineers shall not request, propose, or accept a professional commission on a contingent basis under circumstances in which their professional judgment may be compromised.
   b. Engineers in salaried positions shall accept part-time engineering work only to the extent consistent with policies of the employer and in accordance with ethical considerations.
   c. Engineers shall not use equipment, supplies, laboratory, or office facilities of an employer to carry on outside private practice without consent.
8. Engineers shall not attempt to injure, maliciously or falsely, directly or indirectly, the professional reputation, prospects, practice or employment of other engineers, nor untruthfully criticize other engineers' work. Engineers who believe others are guilty of unethical or illegal practice shall present such information to the proper authority for action.
   a. Engineers in private practice shall not review the work of another engineer for the same client, except with the knowledge of such engineer, or unless the connection of such engineer with the work has been terminated.
   b. Engineers in governmental, industrial or educational employ are entitled to review and evaluate the work of other engineers when so required by their employment duties.
   c. Engineers in sales or industrial employ are entitled to make engineering comparisons of represented products with products of other suppliers.
9. Engineers shall accept personal responsibility for their professional activities; provided, however, that Engineers may seek indemnification for professional services arising out of their practice for other than gross negligence, where the Engineer's interests cannot otherwise be protected.
   a. Engineers shall conform with state registration laws in the practice of engineering.
   b. Engineers shall not use association with a nonengineer, a corporation, or partnership as a "cloak" for unethical acts, but must accept personal responsibility for all professional acts.
10. Engineers shall give credit for engineering work to those to whom credit is due, and will recognize the proprietary interests of others.
    a. Engineers shall, whenever possible, name the person or persons who may be individually responsible for designs, inventions, writings, or other accomplishments.
    b. Engineers using designs supplied by a client recognize that the designs remain the property of the client and may not be duplicated by the Engineer for others without express permission.
    c. Engineers, before undertaking work for others in connection with which the Engineer may make improvements, plans, designs, inventions, or other records which may justify copyrights or patents, should enter into a positive agreement regarding ownership.
    d. Engineers' designs, data, records, and notes referring exclusively to an employer's work are the employer's property.
11. Engineers shall cooperate in extending the effectiveness of the profession by interchanging information and experience with other engineers and students, and

will endeavor to provide opportunity for the professional development and advancement of engineers under their supervision.

a. Engineers shall encourage engineering employee's efforts to improve their education.
b. Engineers shall encourage engineering employees to attend and present papers at professional and technical society meetings.
c. Engineers shall urge engineering employees to become registered at the earliest possible date.
d. Engineers shall assign a professional engineer duties of a nature to utilize full training and experience, insofar as possible, and delegate lesser functions to subprofessionals or to technicians.
e. Engineers shall provide a prospective engineering employee with complete information on working conditions and proposed status of employment, and after employment will keep employees informed of any changes.

Note: In regard to the question of application of the Code to corporations vis-a-vis real persons, business form or type should not negate nor influence conformance of individuals to the Code. The Code deals with professional services, which services must be performed by real persons. Real persons in turn establish and implement policies within business structures. The Code is clearly written to apply to the Engineer and it is incumbent on a member of NSPE to endeavor to live up to its provisions. This applies to all pertinent sections of the Code.

Publication date as revised: July 1993.
Publication # 1102

APPENDIX C

z TABLE

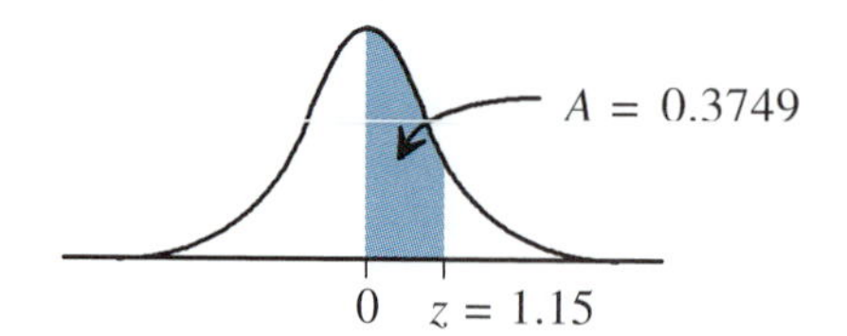

Areas Under the Standard Normal Curve from 0 to z

z	0	1	2	3	4	5	6	7	8	9
0.0	0.0000	0.0040	0.0080	0.0120	0.0160	0.0199	0.0239	0.0279	0.0319	0.0359
0.1	0.0398	0.0438	0.0478	0.0517	0.0557	0.0596	0.0636	0.0675	0.0714	0.0754
0.2	0.0793	0.0832	0.0871	0.0910	0.0948	0.0987	0.1026	0.1064	0.1103	0.1141
0.3	0.1179	0.1217	0.1255	0.1293	0.1331	0.1368	0.1406	0.1443	0.1480	0.1517
0.4	0.1554	0.1591	0.1628	0.1664	0.1700	0.1736	0.1772	0.1808	0.1844	0.1879
0.5	0.1915	0.1950	0.1985	0.2019	0.2054	0.2088	0.2123	0.2157	0.2190	0.2224
0.6	0.2258	0.2291	0.2324	0.2357	0.2389	0.2422	0.2454	0.2486	0.2518	0.2549
0.7	0.2580	0.2612	0.2642	0.2673	0.2704	0.2734	0.2764	0.2794	0.2823	0.2852
0.8	0.2881	0.2910	0.2939	0.2967	0.2996	0.3023	0.3051	0.3078	0.3106	0.3133
0.9	0.3159	0.3186	0.3212	0.3238	0.3264	0.3289	0.3315	0.3340	0.3365	0.3389
1.0	0.3413	0.3438	0.3461	0.3485	0.3508	0.3531	0.3554	0.3577	0.3599	0.3621
1.1	0.3643	0.3665	0.3686	0.3708	0.3729	0.3749	0.3770	0.3790	0.3810	0.3830
1.2	0.3849	0.3869	0.3888	0.3907	0.3925	0.3944	0.3962	0.3980	0.3997	0.4015
1.3	0.4032	0.4049	0.4066	0.4082	0.4099	0.4115	0.4131	0.4147	0.4162	0.4177
1.4	0.4192	0.4207	0.4222	0.4236	0.4251	0.4265	0.4279	0.4292	0.4306	0.4319
1.5	0.4332	0.4345	0.4357	0.4370	0.4382	0.4394	0.4406	0.4418	0.4429	0.4441
1.6	0.4452	0.4463	0.4474	0.4484	0.4495	0.4505	0.4515	0.4525	0.4535	0.4545
1.7	0.4554	0.4564	0.4573	0.4582	0.4591	0.4599	0.4608	0.4616	0.4625	0.4633
1.8	0.4641	0.4649	0.4656	0.4664	0.4671	0.4678	0.4686	0.4693	0.4699	0.4706
1.9	0.4713	0.4719	0.4726	0.4732	0.4738	0.4744	0.4750	0.4756	0.4761	0.4767
2.0	0.4772	0.4778	0.4783	0.4788	0.4793	0.4798	0.4803	0.4808	0.4812	0.4817
2.1	0.4821	0.4826	0.4830	0.4834	0.4838	0.4842	0.4846	0.4850	0.4854	0.4857
2.2	0.4861	0.4864	0.4868	0.4871	0.4875	0.4878	0.4881	0.4884	0.4887	0.4890
2.3	0.4893	0.4896	0.4898	0.4901	0.4904	0.4906	0.4909	0.4991	0.4913	0.4916
2.4	0.4918	0.4920	0.4922	0.4925	0.4927	0.4929	0.4931	0.4932	0.4934	0.4936
2.5	0.4938	0.4940	0.4941	0.4943	0.4945	0.4946	0.4948	0.4949	0.4951	0.4952
2.6	0.4953	0.4955	0.4956	0.4957	0.4959	0.4960	0.4961	0.4962	0.4963	0.4964
2.7	0.4965	0.4966	0.4967	0.4968	0.4969	0.4970	0.4971	0.4972	0.4973	0.4974
2.8	0.4974	0.4975	0.4976	0.4977	0.4977	0.4978	0.4979	0.4979	0.4980	0.4981
2.9	0.4981	0.4982	0.4982	0.4983	0.4984	0.4984	0.4985	0.4985	0.4986	0.4986

z	0	1	2	3	4	5	6	7	8	9
3.0	0.4987	0.4987	0.4987	0.4988	0.4988	0.4989	0.4989	0.4989	0.4990	0.4990
3.1	0.4990	0.4991	0.4991	0.4991	0.4992	0.4992	0.4992	0.4992	0.4993	0.4993
3.2	0.4993	0.4993	0.4994	0.4994	0.4994	0.4994	0.4994	0.4995	0.4995	0.4995
3.3	0.4995	0.4995	0.4995	0.4996	0.4996	0.4996	0.4996	0.4996	0.4996	0.499
3.4	0.4997	0.4997	0.4997	0.4997	0.4997	0.4997	0.4997	0.4997	0.4997	0.4998
3.5	0.4998	0.4998	0.4998	0.4998	0.4998	0.4998	0.4998	0.4998	0.4998	0.4998
3.6	0.4998	0.4998	0.4999	0.4999	0.4999	0.4999	0.4999	0.4999	0.4999	0.4999
3.7	0.4999	0.4999	0.4999	0.4999	0.4999	0.4999	0.4999	0.4999	0.4999	0.4999
3.8	0.4999	0.4999	0.4999	0.4999	0.4999	0.4999	0.4999	0.4999	0.4999	0.4999
3.9	0.5000	0.5000	0.5000	0.5000	0.5000	0.5000	0.5000	0.5000	0.5000	0.5000

APPENDIX D

SUMMARY OF SOME ENGINEERING MILESTONES

Date	Milestone
B.C.	
6000 to 3000	People built permanent houses, cultivated plants, and domesticated animals. Irrigation systems were constructed; plows and animal yokes were used. Wind- and water-powered mills were used to grind grain. Copper ores were mined and transformed into copper and bronze tools. Mathematics was used. Information was written on papyrus, parchment, or clay tablets.
ca. 3050	Earliest evidence of stone masonry in Egypt.
ca. 2930	First pyramid constructed (214 ft).
ca. 2900	Great Pyramid at Gizeh begun. At 481 ft, it was the largest stone building ever erected by ancient humans.
ca. 2000	The Egyptians built irrigation dams and canals.
ca. 1600	The first engineer's handbook, the Rhind Papyrus, was created.
ca. 1500	The palace of Cnossus in Crete was built. It included the first adequate sanitary drains.
ca. 1100	Military engineering was started under the Assyrian king Tiglathpileser I.
ca. 1000	Phoenicians constructed mines. King Solomon's temple was built in Jerusalem.
691	Assyrian aqueduct of Jerwan was constructed.
ca. 600	The Egyptians built a canal connecting the Nile and Red Sea. The Etruscans built the first arch bridge.
484	Mining was a major source of Greek taxes.
ca. 450	The Greek Empedocles of Akragas drained swamps to prevent disease. The Greek Parthenon was constructed.
ca. 300	Appius Claudius constructed the Appian Road in Rome. Appius Claudius constructed the Aqua Appia water supply in Rome. Greek lighthouse near Alexandria was constructed. It lasted 16 centuries before falling.
ca. 250	Greece's Archimedes designed military machines and screw pump.
ca. 200	The Great Wall of China was completed.
ca. 150	Greece's Hero designed military machines, derricks, presses, rotary steam turbine, odometer, and the hand-pump fire engine.
142	First stone arch bridge Pons Aemilius was constructed.
140	High-level aqueduct Marcia was constructed.
ca. 15	Rome's Marcus Vitruvius Pollio wrote *De Architectura,* which was used as a standard engineering reference work until the Renaissance. The book described land-leveling devices, water supply, time measurement by sundials and water clocks, hoists, derricks, pulleys, pumps, water organs, military "engines of war" (e.g., catapults), and ethics.

A.D.

ca. 45 The Romans constructed a 3.5-mile tunnel to drain rich agricultural lands.
79 The Roman surveyor Frontinius described the Roman 250-mile aqueduct system, which could deliver an estimated 300 million gallons per day to Rome.
80 Roman Coliseum was constructed.
ca. 200 Cast iron was used in China.
ca. 300 Romans constructed a water-powered flour mill in Arles, France, to replace scarce slave labor.
ca. 1000 The abacus calculating machine was introduced to Europe from the Orient.
ca. 1100 Construction of medieval stone fortresses began. They were obsolete by ca. 1500 when gunpowder and cannons could destroy them.
Windmills were introduced to mill grain, pump, and grind paint and snuff.
ca. 1150 Spanish papermaking became an industry based upon imported Chinese technology.
Chimneys first appeared in European buildings.
ca. 1200 Black powder was used in Europe.
Locks for canals were developed in Italy.
ca. 1230 A notebook by Frenchman Wilars de Honecourt described surveying, stone cutting, water-powered saws, and a perpetual motion machine.
ca. 1300 Construction of great Gothic cathedrals began in Europe.
Spinning wheels were developed to twist fibers to make thread.
Cast iron was used in Europe.
ca. 1400 Water mills were widely available in European villages.
ca. 1450 Germany's Johann Gutenberg published the first book using a combination of previously known techniques.
ca. 1500 First engineering book, Valturius' *De re militari*, was published.
Current-driven water-wheel pumps were used in Paris and London.
Star-shaped earthen fortifications were developed to resist cannon fire.
ca. 1530 First known horse-powered railway was constructed.
1556 Georgius Agricola published *De Re Metallica,* which described mining methods, ore distribution, pumps, hoists, fans for mine ventilation, mining law, mine surveying, ore processing, and the manufacture of salt, soda, alum, vitriol, sulfur, bitumen, and glass.
ca. 1600 Edmund Gunter developed the "graphicall logarithmic scale," forerunner to the slide rule.
1619 Dud Dudley developed a process to convert coal into coke for cast iron production. Coke replaced charcoal, which was no longer available because forests were decimated.
1642 France's 19-year-old Blaise Pascal devised an adding machine consisting of 10 numbered wheels linked by gears.
1671 Germany's 25-year-old Gottfried von Leibnitz improved upon Pascal's adding machine.
1672 Engineers organized as a separate unit, *Corps du génie*, in the French army.
1698 Britain's Thomas Savery developed the first practical steam engine used to pump water from mines.
1705 Britain's Thomas Newcomen improved the steam engine used to pump water from mines.
1716 French highway department, the *Corps des Ponts et Chaussées*, was organized.
1733 Fly-shuttle loom invented by John Kay in England.
1740 Sulfuric acid production began in England.
1742 America's Benjamin Franklin invented the "Franklin stove," which used fuel more efficiently than fireplaces.
1752 Benjamin Franklin established the similarity between lightning and static electricity in his famous kite-flying experiment.
1759 John Smeaton completed the Eddystone Lighthouse on the treacherous Eddystone Rocks of the English Channel, 14 miles from shore.
1763 Cugnot built a steam locomotive in France.
1770 "Spinning Jenny" was invented by James Hargreaves for making yarn in England.
1775 France's Nicolas LeBlanc developed a process to convert ordinary salt (sodium chloride) to soda (sodium carbonate) for use in glass and soap manufacture. This is regarded as an important milestone in the development of the chemical industry.
1776 First steam engine by Watt and Boulton was installed as a mine pump in England.
1779 First all-metal, cast-iron bridge was constructed in Coalbrookdale, England.
1783 France's Montgolfier brothers flew in hot-air balloons.
1784 First large-scale use of steam for industry was located at Albion Mills in England.
1785 England's Edward Cartwright invented the mechanical loom.
1788 England's William Symington built the first steam-powered boat.

1792	First American canal, only 5 miles long, opened in South Hadley, Massachusetts.
1796	America's Eli Whitney invented the cotton gin for separating cotton from seeds, hulls, etc.
1794	Eli Whitney demonstrated a manufacturing technique based upon interchangeable parts, rather than custom-fitted parts.
ca. 1800	Italy's Volta developed the first battery.
1801	Britain's Sir Humphry Davy developed the electric arc light.
1812	England's 20-year-old Charles Babbage conceived the mechanical "Difference Engine," a calculating machine.
1817	Britain's Henry Cort developed the "puddling process" to transform cast iron to wrought iron.
1818	British Institute of Civil Engineers was founded.
ca. 1820	Analytical mechanics and materials testing were first used for bridge building.
1824	Portland cement, made from lime and clay, was patented by Joseph Aspdin in Britain. This cement improved upon lime mortar known to the ancients, the Greek mixture of lime and Santorin earth, and the Roman mixture of lime and pozzuolana (volcanic ash).
1825	Erie Canal, 363 miles long, joined the Hudson River and Great Lakes.
1829	Britain's George Stephenson built a locomotive called the "Rocket," so named because it could travel at a record speed of 35 miles per hour.
ca. 1830	Britain's William Sturgeon and America's Joseph Henry showed that a magnet is produced when electric current passes through a coiled wire surrounding a metal core.
1831	Britain's Michael Faraday showed that an electric current is induced in a wire when it moves through a magnetic field.
1833	First practical internal combustion engine was developed in England.
	England's Charles Babbage designed the Analytical Engine, the first universal digital computer. It was designed to be programmed using punch cards and could perform logical and arithmetic operations. Unfortunately, it was not built.
1834	America's Cyrus Hall McCormick patented the reaper for harvesting grain.
1836	America's Colt invented the revolver.
1837	Britain's William Cooke and Charles Wheatstone communicated by electric telegraph.
1838	The first Atlantic crossing was made using steam power exclusively. The trip required 15 to 18 days.
1839	Charles Goodyear "vulcanized" rubber by heating rubber latex with sulfur.
1840	There were two engineering schools in the United States.
	Sir William Groves demonstrated an incandescent light by flowing electricity through a platinum wire, but it soon burned out.
1842	The first underwater tunnel was constructed under the Thames River.
1843	America's Samuel Morse commercialized the electric telegraph and sent the first message between Washington and Baltimore.
1845	Guncotton explosive (i.e., cotton treated with nitric and sulfuric acid) was invented. It is more explosive than black powder made of saltpeter (potassium nitrate), sulfur, and charcoal.
1846	America's Elias Howe patented the sewing machine.
	Britain's William Thompson invented pneumatic tires.
1847	James Young patented oil refining by distillation.
ca. 1850	One of the first modern sewage systems was built in Hamburg, Germany.
1852	Henri Giffard powered a dirigible with a steam engine.
1856	Britain's Henry Bessemer invented a steel-making process that allowed steel to be widely produced and ultimately replaced cast and wrought iron in many applications.
1859	The first elevator was developed. It used a steam-powered screw to raise passengers as high as six stories. This invention made skyscrapers possible.
	Edwin Drake's 69-ft-deep oil well came into production, establishing the modern U.S. petroleum industry.
1860	France's Jean-Joseph-Étienne Lenoir made the first practical internal combustion engine.
1861	France's François Coignet demonstrated reinforced concrete by embedding metal bars in *concrete* (Portland cement + sand + stone aggregate), thus improving upon a technology first employed by the ancient Greeks.
1865	Telegraph cable was laid across the Atlantic Ocean, establishing instant communications between America and Europe.
1866	Sweden's Alfred Nobel invented *dynamite* (a mixture of nitroglycerin and diatomaceous earth), an explosive that is safe to handle.
1868	A compressed-air refrigeration plant was built in Paris.
1869	Suez Canal opened.
1870	There were 70 engineering schools in the United States.
ca. 1870	The electric generator was developed using numerous worldwide improvements.
1872	America's John Hyatt opened a factory that produced *celluloid* (guncotton treated with camphor and alcohol), one of the first plastics.

1873	America's Brayton demonstrated an engine that evolved into the jet engine.
	Germany's Carl von Linde developed the first practical ammonia refrigeration machine.
1876	America's Alexander Graham Bell exhibited the telephone at the Philadelphia Centennial.
	Germany's Nikolaus Otto perfected the four-stroke internal combustion engine.
1877	America's Thomas Edison invented the phonograph.
1878	First all-steel bridge was constructed in the United States.
1879	First commercial electric railway was constructed in Berlin.
	First electric power station was installed in San Francisco to power arc lights.
	Thomas Edison invented the lightbulb using a carbonized thread in an evacuated bulb. It lasted almost 2 days.
1882	Thomas Edison started operating the world's first electric generator/electric light system (750 kW) in New York City.
	Von Schroder developed first blood oxygenator machine.
1883	John Roebling's Brooklyn Bridge was completed in New York.
	Sweden's Karl Gustaf Patrick de Laval developed the first practical turbine.
1884	The first American skyscraper (10 stories) was erected in Chicago.
	France's Count Hilaire de Chardonnet patented artificial silk made from nitrated cotton, an explosive.
1885	Germany's Karl Benz built a motorized tricycle.
ca. 1885	America's Frederick Taylor introduced "scientific management" to improve industrial efficiency.
1886	America's Charles Hall developed an electrolytic process to produce aluminum.
1887	Germany's Gottlieb Daimler ran the first motor car.
1888	German physicist Heinrich Hertz built an oscillating circuit that transmitted an electromagnetic wave that induced current in a nearby antenna.
	Alexandre Eiffel constructed the Eiffel Tower in Paris.
	Nikola Tesla patented a multiphase, alternating current, electric motor.
1891	First automobiles were produced in France and Belgium.
1892	Germany's Rudolph Diesel patented an engine using oil as a fuel, rather than gasoline.
	8,000,000 electric lightbulbs were produced.
1894	The turbine-powered (2300-hp) steamship *Turbinia* was launched.
1895	The first large-scale U.S. water power project was completed near Niagara Falls.
1896	America's Samuel B. Langley flew a large steam-powered airplane model.
	Italian inventor Guglielmo Marconi received a patent on wireless radio.
1898	Count Ferdinand von Zeppelin studied rigid, lighter-than-air aircraft.
1900	An electrolytic process was developed to make caustic soda (sodium hydroxide) and chlorine gas from salt (sodium chloride) at Niagara Falls.
	American and British engineers met in Paris to decide whether to standardize on alternating current (AC) or direct current (DC). AC was selected because it can be easily transformed to high voltages for more efficient transmission.
1901	Peter Hewitt developed the mercury-vapor arc lamp, which evolved to the fluorescent tube about 35 years later.
1903	The Wright brothers demonstrated powered flight. In the best flight that year, the plane traveled 852 feet in slightly less than 1 minute.
	Oil-insulated 60,000-volt transformers were developed for efficient electricity transmission.
1904	New York Subway opened.
1905	Albert Einstein proposed the relativity theory, which concluded that $E = mc^2$, i.e., that mass and energy are interchangeable.
1906	The tungsten light filament was introduced, improving lightbulb output 4.7 times and life 27 times.
1907	America's Lee De Forest created the thermionic vacuum tube, the forerunner to the transistor.
1910	Bakelite plastic became a commercial product. It replaced wood, glass, and rubber in many products.
ca. 1910	Germany used the Haber process to fix nitrogen from the air. Although originally used to make explosives for World War I, this process is now used to make fertilizers.
1913	Henry Ford adapted the moving assembly line to automobile production.
1914	The first ship passed through the Panama Canal.
	Robert Goddard began his rocket studies.
1915	X-rays first used for medical imaging.
1920	Spain's Juan de Cierva added an unpowered, horizontal "propeller" to a small-wing airplane to prevent stalling. This was the precursor to the helicopter.
1922	Commercial radio broadcasting was initiated in the United States.
1923	Highly efficient transmission voltages of 220,000 V were used in the western United States.

1925	Electric-powered home refrigerators that used chlorofluorocarbon ("Freon") refrigerants were commercially available.
	Vannevar Bush constructed the "Differential Analyzer," the first analog computer that mechanically solved sets of differential equations.
1927	The first experimental television was demonstrated by transmitting images from Washington to Bell Laboratories in New York.
1930	Empire State Building (102 stories) was completed.
1931	George Washington Bridge was completed.
1932	Britain's John Cockcroft and E. T. S. Walton confirmed Einstein's theory by bombarding lithium with high-energy protons and measuring the resulting changes in mass and energy.
1934	American chemist Wallace Carothers invented "Nylon 66."
	Italy's Enrico Fermi bombarded uranium with neutrons and apparently created a new heavier element called neptunium.
	DeBakey developed the roller pump, which was later used in heart-lung machines.
1936	France's Eugene Houdry developed catalytic oil cracking.
1937	Golden Gate Bridge was completed.
1938	The first commercial fluorescent tubes were sold by General Electric.
	Germany's Otto Hahn and F. Strassman split uranium by bombardment.
1944	Howard Aiken's "Harvard Mark I" electromechanical computer was built by IBM. It performed 200 additions per minute and worked to 23 significant figures.
1945	The United States detonated the first atomic explosion in Alamogordo, New Mexico. It was the result of the 4-year, $2 billion Manhattan Project.
	The ENIAC all-electronic computer was completed at the University of Pennsylvania. It used 18,000 vacuum tubes and 6000 switches to perform 5000 additions per second.
1946	Willem Kolff developed the first artificial kidney machine.
1948	The first transistor was demonstrated at Bell Laboratories.
1952	The United States detonated the first fusion bomb at Eniwetok.
1953	First clinical use of the heart-lung machine.
1956	The first full-scale nuclear power plant was completed at Calder Hall, England.
1957	The Soviet Union launched the Sputnik satellite.
1960	Theodore Maiman demonstrated a laser.
	America's Wilson Greatbatch developed the implantable heart pacemaker.
1961	The Soviet Union placed Yuri Gagarin in orbit.
1969	America's Neil Armstrong walked on the moon.
	Denton Cooley implanted an artificial heart in a patient.
1975	First CAT scanner developed for medical imaging.
1981	The first U.S. space shuttle was launched.
	IBM introduced its first personal computer.
1982	Compact discs were first used to store music.
1994	The "Chunnel" was completed linking England with France by tunneling under the English Channel.
1998	The first components of the International Space Station were launched into orbit.

Note: "ca." is an abbreviation for "circa," meaning "about."

PHOTO CREDITS

3	NASA/NASA Media Services
10	© Richard Passmore/Getty Images
18	© Oliver Benn/Getty Images
20	© Suzanne Murphy/Getty Images
53	NASA/NASA Media Services
55	Photo by Lee Lowery, Jr. Texas A&M University/Lee Lowery c/o Civil Engineering
58	Photo by Lee Lowery, Jr. Texas A&M University/Lee Lowery c/o Civil Engineering
96	© Irving Newman/Archive Photos
96	© Reuters/Archive Photos
97	© Earl Scott/Photo Researchers, Inc.
98	© The Granger Collection Ltd.
127a	Courtesy of Paul MacCready/Mark Holtzapple
127b	Martyn Cowley/Courtesy of Paul MacCready/Mark Holtzapple
127c	Martyn Cowley/Courtesy of Paul MacCready/Mark Holtzapple
127d	Courtesy of Paul MacCready/Mark Holtzapple
127e	Courtesy of Paul MacCready/Mark Holtzapple
143	NASA/NASA Media Services
155	2002 General Motors Corporation. Used with permission GM Media Archives
196	NASA/NASA Media Services
196	NASA/NASA Media Services
248	© Archive Photos
249	© World Perspectives/Getty Images
267`	Credit Missing
270	© Archive Photos
276a	© Archive Photos
276b	© The Granger Collection Ltd.
299	© George Grigoriou/Getty Images
309	Jean Wulfson/Mark Holtzapple c/o Dept. of Chemical Engineering
316	© Science Photo Library/Photo Researchers, Inc.
317	© Science Photo Library/Photo Researchers, Inc.
317	© Science Photo Library/Photo Researchers, Inc.
318	© Archive Photos
336	© Takeshi Takahara/Photo Researchers, Inc.
345	Laboratories/U.S. Department of Commerce
393	© John Sohlden/Visuals Unlimited
402	© Archive Photos
403	© P Crowther/S Carter/Getty Images
453	Courtesy of Lucent Technologies, Inc./Bell Labs
453	Courtesy of Lucent Technologies, Inc./Bell Labs
471	© Tom Pantages

473	Courtesy of Intel
474	Photo courtesy of Texas Instruments
529	© Archive Photos
530	Courtesy *Popular Science*/Los Angeles Times Syndicate
553	© Tony Freeman/PhotoEdit
564	© Doug Densinger/AllSport

INDEX

TOPIC INDEX

Page numbers followed by f indicate figures; t, tables.

A

D

N

BIOGRAPHICAL INDEX

Charles Stark Draper Prize

The Charles Stark Draper Prize is awarded biannually by the National Academy of Engineering to recognize outstanding engineering achievements that contribute to human welfare and freedom. It is the most prestigious award offered to engineers; some people consider it to be the Nobel Prize of engineering.

Year	Recipient(s)	Achievement
2001	Vinton Cerf Robert Kahn Leonard Kleinrock	Inventing the Internet
1999	Charles K. Kao Robert D. Maurer John B. MacChesney	Developing fiber optic technology
1997	Vladimir Haensel	“Platforming” technology used in oil refining
1995	John R. Pierce Harold A. Rosen	Communication satellite technology
1993	John Backus	Development of Fortran computer language
1991	Sir Frank Whittle Hans von Ohain	Turbojet engine
1989	Jack S. Kilby Robert N. Noyce	Monolithic integrated circuit

Charles Stark Draper is the father of intertial guidance.